INDUSTRIAL RAILWAYS and LOCOMOTIVES of TEESSIDE

Cliff Shepherd

INDUSTRIAL RAILWAY SOCIETY

Published by the INDUSTRIAL RAILWAY SOCIETY
at 24, Dulverton Rd, Melton Mowbray, Leicestershire, LE13 0SF

www.irsociety.co.uk

© **INDUSTRIAL RAILWAY SOCIETY 2023**

ISBN 978 1 912995 07 3

British Library Cataloguing-in-Publication Data
A catalogue record for this book is available from The British Library.

The location of the coal drops on the River Tees at Middlesbrough was initially called Port Darlington in the 1830s.

Designed by Karen Robertson of Print Rite, Long Hanborough, Witney, Oxon OX29 8BA

Printed by Artisan Litho Ltd, Kingston Bagpuize, Abingdon, Oxon OX13 5FB

This book is copyright under the Berne Convention. Apart from any fair dealing for the purposes of private study, research, criticism, or review, as permitted under the Copyright Act, 1911, no portion may be reproduced by any process without the written permission of the publisher.

CONTENTS

Introduction	4
Key Maps	12
Locomotive Worked Sites	21
Contractors' Locomotives	357
Locomotive Dealers, Repairers, Hirers and Builders	381
Preserved Locomotives	391
Non-Locomotive Worked Sites	397

Indexes:

Locomotives	419
Locomotive Names	453
Locations & Owners	458

Dedication

Remembering Jenny Shepherd

1946 – 2021

Title page photograph: Locomotive HENRY Black Hawthorn works No. 525 built 1880 for Bolckow Vaughan.
(John Hutchings collection)

Front cover photograph: Dorman Long's number 10 awaits its next duties at the company's Acklam Iron and Steel Works, Middlesbrough on 19th June 1960. It was a popular design of saddle tank constructed by Hawthorn Leslie, maker's number 3352 of 1918, which had been rebuilt by Dorman Long in 1948.
(IRS J.P. Mullett photograph)

Rear cover photograph: A line up of diesel locomotives at the British Steel Corporation's Hartlepool South Iron and Steel Works on 10th May 1978 with the now silent blast furnace dominating the background. The first two locomotives are 454 and 453 (Sentinel 10018 and 10025 of 1960). Behind are six Fowler diesel hydraulics, thought to be the four-coupled 4220027 and then BSC numbers 11, 8, 10, 9 and 5. It was noted that 9 had its Fowler maker's plate 4240008 on one side but a 4240011 plate on the other! *(Adrian Booth photograph)*

INTRODUCTION

Much of modern Teesside – its physical form, character and local communities – is a result of the mining and heavy industry that developed in the latter half of the nineteenth and early twentieth century. It was a way of life based on traditional industry but now, many of those mines and works have gone and knowledge of their existence is rapidly disappearing from the community memory. Although the main line railway networks on Teesside belonged to the North Eastern Railway by 1900, many businesses had their own industrial railways to transport raw materials and goods within their premises.

This book describes the industrial railways that have existed on Teesside, including some that still function, together with the locomotives that worked on them. Teesside is defined as the area covered by the present-day local authorities of Hartlepool, Middlesbrough, Redcar and Cleveland and Stockton-on-Tees. Sections have also been included to deal with some of the often extensive industrial railways that used other forms of motive power, such as rail cranes and haulage engines, as well as dealers and a small number of preservation sites. This publication is the first to provide a comprehensive historical coverage of one the major manufacturing regions of the UK that had been a world leader in iron production and a major supplier of chemicals and ships. It fits into a series of handbooks being prepared by the Industrial Railway Society to cover the whole country; several volumes of which have already been published.

Prior to the arrival of the railways, this was a largely rural area with small market towns, such as Yarm, Stockton and Hartlepool. All three relied on water for some of their trade, but the Tees was a meandering river with many channels and the harbour at Hartlepool required improvement. Attempts were made to straighten some of the bends on the river, but it was the coming of the Stockton and Darlington Railway (SDR) to Stockton in 1825 and its extension to the new town at Middlesbrough in 1830 that was to stimulate industrial development. However, this and other early railways, such as the Clarence and Stockton & Hartlepool on the north bank of the Tees, were mainly concerned with transporting coal from South West Durham to the ports. In addition to the river wharves at Stockton, new docks were established at Middlesbrough (1842) and West Hartlepool (1852).

The event that was to transform the area was the confirmation of the extensive Cleveland ironstone reserves to the south of the river in 1850. This encouraged the construction of more railways to serve the new mines and stimulated the establishment of iron works on both sides of the Tees. These, in turn, led to the setting up of heavy engineering works and the expansion of shipbuilding.

As the twentieth century developed, alternative sources of raw materials and power became available elsewhere; world events meant that countries overseas were better able to compete and road transport became cheaper and more flexible, leaving rail to carry bulk commodities. As a result, there has been a major transformation of industry on Teesside and many of the companies and scenes featured in this book no longer exist.

Many individual people have provided information over the years, particularly members of the Industrial Railway Society and their contributions are gratefully acknowledged. I would also like to thank the staff at Teesside Archives; local authority employees at Hartlepool, Middlesbrough, Stockton-on-Tees and Redcar and Cleveland libraries; the North Eastern Railway Association; Alison at the Ken Hoole Study Centre, Darlington; Alan Betteney for advice on Stockton subjects and Richard Barber of Armstrong Railway Photographic Trust. In particular, I wish to acknowledge the help of those people who assisted to make this book a reality - Alex Betteney who has kept me up to date with current locomotive movements on Teesside: Jane and Andrew Smith who turned my hand written copy into a typed draft; David Monk-Steel for producing the computerised maps from my rough drafts; Geoff Roughton who drew some of the maps; Geoff Cryer for ensuring that the photographs are displayed at their best; Ian Bendall for preparing much of the Locomotive index; Simon Darvill for preparing the Name, Owners and Locations Indexes and, certainly not least, Robin Waywell who transformed the typed text and information into a form suitable for publication. The book includes amendments up to and including Bulletin 1090 (May 2022).

Any comments, corrections and additional information will be most welcome and should be sent to: –

Cliff Shepherd
Creak Hill
1 Easton Lane
Ainthorpe
Whitby
YO21 2LF

EXPLANATORY NOTES

LOCATIONS

Each site/entry is listed alphabetically by the name of the final (or most recent) railway operator. Previous owners or operators are then noted for each location. Every site/entry has a reference number in numerical sequence. Within locations with industrial railway locomotives, these appear as unprefixed numbers. In the following sections, a prefix is applied.

- C   Contract location
- D   Dealers, Repairers, Hirers and Builders
- P   Preserved locomotives or site
- H   Non-locomotive system (Hand, Horse or Haulage)

GAUGE

The gauge of the railway is given at the head of the list. If the gauge is uncertain, then this is stated. Metric measurements are used where the equipment was designed for these units.

NATIONAL GRID REFERENCE

A six-figure grid reference is given in the text, where known, to indicate the location of salient features of the site on Ordnance Survey maps.

LOCOMOTIVE NUMBER, NAME

A number or name formerly carried is shown in brackets ( ). If it is an unofficial name or number, then inverted commas are used " ".

LOCOMOTIVE TYPE

The Whyte system of wheel classification is used wherever possible. When the driving wheels are not connected by outside rods but by chains or cardan shafts, they are shown as 4w, 6w, 8w, etc. For ex-BR diesel locomotives the usual development of the Continental system is adopted. The following abbreviations are employed: -

- BE   Battery powered Electric locomotive.
- BEF   Battery powered Electric Flameproof locomotive.
- CA   Compressed Air locomotive.
- DE   Diesel locomotive; Electric transmission.
- DH   Diesel locomotive; Hydraulic transmission.
- DHF   Diesel locomotive; Hydraulic transmission, Flameproof.
- DM   Diesel locomotive; Mechanical transmission.
- DMF   Diesel locomotive; Mechanical transmission, Flameproof.
- F   Fireless steam locomotive.
- IST   Inverted Saddle Tank.
- PM   Petrol or Paraffin locomotive; Mechanical transmission.
- PT   Pannier Tank; a type of side tank where the tanks are not fastened to the frame.
- R   Railcar; a vehicle primarily designed to carry passengers.
- R/R   Road-Rail motive power.
- ST   Saddle Tank; a round tank which covers the boiler top. This type includes the 'Box' and 'Ogee' versions popular amongst certain manufacturers during the nineteenth century.
- T   Side Tank or similar. The tanks are invariably fastened to the frame.
- Tank   Tank locomotive of unknown type.
- VB   Vertical Boilered locomotive.
- WE   Overhead Wire powered Electric locomotive.
- WT   Well Tank; a tank located between the frames below the level of the boiler.

CYLINDER POSITION

- IC   Inside cylinders
- OC   Outside cylinders
- VC   Vertical cylinders
- G   Geared transmission (used with IC, OC or VC)

STEAM OUTLINE

Diesel or petrol locomotives with a steam locomotive appearance added are shown as S/O.

MAKERS

The normal IRS abbreviations are used to denote makers. If any of these are unfamiliar to the reader, full details can be found in the index of locomotives (which are set out in the alphabetical order of these abbreviations).

MAKER'S NUMBER AND DATE

The first column shows the maker's number, the second shows the date which appeared on the plate or the date the locomotive was built if none appears on the plate.

It should be noted that the ex-works date given in the locomotive index may be for a later year than that recorded as the building date.

Rebuilding details are noted, usually recording significant alterations to the locomotive.

Locomotives built by Black, Hawthorn and its successor, Chapman & Furneaux, were a popular choice by Teesside companies in the latter part of the 19th century. The manufacturer's records, as later recorded, only give the date of first order and not the ex-works date. Black, Hawthorn frequently built locomotives for stock with customers' orders coming later, so there is some uncertainty about the date built and carried on the plate. In 1944, Richard Inness consulted BH records that are no longer available and gives despatch dates and these have been used where no other information is available.

SOURCE OF LOCOMOTIVE

'New' indicates that a locomotive was delivered from the makers to that location. A bracketed letter indicates that a locomotive was transferred to the location from elsewhere. Full details, including the date of arrival, where known, appear in the footnotes below.

DISPOSAL OF LOCOMOTIVE

A locomotive transferred to another location is shown by a bracketed number and footnote, the date of departure being given in the footnote if it is known. In other cases the following abbreviations are used: –

OOU	Locomotive noted to be permanently out of use on the date shown.
Dere	Locomotive noted to be derelict and no longer capable of being used.
Dsm	Locomotive both OOU and incomplete on the date shown.
Scr	Locomotive broken up for scrap on the date shown.
s/s	Locomotive sold or scrapped; disposal unknown.
Wdn	Withdrawn from traffic.

Many sales of locomotives have been effected through dealers and contractors with details given where known. If the dealer's name is followed by location, e.g. Abelson, Sheldon, it is understood that the locomotive went to Sheldon depot before resale. If no location is given, the locomotive either went direct to its new owner or else definite information on this point is lacking. If the direct transfer is known to have been effected by a dealer, the word 'per' is used.

GENERAL ABBREVIATIONS

c	circa; i.e. about the time of the date quoted.

ABBREVIATIONS

ARPT	Armstrong Railway Photographic Trust
BR	British Railways
BSC	British Steel Corporation
CEGB	Central Electricity Generating Board
CIAS	Cleveland Industrial Archaeology Society
DIC	Derwent Iron Company
Haigh	Ted Haigh kept lists of industrial locomotives that he had seen from 1924 to 1958.
Hartlepool	Hartlepool was the original settlement around the Headland and old harbour. West Hartlepool developed to the south based on the creation of new docks. The boundary between the two authorities ran across some of the industrial railways within the docks. The overall area was sometimes referred to as "The Hartlepools". On 1/4/1967 the two authorities merged to become Hartlepool.
ICI	Imperial Chemical Industries Ltd

ILS	Industrial Locomotive Society	
Inness	Richard Inness was born on 17/9/1884, joined the NER in 1898 working as a draughtsman and later locomotive inspector. He kept lists of North East Industrial locomotives from 1911 and was a meticulous recorder. These have been of considerable use in the preparation of this book. He died in 1971.	
LNER	London & North Eastern Railway	
MoM	Ministry of Munitions	
MOR	Middlesbrough Owners Railway	
NCB	National Coal Board	
NER	North Eastern Railway	
NERA	North Eastern Railway Association	
OME	The Owners of Middlesbrough Estate	
RCTS	Railway Correspondence and Travel Society	
SDR	Stockton and Darlington Railway	
SDSI	South Durham Steel & Iron Co Ltd	
Stockton	The full title is Stockton-on-Tees but, for brevity, Stockton is used. Saltburn-by-the-Sea is similarly treated as Saltburn	
TCC	Tees Conservancy Commissioners	
TCP	Teesside Cast Products	
THPA	Tees & Hartlepool Port Authority	
WD	War Department	
w/c	week commencing	
w/e	week ending	
w/e/f	week ending from	

DOUBTFUL INFORMATION

Information which is known to be of a doubtful nature or subject to confirmation is denoted as such by the wording chosen or else printed enclosed in square brackets, sometimes with a question mark, e.g. [1910?].

LOCOMOTIVE OWNERSHIP

The information within this publication does not confirm ownership of any locomotives, only their physical presence at a location. When a locomotive is sold, the transaction may have taken place sometime before the date that we record the locomotive has moved to another location.

INDUSTRIAL/COMMERCIAL LOCATIONS

Some locomotives may be hired or leased from third parties. Where the hirer or leasing company is known, this information is included in a footnote.

PRESERVATION SITES

Locomotives may be owned by individuals, groups or trusts etc and may not be the property of the host site under which they are listed.

THE MAPS

The purpose of the maps is to give an approximate position of each site using their individual codes.

Given the scale of the maps, it is not possible to indicate **precisely** the position of the connection between the public and industrial railways. This becomes even more complicated around Middlesbrough due to the sheer complexity of the different railways and the existence of the Middlesbrough Owners Railway which involved various companies' locomotives running over NER lines resulting in the payment of tolls.

IRONMASTERS DISTRICT

To the west of the site of Middlesbrough, the River Tees formed a broad curve enclosing Newport and West Marsh, an area of mud banks and rough grazing. In 1830 the SDR opened its line from South Stockton, passing a small cluster of buildings at Newport, before heading in a straight line across the marsh to its new coal drops on the river at Port Darlington (later Middlesbrough). This section of line became known as the Old Town Branch. To maximise the development opportunities, some of the people behind the SDR (Joseph Pease and his partners) purchased land and formed The Owners of the Middlesbrough Estate (OME) to establish the new town of Middlesbrough. Later, sites were sold to new businesses but strategic pieces of land were retained by OME, on which it laid sidings connecting these works with the railway company. Industrial locomotives were allowed to travel over the OME lines with traffic, but tolls were charged. In 1884, the Pease family sold parcels of land, including these lines, to the NER. The railway company took over responsibility for maintaining these lines, known as the Middlesbrough Owners Railway (MOR) and received the toll revenue.

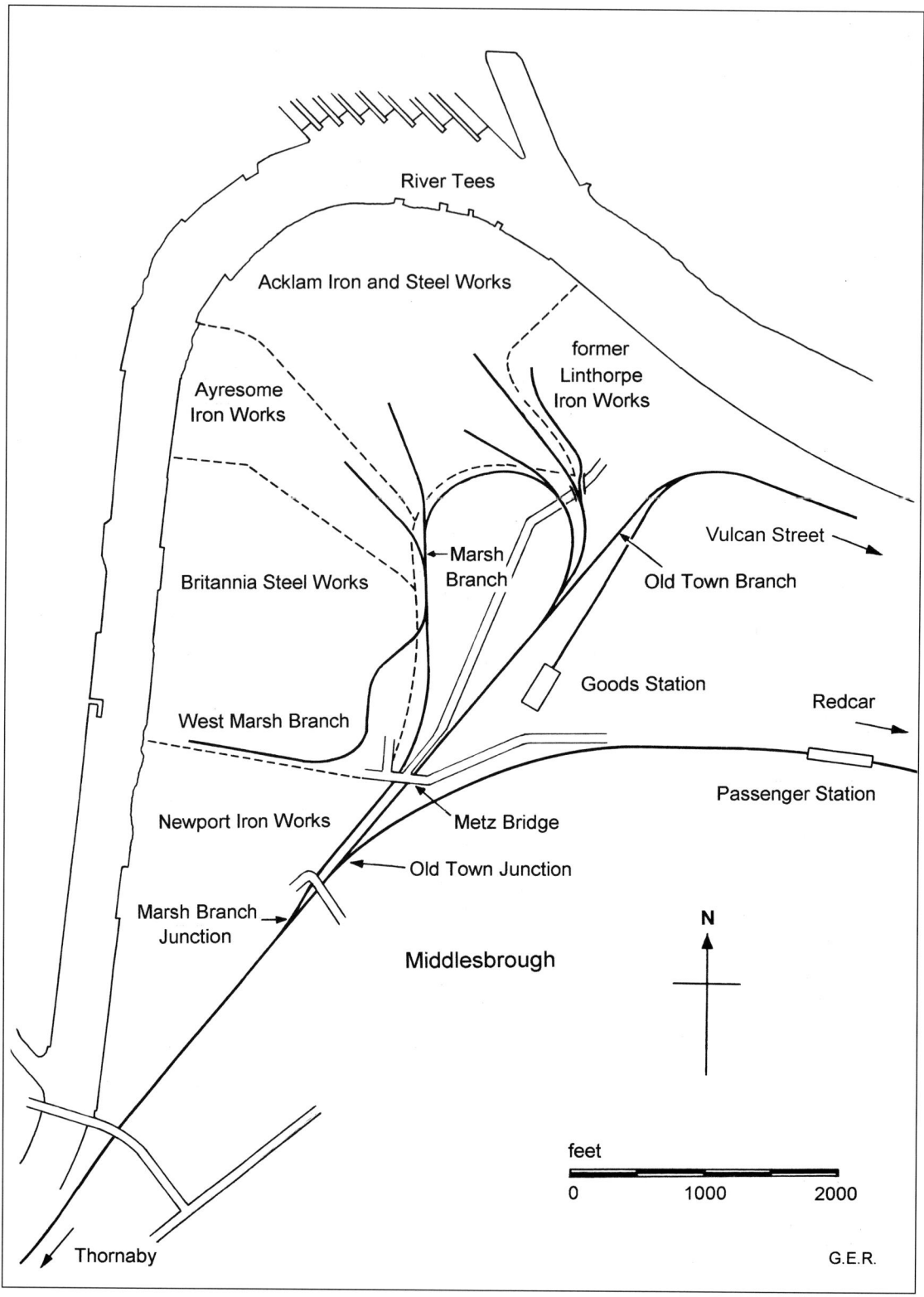

A simplified map of the Ironmasters district in Middlesbrough showing the railway branches serving the main works.

Meanwhile, despite its wet condition, West Marsh was seen as potential sites for the growing iron industry and several works began to be established there. As a result, the area was incorporated into Middlesbrough by the Middlesbrough Improvement and Extension Act of 1866; the land being divided between the existing parishes of Newport, Ayresome, Acklam and Linthorpe – hence the names for some of the works. In return, the owners of the ironworks paid much lower rates than other parts of Middlesbrough. The increase in size and complexity of plant in the 1860s meant that these new companies struggled to raise enough capital and so they initially tended to concentrate on either iron production e.g. Newport, Ayresome, Linthorpe Iron Works or puddling and rolling iron e.g. Britannia, West Marsh.

The earliest blast furnaces were built on wooden piles, but soon large quantities of slag were being used to reclaim the marsh. This reached 15 to 20 feet in depth allowing the whole area to be developed so that it became a tangled web of blast furnaces, mills, railway lines and slag heaps under a pall of smoke and steam. In its heyday, it must have been an impressive, if confusing, spectacle and difficult to distinguish between the various plants and whether the railway lines belonged to several private owners, the NER or were classified as a MOR.

To give some structure, this was the position in 1895:

- Old Town Branch. The branch and its associated sidings were owned by the NER, together with the Middlesbrough Town Goods Station and Mineral Depot.
- On the north west side at the end of the Old Town Branch, lines led to Connal's Wharf, Acklam Iron Works, Linthorpe Iron Works and Tees Side Iron Works.
- A MOR line left the Old Town Branch and ran parallel to the river along Vulcan Street (it was just a narrow strip of land at this time) to reach Middlesbrough Dock and Cargo Fleet.
- The Old Town Branch Loop was a MOR line connecting with the NER's Marsh Branch. It also gave access to a number of private sidings including the Acklam Foundry, Lowood's brick and gannister works, Newport Wire Works, the Wellington Foundry and Ayrton Rolling Mills.

Part of the Ironmasters district at Middlesbrough when it was dominated by iron and steel works. Britannia open hearth plant and mills occupy most of the view with the Cleveland Wire Works to the left and Gjers Mills' blast furnaces behind all bordered by the River Tees.

- The short NER Marsh Branch headed into the central part of Ironmasters to serve the Newport Iron Works, Newport Rolling Mills, Marsh Wire Works, West Marsh Iron Works, Cleveland Wire Works, Britannia Iron Works, Ayresome Iron Works and North Eastern Steel Works.

Eventually some of these works came into the ownership of Dorman Long and, even as late as 1959, *A Technical Survey of Dorman Long (Steel) Developments* summarises the position on railways within the Ironmasters District. "Expansion is impossible and rearrangement difficult; moreover, the constituent parts were in the past in separate ownership and, therefore, had their own rail connections and working arrangements. Add to this the rapid development of the area in its early years, and it will be appreciated that the traffic working is now complex and probably impossible to disentangle short of a complete reconstruction." It went on to say MOR agreements for fees were still "regularly exercised for inter-works movement and for the movement of export traffic to the river wharfs and to Middlesbrough Dock." Road access to the area was poor and severely limited by weight or height restrictions on bridges.

THE MIDDLESBROUGH OWNERS RAILWAYS (MOR)

When the SDR obtained Parliamentary approval to extend its line from Stockton to new coal staithes at Port Darlington in Middlesbrough, Joseph Pease and his partners recognised the resulting development opportunities. They formed a company called The Owners of the Middlesbrough Estate (OME) and purchased 520 acres from a William Chilton of Billingham for £30,000 in 5/1830. The OME began to lay out a grid of new residential streets and encouraged the development of industry near the River Tees. By about 1844, the Pease family had acquired the whole interest in the OME and the latter was to occupy a prominent role in many of the negotiations concerning the use of land. In addition to having direct responsibility for two locomotive-worked industrial railways, the OME also had sections of lines connecting some of the works' sidings with the public railway. This allowed it to charge a wayleave toll for traffic passing over its tracks.

An agreement was signed on 5/9/1884 in which four members of the Pease family sold various parcels of land, including the above lines, to the NER for £125,000. The NER became responsible for the maintenance of the lines, but was allowed to receive the income from the tolls, although the rate remained fixed at the date of transfer for many years. After 1919, it was decided that there could be periodic increases. The principal works included in the £3,875 receipt from tolls in 1933 were as follows: –

Company	Works	£
Pease & Partners Ltd	Normanby Iron Works	389
"	Tees Iron Works	43
Cochranes (M'bro) Ltd	Ormesby Iron Works	99
Dorman, Long & Co Ltd	Middlesbrough Iron Works	62
"	Cleveland Salt Works	33
"	Ayrton Rolling Mills	54
"	Acklam Steel Works	1,583
"	Britannia Works	537
"	Newport Iron Works	156
"	Cleveland Wire Works	104
"	No.3 Mill	57
"	No.4 Mill	250
Tyne Tees S. S. Co	Packet Wharf	281
J. & R. Ritchie	Acklam Foundry	56
W. Shaw & Co	Wellington Foundry	24
R. Hill	Newport Wire Works	22
"	Marsh Wire Works	17
Sundries 32 other companies		108
		Source: LNER

Three businesses were still making payments to BR as late as 1982. Despite the changes of ownership, these lines are referred to as the "Middlesbrough Owners Railway" (MOR) throughout the book. In view of the extensive use of companies' private locomotives over the MOR and the NER lines, the railway company insisted that they must comply with the accompanying set of instructions.

Reference: 'The Middlesbrough Owners Railway: An Abstruse Topic', Cliff Shepherd, *North Eastern Express* 177, NERA, 2005, pp.12-20

NORTH EASTERN RAILWAY

INSTRUCTIONS as to the Working of Private Owners' Locomotive Engines over the late Middlesbro' Owners' Railway and the North Eastern Railway where the latter connects one portion with another of the former, and is allowed by the North Eastern Railway Company (hereinafter called " the Company ") to be worked over.

1st JANUARY, 1902

1. Engines with Wagons or Bogies attached thereto must not work on or over the Middlesbro' Dock Lock Bridge, or travel on or over the Line between the South entrance to the Timber Yard of the Owners of the Middlesbro' Estate, Limited, Cargo Fleet and that Bridge, unless the Wagons or Bogies are in the rear of the Engines.

Except as stated above, a Train of laden or empty Vehicles may, when necessary, be propelled on the Lines, provided a competent man (employed by the Traders) proceeds in front of the Train to signal the Driver.

2. An Engine or a loaded or empty Wagon or Bogie must not work on or over the Dock Lock Bridge, if the weight of the Engine or Wagon or Bogie exceeds one ton per lineal foot of the Vehicle.

3. When Engines work Wagons or Bogies over the Line between the South entrance to the Timber Yard of the Owners of the Middlesbro' Estate, Limited, Cargo Fleet and the Dock Lock Bridge, there must be a man (provided by the Traders) in the rear of the Train during the day.

4. No Engine must be allowed to work over the Lines, until it is certified by the Company's Locomotive Superintendent as fit to travel over them, and the man or men in charge of working on the Engine must be certified by him as suitable, and provided by him with a copy of the Company's Rules and Regulations, to which the man or men must work.

5. No Wagon or Bogie must be worked over the Lines until it is certified by the Company's Locomotive Superintendent as suitable.

Each Wagon or Bogie must have plainly indicated thereon the name of the Owner and its number, tare and registered weight-carrying capacity, and bear the Company's register plate.

Traffic must be loaded on Wagons or Bogies in such a manner as to prevent the falling off of the load or any portion of it during transit, and no Vehicle must be loaded beyond the Company's standard gauge without in each case first obtaining the authority of their Goods Agent in whose district the place of forwarding is situated.

No Wagon or Bogie must be loaded beyond its registered weight-carrying capacity. See Instruction No. 2.

6. The Company reserve the right to stop, for the purpose of inspecting it, any Train, Engine, Wagon or Bogie, at any point on the Lines clear of the Dock Lock Bridge or Public Level Crossings.

7. During night time Engines must carry one or more suitable distinctive head lights, as may be required by the Company, and Trains or Light Engines must have a red light on the rear thereof, the lights to be provided by the Traders, subject to the approval of the Company.

8. The speed of Engines and Trains must not exceed six miles per hour, which must be reduced to not more than three miles per hour when working over Public Level Crossings and the Dock Lock Bridge.

9. Traders and their Servants are subject to the control of the Company's Officials when working over the Lines.

10. Where the Lines pass through or adjacent to the premises of Traders, no materials or structure, temporary or otherwise, must be placed within four feet six inches of the Rails or on the Company's ground, without the permission of the Company.

PHILIP BURTT

York 1st November 1901

General Traffic Manager

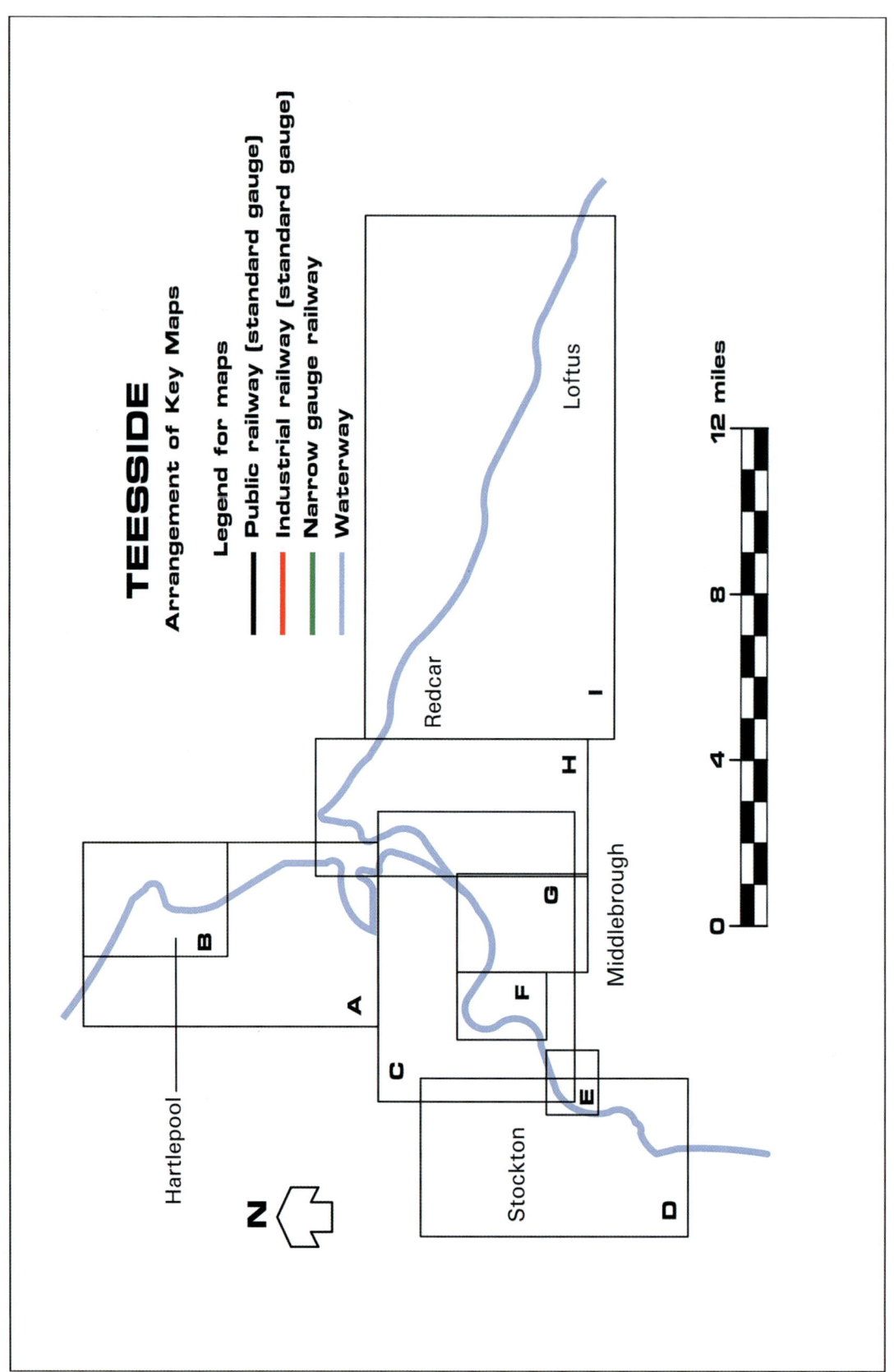

Page 12 – Key Maps

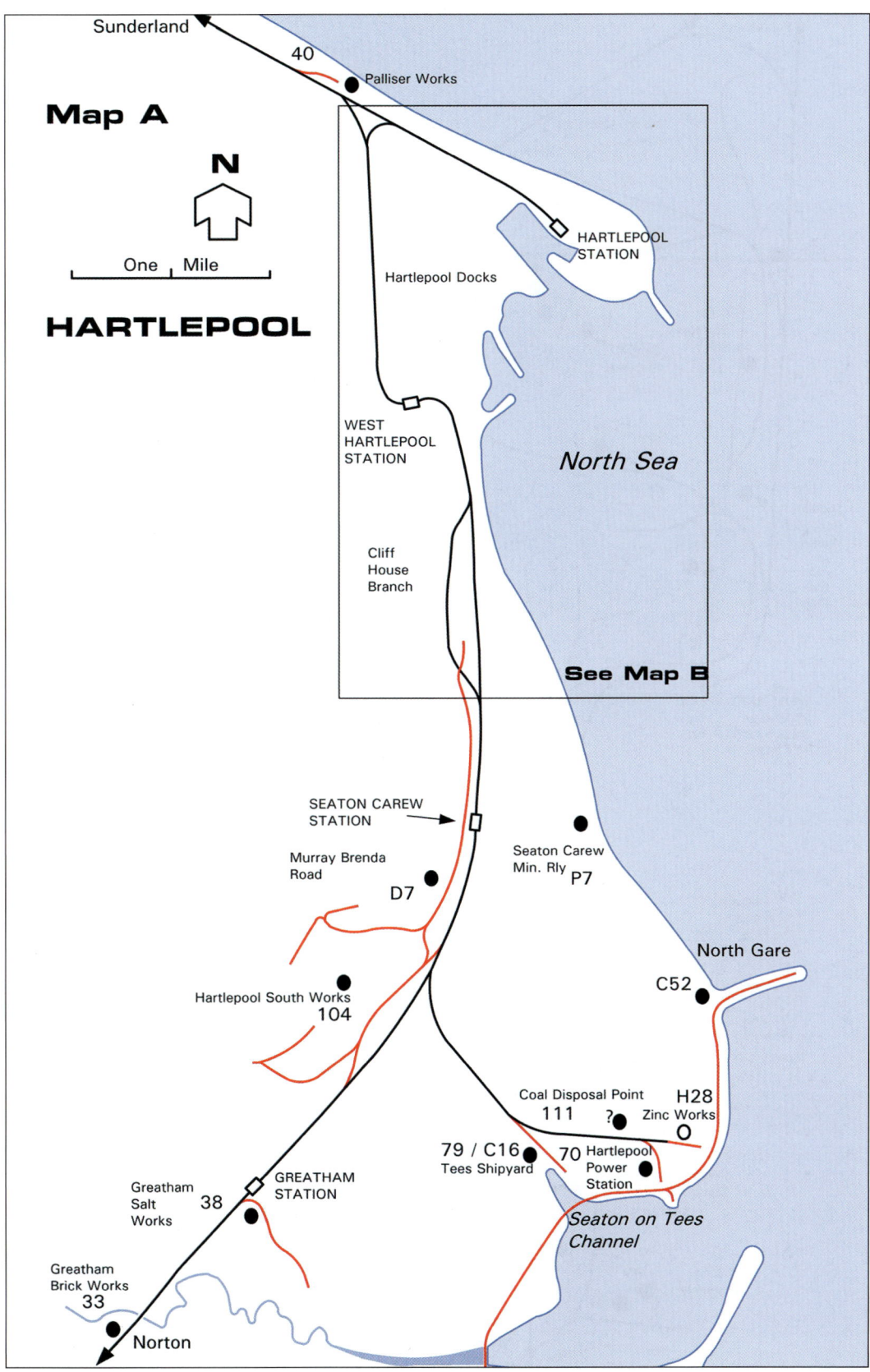

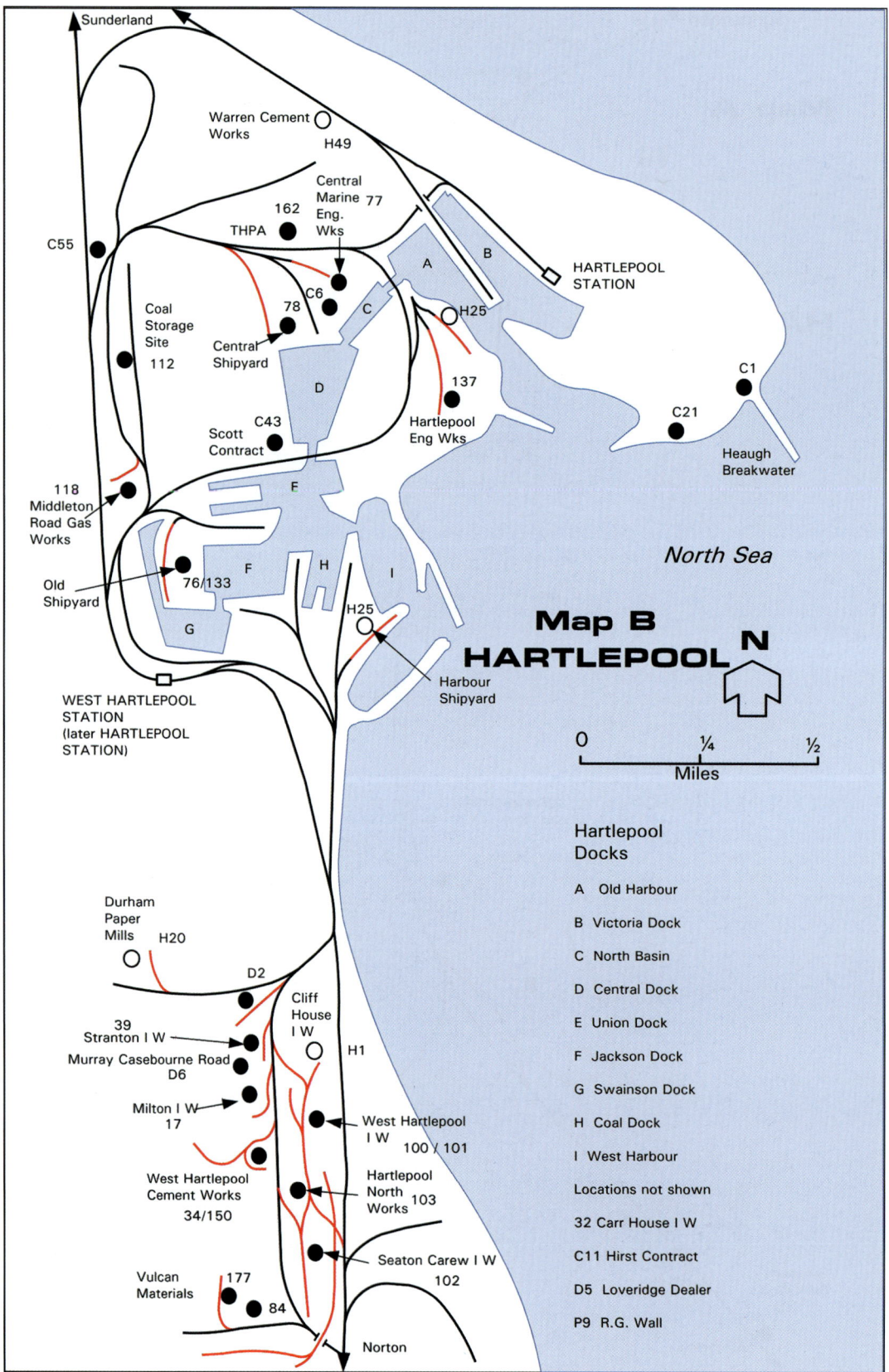

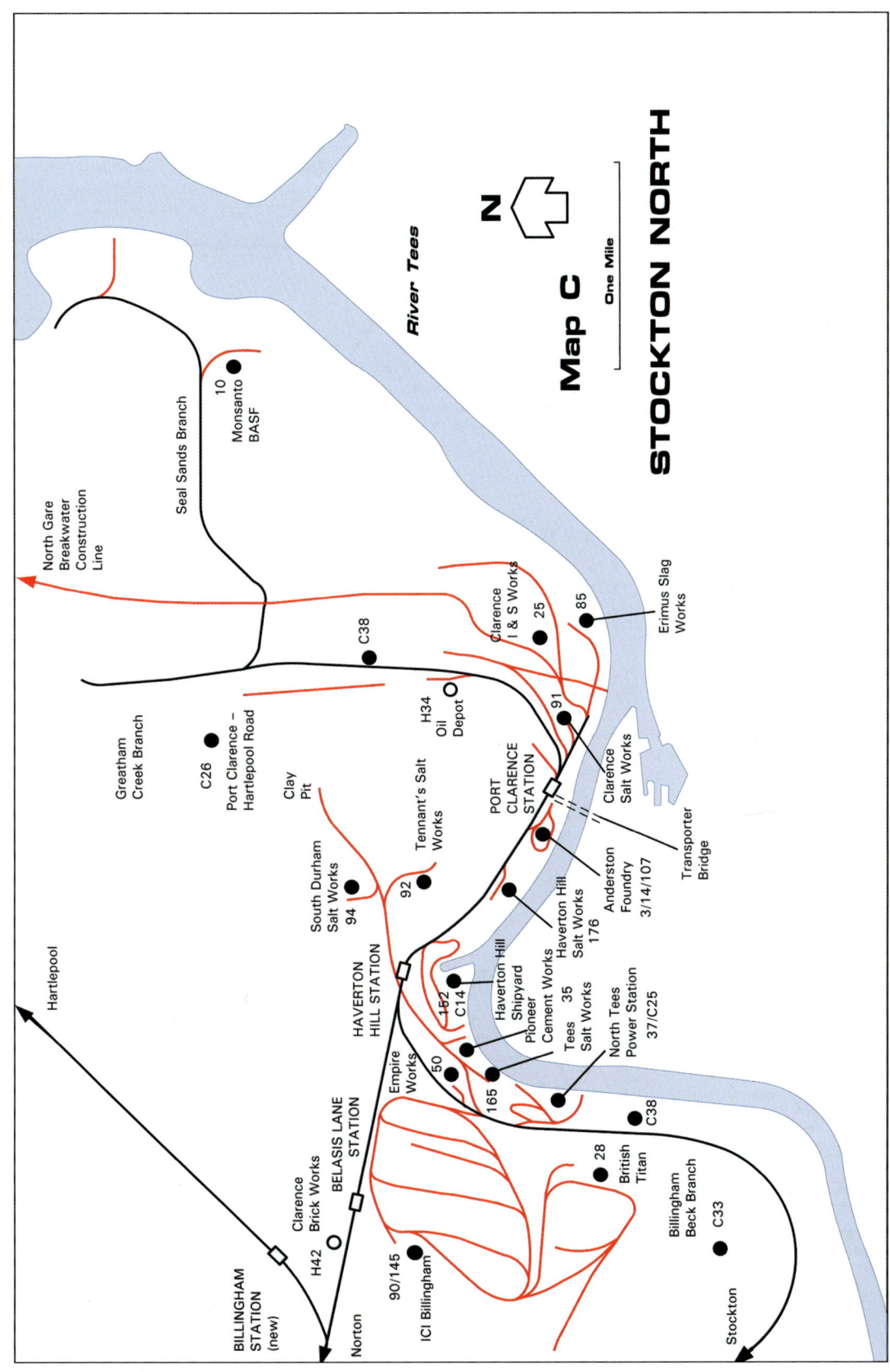

Page 15 – Key Maps

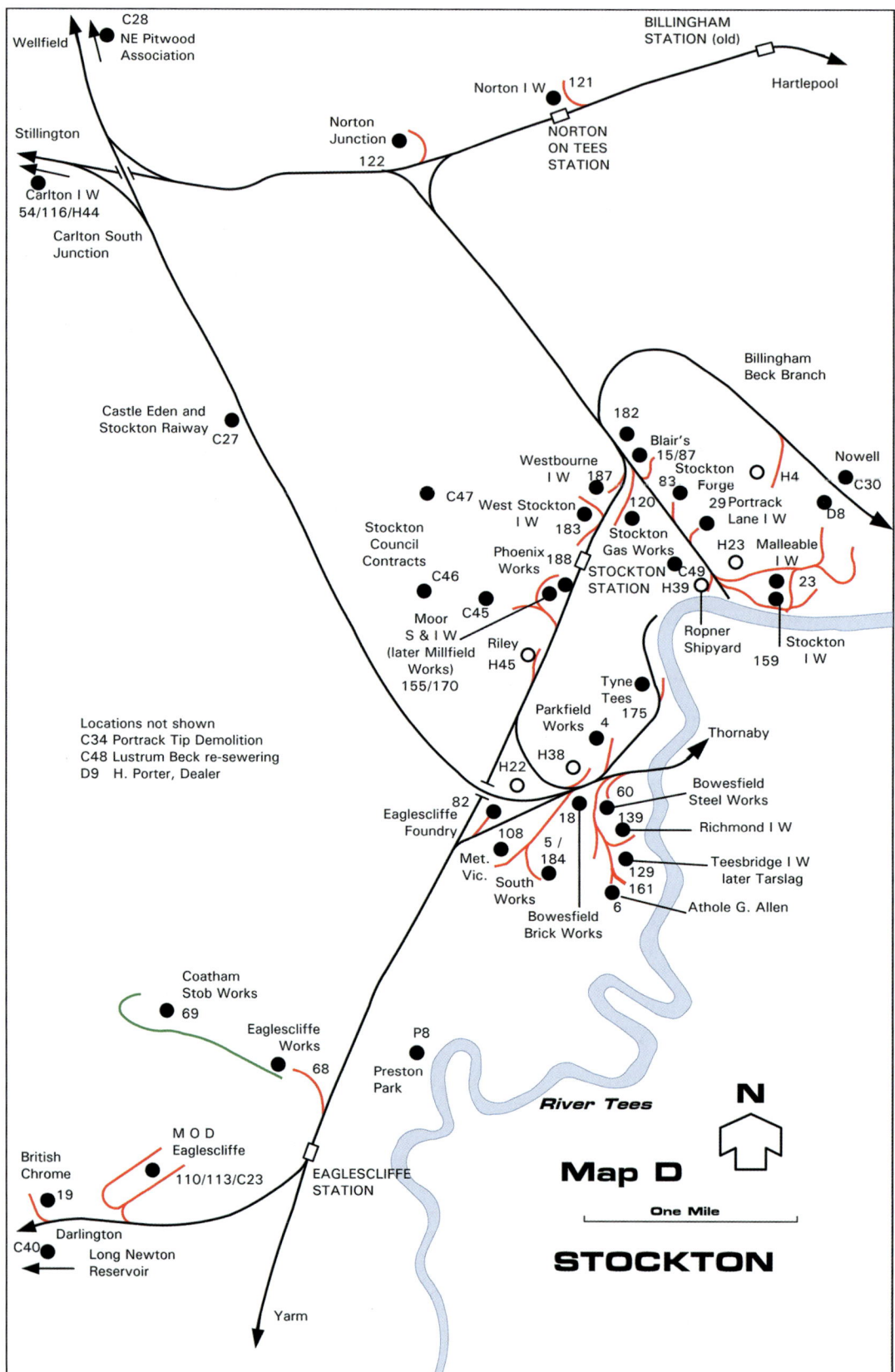

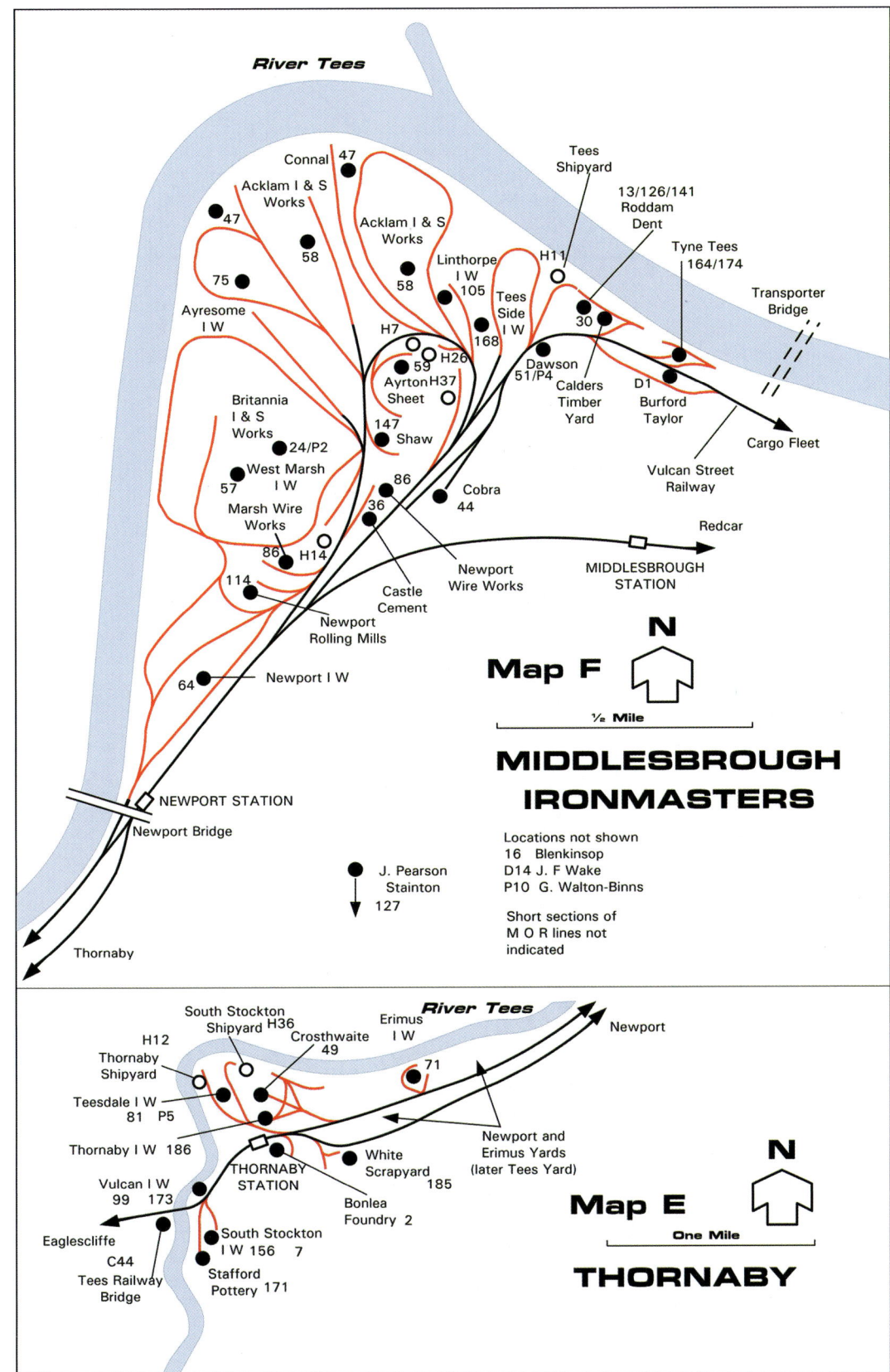

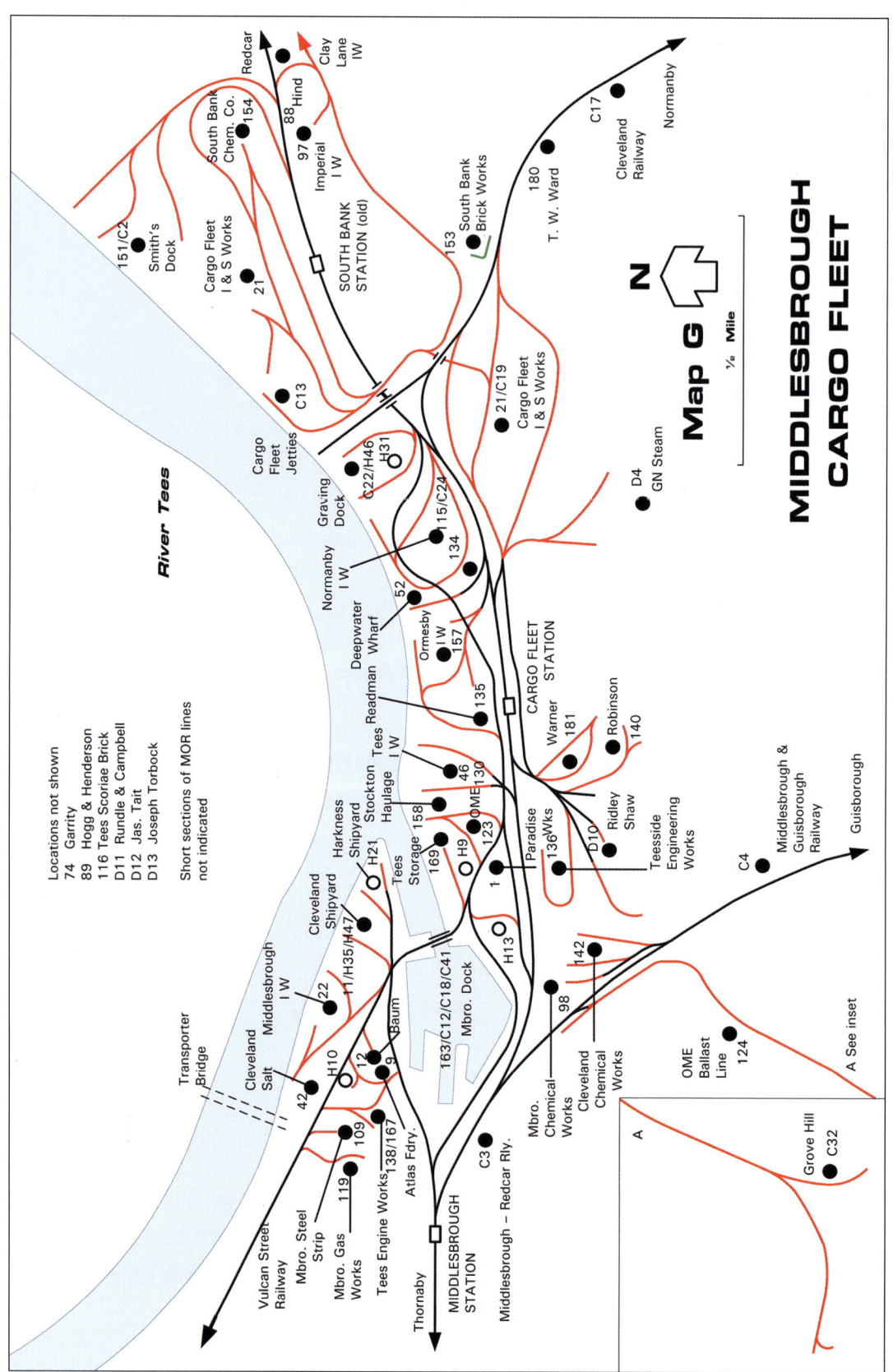

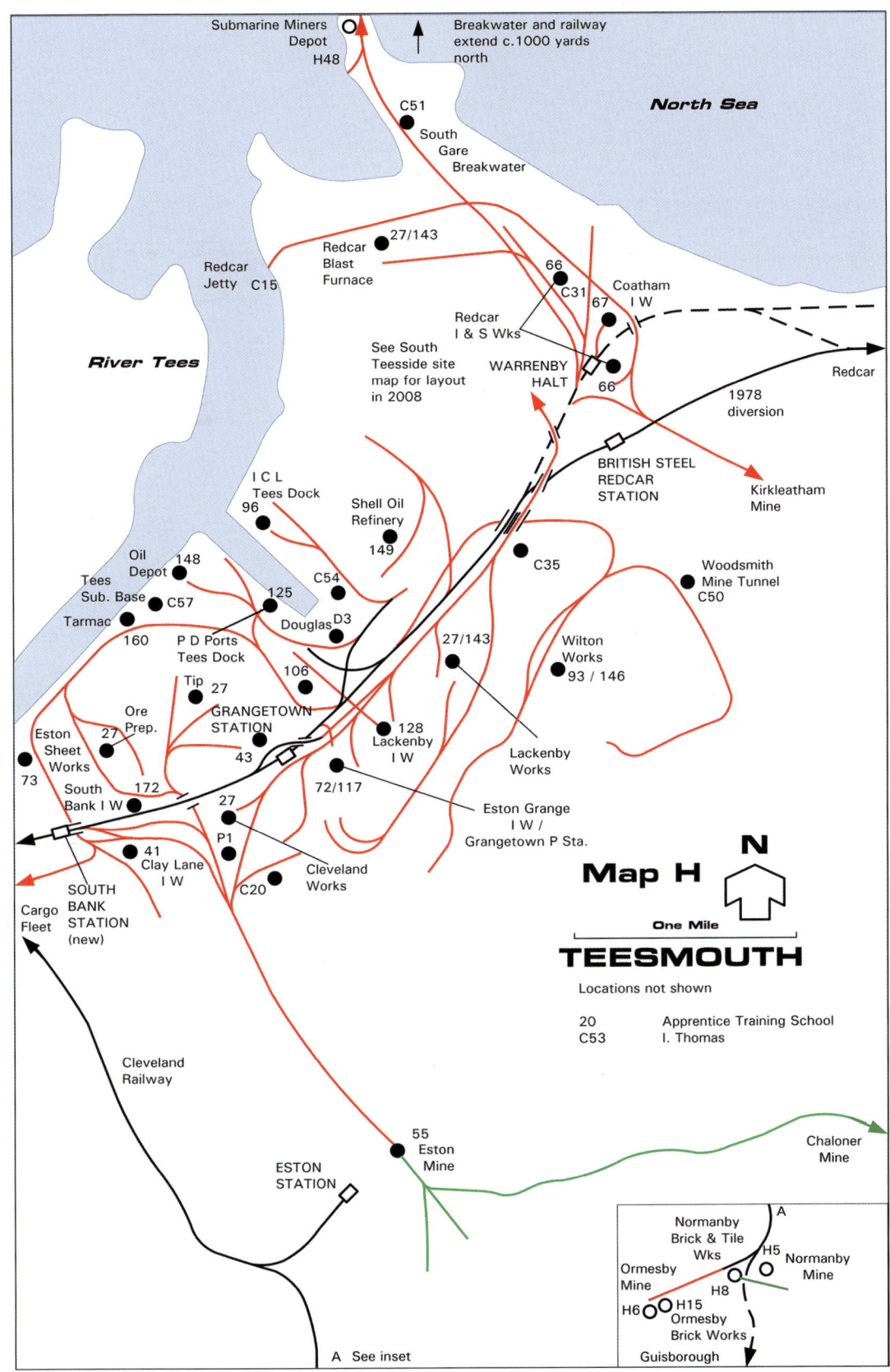

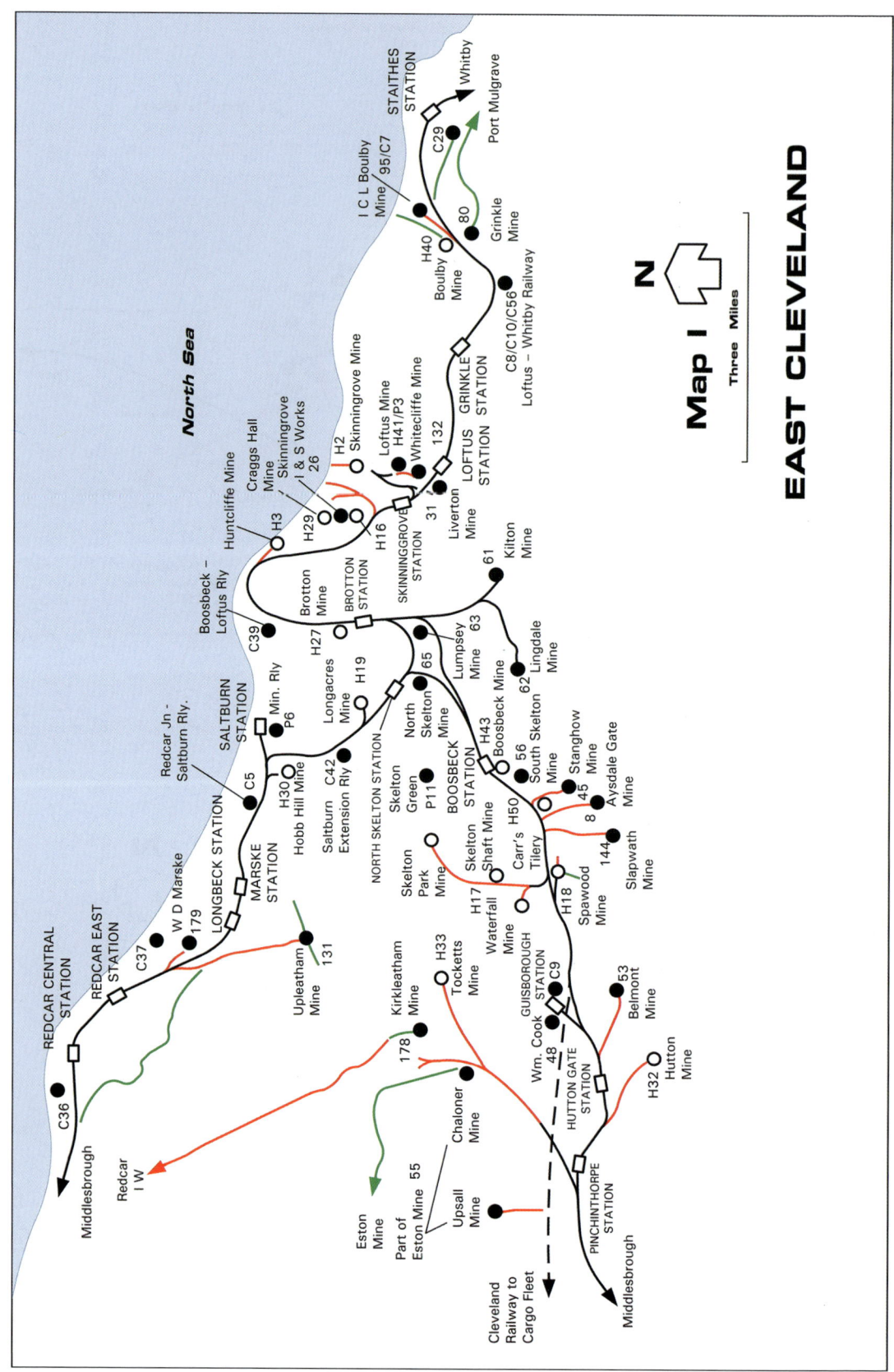

LOCOMOTIVE WORKED SITES

ACCREDITED PROCESSED METALS LTD

Paradise Works, Cargo Fleet   **1**
M. Henderson Clark Ltd until /1977   NZ 508205 – **Map G**

The scrapyard was established on the north side of the Middlesbrough-Redcar railway in the angle between it and the MOR line to Dock Lock Bridge. This area had once been occupied by Warner's and Calder & Co's premises before they were replaced by sidings and buildings associated with the NER's Middlesbrough Dock. The site was leased to William Blenkinsop and Co and an agreement for a new siding into the yard was signed with the NER on 25/6/1909. The lease was then taken on by M. Henderson Clark Ltd beginning on 11/10/1938, although the LNER Estates Department only realised in 1941 that it had forgotten to include provision for paying MOR tolls. It was reported on 17/12/1947 that rail traffic was amounting to 3,000–4,000 tons a month. The omission concerning the tolls was remedied when the site agreement was renewed in 10/1948. A siding reached the site off the MOR line shortly after the latter crossed the railway track into the OME's Cargo Fleet timber yard. On entering Henderson Clark's premises, the sidings split into a number of lines - one equipped with a weighbridge and running alongside a travelling crane gantry, another serving the centre of the yard with two fixed cranes and a shorter line entering what may have been an engine shed. Two sidings at the yard were designated for the LNER to deliver and collect wagons but Henderson Clark was responsible for any shunting. It is not known which locomotives shunted the scrapyard in its early days, but a new Ruston & Hornsby diesel was purchased in 1950 and various other diesels were used before going for scrap. A few industrial steam locomotives have also been scrapped here. A visitor in 10/1974 reported that the working locomotives RH 279599 and RH 411318 were used in alternate weeks and kept in a small engine shed. Accredited Processed Metals Ltd was established in 1977 by the owners of T.J. Thomson & Son Ltd of Stockton to specialise in buying and selling non-ferrous metal scrap and it took over Henderson Clark. The last tolls paid by Henderson Clark were in 1976 and a small amount in 1977. BR commented that the company had ceased using the siding from 26/3/1977. According to reports on 26/7/1979, the remaining two Rustons were not used and kept in the small shed. APM Ltd moved to a new site adjoining Thomson's Millfield Works in Stockton during 1981.

According to the LNER, M. Henderson Clark also leased, from 23/9/1929, a small site (NZ 453185) on the north side of Thornaby Station next to the sidings leading to Head Wrightson's Teesdale Works. Messrs Pettigrew & Co had been an earlier tenant and, in 1915, it was shown as an iron foundry. Henderson Clark operated it as the Victoria Works scrapyard certainly up to the end of the 1960s. RH 279599 was ordered from the Victoria Yard but despatched to the Cargo Fleet site.

Gauge : 4ft 8½in

	name	type	builder	works no	date	origin	disposal
	(PETER)	4wDM	RH	279599	1950	New	s/s c/1981
-		4wDM	RH	187071	1937	(a)	Scr c/1977
-		4wDM	RH	279593	1949	(a)	Scr c/1977
	(RICHARD PEASE)	4wDM	RH	411318	1957	(b)	s/s c/1981
		4wD	FH			(c)	s/s c/1971
2		4wDM	RH	321735	1952	(d)	(1)

(a)   ex North Eastern Iron Refining Co Ltd, Stillington, 11/1968.
(b)   ex OME Ltd, Cargo Fleet Timber Yard, c/1969.
(c)   origin and identity unknown.
(d)   ex Head Wrightson & Co Ltd, Thornaby, on hire, c/1970.
(1)   to Head Wrightson & Co Ltd, Thornaby, ex hire, c/1973.

THOMAS ALLAN & SONS LTD

BONLEA FOUNDRY, Thornaby   **2**
Merged with Allied Ironfounders Ltd /1964   NZ 454185 – **Map E**
(Subsidiary of **Federated Foundries Ltd**, from 9/12/1935)
Thomas Allan & Sons until 30/11/1900

As a consequence of the construction boom in the later nineteenth century, there was a huge demand for ironwork. Thomas Allan had established a works at Springbank, Glasgow in 1848 and went on to build the foundry at Thornaby in 1872; it was named after the former Bonlea Farm on the site. Many of its 200 workers came from Scotland and Allan & Sons built terrace houses for them, Glasgow Street becoming known as

"Scotch Row". The rail connection left the NER running line just to the east of Thornaby Station and extended round the foundry site forming an oval shaped layout with individual sidings serving the various buildings. The foundry produced cast iron gas, water, steam and drainage pipes, together with rainwater goods and stable fittings. From the early 1950s, it concentrated on manufacturing pressure pipes and engineering castings. In 1968 it formed part of the Flanged Pipe Division of Allied Ironfounders, but closed on 30/1/1970 when the latter withdrew from Thornaby. All the railway tracks had been removed from the site by 8/1967.

Gauge : 4ft 8½in

-		0-4-0VBT		Chaplin	1585	1873	New	s/s
No.1	TOM	0-4-0ST	OC	MW	591	1877	New	s/s
	BONLEA No.2	0-4-0ST	OC	MW	1075	1888	(a)	s/s
	BONLEA No.2	0-4-0ST	OC	P	463	1887		
		Rebuilt		Ridley Shaw		1931	(b)	
		Rebuilt		Ridley Shaw		1944		Scr c/1959
No.26	BONLEA No.3	0-4-0T*	OC	HL	2415	1899	(c)	Scr 11/1957
-		4wDM		RH	417894	1959	New	(1)

* formerly 0-4-0CT

(a)   ex Locke & Co (Newland) Ltd, Kippax Colliery, Kippax, Yorks (WR), /1903.
(b)   ex Smith's Dock Co Ltd, South Bank, /1931.
(c)   ex English Steel Corporation Ltd, Manchester, Lancashire. To Walker MacFarlane & Co Ltd, Glasgow, /1950; returned by 3/1951. This was another subsidiary of Federated Foundries Ltd.

(1)   to Warner & Co Ltd, Cargo Fleet c/1969.

ANDERSTON FOUNDRY CO LTD

PORT CLARENCE FOUNDRY, Port Clarence 3
NZ 499216 – **Map C**

The business was initially based in Glasgow and had established a foundry on a constricted site at Anderston in 1869–71. However, in 1875 it purchased a tract of reclaimed mudflats on the north bank of the Tees at Port Clarence. This site offered space for expansion and was in close proximity to the technically advanced Teesside iron works and their abundant raw materials (iron formed 70% of the company's production costs). The works specialised in producing equipment for railway permanent way, excluding signalling. To begin with, this comprised iron sleepers and rail chairs, but it acquired additional land between 1884 and 1894 and expanded to make tiebars, points and crossings, steel sleepers and rail fittings. From the 1930s, it also manufactured tunnel segments for London Underground.

The premises were located on the west side of the ferry crossing (later Transporter Bridge) to Middlesbrough. The NER Port Clarence Branch ran along the northern boundary of the works and a siding left this line, next to Port Clarence Engine Shed, to serve the foundry buildings. The western half of the site was occupied by approximately fifteen parallel sidings, semi-circular in alignment and presumably these were the stockyards. Two other sidings ran across a short causeway over the mudflats to the company's wharf. The latter was extended in 1897 and a second causeway was constructed so that locomotives could enter and leave without reversal by 1913. The wharf was stated to be 300 feet long with four 3 ton cranes in 1926. Ninety per cent of the output was shipped out from the wharf or Middlesbrough Dock in 1913 particularly to the British colonies overseas. Its single road engine shed stood next to the approach road to the Transporter Bridge.

After the 1880s, the company was content to rely on its established contacts and products, while using its reserves to weather economic downturns. Although sales of the Port Clarence foundry were approximately double that of the Glasgow plant by 1884/85, overall control of the company remained in Scotland to the 1930s. There was little thought given to the overall planning of the site so that temporary buildings were thrown up and tended to become permanent. As the reference states "it carried the attitudes and practices of the 1890s into the 1950s". In 1885 the works was reported to have three steam locomotives and four steam cranes and it is symptomatic that it mostly relied on four engines purchased in the 1870–80s from Scottish builders for many years. Some limited development expenditure was sanctioned after World War 2, including the purchase of a new diesel locomotive, but it was too little too late. Railways were no longer an expanding market; it was difficult to compete in former colonial countries and it was unable to manufacture larger products to the required tolerances. Following financial losses in 1961/62, the foundry closed in 5/1962 and sales of stock ceased a few months later. AB 654 of 1889 was noted still in light steam on 12/10/1962. F.R. Evans (Leeds) Ltd acquired the site for its industrial property value (it was later used by the offshore oil industry) and the fixtures were sold by auction on 11–13/6/1963. The tracks on the wharf had gone by 1963 but most of the other sidings on the rest of the site remained in 1967.

Reference : *Competition and Collusion in the British Railway Track Fittings Industry: the Case of Anderston Foundry 1800–1960,* J.F. Hargreaves, Doctoral Thesis Durham University, 1991 (HTTP://ethesis.dur.ac.uk/1984/)

Gauge : 4ft 8½in

No.1		0-4-0ST	OC	B	226	1876	New	(2)
2		0-4-0ST	OC	B	283	1881	New	(4)
No.3		0-4-0ST	OC	B	305	1883	New	(3)
3	(formerly 5)	0-4-0ST	OC	AB	654	1889	New	(7)
No.4		0-4-0ST	OC	BH	306	1875		
		Rebuilt	Ridley Shaw				(a)	(1)
4		4wDM		RH	235514	1945	New	(8)
2	(formerly 3)	0-4-0ST	OC	P	949	1902	(b)	(5)
1		4wDM		RH	305323	1951	New	(6)

(a) ex Hartlepool Gas & Water Co Ltd, c/1947. Ted Haigh was told on 17/11/1948 that it was bought secondhand from the Water Board.
(b) ex T.J. Thomson & Son Ltd, Stockton, per Thos. W. Ward Ltd, 12/4/1948.
(1) ILS has scrapped 31/10/1930, but see footnote (a)
(2) ILS has scrapped 31/10/1930, although Ted Haigh claims to have seen it at the Port Clarence Foundry on 24/5/1947 from the road but he could have been mistaken.
(3) scrapped 1/1946, according to Ted Haigh reporting in 1948.
(4) scrapped /1954, after 16/4/1954.
(5) scrapped c/1961, after 24/6/1961.
(6) to Slater, Pickering, c1/1963; then to Forcett Limestone Co Ltd, Yorkshire (NR), c5/1963.
(7) to J. Hanratty & Co Ltd, scrap merchants, former Browney Colliery Site, Co. Durham, per W. Basey, Langley Moor, Co. Durham, by 21/9/1963.
(8) to Wm Robinson Ltd, Cargo Fleet, per Tomlinson, Hall & Co, Stockton, /1963, after 12/6/1963.

Peckett 949 of 1902 stands next to some steam rail cranes at the Anderston Foundry, Port Clarence on 21st May 1960. *(IRS John Hill photograph)*

ASHMORE, BENSON, PEASE & CO LTD

PARKFIELD WORKS, Stockton   4
(Subsidiary of **The Power-Gas Corporation Ltd** from 12/11/1901)
Ashmore & While until /1885   NZ 444181 NZ 441178 – **Map D**

In 1871 William Ashmore established the Hope Iron Works on the east side of Bowesfield Lane, north of the Darlington–Middlesbrough railway. He was joined by his brother-in-law, S. While and the company became Ashmore and While in 1876. Subsequently R.S. Benson and Edward Lloyd Pease were included in the business and, when S. While retired, it was incorporated as a limited company in 1885 under the name Ashmore, Benson, Pease with William Whitwell as chairman. The works was also extended and renamed Parkfield (NZ 444181). It constructed railway bridges, vertical boilers and wagon hoists, but specialised in the production of gas holders and gas plant; E.L. Pease had invented the telescopic gas holder in 1888.

The company's railway left the Stockton Goods Branch soon after the latter diverged from the main line at Bowesfield Junction. After passing over a weighbridge, the sidings served the main buildings on the west and east sides of the then site in 1899. Three sidings ran to the north end of the premises with wagon turntables connecting with a line across the site. It seems that the company relied on horses or rail cranes for shunting because no locomotive is known until a new Hawthorn Leslie crane tank was purchased in 1902. There was also a narrow gauge railway system equipped with wagon turntables on the western half of the site, including two large complete circles of track; the purpose of which is not known.

On 16/7/1901, Dr Ludwig Mond launched The Power-Gas Corporation to develop his gas ammonia recovery process. He wanted a works to manufacture plant for the growing chemical industry and purchased Ashmore, Benson, Pease & Co, which had suitable equipment and experience of making older type gas plant. This resulted in increased business for the Parkfield Works and, in 1928, it expanded its site to 23 acres by acquiring two premises on the opposite west side of Bowesfield Lane (NZ 441178). These were Robert Roger & Co's Bowesfield Foundry and the Imperial Boiler Works of the late Thomas Sudron & Co Ltd. They already had short sidings off the main line at Bowesfield Junction. A plant was soon installed on the former Imperial site to fit lead linings into tanks and pipes.

Ashmore, Benson, Pease continued to be successful due to the diversity of its products, especially the design and construction of gas producing plant, storage vessels, blast furnaces and ancillary equipment. It built the largest water gas plant in the world at that time for Synthetic Ammonia & Nitrates Ltd at Billingham. Collaboration with the Freyn Engineering Co of Chicago gave it access to the most efficient blast furnace technology. It also sought new processes for corrosion and acid-resistant steels. In 1951 Ashmore, Benson, Pease was converted into a partnership with Power-Gas to simplify accounting procedures. The Parkfield Works was split by Bowesfield Lane and hemmed in by houses and railway lines so, in that year, a new factory called the South Works was constructed to the south of the main line. Ashmore, Benson, Pease often co-operated with Davy & United Engineering Co Ltd of Sheffield on projects and, in 1960, the Davy Ashmore group was formed by Power-Gas and Davy. Also in 1960 the Parkfield Works closed, some of the premises being taken over by Parkfield Foundries (Tees-side) Ltd in 1961. The latter finally finished in 10/1996.

References :   *A Short History of Ashmore, Benson, Pease and Co Ltd*, Manuscript in Middlesbrough Central Library, c/1950

Teesside's Economic Heritage, G.A. North, Cleveland County Council, 1975

Gauge : 4ft 8½in

	PARKFIELD	0-4-0CT	OC	HL	2536	1902	New	Scr /1930
533		0-4-0ST	OC	AB	823	1898	(a)	s/s c/1959
258	No 3	0-4-0ST	OC	AE	1881	1921	(b)	s/s /1960
1492	(MONARCH)	0-4-0ST	OC	P	1210	1910	(c)	s/s /1960
328		0-4-0DM		RSHN	7900	1958	(d)	(1)

(a)   ex Richardsons, Westgarth & Co Ltd, Hartlepool, c/1922.
(b)   ex Thos. W. Ward Ltd, Grays, Essex by 3/1932, previously Bombay Improvement Trust, Bombay, India.
(c)   ex Whessoe Ltd, Darlington, Co. Durham, per Cox & Danks, dealers, by 10/5/1953. A photograph of the locomotive taken by J.T. Clewley on 6/6/1958 was stated to be at the South Works, but it was at Parkfield Works, out of use, on 22/5/1960.
(d)   new. There is some uncertainty whether it came here new or went direct to the South Works.

(1)   to South Works, /1960 (see above comment).

SOUTH WORKS, Stockton

5
NZ 436172 – **Map D**

The Parkfield Works had been extended on numerous occasions over its 80 years and the premises were congested and not suited to the manufacture of the current heavy, more complicated plant. In 1/1949, the decision was taken by Ashmore, Benson, Pease & Co Ltd/The Power-Gas Corporation Ltd to construct the South Works on a 100 acre site half a mile south of the Parkfield Works. The plans envisaged four main blocks of buildings comprising construction and machine shops, jobbing foundries and mechanical foundries all served by railway sidings that looped round and entered the buildings. In the event, only the eastern half of the new works was built and the foundries remained at Parkfield. The construction shop started production on 7/8/1951. The South Works was served by an extension of the Metropolitan-Vickers-Beyer Peacock Ltd's line from Bowesfield Junction, which curved round eastwards dividing into a number of sidings that entered the two main buildings. A small structure on a spur off one of the sidings could have been the engine shed. Towards the end of the 1960s, Ashmore, Benson, Pease withdrew from manufacturing and concentrated on design and contracting services. The South Works closed in 1967 but was taken over by Whessoe Ltd (which see).

Reference :   *The Golden Jubilee of the Power-Gas Corporation Ltd* booklet, 1951

Gauge : 4ft 8½in

| (328) | | 0-4-0DM | RSHN | 7900 | 1958 | (a) | (1) |

(a)   ex Parkfield Works, /1960 or new here in /1958.
(1)   to Whessoe Ltd, with works, /1967.

ATHOLE G. ALLEN (STOCKTON) LTD

BOWESFIELD LANE WORKS, Stockton
Athole G. Allen Ltd until /1934

6
NZ 444171 – **Map D**

At the end of the 19th century, a small cement works was located at the end of the branch that ran south from the Darlington-Middlesbrough railway at Bowesfield Junction. Identified as the South Durham Cement Works operated by Coultas & Co, the branch not only served the premises at the end of Adam Street, off Bowesfield Lane, but swung eastwards to reach a loading stage on the bank of the River Tees. However, by 1923, the site

Hopkins Gilkes 355/1874 had arrived at Athole G. Allen's chemical works in 1948 for use as a stationary boiler.
(Kevin Lane collection)

of the cement works had been covered with slag from Tees Bridge Iron Works and the branch now terminated at the Tees Bridge Brick Co's Works located on the south side of Adam Street. The brick works was probably operated by the Cleveland Magnesite & Refractory Co Ltd during World War I producing silica bricks from ganister, a sandstone rich in silica.

The Stockton-on-Tees Chemical Works was established on five acres of land, south of the Tees Bridge Brick Works, during World War I to produce TNT for the Government. After the war, orders for TNT fell and, despite attempts at diversification, Stockton-on-Tees Chemical Works Ltd closed down in 1927. In the following year, Athole G. Allen took over the works to resume the production of TNT, together with barium chloride, barium carbonate, ferric chloride, iron perchloride and other chemicals. With the approach of World War 2, the production of TNT became a priority. The factory expanded to 25 acres, it employed 500 people during the conflict and this no doubt explains the purchase of a new locomotive in 1939. A total of 86 million pounds of TNT was produced over 16 years.

Government contracts for TNT were cancelled at the end of the war and about 1948 the company concentrated on the processing of barytes and witherite (barium carbonate). Barytes was supplied from the company's own mine near Middleton-in-Teesdale, with supplementary supplies from New Brancepeth Colliery in County Durham. Large quantities of barytes aggregate was sent to hospitals and nuclear power stations for radiation shielding. The company also started to develop surplus land as an industrial estate. The branch from Bowesfield Junction continued to terminate at and serve the works, but IRS records indicate that the remaining rail traffic was worked by a locomotive from Dorman Long's Bowesfield Works. A report dated 8/9/1951 stated that the company had closed down and was for sale by Sanderson, Townsend & Gilbert with the still complete Hopkins Gilkes locomotive as Lot 14.

Gauge : 4ft 8½In

-	0-4-0ST	OC	RSHD	6974	1939	New	(1)
No.5	0-4-0ST	OC	HG	355	1874	(a)	s/s

(a) ex Highways Construction Ltd, Port Clarence, /1948 for use as a stationary boiler.

(1) to English Steel Corporation Ltd, Openshaw, Lancashire, via RSH, /1941. IRS records have a note that Tarslag Ltd would not let Allen use the locomotive over its lines so the engine was sold. Athole G. Allen had gained approval (dated 15/3/1940) for its locomotive to work over LNER sidings but this was terminated as from 28/2/1941.

A. BAINBRIDGE LTD

NORTH YORKSHIRE IRON WORKS, Thornaby

7
NZ 449179 – **Map E**

A. Bainbridge Ltd was established as an iron and steel merchant in 1876 with offices in the Exchange Buildings, Stockton. The *NER Plans Book of Collieries, Works & Sidings* dated 1895 shows a private siding described as "Bainbridge's Metal Yard" associated with the Bowesfield Iron Works so the company may have also become directly involved in processing scrap. A. Bainbridge, Stockton supplied wire rope and scrap to Warner's between 1904 and 1912. It was probably about this time that Bainbridge took over the site of the South Stockton Iron Works (NZ 449179); a siding agreement with the NER was dated 7/5/1909. By 1915, the iron works had been demolished but remnants of the sidings still ran across the site; one to the river, another to the slag tip and the third to the former pottery. The premises had also reverted to its old name, North Yorkshire Iron Works. The railway company delivered and collected rail traffic, but Bainbridge was permitted to use its rail crane for shunting on those sidings owned by the railway company, including that serving the adjacent coal depot.

Acting as a dealer, Bainbridge offered various secondhand locomotives for sale between at least 1884 and 1912 in fourteen separate advertisements. Among these were:

12 inch standard gauge '4-wheel' locomotive with 3ft 6½in wheels by Gilkes & Co (*Machinery Market*, 1/11/1884).

9½ in x 14in outside cylinder 2ft 1in gauge locomotive (*Machinery Market*, 1/1/1892).

12 inch 0-4-0 Hunslet 'tank' built in /1880 with 3ft 0in diameter wheels, 5ft 4in wheelbase and a working weight of 18 tons 7cwt. Overhauled by maker /1890. (*Machinery Market* 1/7/1892).

There are a few records of rolling stock scrapped by Bainbridge. Two 10 ton tank wagons Nos.5 and 6, belonging to the Eaglescliffe Chemical Co, first registered about 1893, were sold to Bainbridge on 18/10/1928. A. Bainbridge Ltd amalgamated with Thomas & Thomas of Middlesbrough in 1960, the resulting company becoming known as Thomas, Thomas & Bainbridge Ltd. The siding agreement with BR was terminated in 2/1966 and the company went into liquidation in 1968.

Gauge : 4ft 8½in

WARRINGTON		0-4-0ST	OC	MW	1027	1887		
		Rebuilt		MW		1907	(a)	s/s
-		4wPM		FH/KC			(b)	(2)
40		0-6-0ST	OC	RSHD	6948	1938		
		Rebuilt		CEW		3/1953	(c)	(1)

(a) ex Tyne-Tees Steam Shipping Co Ltd, Stockton Wharf, Stockton, for scrap, 4/1958.
(b) ex Tyne-Tees Steam Shipping Co Ltd, Stockton Wharf, Stockton, 5/1958; after 4/4/1958, by 1/6/1958. A photograph of it standing at the weighbridge suggests that it may have seen some use.
(c) ex Appleby-Frodingham Steel Co Ltd, Appleby-Frodingham Works, Frodingham, for scrap, /1959.

(1) s/s. It was in Bainbridge's yard on an isolated section of track on 22/5/1960.
(2) s/s. The partially dismantled locomotive was still in the yard on 2/5/1964.

W. BARNINGHAM & CO

AYSDALEGATE IRONSTONE MINE, near Slapewath   8
NZ 652148 – **Map I**

William Barningham was persuaded by Henry Pease to establish his new iron works at Albert Hill in Darlington. By 1864 Barningham, trading as the Darlington Iron Company, was producing 800 tons of wrought iron weekly. He was also looking to acquire his own supply of ironstone and purchased two farms totalling 272 acres at Aysdalegate, Slapewath on 6/4/1864. Progress in establishing the mine was very slow and the shaft was reported in 3/1868 to be only 80 feet deep and had not yet reached the seam. Barningham sold his share of the ironworks in 1872 to the Darlington Iron Co Ltd when the business comprised the Albert Hill and nearby Springfield Works with 192 puddling furnaces and associated rolling mills. He appears to have kept his interest in the mine, but it was not until 1876 that he was seeking approval for a wayleave in order to construct a single track railway running north from the mine for just under a mile to join the Guisborough-Boosbeck line. Aysdalegate Junction Signal Box, 100 yards to the west, controlled the connection (authorised by the NER on 31/8/1876) as well as the separate line to Slapewath Mine. The *Guisborough Exchange* on 14/12/1876 reported that Aysdalegate Mine situated at the foot of Birk Brow hill on the present A171 road was about to open and that the railway was to be locomotive worked. There was only a simple headshunt and siding at the mine. The late 1870s was not a good time to start a new mine and it was no surprise when Aysdalegate Mine closed on 23/10/1880, supposedly because of the poor quality of the ironstone. William Barningham died in 1882. Later the ironstone was worked by the Spawood Mine, with the shafts at Aysdalegate being retained solely for ventilation. Some of the mine buildings were converted to cottages and remain today.

References :   *Guisborough District Mines* Simon Chapman, pub by Peter Tuffs, 2001

                'William Barningham, Railway Engineer', Trevor Lodge, *Industrial Railway Record* 219, IRS,12/2014

Gauge : 4ft 8½in

-	0-4-0WT	Pendleton*		1867	(a)	(1)

* Barningham's Pendleton Works, Manchester

(a) ex Darlington Iron Co Ltd, Albert Hill, Darlington, Co. Durham, c1876/77.
(b) purchased at the sale of plant [at Aysdalegate?] by T. Johnson, 11/1886. The locomotive broke down whilst in transit to Wigan.

BARTON ABRASIVES LTD

ATLAS FOUNDRY, Middlesbrough   9
Harrison Bros (England) Ltd until early 1950s
NZ 501208 – **Map G**

The company, which was founded in 1887 by the Harrison brothers of Massachusetts, USA, took in cast iron scrap and pig iron. This was melted in a cupola furnace (later two from the 1950s) with coke to provide the raw material for producing metallic abrasives in the form of round chilled shot and grit. These were used in the engineering trades, for stone sawing and in well boring for irrigation. The company was sometimes referred to as a "diamond grit manufacturer". The works was situated between Lower Commercial Street and Dock Street. Its offices, emblazoned by the company's name, faced on to Dock Street and the siding was taken off the railway leading from the Middlesbrough Dock lines into the Tees Engine Works site. The two-storey production buildings stretched from the Engine Works siding north to Lower Commercial Street. From the Vulcan Street railway, a single siding ran by Wilson, Copley & Co 's premises to cross Lower Commercial Street, passing through a doorway in the frontage to enter Harrison Bros' internal yard. Here a crossover provided access to a second siding and two rail cranes assisted with unloading the railway wagons.

Production was originally about 1,000 tons pa and this subsequently increased to about 10,000 tons with few additions to the plant. The company had hoped to revamp the works but World War 2 intervened and it was not until 1949 that extensions were planned, with production rising to 18,500 tons by 1955. In 1973 coke was coming mainly from the Derwenthaugh Coking Plant and it was stored in the Top Yard, at the cupolas and in railway wagons. Scrap came from local companies such as Thomson, Cohen, Prosser and Herring.

A handful of railway wagons would be delivered and taken away so shunting requirements were limited (e.g. the number of wagons received each day from 5 – 8/10th May 1948 was 4, 1, 6, 4, 12). Harrison Bros signed a new agreement with the British Transport Commission on 7/7/1950 which allowed its locomotive to run over BR's lines along Vulcan Street as far as North Street, including access to Middlesbrough Gas Works, and eastwards to the Dock Clock Tower where a reversal enabled it to travel alongside Dock Street and enter the former Tees Engine Works siding. There appears to have been a canopy on the south side of Harrison Bros works and it may be that loading out of metallic abrasives could take place here as well. The agreement, backdated to 1/5/1948, specified that one locomotive, with or without wagons could run along these lines. Harrison Bros 'Register of Rail Transport of Materials', 30th April 1949 onwards includes the comment "Own loco started shunting 1/5". It would therefore seem that the intended increase in production would result in more shunting by the company's locomotive in order to avoid hold-ups on the limited length of track within the works. The company was taken over by the Barton Group in the early 1950s after which it was known as Barton Abrasives Ltd.

By 1974 there was little regular railway traffic, although the company's locomotive was still used daily. Rail traffic ceased c1976 and the private siding agreement was terminated from 1/5/1977: "the firm have disposed of their locomotives and that they do not intend to forward or receive traffic by rail in the future."

References : Teesside Archives files U/BA/3/5, U/BA/3/3, U/BA/6/8

Gauge: 4ft 8½in

-	4wDM	RH	183764	1937	(a)	(1)
-	4wDM	RH	402808	1956	New	(2)
-	4wDM	RH	306088	1949	(b)	(3)

(a) ex Thos. W. Ward Ltd, Glasgow, /1948; previously Murex (M.A.P.), Mossend, Bellshill, Lanarkshire. Spares were ordered for the Atlas Foundry 11/6/1948 to 21/10/1955.
(b) ex Michael Baum & Co Ltd, Middlesborough, c11/1973
(1) to G.E. Simm (Machinery) Ltd, Sheffield, Yorkshire WR, /1957; resold to Central Wagon Co. Ltd, Ince, Lancashire, /1958. Harrison Bros had advertised a Ruston & Hornsby 44/46hp diesel locomotive for sale on 7/12/1956.
(2) to George Cohen, Sons & Co Ltd, Cargo Fleet, /1974.
(3) to George Cohen, Sons & Co Ltd, Cargo Fleet, c/1976.

BASF plc

CHEMICAL WORKS, Seal Sands, near Port Clarence   **10**
BASF Chemicals Ltd until /1989   NZ 535241 – **Map C**
Monsanto Textiles Ltd until 2/12/1985

Much of the Greatham Creek Branch had seen little, if any, traffic since its opening in 1901 and, by 18th June 1915, it had been reduced in length to about 2½ miles from Bells Bank Foot Junction. Some of the remaining section was finally brought into use to connect with Phillips-Imperial Petroleum Limited's Refinery Filling Terminal in 1965. Monsanto opened its plant at Seal Sands in 1970 to produce acrylonitrile for use in the manufacture of acrylic and polyamide fibres in clothing and carpets. Under an agreement between BR and Monsanto dated 22/7/1969, BR agreed "to construct a single line railway with run round loops" to connect the plant with the existing line on the Greatham Branch. Monsanto's works was located on the south side of this new Seal Sands Branch and had two deep water jetties near the mouth of the Tees. As more development took place at Seal Sands, the branch was extended past Monsanto to alongside the North Tees Refinery. Here there was a run round loop and a siding into Seal Sands Storage Ltd (agreement signed 14/10/1981).

From the connection with the Seal Sands Branch, a single line ran over the road into Monsanto's works. A siding, shortly after the crossing, passed through the small engine shed. The line continued on to serve a fan of sidings used for stabling and emptying/filling tank wagons, while one of these tracks crossed an internal

BASF's engine shed at Seal Sands on 2nd April 2001 with the company's shunter, EEV 3870/1969 standing outside.   (Cliff Shepherd photograph)

road to a concrete pad. The first timetabled train leaving the plant was BR's Class 37 6893 hauling four empty liquefied ammonia tank wagons to Shellstar, Thameshaven on 26/5/1970; previous movements during the plant's construction had run as specials. The siding agreement was not signed until 31/3/1972! The main duty of the Monsanto locomotive was to move tank wagons between the holding sidings and the filling racks. About six trains a week were being received in 1983. BASF (UK) Ltd and BASF Chemicals Ltd merged to form BASF plc in 1989. BASF* began to hire locomotives when the resident EEV 3870 required repairs and in 1997 it was borrowing from the EWS Thornaby Depot 09005 on 14th and 25th February, 09106 on 5th September and 08813 on 28th October. The last rail traffic was a daily working of highly toxic hydrocyanic acid tanks which ran with barrier wagons. It was one of the shortest traffic flows in the UK travelling just 5½ miles to Billingham but with the locomotive having to run round twice during the journey. In 1997, ICI investigated a safety case to use its own locomotives and a trial took place on 10/10/1997 when 08743 worked the train, but nothing came of it. Eventually a pipeline was opened to transfer the chemical and the final train ran on 25/4/2002. BASF's plant was sold to INEOS Nitrates in 2008.

* Badische Anilin und Soda Fabric

Gauge : 4ft 8½in

25-1 (GM 245)	0-6-0DH	EEV	3870	1969	New	(4)
004 (08785) (D3953) CLARENCE	0-6-0DE	Derby		1960	(a)	(1)
-	0-4-0DH	EEV	D1205	1967	(b)	(2)
No.57 689/167	0-6-0DH	RR	10214	1964	(c)	(3)

(a) ex R.F.S. Engineering Ltd, Doncaster, South Yorkshire, on hire, 9/8/1994.
(b) ex Creative Logistics, Salford, Manchester, property of Harry Needle Railroad Co Ltd, on hire, 8/9/2000.
(c) ex Harry Needle Railroad Co Ltd, Barrow Hill, Derbyshire, on hire, 19/2/2001.
(1) returned to R.F.S. Engineering Ltd, Doncaster, South Yorkshire, ex hire, 25/11/1994.
(2) returned to Creative Logistics Ltd, Salford, Manchester, property of Harry Needle Railroad Co Ltd, ex hire, 28/9/2000.
(3) to European Metal Recycling Ltd, Kingsbury, Warwickshire, property of Harry Needle Railroad Co Ltd, ex hire, 1/5/2001.
(4) to Weardale Railway Society, Wolsingham, Co. Durham, 7/11/2003.

MICHAEL BAUM & CO LTD
CLEVELAND DOCKYARD, Middlesbrough

NZ 504210 – **Map G**

After Raylton Dixon and then Tees Side Bridge & Engineering Works (which see) had vacated the Cleveland Dockyard, the name against the site was Tees Conservancy Commissioners (7/1/1948), followed by OME (30/7/1951). The 12/1951 survey shows no development or sidings on the former shipyard sites. A BR letter dated 15/2/1957 states that Baum had taken over the Cleveland Dockyard. It was generally used as a scrapyard although, during 9/1959, a quantity of Russian pig iron had been transported by rail from here by T.H. Donking on behalf of "B.I.S.C.". A 1960 advertisement for Baum stated they were "iron and steel stockholders, scrap merchants, dismantling and contractors and warehousemen" with a head office at the Cleveland Dockyard. Baum owned two diesel shunters during the 1960s and the 1973 Ordnance Survey plan shows a siding off the Vulcan Street railway running into the scrapyard and ending at a small building with another track serving the rest of the yard.

In 1964 Baum was also using part of the Deepwater Wharf site at Cargo Fleet to break up condemned wagons delivered to the location by BR. Forwarded scrap in wagons from Deepwater Wharf was subject to MOR tolls, as well as the usual carriage rates. Condemned wagons were also being delivered by BR via Vulcan Street to the Cleveland Dockyard and Baum's locomotive was hauling these via Dock Lock Bridge over the former OME line to Deepwater Wharf to be broken up there. This traffic was also subject to MOR tolls at the rate of "empty private wagons". Baum offered diesel locomotives for sale in *Machinery Market* as follows – RH 48DS of 1958 (8/7/1965), a standard gauge RH of 1958 (19/5/1966) and a standard gauge RH 88DS (27/2/1969).

Meanwhile Whessoe Ltd had leased three acres of land from the OME in 1956 between the Cleveland Dockyard and Dock Point. About 1972/73 Baum moved to a site at Commercial Street taking his two locomotives with him. Whessoe subsequently took over the Dockyard to give it, by 1975/76, a 31 acre site with a 2,000ft frontage to the River Tees and a new 100ft load out platform; this formed the base for its Offshore erection activities although these were not served by rail.

In the nature of scrapyards, there would have been other locomotives here for dismantling, in addition to those listed below. IRS records show a 2ft 0in gauge four–wheel Hunslet 2965 of 1944, ex War Department Barlow, here in 8/1964 and two unidentified Hibberd diesels, numbers 1 and 2, in 1/1965. Baum also had three BR Class J94 saddle tanks for cutting up in 8/1964.

Ruston & Hornsby 417891/1959 shunts amongst the remains of the scrapped wagons at Baum's Cleveland Dockyard premises on 8th February 1962. (*Courtesy David Webb*)

Gauge : 4ft 8½in

BEATTY	0-4-0ST	OC	KS	3107	1918	(a)	s/s after 28/7/1958
-	4wPM		FH	1675	1930	(b)	Scr c1/1965
-	4wDM		RH	417891	1959	New	(1)
-	4wDM		RH	306088	1949	(c)	(1)

(a) ex British Insulated Callender's Cables Ltd, Prescot, Lancashire, c/1957.
(b) ex Doncaster Corporation Highways Department, Doncaster, here by 4/11/1958.
(c) ex Rivet Bolt & Nut Co Ltd, Stobcross Rivet Works, Coatbridge, Lanarkshire, after 6/1964.

(1) to Michael Baum & Co Ltd, Commercial Street, Middlesbrough, c1972/73, by 5/6/1973.

COMMERCIAL STREET, Middlesbrough 12
NZ 502209 – **Map G**

This site was yet another in the row of premises on the south side of the Vulcan Street Railway. It was located to the east of Barton Abrasives Ltd's siding, had a couple of short connections linking it to the Vulcan Street railway and was previously a commercial foundry and engineering works owned by Wilson, Copley and Co. This business appears to have been established in the 1890s. Correspondence in the MOR records advises that it had gone by 23/10/1930 and was replaced by Messrs Harrison and Whitfield, which had purchased the site by 4/2/1946 and "private traffic may have begun to pass over the private siding connection". A note in Bob Payne's records states that he saw a four-wheel diesel, Ruston and Hornsby 402928, at Harrison Bros and Whitfield, Middlesbrough on 13/2/1961. Unfortunately a locomotive with this number does not exist. The site was then taken over by M. Baum & Co Ltd's scrapyard, which had moved from the Cleveland Dockyard by 5/6/1973. The stay was of short duration because, in 1974/75, Baum transferred his business to a new site adjoining E. Pearson's scrapyard at Depot Road, Middlesbrough.

Gauge : 4ft 8½in

-	4wDM	RH	306088	1949	(a)	(1)
-	4wDM	RH	417891	1959	(a)	(2)

(a) ex Michael Baum & Co Ltd, Cleveland Dockyard, Middlesbrough, c1972/73, by 5/6/1973.

(1) to Barton Abrasives Ltd, Middlesbrough, c11/1973.
(2) to Michael Baum & Co Ltd, Depot Road, Middlesbrough, c1974/75.

DEPOT ROAD, Middlesbrough 13
NZ 492214 – **Map F**

This was Michael Baum's final site and adjoining E. Pearson's yard at Depot Road. The company took RH 417891 with them but it was not in good condition and so it obtained another Ruston & Hornsby diesel locomotive from Cohen and made one good unit out of the two using 417891's chassis. It was not long before the company started using cranes to move any wagons and the locomotive and remains were scrapped.

Gauge : 4ft 8½in

-	4wDM	RH	417891	1959	(a)	
			Rebuilt 1975/76 with parts of UID loco			Scr 10/1978
-	4wDM	RH	+		(b)	Scr 10/1978

+ One of RH 235514/1945 or RH 402808/1956.

(a) ex Commercial Street premises with company, c1974/75.
(b) ex George Cohen, Sons & Co Ltd, Coborn Works, Cargo Fleet, c1975/76.

BELL & SON (DONCASTER) LTD

ANDERSTON FOUNDRY, Port Clarence 14
NZ 499216 – **Map C**

Following the closure of the Anderston Foundry and sale of stock in 1962-63, the site stood vacant until part was occupied by Bell & Son, although for what purpose is not known. The siding agreement with BR was dated 25/7/1968 and was terminated 31/12/1974. A note in the file reads that "the siding and connection were to be retained in situ at present because they may be required by the Cleveland Bridge & Engineering Co who are now in occupation of Bell's premises". **E. Marson** (which see) also occupied part of the premises.

Gauge : 4ft 8½in

2652		0-4-0DM	HE	2652	1942	(a)	s/s c/1970	
15	ROSEDALE	4wDH	S	10070	1961	(b)	(1)	

(a) ex South Durham Steel and Iron Co., South Works, Hartlepool, 3/1968, by 23/3/1968.
(b) ex ICI Ltd, Billingham Works, Billingham, /1974.

(1) to E. Pearson & Sons Ltd, Middlesbrough, c/1977, by 24/10/1978; then to Fairfield-Mabey Ltd, Chepstow, Gwent.

BLAIR'S (1926) LTD, MARINE ENGINE WORKS

North Shore Branch, Stockton 15
Blair & Co Ltd (taken over by **Gould Steamships and Industrials Ltd** /1919 and liquidated 19/5/1925)
Blair & Co until c/1893 NZ 446203 – **Map D**
Fossick, Blair & Co until/1866
Fossick & Hackworth until/1865

The Clarence Railway (CR) was established to connect the South West Durham collieries to the River Tees by a shorter route than the SDR. The CR's main line ran from a junction with the SDR at Simpasture through Billingham to Samphire Batts (known as Port Clarence from 5/1835) where coal drops were erected and began shipping coal in 1834. A branch line headed south from Norton to North Shore on the Tees at Stockton. Coal was first delivered by the CR along the North Shore Branch to Stockton shipping staithes (later known as "Old Portrack Staith") in 8/1833, although they were subsequently little used as other locations with deeper water down river were preferred. The CR was leased to the Stockton & Hartlepool Railway Company in 1844, but both businesses were controlled by the West Hartlepool Harbour and Railway Co (WHH&R) in 1853. Meanwhile, the Leeds Northern Railway had arrived from the south in 1852, passing through a new Stockton (called North Stockton until 1893) passenger station and joining the former CR branch at North Shore Junction. During the later 1830s and 1840s, the CR allowed the use of private locomotives on its lines and this included some industrial locomotives which hauled mostly coal to the staithes at Stockton and Port Clarence. More information is given in the references but locomotives from the following pits were involved – William Hedley & Son's Crowtrees and West Hetton Collieries, Stephen Walton's Thrislington Colliery and Nicholas Wood & Co's Westerton Colliery.

Just before F. Hills took over the former Blair's Marine Engine Works on the North Shore Branch at Stockton, Andrew Barclay 239/1881 was photographed by B.D. Stoyel on 8th July 1933. (ILS collection)

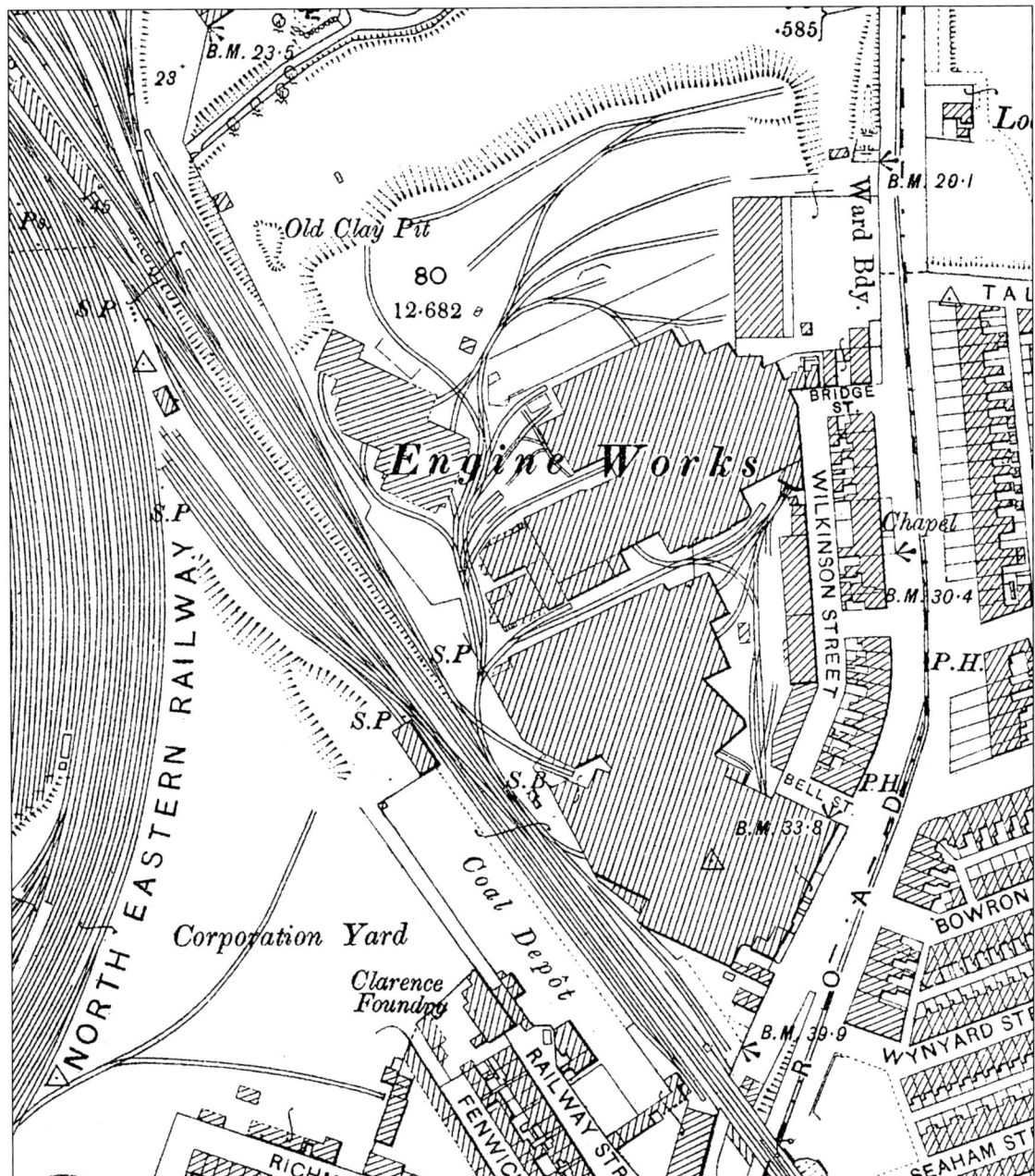

Blair's Marine Engine works alongside the NER's North Shore Branch at Stockton.

(Extract from Ordnance Survey 1:2500 map, 1915 edition)

Fossick & Hackworth's works was located on the east side of the North Shore Branch next to Norton Road. The partners took over the site of a former saw mill and flour mill, erecting a factory to conduct business as engineers and iron founders in 4/1840. Thomas Hackworth, brother of the better-known Timothy, had been manager of the Soho Engine Works at Shildon and George Fossick contributed business and sales skills. The firm soon began building steam locomotives, a 0-6-0 goods engine named STOCKTON was begun for the SDR at the end of 1840 and delivered in 6/1841. Other early locomotives were constructed for the Llanelly Dock & Railway Co and the Hartlepool Dock & Railway Co. These locomotives were similar to Timothy Hackworth's designs, except that they usually featured two outside inclined cylinders. For 10 years from 1844, Fossick and Hackworth had the contract to operate and maintain the Stockton & Hartlepool Railway's locomotives, including constructing all the wagons required. In 1851 the workforce totalled 231 although the 1857 Ordnance Survey reveals that the premises were still relatively small and served by one principal siding off the North Shore Branch. Nevertheless the works went on to construct locomotives for a number of railway companies, including eleven for the WHH&R and others for such as the Londonderry & Lough Swilly Railway and the Great Grimsby & Sheffield Junction Railway, with the last two being built for the Llanelly Dock & Railway Co in 1866. A full list of locomotives manufactured has not survived but George Turner Smith estimates 60 to 70 and one is left to wonder whether any unknown locomotives were constructed for Teesside industrial works.

In the mid-1850s, Fossick and Hackworth began to broaden their output by making stationary engines for the iron works and marine engines. They built the engine for the ADVANCE, the first iron steamer to be launched on the Tees in 1854. This became an increasing part of their output, particularly after the appointment of George Blair as works manager in 1855, who had a background in marine engineering. Thomas Hackworth retired in 1865 and George Fossick departed following his bankruptcy. This left the stage open for George Blair and, under his leadership, the company prospered. It marked the start of a period of considerable expansion with several new buildings, an extensive internal rail system and the two new Andrew Barclay locomotives to shunt it. By 1900, the company's premises covered 23 acres and employed 2,000 people; it was making engines and electrical fittings for 70 steam ships a year. The connections with the North Shore Branch were controlled by a small signal box called "Loco Junction". The other side of the branch was occupied by the Clarence Coal Depots and a short siding which ran south to Blair's depot. Here, the engines were lifted into ships by Blair's sheerlegs, erected in 1887 on two acres of land and served by sidings at the end of the North Shore Branch next to the River Tees. Blair's locomotive operated over 49 chains 12 yards of the NER's metals.

Almost 1,400 marine engines had been built by 1914 but Blair's only made steam engines and, after World War 1, business declined as the local shipyards closed and the demand was for motor vessels. After the liquidation, the works shut for a while until it reopened in 12/1926. The last engines were built in 1/ and 3/1931, with the works finally closing in 1932 and the premises then being taken over by F. Hills (which see).

References: 'Fossick & Hackworth – Haulage Contractor', Russell Wear, *The Industrial Locomotive* 106–07, ILS, 2002–03, pp. 252–55, 270–76

Thomas Hackworth Locomotive Engineer, George Turner Smith, Fonthill, 2015

Shipbuilding in Stockton and Thornaby, Alan Bettany, Tees Valley Heritage Group, 2003, pp. 78–81

'The Locomotives of the Stockton & Hartlepool Railway 1844 Onwards', K. L. Taylor, *The North Eastern Express* 142–44, 146–48, 1996–97

'Some Early Durham Colliery Locomotives – 1', Ken Fleming, *The Industrial Locomotive* 31, ILS, 1983, pp.145-52.

'Private Locomotives on the Clarence Railway', Ken Fleming, *The Industrial Locomotive* 168 and 170, ILS, 2018-19 pp.280-95, 10-28

Gauge : 4ft 8½in

-	0-4-0ST	OC	AB	187	1877	New	Scr /1933
-	0-4-0ST	OC	AB	239	1881	New	(1)

(1)   to F. Hills & Sons Ltd, Stockton, /1933 with premises.

W. BLENKINSOP & CO LTD

MARSH IRON YARD, Middlesbrough

16
Map F

The company was described in 1936 as "buyers of all descriptions of scrap metal, surplus plant and machinery for reuse or scrap." It had an office at various locations in central Middlesbrough. Its Iron Yard was situated on the Ironmaster but there is conflicting information about its specific position. However, it was liable for a 1d per ton MOR toll for all rail traffic entering and leaving its site until this ceased by 1/11/1946. Blenkinsop advertised locomotives for sale on eight occasions between 1891 and 1916.

In 1933 Blenkinsop gained the contract to dismantle the plant at The Wingate Limestone Co Ltd's quarry in County Durham, including three standard gauge locomotives:

HOLMLEA	0-4-0ST	OC	MW	1887	1915
CARMEL	0-4-0ST	OC	MW	1911	1917
-	0-4-0ST	OC	HL	2177	1890
			Rebuilt J.Tait	100	1920

Blenkinsop advertised one of the Manning Wardles for sale in the *Contract Journal* on 1/3/1933. Both Manning Wardles subsequently went to Robert Stephenson & Co Ltd, Darlington for repairs. The Hawthorn Leslie was sold to the Witton Park Slag Co Ltd at Etherley. It is not known whether they were brought to the Marsh Iron Yard or moved direct.

GEORGE BOWER

MILTON IRON WORKS, West Hartlepool 17
NZ 516315 – **Map B**

The premises appear on the 1861 Ordnance Survey and is the only one at this date located on the west side of what was later to become the NER Cliff House Branch. An advertisement by George Bower in 1868 describes the Milton Iron Works as a "pipe and general foundry". In particular, he was "prepared to supply and erect gas works". Slater's 1876–77 directory entry for Bower has "gas and water works engineer and contractor for public works, pipes and general castings". The 1896 Ordnance Survey shows the site on the south side of Greatham Street identified as Milton Works (Engineering) with two sidings entering the premises, one of which crossed over Greatham Street near Steelworks Crossing Signal Box. An 1895 List of Sidings compiled by the NER has an entry for "Milton Iron Works (Bower's Foundry) not working at present".

Shortly after, part of the iron works was occupied by W.H. Loveridge & Co, an iron merchant, which formed the Milton Forge & Engineering Co. A siding agreement between the NER and Loveridge was dated 2/11/1894. However, the northern part of the premises became G. Willson's Milton Saw Mill about 1896 and this used the siding over Greatham Street. The first agreement for use of the siding connection between the NER and George Willson & Sons was dated 22/3/1899. The company was a timber importer and manufactured packing cases. Some sources have the name as Wilson. According to published sources, William Gray & Co took over the Milton Forge & Engineering Co in 1898, although the siding agreement with the NER was dated 31/10/1901. This was to supplement the output of components from its Central Marine Engineering Works. Gray later relied on its Cliff House Foundry for these products and, by 1914, the whole site was occupied by Willson's Milton Saw Mill. Interestingly, in 1909, the siding alongside the Cliff House Branch where the track from the sawmill crossed Greatham Street to join the NER was called "Bower's Siding". The following locomotives are thought to have been used at the Milton Iron Works by George Bower.

Gauge : 4ft 8½in

		0-6-0ST	BH		(a)	s/s
–		0-4-0ST(?)			(a)	s/s

(a) origins and identities unknown.

BOWESFIELD BRICK CO LTD

BOWESFIELD BRICK WORKS, Stockton 18
(Subsidiary of **Crossley Building Products Ltd**, later **Crossley & Sons Ltd**)
NZ 442178 – **Map D**

The brick works was located behind Bowesfield Junction Signal Box abutting the south side of the Darlington-Middlesbrough railway. The reference gives details of some form of brick works in this area from the late 1870s, but Ordnance Survey maps suggest that the major development occurred with the establishment of the Bowesfield Brick Co, which was in existence by 27/2/1911. Earlier, the NER had an agreement on 29/1/1898 with T.E. Atterby, brick manufacturer, for a rail link and commented "A timber staging was erected alongside this siding, connected with the Brick Yard by a plank road". The Bowesfield company was taken over by Crossley's in 1927 but seems to have retained its Bowesfield name. In 1933, there was stated to be a 30 feet bed of clay below 4 feet of loam and the works had two old Newcastle kilns, a 22 chamber Belgian kiln and a down draught Scotch kiln. Six hand diggers were working in the clay pit until a Ruston Bucyrus excavator was purchased and this was in constant use for nearly 30 years, apart from World War 2 when the works was stated to be mothballed but see below. Common bricks and field drainpipes were manufactured from a mixture of loam and clay. A 3 foot gauge railway enabled clay to be transported from the pit to the works. When Ted Haigh visited on 24/4/1934, he found both Black, Hawthorn 1011 and Bagnall 1910 present. The railway was still shown on the 4/1950 Ordnance Survey when it comprised a single line between the pit and the works, apart from a passing loop midway along, possibly to avoid a weighbridge. The brick works closed in 1965.

References: *The Brickworks of the Stockton-on-Tees Area*, Alan Betteney, Tees Valley Heritage Group, 2007

Brickworks of the North East, Peter J. Davison, Gateshead Libraries and Arts Service, 1986

Three foot gauge Bagnall, maker's number 1910 built in 1910, stands out of use at the Bowesfield Brick Works about 1948. Behind can be seen one of the kilns, Bowesfield Signal Box and, on the other side of the main line, the premises of Ashmore, Benson, Pease & Co. (Kevin Lane collection)

Gauge : 3ft 0in

-	0-4-0ST	OC	BH	1009	1890	(a)	Scr
-	0-4-0ST	OC	BH	1011	1890	(a)	Scr
(28)	0-4-0ST	OC	WB	1910	1910	(b)	Scr /1950
-	4wDM		RH	202037	1942	(c)	(1)

(a) ex South Durham Steel & Iron Co Ltd, Malleable Works, Stockton, by 24/4/1934.
(b) ex Armstrong Whitworth & Co Ltd, Scotswood Works, via J.B. Watson & Sons, dealer, Leeds, /1933.
(c) new, ex works 19/2/1942 and first spares ordered 28/2/1942 suggesting that the works was no longer mothballed.

(1) the last spares for Bowesfield were ordered 10/1/1952. A spares order from Wheatley & Whiteley, 99 Kirkstall Road, Leeds, was dated 9/11/1961.

BRITISH CHROME & CHEMICALS LTD

EAGLESCLIFFE WORKS, Urlay Nook   19
Subsidiary of **Harrisons & Crosfield Ltd**, which took over and NZ 402146 – **Map D**
set up a new company using the title of the former owners, 1/11/1973
Albright & Wilson Ltd, Industrial Chemical Division until 1/11/1973
Albright & Wilson Ltd, Associated Chemical Companies Division until 6/1972
Associated Chemical Companies Ltd until /1965
British Chrome & Chemicals Ltd until 29/5/1958
Eaglescliffe Chemical Co Ltd until 4/1953
Egglescliffe Chemical Co Ltd until /1938
Egglescliffe Chemical Co until 3/1910
Wilson & Co until?

This works was unusual because it was situated away from the main focus of industry, 2½ miles north west of Yarm and was one of the earliest factories on Teesside. The SDR had opened in 1825 and, in 1833, a local farmer named Robert Wilson established a chemical works on the north side of the line at Urlay Nook to manufacture fertiliser and later sulphuric acid. The premises are shown on the 1857 First Edition of the Ordnance Survey map with a siding coming off the main line, crossing the road and entering the works. Horses were used initially for shunting. The special edition of the *British Chrome Clarion* recalls that old Tom Bell had been a horse carter delivering carboys of sulphuric acid to the Whessoe Foundry, Darlington and Dorman Long in Middlesbrough. The horse was also used to pull railway wagons across the road to the works.

Over the years, the company diversified into producing various chemicals. Beginning in 1865, it also had linseed oil and cake mills manufacturing animal feeds for 25 years. A plant to make oleum was built in 1890. The production of sulphuric acid was expanded in 1900. The use of cuprous pyrites from Cyprus in the sulphuric acid plant led to byproducts being manufactured, including iron oxide briquettes and copper requiring additional refining. There was much reliance on the railway, although the expanding premises still depended on the siding running through the centre of the works. The connection with the main line and a couple of short reception sidings were controlled by the NER's Urlay Nook Signal Box. The Egglescliffe Chemical Co owned a sizeable fleet of 10, 12 and 14 ton tank wagons registered to travel over the NER; many of them being constructed by Charles Roberts & Co Ltd. The numbers of individual wagons ran up to at least 58. Production of sulphuric acid had expanded during World War I for the manufacture of TNT and, to fully use its capacity, the company entered the chromium industry in 1927–28. The growing demand for chromium products encouraged the company to expand its production. In 1953, it merged with two other businesses to form British Chrome & Chemicals Ltd, which controlled the entire output of chromium chemicals in the UK. After the takeover by Albright & Wilson, fertiliser manufacture ceased in 1967 but an increased range of chromium products made it the sole factory of its kind in Europe. The production of sulphuric acid ended in 4/1983. The company was an early user of diesel locomotives, purchasing a John Fowler four-coupled machine in 1947. Latterly it relied on two Hibberd diesels of 1956 vintage, with one taking wagons to and from the reception sidings and the other spare. The purpose of the 2ft 0in gauge Lister is not known. Rail traffic ceased about 1/1982, but the locomotives were kept for three years, being started up every month or so, in case there was a revival. Elementis Chromium closed the plant in 2009.

References: *British Chrome Clarion*, issue 44, special edition to commemorate 150 years of the works, 9/1983

Private Owner Wagons An Eleventh Collection, Keith Turton, Lightmoor Press, 2012, p.68

Positioned on the siding leading into British Chrome & Chemicals' factory at Urlay Nook are Hibberd 3822 and 3808 of 1956 during an IRS visit on 25th July 1979. (Cliff Shepherd photograph)

Gauge : 4ft 8½in

	"COFFEE POT"	0-4-0T	VC	HG	238	1866	(a)	s/s 1920
	-	0-4-0ST	OC	HCR	139	1874	(b)	(1)
	-	0-4-0ST	OC	AB	266	1884	(c)	(2)
	VICTORIA	0-4-0ST	OC	P	634	1897	(d)	Scr 3/1939
	-	0-4-0ST	OC	AB			(e)	s/s
	PRINCESS ELIZABETH	0-4-0DM		JF	4200020	1947	New	(3)
	PRINCESS MARGARET	4wDM		FH	3492	1951	New	(4)
No.2	FD 62/20	4wDM		FH	3808	1956	(f)	Scr /1985
No.1	FD 62/1	4wDM		FH	3822	1956	(g)	Scr /1985

(a) ex Bell Bros & Co, Clarence Iron Works, Port Clarence. Inness suggests in /1895. A photograph published in *British Chrome Clarion*, issue 44, may depict this locomotive.
(b) ex Hogg & Henderson, Middlesbrough.
(c) ex J.F. Wake, Darlington, Co. Durham, on hire, /1917.
(d) ex Linthorpe-Dinsdale Smelting Co Ltd, Middleton St. George, Co. Durham, by 5/8/1936.
(e) ex Dorman, Long & Co Ltd, Britannia Works, Middlesbrough, /1943.
(f) ex J.& J. White Ltd, Shawfield Works, Rutherglen, Glasgow, 14/10/1966. This was one of the companies that amalgamated to form British Chrome & Chemicals Ltd.
(g) ex Thos. Hill (Rotherham) Ltd, Kilnhurst, South Yorkshire, 25/9/1975; formerly L.B. Holliday & Co Ltd, Huddersfield, West Yorkshire.

(1) to Hudswell, Clarke & Co Ltd, Leeds, /1914; resold to Hugh Symington & Co Ltd, Gretna Contract, Cumberland, 5/1915.
(2) returned to J.F. Wake, Darlington, Co. Durham, ex hire.
(3) to Henderson Clark Ltd, Middlesbrough, for scrap, /1970.
(4) scrapped 11/1975. The company said that a locomotive was "disposed of to Miles Turnbull, Bowesfield Lane" in 1975; possibly this is a reference to Thomas Turnbull at Thornaby.

Gauge : 2ft 0in

-	4wPM		L	34031	1949	New	s/s

BRITISH STEEL CORPORATION,

APPRENTICE TRAINING CENTRE, South Bank   20
Dorman Long (Steel) Ltd until 28/7/1967   NZ 546211 – **Map H**

(Operation not taken over until 1/7/1968)

The locomotives were only used as static examples to train apprentices.

Gauge : 4ft 8½in

No.3		4wVBT	VCG	S	9294	1936	(a)	Scr c/1966
42		0-4-0DH		JF	4220020	1961	(b)	(1)

(a) purchased by Thos Hill Ltd, Kilnhurst, Yorkshire (WR) from Ley's Malleable Castings Co Ltd, Derby, and transported direct to South Bank on 11/2/1963 as a gift from TH for the Apprentice Training Centre. The original bonnet sides from S 9396 were sent to Dorman Long to fit to 9294 to achieve a post war appearance.
(b) ex Cargo Fleet Works, Cargo Fleet, 6/1972.

(1) scrapped c5/1991; after 16/5/1991, by 3/7/1992.

Gauge: 2ft 6in

No.5	4wDM		RH	353484	1953	(a)	Scr /1978

(a) ex North Skelton Ironstone Mine, North Skelton , 6/1974.

CARGO FLEET IRON AND STEEL WORKS
CARGO FLEET

South Durham Steel & Iron Co Ltd until 28/7/1967 NZ 523205 – **Map G**
(Although operation of the works was not taken over until 1/7/1968 and SDSI did not cease to be a separate company until 28/3/1970.)
The Cargo Fleet Iron Co Ltd until 3/10/1953
(Fully controlled subsidiary of SDSI from 28/11/1928)
The Cargo Fleet Iron Co until 29/1/1883
Swan Coates & Co until /1879

For the first 40 years of its life, the business functioned as a traditional iron works using Cleveland ironstone. After the turn-of-the-century, the by now largely obsolete ironworks was replaced by a new integrated iron and steel works – the first on Teesside – which became known for its technical innovations, only to succumb in the "rationalisation" of the 1970-80s.

The company was founded in 1864 with four directors – John Swan, G. Newcomen, J.W. Coates and Colonel Van Stravbenzee. The business changed its name when the latter left in 1865–66. *The Middlesbrough Weekly News and Cleveland Advertiser* on 15/6/1866 announced that the first two blast furnaces had been tapped. The company went into liquidation on 30/6/1876, but continued in business until sold in 9/1879 to Charles Mitchell and H.F. Swan; the latter was described as a Newcastle shipbuilder. Mitchell later sold his share to H.F. Swan who then changed its name to the Cargo Fleet Iron Co (CFI). It was reorganised on a new financial footing when it became a limited company in 1883, John Swan being the managing director.

The next two blast furnaces, blown in during 1870 and 1871, were more advanced and used vertical blowing engines of Gjers design constructed by Cochrane, Grove & Co. A fifth blast furnace had been added by 1875, although only three were in blast in 1877–79. From then on it was usual to have four in blast producing approximately 2,000 tons of pig iron per week with ironstone from Ormesby Mine, Durham coke and flux from Mickleton limestone quarries. The premises in 1874 included the works (44 acres) and a separate stretch of foreshore on the Tees. The NER Normanby Branch ran down to a jetty on the river with a south to west curve connecting it to the Middlesbrough–Redcar line. Cargo Fleet Iron Works was located on the south side of this curve and a siding left the branch, (controlled by Cargo Fleet Inner Junction Signal Box) shortly before the curve began, to serve the calcining kilns and mineral stores on the north side of the blast furnaces. There were

No.4 (YE 289/1877) at the Cargo Fleet Iron and Steel Works near the Works Road level crossing over the main line. The blast furnaces and engine house of the Normanby Iron Works can be seen beyond.
(IRS Noel Needle photograph)

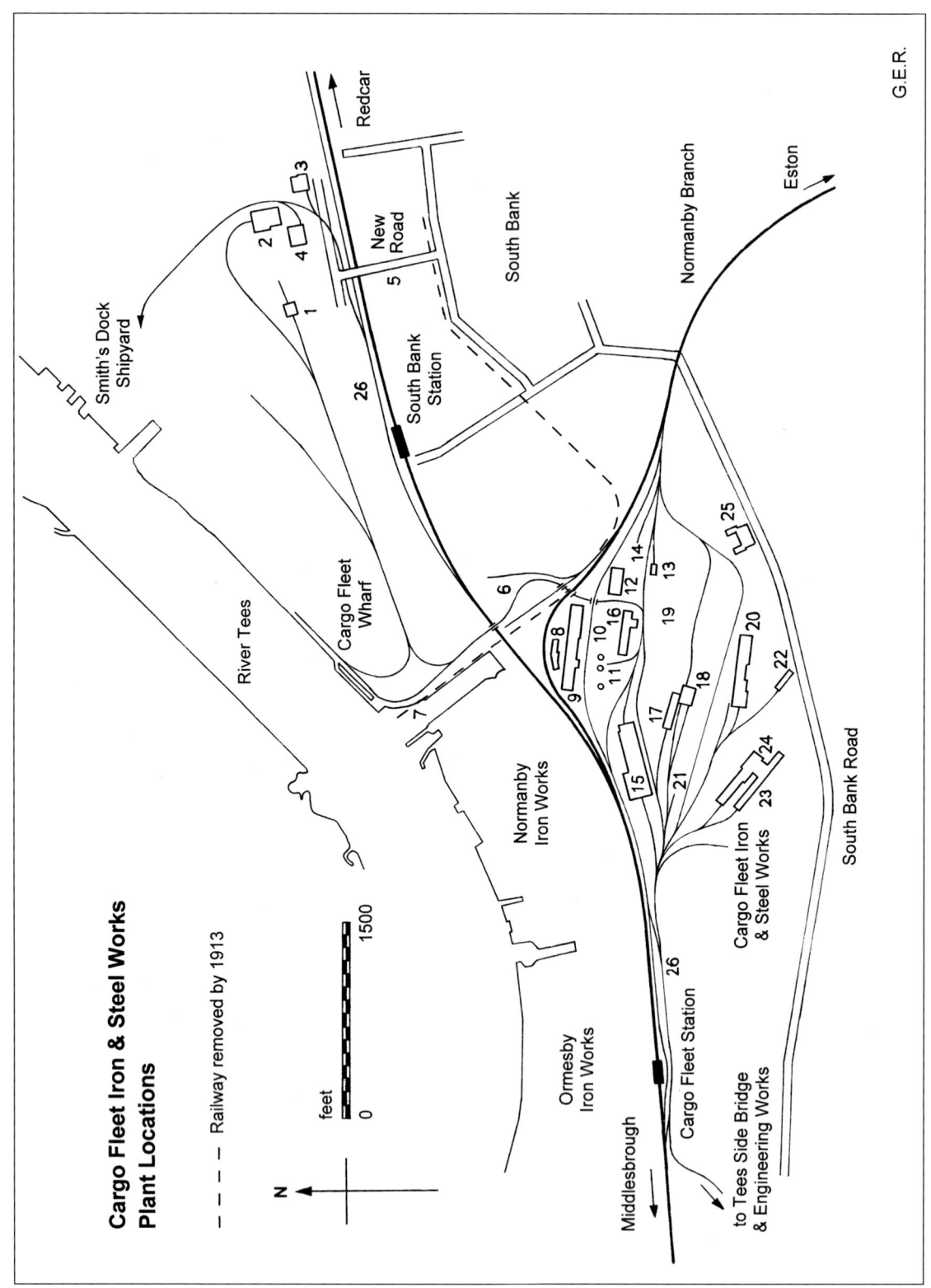

Plant locations

1	South Bank tar macadam works	14	Incoming sidings
2	South Bank basic slag works	15	Steel plant
3	F. Jones slag wool works	16	Gas blowing engine house
4	South Bank Chemical Co's works	17	Ingot mould bay and soaking pits
5	Former Imperial Works site	18	32 inch rolling mill
6	Sinter plant	19	Section and rail banks (later covered in)
7	Former NER jetty	20	Fitting and boiler shops
8	Benzol and sulphate 'houses'	21	Outgoing sidings
9	Kogag coke ovens	22	Wagon shed
10	Elevated railway for minerals	23	11 inch rolling mill
11	Blast furnaces	24	21 inch rolling mill
12	Power station	25	General offices
13	Engine shed	26	Exchange sidings

The simplified map opposite shows the position of the various plants. These did not all co-exist and the building outlines are diagrammatic because there were changes both in layout and extent over time.

ten 35 feet high kilns in 1893 and a raised ramp enabled loaded wagons to be propelled up on to the gantry by a company locomotive. The engine house and boilers were situated at the east end of the blast furnaces. Other lines ran south from the blast furnace discharge chutes and pig beds. A fan of sidings was available west of the blast furnaces for stabling wagons and this, together with other lines, converged where the prominent footbridge carried the 'Black Path' (Sailors' Trod) over these tracks before they joined the Middlesbrough-Redcar railway near Cargo Fleet Junction. This formed the entry point for incoming coke and limestone with pig iron being sent out. Another siding ran south from this strategic location to its open land south of the main works and east of Normanby Beck. Slag was tipped on to the surrounding, still marshy, ground and this may have been the initial purpose of the siding but the 1892/93 Ordnance Survey has it running to the Cargo Fleet Brick Works, which was owned by CFI, manufacturing red bricks and drainage pipes. The plan reveals a number of pools and a separate tramway from the brick works, passing under an access road from the slag tips to the clay pit. The company also owned an increasingly substantial amount of land between the River Tees and the Middlesbrough-Redcar line. By 1875, the company had a railway which left the blast furnace sidings and ran eastwards under the Normanby Branch and then turned northwards to cross over the main line on a bridge to reach its wharf on the river. The land behind the wharf was still mainly occupied by mudflats, even in 1892/93, although CFI had laid a long siding eastward and its locomotives were taking slag there for dumping to reclaim the area. At the same date, another siding left the company's line shortly before the bridge over the main line, headed south parallel to the Normanby Branch before turning north east to cross open ground to reach South Bank. Here it ran behind the houses in North Street to enter the former Imperial Iron Works site of Jackson Gill and Co Ltd. The purpose of this line is not known but it may have been to obtain additional supplies of clay.

William Gray and Christopher Furness were heavily involved in shipbuilding and shipping at the Hartlepools and had seen a need to achieve more economic production of plates, angles and bars. On 28/12/1898, the South Durham Steel & Iron Co Ltd was formed as a private company by the amalgamation of three relatively small businesses (the West Hartlepool Steel & Iron Co Ltd, the Stockton Malleable Iron Co Ltd and the Moor Steel and Iron Co Ltd) and it became a public company in 3/1900 under the chairmanship of Sir Christopher Furness. Meanwhile Furness had purchased the Weardale Company in 1899 and reconstituted it as the Weardale Steel, Coal & Coke Co Ltd. In 1901 control of CFI passed to the Weardale company when John Swan died. Furness recognised the increasing domination of steel and the strategic position of the Cargo Fleet site next to the railway and with over 1,000 feet of river frontage. An exchange of shares between SDSI and the Weardale Company resulted in the Cargo Fleet Works coming under SDSI's control in 1903. The old blast furnaces at Cargo Fleet were demolished and a new integrated iron and steel works erected on the site.

The design of the new works was based on German and American practice. Two mechanically charged blast furnaces were created, each capable of providing 1,400 tons of iron per week and powered by seven gas blowing engines by Richardsons, Westgarth & Co Ltd. A third blast furnace was added during World War I. A total of 100 Koppers by-product coke ovens were constructed in 1904-05, supplemented by another 45/50 in

1914. The coke was pushed out of the ovens on to a sloping bench and a photograph (see reference) shows two coke cars running on a railway track below the bench and connecting with a large hoist which lifted each container for discharge on to a conveyor belt leading to the blast furnaces. These may have been the first combined coke oven locomotive and car used in the UK. The iron was converted into steel in three 150 ton Talbot tilting basic open hearth furnaces. It was transported from the blast furnaces as 'hot metal' in 20 ton ladles on rail 'carriages'. The molten iron could be poured into the inactive mixers without removing the ladles from their carriages. An active mixer was installed in 1922. An impressive office block on South Bank Road was added in 1916 and still stands today.

The basic pattern of railways was retained for the new iron and steel works but with an increase in the number and complexity of sidings. The coke ovens were situated north of the blast furnaces, with the open hearth furnaces to the west near Cargo Fleet Junction and rolling mills to the south. A set of exchange sidings lay alongside the main line next to the final position of Cargo Fleet Station. On 30/5/1910, the NER agreed that the CFI could use two lines south of these sidings and a connection to be constructed by the NER to the Marsh Branch, in order that CFI's locomotives and wagons could transport "channels, angles and joists" to the Tees Side Bridge & Engineering Works Ltd's premises at North Ormesby.

An increasing proportion of the company's land north of the main line had been reclaimed by slag tipping with additional sidings laid, including alongside the Middlesbrough-Redcar railway. An agreement for the sidings was approved by the NER on 12/12/1906, together with the proposed line to a slag wool works. Three small rail connected businesses had been established in the angle between the main line and the branch to Smith's Dock with vehicular access provided by the new road from South Bank across the former Jackson Gill site. The plants were: –

- Slag Wool Works operated by F. Jones & Co
- Slag and Tar Macadam Works producing road making materials from blast furnace slag (South Bank Tar Macadam Co Ltd)
- Basic Slag Works processing slag from the open hearth furnaces to make fertiliser. Established by Alexander Cross & Sons Ltd, it was sold to the CFI on 6/3/1922, which in turn leased it on the same date to its wholly-owned subsidiary, South Bank Basic Slag Co Ltd.

The CFI purchased six new 0-4-0 saddle tanks from Andrew Barclay in 1905–06 to operate its integrated works and these were supplemented by six from Hawthorn Leslie in 1913–19 due to the demands of wartime. Two more tilting open hearth furnaces were added, one in 1908 and the other in 1917.

SENTINEL (6154 of 1926) was a major rebuild of a Dubs crane tank constructed in 1884. It was photographed by Bernard Roberts at the Cargo Fleet Works on 29th September 1951. (IRS collection)

A trial took place on 5th August 1969 to move molten iron in two torpedo wagons from the Cargo Fleet works to Consett. One of the BSC John Fowler diesels is assembling the train near the Cargo Fleet blast furnaces prior to handing it over for collection by British Railways.   (ARPT J.M. Boyes photograph)

An increasing use began to be made of imported iron ore; more railway lines being laid at the wharf which was modernised in 1926. It could handle 10,000 ton ships by 1937. The Koppers coke ovens were replaced at the same location by 64 Gibbons Bros Ltd Kogag ovens in 1930–35 with another 26 added in 1946 and 1952/53. Three 3ft 10in gauge electric locomotives hauled the quenched coke from the benches in bottom door tubs to the blast furnace skip hoists. There was a corrugated iron shed for these locomotives at the west end of the line. By 1937 the three blast furnaces were producing 300,000 tonnes of iron annually, most of which was taken in ladles to the steel plant which contained two 150 ton inactive mixers, a 300 ton active mixer and five open hearth furnaces of 175 to 200 tons capacity. At the rolling mills, the steel ingots were reduced to blooms at a 40 inch cogging mill before going to the section, plate and rail mills. At the end of 1942, Cargo Fleet Works had 21 standard gauge locomotives to shunt its traffic. The two-road engine shed (NZ 524205) was situated immediately south of a fan of sidings and the power station.

Major reconstruction of the blast furnaces took place after World War 2 – No.2 (in 1948), No.1 (1952) and No.3 (1957). The mineral handling facilities were further developed. The LNER had agreed on 29/4/1946 that the CFI could construct seven additional sidings connecting with the north side of the Middlesbrough – Redcar railway for the "ore crushing, grading, screening and sintering plant". These stocking areas were connected by conveyor to the sinter plant (installed 1943 and replaced 1958).

The works began to replace its steam shunting locomotives in 1953. With a few exceptions, it purchased most of its diesels from John Fowler, firstly equipped with mechanical transmission and later hydraulic transmission. Like many iron and steel works, Cargo Fleet employed an extensive fleet of internal wagons and a wagon shop was located on the south side of the works.

On nationalisation of the steel industry in 1967, Cargo Fleet employed 4,097 people. An interesting trial took place in 8/1969 to see if molten iron could be transported from Cargo Fleet to Consett while the latter's blast furnace capacity was reduced. Two specially designed torpedoes made the 63 mile journey behind a pair of BR Class 37 diesels but it was concluded that it was neither practical nor economic.

With the commissioning of the BOS plant at Lackenby, the days of the neighbouring open hearth steel plants were numbered and the iron and steel making plant at Cargo Fleet closed on 29/5/1971. The coke ovens had shut in 6/1961, although the electric locomotives lay around the site for some time. Dismantling of the steel plant began almost immediately and the blast furnaces were toppled during the summer of 1978. It was not quite the end of the story because, although the 40 inch cogging mill finished in 1972 and the 63 years old 21 inch section mill on 28/9/1973, a bloom heating furnace had been installed in 1971 to service the remaining rolling mills with steel from Lackenby. The premises were now concentrated on these mills and stockyard with rail access provided from near Cargo Fleet Station. The engine shed had been converted for use as a garage and the locomotives were stabled in the former fitting shop. For much of the decade, Cargo Fleet continued to roll rails and was supplying 1,500 tonnes per week to BR. When that business was transferred away, it was left with some beams and channels, together with the Larsson piling; 1.5 million tonnes of piling were rolled 1929–59. Production at Cargo Fleet Works finally ceased on 8/5/1984 when 150 tonnes of piling were rolled in the 32 inch Mill.

References:
'Coke Making at Cargo Fleet in 1904', Paul Jackson, *Archive* 47, Lightmoor Press, 2005, p.p. 62–63

History of South Durham Steel & Iron Co Ltd, W.G. Willis, SDSI, 1969

'The Iron, Steel & Engineering Industries of Middlesbrough and District', visit of Iron & Steel Institute, 9/1937

'105 Year Era Ends at Works', G.M. Barnes, *Steel News*, BSC, 17/6/1971

'Cargo Fleet Iron and Steel Works: A remarkable transformation', J. and G. Braddy, *Cleveland Industrial Heritage* 36, pub. Peter Tuffs, 2015

Iron & Steel: A Survey of the Manufacturing Processes & Chief Productions of South Durham Steel & Iron Co Ltd, 1936

'Torpedo Service to B.O.S. is Running Well', Bill Salter, *Steel News*, British Steel Corporation, 15/7/1971

'Cargo Fleet Iron Company', Cliff Shepherd, *Industrial Railway Record* 237, 241, 249, 2019-2022

'Cargo Fleet Iron Company', Roger West, *Industrial Railway Record* 246, 2021

Gauge : 4ft 8½in

	WALKER No.1	0-4-0tank		HG	231	1866	New	(1)
	MOSELEY	0-4-0ST		BH	98	1869	New	(2)
2	UPSALL	0-4-0tank		HG	277	1871	New	s/s
	WHITECLIFFE	0-4-0ST	OC	MW	385	1872	(a)	(4)
1	BEARPARK	0-6-0ST	OC	FW	245	1874	New	s/s
	STOKESLEY	0-4-0ST	OC	MW	84	1863	(b)	(3)
	JESMOND	?		?	?		(c)	(5)
3	SPERO	0-4-0tank		HG	247	1867	(d)	
		Rebuilt		RS		1896		s/s
	NELLY	?		?	?		(e)	(6)
4	IMPERIAL	0-4-0ST	OC	YE	289	1877	(f)	Scr 2/1955
5	LIVERTON	0-4-0ST	IC	Robey			(g)	(7)
6	(MICKLETON)	0-4-0ST	OC	AB	305	1888	(h)	
		Rebuilt		CFI		1945		Scr/1960
	GUY DAWNAY	?		?	?	1886?	(i)	s/s
7		0-4-0ST	OC	Joicey	230	1872		
		Rebuilt		Tudhoe		1903	(j)	(8)
8		0-4-0ST	OC	Joicey	215	1870	(j)	s/s
5		0-4-0ST	OC	N	2280	1878	(j)	
		Rebuilt				1905		Scr c2/1952
No.2		0-4-0CT	OC	D	2051	1884	(j)	
	SENTINEL	Rebuilt 0-4-0VBT	VCG	S	6154	1926		Scr 9/1956
14 (10) (9)		0-4-0ST	OC	AB	1008	1905	New	Scr c/1957
10 (11)		0-4-0ST	OC	AB	1029	1905	New	Scr c/1957
11		0-4-0ST	OC	AB	1047	1905	New	(9)
13		0-4-0ST	OC	AB	1061	1905	New	Scr c5/1968
14 (9) (8)		0-4-0ST	OC	AB	1068	1906	New	Scr /1959
12		0-4-0ST	OC	AB	1095	1906	New	Scr c/1959

15		0-4-0ST	OC	HL	2995	1913	New	Scr /1967
16		0-4-0ST	OC	HL	2997	1913	New	
					Rebuilt	1951		Scr /1968
No.17		0-4-0ST	OC	HL	3212	1916	New	s/s c/1962
18		0-4-0ST	OC	HL	3213	1916	New	Scr c/1964
No.7		0-4-0ST	OC	HL	3420	1919	New	Scr c/1959
No.8		0-4-0ST	OC	HL	3430	1919	New	Scr c/1964
No.19		0-4-0ST	OC	HL	3779	1930	New	Scr /1957
No.22		0-4-0ST	OC	HL	3936	1938	New	Scr 10/1966
23		0-4-0ST	OC	AB	2106	1940	New	Scr /1968
24		0-4-0DM		HE	2653	1943	New*	Scr c/1973
No.25		0-4-0DM		HE	2840	1943	New*	Scr c/1973
20		0-6-0ST	OC	HL	3365	1918	(k)	Scr c/1957
3		0-6-0T	OC	HL	3531	1922	(l)	Scr c/1964
No.5		0-6-0ST	IC	P	877	1901	(m)	Scr c/1958
No.27		0-4-0DM		JF	4210097	1955	New	Scr c/1972
21		0-6-0ST	IC	P	1444	1916	(n)	Scr c/1958
No.29		0-4-0DM		JF	4210106	1955	(o)	Scr c/1973
No.28		0-4-0DM		JF	4160008	1953	(p)	Scr c/1972
No.4		0-4-0ST	OC	HL	2461	1900	(p)	Scr c/1958
No.30		0-4-0DM		JF	4210113	1956	New	Scr c/1973
No.31		0-4-0DM		JF	4210122	1956	New	Scr c/1972
No.32		0-4-0DM		HC	D1013	1957	New	(12)
35		0-4-0DM		JF	4210135	1957	New	Scr c/1973
No.9		0-4-0ST	OC	HL	3935	1937	(q)	s/s c/1962
No.10	COLONEL	0-4-0ST	OC	AB	1501	1917	(r)	Scr c/1964
No.19 (15)		0-4-0ST	OC	RSHN	7045	1942	(s)	Scr /1968
No.36		0-4-0DM		HC	D1032	1957	New	(12)
33		0-4-0DM		HC	D1081	1958	New	(12)
No.34		0-4-0DM		JF	4210139	1958	New	Scr c/1973
No.37		0-4-0DM		JF	4210149	1958	New	Scr c/1972
No.38		0-4-0DH		JF	4220004	1959	New	Scr c/1973
No.39		0-4-0DH		JF	4220005	1959	New	Scr c/1973
No.40		0-4-0DH		JF	4220006	1959	New	Scr c/1973
No.41		0-4-0DH		JF	4220019	1961	New	Scr /1972
No.42		0-4-0DH		JF	4220020	1961	New	(13)
No.26		0-4-0DM		JF	4200006	1946	(t)	Scr c/1972
No.44		0-4-0DM		JF	4210147	1958	(u)	Scr c/1973
No.43		0-4-0DM		JF	4210086	1953	(v)	Scr /1972
6		0-4-0ST	OC	AB	2323	1952	(w)	(10)
7		0-4-0ST	OC	AB	2324	1952	(w)	(11)
No.45		0-4-0DH		JF	4220036	1965	New	Scr /1972
No.46		0-4-0DH		JF	4220035	1965	New	Scr c/1973
No.47		0-4-0DH		JF	4220040	1967	New	Scr 9/1979
No.48		0-4-0DH		JF	4220041	1967	New	Scr c7/1979
No.49		0-4-0DH		JF	4220042	1967	New	Scr c7/1979
No.50		0-4-0DH		JF	4220043	1968	New	s/s /1986
51		4wDH		TH	225V	1970	New	s/s /1986
52		4wDH		TH	226V	1970	New	s/s /1986

* Fitted with new Gardner 4LW engines 6/1952 (2653) and 5/1959 (2840)

The post war 200hp prototype, Sentinel 9538, was demonstrated to CFI at Cargo Fleet c3/1954.

(a) new. Manning Wardle records show the name as WHITE CLIFF but it was WHITECLIFFE after the mine owned by Swan Coates & Co in 1872–76.

(b) ex Smith & Knight, contractors, Wakefield, Yorkshire (WR), by 31/1/1877.

(c) here by 31/1/1877.

(d) ex Clay Lane Iron Co Ltd, Clay Lane Iron Works, South Bank, by 28/1/1880.
(e) here by 28/1/1880.
(f) ex Imperial Puddling Co, by 3/1/1882.
(g) ex Liverton Mine, Loftus, 5/1883.
(h) new? per Joseph Torbock, dealer Middlesbrough. To the Weardale Iron & Coal Co Ltd, Tudhoe Iron Works, Spennymoor, Co. Durham and returned.
(i) here by 30/1/1889.
(j) ex Weardale Steel, Coal & Coke Co Ltd, Tudhoe Iron Works, Spennymoor, Co. Durham.
A new boiler was supplied by Andrew Barclay to the Cargo Fleet Works for N 2280 in 7/1908.
(k) ex SDSI, Malleable Works, Stockton, by 11/12/1942.
(l) ex War Department, via T.J. Thomson & Son Ltd, Stockton, /1949.
(m) ex SDSI, Malleable Works, Stockton, 8/4/1952.
(n) ex SDSI, Malleable Works, Stockton, c/1952.
(o) ex SDSI, Malleable Works, Stockton, /1955.
(p) ex SDSI, Malleable Works, Stockton, c/1955.
(q) ex SDSI, West Hartlepool Works, c1957.
(r) ex SDSI, West Hartlepool Works, 11/1957.
(s) ex SDSI, West Hartlepool Works, 6/1957.
(t) ex South Bank Chemical Works, /1961.
(u) ex SDSI, West Hartlepool Works, 12/1962.
(v) ex SDSI, either from West Hartlepool Works, 12/1962 or Malleable Works, 1964.
(w) ex Dorman Long (Steel) Ltd, Lackenby, /1964.

(1) to The Bearpark Coal & Coke Co Ltd, Bearpark Colliery and Coke Ovens, Bearpark, Co. Durham. This company shared some of the same directors with Swan Coates & Co.
(2) s/s by 31/1/1877.
(3) to J. Torbock, dealer, Middlesbrough, by 4/1/1881; later at Casebourne & Co Ltd, Pioneer Cement Works, Haverton Hill.
(4) to C. Mitchell & Co Ltd, Low Walker Yard and Walker Naval Yard, Walker-on-Tyne, Northumberland, by 13/1/1882; later at Wallsend Slipway & Engineering Co Ltd, Wallsend, Northumberland.
(5) to "NB Burn Coal Co"? by 13/1/1882.
(6) s/s by 13/1/1882.
(7) to Haverton Hill Salt Co, South Durham Salt Works, Haverton Hill, by /1886.
(8) to North Eastern Steel Co Ltd, (subsidiary of Dorman, Long & Co Ltd), Acklam Iron and Steel Works, Middlesbrough ,/1917.
(9) to Irchester Ironstone Co Ltd, Irchester Ironstone Quarries, Irchester, Northamptonshire, /1922.
(10) to SDSI, Irchester Ironstone Quarries, Northamptonshire, 13/4/1964.
(11) to SDSI, Irchester Ironstone Quarries, Northamptonshire, 7/1964.
(12) to BSC, West Hartlepool Works for scrap, 2/1970.
(13) to BSC, Apprentice Training Centre, South Bank, 6/1972.

Gauge : 3ft 10in (Coke Oven Locomotives)

-	4wWE	Scr
-	4wWE	Scr
-	4wWE	s/s

There were also two coke oven locomotives operating by 1905 but no details are known.

DOCK STREET FOUNDRY (formerly MIDDLESBROUGH IRON WORKS), Middlesborough 22
NZ 502211 – **Map G**

Dorman Long (Steel) Ltd until 28/7/1967
(Operation not taken over until 1/7/1968)
Dorman, Long & Co Ltd until 2/10/1954
Bolckow, Vaughan & Co Ltd until 29/11/1929
Bolckow & Vaughan until 19/11/1864

A significant event in the history of Middlesbrough occurred when the financier, Henry Bolckow and engineer, John Vaughan set up in partnership in 1839 and agreed on 18/5/1840 to purchase six acres of marsh and tidal mud alongside the River Tees from the OME in order to construct an iron works. The land to the north east of the new town was reclaimed using clay from the Middlesbrough Dock excavation and ballast from incoming ships. The resulting firm bank to the river accommodated a wharf which had three cranes by 1857. The works contained puddling furnaces to produce wrought iron, together with a forge, foundry and rolling mills. Upwards of 150 workmen were employed when production commenced on 5/8/1841. In 1846, 20,000 tons of finished iron, mainly bars and rails, were manufactured. The existence of a foundry and engineering shop meant that the company also sold substantial amounts of cast iron pipes, especially for London's water mains.

Pig iron was initially shipped in from Scotland but, when this became too expensive, Bolckow & Vaughan erected blast furnaces at Witton Park near Bishop Auckland making use of local coal measure ironstones. The confirmation of Main Seam ironstone in the Eston Hills in 1850 dramatically altered the situation and Bolckow & Vaughan began to construct a new iron works at Eston. They also erected three 42 feet high blast furnaces at the Middlesbrough Works in 1851-1852 and, in the following year, produced 56,000 tons of plates, rails and finished iron.

Although the OME's Vulcan Street railway extended alongside the iron works by 1845–46, plans and an engraving of that date do not appear to show any railways within the site. This had changed by the 1853 Ordnance Survey map. John Dunning's 1856 plan reveals at least three railway lines running along Vulcan Street with two connections to the iron works, one from each direction. All the major parts of the works were served by standard gauge railways, although there was still extensive use of wagon turntables. Looking at the site from east to west - the large foundry building was located near the river bank adjoining the marshland to the east; the three blast furnaces with their vertical hoists stood at right angles to the river with the engine house and boilers nearby; three "moulding sheds" occupied the central area with the puddling furnaces buildings adjoining Vulcan Street and next to the western boundary.

Middlesbrough Iron Works suffered due to of its constricted site. Expansion to the west was not possible because of the railway to the Pottery Wharf. The extension to the east almost to Middlesbrough Dock was hindered by the strip of land giving access to the river bank for Sidney, Sherwood & Smith's foundry and this was not overcome until 1859. Stockyards and a large engineering building were then provided on this land so that, by 1862, the iron works covered 25 acres and employed 1,600 people. Bolckow & Vaughan had functioned as an informal partnership to 1853, when a co-partnership agreement was signed but did not become a company until 1864.

The original blast furnaces were now very antiquated and were rebuilt as taller structures in 1862-66. A combination of the increasing popularity of steel and the depression of the mid-1870s resulted in the closure of 67 puddling furnaces being announced in 1875. The invention to produce good steel from phosphoric ironstones using a 'basic' lining in a 30cwt experimental converter was successfully demonstrated on 4/4/1879

No.1, Black, Hawthorn 525/1880 with a modified cab, was based at Dorman Long's Dock Street Foundry but is seen on the adjoining Vulcan Street railway in 1955. The ornate wall marked the boundary of the former Cleveland Salt Works.   (ILS Frank Jones collection)

139 BEATTY (HL 3240/1917) stands at the entrance to the Dock Street Foundry on 4th April 1956 near an improvised signal; these premises had once been the Middlesbrough Iron Works.

(F. Bell photograph, ARPT)

at the Middlesbrough Iron Works. Soon Bolckow Vaughan was expanding steel production at Eston and the remaining puddling furnaces and rolling mills were dismantled at Middlesbrough Iron Works. The building on the western boundary was probably converted for use by the Cleveland Salt Works (which see).

As the iron works had developed, the two tight curves from the Vulcan Street railway and the wagon turntables within the site proved impractical and a new connection was provided at the south-east corner of the premises linking it to the NER lines alongside Middlesbrough Dock. Two pairs of tracks entered the works, those on the left running by the offices and passing between three 12ft high cast iron gate posts. Within the site, railway lines covered much of the open ground serving two new replacement blast furnaces by 1882; the foundry; the impressive 400ft x 60ft engineering building at the same angle as the incoming tracks (first shown on an 1875 plan and still standing in the third decade of the 21st century); the "gas works" at the east end of the site where there were numerous sidings and the 425 feet long wharf equipped with 2½–3 ton steam cranes on pedestals. According to MOR records in 1897, the iron works' locomotives could travel to the dock, Marsh Branch and Watson's and Tees Union's wharves. Both blast furnaces functioned for several years but only one in 1915 and neither in 1920. *The North-Eastern Daily Gazette* on 21/12/1927 contained a picture of them being demolished and claimed that they had stood idle for ten years.

The foundry building was near the site of the first blast furnaces. It initially supplied the needs of South Bank Iron Works, but, after that at the Clarence Works finished, it became the central foundry for all of Dorman Long's plants, including the collieries and ironstone mines. From 1934, it had excess capacity and, by 1937, about 1,000 tons of castings were made each month, of which one third was for the open market. Raw materials came by rail and wagons were discharged on a raised gantry adjacent to the foundry containing the cupola furnaces. This was roofed over so the sand and coke did not get wet. From then on, the premises functioned as a foundry and engineering workshops, being renamed the Dock Street Foundry. By 1951, the number of sidings within the site had been much reduced, although the two-road engine shed still stood between the foundry and engineering shop. The river wharf was stated to be out of commission circa 1954. Dorman Long's locomotives shunted within the works and placed wagons on the sidings alongside Dock Street for British Railways to collect.

Latterly the foundry mainly produced castings for BSC's other works on Teesside. *Steel News* on 25/4/1974 reported that two overhead cranes at Dock Street Foundry were to be dismantled following the recent closure of the works; they had been installed when the foundry was constructed in the 1930s.

Information on the early locomotives is scarce, but some clues come from published accounts following a boiler explosion at Middlesbrough Iron Works as reported in *The Times* on 18/ and 21/10/1856. This recorded that a locomotive built at the iron works and "only been completed a fortnight ago, and had not been named" exploded after shunting coal and limestone about the yard and on its return to the engine shed for repairs. According to *The Engineer* on 14/11/1856 its boiler was of curious shape with two steam domes connected by an inverted U-shaped saddle. The boiler had split where the saddle was riveted onto the barrel killing four men, including the unfortunate draughtsman responsible for the design. Robert Archer, the locomotive driver, said that there had been "two or three engines built in the works", presumably meaning locomotives. This is not surprising as Bolckow & Vaughan had also constructed engines for other purposes, such as to power THE ENGLISH ROSE, a steam vessel launched at South Stockton in 1843.

The difficulty posed by unknown early locomotives is compounded by the surviving Bolckow Vaughan list not distinguishing between those at the Eston Works (see Teesside Works entry), Middlesbrough Works or, in some cases, its County Durham collieries. For example, in 9/1854, Bolckow & Vaughan purchased SALFORD, a 0-4-2 tender locomotive (built 1843) from the Lancashire & Yorkshire Railway, but it is not known to which of the firm's works it was destined.

References: *The Industrial Heart of Old Middlesbrough*, J.K. Harrison, CIAS, 2010

"Explosion at Middlesbrough", Frank Jux, *The Industrial Locomotive* 115, ILS, 2005

Works on the North-East Coast of Dorman, Long & Company Limited, Reprint from *The Iron & Coal Trades Review*, 1937

Gauge : 4ft 8½in

-		?		BV			New	s/s
-		?		BV			New	s/s
-		?		BV			New	(1)
No.16		0-4-0ST	OC	MW	97	1863	(a)	s/s after 7/11/1926
No.17		0-4-0ST	OC	MW	98	1863	(a)	s/s after 7/11/1926
No.18		0-4-0ST	OC	MW	113	1864	(a)	s/s
	TEES	0-4-0ST	OC	BH	428	1877	(b)	s/s
No.1	(HENRY)	0-4-0ST	OC	BH	525	1880	(c)	(4)
3				BH			(d)	s/s
5				BH			(d)	s/s
140	HAIG	0-4-0ST	OC	HL	3209	1916	New	(2)
139	BEATTY	0-4-0ST	OC	HL	3240	1917	(e)	(6)
185	DAVID PAYNE	0-4-0DM		JF	4110006	1950	(f)	(5)
10		4wDH		S	10010	1959	(g)	(3)
83		4wDM		FH	3883	1958	(h)	(7)
32		4wDM		FH	3832	1957	(i)	(8)

(a) new? MW records state that the customer was "Bolckow & Vaughan, Middlesbrough". A correspondent wrote to Inness on 20/10/1931 stating that, about 1885–90, the Middlesbrough Works had "two small Manning Wardle 0-4-0 saddle tanks Nos.16 and 17."
(b) new. It was stated to be here about 1885–90 according to the above correspondent.
(c) ex Cleveland Works, by /1885. When seen on 6/10/1951, it was stated to have been here at least 40 years.
(d) a note in pencil in IRS records suggests that these were here in /1926.
(e) new. The locomotive appears to have gone on loan to the Cleveland Works and later returned. Ted Haigh saw it there on 3/7/1927 and Inness also lists it at the Cleveland Works. IRS records show it at the Middlesbrough Works at least from 6/1951.
(f) ex Cleveland Works, /1960; returned by 10/1960 and ex Cleveland Works again c/1961, by 15/8/1961.
(g) ex Lackenby Works, /1960.
(h) ex Ayrton Sheet Works, c/1962.
(i) ex Redcar Works, 14/12/1964.

(1) exploded 10/1856 (see text which mentions two possible earlier locomotives built here).
(2) to Cleveland Works; it was rebuilt there in /1955.
(3) to Lackenby Works, c/1961, after 23/4/1961.
(4) scrapped /1965, last reported 1/1965.
(5) to Armstrong Whitworth (Metal Industries) Ltd, Close Works, Gateshead, 5/1965.
(6) to Sharpes Autos (London) Ltd, Gables Service Station, Rawreth, Essex, 7/1972 for preservation.

(7)   to British Steel Corporation, Fullwood Foundry, Holytown, Lanarkshire, /1974.
(8)   stored in yard of H.E.T. Steel Ltd, Snowdon Road, Middlesbrough and then to Frank Berry Ltd, Leicester, 7/1974 per Northern Machine Tools Ltd, Middlesbrough.

Gauge : 3ft 0in

127	0-4-0T	OC	AB	1350	1913	(a)	s/s
128	0-4-0T	OC	AB	1351	1913	(a)	s/s
138	0-4-0T	OC	AB	1464	1916	(a)	s/s
158	0-4-0T	OC	AB	1735	1921	(a)	s/s

(a)   these locomotives went new to the Cleveland Works. When Ted Haigh visited the Middlesbrough Works on 12/7/1934, he noted the locomotives there and it is assumed that they were stored. A Dorman Long schedule dated 3/10/1931 lists the four locomotives commenting that three were in good condition but AB1464 was dismantled and missing its boiler and left-hand cylinder.

STOCKTON WORKS (previously called MALLEABLE WORKS)
Stockton   **23**
South Durham Steel & Iron Co Ltd until 28/7/1967   NZ 456198 – **Map D**
(Operation not taken over until 1/7/1968)
Stockton Malleable Iron Co Ltd until 29/12/1898

The company was established in 5/1861 and developed the Malleable Works on a site to the north of the Stockton Iron Works. It was reported in *The Engineer* on 28/8/1862 that the proprietors of the new venture had previously been involved with an iron works at Consett. Among the shareholders were William Bennington and John Holdsworth who had helped found the Stockton Iron Works. An early photograph of the Malleable Works shows an extensive layout of single-storey buildings and many tall chimneys that were typical of a puddling furnace plant that converted pig iron into wrought iron, prior to rolling into plates and girders.

In the 1860s, there were three works at the end of the NER's North Shore Branch on its eastern side – the three blast furnaces of the Stockton Iron Works, the Stockton Rail Mill, together with the Malleable Works. In 1871 the Rail Mill had 70 puddling furnaces and the Malleable Works 57, although the latter had increased to 64 by 1877. Malleable Works also had plate, rail and angle mills by 1865 and employed about 700 people. The depression in the iron trade at this time resulted in the closure of Stockton Iron Works and the Rail Mill. The blast furnaces were dismantled and both sites were eventually incorporated in the Malleable Works premises. *The Stockton Extension & Improvement Plan* prepared for the 1889 Parliamentary Session shows that the original railway line from the North Shore Branch to the Stockton Iron Works remained, although the latter was described as "Old Works". Shortly before this line turned southwards, two tracks headed north to serve the Malleable Works. A substantial building stood on the Rail Mill site with a rail connection from and running parallel to the North Shore Branch, which terminated in a headshunt serving the jetty before reversing into the building.

In 1888 a large basic open hearth plant for the production of steel had been erected. *The Hartlepool Mail*, in its 12/1/1889 issue, stated "Yesterday Stockton Malleable Iron Co made a successful start with their new steelworks in immediate proximity to the company's iron works." The first of three steel furnaces, with a capacity of 200 tons per week, was tapped and a mill to roll about 600 tons per week was expected to be completed in about six weeks. Unfortunately details of the early fleet of locomotives are sparse. A report in 1871 refers to No.1 locomotive from the Malleable Works helping out on occasional Sundays at the Stockton Iron Works.

Towards the end of the century Christopher Furness and William Creswell Gray wished to rationalise and end wasteful competition in the production of plates, angles and bars, especially for shipbuilding. They brought together three small companies with indifferent financial returns – West Hartlepool Steel & Iron Co, Moor Steel & Iron Co and the Stockton Malleable Iron Co - to form South Durham Steel & Iron Co Ltd; this was a private company under the chairman, Christopher Furness. It became a public company in 1900 and three years later developed into a major iron and steel manufacturer with the acquisition of the Cargo Fleet Iron Co Ltd. The Malleable Works railway system was still based on that of 20 years earlier but with a considerable expansion in the number of sidings. The original connection with the North Shore Branch now fed into a set of exchange sidings. In addition to serving the steel melting shop and rolling mills, other lines terminated end-on at two landing stages on the bank of the Tees, while sidings also ran on to the slag tip occupying the south east part of the site. Later the company's railway also served the Blue House Point Brick Works located between the slag tip and the river.

In order to increase its range of products SDSI, through the East Coast Steel Corporation, purchased Cochrane & Co which produced pipes by the spun iron method at Cargo Fleet; these having the advantage of resisting corrosion. Benjamin Talbot had developed a hydro-carbon lining but this was not wholly satisfactory for iron pipes. As platemaking capacity was in excess of demand, SDSI decided to commence the manufacture of steel pipes at the Malleable Works; the Talbot lining being useful to reduce corrosion of the steel. The spun iron plant at Cochranes was sold to the Stanton Iron Co Ltd and the first 30 feet long welded steel pipes were made in the new 84 inch mill at Stockton in 1926. Pipes ranging from 6in to 96in in diameter were manufactured using plate received from the company's West Hartlepool Works. Other plants on the site included the Sheet Works, which received steel bars and turned them into rectangular sheets for manufacturing into drums and the Tank Works producing plate for making into steel tanks, pylons and lamp posts. Facilities on the river were relatively modest in 1926 with a 90ft long "high wharf" with one electric crane and a 40ft long "low wharf". Bulmer reported "There are no steam cranes fixed on the wharves permanently, but they are served by locomotive cranes, as required".

The ten open hearth furnaces of 50 to 80 tons were closed down in 1928 because of the depressed economy and a major portion of the plant was dismantled but, in 1936, SDSI decided to reopen the shop with five furnaces each of 85 tons capacity. Materials, including cold iron, were brought into the plant on a single railway track for unloading by magnet crane into charging boxes. A series of sidings outside connected with the single track to ensure prompt delivery. The steel was cast into ladles on rail carriages. Three of these had integral electric drive but two were propelled by steam jib cranes working on an adjacent track.

As production of pipes increased, other operations reduced with the closure of the sheet mills in 9/1940 and the steel plant in 8/1951. There was more modernisation in the early 1950s, with an Automatic Submerged Arc Welding Plant opening in 1950 and 42 inch pipe mill on 25/11/1955 based on expanding demand from the oil industry. A fleet of new Fowler diesel mechanical locomotives was purchased, with some of the steam engines being passed to SDSI's other works. Pipes were also sent by railway to Stockton Corporation Quay and Middlesbrough Dock for export. The company's premises had now expanded further and abutted Portrack Lane.

SDSI, Dorman Long and Stewarts & Lloyds agreed to merge as British Steel & Tube Ltd, but this was overtaken by nationalisation on 28/7/1967. The Malleable Works was renamed the Stockton Works and employed 2,821 people. *Steel News* in 3/1969 reported that the 42 inch submerged arc welding mill was working 24 hours a day turning out gas pipelines – "They help to bring natural gas to Britain" – and the 84 inch mill was also on a 3-shift basis producing pipes for water mains, in addition to tubular piling. Following nationalisation, further rationalisation took place within BSC and Stockton's 42 inch mill closed in 1970 with rail traffic ceasing about the same time. In 2005 the 84 inch mill equipment was transferred by Corus to Hartlepool and the Stockton Works closed.

The original manufacturer of SOUTH DURHAM MALLEABLE No.5 is not known but it was reputedly rebuilt by SDSI in 1900. During September 1967, it stands in front of the large buildings that were a feature of the Stockton Malleable Works. The locomotive moved into the care of Beamish Museum in 1971 but remains there largely unrestored to this day. *(E. Lowden collection)*

References: *STOCKTON WORKS A Pictorial History of 150 Years in Iron & Steel*, Corus Tubes Group, 2005

'Iron and Steel', *South Durham Cargo Fleet Steel*, SDSI, 1936

'They Help to Bring Natural Gas to Britain', *Steel News*, British Steel Corporation, 3/1969

'The First Stockton Ironworks', Alan Betteney, *The Cleveland Industrial Archaeologist* 38, CIAS, 2018, p.p.19-34

Gauge : 4ft 8½in

MALLEABLE No.6	0-4-0ST	OC	AB	662	1890	(a)		
Rebuilt			Ridley Shaw		1935			(4)
MALLEABLE No.7	0-4-0ST	OC	AB	673	1891	New		(12)
MALLEABLE No.1	0-4-0ST	OC	HL	2461	1900	New		(10)
MALLEABLE No.2	0-4-0ST	OC	HL	2462	1900	New		(8)
MALLEABLE No.5	0-4-0ST	OC	HC?					
Rebuilt			SDSI		1900	(b)		(16)
No.8	0-4-0ST	OC	HE	1087	1911	(c)		(3)
9	0-6-0ST	IC	P	1444	1916	New		(6)
No.10	0-6-0ST	OC	HL	3365	1918	New		(2)
No.11	0-6-0ST	OC	AE	1787	1917	(d)		(7)
12	0-4-0ST	OC	HE	894	1905	(e)		(1)
SYDNEY	0-6-0ST	IC	P	877	1901	(f)		(5)
No.3	0-6-0ST	OC	AB	1497	1916	(g)		(9)
1	0-4-0DM		JF	4160007	1952	New		(14)
-	0-4-0DM		JF	4160008	1953	(h)		(10)
No.3	0-4-0DM		JF	4160009	1953	(i)		(14)
No.4	0-4-0DM		JF	4210087	1953	New		(17)
No.5	0-4-0DM		JF	4210088	1953	New		(13)
No.6	0-4-0DM		JF	4210116	1956	New		(15)
No.7	0-4-0DM		JF	4210125	1957	New		(15)
No.18	0-4-0DM		JF	4210086	1953	(j)		(11)

(a) new, per James Torbock (dealer), Middlesbrough. To Wensley Lime Co Ltd, Preston-under-Scar Quarry, Wensley, Yorkshire (NR), c1/10/1935 and returned c/1936. Inness has No.6 originally BENNINGTON and No.7 FOWLER, but the IRS only shows No.7 as BENNINGTON.

(b) origin unknown; Inness suggests SDSI Moor Steel and Iron Works but no confirmation.

(c) ex SDSI, Moor Steel and Iron Works, Stockton.

(d) ex Werddu Railway & Colliery Co Ltd, Pontardawe, Glamorgan, by 11/1924.

(e) ex SDSI, Moor Steel and Iron Works, Stockton.

(f) ex Crompton & Shawcross Ltd, Amberswood and Strangeways Hall Collieries, near Wigan, Lancashire, c/1937.

(g) ex Ferens & Love Ltd, Cornsay Colliery, near Waterhouses, Co. Durham, /1943. It had been offered for auction, 7/7/1943.

(h) new, IRS has new to Cargo Fleet Works but JF source shows new to Malleable Works and Cargo Fleet numbering appears to confirm this version.

(i) new, although JF records show as new to West Hartlepool Works. To West Hartlepool Works, /1957 and returned 1/1962.

(j) ex West Hartlepool Works, 9/1963.

(1) to West Hartlepool Works, /1936.
(2) to Cargo Fleet Iron Co Ltd, Cargo Fleet, by 10/1946.
(3) to West Hartlepool Works, Hartlepool, /1949.
(4) to Wensley Lime Co Ltd, Wensley Quarry, Preston-under-Scar, Yorkshire (NR), c/1951, but possibly as early as 1945–46.
(5) to Cargo Fleet Iron Co Ltd, Cargo Fleet, 8/4/1952.
(6) to Cargo Fleet Iron Co Ltd, Cargo Fleet c/1952.
(7) to Irchester Ironstone Co Ltd, Irchester Ironstone Quarries, Northamptonshire, 6/1952
(8) s/s. It was here at the Malleable Works on 20/9/1952.
(9) to SDSI, Storefield Ironstone Quarries, Northamptonshire, 8/1954.
(10) to Cargo Fleet Works, Cargo Fleet, c/1955.

(11) to Cargo Fleet Works, Cargo Fleet, /1964. An alternative suggestion is that the locomotive moved direct from the West Hartlepool Works to the Cargo Fleet Works, 12/1962.
(12) s/s. It was photographed at the Malleable Works, 22/7/1960.
(13) to British Steel Corporation, Imperial Tube Works, Airdrie, Lanarkshire, c9/1970.
(14) to J.D. White Ltd, Thornaby, /1971.
(15) to British Steel Corporation, Bromford Tube Works, Erdington, Birmingham, 8/1971.
(16) to Marley Hill Engine Shed, Co. Durham, for storage, 3/8/1971; then to North of England Open Air Museum, Beamish, County Durham.
(17) to Durham County Council, Dinsdale Residential School, Dinsdale, Co. Durham, 11/1971.

Gauge : 3ft 0in

5		0-4-0ST	OC	BH	1009	1890	New	(1)
6		0-4-0ST	OC	BH	1011	1890	New	(1)

(1) to Bowesfield Brick Co Ltd, Stockton, by 24/4/1934.

BRITISH STEEL CORPORATION/REDPATH DORMAN LONG LTD (subsidiary of BSC)

BRITANNIA STEEL WORKS, MIDDLESBROUGH 24
Dorman Long (Steel) Ltd/Dorman Long (Bridge & Engineering) Ltd NZ 484212 – **Map F**
until 28/7/1967, (Operation not taken over until 1/7/1968)
Dorman, Long & Co Ltd until 2/10/1954
Dorman, Long & Co until 2/11/1889
Britannia Iron Works Co Ltd until 11/1876
Britannia Iron Works Co until 12/8/1872

The works was constructed on 20 acres of land immediately north of the West Marsh Iron Works in the Ironmasters District on a large scale incorporating 120 puddling furnaces. It represented one of the biggest investments in wrought iron production in the area and involved two forge trains (these manipulated the iron while rolling), a blooming mill and a rail mill. It was founded by the Britannia Iron Works Co, but the main proponent was Bernhard Samuelson of the Newport Iron Works who was, no doubt, interested in a market for his pig iron. A private railway connected the two works. Half of the puddling furnaces were operational by 6/1871 and production initially concentrated on the rolling of iron rails. The company's sidings connected with the NER's Marsh Branch and ran from there on the north side of the works before swinging round to connect with its wharf at the river. The first section of the sidings, after leaving the Marsh Branch, formed part of the MOR and so tolls were payable for traffic over them.

Samuelson soon withdrew from active involvement in the works but retained a mortgage on the plant. The depression in the iron trade in the late 1870s and the increasing use of steel for rails caused the plant to struggle. The Britannia Iron Works Co Ltd (company had been formally registered 12/8/1872) decided to close the works and the business was wound up in 11/1876; the premises passing back to Samuelson. After one abortive attempt to sell it (according to *The Engineer* magazine, the auction was to be on 27/9/1876 and the plant for sale included **two** Fletcher Jennings tank locomotives), Samuelson let the works to the "proprietors of the Skerne Ironworks, Darlington" on 25/2/1878. The works reopened in 7/1878 but one month later the Skerne Ironworks Co was in liquidation and its stock was to be sold by auction on 9/9/1879 at the Britannia Iron Works, including a recently new I'Anson 12 inch locomotive. In 12/1879, Dorman & Long leased Britannia for three years with an option to purchase for £50,000. Dorman, Long & Co bought the Britannia Works on 29/4/1882, although still holding a mortgage from Samuelson; the outstanding debt was settled in 1890. It began rolling iron girders in 1883 but it was not long before half of the puddling furnaces were dismantled and seven acid open hearth steel furnaces, each of 15 tons capacity, were installed in 1887 (first three tapped 5/8/1887). The company became the first on Teesside to start rolling steel girders in the same year. To complement this development, a small bridge and construction yard was opened at the works with steel fabrication and erection in 1889. The premises in 1893 formed a triangular shaped relatively constrained site situated between the slag heaps of the West Marsh and Ayresome Iron Works, with the base of the triangle alongside the river. Sidings covered much of the open ground within the site.

The next twenty years witnessed a major expansion of the Britannia Works. In 1893, one of the first fully continuous rod mills was commissioned, based on a design patented by George Bedson at the Bradford Works in Manchester. The plant was started by The Bedson Wire Co Ltd but soon carried on by Arthur Dorman and his son Charles, under the style C. Dorman & Co until it was taken over by Dorman, Long & Co in 1898. Earlier, the NER, had reached agreement with The Bedson Wire Co Ltd on 6/10/1894 for sidings to be constructed to serve the plant. By 1899, it was producing about 400 tons of wire rods from two inch square billets and drawing

200 tons of wire per week. The works was located behind the West Marsh River wharf with a rail connection via the MOR's "West Marsh Wharf Siding" to the NER, as well as numerous lines linking it with Britannia Works, which carried out all of the shunting.

In 1895, Britannia Works had ten open hearth furnaces producing acid steel but a change was made to basic steel in 1901. A 3ft 0in gauge railway system was installed and a fleet of five new 6 inch four-coupled saddle tanks were purchased from Black, Hawthorn and its successor in 1893–99 to operate it. An additional two 8 inch locomotives were later obtained from Peckett. A Guide to the Ports on the Tees & The Hartlepool's (1900) commented that ingots from the steel furnaces "are taken hot along a 3ft gauge railway to vertical heating furnaces and then by roller gear to a 40-in. cogging mill where they are rolled into blooms of suitable sizes". After completion of work on the girders, they could be shipped using the gantries that extended to the wharf or put in railway wagons. Larger sections were rolled at Britannia and smaller sections at West Marsh at this time.

On a plan, dated 19/10/1903, seven sidings comprised the NER's West Marsh Branch. Two of these were reserved for Cleveland Wire Works traffic and one for the West Marsh Wharf siding. Dorman Long had laid a number of sidings for storing iron ore on the west side of the West Marsh Iron Works site with a connecting line to Britannia Wharf. In a letter dated 23/3/1905, Dorman Long was identified as running over the NER sidings with iron ore from Britannia Wharf. Dorman Long argued that this was covered by the £20pa payment to the NER for West Marsh Iron Works, but the NER insisted that an extra £5 be paid to run over its tracks. West Marsh Works was increasingly integrated into the Britannia premises. Nos.1 and 2 rolling mills were modernised in 1904 and two more (Nos.3 and 4) installed between the Cleveland Wire Works and West Marsh Wharf in 1908. As a result, Britannia produced most of the joists, larger sections and structural steel made by Dorman Long until after World War 2. For example, it provided 10,000 tons of steelwork in 1902 for the erection of the British Westinghouse Co's factory at Trafford Park (later Metropolitan Vickers). Dorman Long had relied on a miscellaneous collection of standard gauge saddle tanks to shunt Britannia's railways but, from 1908, it standardised on Hawthorn Leslie locomotives, purchasing ten before the company's fleets on Ironmasters were combined in the early 1930s. The engine shed (NZ 483214) was located on the northern boundary. The average output of steel from the eleven open hearth furnaces was about 3,300 tons per week in 1908.

The bridge and constructional shops at Britannia Works were described as the largest of their kind in Europe and employed 450 people. Dorman Long wished to extend its bridge yard, install a stockyard and enable its locomotives to handle traffic on more of its own tracks. Under the 1914 NER Act and an agreement dated 19/7/1916, the West Marsh Branch and its sidings were abandoned. Instead, Dorman Long transferred some of its land to the NER abutting Richard Hill's Marsh Wire Works and constructed a replacement branch comprising

A Chapman & Furneaux maker's photograph of 1178 of 1899 which worked on the 3ft gauge railway serving the Britannia steel plant. (NERA R. Inness collection)

Many of the Dorman Long locomotives in the Ironmasters district carried large cast number plates to aid identification or did they? Number 43 was originally CF 1199/1900 but much of it was scrapped in 1949 and a new '43' constructed by Dorman Long in the same year; it is seen at the Britannia Works on 29th September 1951. *(IRS Bernard Roberts photograph)*

ten sidings to the satisfaction of the NER's Engineer. These joined near the river connecting with existing lines at Newport Iron Works, West Marsh Wharf and the Cleveland Wire Works.

The yard went on to provide the materials for some iconic bridges, including the Sydney Harbour Bridge. The contract was signed in 1924 for what was then the largest arch bridge in the world and 51,000 tons of constructional steel were rolled at the Britannia before being shipped to Australia where it was fabricated. A "Universal" testing machine was installed in a test house at the works and was first used on the Sydney Harbour bridge materials.

The need to move and process heavy sections around the works called for an extensive fleet of travelling rail cranes and those existing in 1929 at Britannia Works are shown below:

No.	Maker	Year	Capacity	No.	Maker	Year	Capacity
2	Roger	1895	5 tons	14	Priestman	1917	5 tons
3	Roger	1886	5 tons	15	Smith	1900	5 tons
4	Roger	1887	5 tons	17	Dorman Long	1908	10 tons
6	Roger	1889	5 tons	20	Ridley	1895	3 tons
7	Roger	1913	5 tons	21	Isles	1913	5 tons
8	Roger	1901	10 tons	22	Dorman Long	1913	10 tons
9	Smith	1892	5 tons	23	Isles	1919	10 tons
10	Booth	1892	10 tons	31	Smith		5 tons *
12	Smith	1897	5 tons	66	Booth		5 tons *
13	Smith	1900	5 tons	72	Smith		5 tons *

* Belonged to Bridge & Construction Yard.

Following closure of the Newport Iron Works, molten iron was delivered solely from the Acklam blast furnaces in 30 ton ladles hauled by one of Dorman Long's locomotives. In the early 1930s, the locomotive fleets at Acklam, Britannia and Newport Works were combined and renumbered.

The purchase of Bolckow Vaughan by Dorman Long in 1929 brought the former's subsidiary, Redpath Brown & Co Ltd, into the fold. In 1937 Britannia's Bridge and Construction Department occupied eleven acres, employed

1,200 people and had a monthly output of 5,000 tons. In 11/1947 the railway company identified that there was insufficient accommodation for the reception of coal at the Britannia Works and Dorman Long had been using the former Crombie sidings for this purpose. The LNER delivered the coal to the sidings and one of Dorman's locomotives collected the wagons from there. This movement entailed the locomotive working across the MOR diverted West Marsh Branch on to the LNER Marsh Branch and thence into the Britannia Works over the MOR connection. This movement had not been declared by Dorman Long which had hitherto escaped these MOR tolls. A survey by BR for a three week period in January 1948 revealed the number of loaded wagons standing in the former Crombie sidings to be 163 coal and 135 goods. During the same period traffic dealt with at "New Ground", the normal reception point for Britannia Nos.3 and 4 mills, was 1,732 wagons. Dorman Long was paying £50 pa in 1946 for traffic over MOR lines to and from Britannia.

Dorman Long's process of rationalisation gathered momentum after World War 2 to the detriment of Britannia. As T.R. Tighe comments, in the good times its profits helped pay for expansion at Cleveland and Lackenby but, with a downturn in trade, there was no money to invest in Britannia. Completion of the massive Lackenby open hearth plant marked the end of steelmaking at Britannia in 11/1953. The Bridge Yard continued to be involved in major projects, with others, including the Forth, Severn and Humber Road bridges and the Eggborough and Ironbridge Power Stations. In 1959, old Nos.1 and 2 mills were working alternately with Nos.3 and 4 dealing with small orders using steel brought in by BR. The Cleveland Wire Works was still operating. Dorman Long began replacing its steam locomotive fleets with diesels in 1959 concentrating initially on its Cleveland–Lackenby complex. Steam continued on at the Ironmasters for a couple of years and a number of locomotives made redundant at Cleveland–Lackenby moved across to join the Ironmasters fleet. In the meantime, Britannia had acquired a few small Hibberd diesel locomotives. Although too light for heavy iron and steel work, they were ideal for shunting the construction yard. Some of the new Sentinel diesels were allocated to the remaining Acklam iron and steel traffic.

When Dorman Long & Co Ltd became the holding company for a number of trading businesses in 1954, the Britannia rolling mills come under Dorman Long (Steel) Ltd while the Bridge Yard became part of Dorman Long (Bridge & Engineering) Ltd. About the time of nationalisation in 1967, the latter merged with Redpath Brown & Co Ltd to form Redpath Dorman Long Ltd. On 19/3/1971, the rolling mills closed with the loss of 713 jobs and was followed by their demolition. *Steel News* on 12/10/1972 contained the comment "There is not a brick left now of Britannia Works and we took out over 12,000 tons of scrap". This was rather a simplification because the construction yard remained open comprising (from north to south) a fabrication shop (known as the Acklam Fabrication Shop), steel preparation building, stockyard, and Autofabrication Shop. These were located on the southern portion of the Britannia site, part of which had once been the location of the West Marsh Works. The premises were now connected to BR's Middlesbrough Town Goods Yard by a single siding, via the Old Town Branch Loop. On entering the premises, this line divided in two with one leading to the stockyard sidings and the other giving access to the exchange sidings before running between the Autofabrication Shop and Marsh

FH 3942/1960 shunts a bogie bolster wagon carrying girders at the Britannia Works on 26th July 1979.
(Cliff Shepherd photograph)

Wire Works to serve the Cleveland Wire Works and fabrication shop. According to BR, the ownership of the Cleveland Wire Works siding was changed to Rylands & Whitecross Ltd on 23/8/1973. The last year that MOR tolls were paid was 1975.

By 1979, only one working locomotive was required each day and this was used to move individual bogie bolster wagons for loading with steel girders in the stockyard. On 25/1/1980 the Evening Gazette announced that Britannia Works was to close with the loss of 630 jobs. After it finished in 1980, an auction of the plant took place in 2/1981 and some of the internal wagons were being cut up in the following month. The BSC was looking to sell Redpath Dorman Long Ltd to the Trafalgar House Group in 4/1982 and the Autofabrication Shop and test house appear to have seen some intermittent use after the closure. The demolition of the main buildings was well underway in 3/1986, but the Test House remained operational.

References : K. Fleming letter, *The Industrial Locomotive* 16, ILS, 1979

'They span the world with their expertise', *Steel News*, British Steel Corporation, 9/1968

'Britannia Works "Engineers"', Frank Jones, *The Industrial Locomotive* 73, ILS, 1994 and subsequent letters in the following issue.

'More Dorman Long Locomotives', John Fletcher, *The Industrial Locomotive* 103, ILS, 2002

Locomotives at Britannia Steel Works prior to the formation of combined fleet.
Gauge : 4ft 8½in

	SUMUS	0-4-0tank	OC	FJ	90	1872	New	(1)
-		?		FJ		pre1876	?	s/s
-		?		l'Anson		c1878	New	s/s
No.3		0-4-0ST	OC	BH	477	1879	(a)	s/s
No.4		0-4-0ST	OC	BH	478	1880	(b)	s/s
1		0-4-0ST	OC	HL	2133	1889	New	(2)
-		0-4-0ST	OC	BH	977	1889	(c)	s/s
3		0-4-0ST	OC	YE	262	1875	(d)	(3)
7		0-4-0ST	OC	P	462	1887	(e)	(4)
6		0-4-0ST	OC	CF	1199	1900	New	(4)
No.7		0-4-0ST	OC	BH	547	1881	(f)	s/s
8		0-4-0ST	OC	BH	679	1882	(f)	(3)
9		0-4-0ST	OC	HL	2748	1908	New	(4)
10		0-4-0ST	OC	HL	2796	1910	New	(4)
11		0-4-0ST	OC	HL	2905	1911	New	(4)
12		0-4-0ST	OC	HL	3032	1913	New	(4)
No.4		0-4-0ST	OC	HL	3157	1915	New	(4)
5		0-4-0ST	OC	HL	3211	1916	New	(4)
13		0-4-0ST	OC	HL	3334	1918	New	(4)
14		0-4-0ST	OC	HL	3388	1919	New	(4)
27		0-4-0ST	OC	HL	3353	1918	(g)	(4)
2		0-4-0ST	OC	HL	3566	1927	New	(4)

(a) ex West Marsh Iron Works, Middlesbrough and probably absorbed into Dorman Long fleet at Britannia Works when premises merged.
(b) ex West Marsh Iron Works, Middlesbrough. Probably absorbed into Dorman Long fleet at Britannia Works, although IRS suggests that it came via the Ayrton Sheet Works.
(c) ex Ayrton Sheet Works, Middlesbrough.
(d) ex Newport Iron Works, Middlesbrough.
(e) ex Newport Iron Works, Middlesbrough.
(f) ex West Stockton Iron Co Ltd, Stockton.
(g) ex Clarence Works, 9/11/1923.
(1) to B. Samuelson & Co, Newport Iron Works, Middlesbrough.
(2) scrapped by /1929.
(3) scrapped after /1932. Assumed not included in combined list although Inness suggests that it became No.42 and was replaced by AB 1620.
(4) renumbered in early 1930s (after 10/1931) into combined Acklam, Britannia and Newport Works list of steam locomotives (see below).

Gauge : 3ft 0in

11	0-4-0ST	OC	BH	1066	1892	New	Scr
12	0-4-0ST	OC	BH	1067	1892	New	Scr
13	0-4-0ST	OC	BH	1077	1892	New	Scr
14	0-4-0ST	OC	BH	1081	1893	New	Scr
15	0-4-0ST	OC	CF	1178	1899	New	Scr
No.16	0-4-0ST	OC	P	1331	1913	New	Scr
No.17	0-4-0ST	OC	P	1440	1916	New	Scr

Combined Acklam, Britannia & Newport Works steam locomotives

Dorman Long renumbered its steam locomotives on the Ironmasters into one combined fleet in the early 1930s. They continued to operate from one of the three works but might be seen at any of them because of the interconnecting railways. No attempt has been made to trace the change in allocation between the individual sheds. Replacing worn out parts was a normal part of operating industrial locomotives and this might include major items, such as the boiler and firebox, but usually the engine's appearance did not change. This was not the position with Dorman Long's Ironmasters fleet in the early 1950s when more parts were replaced resulting in altered appearances to several locomotives. Previously these have been classified by the IRS as new locomotives but it appears, in some cases, that the original chassis was retained thus representing a rebuild, A further complication is that in one case of rebuilding, the number plates on the locomotive were changed! Therefore a qualifying note has been added to indicated where this may have occurred. In October 1948, Dorman Long's Bridge Department, Middlesbrough advertised a Peckett locomotive with 14in x 20in cylinders for sale but no details are known.

Gauge : 4ft 8½in

1		0-4-0ST	OC	AB	1888	1926	(a)	
		Rebuilt *	DL			c1953		(10)
2		0-4-0ST	OC	AB	1599	1918	(a)	Scr c2/1963
3		0-4-0ST	OC	AB	924	1901		
		Rebuilt				1919	(a)	Scr c2/1963
4		0-4-0ST	OC	AB	1478	1916	(a)	Scr c2/1963
5		0-4-0ST	OC	AB	1515	1917	(a)	
		Rebuilt		AB				(9)
6	(22)	0-4-0ST	OC	AB	1274	1912	(a)	Scr /1961
7		0-4-0ST	OC	AB	1040	1905	(a)	Scr c10/1962
8		0-4-0ST	OC	AB	879	1900	(a)	Scr /1963
9		0-4-0ST	OC	HL	3566	1927	(b)	Scr c3/1963
10		0-4-0ST	OC	HL	3352	1918	(a)	
		Rebuilt				7/1948		(11)
11		0-4-0ST	OC	HL	3353	1918	(b)	Scr 5/1963
12		0-4-0ST	OC	HL	2748	1908	(b)	Scr c3/1963
13		0-4-0ST	OC	HL	3477	1920	(c)	Scr c/1961
14		0-4-0ST	OC	HL	3482	1920	(c)	Scr c10/1962
15		0-4-0ST	OC	HL	3334	1918	(b)	(6)
16		0-4-0ST	OC	HL	3388	1919	(b)	Scr c/1961
17		0-4-0ST	OC	HL	3211	1916	(b)	Scr /1960
18		0-4-0ST	OC	HL	3169	1916	(a)	Scr 6/1961
19		0-4-0ST	OC	HL	3157	1915	(b)	Scr 24/11/1960
20		0-4-0ST	OC	HL	3032	1913	(b)	(8)
21		0-4-0ST	OC	HL	2913	1912	(a)	Scr /1961
22		0-4-0ST	OC	HL	2905	1911	(b)	(1)
22		0-4-0ST	OC	DL*		1953	New	Scr c10/1962
23		0-4-0ST	OC	HL	2796	1910	(b)	(2)
24		0-4-0ST	OC	HL	2712	1907	(a)	Scr 24/11/1960
25		0-4-0WTST	OC	K	4945	1913	(c)	Scr 24/11/1960
26		0-4-0WTST	OC	K	4737	1910	(c)	Scr c/1956
27		0-4-0WT	OC	K	2363	1881	(c)	Scr /1952
27		0-4-0ST	OC	DL*		1952	(k)	(7)
28		0-4-0WTST	OC	K	3882	1899	(c)	

No.	Name	Type	Cyl	Builder	Works No.	Date	Source	Disposal
		Rebuilt		DL		after 1951		(5)
29	+	0-4-0WTST	OC	K	3982	1900	(c)	
		Rebuilt		Britannia		1941		Scr after 19/9/1960
30		0-4-0WTST	OC	K	5114	1914	(c)	Scr /1959
31	+	0-4-0WTST	OC	K	5115	1914	(c)	Scr 24/11/1960
32		0-4-0WT	OC	K	2362	1881	(c)	Scr c/1938
33		0-4-0ST	OC	AB	1059	1905	(a)	(4)
34		0-4-0ST	OC	AB	650	1889	(c)	Scr /1960
35		0-4-0ST	OC	BH	1124	1896	(a)	Scr /1952
35		0-4-0ST	OC	DL*		1952	New	Scr c10/1962
44 (36)	ENGINEERS	0-4-0ST	OC	P	462	1887	(b)	
		Rebuilt		DL		1953		Scr 12/1958
36		0-4-0ST	OC	DL*		1953	New	(7)
37		0-4-0ST	OC	HC	324	1889		
		Rebuilt		Acklam		1926	(a)	Scr c/1948
38		0-4-0ST	OC	HC	323	1888		
		Rebuilt		Acklam		1923	(a)	Scr c/1948
39		0-4-0ST	OC	BH		1888		
		Rebuilt		Acklam		1909		
		Rebuilt		Acklam		1934	(a)	Scr /1951
39		0-4-0ST	OC	DL*		1951	New	(7)
40		0-4-0ST	OC	BH		1888		
		Rebuilt		Acklam		1919	(a)	Scr c/1952
41		0-4-0ST	OC	BH	614	1882	(a)	Scr c/1949
43		0-4-0ST	OC	CF	1199	1900	(d)	Scr /1949
43		0-4-0ST	OC	DL*		1949	New	(9)
42		0-4-0ST	OC	AB	1620	1919	(h)	(10)
45		0-4-0ST	OC	HL	3031	1913	(e)	(9)
46		0-4-0ST	OC	AB	2032	1937	New	
		Rebuilt		DL		c1956-58		Scr c3/1963
47		0-4-0ST	OC	HL	3918	1937	New	Scr /1963
48		0-4-0ST	OC	AB	300	1888	(f)	Scr c/1956
49		0-4-0ST	OC	AB	671	1891	(f)	Scr c/1956
50		0-4-0ST	OC	AB	1102	1907	(g)	Scr /1960
51		0-4-0ST	OC	AB	897	1900	(f)	
		Rebuilt		.		1948		Scr c/1960
52		0-4-0ST	OC	AB			(i)	(3)
52		0-4-0ST	OC	RSHN	7340	1947	New	(11)
53		0-4-0ST	OC	RSHN	7343	1947	New	Scr 5/1963
54		0-4-0ST	OC	RSHN	7346	1947	New	(12)
(T.A. 18/1)	(ACKLAM WORKS)	0-4-0CT	OC	AB	2152	1943	(j)	Scr c10/1962
55		0-4-0ST	OC	HL	3429	1919	(l)	s/s c/1965
20		0-4-0ST	OC	RSHN	7072	1942	(m)	Scr 11/1964
21		0-4-0ST	OC	RSHN	7075	1943	(m)	(13)
23		0-4-0ST	OC	RSHN	7341	1947	(m)	Scr 10/1964
24		0-4-0ST	OC	RSHN	7342	1947	(m)	(12)
25		0-4-0ST	OC	RSHN	7345	1947	(m)	(12)
184	(JOSEPH STIRZAKER)	0-4-0ST	OC	RSHN	7348	1947	(n)	(14)
178	SIR ARTHUR DORMAN	0-4-0ST	OC	RSHD	7037	1941	(o)	Scr 12/1964
180	ROOSEVELT	0-4-0ST	OC	RSHN	7065	1942	(o)	Scr 11/1964
173	ALFRED E. ALCOCK	0-4-0ST	OC	HL	3919	1937	(p)	(11)
177	SIR ELLIS HUNTER	0-4-0ST	OC	RSHD	7036	1941	(p)	(12)
179	CHURCHILL	0-4-0ST	OC	RSHD	7041	1941	(p)	(13)
183	WILLIAM JONES	0-4-0ST	OC	RSHN	7347	1947	(p)	(11)

Note: Some dates of transfers between the Clarence and Britannia Works vary by a couple of days depending on which records are used.

+ According to photographic evidence, by 21/5/1959, the plates on 29 and 31 had been switched so that the rebuilt locomotive was now 31.

* The above locomotives were either substantially rebuilt or were virtually new engines with the reuse of some old parts. The *possible* position is: -

1	rebuild of AB1888
22	new but using HL2905's number
27	new but using K2363's number
35	new but using BH1124's number
36	new but using P462's number
39	new but using BH /1874's number. Completed 12/4/1951.
43	new but using CF1199's number

(a) ex Acklam Works, Middlesbrough. HL3169 and HC324 to Carlton Works Stillington in 9/1933 and returned, HC324 in 3/1934.
(b) ex Britannia Works, Middlesbrough.
(c) ex Newport Works, Middlesbrough. A note in DL records suggests 31 went to "Eaglescliffe" 13/8/1938 and returned 15/9/1938.
(d) ex Britannia Works, Middlesbrough. To Carlton Works, Stillington and returned 8/1/1938.
(e) ex Clarence Works, Port Clarence, 18/3/1935.
(f) ex Linthorpe-Dinsdale Smelting Co Ltd, Linthorpe Iron Works, Middlesbrough, /1937.
(g) ex Linthorpe-Dinsdale Smelting Co Ltd, Linthorpe Iron Works, Middlesbrough, /1937. To Clarence Works, Port Clarence, c/1954 and returned /1955.
(h) ex Cleveland Works, South Bank, 20/8/1941 and re-numbered 1/1942 becoming 42.
(i) origin?
(j) ex Ministry of Supply. It subsequently had its crane removed after /1952, by 18/5/1957.
(k) new. To Bowesfield Works, Stockton, 6/9/1957, although Bob Payne saw it at Bowesfield on 21/11/1955 and at Newport on 5/5/1957. Ex Clarence Works, Port Clarence, by 5/1958.
(l) ex Redcar Works, Redcar, c4/1962.
(m) ex Redcar Works, Redcar, 9/1962.
(n) ex Cleveland Works, South Bank, 24/9/1962.
(o) ex Cleveland Works, South Bank, 5/10/1962.
(p) ex Cleveland Works, South Bank, 11/1962.

(1) to Redcar Works, Redcar, /1938.
(2) to Cleveland Works, South Bank, c/1938.
(3) to Eaglescliffe Chemical Co Ltd, Urlay Nook, /1943.
(4) to Burley Quarry, Rutland, 8/1949.
(5) to Lackenby Works, Grangetown, c9/1954.
(6) to Clarence Works, Port Clarence, c2/1959.
(7) to Clarence Works, Port Clarence, 5/1960.
(8) to Clarence Works, Port Clarence, 6/1960.
(9) to Newfield Brickworks, Newfield, Co. Durham, c11/1960.
(10) to Redcar Works, Redcar, /1962.
(11) to Seaham Harbour Dock Co Ltd, Seaham, Co. Durham, 6/1963.
(12) to Seaham Harbour Dock Co Ltd, Seaham, Co. Durham, 11/1963.
(13) to Gjers Mills & Co Ltd, Middlesbrough, 13/2/1964.
(14) to Gjers Mills & Co Ltd, Middlesbrough, 11/1964.

Combined Acklam, Britannia, Newport diesel locomotives

From 1958 Dorman Long began to dieselise the locomotive duties at the Acklam, Britannia and Newport Works, but the closure of the Acklam Iron & Steel Works in 1963 meant that the remaining locomotives were confined to shunting at Britannia.

Gauge : 4ft 8½in

-	4wDM	FH	3883	1958	(a)	(1)
11	4wDH	S	10011	1959	New	(4)
14	4wDH	S	10014	1960	New	(5)
15	4wDH	S	10015	1960	New	(7)
16	4wDH	S	10016	1960	New	(4)
17	4wDH	S	10017	1960	New	(5)

18		4wDH	S	10018	1960	New	(8)
19		4wDH	S	10019	1960	New	(2)
24		4wDH	S	10024	1960	New	(3)
2	(41)	4wDM	FH	3941	1960	New	(9)
3	(42)	4wDM	FH	3942	1960	New	(10)
4		4wDM	FH	3888	1958	(b)	(9)
25		4wDH	S	10025	1960	(c)	(6)
1	(34)	4wDH	FH	3934	1960	(d)	(13)
	TEES SIDE No.8	4wDH	TH	115V	1962	(e)	(11)
-		4wDM	RH	299107	1950	(f)	(12)

(a) new, delivered to Newport Works, but it appears to have been based at Ayrton Sheet Works.
(b) ex Lackenby Works, Grangetown, 2/6/1963.
(c) ex Redcar Works, Redcar, 27/4/1964.
(d) ex Lackenby Works, Grangetown, before 12/1969.
(e) ex Redpath Dorman Long Ltd, Teesside Engineering Works, Cargo Fleet, c3/1979.
(f) ex Redpath Dorman Long Ltd, Westburn Works, Cambuslang, Strathclyde, /1980.

(1) to Dock Street Foundry, Middlesbrough, c/1962.
(2) to Cleveland Works, South Bank, 21/6/1962.
(3) to Cleveland Works, South Bank, 22/6/1962.
(4) to Cleveland Works, South Bank, 25/6/1962.
(5) to Lackenby Works, Grangetown.
(6) returned to Redcar Works, Redcar, by /1968.
(7) to Lackenby Works, Grangetown, 22/3/1971.
(8) to Hartlepool Works, 19/7/1972.
(9) scrapped on site c6/1980.
(10) to T. Smith & Son, Hartlepool, for scrap, 9/1980.
(11) to Redpath Dorman Long Ltd, Westburn Works, Cambuslang, Scotland, c1/11/1980.
(12) scrapped on site by Mace Metals, Darlington, Co. Durham, 7/1981.
(13) to Bromley Iron & Steel Co Ltd, Lye, Staffordshire, for scrap, 12/1981.

BSC (CHEMICALS) LTD

PORT CLARENCE DISTILLATION WORKS, previously CLARENCE IRON & STEEL WORKS, Port Clarence 25
British Steel Corporation until /1973 NZ 508213 – **Map C**
Dorman Long (Chemicals) Ltd until 28/7/1967 (Operation not taken over until 1/7/1968)
Dorman, Long & Co Ltd until 2/10/1954
Bell Brothers Ltd until 2/5/1923 (subsidiary of Dorman Long from /1902)
Bell Brothers & Co until 27/11/1873
Bell Brothers until /1864

Bell Bros was formed in 1844 by Isaac Lowthian, Thomas and John Bell to lease an iron works at Wylam. They were the sons of Thomas Bell who was a partner in Losh, Wilson & Bell's iron and alkali works at Walker on the River Tyne. Following confirmation of substantial deposits of Main Seam ironstone at Eston in 1850, Bell Bros negotiated a lease to extract ironstone from the Normanby Estate but, in return, Ralph Ward Jackson insisted that the lessees build their iron works on the north bank of the River Tees. Bell Bros purchased 30 acres of land at Port Clarence from the West Hartlepool Dock & Railway Company in 1852. The foundations of the Clarence Iron Works were laid in 3/1853 and it expanded rapidly as a pig iron producer under the direction of Isaac Lothian Bell. A row of blast furnaces was erected at right angles to the river known as the "Old Side": Nos.1 and 2 (blown in 1/1854), 3 (10/1854), 4 and 5 (11/1858) and 6 (10/1861). Two more blast furnaces, numbers 7 and 8, were added at the increased height of 80 feet and were blown in during 5/1865. Subsequently, between 1866 and 1871, the others were rebuilt to also stand 80 feet high. Ironstone from Bell Bros' Normanby Mine had to be transported initially by the SDR over its Guisborough Branch to Middlesbrough where it was ferried downstream to Bell's Clarence Wharf. With the opening of the Cleveland Railway in 1861, the ironstone could be brought across the river from Normanby Jetty in sets of 24 wagons loaded on a barge. To begin with, the calcining kilns were rectangular and placed at right angles to the Old Side blast furnaces but, about the end of the 1860s, they were taken down and replaced by 15 circular kilns. The company also owned collieries in County Durham and limestone quarries in Weardale.

The Clarence Railway had opened its main line from Simpasture on the SDR near Shildon to Samphire Batts (renamed Port Clarence) where coal drops were erected on the north bank of the Tees and coal began to be shipped in 1834. After the Clarence Railway was leased to the Stockton & Hartlepool Railway, the portion of line to Port Clarence became a branch. In 1859 this crossed the marshy ground to reach the five spurs to the coal drops although, by then, most coal shipments had been transferred to West Hartlepool Docks. The branch continued on a short distance to the Clarence Iron Works where a curve and reversal enabled trains to reach the Old Side blast furnaces.

In 1873 Bell Bros designed the "New Side" blast furnaces which stood parallel to and east of the "Old Side" furnaces. Although intended for another eight blast furnaces, only four were built: Nos.9 and 10 (blown in 10/1874) and 11 and 12 (10/1875). By this date, Bell Bros was responsible for about 16% of iron production on Teesside.

Bell Bros identified a 100 feet thick bed of salt underlying its iron works and Salt Holme Farm to the north; it began pumping brine in 1882. The brine was evaporated in pans heated by both coal and waste heat from the blast furnaces. Two panhouses were located next to the blast furnaces while the third and largest stood to the north of the exchange sidings near the later Greatham Creek Branch. A total of 51,871 tons of salt was produced in 1885 from five boreholes. Bell Bros also opened an Ammonium Soda plant (in 1884) and a Chlorine plant. The Clarence Salt Works (which see) was absorbed into The Salt Union Ltd (later ICI) in 1888 and the Ammonia Soda plant was taken over by Brunner Mond in 1896 and closed.

Towards the end of the nineteenth century, Bell Bros ran into difficulties because steel was taking an increasing share of the market and overseas competition was increasing. It was financially restructured in 1895 and in 1/1899 Dorman Long acquired 50% of Bell Bros shares. The remainder were purchased in 1902 and the company then functioned as a wholly owned subsidiary of Dorman Long. The latter made the Clarence Works one of its main production units with increased investment. Most significantly, a new steel plant was authorised on the northern part of the site. On 22/3/1899, Bell Bros' Board approved the erection of four basic open hearth furnaces, washing and cleaning machinery, four new coke ovens, two blowing engines for the "New Side" blast furnaces, four hot blast stoves and four calcining kilns, as well as the purchase of two 16 inch locomotives from Andrew Barclay at a cost of £1,375 each and one 5 ton "loco crane" from Smith of Rodley. Within the next three years, four more open hearth furnaces, a billet mill and mixer had also been approved.

An agreement on 26/9/1908 between the NER and Bell Bros/Dorman Long saw the enlargement of the exchange sidings alongside the north bank of the River Tees; some of which were called the "Low Sidings". There was a significant difference in levels here and the original signal box near station cottages was referred

The Clarence Iron & Steel Works with the Old Bank blast furnaces on the left and the coal shipping gantry running past the New Bank furnaces in the centre to the River Tees. (IRS J.A Peden collection)

to as "Bells Bank Foot" in the agreement. On 4/10/1928, Dorman Long agreed to pay £125 every six months for the LNER to haul inward mineral traffic up Bell's Bank to the High Level Sidings.

The new coking plant comprising 60 Huessener ovens was erected in 1901 and another sixty were added in 1904 after the adjoining byproducts plant had been completed. These ovens were financed by the Coal Distillation Company with its expenditure being recouped from the by-products sales. This company operated a fleet of mainly 12 ton tank wagons registered for use on the main line. Purchases appear to have begun in 1908 when numbers 1 to 9 were recorded by the NER. The wagons were mainly built by Charles Roberts & Co Ltd. Numbers had reached 30 when four 20 ton tank wagons were registered in 12/1911. The Government took over the Coal Distillation Co during World War I because of its German connections and Bell Bros acquired the coke ovens in 1919 after the conflict. They ceased operating 12/7/1921, their place being taken by 72 Collin ovens to the west of the Old Side blast furnaces in 1916 and 84 Otto waste heat ovens (under construction in 1922). Two old blast furnaces were replaced by a modern furnace, designated C, on the "New Side" and this was blown in on 9/10/1923 in connection with a revamped river wharf and ore unloading conveyor. The stocking ground behind the wharf had a capacity of 50,000 tons and received incoming iron ore, which could then be lifted out by an overhead crane for putting into railway wagons. Bell Bros had also purchased one hundred 20 ton steel hopper iron ore wagons (numbers B628–B727) from Tees Side Bridge & Engineering Works in 2-4/1922 registered by the NER to run over the main line. Also, about this time, a "coal shipping plant" was installed comprising twin tracks on a prominent raised gantry running by the "New Side" blast furnaces to a swinging conveyor, supported on two sets of railway lines, which delivered coal for bunkering ships or despatch.

The works had reached its maximum extent by this date. The Port Clarence Branch passed by the single platform of the passenger station (closed 9/11/1939) and ran on as four tracks by Port Clarence Junction Signal Box to enter the exchange sidings on the north bank of the river where a NER goods depot was also located. From here, lines led to the Old or Upper Wharf with four electric and two steam cranes and the New or Lower Wharf with four electric cranes. The blast furnaces stood in two rows with railways alongside to bring in raw materials and other sidings diverged from them to take away iron and slag. In 1913 a two-road engine shed was situated between the rows, but this building later became a boiler shop and was replaced by a three-road shed at the southern end of seven streets of mean terrace houses dating from 1865 and situated amongst the works. Some of the coke ovens stood immediately north of these houses! More railway lines ran northwards to serve the steel plant and mills, while tracks extended eastwards on to the slag tips where the Erimus Slag

WHITWORTH (BH 519 of 1879) stands in front of some of the impressive buildings that were a feature of the Clarence Iron and Steel Works. Lowthian Bell had invited Philip Webb, an architect of national repute, to prepare some of the designs. *(Kevin Lane collection)*

Works was situated. Some of Bell Bros' slag had been used to help construct the North Gare breakwater. The western boundary of the iron and steel works was defined by the Greatham Creek Branch opened by the NER in 1901. In 1929 the Clarence Works had the following rail cranes:

No. of cranes	Maker	Load limit(tons)	Duties
One	Cowans Sheldon	12	Slag
Five	Roger	10	Pit Side
One	Cole	5	Charge crane
One	Booth	5	General
One	Thos. Smith	7	General
One	Roger	5	Hot Blooms
One	Roger	5	Hot Bank
Two	Roger	5	Rail Bank
One	Isles	3	Ore
Two	Cowans Sheldon	2½	Slag
Two	?	5	Smith's Yard

As a result of the amalgamation of Dorman Long and Bolckow Vaughan in 1929 and the continuing economic depression, the stage was set for a major rationalisation of production facilities. Iron and steel manufacture ceased at the Clarence Works in the early 1930s (no blast furnaces were operating 1/1/1931) with the coke ovens finishing in 1935–37. From then on, the works functioned as a tar distillation and benzol refinery producing pitch, road tar, creosote oil, solvents and resin. It processed by-product from all Dorman Long's coke ovens and later for the NCB in County Durham and ICI. With the decline in rail traffic, several of the locomotives were transferred to Dorman Long's works at the Ironmasters and Cleveland during the 1930s. The demolition of much of the plant left large areas of derelict land. This, plus the large slag tips and residues from the by-product plant, gave a fairly depressing appearance to the area. The exchange sidings and railways still connected with the river wharves, by-product plant and tipping sites.

Dorman Long (Chemicals) Ltd took over responsibility for the works in 1955. A programme of improvements was carried out in 1954-58 with the erection of new boiler plant, greater dispersal of facilities and provision of new roads. Much of the crude tar and benzol came by barge to the Clarence wharf although some tar was still arriving in railway tank wagons in 1970. A number of surplus Dorman Long locomotives were transferred to the works to finish their days but, in 1964, a new Thomas. Hill diesel was obtained. The position of the engine shed also changed between 1951 and 1971 and it then occupied a building near the wharf. Rail traffic ceased about 1977 and the works were sold to Bitmac Ltd on 1/4/1983; the plant continuing to function for many more years.

References : 'A Remnant of the Clarence Ironworks', Cliff Shepherd, *Industrial Railway Record* 159 IRS 1999 pp.241 – 54

'A Description of Messrs. Bell Brothers' Blast Furnaces from 1844-1908', *The Journal of the Iron & Steel Institute* 111, 1908

Gauge : 4ft 8½in

No.5	0-4-0tank	VC	GW	106	1860	New	(2)
-	0-4-0tank		GW	109	1861	New	
	Rebuilt		AB		1904		Scr/1935
No.4	0-4-0tank		GW	166	1863	New	
	Rebuilt		RS	3001	1901		(12)
No.7	0-4-0tank	VC	GW	192	1864	New	(5)
No.14	0-4-0tank	VC	HG	238	1866	New	(3)
No.11			FW				
No.12			FW				
No.13	0-6-0ST		BH	307	1874	New	(9)
No.15	0-4-0ST		YE	236	1874	(a)	(4)
-	0-4-0ST		HH		1876	New	(1)
No.5	0-4-0ST	OC	BH	973	1889	(b)	
	Rebuilt?				1908		Scr 7/1959
No.1	0-4-0ST	OC	BH	985	1890	New	
	Rebuilt		AB		1904		(19)
No.8	0-4-0ST	OC	BH	994	1891	New	
	Rebuilt		AB		1905		Scr/1959
11	0-4-0ST	OC	BH	1044	1891	New	

No.	Name	Type	Cyl	Builder	Works No.	Date	Origin	Disposal
		Rebuilt				1903		(14)
6		0-4-0ST	OC	BH		1892?	(c)	
		Rebuilt		P.Clarence		1907		Scr/1959
12		0-4-0ST	OC	HG?			(d)	(11)
13		0-4-0ST	OC	AB	857	1899	New	(18)
14		0-4-0ST	OC	AB	858	1900	New	(28)
7		0-4-0ST	OC	AB	924	1901	New	(6)
15		0-4-0ST	OC	AB	925	1901	New	(10)
No.16		0-4-0ST	OC	AB	926	1902	New	(24)
17		0-4-0ST	OC	AB	998	1903	New	(27)
18		0-4-0ST	OC	HL	2570	1903	New	(25)
10		0-4-0ST	OC	HL	2604	1905	New	(26)
19		0-4-0ST	OC	HL	2615	1905	New	(23)
20		0-4-0ST	OC	HL	2712	1907	(e)	(21)
21		0-4-0ST	OC	HL	2913	1912	New	(21)
22		0-4-0ST	OC	AB	1274	1912	New	(22)
45(23)		0-4-0ST	OC	HL	3031	1913	(f)	Scr c/1960
9		0-4-0ST	OC	HL	3169	1916	(g)	(17)
4		0-4-0ST	OC	HL	3248	1917	New	(15)
24		0-4-0ST	OC	AE	1793	1918	New	(7)
25		0-4-0ST	OC	P	1513	1918	New	(13)
26		0-4-0ST	OC	HL	3352	1918	New	(16)
27		0-4-0ST	OC	HL	3353	1918	New	(8)
28		0-4-0ST		?			(h)	(20)
161	MARTON	0-4-0ST	OC	BH	609	1881	(i)	Scr 8/1961
162	WHITWORTH	0-4-0ST	OC	BH	519	1879	(j)	Scr c/1962
	CLARENCE NO.1	0-4-0DM		JF	22939	1941	(k)	s/s c/1969
50		0-4-0ST	OC	AB	1102	1907	(l)	(29)
"No 2"		0-4-0DM		HE	4630	1956	New	(31)
27		0-4-0ST	OC	DL		1952	(m)	(30)
15		0-4-0ST	OC	HL	3334	1918	(n)	s/s c/1969
147	CLYDE	0-4-0ST	OC	HL	3350	1918	(o)	Scr c/1962
36		0-4-0ST	OC	DL		1953	(p)	s/s c/1969
39		0-4-0ST	OC	DL		1951	(p)	Scr c/1962
20		0-4-0ST	OC	HL	3032	1913	(q)	Scr 8/1961
No.3		4wDH		TH	144V	1964	New	(32)

There are likely to have been three or more earlier locomotives of which there is now no trace. No.2 0-4-0ST OC BH 992 of 1890 was new to Bell Bros' Parson Byers Quarry, thence to Dorman Long's Tursdale Colliery, but there is an unconfirmed suggestion that the locomotive may have visited the Clarence Works.

(a) ex Eston Grange Iron Co / W. Bacon & Co, Grangetown. A sale of plant and materials at the Eston Grange Iron Works for 1-2/3/1876 included a "nearly new locomotive", presumed to have been YE 236.
(b) new. Bell Bros Board minute 5/12/1889 – purchase of a locomotive from Messrs Black, Hawthorn for £1,050 to be confirmed.
(c) origin not known. It carried a plate reading "MESSRS BELL BROS LTD BUILT 1907 PORT CLARENCE".
(d) may have been an old HG locomotive, although Inness suggests Nos.11 and 12 were early Fox Walker locomotives.
(e) new. Bell Bros Board minute 11/6/1907 – purchase of one locomotive reported.
(f) new. To Acklam Works, 18/3/1935. Ex Britannia Works, 6/1960.
(g) new. Clarence Works records suggest an earlier No.9 was built by HH.
(h) ex Carlton Iron Co Ltd, Mainsforth Colliery, Ferryhill Station, Co. Durham, 4/1918. To Dorman Long Newport Bridge contract, /1931 and returned before 22/12/1933.
(i) ex Cleveland Works; after 18/9/1934, by 1/1/1947.
(j) ex Cleveland Works; after 3/10/1931, by 1/1/1947.
(k) ex Abelson (Engineers) Ltd, Sheldon Plant Depot, Birmingham, c/1952; previously MOS Pontrilas Ordnance Storage Depot, Herefordshire.
(l) ex Acklam Works (Ironmasters combined fleet), c/1954.

(m)	ex Bowesfield Works, Stockton, 16/10/1957. To Britannia Works by 5/1958 and returned to Clarence Works 5/1960. A photograph exists of a BR Class Q6 hauling 27, 36 and 39 at Port Clarence en route from Britannia to the Clarence Works.
(n)	ex Acklam Works, c2/1959.
(o)	ex Cleveland Works, c/4/1959.
(p)	ex Britannia Works, 5/1960.
(q)	ex Britannia Works, 6/1960.
(1)	to Tees Conservancy Commissioners, /1877.
(2)	to Bell Brothers Ltd, Browney Colliery, near Meadowfield, Co. Durham.
(3)	to Egglescliffe Chemical Co, Eaglescliffe Works, Urlay Nook; Inness suggests in /1895.
(4)	to T.D. Ridley & Sons, Harton Low Staithes Contract, South Shields, c/1902.
(5)	s/s, but may have been replaced by HL 3248 in 1917.
(6)	to Acklam Works, 18/2/1922.
(7)	to Burley Ironstone Quarries, Rutland, 21/5/1922.
(8)	to Britannia Works, 9/11/1923.
(9)	s/s before /1929.
(10)	scrapped after 2/5/1930 (date of last record).
(11)	scrapped c/1931 (shown in 3/10/1931 schedule as "scrap").
(12)	scrapped c/1931 (according to 3/10/1931 schedule, it was in bad condition with a poor boiler and a broken frame).
(13)	to Upton Colliery, near Doncaster, Yorkshire (WR), 15/9/1931.
(14)	scrapped after 3/10/1931, probably in the 1930s.
(15)	to Dean & Chapter Colliery, Ferryhill, Co. Durham, 16/9/1932.
(16)	to Acklam Works, 5/5/1933.
(17)	to Acklam Works, 24 or 29/8/1933.
(18)	to Burley Ironstone Quarries, Rutland, 18/10/1933.
(19)	to Chilton Colliery, Chilton, Co. Durham, 20/1/1934.
(20)	to Britannia Works, 2/2/1934, en route to Copenhagen (locomotive not shown in Britannia Works entry).
(21)	to Acklam Works, 10/4/1934.
(22)	to Acklam Works, 9/6/1934.
(23)	Clarence Works records have to Cleveland Works, 1/5/1936 but so far, no evidence of this locomotive has been located at Cleveland.
(24)	to Cleveland Works, 8/10/1936.
(25)	to Cleveland Works, 27/10/1936.
(26)	to Cleveland Works, 3/11/1936.
(27)	to Cleveland Works after 4/11/1935. Note: numbers 14 and 17 were exchanged for 161 and 162 from Cleveland presumably on the basis that the smaller Black, Hawthorn locomotives were more suited to the less demanding work at the Clarence Works.
(28)	to Cleveland Works, after 10/1937 (see above note).
(29)	returned to Acklam Works, /1955.
(30)	s/s c/1969, after 8/1969.
(31)	to Hunslet Engine Co Ltd, Leeds, c8/1977.
(32)	to T.J. Thomson & Son Ltd, Stockton, c7/5/1978.

BRITISH STEEL LTD

A subsidiary of **Jingye Group, China** from 9/3/2020
British Steel Ltd from 1/6/2016, a **Greybull Capital LLP Company,**
until it entered Administration from 22/5/2019
Longs Steel UK Ltd (subsidiary of **Tata Steel European Ltd**) until 1/6/2016
Tata Steel Europe Ltd until 27/6/2015
Corus plc until 27/9/2010 (became a subsidiary of **Tata Steel 2/4/2007** and rebranded
Tata Steel Europe 27/9/2010)
British Steel plc until 6/10/1999
British Steel Ltd until 1/4/1989
British Steel Corporation until 5/9/1988

SKINNINGROVE WORKS, Carlin How
(includes North Loftus Mine)
Skinningrove Iron Co Ltd until 28/7/1967

26

NZ 708195 – **Map I**

(Although operation of the Skinningrove Works was not taken over until 1/7/1968 and SICo did not cease to be a separate company until 28/3/1970.)

Loftus Iron Co Ltd until 2/4/1878

The works is situated adjacent to the railway line from Saltburn West Junction to Boulby. It was the opening of the Cleveland Railway and the booming iron trade that encouraged four partners to negotiate access to the ironstone and take over the lease of land on Skinningrove Ridge from the Cragshall Ironstone Co. On 13/6/1872 the Loftus Iron Co Ltd was incorporated to acquire the leases and land from the partners. It began to sink the shafts for its North Loftus Mine, the winding shaft being situated on the south east corner of the iron works site. Work also began on the construction of two blast furnaces with associated pipe stoves and blowing engines. The blast furnaces commenced operations by 8/1874 and the Loftus Iron Co used two new Fox Walker locomotives to carry out shunting. This involved collecting ironstone from the North Loftus pit head, iron from the pig beds and taking slag to the tips on the west side of the works. The boom in the market did not last and the iron works and mine stopped trading about 4/1877; the Loftus Iron Co being wound up on 2/4/1878.

Much of the capital invested was written off and the Skinningrove Iron Co (SICo) was formed on 8/6/1880 to acquire the works and mine. Both blast furnaces were operating again three months later. The railway system at this time was relatively simple. A set of exchange sidings connected with the NER branch near Crag Hall Signal Box. Further east, additional lines left the branch at Carlin How Junction to form four sidings for incoming traffic, with another track swinging round to serve North Loftus pithead. One of the lines ran up an inclined bank and over the top of storage bins and a row of calcining kilns. Other railway lines fanned out over the spoil tips heading towards the cliffs. The works produced foundry iron aimed at the main market around Falkirk and Grangemouth. Transport by rail over that distance was expensive and so the SICo constructed a jetty at the foot of the cliffs with a rope-worked incline powered by a stationary engine at the works to lower the railway bogies with their bundles of pig iron and raise the empties. The first shipment left from the jetty in 1888 but it was not until 1891 that substantial amounts of iron began to leave via this route with a consequent impact on NER traffic receipts. Ironstone initially came from the company's North Loftus Mine but its royalty was relatively small and, from 1887 to 30/11/1894, almost all of the stone from Bell Bros' Carlin How Mine was brought up the North Loftus shaft. Pease & Partners Ltd, with Lord Zetland, offered to finance the erection of two more blast furnaces if the SICo took the bulk of the nearby Loftus Mine's ironstone. A wooden viaduct was constructed from Loftus Mine so that tubs could run across the valley to a new drift in the hillside connecting with the bottom of the North Loftus shaft. The SICo started to take Loftus ironstone from the shaft in 2/1895 but, in the intervening period, it had to be hauled up the zig zag railway by the NER. The two blast furnaces were blown in 10/1895 and 1/1896. A picking belt began work at the North Loftus pithead on 10/9/1901 to improve the proportion of ironstone for the blast furnaces. A fifth blast furnace was blown in during 11/1901.

Steel was increasingly taking over from iron for most constructional purposes and the Skinningrove plant was transformed into an integrated iron and steel works in 1910–14. As a result, there was a major expansion of the railway which was responsible for moving raw materials to the calcining kilns, blast furnace stores and two batteries of 60 by-product coke ovens. Railway sidings linked the liquid iron and slag discharge chutes at the five blast furnaces to the steel works and slag tips. The former comprised the melting shop with its basic open hearth furnaces, the soaking pits for equalising the temperature of the steel ingots, the 36 inch rolling mill and the extensive stockyards. The engine shed in the centre of the works was erected during 1910 and had three tracks. Twelve new locomotives were purchased between 1890 and the end of World War 1. These mainly operated the railway for the next three decades with an extensive fleet of internal wagons.

After initial optimism, trading conditions grew steadily worse and Pease & Partners Ltd leased the works between 1/10/1922 and 30/6/1932, although it continued to operate under the SICo name. The railway also served new industries processing slag, including a scoria block plant, a works producing agricultural fertiliser and a tarmacadam works. The latter was initially located by the LNER line near the former Huntcliff Mine but connected to Skinningrove's tips by an aerial ropeway. Later a new plant was built on the slag tips. The iron and steel works shut from 5/11/1932 to 17/3/1933 to conserve finance and carry out some renewals. The coking plant never reopened and the LNER then delivered supplies of coke. Shipments from the jetty ceased in 1936, although small amounts of sand were still hauled up the incline and this required one of the small vertical boiler locomotives to be based at the jetty. An 18 inch rolling mill was erected in 1937.

With the first nationalisation of the iron and steel industry in 1951, Skinningrove Works became the property of the Iron & Steel Corporation of Great Britain, but this position only lasted ten months when the new Conservative Government reversed the process and the SICo continued to retain its identity. The five blast furnaces had become very outdated and three were demolished to be replaced by a new blast furnace that was lit on 9/6/1952; the other two had ceased operating by 1960. A sinter plant was erected in 1961 to improve the quality of iron ore being fed into the furnaces. Loftus Mine closed 26/9/1958 and all iron ore then came in by

The elusive Black, Hawthorn locomotive near the engine shed at Skinningrove Works on 15th September 1923; was this maker's number 354 or a more recent arrival (see footnote 5)? (NERA R. Inness photograph)

The vertical boiler locomotive built by Cochrane in 1871 stands on Skinningrove jetty in 1954 with wagons of sand destined for the works which is located at the top of the cliffs. (D.G. Charlton photograph)

rail. An additional set of exchange sidings was provided on the north side of the existing tracks near Crag Hall Signal Box and the in-coming mineral sidings within the works were enlarged. New wagon tipplers fed iron ore and coke on to conveyor belts which carried them on a gantry over the railway lines on the inclined bank to the blast furnace stores and sinter plant. Several secondhand steam locomotives were acquired in the 1950s to replace some of the ageing stock. However, in order to comply with the Clean Air Act, ten new Sentinel diesel hydraulic locomotives were purchased in 1963, nine of which arrived in a single cavalcade from Shrewsbury. A new diesel locomotive servicing shed was constructed near the site of the North Loftus pithead.

With the renationalisation of the iron and steel industry on 28/7/1967, Skinningrove Works was vested in the British Steel Corporation. Rationalisation of iron production soon began in the UK with concentration on a limited number of large blast furnace plants located close to deep water facilities. Also the basic oxygen steel (BOS) plant at Lackenby replaced the outdated open hearth plants in the area. Iron making ceased at Skinningrove on 1/10/1971 and steel making on 7/4/1972.

The BSC recognised that there was still a need for the rolling mill at Skinningrove to produce specialist sections for the earth moving, material handling and shipbuilding industries. The 36 inch mill was retained and supplemented by a bloom heating furnace and covered mechanised stock handling banks. The railway system now comprises a single line from Crag Hall Signal Box which enters the works to connect with six exchange sidings, from which one line serves the bloom heating bay and another the stockyard in the former 18 inch mill. Following the cessation of steelmaking, there was less need for the locomotives and by the late 1970s only one or two were required each day. Rail traffic consisted of incoming blooms and the intermittent despatch of steel sections. Latterly only one locomotive is required and is provided by contract hire to shunt steel blooms to the reheating furnace.

Reference : *Skinningrove Iron and Steel Works Its History, Railways and Locomotives*, Cliff Shepherd, IRS, 2012

Gauge : 4ft 8½in

-		0-4-0ST	OC	FW	162	1872	New	Scr c/1916
-		0-4-0ST	OC	FW	252	1874	New	(3)
	MIDGE	0-4-0VBT	OC	?	?		(a)	(2)
	"COFFEE POT"	0-4-0VBT	OC	Cochrane		1871	(b)	Scr 1/1954
	MINNIE	0-6-0ST	OC	FW	358	1877	(c)	(10)
-		0-4-0ST	OC	AB	656	1890	New	(1)
	BLACK PRINCE	0-6-0ST	OC	BH	354	1875	(d)	(5)
	ANNIE	0-4-0ST	OC	P	582	1894	New	Scr 10/1957
	RAKIE	0-4-0ST	OC	P	583	1894	New*	Scr c/1961
	RENNIE	0-4-0ST	OC	BH	1102	1896	New	Scr 6/1957
	HILDA	0-4-0ST	OC	P	736	1899	New	
		Rebuilt		SICo		1950		Scr c5/1961
	LIZZIE	0-4-0ST	OC	P	919	1902	New	Scr 10/1950
	ETHEL	0-4-0ST	OC	P	1084	1906	New	Scr/1948-49
	BRIDGET	0-4-0ST	OC	RS	3415	1910	New	Scr 4/1964
	ABBOT	0-4-0ST	OC	HL	2425	1899	(e)	Scr 6/1957
	GREENBANK	0-4-0ST	OC	AB	721	1894	(f)	Scr /1961
	MARY	0-4-0ST	OC	SICo		1913	New	Scr c/1955
	MARGARET	0-4-0ST	OC	HL	3025	1913	New	Scr 4/1962
	ARTHUR	0-4-0ST	OC	RS	3673	1917	New	Scr c3/1964
	ALFRED	0-4-0ST	OC	RS	3674	1917	New	Scr 3/1964
	EDGAR	0-6-0ST	OC	AE	1815	1918	(g)	Scr /1959
	JEANIE	0-6-0T	IC	HC	694	1904	(h)	(4)
	ROSEBERRY	0-4-0ST	OC	HE	1294	1918	(i)	Scr 3/1962
No.8		0-4-0ST	OC	AB	2045	1937	New	Scr c/1965
No.18		0-4-0ST	OC	RSHN	7059	1942	New	Scr c/1964
	JERSEY	0-4-0T	OC	9 E		1893	(j)	Scr 5/1962
	CATTERSTY	0-4-0F	OC	AB	2269	1950	New	Scr c2/1972
	HUMMERSEA	0-4-0F	OC	AB	2270	1950	New	Scr c2/1972
	ELIZABETH II	0-4-0ST	OC	P	1961	1939	(k)	Scr 4/1964
-		0-4-0DH		JF	4000007	1947	(l)	(6)
	FREEBROUGH	0-4-0ST	OC	AB	2209	1946	(m)	Scr c/1968
	BOULBY	0-4-0ST	OC	AB	2210	1946	(m)	Scr c9/1966

	HUNTCLIFF	0-4-0ST	OC	AB	2254	1948	(m)	(11)
No.26	KILTON	0-4-0ST	OC	YE	2384	1937	(n)	(9)
No.28		0-4-0ST	OC	AB	1987	1930	(o)	Scr 12/1971
107		0-4-0ST	OC	P	2002	1941	(p)	(8)
110		0-4-0ST	OC	P	2020	1942	(p)	(8)
-		0-4-0DH		S	10105	1962	(q)	(7)
3	HUMMERSEA 1	0-4-0DH		S	10127	1963	(r)	(14)
1		0-4-0DH		S	10125	1963	(s)	Scr c1/1994
2	CATTERSTY 2	0-4-0DH		S	10126	1963	New	(15)
4		0-4-0DH		S	10128	1963	New	(12)
5		0-4-0DH		S	10129	1963	New	(13)
6		0-4-0DH		S	10130	1963	New	(12)
7		0-4-0DH		S	10131	1963	(t)	Scr 12/1983
8		0-4-0DH		S	10132	1963	New	(12)
9		0-4-0DH		S	10133	1963	New	Scr /1982
10		0-4-0DH		S	10134	1963	New	(12)
44		0-6-0DH		S	10081	1961	(u)	Scr 12/1983
45		0-6-0DH		S	10080	1961	(u)	Scr 12/1983
-		4wDH		RR	10281	1968	(v)	(16)
-		4wDH		TH	232V	1971	(w)	(18)
	WILLIAM	4wDH		S	10048	1960	(x)	(19)
-		0-6-0DH		RR	10217	1965	(y)	(17)
309		0-6-0DH		YE	2825	1961	(z)	
No.314		0-6-0DH		YE	2832	1962	(aa)	
-		4wDH		RR	10229	1965	(ab)	(22)
	LADY POTTER	0-6-0DH		RR	10214	1964	(ac)	(20)
	HARRY POTTER	0-6-0DH		RR	10220	1965	(ad)	(21)
-		0-6-0DH		TH	150c	1965	(ae)	

*   Inness suggests that it was loaned to the Tees Iron Works at some stage.

A notable event took place when the SICo replaced most of its steam locomotives by ten new Sentinel diesel hydraulics, nine of which came as a single cavalcade from Shrewsbury. They are lined up here after arrival on 26th July 1963 outside the new workshops built specially for them.   (IRS Brian Webb photograph)

(a) MIDGE was described as a "Camelback" locomotive because it had a combined vertical and horizontal boiler. It could have been similar to a locomotive at Cochrane's Works. Another suggestion is that it may have been Gilkes Wilson maker's number 188 of 1864 from the Tees Iron Works.
(b) ex Cochrane & Co Ltd, Cargo Fleet, 1877.
(c) ex John Waddell & Sons, contractor, Whitby, Redcar & Middlesbrough Union Railway, 5/1884.
(d) ex Darlington Forge Ltd, Darlington, 1/1891.
(e) ex J. Abbot & Co Ltd, Gateshead, Co. Durham, per Thos W. Ward Ltd, c/1910.
(f) ex Thos W. Ward Ltd, /1912; previously W. Toulson & Son, contractor.
(g) ex Bloxham & Whiston Ironstone Co Ltd, Bloxham Ironstone Pits, Oxfordshire, per War Stores Disposal Board, Canterbury, 3/1919.
(h) ex Thos Rigley Ltd, Prestwich, Lancashire, on hire, 2/1920.
(i) ex Pease & Partners Ltd, Tees Bridge Iron Works, Stockton, /1937.
(j) ex British Railways (Southern Region), per Messrs Bell, Doncaster, 9/1949.
(k) ex Butterley Co Ltd, Codnor Park Forge, Derbyshire, via Bush, Alfreton, /1952.
(l) ex Abelson (Engineers) Ltd, Sheldon, Birmingham, on hire, /1953.
(m) ex Imperial Chemical Industries Ltd, Winnington, Cheshire, 5/1957.
(n) ex Steel, Peech & Tozer Ltd, Rotherham, Yorkshire (WR), 10/1960.
(o) ex Park Gate Iron & Steel Co Ltd, Rotherham, Yorkshire (WR), c4/1961.
(p) ex E.L. Pitt & Co (Coventry) Ltd, Brackley, Northamptonshire, 1/1962; previously Admiralty, Risley, Lancashire.
(q) ex Sentinel (Shrewsbury) Ltd, Shrewsbury, Shropshire on loan for demonstration, 26/6/1962.
(r) new, ex-works 19/4/1963. To Teesbulk Handling Ltd (subsidiary of Cleveland Potash Ltd), Tees Dock, Grangetown, on hire, 2/1982 and returned after 23/12/1982.
(s) new, ex-works 7/1963. To British Steel Corporation, Lackenby Workshops, Grangetown for repairs, 26/5/1983 and returned 10/1983.
(t) new ex-works 7/1963. To Cleveland Potash Ltd, Boulby Mine, on hire 9–10/1979 and returned by 8/1980.
(u) ex British Steel Corporation, Consett Works, Co. Durham, for spares, 10/7/1981.
(v) ex Staffordshire Locomotive Co Ltd, c/o Telford Horsehay Steam Trust, on hire initially, 2/1998. Locomotive purchased by Corus plc, /2002.
(w) ex Corus plc, Hartlepool Pipe Mill, Hartlepool, /2002.
(x) ex LH Group Services Ltd, Barton-under-Needwood, Staffordshire, on hire, 4/2004. To LH Group Services Ltd for repairs, 1/2005 and returned 24/1/2005.
(y) ex LH Group Services Ltd, Barton-under-Needwood, Staffordshire, on hire, 1/2005.
(z) ex Corus plc, Moss Bay Works, Workington, Cumbria, /2005, by 6/2005. To Cleveland Potash Ltd, Boulby Mine, on hire, 16/10/2009 and returned 1/12/2011. Became property of Ed Murray & Son Ltd, c1/2014 but sale later rescinded.
(aa) ex Corus plc, Moss Bay Works, Workington, Cumbria, 3/4/2007. Became property of Ed Murray & Sons Ltd, c1/2014 but sale later rescinded.
(ab) ex Ed Murray & Sons Ltd, Hartlepool, on hire for trials, 21/11/2011. To Tata Steel Europe Ltd, Hartlepool Pipe Mill, Hartlepool and returned, c25/9/2014. To Liberty Pipes (Hartlepool) Ltd, Hartlepool, 9/12/2019. Ex Ed Murray & Sons Ltd, Brenda Road, Hartlepool, on hire, 26/6/2020.
(ac) ex Ed Murray & Sons Ltd, Brenda Road Depot, Hartlepool, on hire, 1/12/2011; previously Selby Storage & Freight Co Ltd, Selby, North Yorkshire.
(ad) ex Ed Murray & Sons Ltd, Brenda Road Depot, Hartlepool, on hire, 21/12/2011; previously Selby Storage & Freight Co Ltd, Selby, North Yorkshire.
(ae) ex Chasewater Light Railway and Museum Company, Brownhills, Staffordshire, 8/3/2019; unloaded w/c 11/3/2019, property of Ed Murray & Sons Ltd, on hire.

(1) to T.D. Ridley and Sons, Middlesbrough, by 12/1904.
(2) disposed of either before or during World War I.
(3) sold or scrapped; it may have been sold to the Acklam Iron Co Ltd, Middlesbrough at an unknown date and certainly before /1922.
(4) returned to Thos Rigley Ltd, Prestwich, Lancashire, ex hire, by 2/7/1920.
(5) there is confusion about the eventual fate of this locomotive. The company's accounts suggest the possible sale of BH 354 about 7/1920 and the purchase of another Black, Hawthorn locomotive, also called BLACK PRINCE. The loose plant account for 7/1920 has the entry "1 14" cyl locomotive Black Prince £1,035.0.0." on the credit side as if the locomotive was a new acquisition. Also Ted Haigh was told by Skinningrove's traffic department in 9/1952 that BLACK PRINCE was a 0-6-0 saddle tank purchased from a colliery near Newcastle upon Tyne in 1920 but was no use and scrapped in 1924.

A possible coincidence, but the *Contract Journal* on 11/2/1920 advertised for sale from Macdonald, Wilson & Co Ltd's Fulham Plant Depot on 11/2/1920 a six-coupled Black, Hawthorn named BLACK PRINCE. It was not mentioned at a subsequent sale on 3/9/1920. One thing is certain, whether it was BH 354 or a recently acquired Black Hawthorn, Richard Inness took a photograph of it at Skinningrove Works on 15/9/1923. Despite a repaired saddle tank and, with no visible name, it looked well-kept.

(6) returned to Abelson (Engineers) Ltd, Sheldon, Birmingham, ex hire, by 9/1953.
(7) to Dorman Long (Steel) Ltd Lackenby, 29/6/1962.
(8) to John Baker & Bessemer Ltd, Kilnhurst, Yorkshire (WR), 19/10/1963.
(9) dismantled and frame converted to ingot carrier, /1964. Chassis scrapped /1977.
(10) to Industrial Locomotive Preservation Group, Kent & East Sussex Railway, 23/3/1968.
(11) either scrapped or to Higgs & Hill Ltd, contractor, Fiddlers Ferry Power Station, near Widnes, Lancashire, c/1968.
(12) to British Steel Corporation, South Teesside Works, Lackenby, c6/1972.
(13) to British Steel Corporation, South Teesside Works, Lackenby, /1976. Geoff Allen, who worked at Cleveland Potash Ltd, considered that S10129 was also hired temporarily by Boulby Mine from Skinningrove Works to cover for a locomotive failure.
(14) to British Steel plc, Teesside Works, Lackenby, 5/1/1998.
(15) to British Steel plc, Teesside Works, Lackenby, 6/1/1998.
(16) to LH Group Services Ltd, Barton-under-Needwood, Staffordshire for repairs, 4/2004, but scrapped there, 23/9/2005.
(17) to LH Group Services Ltd, Barton-under-Needwood, Staffordshire, ex hire, 24/1/2005.
(18) to LH Group Services Ltd, Barton-under-Needwood, Staffordshire, w/c 6/6/2005, by 10/6/2005 and scrapped there, 10/8/2005.
(19) to LH Group Services Ltd, Barton-under-Needwood, Staffordshire, ex hire, 1/12/2011, and subsequently to EMR, Kingsbury for scrap, 18/7/2012.
(20) to Cleveland Potash Ltd, Boulby Mine, (property of Ed Murray & Sons Ltd, Hartlepool), ex hire, 25/9/2014.
(21) to Chasewater Light Railway & Museum Co, Brownhills, Staffordshire, ex hire, 15/3/2019.
(22) to Ed Murray & Sons Ltd, Casebourne Road Depot, Hartlepool, ex hire, from 25/2/2022.

TEESSIDE WORKS, South Bank, Grangetown, Lackenby, Redcar 27
Dorman Long (Steel) Ltd until 28/7/1967 **Map H**
(Although operation of the works was not taken over until 1/7/1968 and Dorman Long did not cease to be a separate company until 28/3/1970)
Dorman, Long & Co Ltd until 2/10/1954   NZ 535216, NZ 545212, NZ 560223, NZ 565235
Bolckow, Vaughan & Co Ltd until 29/11/1929
Bolckow & Vaughan until 19/11/1864

History of the Company
Bolckow and Vaughan began the development of the iron and steel industry on Teesside when they erected the Middlesbrough Iron Works (which see) in 1841, followed by the blast furnaces at Eston in 1852–53. By 1907, Bolckow Vaughan & Co employed 20,000 people in its iron and steel works, coal and ironstone mines and limestone quarries. However, there were hidden weaknesses – its continued reliance on Bessemer steel, a failure to diversify into more steel products and a lack of innovation by the Board. Growing demand to the end of World War 1 hid these weaknesses but, by 1925, its capital was only half that of Dorman Long. The latter was also in trouble but the appointment of Ellis Hunter to run the company and improved trading conditions saved the business. Bolckow Vaughan was absorbed by Dorman Long in 1929. A major reorganisation took place involving concentrating coke manufacture at a new Cleveland coking plant in 1936, opening light section and bar mills at the Cleveland Works, concentrating rolling of plates at Redcar, heavy sections at Britannia with Port Clarence becoming a central distillation plant. After World War 2, investment was concentrated on plants down river between South Bank and Redcar.

The first nationalisation of the UK iron and steel industry took place in 1951 but was reversed by the incoming Conservative Government within a year and Dorman Long was back in the private sector by 11/1954. Dorman Long & Co Ltd became the holding company in 1954 with Dorman Long (Steel) Ltd responsible for iron and steel production. In 1966 Dorman Long agreed to merge with South Durham Steel & Iron Co Ltd and Stewarts & Lloyds Ltd as British Steel & Tube Ltd but this came into being only a month before nationalisation during the following year.

Vesting day for the British Steel Corporation was 28/7/1967, although operation of the pre-nationalisation businesses was not taken over until 1/7/1968. For the next decade, BSC continued the improvement of the iron and steel plants but trading conditions became very difficult in the 1980s and BSC was privatised in 1988. British Steel continued to struggle and, in 10/1999, it merged with Koninklijke Hoogovens Staal of the Netherlands to form Corus plc. Following heavy losses, Corus created Teesside Cast Products to operate iron and steel making at Teesside Works as a sub business unit, with the Lackenby Universal Beam Mill being managed from Scunthorpe. In 4/2003 Corus decided to concentrate its UK iron and steel production on its Scunthorpe and Port Talbot works and, in 11/2003, TCP became a separate business within Corus selling steel slab to world markets and it began to despatch slab to a consortium of four re-rolling companies on 2/1/2005.

Tata Steel of India purchased Corus in 2007 and the latter continued to function as a subsidiary company until rebranded as Tata Steel Europe Ltd on 27/9/2010. Meanwhile, in the latter part of 2008, world steel prices had fallen in response to a worldwide recession. On 12/5/2009 Corus announced that the consortium had terminated the agreement. After efforts to find alternative markets failed, the iron and steel making plants at Teesside Works were mothballed on 20/2/2010. This was not quite the end of iron and steel production, however, and the restart is covered under the Sahaviriya Steel Industries UK Ltd entry. The Universal Beam Mill remained with Tata Steel Europe as part of its Long Products Division; the latter became a stand-alone business (Longs Steel UK Ltd) from 2/8/2015, although still wholly owned by Tata Steel Europe Ltd and was then purchased by a new company, British Steel Ltd in 2016. The latter went into administration on 22/5/2019 and was acquired by the Chinese Jingye Group on 9/3/2020 but retained its name.

Development of the Teesside Works and its Railways

Teesside Works on the south bank of the River Tees, east of Middlesbrough, became one of the U.K.'s most important iron and steel manufacturing centres, but it had originated as a series of individual plants. Following the discovery in 1850 of large deposits of Cleveland ironstone in the Eston Hills, Bolckow & Vaughan constructed its two mile long Eston Branch Railway northwards to a junction with the Middlesbrough-Redcar line. The first load of ironstone passed over it in 12/1850 and a locomotive was used to pull the wagons on 4/1/1851. Initially the ironstone was transported by the SDR to Bolckow & Vaughan's blast furnaces at Witton Park in County Durham but, in 1852, Bolckow & Vaughan began to establish a new iron works next to the Branch Railway near to its junction with the main line. By 1856 Bolckow & Vaughan had a row of six blast furnaces, 54 feet in height, standing parallel to and approximately 150 yards south of the main line. A rail connection was made with the Branch Railway to service the new plant.

In 1853 John Vaughan's son-in-law, Thomas Light Elwon, began to develop a row of three blast furnaces alongside the Eston Branch Railway. These were the Cleveland furnaces (sometimes known as the Branch

A row of old locomotives at the Cleveland Iron and Steel Works on 5th September 1926 were destined for scrap and included a number of Black, Hawthorn engines led by No.24 (364 of 1876).

(NERA R. Inness photograph)

WILLIAM JONES (RSHN 7347/1947) shunts on the sidings leading to South Bank Wharf with the South Bank Coke Ovens in the background, 30th May 1962. Many of the locomotives at the Cleveland Works had reduced height cabs and boiler mountings because of limited clearances. *(Courtesy David Webb)*

furnaces). Even before they were operational, they were transferred to Bolckow & Vaughan as part of the Eston Works, while Elwon moved on to help establish the Clay Lane Iron Works to the west. By 1856 Bolckow & Vaughan's weekly output of pig iron was 900 tons from the Eston and 430 tons from the Cleveland blast furnaces. Seven out of the nine blast furnaces were operating on 1/5/1859. It is not known how long the Eston furnaces functioned but Bolckow & Vaughan increasingly concentrated on expanding the number of Cleveland furnaces; approval being given by the Board on 16/5/1866 to build four more blast furnaces at Cleveland.

Eventually the Cleveland furnaces comprised a row of five with another three to the south. This is partly explained by the rapid technical development of local blast furnace design which may have made the original Eston structures obsolete. Also, with the increasing height of the furnaces, there was more space to construct the approach ramps for the sidings off the Branch Railway at the Cleveland site in order to run over the top of the mineral stores and calcining kilns. By 1858 Bolckow & Vaughan had also constructed a railway from the Cleveland blast furnaces passing under the Middlesbrough-Redcar line to head along a jetty over the mudflats to its Eston Wharf on the south bank training walls of the River Tees.*

* During much of the latter half of the nineteenth century, the Tees Conservancy Commissioners used slag from the various iron works to construct "training walls" to clearly define the river channel and ensure an adequate depth of water for shipping. By 1906 there were 24 miles of training walls built up to five feet above the Low Water Mark.

Bolckow Vaughan changed from a partnership to a limited company in 1864 when it employed 8,000 men, raised 750,000 tons of ironstone and produced 300,000 tons of pig iron each year. It was content to either sell the pig iron or convert it to wrought iron in its puddling furnaces at the Middlesbrough Iron Works or Witton Park. In 1879 Bolckow Vaughan purchased the South Bank Iron Works with its eight blast furnaces from the Trustees of Thomas Vaughan. This was situated on the north side of the Middlesbrough-Redcar railway and close to its line to Eston Wharf.

Following Henry Bessemer's identification of a process for making bulk steel in a converter during 1856, steel became increasingly popular and was especially specified by the railway companies for their new track. Unfortunately the high phosphorus content of Cleveland ironstone made it unsuitable for this process. Bolckow Vaughan recognised the need to produce steel and had purchased a small plant at Gorton, Manchester in 1870. When this was sold in 1875, the proceeds were put towards a new steelworks on 40 acres of recently purchased land. A set of four 8 ton Bessemer converters were installed in the works situated immediately south of the Middlesbrough-Redcar railway and next to the company's private line to Eston Wharf. High grade haematite iron ore from Cumberland and increasingly Spain, low in phosphorus and high in silica, was required to produce 'acid' steel. According to the 1874 Middlesbrough directory, Bolckow Vaughan had "four steamers trading regularly and are the joint owners of extensive iron ore mines in Spain". Substantial quantities of the mineral were hauled by its locomotives over the company's line from Eston Wharf. Three blast furnaces were erected south of the steelworks to supply the converters with iron and were usually referred to as the "Bessemer Blast Furnaces". Operations at the new plant commenced in 1877 and the first steel rails were rolled at the nearby mills on 17/10/1877. By the following year, 1,000 tons of steel rails were being produced each week and this enabled Bolckow Vaughan to better survive the late 1870s recession, although the iron rail mills at its Middlesbrough and Witton Park Works stood idle for much of 1876–77.

Following experiments at Bolckow Vaughan's works, Sidney Thomas and Percy Gilchrist demonstrated at the Middlesbrough (4/4/1879) and Eston Works (13/5/1879) that steel could be produced from phosphoric iron ores by introducing a 'basic' lining into the converter. As a result, Bolckow Vaughan had added five basic-lined 15 ton converters to its Cleveland Steel Works by 1882. Metal mixers were introduced in 1893 to reduce irregularities in the iron by combining several batches from the blast furnaces. A railway line ran up an incline to the mixer gantry allowing molten iron to be poured from 15 ton ladles into the mixers which, when ready, could be tilted to empty into another rail mounted ladle at ground level.

After the discovery of a thick salt bed under the Middlesbrough Iron Works in 1863, Bolckow Vaughan began to sink two shafts at South Bank towards the end of 1869 with the intention of mining the salt. This project was abandoned in 1872 and it was not until 1886 that the company sank a borehole through which to send water and then pump up the resulting brine. By May of that year, three salt pans were evaporating the brine, prior to the salt being stored ready for despatch. The pan house was situated at the west end of the South Bank blast furnaces; the waste gases of which provided a cheap source of heat. The borehole was located on the mudflats to the north west where slag was also being tipped from rail bogies to reclaim the land. As at the Middlesbrough Iron Works, the lease for the salt deposits and surface plant was let to the Cleveland Salt Co Ltd (incorporated 11/10/1887), although Bolckow Vaughan continued to hold a controlling interest in the business.

By 1893, Bolckow Vaughan operated a very extensive railway system at its Eston Works, which was now increasingly called the Cleveland Works. It is important to realise that for many years, indeed up to the 1950s, there were few roads on the premises and most goods and equipment was moved by rail.

Ironstone came down the Eston Branch Railway from the mine, was crushed and then reloaded into railway wagons for movement to the kilns and stores. Between Bolckow Terrace and the houses of Grangetown,

MC2, a compressed air locomotive used to empty the slag ladles on the Cleveland Works tip, incorporated the chassis and cab of steam locomotive HL 3082/1914. It was photographed on the extensive tip on 6th October 1978. *(Cliff Shepherd photograph)*

numerous sidings diverged from the branch to hold the many mineral wagons. A little further north along the branch, a line ran off to the east passing under the rail ramps to the Bessemer blast furnaces and then dividing to reach the steel works and rolling mills; the latter comprising a 48 inch cogging mill, a 60 inch plate mill and four rolling mills – No.1 heavy sections; No.2 sleepers, angles, rails; No.3 angles, light rails and No.4 fish plates, merchant bars. The Eston Branch Railway continued north dividing into two to run in a curve either side of the Cleveland blast furnaces. These were served by two tracks that climbed a ramp to pass over the calcining kilns and mineral stores on the west side, whilst other sidings gave access to the pig beds and slag discharge spouts on the east side. Steam locomotives, often in pairs, shunted the wagons of raw materials up the ramp. The line under the Middlesbrough-Redcar railway at Holme Beck Bridge had expanded to three tracks and linked up with the South Bank blast furnaces where the pig beds and slag lines were on the north side. Sidings also served the salt stores, together with a separate line to the brine tanks. The railway along the jetty to Eston Wharf was busy with incoming iron ore and outgoing iron and steel products. Substantial amounts of slag had already been tipped on the north side of the main line as far as Eston Grange Station (later Grangetown) using various sidings across the area. The addition of lime to treat phosphoric iron by the basic steel process resulted in slag with significant proportions of phosphorus, which could form the basis of a useful agricultural fertiliser. In 1887 Bolckow Vaughan introduced the first grinding plant for this product in the UK. The Antonian Basic Manure Works (later chemical works) was located on the slag tips and served by railway; H.& E. Albert operated it for some years. Beyond, to the north, was still mainly mudflats up to the river training wall.

In 1898 Bolckow Vaughan expanded even further when it purchased the Clay Lane Iron Works with its six blast furnaces. Subsequently the jetty lines from the Clay Lane and Cleveland/South Bank blast furnaces and their respective wharves on the Tees were combined into one railway serving a longer South Bank Wharf. A siding off this line entered the Eston Sheet & Galvanising Works (which see), whilst on the north side of the Middlesbrough-Redcar railway, where the track from Clay Lane dived under it, Bolckow Vaughan had established a concrete works producing paving slabs, kerbs, cills and artificial stone. Locomotives brought molten slag in ladles from the blast furnaces and these had their contents tipped into long shallow pits to cool. Slag bricks were also manufactured using brick making machines located between the concrete plant and salt works. The momentum for technical innovation had increasingly moved to Germany and the USA so it was not surprising that the row of three blast furnaces at the south end of the Cleveland furnaces were demolished in 1903–06 and replaced by two 100 feet high 'Yankee' furnaces (to be known as the Grangetown furnaces). To sum up, Bolckow Vaughan had the following blast furnaces at the Cleveland Works – Cleveland (Branch) 5, Grangetown 2, Clay Lane 6 (4 in blast), Bessemer 2 (1 in blast) and South Bank 8 (5 in blast). The original three Bessemer blast furnaces had been dismantled in 1900–03 and replaced by No.3 (1903) and No.1, later 4 (1911).

On 20th June 1960, 4 (S 9590/1956), one of a batch of six-coupled vertical boiler locomotives, hauls a rake of hot metal ladles past the boiler house on the Cleveland Works site. (IRS J.P. Mullett photograph)

In 1911 Bolckow Vaughan decided to discontinue making steel in Bessemer converters and replaced them with the North open hearth steel plant. The advantage of the latter was that it took hours rather than minutes to make steel so that there was a greater amount of control and consistency of product. Scrap could also be used in open hearth furnaces resulting in better yields. Some of the new furnaces came into use in 1913 and, two years later, there were seven 60 ton furnaces and two 400 ton mixers. During World War 1 Bolckow Vaughan agreed with the Government to increase its steel output by erecting the South open hearth steel plant, just a few yards north of the terrace houses in Grangetown, although full production was not achieved until after the war when it comprised eight 60 ton furnaces.

In 1925 Bolckow Vaughan stopped supplying blast furnace gas to the South Bank salt works. It was not feasible to convert the pans to coal firing and the works closed after producing a total of 243,430 tons of salt. The Clay Lane blast furnaces ceased operating after World War 1 and, on 1/4/1927, it was announced that Bolckow Vaughan had begun dismantling the Clay Lane Iron Works and two of the blast furnaces had been demolished by 16th October 1928 (*North-Eastern Daily Gazette*). Bolckow Vaughan published *A Romance of Industry* in 1928 with details of its Cleveland Works which now occupied 1,273 acres. The company's South Bank Wharf was 1,750 feet long and equipped with ten electrically operated cranes. Previously, imported iron ore had been loaded into railway wagons, taken to a fan of sidings mentioned above and lifted out on to the ground by grab cranes. When required, it had been reloaded into the wagons and transported to the blast furnaces. This wasteful system had been replaced by a ramp and steel gantry, 1,810 feet long and 50 feet above the ground. Full wagons were hauled up the incline on to the gantry by an electric winder and the ore tipped through bottom discharge doors. A Goliath crane spanned the width of the gantry and its grab could fill two 12 ton railway wagons in two scoops before the train left for the blast furnaces. Ironstone from Eston Mine continued to arrive in hopper wagons and was hauled by a locomotive on to the top of the kilns. According to *A Romance of Industry*, "there are within the works 92 miles of standard gauge railway, operated from signal cabins, 84 locomotives, 46 steam cranes, 3,100 wagons of various types, 350 furnace charging cars, 140 slag ladles. There are 13 blast furnaces in commission in four separate groups". Iron to be made into steel was tapped out of the blast furnaces into 20 ton ladles mounted on bogies and hauled by locomotives to the North and South melting shops. From there, steel ingots were taken to the soaking pits to equalise the temperature and then to the rolling mills. The resulting rails, bars and sections were placed on inspection banks. The rail bank occupied 4½ acres and was intersected by two sidings to facilitate despatches. The slag tips now extended to the banks of the Tees. An "engine house" appears here on the 1927 Ordnance Survey map and a "hauler engine" pulled locomotives and slag ladles up the incline on the tip. In addition to the Basic Slag Works, a Slag Reduction

Hawthorn Leslie maker's no. 3861 was one of a pair of new locomotives supplied to operate the coke cars at Dorman Long's Cleveland Coke Ovens when they opened in 1936. (IRS Brian Webb collection)

The prototype Sentinel rod-drive 0-6-0 diesel hydraulic was tried out at the Lackenby Works on 22nd July 1960. This became Sentinel 10032 CHRIS MOODY and the basis for a large fleet of these locomotives at the Teesside Works; it differed from the production series by having no walkway at the rear.

(Dorman Long photograph)

Works was also rail served. A Tar Macadam Works had been established immediately north of Grangetown Station with a separate rail connection off the main line.

It was ironic that shortly after the publication of *A Romance of Industry*, Bolckow Vaughan's written down capital was exchanged for shares and debentures in Dorman Long, which took over the Cleveland Works. Dorman Long already owned an iron and steel works at Redcar that was separated from the Cleveland Works by the fields and marshes at Lackenby. Bolckow Vaughan had relied for its coke on supplies from ovens at its collieries, but many of the latter were approaching exhaustion and losing money so Dorman Long erected the Cleveland Coking and by-product plant in 1936 at the south end of the Bessemer blast furnaces. Reputed to be the largest single coking unit in the UK when constructed, it had 136 Simon-Carves ovens divided into two batteries and was capable of carbonising 1.12 million tons of coal in a year. Two new 0-4-0 electric locomotives were purchased from Hawthorn Leslie to operate the coke quenching cars. Coal came from Dorman Long's collieries in 10 ton and 20 ton bottom discharge railway wagons and was fed by conveyor to the blending bunkers. The coking plant was reconstructed in 1947–50 resulting in it having two batteries, each of 66 ovens. The North Steel Plant, which had been out of action since 1928, was re-equipped with six 100 ton open hearth furnaces and two 400 ton mixers in 1936–38. At this date, the blast furnaces comprised Cleveland (five), Grangetown (two, A and B), South Bank (four operational 1, 2, 4 and 7) and Bessemer (two). A third blast furnace (No.5) was erected at the Bessemer site in 1938–39. The other two relied on a telpher system which involved cars with two tilting buckets and a small operator's cab travelling on a monorail, with power coming from an overhead conductor rail, to take minerals from the kilns and stores. Later all three furnaces used scale cars. The hot metal was tipped into 25 ton ladles to be conveyed to the steel plants by locomotive with the exception of South Bank. At the latter, all of the iron was cast; two of the furnaces producing ferro-manganese iron and two haematite iron.

Dorman Long had 27,000 employees in 1947 and two years later was reported to be operating iron and steel works at Acklam, Britannia, Cleveland and Redcar with 254 miles of standard gauge track and owned 3,456 internal railway wagons and bogies, 128 locomotives, 50 hot metal ladles, 552 slag ladles and had 400 men maintaining the track and locomotives. In 1946 Dorman Long published a development plan which would see

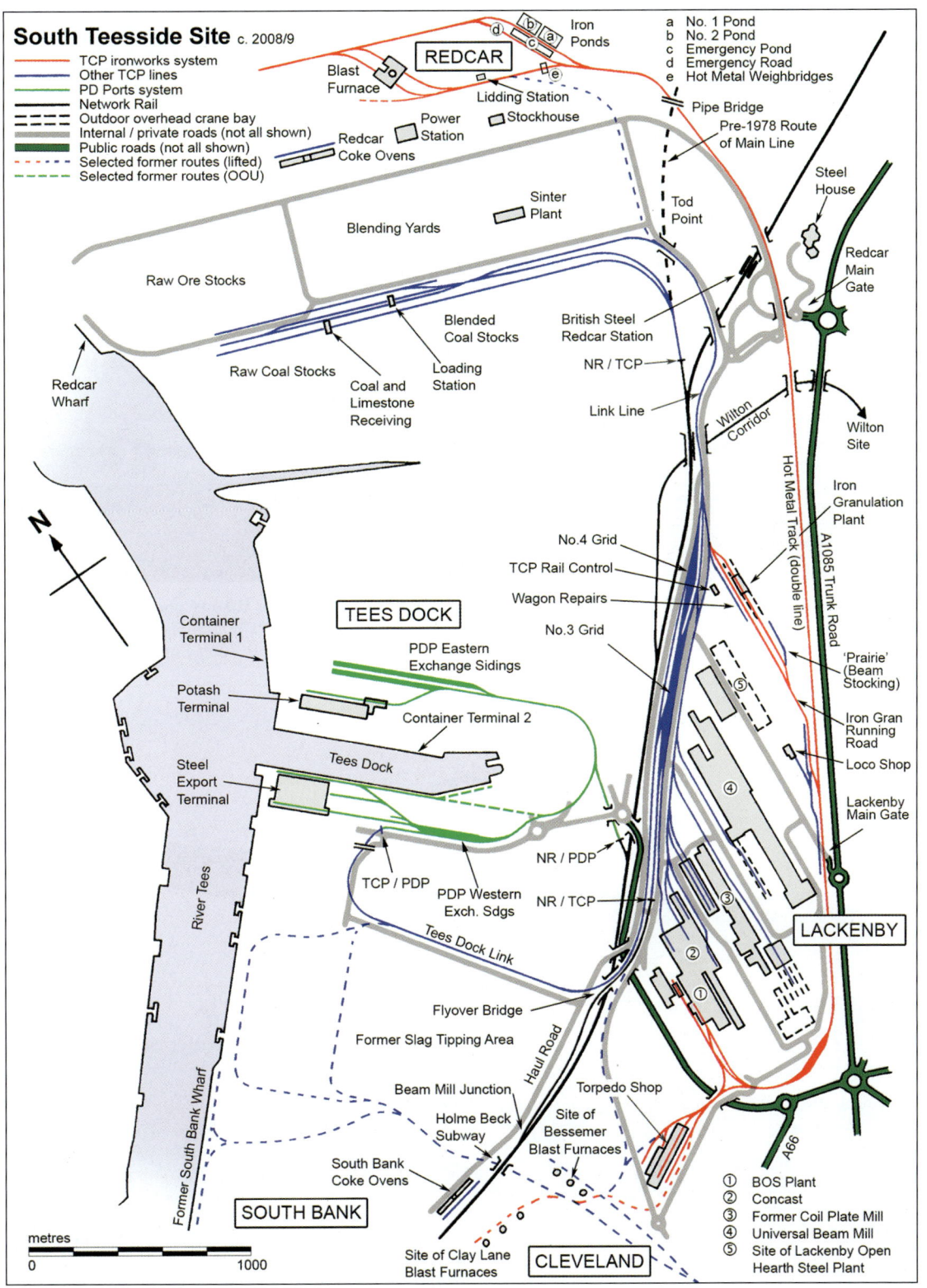

Steam rises from the South Bank Coke Ovens on 3rd November 2009 as hot coke is quenched. The spare coke oven locomotive and car are stabled next to the iconic, but largely disused, coal bunker.
(Cliff Shepherd photograph)

Dorman Long introduced two AB/MV overhead wire electrics at its South Bank Ore Processing Plant in 1948 and this shows No.2 in 1967 with the inclined gantry over the stocking ground behind.
(IRS Peter Excell photograph)

The Thos Hill Tridem locomotives were not a success and most are seen dumped at South Bank Wharf on 24th July 1978. *(Cliff Shepherd photograph)*

iron and steel production, predominantly based on imported iron ore, increasingly concentrated near the mouth of the River Tees. To accomplish this, land comprising 680 acres at Lackenby was purchased in 1946, together with additional areas at Redcar for future reclamation.

The receipt of foreign iron ores was centred on an expanded South Bank Wharf capable of handling 15,000 ton ships and, by the summer of 1949, this was linked by rail to a revamped ore processing plant on the site of the previous Bolckow Vaughan facility. All of the ores were crushed and screened fines sent to a new sinter plant, which was later connected to blast furnaces at Cleveland by conveyor. Railway company wagons bringing in Cleveland ironstone were then used to take ore from the sinter plant to the blast furnaces at Acklam and Redcar.

An internal railway, known as the "Link Line", was constructed in 1949 on the south side of the BR Middlesbrough-Redcar line connecting the Cleveland and Redcar Works and enabling the movement of molten iron, hot ingots, blooms and raw materials between the two sites. Four sets (or "grids") of ten sidings were laid out between BR and the Link Line for the marshalling and exchange of traffic. The connection with the BR line was on the north side at Beam Mill Junction. From here, a double track crossed over the BR Metals via "Flyover Bridge", possibly constructed in 1951 although the plan attached to the agreement between Dorman Long and BR was dated 5/1/1953. BR also abandoned its signal box at Grangetown Station and erected a new Grangetown Signal Box east of the bridge in 1954.

The first major plant built at Lackenby was a very large open hearth melting shop with five 360 ton tilting furnaces and two 600 ton mixers. This opened in 9/1953 replacing that at the Britannia Works. Molten iron in Kling ladles was hauled by locomotives from the blast furnaces at Cleveland and Redcar to the melting shop and hot steel ingots were returned by rail on 'ingot transfer cars' to the primary mills or else they were allowed to cool and placed on main line wagons for delivery to other works. Seven new conventional steam and two fireless locomotives were initially purchased for the new plant; the latter for hauling slag bogies from below the charging stage where smoke could hinder operations.

Erection of the South Bank coke ovens began in 2/1953 on the north side of the Middlesbrough-Redcar railway immediately to the west of the remaining South Bank blast furnaces. The 150 Simon-Carves twin flue ovens in two batteries and the associated by-product plant became fully operational in 1/1957. Initially coal was delivered by BR to the full coal sidings with a capacity of 500 wagons. Batches of 20 wagons were shunted on to the tipping lines and by gravity over a weighbridge before discharge. Two coke oven locomotives by Wellman, Smith, Owen operated the quenching cars.

Stage 3 of the development plan involved construction of two new blast furnaces at Clay Lane not far from where the old Clay Lane Iron Co's furnaces had stood. These were lit on 11/12/1956 and 22/3/1957 allowing Acklam Iron Works to close and the small hand charged Cleveland (Branch) blast furnaces to finish; the latter being demobilised in 1957–58. The remaining South Bank blast furnaces also closed at this time, being demolished in 1956 with the cleared site used for coal stocking. A conveyor delivered coke from the South Bank ovens to Clay Lane.

The Universal Beam Mill was constructed next to the melting shop at Lackenby. Reputed to be the first of its type in the UK, the first beam was rolled on 10/2/1958 and commercial orders began in 7/1958. A new rod, bar and narrow strip mill at Lackenby was completed in the spring of 1959.

For approaching 100 years, the works had relied on a vast fleet of traditional steam locomotives to handle its rail traffic. In 1956 Dorman Long purchased eight oil fired 0-6-0 vertical boiler locomotives from Sentinel, but these were not particularly successful and, in 1959, it bought the first of a standard range of diesel hydraulics again from Sentinel (maker's number 10001). During the next seven years, it replaced the steam fleet with 71 rod or chain driven four or six-wheeled diesels on the Teesside site and at the Ironmasters. Approximately 2,000 internal wagons were owned in 1959, with maintenance taking place at a central repair shop adjacent to the ore handling plant. Six signal boxes controlled busy parts of the system.

South Bank Wharf was further improved in 1961 so that it could deal with ships of up to 25,000 tonnes. A third blast furnace at Clay Lane was commissioned in 6/1962. In 1963 the Link Line was diverted over a combined road and railway bridge across the Middlesbrough-Redcar railway to improve the route for hot metal from Clay Lane to the Redcar steel plant. Trial rolling took place in 9/1965 at a new coil plate mill at Lackenby; the output of which was doubled in 1972. The North Steel plant was transformed into the Cleveland Arc Melting Plant when the open hearth furnaces were replaced in 1965–69 by three 105 tonne capacity electric arc furnaces using cold scrap. The South Steel Plant closed in 1966.

Dorman Long was nationalised again in 1967 but BSC continued the company's plans to adopt Basic Oxygen Steel making (BOS). The BOS plant situated on the west side of the Lackenby site began operating on 9/3/1971 enabling the closure of the open hearth melting shops at Lackenby (last cast 15/9/1971) and Redcar (26/3/1971). A strengthened double hot metal line, comprising 2000 yards of new track, was laid between the Clay Lane blast furnaces and the BOS plant to allow new 250 tonne capacity torpedo ladles containing molten iron to be hauled by diesel locomotives. The building formerly housing the South Steel Plant was converted to maintain the torpedoes and Kling ladles becoming known as the Torpedo Ladle Repair Shop. BSC ordered three 'Tridem' triple unit locomotives from Thos Hill in 1968 to work the steeply graded and sharply curved line. Previously Kling ladles of 60-75 tonnes capacity had carried hot metal to the open hearth furnaces but they could not keep up with the BOS plant, which made a "heat" every 40 minutes, hence the need for the torpedo

Outside the 3ft 0in gauge engine shed at the Cleveland Works on 7th November 1926 were the Black, Hawthorn locomotives, Nos.12 and 17, together with the larger 127 (AB 1350/1913). (NERA R. Inness photograph)

ladles. 'Bessemer' blast furnace No.3 was demolished in 1968 to make space for gas cleaning plant as No.5 was converted to produce ferro-manganese. A sign of things to come occurred when it was decided to remove slag from the BOS plant by road. In 1972–73 the concast (continuous casting of molten steel into blooms and slabs) plant was opened next to the BOS plant reducing the need for some primary mills. The existing South Bank coke ovens ceased operations in 1971–72, but were replaced by 88 Gibbons Wilputte ovens – one battery of 44 started up 6/1971 and the other battery in 8/1972 – on the site of the South Bank blast furnaces. They were able to make use of the coal handling facilities and by-products plant of its predecessor, but later relied on lorries to bring coal from Redcar. The 'Tridem' triple unit locomotives were not a success, because it proved difficult to synchronise the engine speeds. Instead nine large 'Titan' 0-6-0 diesel hydraulics were purchased from Thos Hill in 1972–75.

A new development strategy was unveiled by BSC which aimed to increase UK steel production, one of its main components being to develop plant at Redcar. The first ship arrived at the new Redcar ore terminal on 17/6/1973. The facility can handle ships of 200,000 tons and it replaced the South Bank Wharf, which was then used for exports of screened coke breeze and ferro-manganese. Iron ore and coal were lifted out of ships and moved by conveyor for stocking behind the wharf. There were also two rail loading bunkers that allowed iron ore to be taken out by main line locomotives, although movement of wagons between the ore loading station and the exchange sidings, together with the removal of defective wagons, was carried out by a BSC locomotive. A new fleet of 114 102 tonne G.L.W. rotary tippler wagons was provided to operate in unit trains. For internal iron ore traffic, BSC locomotives hauled fifteen of these wagons, via a new connection with the Link Line, to the Cleveland ore terminal for the Clay Lane blast furnaces. Traditional unfitted four-wheel wagons were still used for iron ore delivered to the Bessemer blast furnaces. Iron ore was also taken in rakes of only nine G.L.W. wagons, due to the gradients, by pairs of Class 37 diesels to Consett; in thirty 27 ton unfitted wagons and a brake van by two Class 31s to Hartlepool and in forty two 26 ton fitted wagons to Workington, pulled by two Class 31's.

New sinter and pellet plants opened at Redcar in 2/ and 7/1978 respectively, although the pellet plant was later mothballed in 1982. In 1978 the first battery of 66 Gibbons Wilputte ovens at the Redcar coking plant was lit, to be followed by 66 more in 1979. However, the method of charging damaged the ovens and they closed 31/7/1982 to be replaced by two batteries, each of 66 Otto-Simon ovens which started up in 7/1984. A joint venture between BSC and THPA resulted in the construction of the steel export terminal at Tees Dock and a rail connection was provided from an existing BSC line leading to South Bank Wharf to join with the Dock railway system about 1976.

The Redcar blast furnace – the largest in Europe at that time – commenced production on 12/10/1979. To create space for these developments, a 5km section of BR's Middlesbrough-Redcar line was diverted on to a more southerly alignment and a new station (British Steel, Redcar), opened within easy walking distance of the Steel House offices on 19/6/1978. Warrenby Halt closed on the same date. BSC constructed a new four mile double track hot metal railway between the Redcar blast furnace and the BOS plant which came into use in 1979. Eighteen 320 tonne capacity hot metal torpedo ladles were manufactured by the Distington Engineering Co and assembled in the Torpedo Shop for working on this railway. As the 'Tridem' and 'Titan' locomotives had not proved as successful as had been hoped and the Sentinel/Rolls-Royce diesels were underpowered for the developing traffic, a joint venture between BSC and GEC Traction resulted in the introduction of a fleet of 27 six-wheel diesel electric locomotives between late 1976 and 9/1978. Initially the GECT's were based at the Cleveland running shed, where repairs to them were also undertaken because the curves to the Cleveland Workshops were too tight. The previous engine shed at Lackenby was being used for crane maintenance in 10/1977. A purpose-designed workshop for them was erected at Lackenby (NZ 564223) and commissioned on 19/3/1981. It was stated in 1979 that the internal railway system at Teesside Works still comprised 162 miles of track with 1,100 turnouts.

However, the increasing concentration on developments at Redcar and Lackenby meant that many of the plants at the former Cleveland Works were no longer required. This was exacerbated by the 1974 energy crisis and severe competition from overseas. The last coke was produced from the Cleveland ovens on 28/9/1977 and the ore preparation plant at South Bank closed in that year. The Clay Lane blast furnaces were decommissioned in 12/1979 and dismantled in 1988. Most of the 250 tonne torpedoes moved to other BSC works and the Titan locomotives transferred to Scotland. The electric arc plant, together with the No.3 Primary Mill and No.7 Billet Mill closed at the end of 1979, to be followed by the No.9 Medium Section Mill in 1980 and the Colliery Arch plant in 1984. The remaining two Bessemer blast furnaces continued to produce foundry iron and ferro-manganese with the railway being used to bring in raw materials and take away molten iron and slag.

There was also an increasing move to use road transport; one example being the cessation of rail traffic over the internal line to Tees Dock in 1987, despite over 1 million tonnes of steel being exported and its removal in 1990–91. In 1993 Teesside Works had 52 miles of railway and 295 turnouts, but 38 miles of road. The number of industrial locomotives had declined from 93 in 1979 to 35 in 1995. The two Bessemer furnaces finished in 12/1989 (No.5) and 27/4/1993 (No.4) with demolition following in 4/1994 and this resulted in the end of regular rail traffic over the Link Line and on the Cleveland site.

In the final week under Corus ownership, 266 (GECT 5463/1977) pulls torpedo ladles 35 and 42 away from the Redcar blast furnace on 16th February 2010 taking molten iron to the BOS steel plant. *(Cliff Shepherd photograph)*

The coil plate mill at Lackenby had been responsible for sending a daily train load of coils to the Corby Works after the latter finished steelmaking but Corus, faced with continuing financial losses, transferred this work to Port Talbot and the mill closed on 26/6/2001. The formation by Corus of Teesside Cast Products (TCP) with a remit to sell most of its steel slab on the world market resulted in increasing amounts of traffic being despatched to Tees Dock. Initially the railway company handled the export steel, but this required a reversal at South Bank and was not very satisfactory, so TCP reinstated the internal rail link (officially reopened 20/7/2006) and introduced specially adapted wagons for its GECT locomotives to work from Lackenby to Tees Dock.

In the latter part of 2008, world steel prices fell in response to a worldwide recession and the consortium taking Teesside steel withdrew from the agreement on 12/5/2009. Efforts to find alternative markets failed and the Redcar blast furnace and BOS steelmaking plant, together with their ancillary plant, were mothballed with the last slab being produced on 20/2/2010. Most of the GECT locomotives and all of the torpedoes was stored in the torpedo shop. Redcar Wharf continued to principally handle main line coal traffic for the power stations. The coke ovens at Redcar and South Bank remained open with some coke being despatched by rail to Scunthorpe. Also, the Universal Beam Mill continued to operate and received a daily train of steel slab from Scunthorpe. As a result, a few GECT locomotives remained operational to handle the small amount of residual traffic. When Sahaviriya Steel Industries UK Ltd (which see) took over operation of the Redcar blast furnace, Lackenby steel plant and the two coke ovens, it introduced hired Di8 locomotives although the GECT locomotives continued to shunt the concast plant and slab bays until 4/2014. When SSI went into receivership, there was a continuing need to shunt the Universal Beam Mill and Longs Steel UK Ltd hired locomotives from Ed Murray to transfer incoming slabs from the Grid to the Universal Beam Mill. To achieve some clarity, the shunting of the Universal Beam Mill after the SSI UK Ltd liquidation, is shown as a separate sub heading.

References :   *Cleveland Iron and Steel*, Ed. C.A. Hempstead, The British Steel Corporation, 1979

Chronology of the Development of the Iron and Steel Industries of Teesside, K.H.R. Edwards, 1954

A Romance of Industry, Bolckow, Vaughan & Co Ltd, 1928

'Cleveland Iron Works', Cliff Shepherd, *Industrial Railway Record* 143, IRS, 12/1995

'South Bank Coking Plant 1 and 2', Cliff Shepherd and John M. Hutchings, *Industrial Railway Record* 167 and 168, IRS, 12/2001 and 3/2002

The Cleveland Industrial Archaeologist 20, The Cleveland Industrial Archaeology Society, 1990.

Various Dorman Long/British Steel Corporation technical surveys dated 1937, 1959 and 1979

Issues of *Steel News* from 10/1969 to 6/1985, British Steel Corporation

Teesside Cast Products Its Railways and their Origins, John Cowburn, IRS, 2016

Sentinel Locomotives and Sentinel-Cammell Railcars Their design and Development, John M. Hutchings, IRS, 2020

Gauge : 4ft 8½in

THOMAS	0-6-0ST	IC	MW	5	1859	New	s/s

In 1863-64, Bolckow Vaughan purchased seven new 0-4-0 "tank" vertical boiler locomotives from Gilkes Wilson. They are all shown below, but it is not known at which works or colliery they operated.

-	0-4-0tank	VB	GW	167	1863	New	s/s
-	0-4-0tank	VB	GW	168	1863	New	s/s
No.15	0-4-0tank	VB	GW	170	1863	New	s/s
No.18	0-4-0tank	VB	GW	171	1864	New	s/s
No.20	0-4-0tank	VB	GW	176	1864	New	s/s
No.21	0-4-0tank	VB	GW	177	1864	New	s/s
No.22	0-4-0tank	VB	GW	190	1864	New	s/s
153 (No.23) PATRIOT (MARTON) +	0-6-0T	IC	MW	153	1865	New	
-	Rebuilt	B. Vaughan			1920		(18)
BOLCKOW	0-4-0ST	OC	MW	352	1872	New	s/s
BESSEMER	0-4-0ST	OC	MW	415	1872	New	s/s after 1911
VAUGHAN	0-4-0ST	OC	MW	416	1872	New	s/s
VICKERS	0-6-0ST	OC	FW	249	1874	New	Scr
-	0-6-0ST	IC	MW	571	1876	(a)	s/s

No.	Name	Type	Cyl	Mfr	Works No.	Date	Source	Disposal
No.24	MUSHET	0-4-0ST	OC	BH	364	1876	New	s/s
	POCHIN	0-4-0ST	OC	BH	396	1877?	New	(5)
	CLEVELAND	0-4-0ST	OC	BH	429	1877	New	s/s
	GRANGETOWN	0-4-0ST	OC	BH	431	1877	New	s/s
	SOUTH BANK	0-4-0ST	OC	BH	487	1879	New	s/s
	MIDDLESBRO'	0-4-0ST	OC	BH	488	1879	New	s/s
	TYNE	0-4-0ST	OC	BH	489	1879	New	s/s
	ESTON	0-4-0ST	OC	BH	491	1879	New	s/s
	VAUGHAN	0-4-0ST	OC	BH	492	1879	New	s/s
	CHALONER	0-4-0ST	OC	BH	493	1879	New	s/s
-		0-4-0ST	OC	MW	199	1866	(b)	(6)
	LAING	0-4-0ST	OC	BH	523	1880	New	(5)
23	CARL	0-6-0ST	IC	MW	701	1880	New	s/s
	WYLLIE	0-4-0ST	OC	BH	524	1880	(c)	(7)
	HENRY	0-4-0ST	OC	BH	525	1880	New	(1)
	JUPITER	0-4-0ST	OC	BH	527	1880	(d)	s/s
	VULCAN	0-4-0ST	OC	BH	543	1880	New	s/s
	COMET	0-4-0ST	OC	BH	544	1881	New	(3)
	EDWARD WILLIAMS	0-4-0ST	OC	BH	605	1881	New	(7)
	JUNO	0-4-0ST	OC	BH	606	1881	New	(8)
	HENRY CORT	0-4-0ST	OC	BH	607	1881	New	(9)
	WINDSOR	0-4-0ST	OC	BH	608	1881	New	s/s
	MARTON	0-4-0ST	OC	BH	609	1881	New	
		Rebuilt	B. Vaughan			1924		(19)
161	NORMANBY	0-4-0ST	OC	BH	611	1881	New	s/s after 10/10/1926
162	WHITWORTH	0-4-0ST	OC	BH	519	1879	(e)	
		Rebuilt	B. Vaughan			1929		(20)
24		0-4-0ST	OC	BH	1039	1891	New	(7)
25		0-4-0ST	OC	BH	1041	1891	New	s/s
No.26	JOHN EVANS	0-6-0ST	IC	P	629	1896	New	(10)
No.27		0-6-0ST	IC	P	652	1896	New	Scr 10/1930
No.28		0-4-0ST	OC	P	667	1897	New	(7)
No.29	ESTON	0-4-0ST	OC	P	668	1897	New	
		Rebuilt	B. Vaughan			1924-25		(30)
No.30		0-4-0ST	OC	P	669	1897	New	(11)
	CHEETHAM	?	?	?	?	1877?	(f)	(2)
103		0-4-0ST	OC	CF	1186	1899	New	(7)
4		0-6-0ST	IC	MW	1367	1897	(g)	s/s
104		0-6-0ST	IC	MW	1469	1900	New	(13)
105	KELVIN	0-4-0ST	OC	CF	1211	1901	(h)	(4)
No.1		0-4-0ST	OC	BH	1094	1895	(i)	s/s after 24/10/1926
No.2		0-4-0ST	OC	BH	1123	1896	(i)	s/s after 5/9/1926
109	ARTHUR	0-6-0ST	IC	MW	1654	1905	New	(21)
111	SIEMENS	0-6-0ST	OC	AB	1100	1906	New	Scr /1962
112	CYCLOPS	0-4-0ST	OC	HL	2711	1907	New	(58)
114	AIREDALE	0-6-0T	IC	K	4624	1908	New*	
	Rebuilt as	0-6-0ST						Scr /1963
115	ILLTYD	0-4-0ST	OC	HL	2797	1910	New	Scr /1963
117	QUEEN MARY	0-4-0ST	OC	HL	2825	1910	New	Scr c/1960
118	(BELMONT)	0-6-0ST	OC	HL	2909	1911	New	(14)
119	TEES	0-4-0ST	OC	HL	2904	1911	New	Scr /1962
121	WOODFIELD	0-4-0ST	OC	AB	1315	1913	New	Scr /1963
124	JOHN TURNER	0-4-0ST	OC	AB	1344	1913	New	Scr /1963

No.	Name	Type		Maker	Works No.	Date	Acquired	Disposal
125	BELMONT	0-4-0WTST	OC	K	5020	1913	New	Scr c/1957
126	ROSEBERRY	0-4-0ST	OC	?	?			
		Rebuilt		AB	5926	1913	(j)	(7)
130	JUBILEE	0-4-0ST	OC	AB	1364	1914	New	(41)
131	FRENCH	0-4-0ST	OC	AB	1384	1914	New	Scr c/1960
132	KITCHENER	0-4-0ST	OC	HL	3082	1914	New	(31)
133	SCOTT	0-4-0ST	OC	AB	1414	1915	New	Scr /1962
134	JELLICOE	0-4-0ST	OC	HL	3131	1915	New	Scr c/1966
135	ALBERT	0-4-0ST	OC	HL	3132	1915	New	Scr c/1966
136	SALTBURN	0-4-0ST	OC	HL	3139	1915	New	(23)
137	ARGYLE	0-4-0ST	OC	HL	3140	1915	New	(17)
140	HAIG	0-4-0ST	OC	HL	3209	1916	(k)	
		Rebuilt		Cleveland Works		1955		Scr /1963
141	JOFFRE	0-4-0ST	OC	HL	3246	1917	New	Scr /1962
142	VICTORY	0-4-0ST	OC	HL	3257	1917	New	Scr 8/1961
139	BEATTY	0-4-0ST	OC	HL	3240	1917	(l)	(12)
143	VERDUN	0-4-0ST	OC	AB	1567	1917	New	Scr /1962
144(145)	BYNG	0-4-0ST	OC	HL	3326	1918	New	Scr /1963
145	SIR HENRY WILSON	0-4-0ST	OC	AB	1589	1918	New	Scr c/1960
	(formerly 144 PRESIDENT WILSON)							
146	FRASER	0-4-0ST	OC	HL	3348	1918	New	(28)
147	CLYDE	0-4-0ST	OC	HL	3350	1918	New	(36)
149	ANGUS	0-4-0ST	OC	HL	3369	1919	New	Scr /1962
150	LESLIE	0-4-0ST	OC	HL	3370	1919	New	Scr /1961
151	WILLIAM ANDERSON	0-4-0ST	OC	AB	1643	1919	New	Scr /1958
152	ALEXANDER (JAMES EVANS)							
		0-6-0ST	IC	P?	?	?		
		Rebuilt		B. Vaughan		1920		Scr /1958
154	WILLIAM HEWITT	0-4-0ST	OC	?	?			
		Rebuilt		B. Vaughan #		1921		Scr /1961
155	VAUGHAN	0-4-0ST	OC	HL	3501	1921	New	Scr /1962
156	BESSEMER	0-4-0ST	OC	HL	3502	1921	New	Scr 8/1961
157	MIDDLESBROUGH	0-4-0ST	OC	HL	3503	1921	New	Scr 8/1961
159	CALEDONIA	0-4-0ST	OC	AB	1736	1921	New	(41)
160	KILMARNOCK	0-4-0ST	OC	AB	1737	1921	New	Scr c/1960
163	CARLTON	0-6-0ST	OC	AB	1404	1915	(m)	Scr /1963
164	KING GEORGE V	0-4-0ST	OC	AB	1077	1906	(n)	Scr /1963
165	NORMANBY	0-4-0ST	OC	AB	1620	1919	(n)	(24)
166	GRANGETOWN	0-4-0ST	OC	AB	1076	1906	(n)	Scr /1962
	NEWLANDSIDE	0-4-0ST	OC	BH	365	1876		
		Rebuilt		Wake	2432		(o)	(15)
	ELSA	0-4-0ST	OC	MW	1328	1898		
		Rebuilt		Ridley Shaw		1925	(p)	(16)
167	MARS	0-4-0ST	OC	AB	1360	1913	(q)	Scr /1961
	HUGO	0-4-0ST	OC	MW	1517	1900	(r)	(22)
168	LAWRENCE ELLIS	0-4-0ST	OC	HL	3914	1937	New	Scr /1963
169	KING EDWARD	0-4-0ST	OC	HL	3915	1937	New	(25)
170	JUPITER	0-4-0ST	OC	HL	2604	1905	(s)	(26)
171	SATURN	0-4-0ST	OC	HL	2570	1903	(t)	Scr 8/1961
172	SOUTH BANK	0-4-0ST	OC	AB	926	1902	(u)	(33)
173	ALFRED E. ALCOCK	0-4-0ST	OC	HL	3919	1937	New	(45)
174	MERCURY	0-4-0ST	OC	HL	2796	1910	(v)	(32)
175	CLARENCE	0-4-0ST	OC	AB	858	1900	(w)	(34)
176	ORMESBY	0-4-0ST	OC	AB	998	1903	(x)	Scr /1961
177	SIR ELLIS HUNTER	0-4-0ST	OC	RSHD	7036	1941	New	(45)
178	SIR ARTHUR DORMAN	0-4-0ST	OC	RSHD	7037	1941	New	(44)

179	CHURCHILL	0-4-0ST	OC	RSHD	7041	1941	New	(45)
180	ROOSEVELT	0-4-0ST	OC	RSHN	7065	1942	New	(44)
181	CLEVELAND	0-6-0ST	IC	HE	3218	1945	(y)	Scr /1962
182	JOHN BIRD	0-4-0ST	OC	RSHN	7344	1947	New	Scr /1962
183	WILLIAM JONES	0-4-0ST	OC	RSHN	7347	1947	New	(45)
184	JOSEPH STIRZAKER	0-4-0ST	OC	RSHN	7348	1947	New	(43)
-		0-4-0ST	OC	AB	1224	1911	(z)	(27)
	DINSDALE No.2	0-4-0ST	OC	P	880	1901	(aa)	s/s c/1951
185	DAVID PAYNE	0-4-0DM	JF	4110006		1950	(ab)	(40)

+ Following its rebuilding, the locomotive carried "Built by Bolckow Vaughan 1920" plates and a plate on the side tank reading "PATRIOT TO THE MEMORY OF THE MEN FROM THESE WORKS WHO GAVE THEIR LIVES DURING THE GREAT WAR 1914-1918"

* According to the Locomotive Magazine on 15th April 1909, this locomotive was ordered by Bolckow & Vaughan for its recently opened coal mine in the Richmond district but it is thought to have come new to the Cleveland Works. Inness suggests that it was used on the Eston Mines traffic.

Although a copy of the company's general arrangement drawing states "Designed & Built by Bolckow Vaughan & Co. Cleveland Works".

The above locomotives operated at the Cleveland Works, including the South Bank and old Clay Lane Iron Works and were mainly based at the Cleveland Works running shed, with the workshops for locomotive repairs located immediately to the north. See the separate entry for those locomotives at the old Redcar Works. In 1949 Dorman Long constructed the Link Line connecting the Cleveland and Redcar Works so locomotives from either works could be seen at the other premises. The massive open hearth melting shop at Lackenby opened in 1953 and, to service it, Dorman Long purchased the following fleet of engines which were based at a new shed on the west side of the melting shop, although they might be seen elsewhere on the overall railway system.

1	0-6-0ST	OC	RSHN	7687	1951	New	(48)
2	0-6-0ST	OC	RSHN	7688	1951	New	(48)
3	0-6-0ST	OC	RSHN	7689	1951	New	(46)
4	0-6-0ST	OC	RSHN	7690	1952	New	(47)
5	0-6-0ST	OC	RSHN	7691	1952	New	(48)
6	0-4-0ST	OC	AB	2323	1952	New	(50)
7	0-4-0ST	OC	AB	2324	1952	New	(50)
8	0-4-0F	OC	AB	2328	1952	New	Scr c11/1967
9	0-6-0F	OC	AB	2329	1952	New	Scr c11/1967

During the mid-1950s, Dorman Long tried various types of Sentinel vertical boiler locomotives. It is likely that the purchase of the first batch was prompted by the commissioning of the Clay Lane blast furnaces starting in 1956.

-		4wVBT	VCG	S	9538	1951	(ac)	(28)
28		0-4-0WTST	OC	K	3882	1899	(ad)	Scr 12/1958
1		0-6-0VBT	VCG	S	9587	1956	New	(53)
2		0-6-0VBT	VCG	S	9588	1956	New	(53)
3		0-6-0VBT	VCG	S	9589	1956	New	(53)
4		0-6-0VBT	VCG	S	9590	1956	New	(53)
5		0-6-0VBT	VCG	S	9591	1956	New	(53)
6		0-6-0VBT	VCG	S	9592	1956	New	(53)
7		0-6-0VBT	VCG	S	9601	1956	New	(53)
8		0-6-0VBT	VCG	S	9602	1956	New	(53)
-		0-6-0+0-6-0VBT	VCG	S	9603	1956	New	(54)
-		0-6-0VBT	VCG	S	9633	1957	(ae)	(35)
-		0-6-0F	VCG	S	9650	1957	(af)	(42)
-		0-6-0F	VCG	S	9651	1957	(af)	(42)
-		0-6-0F	VCG	S	9652	1957	(af)	(42)
-		0-6-0F	VCG	S	9653	1957	(af)	(42)
	VANE	4wVBT	VCG	S	9619	1957	(ag)	(42)
2		0-6-0VBT	VCG	S	9604	1956	(ah)	(53)

7		0-6-0VBT	VCG	S	9605	1956	(ah)	(53)
-		0-6-0VBT	VCG	S	9606	1956	(ah)	(38)
MC1		0-4-0CA	OC	Cleveland		1958	(ai)	Scr /1982
MC2		0-4-0CA	OC	Cleveland		1958	(aj)	Scr /1983
MC3		0-4-0CA	OC	Cleveland		1958	(ak)	Scr c8/1982
No.10		0-4-0F	OC	AB	2376	1961	New	Scr c11/1967

In the late 1950s, Dorman Long decided to replace its steam locomotives with diesels and various manufacturers supplied one of their products for demonstration purposes, including RH 412716/1957 (11/12/1957), JF 4230001/1959 (3/3/1959), WB 3160/1959 (8/1959) and NBL.

Dorman Long also purchased seven small Hibberd diesel locomotives during this period and these are shown under the relevant works.

32 (1)		4wDM	FH	3832	1957	(al)	(51)
No.4		4wDM	FH	3888	1958	New	(49)
No.1 (34)		4wDH	FH	3934	1960	New	(52)
35		4wDH	FH	3935	1960	New	(60)

In 1959, Dorman Long began to purchase a large fleet of diesel hydraulic locomotives from Sentinel via its agent, Thos Hill. The diesel locomotives numbered between 10001 and 10183 were designed and built by Rolls-Royce Ltd at Shrewsbury but fitted with Sentinel maker's plates. Initially these locomotives were distributed around the various works and a code letter has been shown where they were first allocated. However, locomotives were subsequently moved between works depending on traffic and repairs.

Works codes – C Cleveland, CL Clay Lane, L Lackenby, R Redcar

1		4wDH	S	10001	1959	(am)	(67)
2		4wDH	S	10002	1959	New L	(72)
4(3)		4wDH	S	10004	1959	New L	(66)
8		4wDH	S	10008	1959	New L	Scr 5/1976
9		4wDH	S	10009	1959	New L	(71)
10		4wDH	S	10010	1959	New L	(77)
11		4wDH	S	10011	1959	(an)	(69)
14		4wDH	S	10014	1960	(ao)	(79)
15		4wDH	S	10015	1960	(ap)	(71)
16		4wDH	S	10016	1960	(an)	(64)
17		4wDH	S	10017	1960	(ao)	(76)
19		4wDH	S	10019	1960	(aq)	(57)
24		4wDH	S	10024	1960	(ar)	Scr /1976
25		4wDH	S	10025	1960	(as)	(59)
26		4wDH	S	10026	1960	New R	(71)
27		4wDH	S	10027	1960	New R	(62)
28		4wDH	S	10028	1960	New R	(73)
29		4wDH	S	10029	1960	New R	(80)
30		4wDH	S	10030	1960	New R	(63)
31		4wDH	S	10031	1960	New C	Scr c/1973
32	CHRIS MOODY	0-6-0DH	S	10032	1960	New (at) L	(111)
34		4wDH	S	10034	1960	New C	(83)
35		4wDH	S	10035	1960	New L	(78)
36		4wDH	S	10036	1960	New R	(55)
38		4wDH	S	10038	1960	New R	(113)
39		4wDH	S	10039	1960	New C	Scr 5/1976
40		4wDH	S	10040	1960	New C	(77)
41		4wDH	S	10041	1960	New C	(68)
42		4wDH	S	10042	1960	New C	(73)
43		4wDH	S	10043	1960	New C	(64)
44		4wDH	S	10044	1960	New C	Scr 5/1976
50		0-6-0DH	S	10050	1961	New (at) L	(75)
53		0-6-0DH	S	10053	1961	New L	(87)
54		0-6-0DH	S	10054	1961	New L	(109)

55	0-6-0DH	TH	109C	1961	New (at) L	(108)
56	0-6-0DH	S	10056	1961	New L	(109)
57	0-4-0DH	S	10057	1961	New R	(97)
64	0-4-0DH	S	10064	1961	New R	(113)
66	0-4-0DH	S	10066	1961	New R	(85)
67	0-4-0DH	S	10067	1961	New R	(82)
68	0-4-0DH	S	10068	1961	New R	(113)
69	0-4-0DH	S	10069	1961	New R	(88)
73	4wDH	S	10073	1961	New C	(57)
74	4wDH	S	10074	1961	New C	Scr 5/1976
75	4wDH	S	10075	1961	New C	Scr 5/1976
76	4wDH	S	10076	1961	New C	Scr 5/1976
78	0-6-0DH	S	10078	1961	New CL	(108)
84	0-4-0DH	S	10084	1961	New L	(85)
91	0-6-0DH	S	10091	1962	New CL	(89)
92	0-6-0DH	S	10092	1962	New CL	Scr 5/1976
93	0-6-0DH	S	10093	1962	New L	(99)
94	0-6-0DH	S	10094	1962	New CL	(100)
95	0-6-0DH	S	10095	1962	New C	(99)
96	0-4-0DH	S	10096	1962	New C	(95)
98	0-4-0DH	S	10098	1962	New C	(86)
99	0-4-0DH	S	10099	1962	New C	(86)
100	0-4-0DH	S	10100	1962	New C	(98)
101	0-6-0DH	S	10101	1962	New C	(110)
102	0-6-0DH	S	10102	1962	New C	(106)
103	0-4-0DH	S	10103	1962	New C	(93)
104	0-4-0DH	S	10104	1962	New L	(96)
105	0-4-0DH	S	10105	1962	New C	(113)
127	4wDH	TH	127V	1963	New C	(61)
128	4wDH	TH	128V	1963	New L	(56)
166	0-6-0DH	S	10166	1963	New L	(107)
167	0-6-0DH	S	10167	1964	New L	(113)
168	0-6-0DH	S	10168	1964	New L	(119)
210	0-6-0DH	RR	10210	1964	New L	(120)
211	0-6-0DH	RR	10211	1964	New L	(117)
224	0-6-0DH	RR	10224	1965	New C	(120)
225	0-6-0DH	RR	10225	1965	New C	(116)
234	0-6-0DH	RR	10234	1965	New C	(115)

After 1965, the following locomotives were obtained for Teesside Works and based at the Cleveland and Lackenby sheds. Of these, 201 to 211 were tridem articulated locomotives, with 201 to 204 being the drive units, 205 to 211 were the 'slaves' or 'power tenders' and 204 and 211 were spare units to provide cover.

201	4wDH	TH	201V	1969	New	Scr /1982
202	4wDH	TH	202V	1969	New	Scr /1982
203	4wDH	TH	203V	1969	New	Scr /1982
204	4wDH	TH	204V	1969	New	Scr 3/1975
205	4wDH	TH	205V	1969	New	Scr /1982
206	4wDH	TH	206V	1969	New	Scr /1982
207	4wDH	TH	207V	1969	New	Scr /1982
208	4wDH	TH	208V	1969	New	Scr 3/1975
209	4wDH	TH	209V	1969	New	Scr /1982
210	4wDH	TH	210V	1969	New	Scr /1982
211	4wDH	TH	211V	1969	New	Scr /1982
292	0-6-0DH	RR	10292	1971	New	(112)
241	0-4-0DH	TH	241V	1972	New (au)	Scr 7/1975
242	0-6-0DH	TH	242V	1972	New	(105)
243	0-6-0DH	TH	243V	1972	New	(103)

244		0-6-0DH	TH	244V	1972	New	(104)
128	(4)	0-4-0DH	S	10128	1963	(av)	(92)
130	(6)	0-4-0DH	S	10130	1963	(av)	(94)
132	(8)	0-4-0DH	S	10132	1963	(av)	(98)
134	(10)	0-4-0DH	S	10134	1963	(av)	(95)
245		0-6-0DH	TH	245V	1973	New	(102)
CN3539		6wDE	Moyse	3539	1973	(aw)	(65)
48	STANTON No.43	0-4-0DE	YE	2621	1956	(ax)	(84)
No.45	STANTON No.44	0-4-0DE	YE	2622	1956	(ax)	(70)
	STANTON No.46	0-4-0DE	YE	2624	1956	(ax)	(84)
47	STANTON No.47	0-4-0DE	YE	2625	1956	(ax)	(84)
	STANTON No.51	0-4-0DE	YE	2886	1961	(ax)	Scr 8/1984
250		0-6-0DH	TH	252V	1974	New	(101)
246		0-6-0DH	TH	255V	1975	New	(102)
247		0-6-0DH	TH	253V	1975	New	(105)
248		0-6-0DH	TH	254V	1975	New	(105)
249		0-6-0DH	TH	256V	1975	New	(104)
	STANTON No.40	0-4-0DE	YE	2596	1955	(ay)	(74)
	STANTON No.41	0-4-0DE	YE	2597	1955	(az)	(74)
134		0-6-0DH	S	10153	1963	(ba)	(81)
186		0-6-0DH	RR	10186	1964	(ba)	(91)
129	(5)	0-4-0DH	S	10129	1963	(bb)	(90)
251	WALTER URWIN	6wDE	GECT	5414	1976	New	(134)
252	BOULBY	6wDE	GECT	5415	1976	New	(134)
253	ESTON	6wDE	GECT	5416	1976	New	(134)
254	BROTTON	6wDE	GECT	5417	1976	New	(121)
255	LIVERTON	6wDE	GECT	5418	1976	New	(134)
155		0-8-0DH	S	10136	1962	(bc)	Scr 6/1982
256	LINGDALE	6wDE	GECT	5425	1977	(bd)	(127)
257	NORTH SKELTON	6wDE	GECT	5426	1977	New	(134)
258	GRINKLE	6wDE	GECT	5427	1977	New	(135)
259	CARLIN HOW	6wDE	GECT	5428	1977	New	(131)
260	ROSEDALE	6wDE	GECT	5429	1977	(be)	(135)
261	STAITHES	6wDE	GECT	5430	1977	New	(134)
262	LOFTUS	6wDE	GECT	5431	1977	New	(134)
263	LUMPSEY	6wDE	GECT	5432	1977	New	(132)
264	PORT MULGRAVE	6wDE	GECT	5461	1977	New	(114)
265	ROSEBERRY	6wDE	GECT	5462	1977	New	(134)
266	SHERRIFFS	6wDE	GECT	5463	1977	New	(134)
267	SLAPEWATH	6wDE	GECT	5464	1977	(bf)	(134)
268	KIRKLEATHAM	6wDE	GECT	5465	1977	New	(135)
269	LONGACRES	6wDE	GECT	5466	1977	New	(134)
270	CHALONER	6wDE	GECT	5467	1977	New	(136)
271	GLAISDALE	6wDE	GECT	5469	1978	New	(135)
272	GROSMONT	6wDE	GECT	5470	1978	(bg)	(122)
273	KILTON	6wDE	GECT	5471	1978	New	(136)
274	ESKDALESIDE	6wDE	GECT	5472	1978	New	(136)
275	RAITHWAITE	6wDE	GECT	5473	1978	New	(136)
276	SPAWOOD	6wDE	GECT	5474	1978	New	(135)
277	WATERFALL	6wDE	GECT	5475	1978	New	(135)
1	(LM20) LACKENBY	4wDM	Robel	54.12-107 AD184	1980	New	(135)
2	(LM21) REDCAR	4wDM	Robel	54.12-107 AD183	1980	New	(135)
43		6wDH	TH	V317	1987	(bh)	(118)
42		6wDH	TH	V316	1987+	(bi)	(123)

	+plate reads TH P214 1987						
41		6wDH	RR	10277	1968	(bj)	(124)
278		6wDE	GECT	5468	1977	(bk)	(134)
-		6wDE	GECT	5421	1977	(bl)	(136)
-		6wDE	GECT	5479	1979	(bm)	(133)
2	CATTERSTY 2	0-4-0DH	S	10126	1963	(bn)	(122)
3	HUMMERSEA 1	0-4-0DH	S	10127	1963	(bo)	(128)
-		6wDE	GECT	5578	1980	(bp)	(125)
279		6wDE	GECT	5478	1978	(bq)	(136)
-		0-6-0DH	HE	7041	1971	(br)	(129)
	PANDA	4w-4wDH	HAB	776	1981	(bs)	(130)

Miscellaneous observations:

A Mercedes Benz Unimog was present on 28/10/1977. Its rail wheels had been removed by 15/4/1978.

On 19-20/10/1996, Freight Train Productions staged a photographic charter at the Redcar blast furnace using steam locomotives, MIRVALE (HC 1582/1955) and LION (P 1351/1914).

(a) ex Logan & Hemingway, contractor, Sheffield, Yorkshire (WR).
(b) ex The Rosedale & Ferry Hill Iron Co Ltd, Ferry Hill Iron Works, West Cornforth, Co. Durham, c/1880. This company went into liquidation and its plant including locomotives were auctioned on 16/12/1879, although the Durham Handbook suggests that the locomotives were not despatched until a year later.
(c) new. To Darlington Rolling Mills Co Ltd, Darlington, Co. Durham, possibly via Byers Green Colliery, Byers Green, Co. Durham, on hire and later returned, ex hire. The boiler on this locomotive blew up on 26/12/1898 due to a rupture in the firebox. The boiler had been made by Black, Hawthorn in 1880 and fitted to a Bolckow Vaughan locomotive known as "No.1" or "HENRY".
(d) new; Bolckow Vaughan is given as the customer but no destination is stated in the BH records.
(e) ex Witton Park Iron Works, Witton Park, Co. Durham. The iron works closed 19/5/1884 so presumably the locomotive was transferred at some time after that date.
(f) origin unknown. It was stated to be at Eston in the 1885–90 period.
(g) ex Midland Coal, Coke & Iron Co Ltd, Apedale Iron Works, Chesterton, Staffordshire; the MW Locomotive Customer Index shows it under Bolckow Vaughan before MW 1469 so it may have arrived here before 1900.
(h) ex Dean & Chapter Colliery, Ferryhill, Co. Durham.
(i) ex Clay Lane Iron Co Ltd, Clay Lane Iron Works, South Bank, c/1898.
(j) origin unknown. Inness suggested Dick & Stevenson.
(k) ex Middlesbrough Iron Works, Middlesbrough.
(l) ex Middlesbrough Iron Works, Middlesbrough. It was seen at the Cleveland Works on 3/7/1927.
(m) ex Carlton Iron Works, Stillington; the date is unknown although the iron works closed in 3/1930 and it was here 3/10/1931.
(n) ex Glasgow Iron & Steel Co Ltd. This company owned the Wishaw Iron & Steel Works and various collieries. It is not known where the three locomotives worked, nor when they arrived at the Cleveland Works, but the Wishaw Iron Works closed in /1930.
(o) ex Newlandside Quarry, Stanhope, Co. Durham.
(p) ex Newlandside Quarry, Stanhope, Co. Durham after 6/1935.
(q) ex Thos W. Ward Ltd, Sheffield, Yorkshire (WR), sold 6/7/1936; previously hired by Ward to British Celanese Ltd, Spondon, Derbyshire.
(r) ex Newlandside Quarry, Stanhope, Co. Durham, after 5/1932.
(s) ex Clarence Works, Port Clarence, 3/11/1936.
(t) ex Clarence Works, Port Clarence, 27/10/1936.
(u) ex Clarence Works, Port Clarence, 8/10/1936.
(v) ex Britannia Works, Middlesbrough, c/1938.
(w) ex Clarence Works, Port Clarence, after 10/1937.
(x) ex Clarence Works, Port Clarence, after 4/11/1935.
(y) ex War Department, Longmoor Military Railway, Hampshire, 71454, 4/1946.
(z) ex Netherseal Colliery Co Ltd, Netherseal Colliery, Linton, Derbyshire, /1947. The locomotive was not vested in the NCB but was still at Netherseal on 1/1/1947.
(aa) ex Linthorpe–Dinsdale Smelting Co Ltd, Middleton Iron Works, Middleton St George, Co. Durham, after 10/8/1949.

(ab)	new. To Dock Street Foundry, Middlesbrough, /1960 and returned by 10/1960.
(ac)	ex Sentinel (Shrewsbury) Ltd, Shrewsbury, Shropshire, on demonstration, probably 12/2/1954.
(ad)	ex Britannia Works, Middlesbrough, c9/1954.
(ae)	ex Sentinel (Shrewsbury) Ltd, Shrewsbury, Shropshire, on loan 14/5/1957 to cover problems with the Sentinel receiver locomotives.
(af)	new. These were 'steam receiver' locomotives. Only S 9650 was steamed, performing test runs between 5/1958 and 2/1959. They did not achieve the specification and did not become Dorman Long property being returned to Thos Hill.
(ag)	ex Thos Hill Ltd, Kilnhurst, Yorkshire (WR), on loan to cover problems with the Sentinel receiver locomotives.
(ah)	ex Guest Keen Iron & Steel Co Ltd, East Moors Works, Cardiff, Glamorgan, per F. Burrill & Co Ltd, dealer, c8/1959. Probably shortly after arrival, S 9604 received the cab with number 2 from S 9588 and, similarly, S 9605 received the cab and number 7 from S 9601.
(ai)	incorporates frame, cylinders and wheels of Peckett 668/1897. Operated slag bogies on tip.
(aj)	incorporates frame, cylinders and wheels of Hawthorn Leslie 3082/1914. Operated slag bogies on tip.
(ak)	incorporates frame, cylinders and wheels of Hawthorn Leslie 2796/1910. Operated slag bogies on tip.
(al)	new. FH quote "Redcar Works" for the order but no despatch details are given. It is most likely to have come to Lackenby Shed but was later working at Redcar Works.
(am)	new L, although this first of the type was initially used by Sentinel for demonstration purposes. Purchased by Dorman Long and delivered 20/10/1959.
(an)	ex Britannia Works, Middlesbrough, 25/6/1962.
(ao)	ex Britannia Works, Middlesbrough, by 12/1969.
(ap)	ex Britannia Works, Middlesbrough, 22/3/1971.
(aq)	ex Britannia Works, Middlesbrough, 21/6/1962.
(ar)	ex Britannia Works, Middlesbrough, 22/6/1962.
(as)	new R, to Britannia Works, Middlesbrough. 27/4/1964 and returned to Redcar Works by /1968.
(at)	these three locomotives incorporated frames and wheelsets from the Sentinel 0-6-0VBT locomotives. The Dorman Long version was that S 9588 was rebuilt to diesel 50 at Sentinel and 9606 to diesel 32 – the latter commenced at Dorman Long and completed at Sentinel. However it is more likely that exchanges of cabs and, therefore, plates took place so Sentinel 9604–06 were used to supply frames for 32, 50 and 55.
(au)	new, intended as a slave unit for use with 242 to 245
(av)	ex Skinningrove Works, Carlin How, c6/1972.
(aw)	new, ex Moyse for trials, 10/1973.
(ax)	ex Stanton & Staveley Ltd (subsidiary of BSC), Stanton Iron Works, Ilkeston, Derbyshire, 8/1974.
(ay)	ex Stanton & Staveley Ltd (subsidiary of BSC), Stanton Iron Works, Ilkeston, Derbyshire, /1976.
(az)	ex Stanton & Staveley Ltd (subsidiary of BSC), Stanton Iron Works, Ilkeston, Derbyshire, 4/1976.
(ba)	ex Ebbw Vale Works, Ebbw Vale, Gwent, /1976.
(bb)	ex Skinningrove Works, Carlin How, /1976.
(bc)	ex Ebbw Vale Works, Ebbw Vale, Gwent, for spares, c5/1977.
(bd)	new. To Appleby-Frodingham Works, Scunthorpe, Lincolnshire, /1999, by 30/5/1999. Returned to Teesside Works, c/2000; after 27/11/1999, by 4/9/2000.
(be)	new. To Shapfell limestone quarries and lime works, Shap, Cumbria, 7/4/2010. Returned to Teesside Works 8/8/2011 and then stored with other locomotives.
(bf)	new. To Shapfell limestone quarries and lime works, Shap, Cumbria, /1994; after 15/1/1994, by 8/6/1994. Returned to Teesside Works, c24/5/1996.
(bg)	new. To Shapfell limestone quarries and lime works, Shap, Cumbria, after 23/4/1993, by 17/6/1993. Returned to Teesside Works after 15/1/1994, by 8/6/1994.
(bh)	ex Ravenscraig Works, Motherwell, Strathclyde, here for repairs, 13/7/1995.
(bi)	ex Ravenscraig Works, Motherwell, Strathclyde, here for repairs, 17/7/1995. Left Ravenscraig 14/7/1995.
(bj)	ex Ravenscraig Works, Motherwell, Strathclyde, here for repairs, 4/9/1995.
(bk)	ex Staffordshire Locomotive Co Ltd, c/o Yorkshire Engine Co Ltd, Long Marston, Warwickshire, 14/12/1995; previously British Coal Corporation, Littleton Colliery, Huntington, Staffordshire. Overhauled at Teesside Works and sent to Shapfell limestone quarries and lime works, Shap, Cumbria, 24/5/1996. Returned to Teesside Works, 7/4/2010.
(bl)	ex Staffordshire Locomotive Co Ltd, c/o Yorkshire Engine Co Ltd, Long Marston, Warwickshire, 29/4/1996 (for spares); previously British Coal Corporation, Littleton Colliery, Huntington, Staffordshire.

(bm) ex Staffordshire Locomotive Co Ltd, c/o Yorkshire Engine Co Ltd, Long Marston, Warwickshire, 28/5/1996 (for spares); previously British Coal Corporation, Littleton Colliery, Huntington, Staffordshire.
(bn) ex Skinningrove Works, Carlin How, 5/1/1998.
(bo) ex Skinningrove Works, Carlin How, 6/1/1998.
(bp) ex Brunner-Mond, Winnington Works, Northwich, Cheshire (property of Staffordshire Locomotive Co Ltd), 3/8/1999.
(bq) ex T.J. Thomson & Son Ltd, Millfield Works, Stockton, c/2000; previously Wilmott Bros (Plant Services) Ltd, Ilkeston, Derbyshire where had been used as a hire locomotive at Allied Steel & Wire Ltd, Cardiff. To Hartlepool Pipe Mill, Hartlepool by 4/3/2000 and returned by 2/4/2001.
(br) ex BP Amoco Chemicals Ltd, Salt End Refinery, Kingston upon Hull, East Yorkshire, (property of Staffordshire Locomotive Co Ltd), for storage 8/2002. Lackenby Workshop had been set up as an 'autonomous' concern to service Corus locomotives and to carry out contract work on other locomotives. Staffordshire Locomotive Co Ltd acted as agents. Some Freightliner class 66 main line diesels received attention at the workshop.
(bs) ex Lindsey Oil Refinery Ltd, South Killingholme, Lincolnshire. The locomotive was sent to Hunslet-Barclay Ltd, Kilmarnock, East Ayrshire for repairs, /2002. These were not carried out and the locomotive was purchased at Kilmarnock by Staffordshire Locomotive Co Ltd and moved to Teesside Works for storage, 4/2/2003.

(1) to Middlesbrough Iron Works, Middlesbrough, c/1885.
(2) to Byers Green Colliery, Byers Green, Co. Durham, 9/1899.
(3) to Newlandside Quarry, Stanhope, Co. Durham, by 5/1908.
(4) to Byers Green Colliery, Byers Green, Co. Durham.
(5) s/s after 24/10/1926, by 3/10/1931.
(6) s/s after 11/1929.
(7) s/s after 3/7/1927, by 3/10/1931. Inness saw a line of Bolckow Vaughan locomotives on 5/9/1926 at Cleveland Works "shunted out for scrap" presumably due to a decline in business after World War 1. It comprised Nos. 24, 2, 28, JUNO, HENRY CORT, No.3, LAING, WYLLIE and No.1.
(8) to Darlington Rolling Mills Co Ltd, Darlington, Co. Durham, after 3/7/1927, by 3/10/1931.
(9) the movements of this locomotive are uncertain. It may have gone to Dean & Chapter Colliery, Ferryhill, Co. Durham and returned or it could have moved to Binchester and Westerton Collieries, near Westerton, Co. Durham from the Cleveland Works direct or via Dean & Chapter Colliery. It was at Cleveland Works on 3/7/1927 but had moved by 3/10/1931.
(10) to Dean & Chapter Colliery, Ferryhill, Co. Durham, after 3/7/1927, by 3/10/1931.
(11) to Tursdale Colliery, near Cornforth, Co. Durham, after 1/11/1929. It was rebuilt by Ridley Shaw in /1930 and may have gone there first. It had left Cleveland Works by 3/10/1931.
(12) returned to Middlesbrough Iron Works, Middlesbrough.
(13) to Leasingthorne Colliery, near Coundon, Co. Durham, by 3/10/1931.
(14) to Byers Green Colliery, Byers Green, Co. Durham, after 3/7/1927, by 3/10/1931.
(15) to Newlandside Quarry, Stanhope, Co. Durham, by 5/1939. It was not at the Cleveland Works on 3/10/1931.
(16) to Newlandside Quarry, Stanhope, Co. Durham, after 6/1935.
(17) to Parson Byers Quarry, near Stanhope, Co. Durham, c/1934-35.
(18) to Leasingthorne Colliery, near Coundon, Co. Durham, after 3/10/1931, by 24/3/1940.
(19) to Port Clarence Works, after 18/9/1934. It was there by 1/1/1947.
(20) to Port Clarence Works, after 3/10/1931. It was there by 1/1/1947.
(21) scrapped after 3/10/1931.
(22) to Newlandside Quarry, Stanhope, Co. Durham, after 5/1932.
(23) to Parson Byers Quarry, near Stanhope, Co. Durham, /1941.
(24) to Acklam/Britannia/Newport Works (combined list), 20/8/1941.
(25) to Redcar Works, before 1/8/1954.
(26) to Burley Ironstone Quarries, Rutland, 9/1940.
(27) to Burley Ironstone Quarries, Rutland, 9/1950.
(28) to Redcar Works, c/1954.
(29) to NCB Derwenthaugh, Swalwell, Co. Durham, ex demonstration, by 22/8/1954.
(30) dsm, chassis and cab used by Cleveland Works to construct compressed air locomotive MC1, /1958.
(31) at some time carried HL 3132/1915 plate? Dsm, chassis and cab used by Cleveland Works to construct compressed air locomotive MC2, /1958.
(32) dsm, chassis and cab used by Cleveland Works to construct compressed air locomotive MC3, /1958.

(33) to Burley Ironstone Quarries, Rutland, /1937. Inness mistakenly suggested the maker as HL and that it went to Middlesborough Iron Works but the maker's number and dimensions are those for the AB.
(34) s/s, after 9/1950, by 1/8/1954.
(35) returned to Thos Hill Ltd, Kilnhurst, Yorkshire (WR), after loan, c6/1958.
(36) to Port Clarence Works, c4/1959.
(37) Not Used
(38) No record of use at Cleveland Works and its frame and wheelsets were prepared by Dorman Long in /1960 for incorporation in S10032/1960
(39) returned to Thos. Hill Ltd, Kilnhurst, Yorkshire (WR), ex loan, 5/1960.
(40) to Dock St Foundry, Middlesbrough, c/1961
(41) to Redcar Works, Redcar, /1962.
(42) to Thos Hill Ltd, Kilnhurst, Yorkshire (WR), for conversion to diesel hydraulic locomotives: S 9616 to Kilnhurst c6-7/1961 and S 9650-53 in 1961-64 with S 9650 being the last in 11/1964.
(43) to Britannia Works combined fleet, Middlesbrough, 24/9/1962.
(44) to Britannia Works combined fleet, Middlesbrough, 5/10/1962.
(45) to Britannia Works combined fleet, Middlesbrough, 11/1962.
(46) to NCB, Wearmouth Colliery, Monkwearmouth, Co. Durham, 8/12/1962.
(47) to NCB, Wearmouth Colliery, Monkwearmouth, Co. Durham, 12/1962
(48) to NCB, Lambton Railway, Co. Durham, 12/1962 or 1/1963.
(49) to Britannia Works combined fleet, Middlesbrough, 2/6/1963.
(50) to South Durham Steel & Iron Co Ltd, Cargo Fleet Iron & Steel Works, Cargo Fleet, /1964.
(51) to Dock St Foundry, Middlesbrough, 14/12/1964.
(52) to Britannia Works, Middlesbrough, before 12/1969.
(53) all taken out of use probably by end of /1960. The frame of 2 (S 9588) was sent to Thos Hill at Kilnhurst for incorporation in TH109c/1961; the cab with number 2 passing to S 9604. The frame of 7 (S 9601) was sent to Sentinel at Shrewsbury for incorporation in S 10050/1960; the cab with number 7 passing to S 9605, 8 (S 9602) was reported as having been converted to a brake tender at Dorman Long c/1963. By 6/1964, a row of seven rod-drive Sentinels (7 S 9605, 5, 4, 1, 2, S 9604, 3 and 6) stood out of use at the Cleveland Works and were later moved to Redcar Works for scrapping. A visit here on 6/5/1967 found one still complete, four frames after dismantling and one frame and chassis converted to "Carrier Wagon No.3".
(54) stored at South Bank by 12/1962 and scrapped at Redcar Works, Redcar, /1966.
(55) to Stanton & Staveley Ltd, Ormesby Iron Works, Cargo Fleet, 29/3/1966.
(56) rebuilt into 4wWE at Cleveland Works, c/1967. See Cleveland No.3 Billet Mill.
(57) to Stanton & Staveley Ltd, Ormesby Iron Works, Cargo Fleet, 6/3/1968.
(58) to Apprentice Training Centre, South Bank for some restoration, c1/1970.
(59) to Hartlepool South Works, Hartlepool, 15/10/1971.
(60) to BSC, Teesside Bridge & Engineering Works, Cargo Fleet, c11/1971. (A Locomotive Schedule has the comment "Transferred ex Furness Shipbuilding 2/12/1968".)
(61) to Thos Hill Ltd, Kilnhurst, Yorkshire (WR), /1971.
(62) converted to Redcar Ore Terminal "BARRIER WAGON No.1", /1972. Equipped with rotary couplings at one end in case one of the 102 tonne ore wagons was defective.
(63) converted to Redcar Ore Terminal "BARRIER WAGON No.2", /1972.
(64) scrapped c/1973 due to collision damage.
(65) to Voest Alpine (Austria State Steel Works), Austria, c10/1973.
(66) rebuilt into 4wWE at Cleveland Works, 5/1974. See Cleveland No.3 Billet Mill.
(67) to Thos Hill Ltd, Kilnhurst, South Yorkshire, 9/1974 where rebuilt to TH 258C /1975.
(68) to Thos Hill Ltd, Kilnhurst, South Yorkshire, 9/1974 where rebuilt to TH 259C /1975.
(69) to Thos Hill Ltd, Kilnhurst, South Yorkshire, 6/1975 where rebuilt to TH 260 c/1975.
(70) to North Yorkshire Moors Railway, Grosmont, North Yorkshire, 7/1975.
(71) to River Don Works, Sheffield, South Yorkshire, 12/1975.
(72) to River Don Works, Sheffield, South Yorkshire, c2/1976.
(73) to Hartlepool Works, Hartlepool, 11/1976 (11/1975 is an alternative).
(74) to Stanton & Staveley Ltd (subsidiary of BSC) Stanton Iron Works, Ilkeston, Derbyshire, /1977.
(75) to Consett Works, Co. Durham, 4/1/1977.
(76) to George Cohen, Sons & Co Ltd, Cargo Fleet 17/11/1977.
(77) to Dock Street Foundry, Middlesbrough /1960 and returned c/1961, after 23/4/1961. To Thos Hill Ltd, Kilnhurst, South Yorkshire, 15/12/1977; resold to Rugby Portland Cement Co Ltd, Rochester, Kent, 12/1978.

(78) to Thos Hill Ltd, Kilnhurst, South Yorkshire, 21/12/1977; resold to Rugby Portland Cement Co Ltd, Rochester, Kent, 12/1978.
(79) to Thos W. Ward Ltd, Templeborough Works, Sheffield, South Yorkshire, for resale, 15/2/1978.
(80) to Thos W. Ward Ltd, Templeborough Works, Sheffield, South Yorkshire, for resale, 3/1978.
(81) to Consett Works, Consett, Co. Durham, for spares, 11/1977.
(82) to Consett Works, Consett, Co. Durham, 17/2/1978.
(83) to Thos W. Ward Ltd, Templeborough Works, Sheffield, South Yorkshire, for resale, 2/3/1978.
(84) to T.J. Thomson & Son Ltd, Stockton, 5/1978.
(85) to Consett Works, Consett, Co. Durham, 10/7/1978.
(86) to Tinsley Park Works, Sheffield, South Yorkshire, 19/9/1978.
(87) to Thos Hill Ltd, Kilnhurst, South Yorkshire for overhaul, 21/9/1978, prior to transfer to Tinsley Park Works, Sheffield, South Yorkshire.
(88) to Consett Works, Consett, Co. Durham, 16/2/1978.
(89) to Tinsley Park Works, Sheffield, South Yorkshire, 23/9/1978.
(90) to Falmouth Docks & Engineering Co, Falmouth, Cornwall, 4/10/1978.
(91) to Workington Works, Workington, Cumbria, 10/1978.
(92) to Fairfield-Mabey Ltd, Chepstow, Gwent, c6/1980.
(93) to Andrew Barclay, Sons & Co Ltd, Kilmarnock, Ayrshire, 14/7/1980.
(94) to Andrew Barclay, Sons & Co Ltd, Kilmarnock, Ayrshire, 16/7/1980.
(95) to Andrew Barclay, Sons & Co Ltd, Kilmarnock Ayrshire, 21/7/1980.
(96) to Andrew Barclay, Sons & Co Ltd, Kilmarnock, Ayrshire, 23/7/1980
(97) to Andrew Barclay, Sons & Co Ltd, Kilmarnock, Ayrshire, 28/7/1980.
(98) to Andrew Barclay, Sons & Co Ltd, Kilmarnock Ayrshire, 30/7/1980.
(99) to Andrew Barclay, Sons & Co Ltd, Kilmarnock, Ayrshire, 4/8/1980.
(100) to Andrew Barclay, Sons & Co Ltd, Kilmarnock, Ayrshire, 7/8/1980.
(101) to Ravenscraig Works, Motherwell, Strathclyde, 18/8/1980.
(102) to Ravenscraig Works, Motherwell, Strathclyde, 19/8/1980.
(103) to Ravenscraig Works, Motherwell, Strathclyde, 22/8/1980.
(104) to Ravenscraig Works, Motherwell, Strathclyde, 9/1980.
(105) to Ravenscraig Works, Motherwell, Strathclyde, c10/1980.
(106) to Tees Bulk Handling Ltd, Tees Dock, Grangetown, c5/1982.
(107) to New Holland Bulk Services Ltd, New Holland, Lincolnshire, 1/6/1984.
(108) scrapped after 12/3/1985, by 4/1991.
(109) scrapped after 5/8/1986, by 4/1991.
(110) scrapped after 27/10/1986, by 4/1991.
(111) to Ravenscraig Works, Motherwell, Strathclyde, c/1990.
(112) to Ravenscraig Works, Motherwell, Strathclyde, c11/1990.
(113) scrapped after 31/7/1991, by c1/1993.
(114) to Hartlepool Pipe Mill, Hartlepool, 3/1992.
(115) to Cleveland Potash Ltd, Boulby Mine, Easington, 27/4/1992.
(116) to Cleveland Potash Ltd Boulby Mine, Easington, 28/4/1992
(117) scrapped after 3/7/1992, by c1/1993.
(118) to Shelton Works, Etruria, Stoke-on-Trent, Staffordshire, 18/12/1995.
(119) to Staffordshire Locomotive Co Ltd, 22/12/1995 and scrapped 23/2/1996.
(120) to T.J. Thomson & Son Ltd, Stockton, for scrap, w/e 22/12/1995.
(121) scrapped c5/1996; after 13/3/1996, by 20/5/1996.
(122) to Shapfell limestone quarries and lime works, Shap, Cumbria. The low loader taking it from Lackenby had problems gaining access to Shapfell and left the locomotive at Harrison's Sidings; Class 47 47816 then pulled 272 to Shapfell along the main line on 4/7/1996.
(123) to Shelton Works, Etruria, Stoke-on-Trent, Staffordshire, 16/9/1996.
(124) to Shelton Works, Etruria, Stoke-on-Trent, Staffordshire, 24/4/1997.
(125) Corus decided that the locomotive was not suitable and it was sent to Staffordshire Locomotive Co Ltd, c/o Cambrian Railways Society Ltd, Oswestry, Shropshire, soon after its arrival, c9/1999.
(126) to T.J. Thomson & Son Ltd, Stockton, for scrap, 5/2000.
(127) to Hartlepool Pipe Mill, Hartlepool, after 18/7/2000, by 2/4/2001.
(128) to Wilmot Bros (Plant Services) Ltd, Ilkeston, Derbyshire, via T.J. Thomson & Son Ltd, 6/4/2001.

(129) to Manchester Ship Canal Co Ltd, Barton Dock, Manchester (property of Staffordshire Locomotive Co Ltd), 17/7/2003.
(130) to LH Group, Barton under Needwood, Staffordshire (property of Staffordshire Locomotive Co Ltd), 12/8/2003.
(131) scrapped, after 24/7/2005, by 13/11/2005.
(132) scrapped, after 24/7/2005, by 4/2/2007.
(133) out of use since arrival; it was scrapped after 24/7/2005, by 4/2/2007.
(134) mothballed 2/2010, 278 in 4/2010. To Sahaviriya Steel Industries UK Ltd with part of site, 25/3/2011.
(135) to Sahaviriya Steel Industries UK Ltd with part of site 25/3/2011.
(136) out of use. To Sahaviriya Steel Industries UK Ltd with part of site, 25/3/2011.

Gauge : 3ft 0in

Bolckow Vaughan purchased 20 new 3ft gauge locomotives and a crane tank from Black, Hawthorn between 1876 and 1881. Although the orders mostly specified "Middlesbrough" or one of its works on Teesside, Bolckow Vaughan also used narrow gauge locomotives to propel wagons of coal for charging the beehive coke ovens at some of its County Durham collieries and it is almost certain that some engines went directly or subsequently to these pits. One recollection is that there were about a dozen narrow gauge locomotives at Cleveland Works. They were used on the 3ft gauge to connect the Bessemer converters at the steel plant to the soaking pits where the ingots were reheated, prior to rolling in the mills. The converters began operating in 1877 with an expansion about 1880. Two of the 3ft gauge locomotives also worked in the mills. Although Bolckow Vaughan's Bessemer plant was replaced by the North Open Hearth Steel furnaces, immediately prior to World War 1, the narrow gauge railway continued to function and four new 3ft gauge locomotives were purchased from Andrew Barclay in 1913-21.

There are other complications in deciding on the history of these locomotives. Black, Hawthorn often built locomotives for stock with customers' orders coming in later resulting in doubts about the dates carried on the plates. The original order book was copied by two separate people (John Dawson and Richard Inness) in the 1940s and there are variations between them leading to doubts about which version is correct, particularly as little is known about the later history of some of the locomotives. Therefore all 21 new locomotives are listed below with what information we have about them.

When Richard Inness visited the narrow gauge shed at the Cleveland Works on 7/11/1926, he photographed numbers 12, 16 and 17 (probably Black, Hawthorn locomotives) and 127 (Andrew Barclay), although some appeared out of use. A letter from one of Inness' correspondents dated 20/10/1931 states that Bolckow Vaughan had some narrow gauge locomotives, Nos. 7, 11, 12, 15, 17, 8 and 16, the first five of which had been scrapped. This accords with a Dorman Long schedule dated 3/10/1931 which shows two narrow gauge locomotives in existence from this batch – No.16 Bolckow Vaughan /1888 9in cylinders, 150lb/sq in and No.8 Black, Hawthorn /1881 9in cylinders 150lb/sq in – suggesting that some of the locomotives had been rebuilt.

-	0-4-0ST	OC	BH	402	1877	(a)
-	0-4-0ST	OC	BH	403	1877	(b)
-	0-4-0ST	OC	BH	435	1877	(c)
-	0-4-0ST	OC	BH	436	1877	(c)
-	0-4-0ST	OC	BH	442	1877	(d)
-	0-4-0ST	OC	BH	446	1878	(e)
-	0-4-0ST	OC	BH	447	1878	(f)
-	0-4-0CT	OC	BH	455	1878	(g)
-	0-4-0ST	OC	BH	459	1878	(h)
-	0-4-0ST	OC	BH	494	1879	(c)
-	0-4-0ST	OC	BH	495	1879	(c)
-	0-4-0ST	OC	BH	557	1880	(c)
-	0-4-0ST	OC	BH	558	1880	(c)
-	0-4-0ST	OC	BH	559	1880	(i)
-	0-4-0ST	OC	BH	561	1880	(j)
-	0-4-0ST	OC	BH	595	1881	(c)
-	0-4-0ST	OC	BH	596	1881	(c)
-	0-4-0ST	OC	BH	597	1881	(c)
-	0-4-0ST	OC	BH	598	1881	(c)
-	0-4-0ST	OC	BH	599	1881	(c)
-	0-4-0ST	OC	BH	601	1881	(c)
127	0-4-0T	OC	AB	1350	1913	(k)

128		0-4-0T	OC	AB	1351	1913	(k)
138		0-4-0T	OC	AB	1464	1916	(k)
158		0-4-0T	OC	AB	1735	1921	(k)
-		4wDM		HE	1720	1933	(l) s/s
-		4wDM		HE	1813	1936	(l) s/s

(a) new to BV, Middlesborough but Inness has to Binchester Colliery, Co. Durham, ex works 12/1876.
(b) new to BV, Middlesborough but Inness has to Binchester Colliery and later Auckland Park Colliery, although the gauge at the latter may have been different.
(c) new to BV Cleveland Works (sometimes Eston Steel Works or South Bank Works but the location is the same).
(d) new to BV for Binchester Colliery
(e) new to BV Middlesborough. Inness has Binchester Colliery.
(f) new, it did not arrive at South Bank until 12/1/1880 after display at the Paris Exhibition in 1878, where it was named MIGNONNE and awarded a silver medal, to be followed by a visit to South Kensington. Part 3 of the Durham Handbook suggests that it may have been purchased by Lingford Gardiner c/1900 and resold to Pease & Partners Ltd, Bank Foot Coke Ovens, Pease's West by 4/1901.
(g) new to BV Eston Steel Works. Shown in Order Book as a four wheel coupled tank loco with 4 wheel crane. ILS suggests 0-4-0ST+4c but could be 0-4-4CST.
(h) new to BV Eston Steel Works. A temporary replacement for 447 but named NEWFIELD No.1 so probably went to Newfield Colliery, Co. Durham at some stage.
(i) new to BV Eston Steel Works. Inness has Hutton Henry Collieries. Exhibited at Tynemouth Exhibition. It probably went to the Hutton Henry Coal Co Ltd secondhand.
(j) new to BV Eston Steel Works. Inness has new to Mersey Iron & Steel Co, as No.2. He was probably referring to the Mersey Steel & Iron Co but this sale presumably fell through as this company went bankrupt in 1880.
(k) new to BV Cleveland Works. To Dorman Long, Middlesborough Iron Works for storage, by 12/7/1934.
(l) new to Dorman Long (see *Railway Bylines* 3/2011, page 192). Possibly they were used on the slag tips.

Cleveland Coke Ovens, South Bank   NZ 545210
Gauge : 4ft 8½in

	0-4-0WE	HL	3861	1936	New	(1)
	0-4-0WE	HL	3862	1936	New	(1)
	0-4-0WE	DL/BSC			New	(1)

(1) one of the HL locomotives went to the workshops for overhaul, but was mistakenly scrapped so the workshops constructed a replacement of broadly similar design but incorporating a heavier chassis. The coke works closed in 1977 and the remaining two locomotives were scrapped in 1979.

South Bank Coke Ovens, South Bank   NZ 536214
Gauge : 4ft 8½in

-		0-4-0WE	WSO	5998/1	1953	New	(1)
-		0-4-0WE	WSO	5998/2	1953	New	(1)
1		4wWE	WC	1504	1968	New	(2)
-		4wWE	WC	1505	1968	New	(2)
1	(BSC No.1)	4wWE	BSC Hartepool		1986	New	(3)
2	(BSC No.2)	4wWE	BSC Hartepool		1986	New	(3)

(1) s/s by 1971/72.
(2) out of use by 1986 and scrapped /1989.
(3) to Sahaviriya Steel Industries UK Ltd with coke ovens, 25/3/2011.

Redcar Coke Ovens, Redcar   NZ 562257
Gauge : 4ft 8½in

1	4wWE	GB	420355/1	1976	(a)	(1)
2	4wWE	GB	420355/2	1976	(a)	(1)
3	4wWE	GB	420408	1977	(a)	(1)

(a) new. Rebuilt at SBV Fabrications & Site Services Ltd, South Bank, c4/2009
(b) to Sahaviriya Steel Industries UK Ltd with coke ovens, 25/3/2011.

Cleveland No.3 Billet Mill

The two 1940s 4wWE were replaced by TH 128v, the 4-wheel diesel hydraulic frame being rebuilt as an electric with power collected by a side arm. It was subsequently superseded by the rebuilt S 10004. The Billet Mill had closed by 1/1/1982.

Gauge : 4ft 8½in

-		4wWE	Cleveland		c/1940s	s/s
-		4wWE	Cleveland		c/1940s	s/s
1		4wWE	Cleveland		c/1967	
	Rebuild of	4wDH	TH	128v	1963	Scr c/1983
2		4wWE	Cleveland		1974	
	Rebuild of	4wDH	S	10004	1959	Scr c/1983

Cleveland Branch Blast Furnaces   NZ 542214

The furnaces were situated on the outside of a curve leading to the main line and served by a three track gantry carrying raw materials to the storage bins. At some date, one of the tracks was equipped with overhead electric wires and worked by a small four-wheel locomotive fitted with trolley style poles. The blast furnaces were demolished in 1957.

Gauge : 4ft 8½in

-	4wWE	?		?	?	(a)	s/s /1957

(a)   origin unknown, but possibly built at Cleveland Works.

South Bank Ore Preparation Plant   NZ 534220

Dorman Long installed an overhead electric wire on some of the railways at the plant operated by two electric locomotives in 1948. The locomotives hauled sets of six 45 ton capacity bottom discharge wagons from South Bank Wharf to the foot of a gantry over the stocking ground, up which single wagons were raised by a 'mule'. When required, ore was lifted from the stockpile by two Goliath cranes and placed in wagons for the electric locomotives to take to the ground bunkers. After any crushing and sintering, the ore was carried by conveyors to the blast furnaces. The opening of the Redcar Ore Terminal in 1973 resulted in the electric railway becoming redundant.

Reference: https://player.bfi.org.uk/free/film/watch-steel-strides-ahead-1961-online

Gauge : 4ft 8½in

No.1	4wWE	AB/MV		1948	(a)	(1)
No.2	4wWE	AB/MV		1948	(a)	(2)

(a)   new. Although constructed by Andrew Barclay using parts by Metropolitan Vickers (order number 19349 of 1946), they did not receive that company's maker's numbers. A third electric locomotive was constructed by AB/MV for Dorman Long in 1948 but no destination was given (see below).

(1)   to Vehicle Repair Shop, Cleveland Works with collision damage and stored oou c/1973. Scr c6/1982.

(2)   to Bessemer Blast Furnaces, c/1973.

Bessemer Blast Furnaces   NZ 544212

The blast furnaces were served by two raised gantries, each with a pair of railway lines, for delivering raw materials to the storage bins and shunted by steam, later diesel, locomotives. A diagram in the 1959 Technical Survey shows a conveyor from the Ore Preparation Plant leading to a bunker at the north end of the gantries with a position of an "electric loco" marked a little further south along one of the gantries. The only report we have is dated 2/7/1974 when No.2 (AB/MV 1948) was observed working on 40 yards of electrified track on one of the gantries.

There are two other possibilities that may, or may not, have been involved. Dorman Long ordered a third electric locomotive from AB, job no. 5534 (MV 19356 of 1946). No specific location was recorded but a despatch date of 20/9/1948 suggests that it was built. IRS records show a Head Wrightson battery locomotive allocated to the Cleveland Coke Ovens which is an unlikely location but perhaps it was observed in the vicinity. The locomotive was derelict in 1960 and last noted in 7/1968.

Gauge : 4ft 8½in

No.2	4wWE	AB/MV	1948	(a)	Scr 6/1982

Other possibilities:

-	4wWE	AB/MV	1948	New	s/s
No.1	4wBE	HW	1928		s/s /1970

Teesside Beam Mill, Lackenby   NZ 860223

The Universal Beam Mill continues to receive steel slab from Scunthorpe and this is delivered by rail to the Lackenby Grids. Following the failure of SSI, Ed Murray & Sons Ltd gained the contract to shunt wagons between the Grids and Beam Mill. Most steel sections are despatched by road. After Sahaviriya Industries UK Ltd's liquidation, GECTs 5430, 5465, 5475 and Robel AD184 remained on the Teesside Beam Mill site but became the property of Ed Murray & Sons Ltd, 17/11/2015 and therefore available for hire.

Gauge: 4ft 8½in

(MURR2)		4wDE	Moyse	1464	1979	(a)
261		6wDE	GECT	5430	1977	(b) (2)
268		6wDE	GECT	5465	1977	(c) (3)
277		6wDE	GECT	5475	1978	(d) (1)
1		4wDM	Robel 54.12-107	AD184	1980	
		Rebuilt	Lackenby		2010	(d) (4)
(01569) EMMA		4wDH	TH	281V	1978	(e)
-		0-6-0DH	RR	10255	1966	(f)
	LADY POTTER	0-6-0DH	RR	10214	1964	(g)
-		0-6-0DH	S	10166	1963	(h)

(a) ex Tata Steel Europe Ltd, North East Pipe Mills, Hartlepool, (property of Ed Murray & Sons Ltd), on hire, 10/2015. To Liberty Pipes (Hartlepool) Ltd, 42 and 84 inch Pipe Mills, Hartlepool, (property of Ed Murray & Sons Ltd), w/c 5/8/2019. Ex Liberty Pipes (Hartlepool) Ltd, 42 and 84 inch Pipe Mills, Hartlepool (property of Ed Murray & Sons Ltd, on hire, 1/5/2021.
(b) on hire, (property of Ed Murray & Sons Ltd), 17/11/2015. To Chasewater Light Railway and Museum Co, Brownhills, Staffordshire for overhaul, ex hire 21/12/2016. Ex ICL UK (Cleveland) Ltd, Tees Dock Terminal, Tees Dock, Grangetown, on hire 7/11/2018.
(c) on hire, (property of Ed Murray & Sons Ltd), 17/11/2015. To Chasewater Light Railway and Museum Co, Brownhills, Staffordshire, 30/6/2016. Ex ICL UK (Cleveland) Ltd, Tees Dock Terminal, Tees Dock, Grangetown, on hire, 1/2019; after 18/1/2019, by 28/1/2019.
(d) on hire, (property of Ed Murray & Sons Ltd), 17/11/2015.
(e) ex T.J. Thomson & Son Ltd, Stockton, (property of Ed Murray & Sons Ltd), on hire, 8/2/2017; initially delivered to former SSI Lackenby Workshop.
(f) ex Ed Murray & Sons Ltd, Casebourne Road, Hartlepool, on hire, w/c 5/8/2019; after 3/8/2019 by 9/8/2019.
(g) ex Ed Murray & Sons Ltd, Casebourne Road, Hartlepool, on hire, c26/3/2021.
(h) ex ICL UK (Cleveland) Ltd, Boulby Mine, Easington, (property of Ed Murray & Sons Ltd), on hire, 1/2/2022.
(1) to ICL UK (Cleveland) Ltd, Tees Dock Terminal, Grangetown, travelling via the former SSI line to the dock, ex hire 11/8/2016.
(2) to Tata Steel Europe Ltd, Port Talbot Works, Port Talbot, South Wales, ex hire 10/5/2019.
(3) to Tata Steel Europe Ltd, Port Talbot Works, Port Talbot, South Wales, ex hire, departed 24/5/2019 via Paul Archer Transport, Wishaw, Warwickshire and arrived 30/5/2019.
(4) to Ed Murray & Sons Ltd, Casebourne Road, Hartlepool, ex hire, w/c 21/2/2022

Stored Locomotives

253		6wDE	GECT	5416	1976	(a)	
257		6wDE	GECT	5426	1977	(a)	
260		6wDE	GECT	5429	1977	(a)	
265		6wDE	GECT	5462	1977	(a)	
266		6wDE	GECT	5463	1977	(a)	
269		6wDE	GECT	5466	1977	(a)	
276		6wDE	GECT	5474	1978	(a)	
2	REDCAR 22	4wDM	Robel 54.12-107	AD183	1980	(a)	(1)

(a)	ex Sahaviriya Steel Industries UK Ltd, Teesside Works, Lackenby, (property of Ed Murray & Sons Ltd), 25/3/2022.
(1)	to Ed Murray & Sons Ltd, Casebourne Road, Hartlepool, 4/2022; after 8/4/2022, by 2/5/2022.

BRITISH TITAN PRODUCTS CO LTD

BILLINGHAM WORKS, Billingham   28
NZ 473217 – **Map C**

The company was incorporated 26/7/1930, initially as a subsidiary of the Imperial Smelting Corporation to manufacture titanium dioxide. However, it was not until 16/2/1933 that an agreement was signed involving four partners - Imperial Smelting Corporation Ltd, ICI Ltd, Goodlass Wall & Lead Industries Ltd and Titan Company Incorporated - to proceed with construction. Seven acres of a triangular shaped piece of land between Haverton Hill Road and New Road was purchased from ICI Fertilizers & Synthetic Products Ltd. The first sod was cut on 10/7/1933 and titanium dioxide began to be produced on 9/7/1934. This white opaque powder is used as a paint pigment and in plastics, fibres and paper. Most of the incoming raw materials initially did not require rail transport. Norwegian ilmenite came by ship to a wharf on the Tees and a band conveyor moved it to the silo. Sulphuric acid was supplied by pipeline from ICI's Billingham Works. The Titan plant was located next to ICI's Cassel Works and a couple of sidings of the latter's railway system enabled fuel to be brought in and product sent out; presumably any shunting being carried out by ICI or some form of tractor. During World War 2, titanium tetrachloride was produced for use as smoke screens. In 1953 a decision was taken to make titanium tetrachloride in order for ICI to produce titanium metal at Wilton. The plant at Titan Products opened in 1954 and, although demand was less than expected, the chlorine arrived from ICI in tankers; 60% by rail and 40% by road. This continued until 5/1965 when it was transferred by pipeline. It may be a reason why the Ruston & Hornsby diesel was sent here from the Grimsby Works. By 1971 the Titan site at Billingham had expanded to 18 acres with a production capacity of 27,000 tons pa and other plants had been opened at Grimsby in 1949 and Greatham in 1971. Due to site constraints and environmental consideration, the production plant at Billingham closed in 1981 and was mostly demolished, although the offices became the technical headquarters of Huntsman Tioxide Group.

Reference : *British Titan Products to Huntsman Tioxide*. A Company History 1930 to 2000, John M. Graham, Huntsman Tioxide, 2002

Gauge : 4ft 8½in

	4wDM	RH	237928	1946	(a)	(1)

(a)	ex British Titan Products Co Ltd, Grimsby, Lincolnshire, c5/1937.
(1)	to Yorkshire Tar Distilleries Ltd, Killamarsh, Derbyshire, /1963.

BROWN G. & BROS LTD

PORTRACK LANE IRON WORKS, North Shore Branch, Stockton   29
Brown G. & Bros until c/1924   NZ 450197 – **Map D**
Nesham & Welch until /1844

John Douthwaite Nesham owned a number of pits comprising Newbottle Colliery in County Durham linked by a waggonway to staithes on the River Wear at Sunderland. This was noteworthy for seeing the use of William Brunton's 1813 locomotive. Due to financial difficulties and also because of interest in the property from nearby coal owners, Nesham sold the colliery and waggonway in 1822 and moved to Stockton where the family owned more land. His son, David Nesham (1805-89), founded the Portrack Lane Iron Works with Humphrey Welch and Daniel Hawthorn, although the latter left the partnership on 28/9/1833. In 1838 Portrack Lane Iron Works with its forge and workshops occupied a site of about one acre and was situated on the east side of the Clarence Railway's North Shore Branch (opened 1833). In 1837-40 the partnership supplied engineering equipment to the Stockton & Darlington Railway and the ironwork for a lifting railway bridge at Hartlepool Docks. The business was described in 1841 as iron founders, millwrights, engine and locomotive builders. Nesham & Welch constructed a small number of locomotives in 1837-40 at the Portrack Lane Iron Works. The following standard gauge locomotives are considered to be the most likely:

No.5	GORDON	0-6-0	1837	Clarence Railway
	EXILE	0-6-0	1837	
-		0-6-0	1839	Hartlepool Dock & Railway Co, later YN&B/NER 125
-		0-6-0	1840	Hartlepool Dock & Railway Co, later YN&B/NER 122
-		0-4-0	1840	Hartlepool Dock & Railway Co, later YN&B/NER 126

-		0-4-0	1840	Hartlepool Dock & Railway Co, later YN&B/NER 127
No.27	WITTON CASTLE	0-6-0	1840	Stockton & Darlington Railway

In 1/1837 Nesham & Welch was building a 0-6-0 named GORDON for the Clarence Railway. Later in 2/1839 Nesham & Welch asked the Clarence Railway if it wished to purchase the locomotive EXILE (sometimes quoted as EXILO), but this was refused. Instead Nesham & Welch took on a one year contract in 2/1839 to shunt coal wagons from the Port Clarence sidings on to the shipping staithes. This lasted until 8/1839 when the contract was relinquished and the Clarence Railway took over. The locomotive was subsequently sold to William Hedley & Sons, which seems to have traded as the Crowtrees Coal Co and was used on the Hartlepool Dock & Railway Co's line, being noted in a minute of the latter in December 1841.

In the early 1840s, Fossick & Hackworth (which see) began constructing locomotives at its works to the north of Nesham & Welch and, in 1843, the Portrack Lane Iron Works was offered for sale. The premises were auctioned on 9/5/1844 and purchased by Andrew Brown [& Sons], with David Nesham moving to Haughton-le-Skerne to become a farmer and land agent.

Brown's continued to function as an iron and brass foundry until it closed in 1978, by which time the site contained numerous items of historic equipment. There were efforts to turn it into a museum, but these came to nothing and it was demolished. During Brown's tenure, the premises were served by a short siding which climbed from the North Shore Branch into the works yard. A steam winch pulled the loaded wagons on to a gantry where the coal could be dropped between the rails adjacent to a Lancashire boiler before continuing by gravity past the forge and foundry to the end of the siding near the cupola furnaces. Two separate internal railway tracks ran out of the fitting shop to the yard. Latterly there were two 'wagons', comprising four wheels and a solid rough timber frame, in the fitting shop for moving machined parts by hand to the crane in the yard.

References : 'Some Early Durham Colliery Locomotives – 1', Ken Fleming, *The Industrial Locomotive* No.31, ILS, Autumn 1983

'Nesham and Welch Ironfounders and Engineers of Portrack Lane Iron Works Stockton on Tees', Alan Betteney, *The Cleveland Industrial Archaeologist* 40, CIAS, 2020 pp.13-36

'Richard Parr, Contractor Part 3', Ken Fleming, *The Industrial Locomotive* 180, ILS, 2021

The siding into the Portrack Lane Iron Works on 23rd June 1979 following its closure the previous year. This had once been the site of Nesham & Welch's locomotive manufactory. (Cliff Shepherd photograph)

CALDERS LTD

TIMBER YARD and WEST WHARF, Middlesbrough   30
Calder, Dixon & Co Ltd until c/1913   NZ 496213, NZ 495214 – **Map F**
Calder & Co until c/1898

Initially Calder & Co occupied a small timber yard (NZ 496213) on the north side of Commercial Street. This had previously been the site of the second SDR Middlesbrough passenger station. When the main line was extended to Redcar, a new permanent passenger station was provided at Sussex Street in 1847 and that at Commercial Street became a goods station. In 1870 the NER agreed to purchase a large tract of land south of Depot Road for its new Middlesbrough Town Goods Station and the site at Commercial Street became a saw mill and timber yard. The North Street siding left the Vulcan Street railway and ran parallel to it on the south side passing through Calder's timber yard, crossing an extension of North Street and continuing via Hallam's Tees Saw Mill to rejoin Vulcan Street. Although Calder continued to make some use of this timber yard until it was taken over by various engineering businesses, its principal premises were at West Wharf (NZ 495214) next to the river.

The plan accompanying the 1884 NER Agreement for the MOR indicates that the eastern half of the former Port Darlington coal drops site had been divided into two. The part to the west still belonged to the NER but that to the east was identified as a Timber Yard occupied by Messrs Calder & Co. Four sidings off the Vulcan Street railway curved into the yard with another running alongside the River Tees. Subsequently the NER portion was rented by the Wharfinger, George Watson and eventually both sites became part of T. Roddam Dent's premises, but with Calder, Dixon's saw mill and timber yard still present. *The Guide to the Ports on the Tees and the Hartlepools*, published in 1900, stated that "within the past year or so," Calder & Co had been reconstructed to become Calder, Dixon & Co Ltd with William Dixon as managing director. The yard was filled with huge stacks of home-grown and imported timber, with specific mention given to wood for ironstone mines and sleepers for railway companies.

A letter from the NER Middlesbrough Goods Agent dated 5/3/1897 stated that Calder was one of four companies with private locomotives running over the NER and MOR lines for which no charge was made. These lines were described as "Vulcan St and Dock, Marsh Branch. To the Steel Strip and Nail Co's works in Vulcan Street and to Calder & Co's store yard in Vulcan St and to Calder & Co's store yard at Cargo Fleet for shunting purposes"; a nominal charge was suggested. A year later, the NER agreed to permit Calder's locomotive to run over its lines to R. Craggs & Sons' Tees Dockyard, the annual charge being 10 shillings and was "To cease running the said Locomotive Engine over the Railway Company's Railway" if one calendar month notice was given. Calder Dixon advertised a secondhand standard gauge locomotive, with four wheels and 10 inch cylinders, for sale at Middlesbrough in the *Glasgow Herald* on 10th August 1918. In 1929 LNER gave approval for Calders' "steam travelling crane" to haul empty and loaded wagons over various sidings to T. Roddam Dent. The LNER deleted the Calders Ltd siding entry on 1/11/1946 and included it in the Roddam Dent entries.

Gauge : 4ft 8½in

No.1		0-4-0ST	OC	AB	24	1863	(a)	(1)
		Rebuilt		AB		1880		
No.2		0-4-0ST	OC	AB	775	1896	New	(2)

(a)   ex William Baird & Co, Eglinton Iron Works, Kilwinning, Ayrshire

(1)   it could be the locomotive advertised for sale in 1918.

(2)   to W. Robinson & Co Ltd, Cleveland Timber Yard, North Ormesby.

CARGO FLEET IRON CO LTD, THE

LIVERTON IRONSTONE MINE, Liverton Mines   31
Cargo Fleet Iron Co, The until 1/1883   NZ 710180 – **Map I**
Liverton Co Ltd until 7/1882
Liverton Ironstone Co Ltd until 23/7/1877

The mine was situated on the south side of the Carlin How-Loftus railway at the east end of the Kilton Viaduct. There were substantial buildings on the surface, but the mine suffered from the Main Seam ironstone being divided into two by a shale band and the quality reduced as the workings moved south. Although formed in 1865, the Liverton Ironstone Co did not commence production until 1871 when 290,000 tons of ironstone were despatched. Two shafts descended 480 feet to the ironstone and both cages could hold two tubs, each with a capacity of 25cwt to 28cwt.

A railway curved round from the NER branch to a headshunt south of the mine buildings. From here, sidings descended back past the mine and its tippler to another set of sidings before reconnecting with the curve to the branch. The copious amounts of shale encouraged the owners to provide the first picking belt at a Cleveland mine. Shale from the picking belt was loaded into wooden tubs and hauled up the spoil heap on a 2ft 6in gauge railway powered by a 40hp stationary steam engine. The tubs were adapted for end or side tipping. This system was replaced about 1914 by an aerial ropeway with 19 buckets supplied by Messrs Ropeways Ltd. The distance from the blast furnaces on Teesside placed the mine at a disadvantage and a batch of seven (later ten) calcining kilns was erected; it was more usual practice for these to be located at the iron works. Railway wagons filled with ironstone and coal were raised by steam hoist and the contents tipped into the top of the kilns. The transport of the still warm calcined ironstone encouraged the NER to introduce "iron trucks" for its trains to the blast furnaces. As a result of the late 1870s economic depression, Liverton Mine closed on 28/7/1877. The existence of the kilns had resulted in the need for more shunting and the sale details for the mine, dated 12/11/1878, listed two locomotives, together with an engine shed and 500 tubs.

W. Graham purchased the mine and formed the Liverton Company Ltd on 1/1/1879. Mining restarted a year later, but the Loftus Advertiser announced on 12/8/1882 that the Liverton Co Ltd was to be voluntarily wound up. The mine was then purchased by H.F. Swan, a Newcastle shipbuilder and J.G. Swan, a founder member of Swan, Coates & Co. A new mining lease was taken out on 9/11/1882 and the two Swans assigned their newly acquired mine to the Cargo Fleet Iron Co Ltd on 15/8/1883. The new owners made extensive alterations to the mine before a NER train departed with the company's first stone on 16/8/1883. These changes made the mine's locomotives redundant as a report on the Iron and Steel Institute's visit on 21/9/1883 makes clear: "A thoroughly ingenious method of conveying the stone from the pit mouth to the cleaning belt and on to the siding, of supplying the calcining furnaces and returning the empty wagons to the pit mouth by means of gradients, and without the aid of steam or horse was viewed with universal admiration."

At Cargo Fleet Iron Co's Board meeting on 18/6/1901, concern was expressed about the cost and quality of ironstone from Liverton Mine. However it was decided to persevere with the mine and it was modernised with three new Lancashire boilers, four new kilns, electric pumps and a cleaning belt. The remodelling was completed in 1/1905. Subsidence from Bell Bros mining began to threaten the stability of the viaduct and the NER decided to bury it in an embankment with shale from the mine tip. Plans were submitted in 11/1906 and the Cargo Fleet Iron Co agreed to provide 2,250 tons of shale a week from Liverton Mine, although NER locomotives and wagons would take the shale to the viaduct and drop it from the truck bottoms through openings created in the track. Liverton never achieved its target mainly because its mostly used shale from the picking belt, although a narrow gauge incline may have been subsequently employed to take some shale from the tip directly to the viaduct. Shale also came from the North Loftus Mine and the work was completed in 8/1912.

Liverton Ironstone Mine *(IRS J.A. Peden collection)*

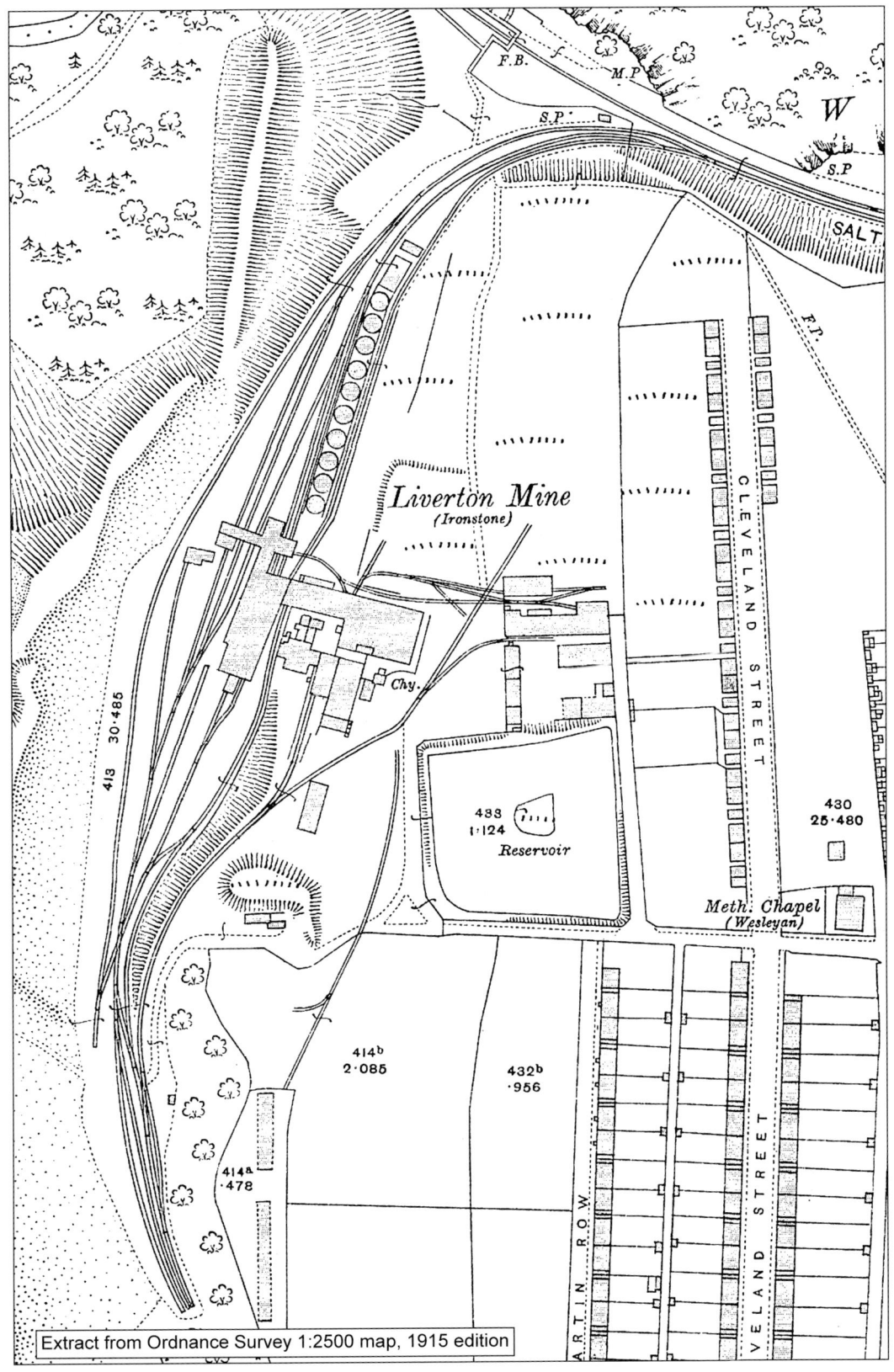

Liverton Ironstone Mine

The quality of the ironstone sent away from the mine progressively became poorer. As a result, the mine closed about 11/1921, although some of its ironstone was later extracted by Dorman Long from its Kilton Mine.

Gauge : 4ft 8½in

LIVERTON	0-4-0ST	IC	Robey	(a)	(1)
"4-wheeled tank locomotive"			DS	(a)	s/s

(a) origin unknown, here by 12/11/1878. Robey was an unusual manufacturer to choose for a locomotive, but the Liverton Ironstone Co also owned a pair of "Horizontal Engines" by the same firm.

(1) to Cargo Fleet Iron Works, Cargo Fleet, 5/1883.

CARR HOUSE IRON CO

CARR HOUSE IRON WORKS, West Hartlepool
Dunlop, Meredith & Co Ltd until?

32
Map B

Further research is required to identify the location of this works. According to the IRS, the Carr House Iron Co acquired the premises about 1878, although note that Dunlop, Meredith & Co Ltd's nut and bolt business at the Cliff House Iron Works had gone into liquidation in 4/1875. A list of puddling furnace owners in 1877 shows nine belonging to Dunlop, Meredith & Co. The Carr House Iron Works was offered for sale in 3/1880. However, Robert Wood states that Blake and Tomkyns, proprietors of the Carr House Iron Works, took over the "top end" of the closed West Hartlepool Iron Works comprising 37 puddling furnaces with hammers and boilers capable of producing a weekly output of 400 tons in 1/1880. Notices appeared in *The Colliery Guardian* on 29/4/1887 and the Manchester Guardian on 30/4/1887, re the dismantling of the Carr House Iron Works, that A.T. Crow was to auction on 4/5/1887 plant including the two locomotives below with 9in x 16in cylinders.

Gauge : 4ft 8½in

-	0-4-0ST	OC	BH	(a)	s/s
-	0-4-0T?	OC	HH	(a)	(1)

(a) origin unknown

(1) this may be the locomotive later at Thomas Richardson & Sons Ltd, Hartlepool Engine Works, Hartlepool (which see).

G. & W.H. CARTER LTD

GREATHAM BRICK WORKS, Cowpen Bewley
C. Casebourne & Co Ltd until c/1904

33
NZ 486257 – **Map A**

The brick works was located on the north west side of the NER Norton-Hartlepool railway approximately half a mile north of Cowpen Bewley village. Brickmaking had commenced from about 1855 – a "brick and tile yard" appears on the 1857 Ordnance Survey map and was operated by the Farrow family. However, on 29/6/1882, the works was taken over by Casebourne, a cement manufacturer based in West Hartlepool when it was known as the Viaduct Steam Brick Works. By 1897, the principal building was located end-on to the main line and a standard gauge siding connected with and ran alongside the NER. A small signal box, "Cowpen Brick Yard" on the opposite side of the line, controlled the connection. According to the Ordnance Survey, the premises were called Cowpen Brick Works but had changed their name to Greatham Brick Works by 1914. Some of the clay pits to the north of the works had been worked out by 1896 and a large pit to the south west of the works had been opened up served by a narrow gauge tramway that ran into the main building and also approached the standard gauge siding.

By 1904, the works was taken over by G.& W.H. Carter Ltd which had another brick works at Wingate in County Durham. The 1914 Ordnance Survey shows that the Greatham Works had increased in size and clay was having to be worked from a large pit on the south east side of the main line. The narrow gauge tramway now ran from this pit under the main line to reach the brick works and standard gauge siding; possibly using rope worked inclines to reach these destinations. An additional standard gauge siding had also been laid into the premises. As clay working extended eastwards, the tramway also had to be lengthened and two new 1ft 5¾in gauge Hunslet diesels were purchased in 1934 to operate it. During World War 2, the works was mothballed but it started up again with 12 kilns when the conflict ended. A dragline excavator was used in the clay pit. On 13/9/1946 Carter had advertised for a 20hp 18 inch gauge internal combustion locomotive but later owned a 2ft 0in gauge secondhand Motor Rail 20/28hp diesel mechanical in 1949 to 1956. It is not known whether the gauge of the tramway had been altered but the 1953 Ordnance Survey 1:2500 map shows that the line at the clay pit had been lengthened with little sign of narrow gauge tracks at the works. On 29/9/1956 the company announced that it was closing the brick works because of a fall in demand, although bricks would be supplied for some time from existing stocks. The premises were sold by auction on 5/12/1957 and, following much landfill, the site on the north east side of the main line now forms part of the Cowpen Bewley country park with a large pond as a remnant of one of the clay pits.

Reference : *The Brickworks of the Stockton-on-Tees Area*, Alan Betteney, Tees Valley Heritage Group, 2007

Gauge : 1ft 5¾in

-	4wDM	HE	1742	1934	New	s/s
-	4wDM	HE	1748	1934	New	s/s

Gauge : 2ft 0in

	4wDM	MR	7207	1937	(a)	s/s

(a) new to Sir Lindsay Parkinson & Co Ltd, Euxton Central Stores, Chorley, Lancashire. According to a letter dated 23/7/1949, it was in service with Carter 27/7/1949 and spares were supplied to Carter until 18/8/1956 (Walters and Waywell, 2006)

CASEBOURNE & CO LTD

WEST HARTLEPOOL CEMENT WORKS, BRICK WORKS AND COAL DEPOT
West Hartlepool   **34**
Casebourne & Lucas until 29/6/1882   NZ 515313 – **Map B**

Charles Casebourne and Albert Lucas operated a small cement works to the north of Middleton Road, Hartlepool near the timber ponds in 1863 making use of the cheap chalk ballast dumped from incoming ships involved in the London coal trade. Two years later, the partners purchased a 16 acre site at Longhill on the west side of what became the NER Cliff House Branch. This was in production by 1866 and the old works had closed. Clay was dug from the western part of the site and also obtained from Cowpen Brick Works. Supplies of chalk came by rail from the harbour, initially from stock-piled waste and then from freshly tipped ballast. In 1891 Casebourne acquired additional land behind the cement works and established the brick works. The clay was dried, ground to a powder and placed in a rotary table press capable of making 12,000 bricks a day. There were six kilns in the yard with a 100 feet high chimney. A cement works in West Hartlepool offered for sale in *Machinery Market* on 1/3/1897 a small Manning Wardle locomotive with 7½ inch cylinders but it is not known whether this was Casebourne.

An agreement between the NER and Casebourne on 3/11/1905 gave approval for the latter's locomotives and wagons to convey slag from the Seaton Carew Iron Co's Works (entrance near Longhill subway) over the NER's Cliff House Branch to Casebourne's cement works, with a payment of one shilling for every loaded or partly loaded wagon hauled. A siding off the Cliff House Branch ran into Casebourne's premises serving the cement works and continuing by the brick works on a low embankment to reach the coal depot. In the NER 1909 operating instructions for the Cliff House Branch, the NER's locomotive was to have exclusive use of the sidings and down line between Cliff House South Junction and Steel Works Box as shunting ground unless a locomotive was delivering ore or "Messrs. Casebourne & Co's Engine or the Blast Furnace Engine have shunting to do". Casebourne decided in 1901 to develop a new cement works at Haverton Hill (which see) and the Longhill plant closed later. In a letter dated 7/9/1921, Casebourne belatedly stated that it no longer required use of the NER's line for its locomotive. In fact, the Seaton Carew Iron Co had purchased the 16.6 acres of Casebourne's "Diamond Cement Works" on 17/8/1917 as an ore stocking yard and the 1940 Ordnance Survey map shows the site served by six pairs of parallel sidings.

Gauge : 4ft 8½in

-	0-4-0ST	OC	MW	84	1863	(a)	(1)

(a) ex J. Torbock, dealer, Middlesborough, after /1881; previously The Cargo Fleet Iron Co. This was a small locomotive built with 6in x 12in cylinders, the first of Manning Wardle's Class B design. A diagram of it appears on page 6 of *The Locomotives Built by Manning Wardle & Company* Volume 2 by Fred Harman. IRS records show it at the West Hartlepool Works.

(1) to Casebourne & Co's Pioneer Cement Works, Haverton Hill.

CASEBOURNE & CO (1926) LTD

PIONEER CEMENT WORKS, Haverton Hill
Casebourne & Co Ltd until 25/5/1926

35
NZ 482222 – **Map C**

The Casebourne Company had been producing cement at Hartlepool for some years and, in 1901, it purchased a 10½ acre site on the north west bank of the river at Haverton Hill between the Tees Salt Co's works and vacant land that later became the Furness Shipyard. A prospectus to raise fresh capital to build the works was issued on 25/11/1902. Four rotary kilns were installed capable of producing 50,000 tons of Portland cement annually. A rail connection was made with The Salt Union's South Durham private railway and, from it, Casebourne's line ran past its own engine shed to serve both sides of the works before terminating at the company's river wharf. This enabled chalk from quarries bordering the Thames to be shipped in by sea and unloaded at the wharf by a steam crane; it was then taken by conveyor to a stockpile. Globe Portland Cement & Whiting Ltd contracted in 1912-13 to supply it with 48,000 tons. In 1926 the wharf was stated to be 220 feet long with two 5 ton steam cranes and one 5 ton electric crane. When required, the chalk was put into "bogies" and hauled to the wash mills where clay and water were added.

The South Durham line could give access to the Port Clarence Branch, but the northern part of what later became the Billingham Beck Branch had been opened in 1901 by the NER. Under an agreement between the railway company and Casebourne dated 17/3/1903, the NER constructed, at the trader's expense, a set of five sidings connected to its branch with stabling space for 28, 28, 26, 25 and 31 wagons respectively. From here, a line ran across Chilton's Lane (later Haverton Hill Road) to join the South Durham railway and the cement works. The Tees Salt Co was permitted to make its own arrangements with Casebourne to use the sidings. John Wilson Watson owned the Haverton Hill Brick Works on the north side of the Billingham Beck Branch and, when this line opened, a siding was constructed to the brick works. The latter had clay pits, both to the north of its works and next to the site of the five exchange sidings and it was probably this latter pit that was leased by Casebourne in 1901 to provide clay for the cement works. Interestingly, the NER Line Diagram also shows that a subway constructed of ferro-concrete existed on the Billingham Beck Branch, north of the British Chilled Roll's siding, with a "tramway under".

Deliveries of chalk by sea became irregular during World War I and alternative supplies had to be brought in by rail from Cambridge and Hitchen. A quarry was then leased at Wharram in the Yorkshire Wolds and this supplied all the company's chalk in 1919-24, with decreasing amounts to 1928. A daily train ran from Wharram to Haverton Hill. However, trials had been conducted in 1924 using a chalk by-product from Synthetic Ammonia

Hopkins Gilkes 276 of 1870 was acquired from Lingford Gardiner and operated at Casebourne's Haverton Hill cement works before being sent to its Wharram Quarry about 1919 where this photograph was taken in 1929. (IRS collection)

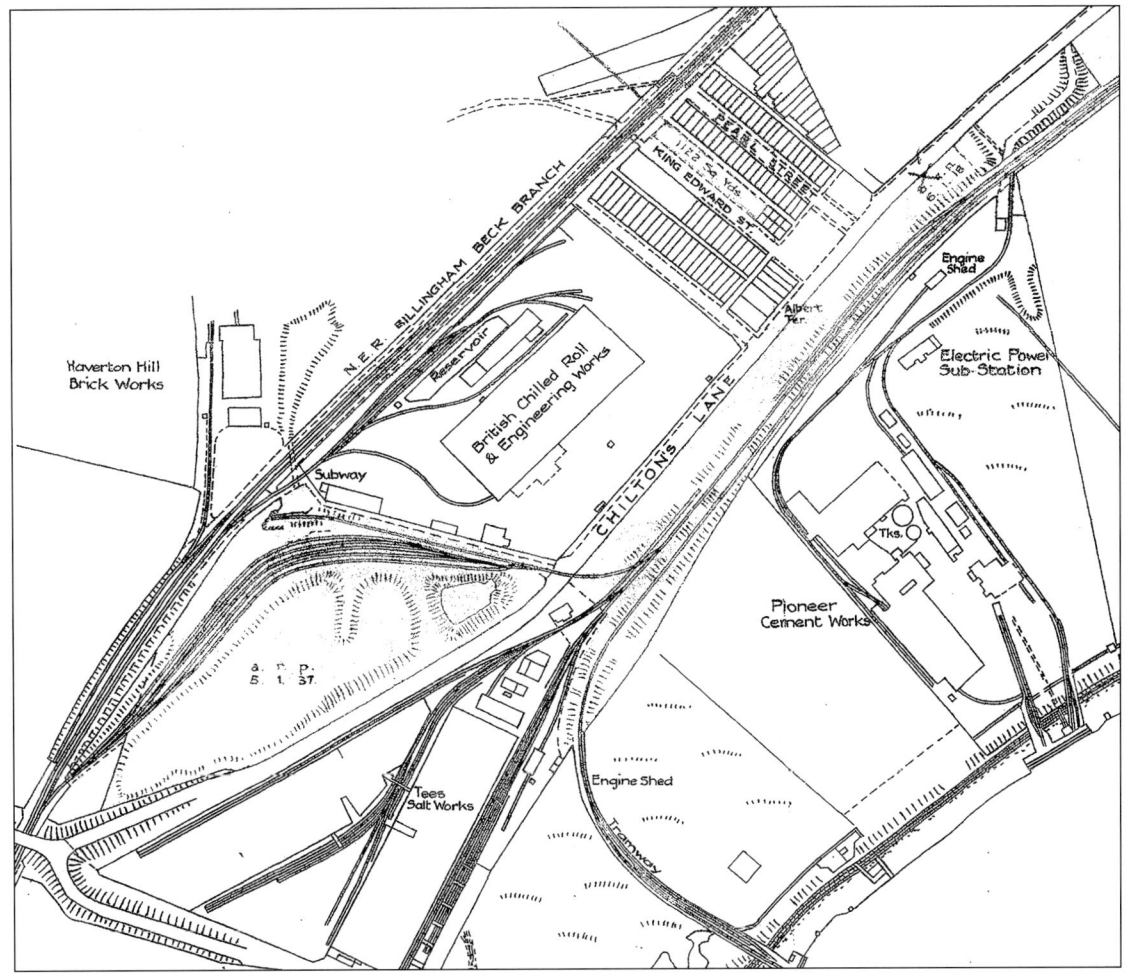

Pioneer Cement Works

& Nitrates Ltd.'s ammonium sulphate plant at Billingham. The resulting cement proved to be of high quality and a new company was formed in 1926 to raise the necessary capital to construct a 280 feet long rotary kiln (first lit 20/10/1927), together with additional silos and storage towers. Annual output increased to 135,000 tons of cement. An agreement between the NER and Casebourne dated 15/12/1922 allowed the latter to convey cement over the up line of the Billingham Beck Branch to the Cleveland & Durham County Electric Power Co's North Tees Power Station using its own locomotive and wagons. A similar agreement had been signed on 16/3/1922 permitting Casebourne's locomotive and wagons to take cement along the branch to Synthetic Ammonia & Nitrates Ltd (then under construction).

In 2/1928, Casebourne & Co (1926) Ltd was acquired by ICI and the subsequent history of the cement works will be covered under the ICI Billingham Works entry. Three of Casebourne's locomotives were absorbed into the ICI fleet.

References : see ICI Ltd Billingham Works entry.

Gauge : 4ft 8½in

-		0-4-0ST	OC	MW	84	1863	(a)	s/s
No.2	DUKE	0-4-0T		HG	276	1870	(b)	(2)
-		0-4-0ST	OC	MW	498	1874	(c)	(1)
	PIONEER No.1	0-4-0ST	OC	BH	298	1875	(d)	
		Rebuilt	LG			1925		(3)
	PIONEER No.4	0-4-0ST	OC	BH			(e)	
		Rebuilt	Ridley Shaw			1927		(3)
	PIONEER No.5	0-4-0ST	OC	AB	1800	1923	(f)	(3)

(a) ex Casebourne & Co, West Hartlepool Cement Works, West Hartlepool.
(b) originally Linthorpe Iron Works, Middlesbrough; it was probably offered for sale at Hutton Henry Colliery, County Durham 19-23/9/1898 and acquired by dealer, Lingford, Gardiner & Co Ltd, Bishop Auckland which sold it to Casebourne & Co Ltd.
(c) there is some uncertainty about the history of this locomotive. It was new to J. Whitham & Sons, Leeds where it was the only locomotive. This works had closed and was up for auction in 6/1901. It moved to Pease & Partners Ltd, Tuthill Quarry, near Haswell, Co. Durham at an unknown date before coming from there to the Pioneer Cement Works again at an unknown date.
(d) ex Lingford, Gardiner & Co Ltd, Bishop Auckland, Co. Durham; previously NER number 996 (898 class) until replaced /1889.
(e) may have come in /1919 via a dealer, R.E. Cowell.
(f) ex Sir William Arrol & Sons Ltd, contractor, possibly ex contract at South Dock, Sunderland, c/1925.

(1) to North Beechburn Coal Co Ltd, Rough Lea Colliery and Brick Works, New Hunwick, Co. Durham.
(2) to Casebourne & Co, Wharram Quarry, near North Grimston, Yorkshire (NR) c/1919.
(3) to ICI Ltd, Billingham Works, Billingham, /1929.

Gauge : 2ft 0in

| - | 4wPM | MR | 1116 | 1918 | (a) | (1) |

(a) ex WDLR 2837

(1) spares for this locomotive were ordered by A. Monk, Warrington, 3/3/1938.

It is not known where the 2ft 0in railway operated but Casebourne had advertised for a 2ft 0in gauge Simplex in the 15/6/1927 *Contract Journal*.

CASTLE CEMENT (RIBBLESDALE) LTD

MIDDLESBROUGH DEPOT, Middlesbrough.   **36**
Ribblesdale Cement Ltd until 1/11/1986   NZ 487207 – **Map F**

The cement depot was established in 1977 next to Forty Foot Road with a rail connection to the adjacent Middlesbrough Town Goods Yard. Trains of 20 cement wagons arrived and were split into two for unloading.

Middlesbrough Town Goods yard on 20th September 1978 with Ruston & Hornsby 312989/1952 shunting wagons at Ribblesdale Cement's depot.   *(Cliff Shepherd photograph)*

The Ruston & Hornsby locomotive was used daily to propel the Prestflo wagons on the siding to the discharge point. When not required, it was stabled in the compound next to the three silos. Cement was then distributed by road vehicles. Rail traffic ceased c6/1992 but the depot still functions using road transport.

Gauge : 4ft 8½in.

-	0-4-0DE	RH	312989	1952	(a)	(2)
-	0-4-0DM	RH	327971	1954	(b)	(1)

(a)   ex Ribblesdale Cement Ltd, Clitheroe Works, Lancashire, /1977.
(b)   ex Clay Cross (Iron & Foundries) Ltd, Clay Cross, Derbyshire, by 11/12/1978.
(1)   to GECT, Newton le Willows, Lancashire, for repairs by 2/7/1979; then to Ribblesdale Cement Ltd, Clitheroe Works, Lancashire, c10/1979.
(2)   to Tanfield Railway Preservation Society, Marley Hill, Co. Durham, 20/6/1994.

CENTRAL ELECTRICITY GENERATING BOARD

NORTH TEES GENERATING STATION, Haverton Hill.   37
Central Electricity Authority until 1/1/1958   NZ 478214 – **Map C**
British Electricity Authority until 1/4/1955
North-Eastern Electric Supply Co Ltd until 1/4/1948
Cleveland & Durham County Electric Power Co Ltd until 30/9/1932

In 1917 the Newcastle upon Tyne Electric Supply Co Ltd gained control of the principal electricity generators in South Durham and North Yorkshire, including the Cleveland & Durham County Electric Power Co which continued as a wholly-owned subsidiary. It commissioned the design of a new base load power station located on the west bank of the River Tees at Haverton Hill. North Tees 'A' was designed to achieve the highest possible thermal efficiency. Steam was raised in ten Babcock & Wilcox boilers to drive two 20,000 kW turbines. The station was opened in 3/1921 and created considerable interest amongst engineers because of its technical innovations. The North-Eastern Electric Supply Co absorbed the Newcastle company in 1932 and commissioned 'B' station at North Tees in 1934 using pulverised fuel firing, which was an advanced technique for the time. 'A' station closed in 1959 after serving for a number of years in reserve to cater for periods of high demand. 'B' station was demolished in the late 1960s.

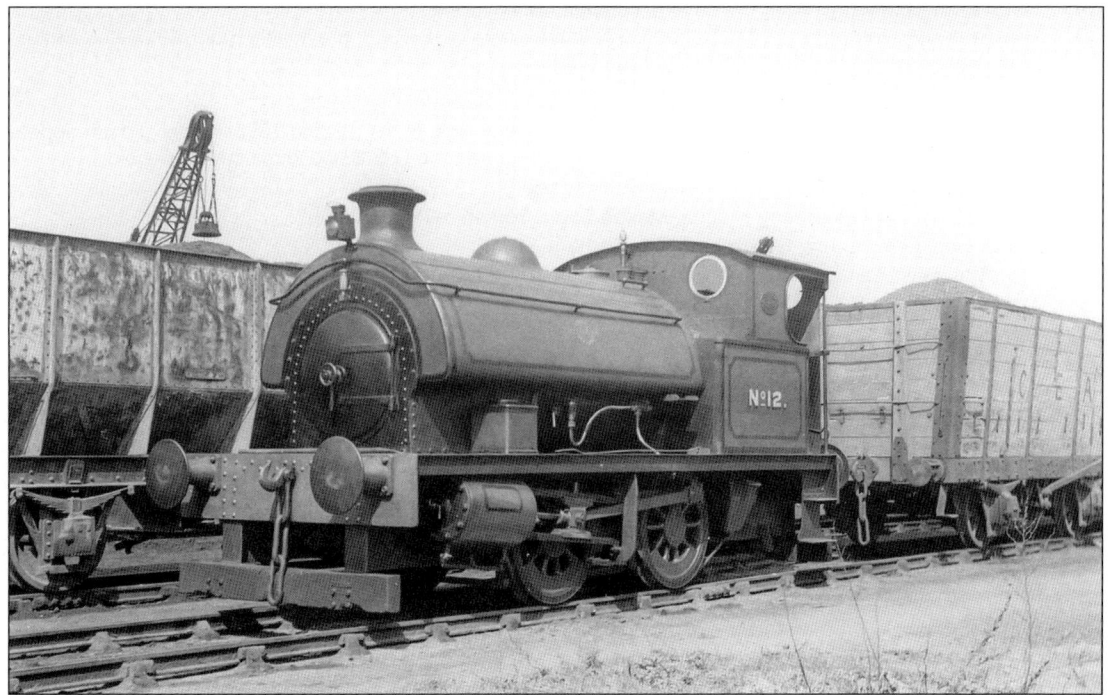

No.12 (HL 3651/1926) in the coal sidings at North Tees Generating Station on 19th April 1960.
(IRS B. Roberts photograph)

The North-Eastern Electric Supply Company proposed the construction of a larger 'C' station in 1945 but, with nationalisation, the project was taken over by the British Electricity Authority and completed in 1949. It used two 60,000 kW turbo-alternators to produce 120,000 MW of electricity. The prominent building occupied the south end of the site and its exterior was designed by Giles Gilbert Scott. North Tees Generating Station lay alongside the Billingham Beck Branch between the Haverton Hill road bridge and the ICI storage tanks. An agreement dated 12/7/1920 was signed between the NER, Newcastle upon Tyne Electric Supply Co Ltd and the Tees Power Station Co Ltd as to a siding at Haverton Hill. A set of sidings next to the branch, together with a coal tippler, enabled incoming trains of coal to be unloaded, while other lines served the remainder of the site.

In 1918 the Newcastle upon Tyne Electric Supply Co had ordered five sets of equipment, each comprising two motors, from British Westinghouse which were subcontracted to Dick, Kerr & Co Ltd. Each set was to be mounted on a "26 Ton Battery Locomotive constructed by Kilmarnock Works" (order no.C1690, 9th September 1918). The North-Eastern Electric Supply Co introduced a single numbering system, based on their age, for all of the locomotives at its power stations which explains the lack of early numbers for those at North Tees. The latter then obtained two new Hawthorn Leslie saddle tanks in 1924-26 to carry out shunting. Rail traffic to North Tees Generating Station ceased in early 1982 and it was decommissioned on 31/10/1983. A visit on 17/3/1983 had found the yellow Sentinel shut in the shed; the locomotive carried BR registration plate, 2591/58. Demolition of the main building took place in 1989.

Gauge : 4ft 8½in

No.	Type		Builder	Works No.	Date		
No.5	4wBE		DK		1920	New	(1)
No.6	4wBE		DK		1920	New	(2)
No.7 (originally No.1)	4wBE					(a)	Scr
No.10	0-4-0ST	OC	HL	3586	1924	New	(3)
No.12	0-4-0ST	OC	HL	3651	1926	(b)	Scr c/1968
No.16	0-4-0ST	OC	RSHN	7359	1947	New	(4)
No.26	0-4-0ST	OC	RSHN	7798	1954	(c)	Scr 3/1970
(No.5)	0-6-0DM		HC	D835	1954	(d)	(5)
2	4wDH		S	10003	1959	(e)	(6)

(a) this was possibly another new locomotive like Nos.5 and 6 or it may have been the battery electric transferred from the Grangetown Power Station (see page 249) when that closed in 1937.
(b) new. To CEGB Dunston Generating Station, Dunston, Co. Durham c/1962 and returned c6/1963.
(c) ex CEGB, Carville Generating Station, Newcastle upon Tyne, 9/1960.
(d) ex CEGB, Skelton Grange Generating Station, Leeds, 14/4/1967.
(e) ex Thos Hill (Rotherham) Ltd, Kilnhurst, Yorkshire (WR), 3/10/1969; previously Bass, Mitchell & Butlers Ltd, Burton on Trent, Staffordshire, until 11/1968.

(1) to CEGB, Dunston Generating Station, Dunston, Co. Durham, by 23/9/1949.
(2) s/s. Ted Haigh visited North Tees on 9/7/1950 and saw Nos. 6 ,10, 12 and 16; the latter did not appear to have been used. All were painted green, lined and had the N.E.S. Co scroll in yellow-gold.
(3) to CEGB, Skelton Grange Generating Station, Leeds, /1959.
(4) to Mexborough Generating Station, Mexborough, Yorkshire (WR), 9/1967.
(5) to D. Sep. Bowran Ltd, Gateshead, Co. Durham, /1974.
(6) to Standard Railway Wagon Co Ltd, Heywood, Greater Manchester, 3/11/1983, via/per Resco (Railways) Ltd, Woolwich, London.

CEREBOS FOODS LTD

GREATHAM SALT WORKS, Greatham   **38**
Cerebos Ltd until 1/1/1963   NZ 499267 – **Map A**
Cerebos (1903) Ltd until c/1906
Hartlepool Salt & Brine Co until /1902

Salt was found at Greatham in 1887 and, two years later, the Hartlepool Salt & Brine Company had sunk nine boreholes on the marshes to the north of Greatham Creek. The salt works was erected on the south side of the NER Norton-Hartlepool railway next to the Marsh House level crossing and Greatham Station. Two brine reservoirs were constructed by the road north of the crossing. The connection with the NER ran into the salt works dividing into several sidings which served the three principal buildings. One of the sidings continued out of the premises across Marsh House Lane and headed south as a single line alongside the road past Marsh House Farm to reach the brine wells. With a couple of exceptions, most of the boreholes were in a row which the line ran along to end at the last one. The NER Private Owner Wagon Register shows that, by

Andrew Barclay 1768 of 1922 stands near the main line at the Greatham Salt Works with Cerebos Foods Ltd's name proudly displayed. (ILS F. Jones photograph)

1891, the Hartlepool Salt & Brine Co at Greatham owned thirty eight 10 ton railway salt vans (its numbers 1-38) for operating on the main line, with another eighty (numbers 39-118) being acquired by 1894; these were constructed by the Darlington Wagon & Engineering Co Ltd. Also, in 1895, the HIGHFIELD ship left Hartlepool Docks with 4,000 tons of salt from the Hartlepool Salt & Brine Co for Calcutta.

By 1901, the company was in receivership and was dissolved by a notice dated 9/12/1902. Cerebos was already based on Tyneside and its salt was claimed to be free running, a useful asset in a damp climate. In 1903 Cerebos was restructured as a company and purchased the works at Greatham. The salt mixing and packing operations were transferred from Newcastle to Greatham in 1906. The company also introduced Bisto gravy salt in 1908. The connection between the main line and the salt works sidings had been originally agreed between the NER and Christopher Furness on 16/9/1889 and it was transferred to Cerebos (1903) Ltd on 3/6/1904. An extension to the agreement on 30/10/1908 allowed Cerebos Ltd's locomotives to cross the main line to the NER's goods siding on the north side. A setback occurred in 1912 when fire destroyed much of the works putting 300 people out of work. However, it was rebuilt in the same year and, shortly after, an economic vacuum evaporating plant was installed enabling the works to survive the Depression based on its well-known brands of "Saxa" and "Cerebos". There is a shortage of information on the first locomotives but, by 1914, the branch to the brine wells had been extended about 100 yards closer to Greatham Creek where a tip was established, presumably for ash from the boilers. Following further modernisation after World War 2, this was the only plant still producing refined salt on Teesside by the late 1950s when it employed about 500 people. Approximately 1,500 tons of salt per week were made; most went out by road but a small quantity still travelled by rail. Cerebos was taken over by Rank Hovis McDougall Ltd in the 1960s, by which time it was making a range of foodstuffs. Rail traffic ceased in 1968, the siding agreement being formally terminated on 19/4/1968. A year later, a new salt works opened at Middlewich and salt production ceased at Greatham in 1970 with the salt works block being demolished in 11/1970. The factory subsequently became Sharwoods and then Centura Foods prior to closure in 2003.

Reference : *The Salt Industry of the River Tees*, David M. Tomlin, De Archaeologische Pers, 1982

Gauge : 4ft 8½in

No.1	CEREBOS				?		(a)	s/s
	CEREBOS	0-4-0ST	OC	AB	773	1897	(b)	s/s
(No.1	MIDDLESBROUGH until /1920)							

8	CEREBOS	0-4-0ST	OC	AB	1768	1922	New		(3)
-		4wPM		KC	1470	1926	(New?)		s/s
	CECIL	0-4-0ST	OC	RWH	1847	1881	(c)		s/s
	BESSIE	0-4-0ST	OC	HE	205	1878	(d)		(1)
	TEES SIDE No.2	0-4-0ST	OC	MW	1327	1897	(e)		(2)
	CEREBOS	0-4-0DM		JF	4100012	1948	New		(4)

(a)   ex ? , by /1910.
(b)   ex Sir W. G. Armstrong Whitworth & Co Ltd, Elswick Works, Newcastle upon Tyne, /1919.
(c)   ex Sir S.A. Sadler Ltd, Malton Colliery, near Lanchester, Co. Durham, c/1933.
(d)   ex Warner & Co Ltd, Cargo Fleet, on hire.
(e)   ex Tees Side Bridge & Engineering Works Ltd, Cargo Fleet, on hire, 1/1947.

(1)   returned to Warner & Co Ltd, Cargo Fleet, ex hire.
(2)   returned to Tees Side Bridge & Engineering Works Ltd, Cargo Fleet, ex hire, 5/1947.
(3)   to C. Herring & Son Ltd, Hartlepool, for scrap, 3/1968.
(4)   to Batchelor, Robinson & Co Ltd, Hartlepool, 8/1969.

R.H. CHARLTON & CO

STRANTON IRON WORKS, West Hartlepool   **39**
Stranton Iron & Steel Co Ltd until 1/1874   NZ 514317 – **Map B**

The works was located on the west side of the NER's Cliff House Branch, north of Greatham Street. It was acquired from R.D. Buckey of London by the Stranton Iron & Steel Co Ltd on 6/12/1871 but this business went into chancery in 8/1873; a year when it had 20 puddling furnaces and two furnaces/mills. It was then purchased by R.H. Charlton & Co on 30/4/1874 for approximately £20,000, compared to the original construction cost of about £100,000. R.H. Charlton was the son of Thomas Charlton who, as T. Charlton & Co, possessed collieries in County Durham and the Slapewath ironstone mine near Guisborough. The depression in the iron trade during the late 1870s meant that the Stranton Works soon ran into difficulties and was in the hands of R.H. Charlton's trustees by 4/7/1878 when an advertisement appeared in the *Hartlepool Mail* inviting tenders for the 12 inch locomotive, MILLIE, built by Hopkins Gilkes in 1874. The engine was stated to be in good condition and it could be viewed at the Stranton Iron Works. Interestingly, this locomotive does not appear in Fred W. Harman's book on the Tees Engine Works which may indicate that his list of Hopkins Gilkes locomotives is incomplete.

The site was the same as the "Stranton Works (Expanded Metal)" premises that appears on the 1896 Ordnance Survey, with a siding off the Cliff House Branch which divided into three on entering the works. The British Expanded Metal Co was established in 1889, having secured access to a patent from the inventor in America. The process involved transforming steel sheet into various types of lattice work. Principal proprietors in 6/1890, when the works in Greatham Street opened, were Christopher Furness, Matthew Gray and R. Irvine. In 1894, the business purchased the Expanded Metal Co Ltd of London and took its name. The Stranton Works produced a comprehensive range of expanded metal products for the automotive, building and construction, security and engineering markets.

Gauge : 4ft 8½in

MILLIE	0-4-0ST	OC	HG		1874	New	s/s

CJC CHEMICALS AND MAGNESIA LTD

PALLISER WORKS, Hartlepool.   **40**
Britmag Ltd until 3/2002 (subsidiary of **KHSL Industries**)   NZ 507352 – **Map A**
Redland Magnesia Ltd until c28/5/1997
Steetley Magnesite Co Ltd until /1992 *
The Palliser Works (Ministry of Supply managed by British Periclase) until 27/10/1952
British Periclase Co Ltd until /1941 (subsidiary of **Steetley Company**)

* Responsibility for the works changed frequently between various Steetley subsidiaries 1969-92.

To avoid confusion on the terminology:
Dolomite is the natural quarried stone (calcium magnesium carbonate)
Dolime is calcined (not deadburnt) dolomite (calcium magnesium oxide)
Magnesia occurs naturally in some parts of the world (magnesium hydroxide)

Steetley quarried dolomite at Coxhoe and later Thrislington in Co. Durham where the stone was converted to dolime using initially shaft kilns and later rotary kilns. At the Palliser Works, the dolime was hydrated to produce a calcium magnesium hydroxide slurry. This was then reacted with seawater to make a suspension of magnesium oxide and magnesium hydroxide which was consolidated and washed in large setting tanks. The magnesium oxide hydrated gradually to increase the level of magnesium hydroxide in the slurry. The magnesia slurry was filtered and the resultant paste deadburnt in large rotary kilns to make Britmag, a refractory grade magnesium oxide, which was used for refractory linings in various industries. If the paste was calcined, a chemical grade of magnesium oxide powder was produced.

The Palliser Works was the first plant in the world to commercially manufacture magnesia from dolomite and seawater. Steetley's only interest in making Britmag was to further the company's development in refractory sales and minimise the reliance on imported natural magnesites from Austria. The Magnesium Elektron Company at Clifton Junction in Lancashire used imported magnesites to manufacture magnesium metal and, with war approaching, the Palliser Works also began producing reactive calcined magnesia (magnesium oxide) to replace the probable loss of supply from Austria. Magnesium metal was vital for magnesium alloys (aeroplane frames) and munitions (incendiaries).

Originally called the Palliser Works, the pilot plant was established at Hartlepool in 1937 funded by the Refractory Brick Co of England, a Steetley subsidiary. On its success, Steetley then formed British Periclase Ltd (incorporated 3/8/1937) to build and operate the works which was in production by the end of 1938. In view of its strategic importance, the works was taken over by the Ministry of Supply in 1941, but was still operated by British Periclase for a management fee. It became the responsibility of the Ministry of Materials in 1951 and was sold back to Steetley. With peacetime, demand for Britmag increased as the economy recovered and the steel industry expanded. The works was situated on the north east side of the Hartlepool-Seaham railway where a set of sidings was located controlled by Cemetery North Signal Box. These sidings were very busy in the 1950s and a locomotive servicing point, with a water tank, was located here. Later this yard functioned solely as exchange sidings for Palliser Works traffic. There were two connections leading from the exchange sidings across the adjacent Old Cemetery Road into the works, although the more southerly line was later removed. Within the premises, a small three-road yard parallel to the perimeter fence served the northern weighbridge, the dolime unloading shed and a short head shunt beyond. Another line led further into the works to reach the Lancashire boiler plant, while a two-road engine shed and repair shop was located near the south

A publicity photograph showing new Hunslet 7425 and some of the PAA wagons received by Steetley at its Palliser Works in 1981. (Bob Dunn collection)

eastern boundary. A dolomite burning shaft kiln was also erected at the Palliser Works during World War 2 as a safeguarding measure, but this required the transport of dolomite from Steetley's Coxhoe quarry before it was converted into dolime. Later it was shut and all the dolime was produced at Coxhoe and, from 1957, at Thrislington.

Coal came by rail for both the Lancashire boilers and kilns, but the former was no longer required when process developments in 1947 removed the necessity to steam heat the magnesia slurries prior to filtration saving 25,000 tons of coal a year. For comparison, coal used to fire the rotary kilns during the same year was about 14,000 tons. Palliser's complement of rotary kilns had expanded to six by 1961 when a change was made to firing them with heavy fuel oil. Dolime travelled to Hartlepool initially in wooden and later 21 ton steel hopper wagons covered with tarpaulins. BR then allocated over one hundred 24 ton Covhop covered hopper wagons for this traffic. Output from the Palliser Works rose from 150,000 tons pa in 1957 to a peak of 236,058 tons in 1972. Magnesium was sent out in small containers on 14 ton Conflat L wagons to steelworks and foundries. Britmag was put in sacks and loaded into coal wagons, which were sheeted. Subsequently loose Britmag was dropped into covered grain vans. However, there was an increasing move to using road transport for outward deliveries. In 1976/77 most of this rail traffic ceased, the final loads of fettling magnesia going in the small containers to the British Steel Corporation at Skelton and Bilston, with just the occasional Transfesa ferry wagon containing 25kg sacks of chemical magnesium powder for a Portuguese paper manufacturer. In 1980 Steetley was awarded a section B grant to purchase its own fleet of wagons and thirty 51 ton airbraked covered hoppers, code PAA, were built by the Standard Wagon Co in 1981 to transport dolime from Thrislington. A new Hunslet four-coupled diesel locomotive was also acquired at the same time. The BR Covhops survived in traffic until 1983 when Steetley hired some secondhand PAB wagons to carry lime from Thrislington. The two types were then used in a daily train of 15-20 wagons to Hartlepool.

The market for refractory materials was declining because the amount required to produce a ton of steel had reduced by one third in 40 years. Also a number of UK steel works had closed. China was able to supply cheap deadburnt natural magnesite. In 1992, the Thrislington-Hartlepool trains were reduced to three a week and the PAB wagons withdrawn. Annual production amounted to only 60,000 tons in 1998. Redland had taken over from Steetley in 1992 but it then sold the plant to KHSL Industries in 1997, which formed Britmag Ltd. The latter went into administration at the end of 2001 and stopped trading in March 2002. CJC Chemicals and Magnesium Ltd was created by a management buyout to continue some processes. Production of refractory grade magnesia ceased and the train from Thrislington only ran once a week until it finished in 2004. The works closed on 1/6/2005 and was demolished in 2012.

References : 'Steetley Rail Freight Remembered', David Ratcliffe, *Rail Express*, 2006, pp.20-24
Steetley Dolomite and Seawater Operations in the North of England, Volumes 1-3, Robert Dunn and John Smailes, 2011–15

Gauge : 4ft 8½in

	MARS	0-4-0T	OC	AE	1701	1915	(a)	s/s
	P.W. 1	0-4-0ST	OC	HC	1735	1942	New	(1)
	-	0-4-0DM		RH	327966	1954	New	(3)
	(DL 2)	0-4-0DH		HC	D1346	1965	(b)	(6)
	03154 (D2154)	0-6-0DM		Sdn		1960	(c)	(2)
	-	0-4-0DH		HE	7425	1981	New	(5)
	HO18	0-4-0DH		HC	D1279	1963	(d)	(4)

- 0-4-0ST OC HC 1734 of 1942 from Steetley Lime & Basic Co Ltd, Coxhoe, Co. Durham, may have worked here at sometime between 1943 and 1945.

(a) ex Sharp, Jones & Co Ltd, Bourne Valley Potteries, Branksome, Dorset, per Ministry of Supply, /1942.
(b) ex Scottish Gas Board, Provan Gas Works, Glasgow, Lanarkshire, 7/1971. The machinery merchants, Thomas Mitchell & Sons of Bolton, purchased the locomotive at Provan in 5/1971 from scrap dealers, Abercrombie & Co, which was presumably dismantling the Gas Works and went on to resell the locomotive to Steetley (Mfg) Ltd in 6/1971.
(c) ex BR, Thornaby Depot on hire, 2/1974.
(d) ex Booth Roe Metals Ltd, Rotherham, South Yorkshire, (property of RMS Locotec), on hire, 23/8/1995.

(1) to Steetley Dolomite (Quarries) Ltd, Coxhoe, Co. Durham, /1954, by 18/4/1954.
(2) returned to BR, Thornaby Depot, ex hire, 3/1974.
(3) to Hunslet Engine Co Ltd, Leeds in exchange for HE7425, 10/3/1981. The lowloader, bringing the new locomotive, took away the RH.
(4) to RMS Locotec, Dewsbury, West Yorkshire, ex hire, 28/11/1995.
(5) to Ed Murray & Sons Ltd, Hartlepool, 5/2005.
(6) to Ed Murray & Sons Ltd, Hartlepool, 25/5/2005.

CLAY LANE IRON CO LTD

CLAY LANE IRON WORKS, South Bank   **41**
Owners of Clay Lane Iron Works until 12/1/1882   NZ 533213 – **Map H**
Trustees of Thomas Vaughan & Co until /1878
Thomas Vaughan and Co until 10/1876
Clay Lane Iron Co until /1870
Elwon, Malcolm & Co until /1859

Thomas Light Elwon (son-in-law of John Vaughan) initially develop three blast furnaces at the Cleveland Iron Works near to Bolckow & Vaughan's Eston Works. However, these blast furnaces were transferred to Bolckow & Vaughan even before they were brought into full operation. Instead Elwon went into partnership with Malcolm & Co to establish the Clay Lane Iron Works, with blast furnaces parallel to and on the south side of the Middlesbrough-Redcar railway. Pig iron production commenced from the first of its blast furnaces towards the end of 1858, with a second blown in during the following year. By 1860 the plant comprised three blast furnaces with cylindrical brick stacks standing on large sandstone blocks.

The Clay Lane Iron Co was formed in 1859 and three more blast furnaces were added in line with the original structures, beginning in 1863. All six were reported to be in blast between 1864 and 1869. Thomas Vaughan, the son of John Vaughan (joint founder of Bolckow & Vaughan), had been associated with Elwon and Malcolm but, when his father died in 1868 he acquired a fortune of about £½ million. Based on his new wealth, he separated from his previous partners and established Thomas Vaughan & Co in 1870 taking over the Clay Lane Iron Works and starting his own ironstone mine at South Skelton in 1872. It may have been a coincidence but, from 1/5/1868, the Clay Lane Iron Co's engines and traffic were no longer allowed to use Bolckow Vaughan's lines at Eston. Unfortunately Thomas Vaughan got into financial difficulties with the economic depression of the mid to late 1870s. A decision was taken on 27/7/1876 to turn the business into a limited liability company, but a newspaper on 2/9/1876 reported on the failure of Thomas Vaughan & Co and the *Guisborough Exchange* on 12/10/1876 stated that it was to be liquidated. The Trustees of Thomas Vaughan & Co continued to run the ironworks, although only three of the six blast furnaces were operating. However, in 1878, the Trustees gave up South Skelton Mine and this probably applied to the Iron Works as well because Riden and Owen show the blast furnaces being operated by the Owners of Clay Lane Iron Works.

On 12/1/1882 the Clay Lane Iron Co Ltd was registered to operate the works. The main protagonists behind the business were James Kitson, the younger of Leeds, the Yorkshire Banking Co and landowners seeking continuation of their ironstone royalties – J.T. Wharton and D.T. Petch. The ironworks traded successfully with most of the blast furnaces operational until Bolckow Vaughan acquired it in 1898. *(The remaining history of the plant will be considered under the composite Teesside Works entry.)*

In 1893 Clay Lane Iron Works had covered an area of 21 acres. Output from the larger blast furnaces was 600 tons per week and from the smaller ones 450 tons. A connection with the Middlesbrough-Redcar railway ran past the termination of Normanby Road at the west end of the site, whilst a link was also made with the NER's Eston Branch on the eastern boundary. Minerals were taken by rail to the north side of the blast furnaces where the stores were located. The boilers and blowing engine house were situated at the eastern end of the blast furnaces. The sidings serving the pig beds and the slag spouts ran south from the blast furnaces. Most of the remainder of the site as far as Middlesbrough Road East was dominated by the slag tips, so much so that the adjacent terraced houses at South Bank endured the nickname "Slaggy Island". Railway lines ran over the tips as well as also serving Brand's Slag Brick Works and a lime kiln. The Clay Lane Iron Co extended its line southwards beyond these tips to cross over Middlesbrough Road East and the lane from Redcar Road East by substantial bridges to enable its locomotives and their slag ladles to reach another extensive tip. It also operated a single railway track which burrowed under the main line to emerge on the reclaimed land to the north. At about the then High Water Mark, it was joined by a siding from the South Bank Iron Works before running along the Clay Lane jetty over the mudflats to Clay Lane Wharf on the bank of the River Tees. Another siding headed west from the works across Normanby Road to a coal depot (identified as Imperial Depots on an 1895 NER plan) at North Street, South Bank.

The company employed a mixed fleet of four-coupled locomotives to shunt its works; the number being increased when the blast furnace plant was extended. As our concern here is primarily with the latter part of the nineteenth century, records of the motive power are not complete. An added complication is that for some years in the 1870s, Thomas Vaughan & Co also owned the South Bank Iron Works so it is possible that some locomotives may have moved between the two premises. In *Machinery Market* on 1/12/1884, dealer Joseph Torbock offered two locomotives for sale "lying at Clay Lane ironworks"; these were a four-wheel Black, Hawthorn with 9in x 16in cylinders and 2ft 8in diameter wheels (£425) and a four-wheel Kitson, built in 1881, with 14in x 21in cylinders and 3ft 2in diameter wheels (£850). Barclays 208 of 1873 suffered a boiler explosion at Woodland Colliery, Co. Durham in 1886 after being recently overhauled by the Clay Lane Iron Co on behalf of Joseph Torbock.

The Clay Lane Iron Works in the 1870s with three older shorter blast furnaces standing on sandstone blocks and the newer taller furnaces on metal columns beyond. The pipe stoves in between the stacks were replaced by eight pairs of Cowper regenerative stoves when the plant was modernised in the 1880s. The pig beds are in front of the blast furnaces and rakes of bogies, some carrying tiny slag ladles, are present as well as an open wagon still labelled "S&D" (SDR). The end of the engine shed is just visible and three of the company's locomotives are at work.
(Jim Stancliffe collection)

Gauge : 4ft 8½in

	GLENMORAG	0-4-0 tank		GW	173	1863	New	(1)
No.3	TOM	0-4-0VB tank		GW	191	1864	New	(1)
-		0-4-0 tank		HG	196	1865	New	(1)
	BALNABOTH	0-4-0 tank		HG	197	1865	New	(1)
-		0-4-0ST	OC	RP	969	1866	New	s/s
	SPERO	0-4-0 tank		HG	247	1867	New	(2)
	ROXBY	0-4-0ST	OC	K	1788	1873	New	(3)
No.3		0-4-0ST	OC	B	294	1882	(a)	(3)
No.5		0-4-0ST	OC	B	295	1882	(a)	(3)
No.1		0-4-0ST	OC	BH	1094	1895	(b)	(4)
No.2		0-4-0ST	OC	BH	1123	1896	New	(4)
No.7		0-4-0ST	OC	MW	465	1873	(c)	s/s

Inness suggests that there was a No.4, possibly a Peckett 0-4-0ST with 12in x 18in cylinders.

(a) new to Clay Lane Iron Co, Saltburn. The reference to Saltburn has led to a suggestion that they may have gone first to South Skelton Mine but this is unlikely.

(b) new. The locomotive was painted "dark chocolate" in colour with "CLICO" on the tank in gilt letters.

(c) ex Stanghow Ironstone Co Ltd, Stanghow Ironstone Mine, Margrove Park by /1900.

(1) s/s, except one of these locomotives went to Preston Coal Co, Preston Colliery, North Shields, Northumberland, 1/1897.

(2) to Cargo Fleet Iron Co Ltd, Cargo Fleet.

(3) to Bolckow Vaughan & Co Ltd, /1898, according to Inness, but it is not known which works. B 294 is stated to have become Bolckow Vaughan No.3 and B 295 to have been still working in 1907.

(4) to Bolckow Vaughan & Co Ltd, /1898. Nos.1 and 2 (assumed to be these locomotives) stood in a line of locomotives awaiting disposal at the Cleveland Works on 5/9/1926.

ROXBY (Kitson 1788/1873) was one of the Clay Lane Iron Works locomotives.

CLEVELAND SALT CO LTD

CLEVELAND SALT WORKS, Middlesbrough
Bolckow, Vaughan & Co Ltd, until 11/10/1887

42
NZ 501211 – Map G

In 1859 Bolckow & Vaughan began sinking a shaft at the Middlesbrough Iron Works allegedly in a search for clean water. This was unsuccessful, but a borehole from the bottom of the shaft located a 99ft thick bed of salt at 1,206 feet in 1863. No action was taken until the process for pumping water down to extract brine, which could be then evaporated using waste heat from the blast furnaces, was demonstrated. The No.1 borehole at Middlesbrough Iron Works commenced raising brine on 17/8/1886. There were eventually four wells – No.2 (started 1888), No.3 (1893) and No.4 (1941). Six brine evaporating pans were constructed in 1886 at the western end of the iron works and another seven were added in 1889, some using space previously occupied by the puddling furnaces. The Cleveland Salt Co Ltd was incorporated on 11/10/1887 to purchase the lease for the salt deposits and the surface plant, although Bolckow Vaughan held a controlling stake in the business for 38 years. The salt evaporating pans were heated by blast furnace gas until the latter stopped operating and then coal was used.

The salt works relied on the Iron Works' railway and locomotives to handle its traffic which was taken to Vulcan Street; it paid £33 in MOR tolls in 1933. As activity at the ironworks ran down, Cleveland Salt obtained its own locomotive in 1937. This was of interest as it was the first of three Fowler standard gauge diesels built with chain drive, as opposed to rod drive transmission. The works ceased operations in 1947 having produced 879,972 tons of salt and the company was wound up on 2/4/1947. The site was then occupied by Thos W. Ward Ltd between 5/7/1947 to 20/8/1952, according to MOR records, to dismantle the plant which had been cleared by 10/1951. A reminder of the salt works remains with a portion of the tall ornate red brick boundary wall, erected in 1887, standing alongside Vulcan Street.

References : *The Salt Industry of the River Tees*, David M. Tomlin, De Archaeologische, 1982

A Brief History of the Cleveland Salt Co Ltd 1887-1947, N.W. Pallister

Gauge : 4ft 8½in

FRANK PALLISTER	4wDM	JF	21750	1937	New	(1)

(1) to Rotherham Corporation Electricity Department, Rotherham Power Station, Yorkshire (WR), per JF, c6/1947, but was quickly returned to JF for repairs and then to Cravens Railway Carriage & Wagon Co Ltd, Darnall, Sheffield, 18/11/1947.

CLEVELAND SLAG ROADS LTD

Grangetown

43
approx NZ 549219 – Map H

The company possessed a crusher behind Grangetown Station for breaking up slag from the large tip associated with Bolckow & Vaughan's (later Dorman Long's) Cleveland Works. Its date of establishment is not known but an entry in the NER *Private Owner Wagons Register* states that a 10 ton coke wagon, by Stabler & Co, Darlington sold by Rolling Stock Co to "Cleveland Slag Roads, Grangetown for tarmac traffic" on 12/4/1923. The name Cleveland Slag Roads Ltd appears in directories for 1934 and 1958 but IRS records indicate that the site was later taken by Tarslag Ltd and the works dismantled in 1960.

Gauge : 2ft 0in

-	4wDM	HE	1863	1937	New	s/s

COBRA (MIDDLESBROUGH) LTD

Middlesbrough
COBRA Railfreight Ltd until /2012

44
NZ 488209 – Map F

Cobra Railfreight Ltd was founded in 1968 and specialised in the movement of goods by rail. The main terminal was at Wakefield. In 1986 it opened a new £750,000 depot on the site of the former BR Middlesbrough Town Goods Station to handle potash, fertiliser, salt and palleted goods. Much of the traffic came from Cleveland Potash Ltd's Boulby Mine as rock salt for storage and onward distribution by road. The depot consists of a warehouse with a covered discharge point on its west side, together with an open storage area for rock salt next to North Road. Trains arrive and depart from the Middlesborough Goods sidings (BR, Railtrack and now A.V. Dawson Ltd) alongside the depot, from where the diesel shunter moves the wagons to the discharge point. Responsibility for shunting has been contracted out for some years. The depot is now operated by a subsidiary company, although its headquarters is still at Wakefield.

Gauge: 4ft 8½in

ID	Name	Type	Builder	Works No.	Year	From	To
(NC23) (08816) (D3984)		0-6-0DE	Derby		1960	(a)	(1)
	COBRA	0-6-0DH	GECT	5378	1972	(b)	(3)
08774 (D3942) ARTHUR VERNON DAWSON		0-6-0DE	Derby		1960	(c)	(2)
H005		0-6-0DH	HE	6295	1965	(d)	(4)
H025	R.A. LAWDAY	0-6-0DE	YE	2878	1963	(e)	(5)
H032	PETE GANNON LOCOMOTIVE ENGINEER 1948-2001	0-6-0DH	HE	7541	1976	(f)	(6)
H037	7	0-6-0DH	EEV	D1201	1967	(g)	(8)
H001		4wDH	S	10003	1959	(h)	(7)
11		0-6-0DH	EE-AEI	3994	1970		
		Rebuilt	YEC	L180	2000	(i)	(9)
H006 15 (01573)		0-6-0DH	HE	6294	1965	(j)	(9)
(01567)	ELIZABETH	4wDH	TH	276V	1977	(k)	

According to the then engineer responsible for the plant, the company stopped using 08 816 on 4/1/1994 so another locomotive must have been hired before GECT 5378 arrived.

(a) ex BR, Thornaby Depot, 15/2/1986.
(b) ex Cobra Railfreight Ltd, Wakefield, West Yorkshire, 27/5/1994 and unloaded here 31/5/1994.
(c) ex A.V. Dawson Ltd, Middlesbrough, on hire, 10/1998.
(d) ex RMS Locotec, Dewsbury, West Yorkshire, on hire, 20/2/1999.
(e) ex RMS Locotec, Dewsbury, West Yorkshire, on hire, 11/2000.
(f) ex RMS Locotec, Dewsbury, West Yorkshire, on hire, by 22/2/2002. It was here 22/2/2002 but stated to have arrived 11/2001.
(g) ex Aggregate Industries UK Ltd, Croft, Leicestershire, (property of RMS Locotec, Dewsbury, Wakefield), West Yorkshire, on hire, 1/8/2005.
(h) ex Mostyn Docks Ltd, Mostyn, Flintshire (property of RMS Locotec, Wakefield, West Yorkshire), on hire, 19/12/2006.

The rail discharge point at Cobra's Middlesbrough goods depot on 3rd November 2009 with HE 6294/1964 hired from British American Railway Services Ltd coupled to wagons which have arrived from Boulby Mine.
(Cliff Shepherd photograph)

(i)   ex Cleveland Potash Ltd, Tees Bulk Terminal, Tees Dock, Grangetown (property of RMS Locotec, Wakefield, West Yorkshire), on hire, 13/12/2007.
(j)   ex PD Ports plc, Teesport, Grangetown, (property of British American Railway Services Ltd, Walsingham, Co. Durham), on hire, 3/4/2009.
(k)   ex T.J. Thomson & Son Ltd, Stockton, (property of Ed Murray & Sons Ltd), 9/2/2017 via Tata Steel Europe, Hartlepool Pipe Mill, Hartlepool, on hire, 11/2/2017.

(1)   purchased by Harry Needle Railroad Company, /1995; then to Johnson (Chopwell) Ltd, Widdrington Disposal Point, Northumberland, 19/8/1995.
(2)   returned to A.V. Dawson Ltd, Middlesbrough, ex hire, /1998.
(3)   to RMS Locotec, Dewsbury, West Yorkshire, by 9/1/1999.
(4)   to RMS Locotec, Dewsbury, West Yorkshire, ex hire, 11/2000.
(5)   to RMS Locotec, Dewsbury, West Yorkshire, ex hire, c/2001, after 2/4/2001.
(6)   to RMS Locotec, Dewsbury, West Yorkshire, ex hire, 5/8/2005.
(7)   to Weardale Railway Trust, Weardale Railways Ltd, Wolsingham, Co. Durham, 13/12/2007.
(8)   to Wyvern Rail plc, Ecclesbourne Valley Railway, Wirksworth, Derbyshire, 23/1/2008.
(9)   to British American Railway Services Ltd, Wolsingham, Co. Durham, ex hire 8/8/2018.

COCHRANE & CO LTD

STANGHOW IRONSTONE MINE, Margrove Park 45
Stanghow Ironstone Co until 1/1892 NZ 654156 – **Map I**

An agreement was signed by G. Wythes and J. Cochrane of the Stanghow Ironstone Co in 1870 to extract ironstone from J. Wharton's Skelton Estate and also to construct houses for the miners. A lease of the ironstone was taken as from 1/7/1871 and the sinking of two shafts commenced in 1872 penetrating 123 feet to the ironstone on the south side of the Margrove Valley. The ironstone was being worked by 8/1872. A total of 100 terrace houses were constructed in the following year forming Margrove Park village immediately east of the mine. It is likely that the bricks of these houses, the mine shafts and other buildings came from the nearby Carrs Tilery. The mine was often known locally as Margrove Park or "Magra Park".

The mine was served by a short branch line which diverged as a single track from the former Cleveland Railway at Stanghow Signal Box, next to Carr's Tilery. It then became the mine company's property and formed a double track which curved round crossing the access road to the Tilery on the level. Immediately after, a loop line served a building from which a narrow gauge tramway ran to a nearby drift entrance and was used for taking pit props into the mine. The standard gauge branch climbed over the Charlton-Boosbeck Road to pass under the chutes from the picking belt next to the shafts. The Main Seam was 5 to 6 feet thick, but had a shale band running through it. Narrow gauge tramways ran from the winding shaft and the picking belt carrying tubs of shale to the spoil heap between the mine and the road. By 1913 this area was full and another tip existed on the north west side of the road with the narrow gauge tramway being carried over the highway on a bridge. In *The Colliery Guardian* on 5/7/1872, a W. Walker advertised wanting two new or secondhand 0-4-0 tank locomotives with 8in x 16in cylinders, 2ft 6in diameter wheels on a 6ft wheelbase for the 3ft 0in gauge. The address was given as "Skinningrove, near Saltburn." William Walker was manager of the Stanghow Mine in 1875 but had interests elsewhere. It is not known whether anything came of his request or where they were intended to work.

The Stanghow Ironstone Co went into liquidation in 1892 and, in the following year, Cochrane & Co Ltd leased the mine. It began sending out ironstone again in 6/1893 and subsequently large tonnages went out by rail to its blast furnaces at the Ormesby Works until the latter were taken out of blast in the early 1920s. The last ironstone was worked in 2/1921 and the mine had effectively finished by 6/1924, although it was not officially abandoned until 11/1928. There is considerable uncertainty whether Stanghow Mine employed a locomotive and the reader Is referred to the discussion on page 333 of *The Industrial Railways & Locomotives of County Durham Part 1* by Mountford and Holroyde (IRS, 2006) concerning Manning Wardle 465.

Gauge : 4ft 8½in

| | 0-4-0ST | OC | MW | 465 | 1873 | (a) | (1) |

(a)   Manning Wardle records show this locomotive as being purchased by "Cassop Colliery Co, Redcar, Durham" and named "CASSOP No.1". Interestingly a Henry Cochrane and then George Wythes were listed as owners of Cassop Colliery in Co. Durham in the mid 1870s. The Stanghow Iron (sic) Co Ltd appears as a subsequent user of the engine in MW records.

(1)   to Clay Lane Iron Co Ltd, South Bank, by /1900.

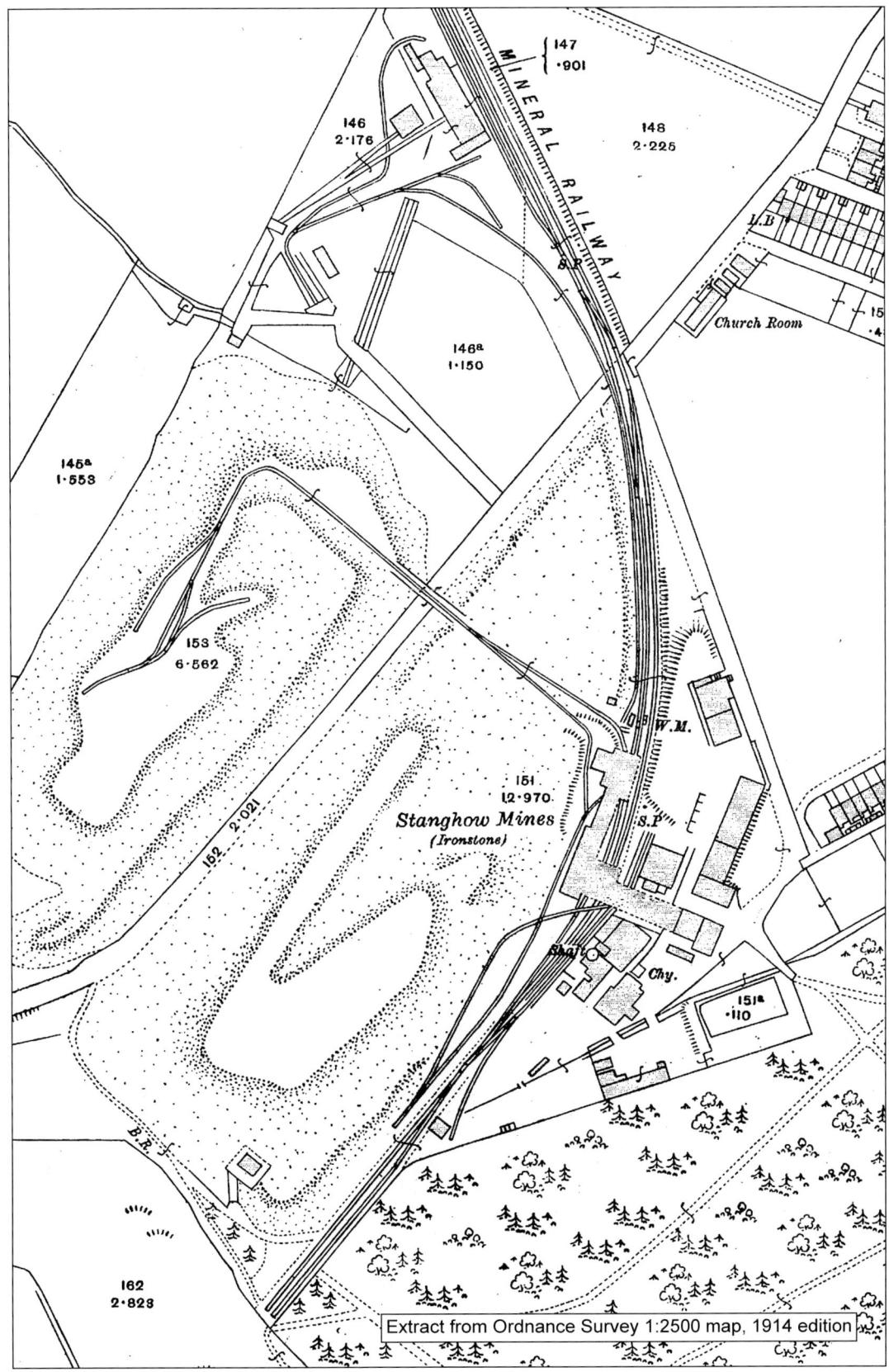

Extract from Ordnance Survey 1:2500 map, 1914 edition

GEORGE COHEN, SONS & CO LTD

COBORN WORKS, Cargo Fleet
46
NZ 511206 – **Map G**

This well known company of machinery merchants and scrap dealers was responsible for dismantling various large works around the UK. Following the closure of the Tees Iron Works in 1959, Cohen purchased the premises in 1960. It was negotiating with BR on 5/9/1960 about the likely rates for scrap being forwarded to Cochranes and the MOR tolls; an agreement operated from 22/11/1960. During the next year Cohen demolished some of the Tees Iron Works and was stated in 6/1961 to be using a diesel locomotive. Following the removal of the main plant, Cohen retained the rail connection to the main line and some buildings. A weighbridge checked wagon loads leaving the site. A series of separate lines then ran from the weighbridge towards the river. Cohen cut up some BR steam locomotives at its yard; the most notable being the A4 class Pacific 60002 SIR MURROUGH WILSON which arrived from store at Heaton on 1/10/1964 and was dismantled 10/4//1965. Also scrapped were ten BR Class J94 Austerity saddle tanks in 1965. A number of local industrial diesels finished their careers at the yard and these are listed below, although some would have seen occasional use until expenditure was required on them. Examples of payments received from Cohen for MOR tolls on its traffic for the quarter ending 31st March were £135.78 (1973), £297.23 (1976) and £257.30 (1977). When amounts of scrap declined in the Teesside area, the yard closed in early 1981. A BR note dated 3/2/1981 stated that, in view of the closure, no further charges were to be raised. The site was up for sale by 21/8/1981 and had been cleared of most scrap by the end of the year.

Gauge : 4ft 8½in

-	4wDM	RH	386871	1955	(a)	(1)
-	4wDM	FH	4011	1966	New	(2)
-	4wDM	RH	235514	1945	(b)	s/s c/1976*
-	4wDM	RH	402808	1956	(c)	s/s c/1976*
RICHARD HILL LIMITED No.4	0-4-0DM	JF	4210016	1949	(d)	Scr c8/1980
-	4wDM	RH	306088	1949	(e)	Scr c5/1979
-	4wDH	S	10017	1960	(f)	(2)

* One of these locomotives moved to Michael Baum & Co Ltd, Middlesbrough, c1975/76.

Following closure of Cohen's Coborn Works scrapyard, S 10017/1960 and FH 4011/1966 await removal to Readman Ltd's adjacent premises on 31/10/1981. Some of the former Tees Iron Works buildings visible here had been used by Cohen.
(IRS Douglas Johnson photograph)

(a) ex George Cohen, Sons & Co Ltd, Kingsbury Scrapyard, Warwickshire, c/1962, (spares order 2/10/1962 to Coborn Works).
(b) ex W. Robinson & Co Ltd, North Ormesby, c/1968.
(c) ex Barton Abrasives Ltd, Middlesbrough, c/1974.
(d) ex Richard Hill & Co Ltd, Middlesbrough, /1974.
(e) ex Barton Abrasives Ltd, Middlesbrough, c/1976.
(f) ex British Steel Corporation, South Teesside Works, 11/1977.
(1) to George Cohen, Sons & Co Ltd, Cransley, Northamptonshire, c4/1966.
(2) to W.G. Readman Ltd, Cargo Fleet, 11/1981; after 31/10/1981, by 19/12/1981.

CONNAL & CO LTD

MIDDLESBROUGH WHARF, Middlesbrough 47
Connal & Co until c/1904   NZ 486219 NZ 482217 – **Map F**

Connal & Co was closely linked with meeting the needs of the iron and steel industry. An advertisement in 1928 describes its activities as "Iron warehouse keepers, wharfingers and warehousemen. Goods stored in covered warehouses. Railway sidings along buildings". It occupied two separate strips of land in the Ironmasters district, each abutting the River Tees. The first map evidence appears in 1875 when an area of land, becoming wider as it reached the south bank of the Tees, was labelled "Cleveland Iron Store" although no railway lines were shown yet. This site (NZ 486219) lay alongside the western boundary of the Acklam Iron Works in the Ironmasters District. At this time, Glasgow's merchants were said to handle the most extensive iron trade in the world. William Connal & Co was based in Glasgow and, according to the reference, a Connal's warrant was "as safe for conversion from paper….as a Bank of England note". Connal purchased the Cleveland Iron Store in 1877 and, within three years, over 100,000 tons of iron was kept there. It was called Cleveland Iron Store Yard (No.1) or Acklam Iron Stores by 1892–93. A railway line from the NER's Old Town Branch entered the yard and divided into six approximately parallel sidings with space in between to store materials in the open. A single line also connected a couple of the sidings with the North-Eastern Steel Works to the west.

In 1881 Connal purchased a separate eight acre site (NZ 482217) from the OME, which was located between the North-Eastern Steel Works and Gjers Mills' Ayresome Iron Works. It was called either Cleveland Iron Stores (No.2) or the Ayresome Iron Stores in 1892-93. This yard was narrower, apart from a widening next to the river where Connal's wharf was situated. Three or four sidings ran along the length of the site and were linked

NANCY (AB1059/1905) began life at Connal's iron stores in the Ironmasters district.
(G. Alliez photograph, NRM collection)

to the NER Marsh Branch, but a six chain section of the connection was classed as MOR meaning that fees were payable for rail traffic worked over it. Outgoing traffic was taken by Connal's locomotives to the NER's sidings. Connal offered wharfage for businesses lacking their own facilities, for example, Warner paid Connal for "wharfage between 1901 and 1910". In 1887 Connal also installed a coal shipping spout at its wharf for bunkering ships, although it was then being used to load slag from the North-Eastern Steel Works. When faced with falling prices and to 'steady' the market, Teesside's iron producers put Cleveland pig iron into both their own and Connal's stores until trade improved; Connal's stock of iron amounted to 323,923 tons in 4/1887. In 1905, a seagoing steamer was used to transport iron from the iron producers' wharves to Connal's wharf which had been extended in 1900.

A letter dated 26/9/1917 confirms that Thos W. Ward Ltd had signed a Third Party Agreement for the use of Connal's Acklam Iron Stores site. It agreed to pay MOR tolls on all traffic, except on the "bear or skull" scrap being hauled by Dorman Long's engines in that company's wagons between Britannia and the Acklam Stores Site "after the traffic has been broken up". Connal's Ayresome Wharf had a length of approximately 410 feet and was equipped with one 7 ton, three 5 ton and one 3 ton electric cranes in 1926. According to the 1929 Dorman Long survey, Connal's premises were excluded from the Acklam Iron and Steel Works but the wharf was now called "Acklam Upper Wharf". A letter dated 23/10/1930 in MOR records states that the Ayresome Yard and wharf were now Dorman Long and they were subsequently absorbed into the Acklam Works site.

Reference : 'Middlesbrough's Wharves', Jenny and Geoff Braddy, *Cleveland Industrial Heritage* 41, pub. Peter Tuffs, 2017

Gauge : 4ft 8½in

Name	Type			No.	Year		
POLLY	0-4-0ST	OC	BH	421	1877	(a)	(1)
MAGGIE	0-4-0ST	OC	BH	528	1880	New	s/s
JEANNIE	0-4-0ST	OC	AB	232	1881	New	s/s
NANCY	0-4-0ST	OC	AB	1059	1905	New	(2)
KATIE	0-4-0ST	OC	MW	399	1872	(b)	s/s

Inness gives numbers for the locomotives – Nos.6/7/5/11/8 respectively – but it is not known if these were carried.

(a) new, painted black with gold lettering. To Black, Hawthorn & Co Ltd, Gateshead for repair and returned, /1883.
(b) ex T.J. Thomson & Son Ltd, Stockton; previously AYRESOME No.4 at Gjers Mills & Co Ltd.
(1) to T. Roddam Dent & Son Ltd, Dent's Wharf, Middlesbrough.
(2) to Dorman, Long & Co Ltd, Acklam Iron & Steel Works, Middlesbrough.

WILLIAM COOK, BLACKETT, HUTTON LTD

FOUNDRY, Guisborough   48
Blackett, Hutton & Co Ltd until /1989   NZ 612155 – **Map I**
Blackett, Hutton & Co until 18/12/1914
Messrs Blackett and Foster c/1895
Messrs Sutherst & Southron until /1895

The foundry was located adjacent to the west side of Guisborough Station goods yard. Passenger trains began to run along the branch from Middlesbrough to the terminus at Guisborough Station in 1854. Land between Providence Street and the railway was leased by John Sutherst from the Chaloner Estate in 1861 and, by 1867, a foundry and engineering works was operating on the site, no doubt as a response to the needs of the nearby ironstone mines. In the early 1890s Sutherst & Southron at its "Cleveland Steel & Iron Works" was manufacturing a range of equipment for both ironstone mines and collieries, including steel wheels, rollers, points etc, but both partners were declared bankrupt in 1895. The premises were taken over, initially by Blackett and Foster, to be followed by Blackett and Hutton. A NER siding ran by the works in 1914, while a single standard gauge line had been laid within the premises connecting the north and south ends. Blackett Hutton acquired a Robert Roger & Co steam rail crane in settlement of a bad debt in late 1915 and this was used to shunt wagons of scrap for the furnaces. Belgian yellow foundry sand also came in wagons which were taken on a siding to the back of the yard where there were sand and coal hoppers. A secondhand 5 ton steam rail crane by T. Smith & Sons (Rodley) Ltd, maker's number 11800 of 1933/34, was purchased in 1938 probably from Cleveland Bridge & Engineering Co Ltd.

Major improvements took place following the acquisition of Blackett Hutton by Glenfield & Kennedy Ltd in 1944, with the Light Foundry being constructed towards the southern end of the site; the Roger crane reportedly being recommissioned to assist with the construction. A new Ruston & Hornsby class 4BDS diesel shunter was

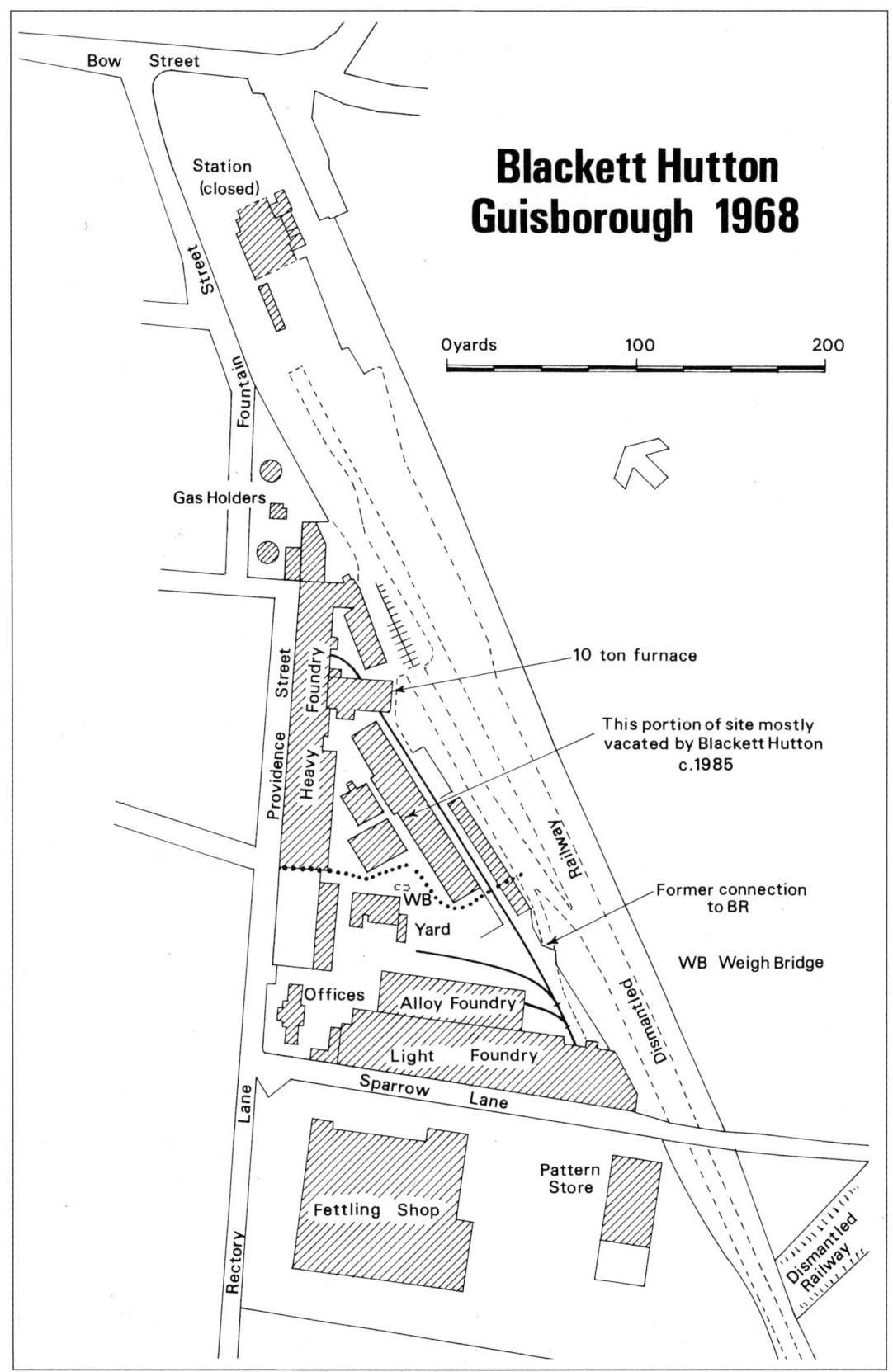

Loading 48DS class RH 265617/1948 on to a lorry at Blackett Hutton's foundry, Guisborough, prior to its move to the Aln Valley Railway for preservation.   (Courtesy Evening Gazette, Middlesbrough)

purchased in 1948. The old crane was disposed of, but the Smith Crane lasted in the stockyard until the 1960s when it was replaced by a road crane. The main tasks of the locomotive were to shunt a single flat wagon, with a cradle on a bed of sand, either transferring a 10 ton ladle of molten metal from the Heavy Foundry to the Light Foundry or a 3 ton ladle in the reverse direction. The locomotive was stabled in the Light Foundry. A fettling shop was constructed on land west of Sparrow Lane in 1960 and Crane Co Ltd of the USA took over Glenfield & Kennedy in 1966. Although the siding into the works from BR had shut years before the Guisborough Branch finally closed on 26/3/1965, the approximately 200 yards of single line between the two foundries continued to be used on a daily basis for more than three decades! In addition, there was a continuing need to move heavy castings and moulding boxes within the Light Foundry and Adrian Booth, during a 1985 visit, noted an 800mm gauge 4-wheel vehicle with an overhead wire pickup; three short lengths of 900mm gauge track; several lengths of 600mm track with 36 flat wagons and a small isolated 400mm system. In 1989 Blackett Hutton became part of the William Cook group. The locomotive was still working in 1996, the last active standard gauge 48DS at an industrial site in the UK, although three were still operated by track maintenance contractors. The erection of a new furnace west of Sparrow Lane resulted in the abandonment of much of the original site. The last charge into the old furnace took place on 15/3/1997 and this marked the final duty of the locomotive.

References :   'The Blackett Hutton Foundry' J. McConnell, *The Cleveland Industrial Archaeologist* 12, CIAS, 1980, pp.23–27

'Last Outpost of the Ruston 48 DS Class', Cliff Shepherd, *Industrial Railway Record* 149, IRS, 1997 pp. 323–33

Gauge : 4ft 8½in

-	4wDM	RH	265617	1948	New	(1)

(1)   to Aln Valley Railway, Alnwick, Northumberland, 13/2/1998. It was loaded on to the lorry at Blackett Hutton on 10/2/1998 (a photograph appeared in the *Evening Gazette* on 11/2/1998).

R.W. CROSTHWAITE LTD

UNION FOUNDRY, Thornaby
(subsidiary of **Allied Ironfounders Ltd** from 21/3/1929)

49
NZ 452189 – **Map E**

Crosthwaite's business was first established at Falkirk in 1849 and it was not until 1878–79 that the Union Foundry was built at Thornaby. This occupied a rather constrained site between Head Wrightson's Teesdale Works and Whitwell's Thornaby Iron Works, north of the Darlington-Middlesbrough railway. It manufactured a range of domestic fittings, such as stoves, heating apparatus and its "Thornaby" patent grate, together with troughs, railings, rainwater goods, stable fixtures and architectural castings. The iron was melted in a cupola furnace and then taken in "bogie–ladles" on railway lines to various parts of the works. In the 1890s the company enjoyed a strong export trade, had London showrooms and employed over 300 people. A single siding ran off the Thornaby Iron Works railway over an area tipped with slag before crossing at right angles Whitwell's line from its blast furnaces. The siding divided into two to enter the foundry site. Much of the three acre premises was occupied by "several ranges of well-constructed buildings of one and two storeys". Crosthwaite held a perpetual wayleave over Whitwell's railway paying 1d per ton on all in and out traffic. According to a *Guide to the Ports on the Tees & the Hartlepools*, published in 1900, the Union Foundry then occupied an area of about 10 acres and had a siding connecting the NER "direct with the works, the firm owning the powerful locomotive used for drawing materials and goods in all parts of the premises". The extent of the works had expanded by 1915, although with only a few additional lengths of sidings. By the 1930s, it was making cylinders for motor car engines. Despite the subsequent demolition of much of the Thornaby Iron Works leaving a vacant space to the east, the layout of the Crosthwaite's buildings and sidings in 1963 remained much the same as 1915, with a single track running from the Tees Marshalling Yard past Whitwell's derelict coal washing plant and into the premises. However, the 1951 agreement specifying BR's lines that could be used by Crosthwaite's locomotive was cancelled w/e/f 6/10/1965. The foundry closed in 1966 with work being transferred to the Bonlea Foundry.

Gauge : 4ft 8½in

R.W.C.	0-4-0ST	OC	I'Anson		1875	(a)	(1)
CROSTHWAITE	0-4-0ST	OC	P	1605	1922	(b)	
	Rebuilt	Ridley Shaw			1947		Scr c/1965

Charles I'Anson of Darlington only built two or three locomotives and one of them came to Crosthwaite's Union Foundry in Thornaby. It left for Thomson's scrapyard in 1927 and was photographed at Stockton by Bernard Stoyel on 18th July 1933. (ILS collection)

(a)   ex Charles I'Anson & Co Ltd, Rise Carr Iron Works, Darlington, Co. Durham.
(b)   ex Charles Roberts & Co Ltd, Horbury, Yorkshire (WR). It was noted at the Union Foundry on 30/4/1934.
(1)   to T.J. Thomson & Son Ltd, Stockton, /1927.

DAVY & UNITED ROLL FOUNDRY LTD
EMPIRE WORKS, Haverton Hill   50
(Subsidiary of **Davy & United Engineering Co Ltd**, later **Davy-Ashmore Ltd**)
British Chilled Roll & Engineering Co Ltd until /1937   NZ 480222 – **Map C**
Hemingway's Chilled Rolls Ltd until /1904

The works was located between Chilton's Lane (later Haverton Hill Road) and the NER Billingham Beck Branch, the northern portion of which had opened in 1901. Hemingway's purchased a secondhand Thornewill & Warham locomotive from Bass, Ratcliffe & Gretton's brewery at Burton on Trent to shunt the foundry, but soon ran into difficulties and the works closed about 1904 with the engine subsequently being sold. *The Glasgow Herald* on 5/5/1906 announced an auction on 15–17/5/1906 of plant including a 14 inch locomotive re Hemingways Ltd (in liquidation), bridge builders, forging mills, at Haverton Hill.

The premises were taken over by the British Chilled Roll & Engineering Company in 1907, which specialised in manufacturing rolls and rolling mill machinery for iron and steel works. It had a small number of sidings serving the main factory building and these joined with the NER branch near the connections to the Haverton Hill Brick Works and, a little further south, the Casebourne Cement Works. The company does not appear to have owned a locomotive for some time because Ted Haigh was told that the Hudswell Clarke was the first. In 1937 Davy & United Engineering Co Ltd of Sheffield acquired the works and a special steel was substituted for iron in the manufacture of rolls, with steel for casting being produced from a small open hearth furnace. Later the foundry made carbon and alloy steel castings, electrically melted, up to 12 tons finished weight. Rolls were also produced for structural plate and continuous strip mills. By 1962, the buildings occupied most of the site but there had been little change to the internal railway. The last locomotive was Manning Wardle 1969 from the Haverton Hill Shipyard. This came in a green livery with yellow lining and the number 2 painted on the back of the cab. Rail traffic probably ceased about 1967 although the sidings were still there in 1972–74.

Manning Wardle maker's no.1969 in steam at Davy & United Roll's Foundry on 24th June 1961. The locomotive had started life as a new engine in 1918 at the Furness Shipyard, Haverton Hill.
                                                                    (IRS C.B. Golding photograph)

Gauge : 4ft 8½in

-	0-4-0ST	OC	TW			(a)	(1)
-	0-4-0ST	OC	HC	656	1903	(b)	(3)
W. SHAW & Co LTD No.3	0-4-0ST	OC	RS	3056	1904		
		Rebuilt	Ridley Shaw		1917,1940	(c)	(2)
No.2	0-4-0ST	OC	MW	1969	1918	(d)	(4)

(a) ex Bass, Ratcliffe & Gretton Ltd, Burton on Trent, Staffordshire. The only 14in Bass locomotive, whose fate is not known, is the company's No.9 (TW 425 of 1877) which was probably replaced in 1901.

(b) ex Joseph Pugsley & Sons Ltd, dealer, Bristol, by 4/1939. Previously owned by C.J. Wills & Sons Ltd and used on its Becontree Contract for London County Council (1920-34) before being disposed of to George Cohen, Sons & Co Ltd at Canning Town. To ? for repairs, including new tubes, c/1947 and returned. It was noted back at the Empire Works on 12/6/1948.

(c) ex W. Shaw & Co Ltd, Middlesbrough, on hire, c/1947, by 24/5/1947.

(d) ex Furness Shipbuilding Co Ltd, Haverton Hill, via Ridley Shaw, /1950, by 9/7/1950.

(1) sold c/1906.
(2) returned to W. Shaw & Co Ltd, Middlesbrough, ex hire, c/1947.
(3) to T.J. Thomson & Son Ltd, Stockton, for scrap, /1950.
(4) to T.J. Thomson & Son Ltd, Stockton, for scrap, /1967, by 2/1967.

A.V. DAWSON LTD

PORT OF MIDDLESBROUGH

51

NZ 491213, NZ 488213, NZ 493215 – **Map F**

Vernie Dawson began as a coal merchant in 1938 using a horse and cart but, in 1944, his first lorry was purchased and the business diversified into road transport. It became A.V. Dawson Ltd in 1954 and a two acre site in Lloyd Street, Middlesbrough was purchased in 1965 to house the growing lorry fleet. By the 1970s, the company owned 14 acres of land next to the River Tees and rented an adjoining five acres of Middlesbrough

Another Branch Line Society's tour of Dawson's railway system took place on 21st July 2018 and 08774 pulls away from Middlesbrough Wharf and crosses Depot Road. This is approximately where the original SDR line had run to reach the Port Darlington coal drops in 1830. *(Cliff Shepherd photograph)*

Dawsons' 08410, built by Derby Works for BR in 1958, pulls loaded wagons from Boulby Mine across Forty Foot Road into the company's Ayrton Railfreight Terminal on 17th May 2022. The dark green livery was a reminder that it had been purchased from First Great Western's Penzance depot in 2020.

(Cliff Shepherd photograph)

Wharf from the THPA; the wharf was purchased in 2002. Initially Dawson concentrated on converting some of the redundant buildings into industrial units but, in 1985, it reopened Middlesbrough Wharf (NZ 493215) with a view to operating road/rail/sea interchange facilities. This involved bringing the single line extending from Middlesbrough Town Goods over Depot Road back into use. The small secondhand Husky diesel was purchased in 10/1986 to replace a road tractor.

Dawson could see a greater market for this activity and in 1986 purchased the 60 acre Ayrton site. This had been vacated by BSC in the previous year following its use to produce fabricated steel goods; earlier it had been Dorman Long's Ayrton Sheet Works. Dawson renovated the buildings to act as a railhead for northern markets and, in 1987, relayed the railway line from Middlesbrough Town Goods across Forty Foot Road to serve the Ayrton Rail Freight Terminal (NZ 488213). Within the premises, the line passed over a weighbridge before dividing into two – one track ran into a warehouse and the other went through a small discharge shed, with a loop to bypass it, before terminating at the rear of the warehouse where a conveyor allowed goods to be transferred from road to rail. The official opening took place on 26/10/1988. A more powerful locomotive was required and a Class 08 diesel electric shunter was purchased from BR's Thornaby Depot and repainted in Dawson's house colours of red with white lining. The Store received potash from Boulby Mine, replaced by polyhalite in 2018 which is also used as a fertiliser.

The 1990s saw further development of the business. The refurbished Dawson No.1 Quay provided a 120 metre berth from which long steel sections were shipped to Antwerp and Vlissingen. The No.2 Wharf came into operation in early 1992. The adjacent 22 acre Linthorpe Dinsdale Offshore Fabrication Yard, with its 45 metre high fabrication hall, was purchased from Kaverna in 1998 to become the North Sea Supply Base Wharf (NZ 491216). With the retrenchment in the steel industry, there has been only limited rail traffic to the river wharves in recent years with most activity concentrated on the other facilities.

A later rail development by Dawson has been on land between Forty Foot Road and Middlesbrough Town Goods Yard. A rail terminal was opened in Summer 2012 called Tees Riverside Intermodal Park (TRIP) and its two new sidings can deal with trains carrying up to 80 containers; there is space for on-site storage of 1,200 containers. A new £6 million Automotive Coil Store building (NZ 491213) also opened about 9/2014. The huge 300m x 60m structure has the capacity to store 32,000 tonnes of steel coil and is supplied by a daily train from Tata's Port Talbot Works. Much of the steel is destined for Nissan's Sunderland car plant. The building is located at the north end of the Town Goods yard. Both developments are usually shunted by main line train locomotives. On 19/11/2014 RSHN 7900 was moved to Middlesbrough Wharf for repainting and was placed on a plinth outside the then new offices south of Depot Road on 6-7/1/2015. Dawson signed a 99 year lease with Network Rail in 2015 to manage the seven acre Middlesbrough Town Goods Yard (now known as Middlesbrough Goods Yard) that connects with its various facilities and some track relaying was carried out. In late 8/2015, Bg/DC 2725 was repainted and placed on a plinth (NZ 491214) in memory of Eleanor Dawson, one of the founders of the company.

The Branch Line Society ran a tour over Dawson's railway on 28/6/2017, topped and tailed by 08598 and 08600. This included the line to the quayside; possibly the first passengers to be conveyed over this route since 5/6/1846 when the SDR station at Commercial Street was replaced by a temporary station on the SDR's line to Redcar. Dawson's increasing range of facilities led to the construction of larger offices north of Depot Road in 2021 and its various facilities being named "Port of Middlesbrough" from 17/9/2020.

Gauge : 4ft 8½in

(7900)	7900/58	0-4-0DM	RSHN	7900	1958	(a)	
(08774)	(D3942)						
ARTHUR VERNON DAWSON		0-6-0DE	Derby		1960	(b)	
	ELEANOR DAWSON	0-4-0DM	Bg/DC	2725	1963	(c)	
	LOC 1	4wDM	RH	421419	1958	(d)	(1)
08600	(D3767)	0-6-0DE	Derby		1959	(e)	
08807	(D3975)	0-6-0DE	Derby		1960*	(f)	(2)
08912	(D4142)	0-6-0DE	Horwich		1962*	(g)	
08598	(D3765)	0-6-0DE	Derby		1959	(h)	
08410	(D3525)	0-6-0DE	Derby		1958	(i)	
1	TWEETY	4wBE	Niteq	M2013	2007	(j)	

* obtained for spares

(a) ex Vulcan Materials Co (UK) Ltd, Hartlepool, 10/1986.
(b) ex BR, Thornaby Depot, 9/1988. To Cobra Railfreight, Middlesbrough, on hire,10/1998 and returned /1998. To Wabtec Rail Ltd, Doncaster, South Yorkshire for overhaul, late /2001 and returned 22/2/2002. To Moveright International/Rail Support Services Ltd, Wishaw, for repairs 26/10/2017 and returned 13/7/2018 via GCR (N), Ruddington, Nottinghamshire.
(c) ex Rea Bulk Handling Ltd, Bidston Ore Dock, Wallasey, Merseyside, c/1990, by 25/9/1990.
(d) ex North Yorkshire Moors Railway, Grosmont, North Yorkshire, by 20/7/1989.
(e) ex EWS Eastleigh, Hampshire, 17/11/1997. To EWS Thornaby Depot for tyre reprofiling, 14/6/2004 and returned 1/7/2004. To LH Group Services Ltd, Barton-under-Needwood, Staffordshire for repairs, 14/10/2008 and returned 23/3/2009.
(f) ex T.J. Thomson & Son Ltd, Stockton, 30/10/2007 via EWS Thornaby Depot for tyre reprofiling and arrived 23/11/2007; previously EWS Motherwell Depot until 26/4/2007.
(g) ex T.J. Thomson & Son Ltd, Stockton, 1/2/2008 via EWS Thornaby Depot for tyre reprofiling and arrived 20/3/2008; previously EWS Toton Depot until 16/3/2007.
(h) ex Chasewater Light Railway & Museum Company, Brownhills, Staffordshire (where property of Ed Murray & Sons Ltd and purchased by A.V. Dawson), 4/5/2017. To R.S.S. Ltd, Wishaw, Warwickshire, for repairs, from 19/7/2019 and returned via Tyseley Diesel Depot, Birmingham for tyre turning, 17/1/2020.
(i) ex First Great Western Ltd, Penzance Long Rock Depot, Penzance, Cornwall, 20/2/2020, arrived 27/2/2020.
(j) ex BEAZ Solutions, Netherlands via Multi-Movers UK (agent) on demonstration, 10/3/2022.
(1) returned to North Yorkshire Moors Railway, Grosmont, North Yorkshire, c8/1989.
(2) to C.L. Prosser & Co Ltd, Cargo Fleet for scrap, 7/1/2017.

THE DEEPWATER WHARF LTD

Cargo Fleet 52
(Subsidiary of **Pease & Partners Ltd** by /1946 and **T. Roddam Dent & Son Ltd** by 1/1957)
The Harris Deepwater Wharf Co Ltd until 1/9/1917   NZ 516207 – **Map C**

The wharf and its associated sidings were situated between the Middlesbrough–Redcar line and the River Tees, with Ormesby Beck and Normanby Beck forming its west and east boundaries respectively. Prior to the wharf being established, this land belonged to John Brown, owner or part owner of the Ormesby estate. At some date, it passed to Miss Elizabeth Caroline Brown. By 1875 a single siding entered the southern half of the site serving a warehouse. The siding joined the main line immediately east of the old Cargo Fleet Station. The latter had opened in 4/1847 and was originally called "Cleveland Port". It was replaced by a new Cargo Fleet Station approximately ¼ mile to the west on 9/11/1885. "Brown's Wharf" is shown on the bank of the River Tees in 1882, but there had been little development of sidings. According to the 1892–93 Ordnance Survey of the site, the river wharf was shown as disused and Cochranes had constructed a siding over Ormesby Beck to tip slag on part of the site next to the main line.

There was, however, one other railway that crossed over the site. The MOR line from Middlesbrough Dock Lock Bridge ran past the Ormesby Iron Works to terminate at the western boundary (Ormesby Beck). In order to traverse the site and reach the Normanby Iron Works, the OME agreed to pay a wayleave of ½d per ton on all rail traffic crossing the property unless it was going to the wharf. It was reported to the OME directors on 13/5/1874 that J.W. Pease had agreed with Miss Brown on straightening the east and west boundaries by land exchanges and approval for the railway passing over her land. In 1948 an approach to increase the ½d rate for passing over "Brown's Committee land" was rejected by the railway company. Nevertheless an average sum of £61 was paid annually for traffic that had passed over the line in the seven years up to the end of 1945.

About the turn of the century (the Siding Agreements with the NER were dated 10 and 15/8/1900), The Harris Deepwater Wharf Co Ltd transformed the site by constructing a 300 feet wharf on the bank of the River Tees – in 1926 this had two 5 ton and two 7 ton electric cranes. Behind the wharf, seven railway lines terminated at right angles to the crane track. Moving south, each of these lines divided into groups of 3-4 sidings before coming together to connect to the main line without incurring MOR tolls. This open storage was mainly for iron ores, plates, angles and timber; there was standing room for 200 wagons in 1926. Initially it appears that the NER worked traffic from the sidings to the wharf for a fee. A wooden single road engine shed was provided next to Normanby Beck for the locomotives later owned by the company, which were recorded as being painted black with red motion in 1950. Presumably this development had been stimulated by the increasing use of imported iron ore by the local iron works and was operational by 6/11/1901 when the Managing Director of the company wrote to the Tees Union Shipping Company encouraging resistance to any increase in MOR tolls. The NER District Goods Manager, Darlington had also proposed from 1/1/1902, under the MOR regulations, that locomotives or wagons exceeding one ton per linear foot should not cross Dock Lock Bridge. The Harris Deepwater Wharf Co objected because it would mean private locomotives that had been using the line for many years would be barred. It wrote to the Tees Union Shipping Co seeking support claiming it was a contravention of the 1884 agreement. Private locomotives and traffic continued to use the line for a considerable time.

According to a NER letter dated 10/9/1917, the name of the trader "should be altered to read Lord Furness, the wharf having been taken over by him from the 1st inst.". This prompted the change in name to The Deepwater Wharf Ltd (noted by the NER 11/4/1919). After World War 2 and certainly by 12/1/1946, Pease & Partners Ltd had purchased the wharf, but retained its existing business name; Pease & Partners acting as a holding company. In the quarter ending 30/9/1946, 6,621 tons of iron ore was unloaded at the wharf and despatched to the Tees and Normanby Iron Works. Some was also taken by the LNER to the Skinningrove Iron and Steel Works. On 9/1/1957 Ted Haigh wrote that Pease & Partners Ltd had sold the wharf to T. Roddam Dent & Son Ltd; BR noted the change on the following day commenting that the wharf would still trade as The Deepwater

Hawthorn Leslie 2730/1907 was outside its rather dilapidated wooden engine shed at the Deepwater Wharf on 19th August 1956. *(RCTS J. Faithfull photograph)*

Wharf Ltd title, but was under Roddam Dent's personal management. It is likely that the site had been acquired to reduce the number of competing wharves on the Tees and it is no surprise that the two locomotives were transferred to Roddam Dent's main wharf in 1958.

However, it is likely that some final work for a locomotive was still required at Deepwater Wharf and it was only a short journey by rail between the two wharves. The IRS has a note that HL 2662/1906 was loaned from Dent's Wharf in 8/1958 and J.T. Clewley photographed HL 2730 now lettered "RODDAM DENT & SON LTD No.8" at the Deepwater Wharf shed on 8/5/1960.

On 30/11/1964 it was stated that M. Baum & Co Ltd was using the Deepwater site on a temporary basis to break up condemned wagons. Some of these had been hauled by one of Baum's locomotives over the MOR from his Cleveland Dockyard site. Scrap was also being despatched by rail from the Deepwater site. Baum then wrote to BR saying that he expected to vacate the site in 3-4/1965. With the formation of the Tees & Hartlepool Port Authority on 1/1/1967, the wharf became its responsibility and BR recorded that the MOR tolls payable by the authority would be increased from 1/2/1968, but it is unlikely that there was any traffic to incur such payments.

Gauge : 4ft 8½in

No.1	NUNTHORPE	0-4-0ST	OC	HG	252	1869	(a)	Scr
No.2		0-4-0ST	OC	JF		1880	(b)	s/s
No.1		0-4-0ST	OC	HL	2730	1907	New	(1)
No.2		0-4-0ST	OC	AB	1196	1912	New	(1)

(a) ex Tees Side Iron & Engine Works Co Ltd, Tees Side Iron Works, Middlesbrough.
(b) origin unknown.
(1) to T. Roddam Dent & Son Ltd, Dent's Wharf, Middlesbrough, /1958.

DORMAN, LONG & CO LTD

Arthur John Dorman (1848–1931) was born in Kent and had set up as an iron merchant in 1872. He became managing director of the small West Marsh Iron Co in 1874. Meanwhile, Albert de Lande Long was employed at Whitwell's Iron Works. The two men joined together to lease the West Marsh Works in 1876. The partners then leased the much larger Britannia Works from Samuelson in 1879. Arthur Dorman was one of the people involved in establishing the North Eastern Steel Co Works in 1881 to produce basic steel. Middlesbrough's Ironmasters District continued to be the principal focus of Dorman Long's activities. On 2/11/1889, Dorman, Long & Co Ltd was formed with the balance of shares offered on the following 8th. Arthur Dorman's brother had taken over the Ayrton Sheet Works but, on Robert's death, this also came to Dorman Long. More shares were issued in 1899 enabling Dorman Long to acquire a 50% interest in Bell Bros which, in addition to the Clarence Works, owned collieries and quarries in Co. Durham. The balance of the company was acquired in 1902 and Bell Bros then functioned as a Dorman Long subsidiary. Further acquisitions followed – the Acklam Works, including the North Eastern Steel plant, in 1903, the Redcar Iron Works (1916), Newport Iron Works (1917) and the Carlton Iron Works at Stillington in 1920. Dorman Long took over Bolckow Vaughan in 1929 creating the largest industrial concern in North East England.

The subsequent history of Dorman Long is covered in the British Steel Ltd entry, but to understand those sites which ceased using railways or closed during the Dorman Long tenure, they are divided into two groups pre 1954 and 1954 to 1967. Dorman Long & Co Ltd became the holding company in 1954 with Dorman Long (Steel) Ltd being responsible for iron and steel production until nationalisation in 1967.

BELMONT IRONSTONE MINES, Guisborough   **53**
NZ 625148 – **Map l**

There were three different workings taking ironstone from the Chaloner royalty below the hills south of Guisborough.

Belmont Mine
Weardale Iron & Coal Ltd, The until 11/11/1886
Weardale Iron Co, The until 23/7/1863

The Middlesbrough & Guisborough Railway was promoted by the SDR to reach the ironstone reserves south of the town. The line was authorised in 1852 and opened for mineral traffic on 11/11/1853. The Weardale Iron Co leased a substantial area of ironstone in 11/11/1853 with the intention of supplying its iron works in Co. Durham. The mine was established on the hill slope about 200 feet above the town and comprised quarries where the ironstone outcropped and drifts that followed the Main Seam underground. The first load of raw ironstone was despatched in 5/1855, although much of the stone was later calcined in open clamps near the mine to reduce the cost of transporting it to Co. Durham. The Act for the Middlesbrough & Guisborough Railway included provision for a private railway from the mine down a self acting incline to Guisborough Station Yard; published accounts differ as to whether it was standard or narrow gauge. When the Cleveland Railway was built, it crossed over this incline and a cart track from Belmangate by a three arch stone bridge. The Weardale Iron & Coal Co also commenced taking ironstone from the east end of the royalty at Spawood Mine (which see) in 3/1864. According to the *Northern Echo* on 8/12/1880, Belmont Mine was being worked by two main drifts – Belmont West extending 1½ miles in a south west direction and Belmont East, about ½ mile long. Belmont Mine suffered various periods of closure and abandonment took place on 11/11/1886.

South Belmont Mine
Weardale Iron & Coal Co Ltd, The until 7/1876
The North of England Industrial Iron & Coal Co Ltd until 11/1875
Henry King Spark and others until 4/1870

As the size of the original lease proved too great a strain on the Weardale Iron Co, it sublet 400 acres to Henry King Spark and others on 8/11/1861 and output from South Belmont commenced in 3/1862. Spark was an entrepreneur and he sold the lease to Henry and Archibald Briggs, West Riding colliery owners, who formed The North of England Industrial Iron & Coal Co Ltd on 13/4/1870. The latter also purchased the Carlton Iron Works at Stillington where two blast furnaces were blown in on 16/6/1871. By 11/1874 the Briggs were looking to surrender the lease on the ironstone because its quality did not justify the royalty rent. In 11/1875 the sublease expired and the mine reverted to the Weardale Iron & Coal Co which continued to provide a separate outlet from the mine until 7/1876.

Later Belmont Mine NZ 616145
Bolckow Vaughan & Co Ltd until 29/11/1929

Bolckow Vaughan was looking for additional supplies of ironstone at the beginning of the twentieth century. The lease of ironstone from Richard Chaloner was dated 18/3/1908 and in 1908-09 Bolckow Vaughan constructed a railway from Hutton Gate Station on the Middlesbrough-Guisborough line to the new mine. The single line branch ran alongside the NER for 300 yards before swinging south east mostly on an embankment; the junction being controlled by the three levers of Belmont Mines Ground Frame (dated 1908). The red brick mine buildings were constructed at the foot of the hills immediately south of Hunter Hill Farm. Near the mine, the branch divided into three tracks, one of which passed under the tippler house. An agreement was signed with the NER on 23/7/1909 for the sidings at Hutton Gate. The first ironstone was despatched in the week ending 7/8/1909. The Main Seam outcropped about 150 yards up the hillside and a drift was put in here with a 400 yards long self-acting narrow gauge railway incline moving sets of 18 tubs to the tipping point over the standard gauge siding. The drift ascended underground nearly half a mile to meet the main engine plane of the old drift. Initially the shale was mostly stowed underground but, with increasing amounts as the workings moved south, a picking belt (constructed by Head Wrightson) and an aerial ropeway (by White & Sons, Widnes) were installed in 1914–15; the ropeway heading westwards to a tip.

Bolckow Vaughan continued to improve the mine particularly extending the use of electricity. It agreed on 28/10/1919 to purchase an electric storage battery for an underground locomotive. *The Iron and Coal Trades Review* on 11/6/1920 contained an illustration of an electric battery locomotive developed by British Electric Vehicles Ltd. They were called 'pony locomotives' because they were intended to replace pit ponies. Eight were manufactured in 1919–20 for the Ebbw Vale Co's Irthlingborough Mines in Northamptonshire. Each was designed to haul 10-15 tons on the level and about 3 tons up a 1 in 30 gradient. Bolckow Vaughan purchased a Type No.1 from BEV on 18/8/1920 and it was probably intended for face work because the charging station was situated well underground. It was not successful reportedly derailing too easily on tight curves. Belmont Mine closed on 31/1/1921 when some work had been taking place to widen the 'roads' in order to increase the railway track gauge by 6 inches. Also a specification survives for new tubs for Belmont with tapered steel sides, a track gauge of 2ft 9½in and a capacity of 36cwt. However, Belmont Mine never reopened, Dorman Long terminated the lease on 30/9/1932 and abandoned the mine on 2/1933. The aerial ropeway was transferred to the Cleveland Works and the picking belt to South Skelton Mine.

Reference : *Guisborough District Mines*, Simon Chapman, pub. by Peter Tuffs, 2001

Gauge : Narrow

-	4wBE	BEV	257	1920	(a)	s/s

(a) new, BEV/WR maker's list says supplied to Bolckow Vaughan & Co Ltd, Middlesbrough, 18/9/1920.

CARLTON IRON WORKS, Stillington
Carlton Iron Co Ltd until 2/5/1923
(Company formally wound up 15/5/1935)
The North of England Industrial Iron & Coal Co Ltd until 1/5/1877
S. Bastow & Co until 4/1870

54
NZ 373237 – **Map D**

The works was established on the north side of the Ferryhill–Stockton railway line, which had recently become part of the NER, near the small village of Stillington. Samuel Bastow already owned the Cliff House Iron Works at West Hartlepool and he erected two blast furnaces in 1866-67 on a 38 acre site. Operations began in 6/1866 by S. Bastow & Co, but financial difficulties resulted in the furnaces being taken out of blast in the following year. Bastow was declared bankrupt in 1869. Henry King Spark, a Darlington entrepreneur, with others purchased the rather dilapidated iron works at a knockdown price. He went on to sell it, together with the South Belmont ironstone mine, near Guisborough, to Henry and Archibald Briggs, West Riding colliery owners, who formed The North of England Industrial Iron & Coal Co Ltd on 13/4/1870. Reconstruction of the works commenced in 7/1870. The blast furnaces were increased in height from 60 to 80 feet, their brickwork was renewed, a second blast engine ordered from Cochrane, Grove & Co and eight Whitwell hot blast stoves were erected. The blast furnaces were blown in on 16/6/1871 and a third was added in 12/1872. A building housing the Danks rotary puddling furnaces was constructed in 1873, to be followed by two adjoining plate mills in 1875. The name of the business changed to the less cumbersome title of the Carlton Iron Company in 1877. It acquired East Howle Colliery in 1872 and this was superceded by Mainsforth Colliery at Ferryhill in 1905. A briquetting plant using slag was set up in 1886 and coke ovens were erected immediately north of the blast furnaces in 1888, although these were renewed to Semet-Solvay ovens (25 in 1896 and another 35 in 1899).

The private railway trailed off from the exchange sidings into the premises east of Stillington Station (this was originally a halt built to serve the works in 1870). The NER running lines were increased from two to four in 1884, a new siding agreement being signed with the Carlton Iron Co on 17/9/1884 and the exchange sidings were enlarged in 1915–16. On entering the iron works, the railway divided into several sidings, two of which climbed a ramp to run alongside the blast furnaces in order to bring in minerals. Other lines ran eastwards, either side of the engine shed, taking pig iron and slag from the base of the furnaces. A fan of sidings served the massive slag heaps on the eastern half of the site and there were lines to the wrought iron works and coke ovens.

Manning Wardle maker's no.60 had once been on display at the 1862 Great Exhibition but is seen towards the end of its days at the Carlton Iron Works on 2nd November 1924. *(NERA R. Inness photograph)*

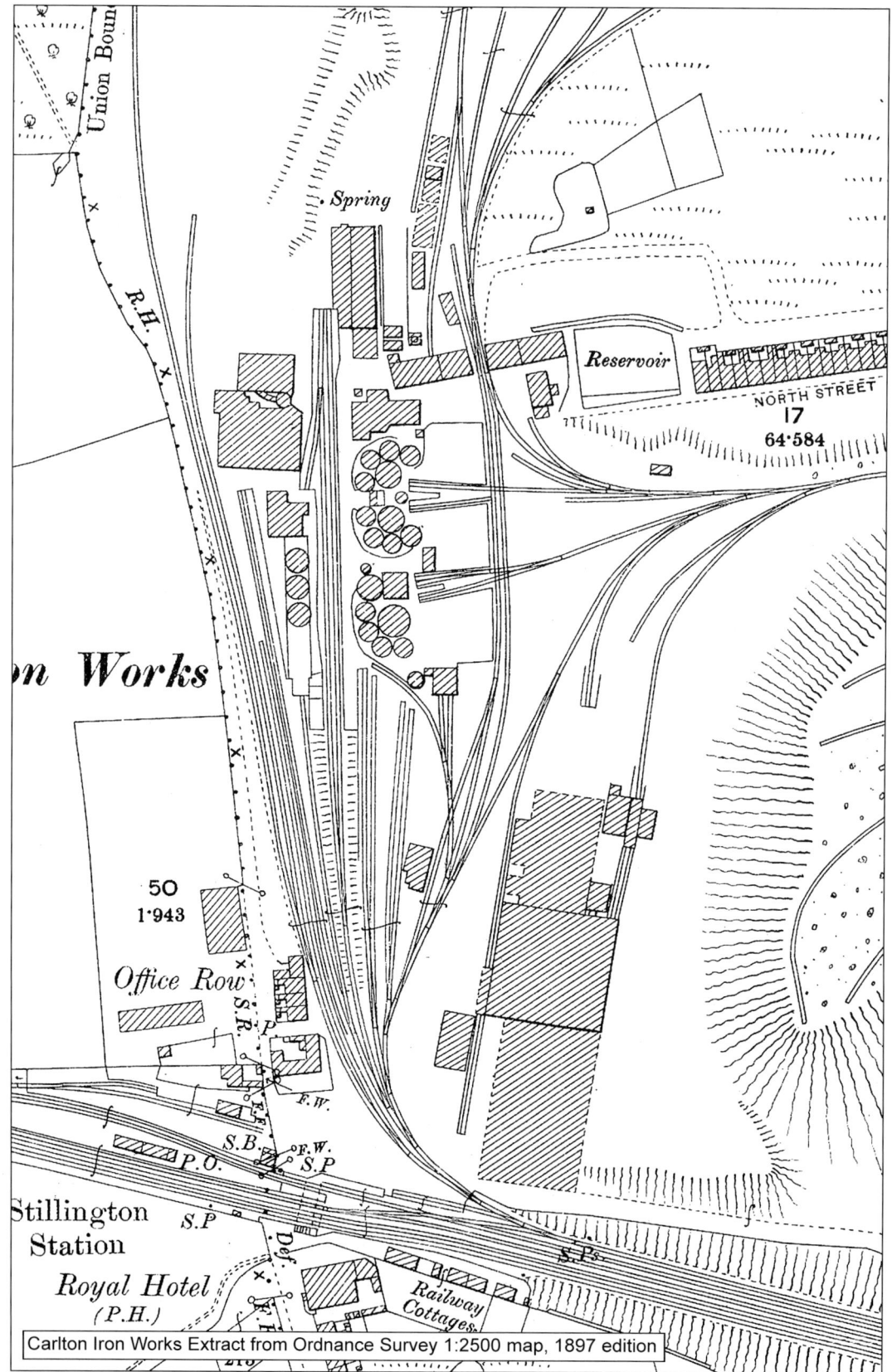

Carlton Iron Works Extract from Ordnance Survey 1:2500 map, 1897 edition

The business was restructured financially in 1914, going into voluntary liquidation on 25/6/1914; a new company with the same name being registered on 7/7/1914. It was taken over by Dorman, Long & Co Ltd on 1/1/1920 and two Dorman Long directors sat on the Board; the attraction being the coal from Mainsforth Colliery rather than the iron making capacity. With the difficult trading conditions of the inter-war years, Dorman Long transferred ferro-manganese production to its South Bank Works and the last furnace was blown out in 1/1930, with the Carlton Iron Works closing in 3/1930 and mostly being dismantled by Hughes, Bolckow Co Ltd of Blyth by 1931.

A number of other companies were involved on the site and it is likely that any rail traffic for them was shunted by the iron works locomotives until 1937. The wrought iron plant and plate mills had not been successful and, in 1887, the Basic Phosphate Co occupied part of the buildings to produce fertiliser using finely ground slag from the tip. It ceased operating in 7/1911 and Thomas Wilkinson leased these buildings in 12/1912 to begin manufacturing foundry blacking. In 9/1933 Dorman Long was charging Thomas Wilkinson & Sons for use of its railway sidings and for Dorman Long to shunt the company's wagons to and from the exchange sidings. The Stillington Slag Co operated a crushing plant alongside the briquetting building in 1914-37 producing road materials. In 1926 the North Eastern Iron Refining Co Ltd (which see) established its works at the former briquetting building and, from 1937, operated its own locomotive. It was paying Dorman Long in 1938 for the right to use the railway sidings to tip at the north end of the slag heap. In 1939 Stillite Products Ltd opened a factory, approximately where the coke ovens had stood, to produce slag wool. This functioned until 1965 when production moved to South Bank. The final duty of the last three Dorman Long locomotives was to move slag from the tip, but Stillite (which see) also operated a hand-worked railway.

References :   *Mining Journal*, 18/10/1873 p.1149

There was a Green Hill, The History of Stillington from its Beginnings until 1950, J.D. Tuffs, pub. by the author, 1999

Stillington in the Borough of Stockton-on-Tees, Industry, Church and School 1860 to 1991, Parish Map Community Programme Project, 1991

Gauge : 4ft 8½in

No.	Name	Type		Builder	Works No.	Date	Notes	Disposal
No.1		0-4-0	VCG	HG		?		Scr
No.2		0-4-0ST		Hopper		1870		s/s
No.5	COMPTE DE PARIS	0-6-0ST	OC	BH	166	1873	New	Scr after 21/8/1924
No.3		0-4-0ST	OC	MW	60	1862	(a)	Scr after 2/11/1924
-		0-4-0ST		HH		?		Scr
No.1	LOTTIE	0-4-0ST	OC	HL	2110	1888	New	
		Rebuilt				1906		(1)
	CARLTON No.7	0-6-0ST	OC	HL	2607	1905	New	(6)
	CARLTON No.1	0-6-0ST	OC	HL	2732	1907	New	(7)
	CARLTON No.4	0-6-0ST	OC	VF	422	1858		
		Rebuilt		Caerphilly		1884,1903	(b)	(2)
	CARLTON No.2	0-6-0ST	OC	AB	1404	1915	New	(3)
6		0-6-0ST	OC	?			(c)	Scr
18		0-4-0ST	OC	HL	3169	1916	(d)	(4)
37		0-4-0ST	OC	HC	324	1889		
		Rebuilt		Acklam		1926	(e)	(5)
43		0-4-0ST	OC	CF	1199	1900	(f)	(8)

(a)   ex Samuel Allsopp & Sons Ltd, Burton on Trent, Staffordshire. Previously appeared at 1862 International Exhibition, South Kensington. MW records state sent to "North of England Industrial Iron & Coal Co Ltd, Middlesbrough" which suggests that it arrived between 1870 and 1877. The locomotive is believed to have worked at Mainsforth Colliery, Ferryhill at some period.

(b)   ex Thos W. Ward Ltd, Sheffield, Yorkshire (WR), c/1911; previously Rhymney Railway No.10, until 5/1911.

(c)   origin unknown.

(d)   ex Acklam Works, Middlesbrough, 9/1933.

(e)   ex Acklam Works, Middlesbrough, 9/1933.

(f)   ex Britannia Works, Middlesbrough.

(1)   to Carlton Iron Co Ltd, Mainsforth Colliery, Ferryhill, Co. Durham. It passed to Dorman Long with the colliery, 2/5/1923 and was still there in 1929.

(2)   to Burley Ironstone Quarries, Rutland, c/1926.
(3)   to Cleveland Works, South Bank, by 3/10/1931.
(4)   returned to Acklam Works, Middlesbrough.
(5)   returned to Acklam Works, Middlesbrough, 3/1934.
(6)   to Chilton Colliery and Coke Ovens, Co. Durham, /1934.
(7)   to Chilton Colliery and Coke Ovens, Co. Durham, 8/1934.
(8)   to Britannia Works, Middlesbrough, 8/1/1938.

ESTON, UPSALL AND CHALONER IRONSTONE MINES   55
Eston Mine NZ 564183, Upsall Mine NZ 574173, Chaloner Mine NZ 604175 – **Map I**
Bolckow, Vaughan & Co Ltd until 29/11/1929
Bolckow & Vaughan until 19/11/1864

Eston was the greatest of the Cleveland ironstone mines; the mineral bed was so thick that workings were 14 to 16 feet high and its peak annual production (in 1883) was 1,372,778 tons. On 8/6/1850 John Vaughan (of Bolckow & Vaughan) and his mining engineer, John Marley, established the existence of a substantial exposure of ironstone on the north face of the Eston Hills, following earlier investigations in the area for the Skinningrove (later Main) Seam ironstone. Bolckow & Vaughan wasted no time in opening up a quarry called the Bold Venture on 13/8/1850. A total of 4,040 tons of ironstone was extracted during the last 17 weeks of 1850. A temporary tramway was laid to carry the tubs down to the roadside where horse drawn carts transported it to Cargo Fleet for onward movement by the SDR to Bolckow & Vaughan's Witton Park Iron Works. Meanwhile the first surveys for Bolckow & Vaughan's standard gauge railway from Eston Junction on the Middlesbrough-Redcar line as far as the Eston-Redcar Road had been completed by 17/8/1850. Work commenced on building the railway in 10/1850 and it was sufficiently completed for 136 tons of ironstone to be taken over it in 12/1850. The first locomotive travelled on the branch on 4/1/1851 and Eston Mine officially opened two days later.

Agreement was reached on 17/2/1851 for the SDR to operate a workmen's service between Eston and Middlesbrough. Also, under an agreement of 10/5/1852, a 'market train' for passengers ran to Middlesbrough on Saturday evenings. This occasionally went up to Eston with Bolckow Vaughan using its locomotive and borrowed NER carriages over its own metals. The shop keepers of Eston requested Bolckow Vaughan to stop running the market trains but the Board declined to interfere stating the great convenience for its work people. This service probably ceased in the 1870s.

The Old Bank drum house and three foot gauge ironstone tubs at the Eston Mine about 1900.
(Kirkleatham Museum, Redcar & Cleveland Council collection)

Bolckow Vaughan's No.6 locomotive stands in the timber sidings, opposite the Miners' Cottage Hospital, west of the Tip Yard at Eston
(R. Goad collection)

The ironstone was moved in narrow gauge railway wagons from the quarries on the side of the Eston Hills down the Old Bank self-acting rope worked incline to California, next to the village of Eston. This location was known locally as "Low Drum", a reference to the brake drum that controlled the descent of wagons. From here, another self-acting incline – 300 yards in length on a 1 in 7 gradient – delivered the wagons down to the Tip Yard at Eston where they were emptied into standard gauge wagons for haulage by Bolckow & Vaughan's iron works locomotives to Eston Junction and also to the company's new iron works being developed there. Amounts of ironstone were so large that Bolckow & Vaughan had to install a second rope-worked incline known as New Bank by 1853. This descended the hillside at an angle east of the original incline but also terminated at California, where houses were constructed for the miners. As the size of the overburden increased, drifts were driven into the hillside following the ironstone underground. They were developed at the top of the Old Bank and New Bank inclines, to be followed by a third when an incline was constructed to Lady Hewley's royalty. This latter incline was installed to the west of the other two and descended diagonally down to California. The railway at the mine was of 3ft gauge and used 4-wheel tubs with sloping sides made of wood.

Bolckow & Vaughan obtained a lease on more ironstone to the east in 1869 and, while some was taken out via the Old Bank Drift, the surface narrow gauge tramway that serve the quarries was relaid and extended beyond the Lazenby to Guisborough road; the sandstone bridge carrying a date stone of 1871. East of the bridge, quarries and the North Drift were employed to extract the ironstone. The tubs were pulled along the tramway by two steam haulage engines. Ironstone had ceased to be removed from Lady Hewley's Drift in 1867 resulting in the probable closure of that incline. However, in the early 1870s the Trustee Drift opened and a steam haulage engine was erected halfway up the original incline to Lady Hewley's and this section reopened down to California.

Upsall Mine

Bolckow & Vaughan sank two 564 feet deep shafts at Upsall on Barnaby Moor towards the south side of the Eston Hills. These were sited at the bottom of the syncline and so were at the lowest part of this section of the Eston workings. Although a lease of the ironstone had been signed in 1853, the mine was not operational until 1869. There was a standard gauge railway constructed from the mine south to the Cleveland Railway, but there are doubts about how much this was used because the ironstone was generally taken underground through the Eston Mine workings to the Old Bank and Trustee Drifts. Later a haulage engine was installed at the foot of the shafts to pull wagons from further east.

Chaloner Mine

The mine was established on the west side of the Lazenby to Guisborough Road (Wilton Lane) and a standard gauge railway branch connected it with the Middlesborough–Guisborough line. The ironstone was some 12-13 feet thick and only 42 feet below the surface. Chaloner Mine opened in 6/1872 with two 56 feet deep shafts

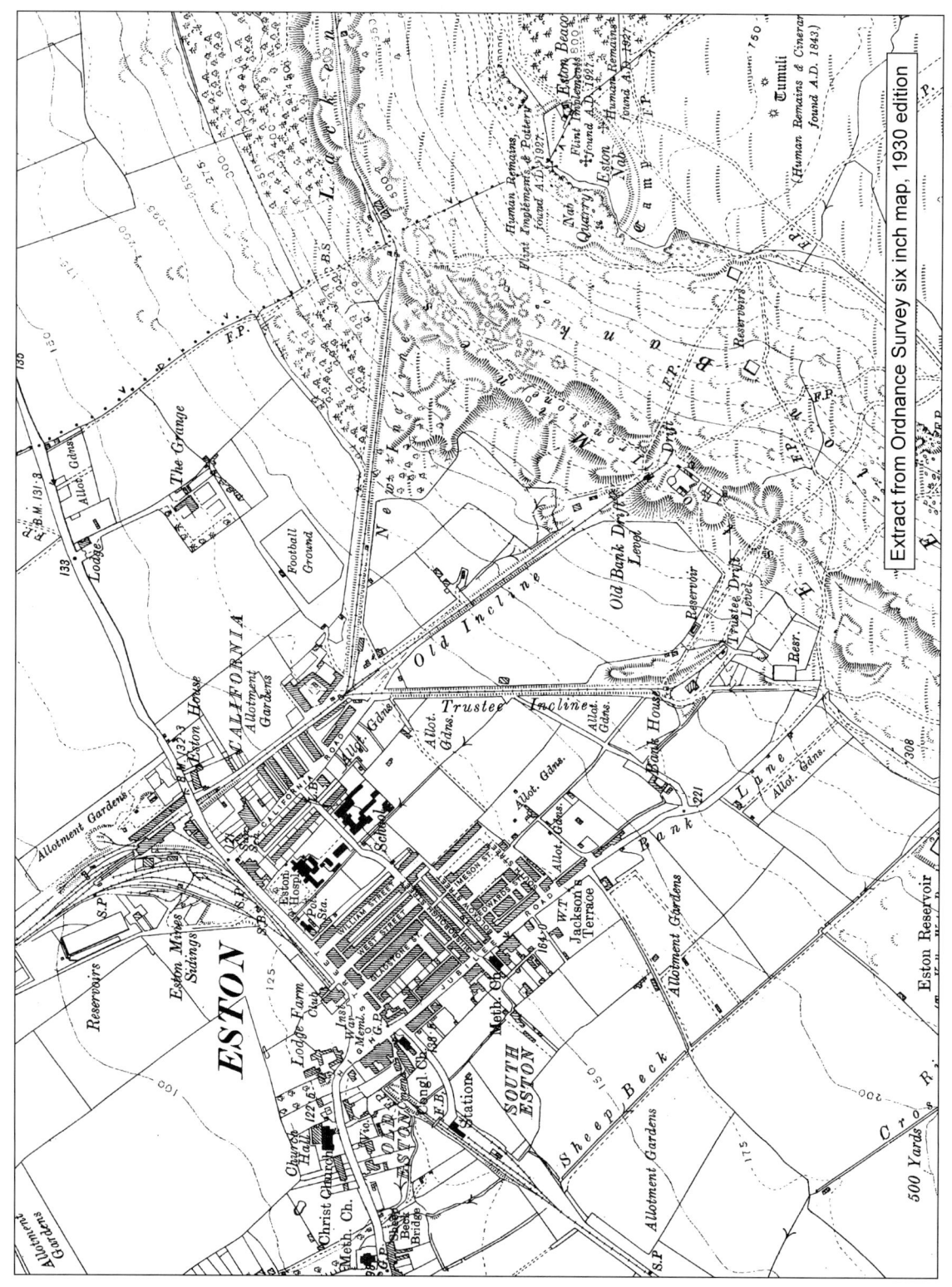

Extract from Ordnance Survey six inch map, 1930 edition

Page 142 – Locomotive Worked Sites

and the first ironstone was despatched on 23/11/1872 presumably along the new railway branch. The initial one mile 45 chains of the mainly single track railway belonged to Bolckow Vaughan, with the remaining 58 chains in the NER's ownership. It ran south west from the mine to Chaloner Junction where there was a signal box just west of Pinchinthorpe Station. Part way along, it occupied the track bed of the former Cleveland Railway for a short distance. Whilst it would have been usual for the NER to have worked the whole branch, Bolckow Vaughan must have had one of its locomotives there in the early days as it was castigated for leaving its locomotive standing next to the Guisborough road. The workings underground headed west to connect with those at Upsall Mine and after 1879 all ironstone went out this way resulting in little traffic being left on the standard gauge branch. Nevertheless, an instruction for Chaloner's Junction Signal Box states that from 1/3/1922 it only opened when required for goods traffic. By 1913 a surface tramway extended from the Chaloner Mine northwards before crossing Wilton Lane to reach extensive ironstone quarries around Stonebridge Farm. Eventually these workings went underground to reach as far as the old Kirkleatham Mine at Dunsdale. In order to remove the remaining pillars of ironstone to the west of Chaloner Mine, a surface 3 feet gauge rope-hauled tramway was constructed northwards (opened mid 12/1915) to connect with the line to North Drift. Three electric haulage engines at Moordale Beck (100hp), Wilton Bank top (220hp) and New Bank Top (320hp) provided the motive power for moving the tubs.

With exhaustion of the reserves at the three mines, Chaloner closed on 12/8/1939 and the Upsall shafts in 1945. On 16/9/1949, after 99 years and more than 63 million tons of ironstone being removed from below the Eston Hills, the final set of tubs was hauled out of Trustees Drift.

Bolckow Vaughan's Eston Mine Branch involved a longer run and sometimes may have had its own dedicated locomotive. On 16/3/1865, the Board had agreed that "management of the Eston Branch Railway and jetty be placed under John Marley" the Mines Manager and that "steps be taken to buy or hire a locomotive if necessary for use of the Eston Branch Railway". A decision for him to hire a locomotive was made on 21/2/1866 although, on 17/10/1866, it was decided to amalgamate the railway with the Cleveland Works due to a financial loss being made on the railway's operation. Later, a former driver, who had started work at Bolckow Vaughan in 1899 recalled that No.6 was in charge of the Eston Mines traffic. It was later rebuilt as a side tank by Bolckow Vaughan. When it was scrapped, the duty was taken on by 111 SIEMENS (AB 1100/1906) of the Cleveland Works locomotives. IRS records also suggest that a four-coupled locomotive may have been based at Eston to shunt the Tip Yard.

References : 'Eston and Normanby Ironstone Mines', Richard Pepper, *South Bank and Eston in Times Past,* C. Books, 1987

Ironstone Mining in Eston A Personal Account, W.E. Brighton, pub. Peter Tuffs, 1996

Catalogue of Cleveland Ironstone Mines, Peter Tuffs, pub. by author, 1997

'Chaloner Mine up to 1879', Richard Pepper, *Cleveland Industrial Heritage* 32, pub. Peter Tuffs, 2013, pp.16-25

Gauge : 4ft 8½in

No.6		0-6-0ST	IC		?	(a)	Scr
	HARE	0-4-0ST	OC	GH	1908	(b)	(1)

(a) the former driver recalled that No.6 had once been NER 571. If correct, this would have been a Class 93 0-6-0 tender locomotive but with double frames and a short wheelbase, probably RWH 1345/1866. It was replaced by the NER in 1884 but would have then been rebuilt into the form in the accompanying photograph.

(b) new. To Bolckow , Vaughan & Co Ltd, Newlandside Quarry, Stanhope, Co. Durham by 9/1914 and returned c/1932, after 5/1932.

(1) to Dorman, Long & Co Ltd, Newlandside Quarry, Stanhope, Co. Durham by 6/1935.

SOUTH SKELTON IRONSTONE MINE 56
Bolckow Vaughan & Co Ltd until 29/11/1929 NZ 655165 – **Map I**
Clay Lane Iron Co Ltd until 31/12/1899
Owners of Clay Lane Iron Works until 12/1/1882
Trustees of Thomas Vaughan & Co until 1/7/1878
Thomas Vaughan & Co until 10/1876

The mine, opened in late 1872 by Thomas Vaughan & Co, was situated alongside the former Cleveland Railway, south west of the later Boosbeck Station. It supplied ironstone to the company's Clay Lane Iron Works. The wooden and glazed heapstead extended from the vertical winding engine house under the wooden headgear of the main shaft to above the siding allowing NER wagons to be filled with ironstone. Additional sidings to the north and south of the mine and parallel to the NER branch were available to stable wagons. There were also lines to a coal depot and to the shale tip south of the mine. The two shafts at South Skelton were sunk 216 feet to the Main Seam of Cleveland ironstone, although this was affected by a shale band as the workings moved south.

The iron trade was in depression for the latter part of the 1870s and Thomas Vaughan & Co went into liquidation in 1876. Mining was continued by the trustees until 1/7/1878 when they decided to sell the stores and movable equipment by auction on 9/10/1878. Generally Cleveland ironstone mines relied on the use of gravity and visiting railway company locomotives to move the standard gauge wagons. *The Daily Gazette, Middlesbrough*, on 25/9/1878, contained a list of plant for sale including a "most superior six wheeled coupled locomotive, double cylinder, 14 inch stroke, 20 inch diameter of wheel 3ft 6in. Copper Fire Box and Brass Tubes, Injector, Boiler Pump, &c;". Strangely the same list in the *Darlington & Stockton Times* on 14/9/1878 had omitted mention of the locomotive.

The loss of income to the landowners prompted early action and mining was resumed at South Skelton (reported 2/11/1878) by new owners of the Clay Lane Iron Works. The mine eventually became the property of Bolckow Vaughan and then Dorman Long, with improvements being made including a new steel headgear, electric winding and picking belt. The years of depression resulted in the mine being shut 1930–36. Underground in 1937, the tubs travelled on a 2ft 6in gauge railway. Motive power was provided by two 320hp electric "main and tail haulages" and supplied by five auxiliary main and tail haulages, with horses moving tubs from the faces. Earlier, the tubs had been constructed with a wooden under frame and body of wrought iron vertical sides that could hold approximately 30cwt. A batch of 50 new steel tubs was purchased in 10/1937 and these could each carry about two tons of ironstone. By the early 1950s, most of the best ironstone had gone and the remaining reserves to the south had reduced iron content with large amounts of shale. The mine finally closed on 6/8/1954.

Reference :   *South Skelton Mine. The story of mining at Boosbeck*, Simon Chapman, pub. Peter Tuffs, 2010

Gauge : 4ft 8½in

-		0-6-0	RWH	?	?	(a)	s/s

(a)   origin and identity unknown.

WEST MARSH IRON WORKS, Middlesbrough 57
Dorman, Long & Co until 2/11/1889 NZ 483211 – **Map F**
West Marsh Iron Co until /1876

The works was located in the centre of the Ironmasters district. It was developed about 1870 by the West Marsh Iron Co, which combined the interests of J.E. Swan & Brothers and Messrs Smith and Thomson (ironfounders) from Stockton. T. James Thomson was the managing director. The eight acre site contained 20 puddling furnaces designed to produce 240 tons of bar iron per week. In 1874 Arthur John Dorman took over as managing director of the West Marsh Iron Co. Two years later, he was joined by Albert de Lande Long and they leased the West Marsh Works; thus it became the first part of the Dorman Long empire. In addition to the 20 puddling furnaces, the premises also had a shingling hammer, an iron rolling mill and a 14 inch finishing mill. A 10 inch rolling mill was then added and the works manufactured iron bars and angles for the shipbuilding industry.

In 1875 the works was surrounded by open land belonging to the OME. It was separated from the Britannia Works to the north by a NER siding from the Marsh Branch that ran to an area marked "NE Railway Wharf", although the latter never came to fruition. A separate siding curved off the Marsh Branch and traversed the open ground to reach the West Marsh Works, where tracks served either side of the main building. Another line off this siding swept round to connect with the company's West Marsh Wharf on the river. It also had a connection with a private line heading north from the Newport Iron Works to the Britannia Works. This was the position in 1882, by which time, Dorman Long had purchased two 12 inch saddle tank locomotives from Black, Hawthorn for the works.

No MOR toll was charged on traffic between West Marsh Iron Works and the wharf because, as the OME explained, "at the time the works site was purchased the wharf did not exist and as a matter of fact was not constructed for a number of years but there was an undertaking to build the wharf which amongst other things should be free of wayleave to them…".

On 8/5/1884 the lease to Dorman Long & Co was renewed from 1/7/1883 for a further seven years with an option to purchase for £35,000. The premises were finally conveyed from the West Marsh Iron Co Ltd to Dorman, Long & Co Ltd in January 1890. Three years later, the works still retained its 20 puddling furnaces and 2 rolling mills, however, there had been some significant changes to the surrounding railways. West Marsh had taken over some of the NER siding to the north and appears to have used the OME land to the east of the works for tipping. In contrast, the line to the south serving the works railway had expanded to 3 or 4 tracks and was identified as West Marsh Wharf Siding (sometimes referred to as the West Marsh Branch). It was part of the MOR and so tolls were payable to the NER for traffic passing over these lines. The MOR line continued beyond the sidings to provide connections with the West Marsh Wharf, Cleveland Wire Works and Ord & Maddison's West Marsh slag works (the latter's closure was recorded on 7/2/1901). In 9/1902 the NER was charging ¼d per ton for traffic passing over a short portion of its line near the Acklam Foundry. This included "Puddler's Tap in Owners' Bogies hauled by Private Loco Engine from West Marsh Iron Works to Acklam Iron Works". However, according to a letter dated 16/4/1903 from A.N. Wilson, District Goods Manager at Darlington, the West Marsh Iron Works had practically been dismantled "but the traffic between Britannia and West Marsh Wharf, using the NER's West Marsh Wharf sidings, would be charged wayleave".

With Dorman Long's ownership of the adjoining land, it is more appropriate to consider the later history of the West Marsh site in the section on the Britannia Works (which see). Dorman Long also controlled a number of wharves on this stretch of the river so it is no surprise that West Marsh Wharf had been occupied by the Tyne-Tees Steam Shipping Co Ltd in 1914 and, by 1926, the 375 feet long wharf was reported to be out of commission. A letter in MOR records dated 10/9/1927 states that it had been taken over by Dorman Long but was not in use.

Gauge : 4ft 8½in

DL & Co No.3	0-4-0ST	OC	BH	477	1879	(a)	(1)
DL & Co No.4	0-4-0ST	OC	BH	478	1880	New	(1)

(a)   new. Order date was 3/9/1878.

(1)   to Dorman, Long & Co, Britannia Works Middlesbrough.

DORMAN LONG (STEEL) LTD

ACKLAM IRON & STEEL WORKS, Middlesbrough   58
NZ 488216, NZ 484218 – **Map F**

This works situated in the Ironmasters District consisted of an iron plant and steel plant which, to begin with, had separate histories.

Acklam Iron Works
Acklam Iron Co Ltd until 1/1/1896
Stevenson, Jaques & Co until 14/8/1888

The ironworks (NZ 488216) was developed by John Stevenson of the Canal Foundry, Preston and Richard Jaques. The three 70 feet high blast furnaces were constructed parallel to and a short distance to the north west of those on the adjoining Linthorpe Iron Works site. They were blown in during 12/1865 and were powered by blowing engines designed by J.G. Beckton of Whitby, although the latter proved unsuccessful and had to be rebuilt. A fourth blast furnace was added about 1869 and all four were in blast in 1873 on the 40 acre site.

A railway connection was provided from the NER Old Town Branch and this ran next to that serving the Linthorpe Iron Works. The railway gantry over the calcining kilns and mineral stores was located on the north east side of the blast furnaces with the pig beds and slag spouts, with their sidings, on the south west side. A line ran round the north end of the engine house crossing over the sidings from the kiln gantry to reach Acklam wharf on the River Tees. The number of railway tracks owned by the company had increased by 1892 with more connecting to the Old Town Branch for the receipt of minerals and the despatch of iron. A large slag heap had been established between the blast furnaces and the river with a railway line climbing up on to it and serving several short temporary sidings for tipping. In 1884–85 the Tees Scoriae Brick Co Ltd, secretary Thomas Roddam Dent, had brick making plants at Acklam Iron Works, as well as Lackenby, Eston and Tees Bridge, Stockton. The Acklam Iron Works was acquired in 1896 by the North-Eastern Steel Co Ltd (NESCL).

North-Eastern Steel Works
North-Eastern Steel Co Ltd until 1/1/1896

The company was incorporated on 9/7/1881 by Sydney Gilchrist Thomas and associates, including Arthur J. Dorman, to produce basic steel from Cleveland iron. It established its works (NZ 484218) on an approximately 22 acre site with four 10 ton basic Bessemer converters standing in line 22 feet above the ground, together with seven adjacent cupolas for melting the pig iron. Two 20 ton hydraulic lifts raised the iron to the staging. Rolling mills were provided to produce steel rails, blooms, billets and bars. The first cast of steel took place on 31/5/1883 and, by the end of the year, the company employed 500 people and was rolling 1,000 tons of rails per week. The main buildings were laid out at right angles to the bend in the river so that the railway lines could run into and alongside them; the sidings connecting to the NER's Marsh Branch. The works was initially separated by Connal & Co Ltd's "iron stores" sidings from the Acklam Iron Works to the east and the Ayresome Iron Works to the south.

The works relied on supplies of iron from other companies and, in view of their proximity, the intention was to transport the molten iron in ladles by rail to save on reheating costs, as had been demonstrated in West Cumberland, but trials in 1883 involving ladles hauled by locomotives from Newport Iron Works were abandoned after a few months because of the molten iron chilling. A significant development was the construction of two 'mixers' close to the converters in 1893 which allowed several batches of iron to be combined equalising irregularities in composition and temperature. Ladles of molten iron were then brought to the mixers by an inclined railway. As a result 2,000 tons of blast furnace iron and 1,000-2,000 tons of cupola iron passed through the mixers each week. Producing steel in the basic converters resulted in large amounts of slag rich in phosphorus and, when ground down, this was a good fertiliser. Most of its 60,000 tons of slag produced each year was exported mainly to Germany but, in 1893, a basic slag works operated by H. & E. Albert opened near the connection with the Marsh Branch. Initially the major market for ground slag was Scandinavia. Output of finished steel had reached 130,000 tons pa by 1894.

Acklam Iron & Steel Works
Dorman, Long & Co Ltd until 2/10/1954
North-Eastern Steel Co Ltd until 2/3/1923
(Subsidiary of **Dorman Long** from /1903)

Following the acquisition of the Acklam Iron Works by the North-Eastern Steel Co Ltd, the blast furnaces were remodelled. Mainly Cleveland ironstone was brought in by rail and calcined in very large kilns. Substantial amounts of coal were consumed by the boiler plant (eleven horizontal and eight water tube boilers in 1908) and the 50 Semet-Solvay coke ovens erected in 1900 between the kilns and the slag heap. The ovens produced approximately 1,500 tons of coke per week.

In 1903 the North-Eastern Steel Co Ltd was taken over by Dorman Long and it then operated as a subsidiary. The NESCL was allowed by the NER "by means of their own locomotives, wagons and ladles convey between their said Iron Works and their said Steel Works molten metal and other traffic approved by the company [NER]" over various lines of the NER. The NESCL was to pay £5 on traffic not exceeding 1,000 tons during the year and ¾d for every extra ton. It also had to provide monthly returns to the NER showing the numbers of wagons and passing times. All of the locomotives, wagons and ladles had to be approved by the NER which also examined the competence of the NESCL's enginemen. In 1908 it was reported that molten iron was being taken from the Acklam blast furnaces in 30 ton ladles to 400 ton mixers that had just been erected at the steel

A postcard showing the Acklam blast furnaces which was posted on 7th September 1909 to Margate. Part of the message on the back reads "I am up amongst the smoke & sulphur, we are having fairly good weather and the change has done my wife a wonderful lot of good, the boy also." (Cliff Shepherd collection)

Hudswell Clarke locomotives were not especially common at Teesside iron and steel works. This 11th June 1939 view of maker's number 324 of 1889 is of interest in showing the Dorman Long lettering and that it was allocated to the Acklam Works. *(G. Alliez photograph, NRM collection)*

works. The four Bessemer converters were each of 12 tons capacity; it was "just like Dante's inferno, smoke and flames everywhere" according to Joe Murphy who had started there in 1911 aged 16. The converters were producing 4,000 tons per week, with the hot slag being transported in 7 ton slag bogies and "dumped in the open", before it was moved into the slag mill and ground into a fine powder known as "Thomas phosphate powder" or "basic slag"; about 1,200 tons of this byproduct was produced each week. The steel then passed to either No.1 or No.2 mills and, after rolling, the steel sections were skidded from the hot banks to the cold banks for straightening, grinding or drilling before being taken by "electric cars" to the inspection and stocking benches. The slag heap near the Acklam Wharf became higher so that the railway line had to climb round and round it to reach the top.

By 1919 the demand for basic Bessemer steel was declining and the converters were taken out of use. One of the mixers was converted to an open hearth tilting furnace. Ten years later, the works had three open hearth furnaces of 140 tons, 160 tons and 180 tons respectively. In 1920 the Semet-Solvay coke ovens were shut down and replaced on approximately the same site by two batteries of 60 Simon-Carves coke ovens in 1920–22 with an output of 4,000 tons of coke per week. Coke was moved from the ovens to the blast furnaces on a largely self-contained railway operated by the Motor Rail petrol locomotives. Benzol and tar were transported by barge across the river to Port Clarence. The position concerning the river wharves is unclear but it appears that the original Acklam or Lower Wharf was of less importance and most imported iron ore came via Connal & Co Ltd's wharf. However, by the time of Dorman Long's 1929 survey, the latter was described as the Acklam Upper Wharf; 408¾ feet long and equipped with 1 x 7 ton, 3 x 5 ton and 1 x 3 ton electric cranes and regarded as part of Dorman Long's works.

An iron ore storage site with a 10 ton overhead grab crane had been erected between the slag grinding plant and the blast furnaces by 1929. The reinforced concrete gantry was 400 feet long with a 117 feet long 1 in 3 incline; a 200hp single drum electric hauler pulled loaded wagons up the railway track allowing the ironstone to be dropped into an excavated hole in the ground. Two railway lines ran under the crane alongside the store of ore for wagons to be filled and taken to the furnaces. Up to 50,000 tons of ore could be kept here.

By now there were 25 miles of railway at the Acklam Iron & Steel Works. A fleet of approximately nineteen locomotives was required to handle the heavy rail traffic, with an initial use of those engines built by Black, Hawthorn and, after 1900, from Andrew Barclay and Hawthorn Leslie. In the early 1930s these were combined with the locomotives from the Britannia and Newport Works to form one fleet (see separate stock list). There appear to have been two separate engine sheds. The one at the iron works was situated to the west of the

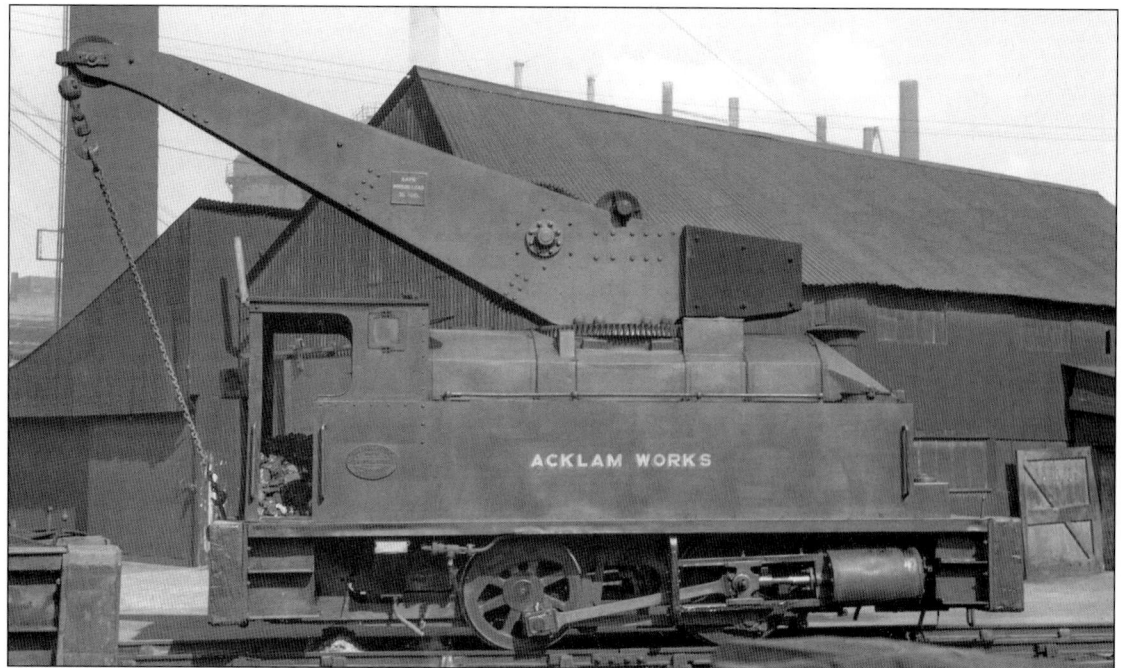

Some 146 crane locomotives were constructed for British industrial customers and AB 2152/1943 was one of the last built. It was seen at the Acklam Works in June 1952 and has the 5 ton crane mounted above the boiler which was the usual arrangement. The crane was removed in the mid-1950s and it then functioned as a traditional locomotive. (ILS Frank Jones photograph)

northernmost blast furnace at NZ 485216. The other engine shed (NZ 484217) was located at the steel works approximately half way along and next to the building housing the rolling mills. The number of jib cranes at the Acklam Works in 1929 was given as four Booth type (2½ tons) Ingots to Mills; one Booth type (2½ tons) Yard Iron Works; three Booth type (2½ tons) Yard Steel Plant; three Priestman grab (5 tons) Yard Steel Plant and one Dunlop Bell magnet (5 tons) Yard Steel Plant.

After World War 2, Acklam had four hand-charged blast furnaces, of which three were normally in blast producing about 6,000 tons of iron per week. Dorman Long was still paying 1d per ton MOR toll on molten metal forwarded from the Acklam Iron Works via the Acklam Steel Works connection in 1948. Iron ore, scrap and sand came in by Acklam Upper Wharf for movement around the site by the Dorman Long locomotives. However, when the company centralised the import of iron ore and processing at its Cleveland Works, supplies for Acklam were then transported by BR over the four miles from South Bank. Production at Acklam was now confined to semi-finished steel products including billets, slabs and sheet bars, however, a new 250 ton tilting furnace was constructed there in 1951.

Acklam Lower Wharf was leased to Tarmac Ltd for it to ship slag and dust along the coast. A tarmacadam plant was located on top of the slag heap and also conveyors led to slag chutes on the river bank for loading vessels. Dorman Long was now implementing its development plan which would see iron and steel production concentrated further down the river. When steelmaking ceased at the Britannia Works in 1953, Acklam was reduced to two working furnaces. The commissioning of the first two Clay Lane blast furnaces and the South Bank coke ovens at the Cleveland Works in 1956–57 resulted in the closure of the Acklam coke ovens on 2/10/1956. Acklam's two working blast furnaces continued to function mainly because their large bells, designed originally for Cleveland ironstone, made them ideal for handling low-grade scrap. According to the 1959 survey any iron, surplus to the steel plant, was cast on a pig bed and the slag was cooled in the ladles to produce dense lump slag suitable for road making. The steel plant now comprised four open hearth tilting furnaces each of 250 tons capacity and a 450 ton mixer, with its slag sent to the Lackenby slag mill. Acklam Iron & Steel Works closed on 26/7/1963 "in the interests of overall production efficiency"; today only the remains of the slag heap, in the form of Teesaurus Park, is a reminder of this once important plant.

References :   *Schedule of Properties*, Dorman Long & Co Ltd, 1929

'Memories of Acklam Works', John L. Brown, *Industrial Railway Record* 245, IRS, June 2021

Acklam Iron Works. This operated its own fleet of locomotives until 1896 when they were incorporated into the following North-Eastern Steel Co Ltd's stock.

Gauge : 4ft 8½in

?								s/s
	ACKLAM No.2	0-4-0ST	OC	BH	21	1867	New	s/s
	ACKLAM No.3	0-4-0ST	OC	BH	23	1868	New	(1)
	ACKLAM No.2	0-4-0ST	OC	HCR	151	1874	New	(3)
	ACKLAM No.3	0-4-0ST	OC	HC	205	1880	New	(2)
	GLADYS	0-4-0ST	OC	HC	292	1888	New	(8)
10	(JENNIE)	0-4-0ST	OC	HC	323	1888	New	
		Rebuilt		Acklam		1923		(9)
11	(NANCY)	0-4-0ST	OC	HC	324	1889	(a)	
		Rebuilt		Acklam		1926		(9)

North-Eastern Steel Co Ltd (later Dorman Long)

Gauge : 4ft 8½in

N.E. STEEL Co No.1	0-4-0ST	OC	BH	614	1882	New	(9)
-	0-4-0ST	OC	BH	689	1882	New	s/s
-	0-4-0ST	OC	BH	691	1882	New	(7)
N.E. STEEL Co No.4	0-4-0ST	OC	BH	764	1885	New	(9)
1 (5)	0-4-0ST	OC	BH	852	1886	New	(5)
No.6	0-4-0ST	OC	BH	991	1890	New	(6)
7	0-4-0ST	OC	BH		1888	(b)	
	Rebuilt		Acklam		1919		(9)
8	0-4-0ST	OC	BH	1124	1896	New	(9)
9	0-4-0ST	OC	AB	879	1900	New	(9)
13	0-4-0ST	OC	AB	1040	1905	New	(9)
3	0-4-0ST	OC	BH		1888	(c)	
	Rebuilt		Acklam		1909		(9)
12	0-4-0ST	OC	AB	1478	1916	New	(9)
15	0-4-0ST	OC	AB	1515	1917	New	(9)
16	0-4-0ST	OC	Joicey	230	1872	(d)	Scr
2	0-4-0ST	OC	AB	924	1901	(e)	
	Rebuilt				1919		(9)
3	0-4-0ST	OC	AB	1888	1926	New	(9)
4	0-4-0ST	OC	AB	1599	1918	(f)	(9)
5	0-4-0ST	OC	AB	1059	1905	(g)	(9)
14	0-4-0ST	OC	HL	3169	1916	(h)	(9)
6	0-4-0ST	OC	HL	3352	1918	(i)	(9)
20	0-4-0ST	OC	HL	2712	1907	(j)	(9)
21	0-4-0ST	OC	HL	2913	1912	(j)	(9)
22	0-4-0ST	OC	AB	1274	1912	(k)	(9)

Note: A Dorman Long list of locomotives at Acklam Works on 3/10/1931 does not include BH 614 of 1882. The list has HC 324 as number 14, not 11 and that BH /1888 was rebuilt Acklam /1921, not 1919. It is not known whether these variations are accurate.

(a) new. To Carlton Iron Works, 9/1933 and returned to Acklam Works, 3/1934.
(b) origin unknown, It was at the Acklam Works in /1929.
(c) ex Ayrton Sheet Works. It was not recorded at Acklam /1929.
(d) ex Cargo Fleet Iron Co Ltd, Cargo Fleet Iron and Steel Works, Middlesbrough, /1917. It was at Acklam Works in /1929.
(e) ex Clarence Works by /1929.
(f) ex Bradford Corporation, Esholt Sewage Works, near Bradford, via Harry Ackroyd, Drighlington, Bradford, 9/1927.
(g) ex Connal & Co Ltd, Middlesbrough. It was not recorded at the Acklam Works in /1929.
(h) ex Clarence Works, 24 or 29/8/1933. To Carlton Iron Works at a date unknown and returned.

(i) ex Clarence Works, 5/5/1933.
(j) ex Clarence Works, 10/4/1934.
(k) ex Clarence Works, 9/6/1934.

(1) s/s. Inness has "to Locke Newlands Wakefield" via Hudswell Clarke; possibly a reference to Locke & Co (Newland) Ltd?
(2) s/s. Inness has "to Brooks & Brooks, Bacup".
(3) to W. Shaw & Co Ltd, Wellington Foundry, Middlesbrough.
(4) to The Salt Union Ltd, South Durham Salt Works, Haverton Hill.
(5) to Lingford, Gardiner & Co Ltd, Bishop Auckland, Co. Durham; resold to J. Spencer & Sons Ltd, Newburn-on-Tyne, Northumberland.
(6) to North Bitchburn Coal Co Ltd, Rough Lea Colliery and Brick Works, Hunwick, Co. Durham.
(7) to W. Blenkinsop & Co Ltd, Middlesbrough for scrap.
(8) s/s. Inness states that this locomotive was originally No.2 GLADYS, became ACKLAM No.7 and was renumbered 40 in 1937 but it does not appear in the 'Combined List'.
(9) renumbered in early 1930s (after 10/1931) into the combined Acklam, Britannia and Newport Works list of steam locomotives (details of the 'Combined List' are located in the Britannia Steel Works entry). However, BH 764/1885 may have been sold to The Salt Union Ltd, South Durham Salt Works, Haverton Hill according to R. Inness.

Acklam Coke Ovens
Gauge : 4ft 8½in

1	4wPM	MR	1926	1919	New	(1)
2	4wPM	MR	1945	1919	New	(1)
-	4wPM	MR			?	Scr

(1) the locomotives were derelict in 6/1960 and scrapped c/1961.

Hawthorn Leslie 2748 of 1908 and a hot metal ladle are pictured at Dorman Long's Acklam Iron and Steel Works on 19th June 1960. (IRS J.P. Mullett photograph)

AYRTON SHEET WORKS, Middlesbrough
Dorman, Long & Co Ltd until 2/10/1954
R.P. Dorman & Co until / 1899
Jones Bros & Co Ltd until /1890
Jones Brothers & Co until c/1876

59
NZ 488212 – **Map F**

Ayrton Rolling Mills (later Sheet Works) were established near the north end of Forty Feet (later Forty Foot) Road in the Ironmasters District, with sidings off the NER Marsh Branch. The business was founded by Jones Brothers & Co in 5/1870 to produce sheet, hoops, billets and rods in the works which was named after Judge W.S. Ayrton, a substantial backer of the project. There were 16 puddling furnaces on the premises, together with rolling mills and production commenced in 3/1871; a speciality being the manufacture of thin iron sheets. By 1880 the works contained 29 puddling furnaces and two rolling mills. The link from the Marsh Branch served a couple of sidings on the western boundary of the works before terminating in a head shunt and 3-4 short sidings entered the main part of the premises.

In 1890 Robert Page Dorman, younger brother of Arthur Dorman, took over the works. Unfortunately he died in 1898, following which Dorman Long gained control, although it continued to be known as "R.P. Dorman" and had a trademark to that effect. At this date, the plant consisted of six sheet mills, 22 puddling furnaces, two shingling hammers and a "puddled-bar rolling train". It was capable of producing 400 tons per week, the bulk of which was galvanised corrugated sheets for roofs and buildings. There appears to have been a small single road engine shed on a short spur from the railway tracks on the western boundary. A new Black, Hawthorn 12 inch four-coupled saddle tank had been acquired in 1890.

On the northern boundary, in the area defined by the Old Town Branch Loop, were the Wharncliffe Ganister Brick Works and a cast steel foundry, although the latter was to be subsequently incorporated into the Ayrton Sheet Works site. In 1908, Ayrton had six sheet mills, four galvanising baths, 16 puddling and two ball furnaces, with 470 people employed. It had an annual capacity of 20,000 tons of mostly corrugated galvanised iron sheets. This had become its staple product for which it was very well known. However, the change to steel was inevitable and, in 1922, the last of the puddling furnaces finished, the existing plant was dismantled and new furnaces and mills were installed to produce sheets using steel supplied by the Britannia Works. At some stage responsibility for shunting Ayrton was taken on by Britannia Works locomotives. The 1929 Dorman Long survey revealed that it had the use of two Hawthorn Leslie locomotives with 12in cylinders, a 10hp petrol tractor by Trackless Rail Ltd of Bedford and a Grafton 2½ ton jib crane.

The Iron and Steel Institute visit to the works in 1937 was informed that the steel bars were delivered from the Acklam Works to the stocking ground at the rear of the mills "and are loaded by a locomotive crane into trucks for delivery to the bar bay which is served with a 10-ton crane". The bars were cut to length and fed into one of four furnaces heated by blast furnace gas from Acklam Works. The steel then passed through seven pairs of hot rolls and three pairs of cold rolls. Galvanising involved a pickling tank to clean the steel, three galvanising machines and then two corrugating machines. A new Lewis Mill commenced production in 1937–38 and this made many of the sheets used in Anderson shelters during World War 2. These sheets went to the Cleveland Works boiler yard for bending and punching because it had the only hydraulic press in the area to do this work. Virtually all of these sheets were despatched in railway wagons.

About the mid 1960s the rolling mills at Ayrton finished, by which time, there was probably little if any rail traffic although much of the siding layout remained. In 1968 the Light Plate Mill (opened 1937) closed and the work transferred to Lackenby which could produce in two hours what had occupied Ayrton for a week. From then on, the works concentrated on 'cold forming processes' producing items such as steel lintels and motorway safety barriers. Operations at Ayrton ceased in 1/1985 when this work was transferred to BSC's Whitehead factory in South Wales. Subsequently the Ayrton site and some of the buildings were taken over by A.V. Dawson Ltd (which see).

The following list of locomotives is probably incomplete, for example, the IRS suggests that BH 478 of 1880 may have moved here from West Marsh Works before going to Britannia. Locomotives used at Ayrton mostly came from one of the other nearby Dorman Long works and are included in their lists.

Gauge : 4ft 8½in

R.P.D. & Co No.1	0-4-0ST	OC	BH	977	1889	New	(2)
SHEET WORKS No.3	0-4-0ST	OC	BH		1888	(a)	(1)

(a) origin & identity unknown.

(1) to Acklam Works, Middlesbrough.

(2) to Britannia Works, Middlesbrough.

Also see the Combined Acklam, Britannia, Newport steam locomotive fleet located in the Britannia Steel Works entry.

Gauge : 2ft 0in

		4wPM	MR		(a)	s/s
-						

(a)   ex Thos. W. Ward Ltd, 30/4/1929. This Simplex 20hp locomotive, equipped with a 2-cylinder Dorman engine, had been purchased by Ward on 16/1/1928 at a sale of plant lying at Wallbridge following completion of the Kirkheaton Colliery Co Ltd's railway from the LNER Ponteland Branch by the contractor, James Goldie & Sons Ltd. No information is available on its use at Ayrton Rolling Mills.

BOWESFIELD STEEL WORKS, Stockton. 60
Bowesfield Steel Co Ltd (later subsidiary of **Dorman, Long & Co Ltd** from 6/1936)
Bowesfield Iron Co Ltd until 22/6/1896   NZ 443177 – **Map D**
Bowesfield Iron Co until /1875

The Bowesfield Iron Works was established by Messrs Jaques and Stevenson on a 21 acre site located on the south side of the Darlington–Middlesbrough railway between Bowesfield Junction and the Tees river bridge. The first iron plate was rolled in 11/1871 when the works had 30 puddling furnaces and two mills. In 1890 the works comprised 51 puddling furnaces, a sheet mill and an extensive plate mill measuring 300ft by 192ft. The layout was very distinctive with the furnaces set out in a semi-circle at the north end of the main building with the hammers and other processing equipment in the centre. Upwards of 600 people were employed producing approximately 800 tons of plates and sheets and 800 tons of bars, which were used in the boiler making, ship and bridge construction industries. Tom Sowler recorded that "on a still summer night, the clanking of the sheets as they passed through the rolls could be heard in the centre of Stockton".

A connection was made with the NER at Bowesfield Junction and, by 1895, this led into a fan of sidings owned by the iron company alongside the main line. From here, railway tracks swept round to serve both sides of the main works building. In addition, another line continued southwards next to Bowesfield Lane to reach the Richmond and Tees Bridge Works. There was also scope for tipping spoil on the east side of the premises.

The works closed in 1890 but the Bowesfield Steel Co Ltd was incorporated on 22/6/1896, the Board of Directors having close links with Head Wrightson. After the takeover, extensive improvements were undertaken with new sheet mills, furnaces and engines installed. It was one of the first works to adopt electric drive for the sheet steel mills and it increasingly concentrated on producing black and galvanised sheeting. Dorman Long began to buy shares in the company in 1912. There were 15 sheet mills in 1930, together with annealing furnaces, galvanising baths, cold rolls and other finishing plant. Additional sidings had been added to give access to the tips which now reached to the bank of the River Tees. Dorman Long eventually owned a considerable shareholding in the Bowesfield Steel Co and the *Leeds Mercury* reported on 30/6/1936 that the company had purchased the whole of the remaining shares, although the steelworks would continue to retain its separate identity. The Bowesfield Steel Co was liquidated in mid-1938 but relaunched with the same name on 27/10/1939 as a wholly owned subsidiary of Dorman Long.

A report from the Iron and Steel Institute visit to Middlesbrough in 9/1937 stated that "the complete works are housed in one building 700 feet long by 80 feet wide." Of the 15 mills, the eight older ones had coal fired furnaces and the newer mills used gas from two 'producers' designed by the Power Gas Corporation. The black sheets passed to the warehouse where they were examined and packed for despatch, "railway sidings extending into the warehouse". Other sheets went to the galvanising plant. The stocking ground extended along one side of the mill building immediately at the rear of the mill furnaces. The works closed in 1958 and stood derelict in 1962.

Reference :   *Iron and Steel Institute Visit to Middlesbrough*, 9/1937, illustrated booklet

Gauge : 4ft 8½in.

	BOWESFIELD No.1	0-4-0ST	OC	BH	163	1871	New	s/s
	BOWESFIELD No.2	0-4-0ST	OC	BH	175	1872	New	s/s
-		0-4-0ST	OC	MW	447	1873	(a)	(1)
	BOWESFIELD No.1	0-4-0ST	OC	CF	1146	1897	New	s/s c/1956
No.2		0-4-0ST	OC	HL	2891	1911	New	s/s c/1960
27		0-4-0ST	OC	DL		1952	(b)	(2)

(a)   ex South Durham Iron Co, Darlington, Co. Durham.
(b)   ex Dorman Long (Steel) Ltd, Britannia Works, Middlesbrough, 6/9/1957 although Bob Payne claims he saw it here on 21/11/1955 and at Newport Works on 5/5/1957.
(1)   to Thornaby Pottery Co Ltd, Thornaby.
(2)   to Dorman Long (Chemicals) Ltd, Port Clarence Distillation Works, Port Clarence, 16/10/1957.

KILTON IRONSTONE MINE, Brotton
Dorman, Long & Co Ltd until 2/10/1954
Walker, Maynard & Co Ltd until 2/1/1916
Walker, Maynard & Co until c12/1900
Kilton Ironstone Co until 2/9/1876

61
NZ 695169 – **Map I**

The mine was situated at the end of the 1 mile 21 chains Kilton Thorpe Branch, approximately one mile south of Brotton. The branch was constructed by J.T. Wharton (owner of the land) and Messrs Robson and Cochrane who formed the Kilton Ironstone Co. As early as 9/1872, J.T. Wharton offered to sell the line to the NER. It opened on 11/6/1873 and the NER took it over under its 16/7/1874 Act. The royalty on the ironstone had been leased by the Kilton Ironstone Co from 1/7/1870 and sinking of the two shafts began on 21/2/1871 but the first stone was not raised until 20/3/1875, by which time a trade depression had set in. The quality of the Main Seam ironstone was only moderate and it had a band of shale in the middle. Only small amounts were raised and taken by the NER to Middleton Iron Works at Fighting Cocks. The mine closed on 2/9/1876.

Walker, Maynard & Co agreed with Wharton on 27/9/1894 to take over the mine as from 1/7/1894; production commenced 7/1898 to supply the company's blast furnaces at Redcar. It transferred to Dorman Long when the latter acquired Walker Maynard in 1916. Electricity had been supplied to the area since 1908 from the Grangetown Power Station and Dorman Long installed an electric winder in 10/1916. The mine worked only intermittently between the wars and was closed for much of the 1930s before it reopened on 2/1/1940. The railway system on the surface was relatively simple with two lines running beyond the winding shaft to a headshunt. Ironstone came up the shaft, was sorted on a picking belt before being tipped into the railway company wagons which descended from the headshunt by gravity. Meanwhile, shale was transported in tubs up the tip north of the mine using, in 1929, a Metropolitan Vickers 30hp motor operating a single drum. Underground the main engine plane had a Metropolitan Vickers 225hp motor working a double drum main and tail rope system over 1,600 yards. A second haulage used a 50hp motor and a single drum. Compared to the 2ft 0in gauge found in many Co. Durham pits, the ironstone mines dealt with thicker seams and larger lumps of rock, so preferred to use wider gauge railways underground. There had to be compromise because this meant greater expense in sinking a wider shaft, therefore, it was usual in Cleveland to adopt a 2ft 6in or 3ft 0in gauge.

After the closure of Eston Mine in 1949, Dorman Long relied on mines in East Cleveland for its locally produced ironstone and, following 1954, these consisted of Kilton, Lingdale and North Skelton. In 1950 Kilton Mine had

No.24 (RH 433388/1959) was a three foot gauge locomotive at Kilton Ironstone Mine on 26th July 1963. It had come from Lingdale Mine for repairs and went to Irthlingborough in Northamptonshire during 1964 after closure of the local mines. *(IRS Brian Webb photograph)*

32 horses, 170 men and an output of 1,500 tons per week. Dorman Long then took the decision to introduce greater mechanisation to these three pits.

At Kilton, the tippler dropped the contents of the tubs on to a steel plate picking belt, where the shale was taken out by hand and moved to the tip by Muir-Hill dumpers, before the ironstone could be put into the BR wagons. Two Eimco compressed air loading machines began work at Kilton on 9/7/1951. The *Cleveland Standard* on 7/11/1952 reported that "Sidney Lightfoot of Liverton changed his job in the Kilton Mine on Wednesday morning… for, from being a horse driver, he changed onto locomotive driver – on the first [sic] locomotive ever to be used for underground haulage in the Cleveland mines." Previously horses had pulled wagons from the face and it was expected that the locomotive would replace four horses. Some 400 yards of track had been laid comprised of five yard prefabricated sections welded to steel sleepers. RH 318748 was the prototype Class LBU locomotive. It was ex works on 17/1/1952 and initially sent on loan to John Lysaght's Nettleton Ironstone Mines in Lincolnshire, presumably on trials, before returning to Ruston's Lincoln Works on 7/3/1952. It was despatched to Kilton Mine on 31/10/1952.

By 1954 Kilton Mine had several diesel locomotives and output had increased to 6,350 tons per week. Late in its life it also received two new 40hp Hunslet diesels. This late change of supplier is explained by the fact that these were three stock locomotives, maker's numbers 5680–82 intended for Hunslet, Canada. When this order was cancelled, construction probably stopped and the partially completed assemblies placed in store. Dorman Long's use of 2ft 6in gauge locomotives offered Hunslet the opportunity of a sale and presumably as 5682 was the least advanced, it was easiest to meet the Dorman Long order specification of 10/5/1960 (delivered 7/1960). No doubt this locomotive proved satisfactory and 5680 was ordered on 23/3/1961 (delivered 7/1961).

However, mining Cleveland ironstone was increasingly uneconomic and its large slag-forming constituents not suited to modern iron-making practice. The mine closed on 2/2/1963 and was formally abandoned 7/12/1963 resulting in the end of BR's Kilton Thorpe Branch; the 'Cleveland Limited' rail tour had travelled over the branch on 8/6/1963. A number of ruined buildings remain including the locomotive garage built in 1951, but the most noticeable feature is the prominent conical spoil heap acting as a tangible reminder of the industry.

Reference : *Kilton Ironstone Mine*, Simon Chapman, pub. Peter Tuffs, 2000

Gauge : 3ft 0in

No.24		4wDM	RH	433388	1959	(a)	(1)

(a) ex Lingdale Mine, for repair only, c/1962, after 16/3/1961.

(1) to Richard Thomas & Baldwins Ltd, Irthlingborough Mines, Northamptonshire, c1/1964.

Gauge : 2ft 6in

	4wDM	RH	318748	1952	New	(2)
-	4wDM	RH	338438	1953	(a)	(4)
No.5	4wDM	RH	353484	1953	(b)	(3)
-	4wDM	RH	353486	1953	New	(4)
No.3	4wDM	RH	353492	1954	New	s/s c/1964
-	4wDM	RH	375329	1954	New	(4)
No.15	4wDM	RH	375693	1954	(c)	(1)
-	4wDM	RH	392119	1956	New	s/s c/1964
-	4wDM	RH	418765	1957	New	(4)
-	4wDM	HE	5682	1960	New	s/s c/1964
-	4wDM	HE	5680	1960	New	s/s c/1964

(a) new. On 5/2/1954 and 10/3/1954 spares orders for this locomotive came from Lingdale, but it was probably still at Kilton Mine.
(b) ex Lumpsey Mine after 13/4/1954.
(c) ex North Skelton Mine, 26/8/1954.

(1) to North Skelton Mine, by 16/8/1955.
(2) to Lingdale Mine for storage only, after 24/4/1956.
(3) to North Skelton Mine, possibly in 1958-61 period (recorded in W.C. Allan's notebook, Mines Engineer at Lumpsey and North Skelton).
(4) to Richard Thomas & Baldwins Ltd, Irthlingborough Mines, Northamptonshire, after 7/1963 (when seven locomotives still reported to be underground at Kilton), by 1964. Purchased for spares only, as all in poor condition and some incomplete.

LINGDALE IRONSTONE MINE, Lingdale
Dorman, Long & Co Ltd until 2/10/1954
The Lingdale Ironstone Mine Co Ltd until 1/1952
Pease & Partners until /1947
J.W. Pease & Co until 19/8/1882

NZ 676165 – **Map I**

J.W. Pease & Co leased the ironstone in the Lingdale locality from 1/1/1871, but it was not until 12/8/1876 that the Main Seam was reached by the shafts. Production began about 3/1877. The pit was situated in the village of Lingdale which had been constructed to house the miners. Simon Chapman records that a quarry of distinctive white freestone was developed to the west of Stanghow and a narrow gauge railway laid to bring the building stone 650 yards north eastwards to the mine and village. The quarry was located near the later mine manager's residence, Seatons Hill House.

The mine was linked to the Kilton Thorpe Branch that served Kilton Mine by a 1 mile 18 chain railway line. This was constructed by Pease with wayleaves paid to J.T. Wharton, owner of the land. The line was leased by the NER on 4/7/1896 although it appears that the railway company always worked it. The line from Lingdale Junction (controlled by Lingdale Branch Junction Signal Box) was difficult to work involving a steep climb, much of it at a gradient of 1 in 36. Brakes on the loaded wagons had to be pinned down for the descent and more than one driver lost control of his train. One such incident reported in the *Loftus Advertiser* happened on 5/11/1900 when NER locomotive, 1248, hauling 22 loaded ironstone and three goods wagons from Lingdale Mine ran away and fell down the embankment near Kiltonthorpe Junction. Nearer Lingdale, where the railway passed under Kilton Lane, two men excavating the cutting were killed in 10/1875 when one side gave way. From this bridge, the line curved round to the mine where it divided into two, with twin tracks passing under the tippler and then terminating in a headshunt and sidings for the empties beyond yet another Kilton Lane bridge. The other line ran by the boiler house in order to deliver coal and stores. In order to not make the gradients on the line any worse, the twin tracks were laid in a deep cutting with the heapstead constructed at ground level. Ironstone was wound up the shaft and subsequently passed over a picking belt, because a band of shale ran through the Main Seam at Lingdale, before being tipped into the wagons. The shale was taken out of the west side of the mine buildings on a narrow gauge railway to be initially dumped at the nearby Nova Scotia Wood and later carried over the road on a wooden bridge erected in 1896 on to the falling ground north of the village. Lingdale Mine was constructed on a lavish scale and employed 3ft 0in gauge railways underground and larger mostly wooden tubs. The latter could hold 56cwt of ironstone and had a tare weight of 14cwt. During the first decade of the twentieth century the mine operated a small brick works using crushed shale with a capacity of 50,000 bricks per week. A stationary steam engine drove an endless chain taking tubs to and from the shale heap and also powered the grinding mill and other machinery.

Like various other nearby mines, Lingdale experienced periods of closure during the interwar years and it was shut between 5/9/1931 and 19/8/1940. An attempt was made to increase the iron content of the stone delivered to the iron works by calcining. Two metal vertical cylindrical kilns were obtained from the Isle of Raasay and erected with a skip hoist in between. They were located at the end of the picking belt so that clean ironstone could be fed into the top of the kilns. After calcining, the iron content increased from 26–29% to about 36%. The treated ore was then removed from the base of the kilns and tipped into the railway wagons. Average output in May 1941 was 1,969 tons per week. The kilns operated until 31/7/1949, but later restarted and continued to about 1959.

When Pease & Partners withdrew from mining following nationalisation of the coal industry, it formed a subsidiary, The Lingdale Ironstone Mine Co Ltd, to facilitate the sale of the mine. Dorman Long acquired the full share capital of the subsidiary and immediately embarked on doubling the output per man shift by the introduction of rotary drills, Eimco 21 rocker shovels and a fleet of Ruston & Hornsby diesel locomotives underground; initially class LBU of 31½hp but later three more powerful 44hp class DLU. This necessitated replacement of the 27lb/yard rails currently in use by heavier track at the faces. By 8/1954 there were no horses underground. Lingdale Mine closed on 23/2/1962 and the lease of the railway branch was terminated by BR in 11/1963.

Reference :   *Lingdale Mine,* Simon Chapman, pub. Peter Tuffs, 2006

Gauge : 3ft 0in

No.	Type	Builder	Works No.	Year		
No.3	4wDM	RH	338439	1953	New	(2)
No.7	4wDM	RH	353491	1953	New	(2)
No.10	4wDM	RH	371938	1954	New	s/s
No.13	4wDM	RH	375694	1954	(a)	(2)
No.20	4wDM	RH	418764	1957	New	(2)
No.21	4wDM	RH	418803	1957	New	(2)
No.22	4wDM	RH	427802	1958	New	(2)

No 24	4wDM	RH	433388	1959	(b)	(1)
-	4wDM	RH	451900	1961	New	(2)
-	4wDM	RH	466578	1961	New	s/s

(a) new. RH received an order dated 6/12/1954 for a gearbox for this loco to be sent to Kilton Mines, but it is not known whether the loco went there as well. The original gearbox was repaired at RH, Grantham in 1955. The locomotive was at Lingdale Mine 7/12/1956.

(b) new. A spares order delivered to Lingdale Mine 16/3/1961 for "1 set of axles and rail wheels to convert the loco to slow speed model".

(1) to Kilton Mine for repair only, c/1962, after 16/3/1961.
(2) to Richard Thomas & Baldwins Ltd, Irthlingborough, Northamptonshire, /1962.

Gauge : 2ft 6in

-	4wDM	RH	318748	1952	(a)	(1)
-	4wDM	RH	353494	1954	(b)	(1)
29	4wDM	RH	466579	1961	(b)	(1)

(a) ex Kilton Mine for storage only, after 24/4/1956.
(b) ex North Skelton Mine for storage only.

(1) s/s. A photograph in the L.G. Charlton collection shows at least two Rustons stored on temporary track under tarpaulin on 17/9/1964.

LUMPSEY IRONSTONE MINE, Brotton 63
Dorman, Long & Co Ltd until 2/10/1954 NZ 686187 – **Map I**
Bell Brothers Ltd until 2/5/1923 (subsidiary of **Dorman, Long & Co Ltd** from /1902)

Shaft sinking at Lumpsey commenced on 26/4/1880 and the Main Seam ironstone was reached on 3/11/1881. There were two shafts and this proved to be the last Cleveland ironstone mine to be sunk, production commencing in 1882. As such, it was the most technically advanced. The mine was located south of Brotton

Bell Bros' Lumpsey Ironstone Mine about 1904 before the old chimney had been demolished. The standard gauge tracks are on the left and the 2ft 6in tub line is in the centre.

in the vee of the junction between the NER lines to Saltburn and Guisborough. A loop off the Saltburn line passed as a single track under the tippler at the heapstead. There was no picking belt so everything out of the mine went directly into the railway company's wagons and down to the iron works because, by this late date, a number of the ironworks had their own calcining kilns. Four sidings on each side of the tippler were available for holding wagons. Empties could be let down by gravity from the north to the south side where they were ready for the journey to Teesside. According to T. Kitching, the tippler was designed to have sufficient height to cater for the larger wagons, including those originally used to carry hot calcined ore from Liverton Mine. Unusually for a Cleveland Mine, there was a second rail connection this time to the Guisborough line which was used to deliver coal to the boilers. The connections with the railway company's lines were controlled by Lumpsey Mines Signal Box (with 20 levers) and Lumpsey Siding Ground Frame (with 3 levers).

The mine was worked on the usual bord and pillar method with haulage underground being provided by a mile long 'endless rope' railway driven by an engine on the surface using steam from three Cornish (later Lancashire) boilers. Electricity was then introduced and a 60hp Metropolitan Vickers motor was installed in the mine to power the endless rope until replaced by a 100hp English Electric motor; the length of the rope-worked railway having been extended to 2,450 yards. Self-acting inclines on the rise side and main rope haulages on the down side delivered tubs of stone to the endless rope. The engine plane intercepted that at Carlin How Mine in 1904 allowing both pits' reserves to be sent out via Lumpsey. In 1917 Lumpsey, including Carlin How, had the largest output of the Cleveland mines, even marginally more than the mighty Eston! Horses were used to move tubs between the faces and "landings" or reception sidings for the ropeways. The largest number of horses employed was 72 and 20–30 remained at the finish.

The cage taking ironstone up the shaft could carry two full tubs. At the pit top, the tubs ran from the cage down an incline to a weighing machine where two men stopped each wagon using sprags. After the weight had been recorded and the token removed indicating the two miners responsible for the ironstone, the tub was pushed into the tippler for unloading. Several hundred tubs were used at Lumpsey and Carlin How. They were subject to heavy wear so tub construction and repair was a full-time activity, with ⅜in steel plate being purchased for the bottom, sides and ends. All the plates were held together with rivets. The oak wooden support frames were made in the joiners' shop, while the wheels, axles and couplings were bought in.

The mine was closed for periods of approximately 3 years, 6 months and 2¼ years in the 1920s–30s. A report in *Iron and Coal Traders Review* 25/4/1947 stated that tubs were still being loaded by hand, but three Eimco rocker shovels were on order. Tubs carried an average of 30cwts of ironstone. On the main tracks 40lb per yard rails were used with lighter rails at the faces. Output in the week prior to 12/3/1947 was 4,055 tons. The first diesel locomotive arrived at Lumpsey to work underground on 14/10/1953; this was Ruston & Hornsby maker's number 353484. However, a scheme had been agreed in 7/1953 to connect the workings to North Skelton Mine in order to reduce costs. The breakthrough came on 8/10/1954 and locomotive working was then mainly in Lower Crosscuts delivering stone to North Skelton. The last shift at Lumpsey took place on 27/11/1954, although pumping and ventilation continued.

References : 'Lumpsey Ironstone Mine', N.A. Chapman and S.A. Chapman, *British Mining* 28, The Northern Mine Research Society, 1985

Lumpsey Mine 'Flower of Cleveland' The History of the Lumpsey and Carlin How Mines, Simon Chapman, pub. P. Tuffs,1997

Cleveland Ironstone from Lumpsey, T. Kitching, Sotheran, 1995

Gauge : 2ft 6in

| No.5 | 4wDM | RH | 353484 | 1953 | (a) | (1) |
| No.9 | 4wDM | RH | 353494 | 1954 | (b) | (2) |

(a) new. Prior to delivery, it was exhibited at *The Engineering & Marine Exhibition* at Olympia, London leaving RH Lincoln on 26/8/1953 and returning in 9/1953. It was then sent to Lumpsey on 13/10/1953 (the Mine Engineer's desk diary records 14/10/1953). On the following Tuesday, a RH representative arrived to try it out in the pit yard only to discover there was a fault with the gearbox.

(b) new. RH records suggest that it was delivered to South Skelton Mine. However the latter was due to close soon and Lumpsey's Mine Engineer recorded that it arrived at Lumpsey 9/3/1954 and was lowered underground to work for "Carlin How" while No.5 was brought out for attention.

(1) to Kilton Mine, after 13/4/1954.
(2) to North Skelton Mine, by 7/12/1955.

NEWPORT IRON WORKS, Middlesbrough

Dorman, Long & Co Ltd until 2/10/1954
Sir Bernhard Samuelson & Co Ltd until 2/5/1923
B. Samuelson & Co until 10/1887

64
NZ 481203 – **Map F**

Newport Iron Works occupied a triangular site at the south end of the Ironmasters district between the River Tees and the Thornaby–Middlesbrough main line. Bernhard Samuelson had sold his South Bank Iron Works in 1863 and, with the proceeds, began to erect three 69 feet high blast furnaces at Newport. By 1864, heaps of ironstone mixed with coal stood ready for calcining and charging into the new furnaces which were blown in during August. These were quickly followed by two more in 1865 and 1867 making a row of five located parallel to the NER main line. The rail gantry over the mineral stores and calcining kilns was situated between the furnaces and NER with the pig beds and slag discharge points on the opposite side.

A separate row of three blast furnaces, constructed in 1869-71, were located further north and at a slight angle to the NER. Here the positions of the rail gantry over the kilns and the pig beds were reversed. The blast furnaces were of typical Cleveland design later using ironstone from the company's Slapewath Mine. The costs associated with erecting the three blast furnaces included the following purchases:

	£	s	d
Two locomotives at £1,000 and £1,250	2,250	0	0
One locomotive for slagging	750	0	0
Thirty metal bogies	870	0	0
Forty five slag bogies	1,055	11	7
About 1½ miles of single track railway	3,034	9	11
Two new iron wagons, 27 wooden hopper wagons	941	3	0

Samuelson favoured a distinctive design of Kitson locomotive incorporating both a saddle and well tank; there were seven of them at the Newport Works. A Kitson gear was used with the valves on top of the cylinders to avoid lineside obstructions and protected by side panels. The design offered greater adhesion in a four-coupled design which could cope with the sharp curves of an industrial railway. Two of the locomotives stand near the engine shed at the Britannia Works awaiting scrapping and illustrate the difficulties of identity on the Ironmasters. The unrebuilt engine seen here on 19th June 1960 appears to be 29 (Kitson 3982/1900) with 31 (Kitson 5115/1914) behind, the latter having received a replacement saddle tank and chimney. However, up to a couple of years previously, the numbers were reversed and 29 was the rebuilt engine!

(IRS J.P. Mullett photograph)

It was reported in 1887–88 that seven of the eight blast furnaces had been rebuilt in the last few years on the overall site of about 60 acres. Their location near the NER gave ready access to the main line. The NER referred to the two sets of blast furnaces as 'Old Side' and 'New Side'. Samuelson's first wharf was the Upper Newport Wharf and this was used to bring in Spanish ore and ship out slag in barges either for dumping at sea in the form of "shingles" or to assist with the construction of retaining walls and the Gares. The Lower Newport Wharf was also being built in 1887 to handle slag disposal.

Incoming minerals entered the premises from the Marsh Branch, while outgoing goods and empty mineral wagons were despatched from the southern end of the site joining the NER's main line south of Newport Station. There were various sets of sidings and a railway line from the two groups of blast furnaces led to an extensive slag tipping area in the north-western 'quarter' of the premises, including behind the Newport Rolling Mills. At the turn of the century, the slag heap stood 140 feet high and was said to contain three million tons of slag.

Initially the works relied on coke supplied from Co. Durham but, in 8-9/1896, 70 Simon-Carves coke ovens commenced production to be followed by 50 (mid 1898) and 80 (1900-02) Otto-Hilgenstock waste heat ovens. Weekly coke production was 4,500 tons in 1908. Samuelson was one of the earliest to install a by-product coking plant. The Simon-Carves plant closed partly in 1913/14 and the remainder in 1918/19. The Otto-Hilgenstock coke ovens shut down in 1919-21. They were replaced by 46 Otto Regenerative ovens in 1914, to be followed by another 65 in 1922. It was in 1914 that Samuelson began supplying gas from the ovens to Middlesbrough Gas Works for town use. The coking plants were located between the two sets of blast furnaces and the river at the southern end of the site.

The number of blast furnaces operating between 1887 and 1913 varied but, latterly, was usually four or five because of the limited capacity of their blowing engines; Newport Works employed 1,500 people in 1889. Bernhard Samuelson had previously been involved in the establishment of the Britannia Works further north; its puddling furnaces and rolling mills provided a useful market for Newport's pig iron and a private railway line connected the two works. In the early 1900s Samuelson entered into an agreement with Dorman Long, now owners of Britannia, to supply it with molten iron. The iron was taken by rail in ladles between the two works – an early example of this practice which offered economies by reducing the amount of reheating required. In 1908 only five blast furnaces were working, because the ancillary plant could only service this number, three making Cleveland iron and two haematite iron. During the same year, a foreign ore gantry was built next to Newport Lower Wharf incorporating an underground longitudinal tunnel and mostly constructed of reinforced concrete. Up to 40,000 tons of incoming ore could be stored and then fed into bogie trucks on a merry-go-round narrow gauge track which was then discharged into standard gauge wagons for transport to the blast furnaces. Four 4-ton electric cranes were installed at the Lower Wharf which could accommodate ships of 7,000 tons. The Old or Upper Wharf at the southern end of the site, near the ancient ferry crossing, was then only used by smaller ships.

The railway system was extensive in 1913 serving both sets of blast furnaces and coke ovens. The area south of Newport Rolling Mills was now laid out as metal stocking grounds with sidings. A large slag heap remained between the Rolling Mills and the river but some of this was being processed in a slag crushing mill and a "Newport Dry Colour Works"; both buildings being served by sidings. The NER agreed on 24/1/1905 to provide additional sidings alongside the Marsh Branch for use by the Middlesbrough Slag Co Ltd, which had its slag breaker at the Newport Iron Works. A fleet of small elderly locomotives had shunted round the works but, from 1899, Samuelson began to standardise on a distinctive design of Kitson four-coupled locomotives with both saddle and well tanks.

The relationship with Dorman Long became more formal when a long-term supply agreement was signed in 1910. Dorman Long took over Samuelson in 1917, with the latter remaining as a subsidiary company until absorbed in 1923. In 1917, Newport Iron Works was being charged £50pa for its locomotives and wagons working over NER lines from various parts of its plant to its wharf. The difficult trading conditions soon called for rationalisation of plant and the three 'New Side' blast furnaces were demolished. On the 'Old Side', No.5 blast furnace had been rebuilt as a revolutionary oval (rather than circular) design in 1907, but it appears to have been worked intermittently, if at all, after relining in 1920. Iron works owned a substantial stock of mainly internal railway wagons as the 1929 schedule for Newport Works reveals;

2	8-ton acid tank wagons	12	"metal crates"
6	8-ton tarmacadam wagons	12	8-ton ballast wagons
75	10-ton tarmacadam wagons	1	10-ton bolster wagon
12	30-ton ore trucks	32	iron slag trucks
86	20-ton ore trucks	167	metal bogies
32	10-ton flat wagons	4	10-ton slag ladles
2	10-ton coke trucks	20	5-ton slag ladles
26	30-ton breaker wagons	14	12-ton hopper bottom door wagons
14	8-ton breeze trucks		

The great majority of wagons for external traffic were provided by the railway companies.

The Newport Iron Works site totalled 65 acres in 1929, the increase being probably due to the inclusion of the Newport Rolling Mills premises and had 13½ miles of railway sidings. The iron making plant closed in 1931. Not long after, the Dorman Long locomotives on Ironmasters were combined into one fleet. The Otto-Regenerative coke ovens continued operating until Dorman Long's large centralised Cleveland Coke Ovens came on-stream at South Bank and then closed in 1938/39. The Lower Wharf, with its ore handling facilities, continued to operate with Dorman Long's locomotives delivering iron ore to the Acklam blast furnaces. In the 1950s, the wharf was reported to be regularly used for the receipt of iron ore and scrap, but probably finished with the closure of the Acklam Iron and Steel Works in 1963.

An area of land at the Newport Works, formerly occupied by the Newport Rolling Mills, was known as the "Sydney Dump". This was used by Dorman Long from time to time for the storage of surplus steelwork and ingots etc. Traffic from Britannia's Nos.1 and 2 Mills and Bridge and Construction Department could travel here over either private lines or the diverted West Marsh Branch. At a LNER meeting on 17/10/1946, it was noted "a considerable traffic passing by private loco between Newport Ironworks and Britannia Works". It was clear that the traffic was travelling via the diverted MOR line and retrospective payment of tolls by Dorman Long was to be considered.

References :   'The Newport Oval Iron Blast Furnace', A.T. Ledgard, *Cleveland Industrial Archaeologist,* CIAS, 1979

'Kitson's Innovative Two-Tank Shunters', Trevor Lodge, *Industrial Railway Record* 220, IRS, 2015

'Obituary Sir Bernhard Samuelson', *The Engineer,* 12/5/1905

Gauge : 4ft 8½in.

	NEWPORT No.1	0-4-0tank		GW	174	1864	New	s/s
	NEWPORT No.2	0-4-0tank	VB	GW	189	1864	New	s/s
	BANBURY	0-4-0tank		HG	233	1866	New	s/s
	BODICOTE	0-4-0WT	OC	FJ	75	1868	New	s/s
	CHELTENHAM	0-4-0WT	OC	FJ	82	1869	New	s/s
	ERIMUS	0-4-0T	OC	FJ	79	1871	New	s/s
	SUMUS	0-4-0T	OC	FJ	90	1872	(a)	s/s
2	(BANBURY)	0-6-0ST	OC	FW	133	1871	New	s/s
3	(NEWPORT)	0-6-0ST	OC	FW	169	1872	New	(2)
	SOUTHFIELD	0-4-0T	OC	FJ	134	1874	New	s/s
4	(CHELSTON)	0-4-0ST	OC	YE	262	1875	(b)	(3)
5	(COMPTON)	0-4-0WT	OC	K	2362	1881	New	(7)
6	NEWPORT (BODICOTE)	0-4-0WT	OC	K	2363	1881	(c)	(7)
7	(SOUTHEND)	0-4-0ST	OC	P	462	1887	New	(3)
8	(STOCKBRIDGE)	0-4-0ST	OC	AB	650	1889	(d)	(7)
10		0-6-0ST	IC	Seaham		1883	(e)	(5)
9		0-4-0WTST	OC	K	3882	1899	New	(7)
11		0-4-0WTST	OC	K	3982	1900	New	(7)
	WHITEHAVEN	0-4-0WT	OC	FJ	76	1867	(f)	(1)
-		0-6-0	IC	RS	1073	1856		
	Rebuilt as 0-6-0ST Seaham					?	(g)	s/s
12		0-4-0WTST	OC	K	4382	1905	New	(6)
2		0-4-0ST	OC	RWH	1847	1881	(h)	(4)
1		0-4-0WTST	OC	K	4737	1910	New	(7)
3		0-4-0WTST	OC	K	4945	1913	New	(7)
13		0-4-0WTST	OC	K	5114	1914	New	(7)
No.14	NEWPORT	0-4-0WTST	OC	K	5115	1914	(i)	(7)
10		0-4-0ST	OC	HL	3482	1920	New	(7)
2		0-4-0ST	OC	HL	3477	1920	(j)	(7)

Several of the locomotives at Newport Iron Works carried a plate with their number and the word NEWPORT underneath. It is possible that others in the above schedule also had this name. Sir Bernhard Samuelson at Newport Iron Works in *Machinery Market* on 30/4/1909 advertised a Fletcher Jennings locomotive with 12in x 20in cylinders for sale. The only one at Newport Iron Works built with 12 inch cylinders was maker's number 82.

Andrew Barclay 650 of 1889 at the south end of the Newport Works in April 1955 with the lifting bridge built by Dorman Long over the Tees behind it. (G. Alliez photograph, NRM collection)

(a)   ex Britannia Iron Works Co Ltd, Middlesbrough.

(b)   locomotive was new to T. Charlton. T. Charlton & Co owned Slapewath Ironstone Mine and had interests in some Co. Durham collieries. The mine and collieries were offered for sale on 20/1/1880. It is known that Slapewath Mine had locomotives – they were reported sold by 22/12/1879 – and it is possible that YE 262 was one of them. Samuelson purchased Slapewath Mine in 1880.

(c)   new? Official Kitson records suggest that 2363 may have gone first to the contractor, Lucas & Aird, named "BUDDYCORT" before coming to Newport Works. Inness has the name as BODICOTE.

(d)   new; it was on loan to the Britannia Works when Dorman Long's locomotive fleets at the Ironmasters were combined.

(e)   ex the Londonderry Railway, Co. Durham, either in 12/1896 or 3/1897. It was obviously not well regarded as Samuelson offered it for sale in *The Colliery Guardian* on 11/11/1898.

(f)   ex Forcett Limestone Quarry Co Ltd, Forcett Quarry, Yorkshire (NR), /1900.

(g)   ex the Londonderry Railway, Co. Durham. It was initially bought by J.H. Denton, Middlesbrough, 3/1897, but the sale fell through and Samuelson purchased it, c/1901.

(h)   ex Erimus Iron Co Ltd, Thornaby.

(i)   new. To Dorman, Long & Co Ltd, Newlandside Quarry, Stanhope, Co. Durham, on loan, by 10/1930 and returned by 3/10/1931.

(j)   ex Thos W. Ward Ltd, Grays, Essex, /1933; previously Bombay Harbour Improvement Trust, Bombay, India.

(1)   to Tyne-Tees Steam Shipping Co Ltd, Stockton Wharf, Stockton, by 5/9/1923.

(2)   to Crown Coke Co Ltd, Consett, Co. Durham; it was subsequently advertised for sale there 1/1912.

(3)   to Britannia Works, Middlesbrough.

(4)   to Sir S.A. Sadler Ltd, Malton Colliery and Coking Plant, Co. Durham, c/1916.

(5)   to James Oakes & Co Ltd, Cotes Park Colliery, Alfreton, Derbyshire, via J.F. Wake c/1919.

(6)   to Redcar Works, Redcar, /1922.

(7)   renumbered in early 1930s (after 10/1931) into combined Acklam, Britannia and Newport Works list of steam locomotives (which see)

The Combined Acklam, Britannia, Newport steam and diesel locomotive fleets are located in the Britannia Steel Works entry.

NORTH SKELTON IRONSTONE MINE, North Skelton

Dorman, Long & Co Ltd until 2/10/1954
Bolckow, Vaughan & Co Ltd until 29/11/1929

65
NZ 675184 – Map I

North Skelton was the last Cleveland ironstone mine to close on 17/1/1964. Following its development of Eston Mine, Bolckow Vaughan looked further south for additional supplies of ironstone and it agreed a lease with J.T. Wharton of Skelton Castle to run from 1/1/1865. The bed of Main Seam ironstone proved to be at a much greater depth than expected and also in the form of a 'saucer-shaped' syncline with a significant inflow of water encountered during sinking, so it was not until 18/12/1873 that the winding shaft reached the ironstone at 720 feet. Meanwhile, a temporary railway line had been provided by the NER from the ex Cleveland Railway near Stanghow Lane in 6/1870. The first ironstone left for Bolckow Vaughan's works in 5/1874.

The NER opened its long loop from Priestfields Junction on the ex Cleveland Railway to North Skelton Junction on the Saltburn line in 1878. The permanent railway connection to the mine diverged from this loop line, south of the Vaughan Street bridge and curved round by the ends of the terrace houses in the appropriately named Williams, Wharton and Bolckow Streets to pass under the tippler building and reach the headshunt. There were sidings either side of the tippler building on which to stable the empty and full wagons. A separate siding ran back from the headshunt to serve the boiler house and mine yard. A total of 4,000 tons of coal were consumed in a year when ironstone production was 360,000 tons. The location of the shafts above the lowest part of the basin meant that a series of self-acting inclines could transport the stone underground to the principal 'roads' where it was handled by main and tail haulages. Power was provided from 1890 by a fixed Robey compound engine and later by 350hp and 200hp electric motors located near the shaft bottom. After World War I, a 1,650 yard underground self-acting incline carried ironstone from Longacres Mine to be raised at North Skelton.

There were four closures of 1-2 years in the 1920s–30s, but North Skelton was more fortunate than many of the other East Cleveland mines. North Skelton was unusual in that the pumping rods reduced the available space in the shaft and therefore the cages had two decks, each of which could take a single tub. The *Whitby Gazette* of 20/12/1946 described the arrangements at the top of the shaft: "… head of the shaft where twin cages were, in turn, ascending and descending. As the ascending cage reached the top it was seen to carry a loaded wagon which was pushed on to a weigh board, and run into another kind of cage where the wagon was somersaulted, thus losing its load of stone, and when righted was run on rails in anti-clockwise fashion until it reached the head of the shaft". There was no picking belt as North Skelton did not have a shale band in its ironstone. Any shale was stowed underground or dumped from the narrow gauge wagons between the mine and the village.

Output declined after World War 2 as a result of the loss of skilled manpower and Dorman Long embarked on a programme of mechanisation. The last steam winder at a Cleveland mine was replaced here in 1951 by an electric winder. Rotary drills, Eimco rocker shovels and a fleet of Ruston & Hornsby diesel locomotives were introduced underground and the last horses came to the surface in 9/1954. Also, shortly after, all of the Lumpsey ironstone began to be raised at North Skelton. During its life North Skelton Mine produced 25 million tons of ironstone; after closure, the last train load departed on 20/1/1964.

Reference : *Hope to Prosper A History of Ironstone Mining at North Skelton*, Simon Chapman, pub. P. Tuffs, 1997

Gauge : 2ft 6in

No.8	4wDM	RH	353493	1954	New	s/s
No.11	4wDM	RH	371942	1954	(a)	s/s
No.12	4wDM	RH	371947	1954	New	s/s
No.14	4wDM	RH	371949	1954	New	s/s
No.15	4wDM	RH	375693	1954	(b)	(2)
No.16	4wDM	RH	392100	1955	New	s/s
No.9	4wDM	RH	353494	1954	(c)	(1)
No.5	4wDM	RH	353484	1953	(d)	(3)
No.23	4wDM	RH	432655	1959	New	s/s
29	4wDM	RH	466579	1961	New	(1)

(a) new, it was delivered to North Skelton, 10/5/1954. A spares order was sent from Kilton Mine for this locomotive on 24/8/1954, but subsequent spares orders (3/1/1956 – 4/12/1957) came from North Skelton.

(b) new to North Skelton, 25/8/1954; to Kilton Mine 26/8/1954 (spares orders for this date state Kilton Mine, also Mines Engineer has "Kilton" written against this locomotive); returned by 16/8/1955 (date of spares order for North Skelton).

(c) ex Lumpsey Mine, by 7/12/1955.

(d) ex Kilton Mine, possibly in 1958-61 (recorded in W.C. Allan's notebook, Mines Engineer at Lumpsey and North Skelton).

(1)   to Lingdale Mine for storage only.
(2)   to Richard Thomas & Baldwins Ltd, Irthlingborough Mines, Northamptonshire, /1964, (purchased for spares only).
(3)   to British Steel Corporation, Apprentice Training Centre, South Bank, 6/1974.

REDCAR IRON AND STEEL WORKS, Redcar.   66
Dorman, Long & Co Ltd until 2/10/1954   NZ 576244, NZ 572253 – **Map H**
Walker, Maynard & Co Ltd until 2/1/1916
Walker, Maynard & Co until 5/12/1900
Robson, Maynard & Co until 1/7/1880

Pre 1916

Robson, Maynard & Co was formed in 1872 to build an iron works on the south east side of the NER Middlesbrough–Redcar railway opposite the site of Coatham Iron Works near Tod Point. Edward Robson's involvement with The Kirkleatham Ironstone Co (which see) meant that ironstone could be carried over the latter's private railway to the Redcar Iron Works. Houses for its workers were built on an extensive rabbit warren starting in 1873. First named Warrentown, later Warrenby, it had a population of 700 in 1881. Two blast furnaces began operating in 1874 and a start made on constructing two more but, almost immediately, there was a serious trade depression and Edward Robson withdrew from the ironstone company in 1878 and the iron works in 1880. Robson, Maynard & Co was dissolved and the business reformed as Walker, Maynard & Co in 1880.

The row of four blast furnaces were constructed at right angles to the NER, with the railway sidings swinging off the main line to curve round past both sides of the furnaces. In addition, the branch from Kirkleatham Mine joined these sidings from the south east; Walker, Maynard had taken over the mine in 1881, but it closed in 1886. A siding still ran south from the iron works in 1893 across Coatham Marsh, with its medieval 'salt mounds', to Wiley Bridge plantation where there had been a brick works and clay pit. At the Iron Works, incoming mineral wagons were raised up a hoist on to a gantry where there were two lines running over the top of the calcining kilns and storage bunkers. At the other end was a 'balanced drop'. On the opposite side of the blast furnaces, railway lines ran past two pig beds allowing wagons to be loaded, weighed, and delivered by the company's locomotives to the exchange sidings with the NER. Other lines led to the slag tipping sites. At the south east

MARY IRVINE, constructed by the local company of Hopkins Gilkes (298 of 1873), was rebuilt by Hudswell Clarke in 1902 before coming to the Redcar Iron and Steel Works.   *(NERA R. Inness photograph)*

A typical Hawthorn Leslie locomotive (3344 of 1918) employed at the Redcar Iron and Steel Works was coupled to a slag ladle on 23rd June 1961; note the chain to facilitate tipping. (IRS C.B. Golding photograph)

end of the blast furnaces were the boilers and blowing engine house with, a little further beyond, the two road engine shed (NZ 576244) and then a slag bank on Coatham Marsh. In 1885 a private railway was constructed by Walker, Maynard north from the works over the NER, to the west of Fishermen's Crossing. It then ran past the Coatham Iron Works site before turning west to cross the Tees Conservancy Commissioners' South Gare Breakwater line at right angles and on the level – a curve was put in to connect with it – and headed over Bran Sands to reach the company's Redcar Jetty, then under construction, on the River Tees. According to a 1900 prospectus, the wharf could accommodate the largest vessels then loading at any of the river wharves and had the capacity to deal with the shipment of all Walker Maynard's pig iron production if required. The opportunity was also taken to lay sidings on the east side of the jetty line providing additional areas for tipping slag on the marshes behind Coatham Sands.

A disastrous boiler explosion occurred at the Redcar Iron Works on 14/6/1895 resulting in the deaths of twelve men and the injury of at least nine others. The plant comprised a row of 15 'long' boilers supplying steam to the blowing engines and blast furnace lift engines. There was insufficient support at the mid-point of the 66 feet long boiler barrels causing them to sag and seams to rip. At around 9:20 pm, fortunately after the blast furnaces had been tapped, No.5 boiler exploded resulting in eleven other boilers bursting and dislodging the remaining three. Some of the boilers landed alongside the locomotive shed breaking all of the windows and destroying a six ton rail crane. The boiler plant was speedily replaced and it was reported that, since 1896, annual pig iron production had exceeded 115,000 tons during the following four years. A total of 140,000 tons had been shipped to Japan, China, India and Australia during this period.

In 1900 Walker, Maynard purchased the closed Coatham Iron Works and, later in the year, it became a limited liability company. Improvements were carried out to the Coatham blast furnaces and they were blown in during 1903–04. From then on, the two works operated as one with the Coatham furnaces as Nos.5 and 6. Up to 1913, of the six blast furnaces, four or five were in blast. The Coatham plant was connected to the Redcar Jetty line and also crossed over it on the level to reach the slag tips on the east side. Ancillary industries had been established on the slag tips, including J.C. Broadbent & Co Ltd's slag wool works, brick kilns, Messrs Brands slag works and a tarmacadam plant.

Post 1916

During World War 1, the MoM continually exhorted manufacturers to increase steel production for shells and ships. Discussions began with Walker, Maynard in 1915 for it to commence making steel, but it was reluctant to do so and sold the business to Dorman Long on 2/1/1916. No doubt with Government assistance, Dorman Long undertook a major expansion of Redcar Works. A battery of 65 Huessener waste heat coke ovens had been commissioned just prior to the takeover. These were located in line with the Redcar blast furnaces and were later supplemented by six 'Cleveland' experimental ovens of the regenerative type. In 1928, 2,500 tons of coke were being produced each week, with the excess gas sold to Redcar Council; a byproducts plant was located to the east. Following the opening of Dorman Long's Cleveland Coke Ovens, the decision was taken in 1938 to shut the Redcar coking plant.

The construction of the open hearth steel plant began in 2/1916 on previously tipped land adjacent to Redcar Jetty. It comprised three tilting and seven fixed furnaces, with a semi active mixer in the centre, and produced about 550,000 tons of steel annually, increasing Dorman Long's pre-war capacity by about 50%. The first steel ingots were produced on 9/2/1917 which was quite an achievement as the ground conditions had been sand and slag to a depth of 17 feet. An extensive network of sidings served the melting shop, stripping shed and adjacent machine shop. There was delay before the rolling mills were completed to the north west in line with the melting shop. The 9ft 6in plate mill began operating in 1920 and the innovative universal plate mill in 1922. Dorman Long had an agreement with the NER dated 21/9/1917 for temporary sidings off the main line into Coatham Works, but these were replaced by the permanent sidings on completion of the new plant (agreement dated 9/5/1921) which required adjustment of the Tees Conservancy Commissioners' Breakwater line and the extension of Tod Point Signal Box. Significant improvements were also made by the contractor, Holloway Bros, to Redcar Wharf about 1917–18. The wharf was 500ft 6in long with three electric 10 ton derrick jib cranes in 1937 and four standard gauge railway tracks ran along its full length. Incoming iron ore and scrap with outgoing pig iron, steel plate and crushed slag were dealt with, all moved by Dorman Long's locomotives. It could accommodate ships of up to 8,000 tons. Fourteen new four-coupled saddle tanks were purchased from Hawthorn Leslie in 1916–19. A new engine shed (NZ 572253) was provided for the steel works fleet next to the melting shop and quaintly labelled in the 1928 article "PUG-ENGINESHEDS".

There was an urgent need for houses for the new workers and Dorman Long developed a new settlement called Dormanstown. The project was overseen by Dormanstown Tenants Ltd, registered on 16/12/1919 and chaired by one of the Dormans. By the end of 1920, 76 houses were occupied and another 95 under construction. A standard gauge temporary railway line ran from near the Redcar blast furnaces past Wiley Bridge Plantation to the construction site in order to bring in materials. The NER had opened Warrenby Halt on its Middlesbrough–Redcar line by 10/1920.

Nos.1 and 4 Redcar blast furnaces had been rebuilt by 1928 with steel shells and larger 14ft 6in hearths; each could produce about 2,500 tons of iron per week. In contrast, the Coatham furnaces were taken out of blast during the 1921 coal miners' strike, never operated again and had been demolished by 1937. Nos.2 and 3 Redcar furnaces were dismantled about 1943–44. A new more extensive ore stocking ground was provided parallel to the main line and full railway wagons were pulled up the 1 in 3 incline to the 820 feet long gantry above the 16 storage bins by a 160hp electric motor. Empty wagons still returned to ground level by the 'balance drop' Most of the iron from the blast furnaces was tapped into 34 ton ladles and taken by locomotives to the steel plant, with only a small amount being put into the pig beds at weekends. Slag was transported in ladles and was being tipped to a height of 22 feet on land "which is still waterlogged".

Dorman Long purchased the land between its Cleveland and Redcar sites and constructed the Link Line in 1949 on the south side of the Middlesbrough–Redcar railway connecting the two works. The remaining two blast furnaces at Redcar were rebuilt with 18ft hearths, although their capacity was limited by the existing ancillary equipment and came on blast in 1953. Each could produce 3,600 tons of iron per week. Six Kling 55 ton or 70 ton ladles were sufficient to carry molten iron to the melting shop. For slag, eight 300cu ft ladles were used. Surplus molten iron could be hauled in Kling ladles from Redcar over the Link Line to the new open hearth melting shop that had been completed at Lackenby and steel ingots made the reverse journey in "ingot transfer cars" to the Redcar plate mills. Iron ore and coke increasingly came from the centralised reception facilities at South Bank and the 1959 Review states that these were moved by BR over its main line to the seven Redcar reception sidings. Use of Redcar Wharf was then mainly confined to iron ore and some scrap for the steelworks, with only an occasional outgoing cargo of steel plate.

When the third new blast furnace came on stream at Clay Lane, the ironworks at Redcar finished in 1962–63 and the two blast furnaces were demolished in 1968. The steel plant and mills now relied on molten iron from Lackenby and the Link Line was diverted in 1963 via a new bridge over the main line at Tod Point to reach them. Continuing developments at Lackenby included new plate mills in 1965–68 resulting in the closure of the Redcar Works' mills. Finally the commissioning of the Basic Oxygen Steel making plant at Lackenby in 1971 meant the end of the Redcar steel plant in that year. Soon the whole site was cleared for the erection of the new Redcar blast furnace. (see British Steel Ltd entry).

Until 1931 and possibly later, the locomotives at Redcar Works were divided into two fleets. A Dorman Long list dated 3/10/1931 shows 1,2,3,4,5,6 (HL/1918), 7 (HL/1918), 8 (HL/1919), 9 (HL/1919) and 10 (K/1905) as steel works locomotives and 6 (P/1896), 9 (rebuilt HC/1902), 7 (HC/1903), 8 (rebuilt HC/1903), 10 (HC/1906), 11, 12, 13, 14, 15 as iron works locomotives. As late as 1951, a few still carried plates showing their allocation e.g. "No.5 DORMAN LONG & CO LTD, REDCAR IRON & STEEL WORKS, STEEL WORKS DEPT". Occasionally another Dorman Long locomotive might be loaned to Redcar. Andrew Barclay 1224/1911 arrived secondhand at the Cleveland Works in 1947, was noted at Redcar on 26/3/1948 but was transferred from Cleveland to Burley Quarries in 9/1950.

From 1959, Dorman Long began to replace its steam locomotives with new Sentinel diesels. These were delivered to and serviced by Cleveland Works and were numbered in one combined list which can be found in the British Steel Ltd Teesside Works section. The initial allocation to the Redcar Works involved S10025–30/1960, S10036 and S10038/1960 and S10057/64/66–69/1961, but these could change subsequently based on traffic levels and overhauls. There was, of course, considerable exchange of traffic between Redcar and Cleveland/Lackenby Works.

References : *Historic Steam Boiler Explosions*, Alan McEwan, Sledgehammer Engineering Press Ltd, 2009, pp 133–44

'Blast Furnaces at Redcar 1870–2010', Charles H. Morris, *The Cleveland Industrial Archaeologist* 33, CIAS, 2012, pp 11–18

'The Redcar Works of Dorman Long & Co Ltd', The Iron & Steel Industry, 1928

The Manufacture of Steel Plates at Redcar Works, Dorman Long and Company Ltd

A Technical Survey of Dorman Long (Steel) Developments, Iron & Coal Trades Review, 1959

Dorman, Long & Co Ltd, Schedule of Properties, Volume 1, 1929

Gauge : 4ft 8½in.

1	(REDCAR No.1)	0-4-0ST	OC	HCR	123	1872	(a)		(3)
2	(REDCAR No.3)	0-4-0ST	OC	BH	333	1876	New		(5)
	REDCAR No.4	0-4-0ST	OC	BH	371	1877	New		(2)
3	(REDCAR No.5)	0-4-0ST	OC	RWH	1819	1880	New		(4)
	REDCAR?	0-4-0ST	OC	JF	2625	1876	(b)		(1)
4	(REDCAR No.6)	0-4-0ST	OC	BH	883	1888	New		(4)
5	(REDCAR No.7)	0-4-0ST	OC	HC	330	1889	New		(5)
6		0-4-0ST	OC	P	640	1896	New		(6)
9	MARY IRVING	0-4-0ST	OC	HG	298	1873			
		Rebuilt	HC			1902	(c)		s/s after /1931
16	(7) REDCAR	0-4-0ST	OC	HC	636	1903	New		Scr c/1953
8		0-4-0ST	OC	HC	405	1894			
		Rebuilt	HC			1903	(d)		Scr after /1931
17	(10) COATHAM	0-4-0ST	OC	HC	754	1906	New		Scr c/1958
1		0-4-0ST	OC	HL	3188	1916	New		Scr c7/1962
2		0-4-0ST	OC	HL	3189	1916	New		Scr /1961
3		0-4-0ST	OC	HL	3210	1916	New		Scr 3/1961
4		0-4-0ST	OC	HL	3255	1917	New		Scr /1962
11		0-4-0ST	OC	HL	3256	1917	New		s/s c/1963
12		0-4-0ST	OC	HL	3267	1917	New		Scr /1962
13		0-4-0ST	OC	HL	3268	1917	New		Scr /1961
5		0-4-0ST	OC	HL	3342	1918	New		Scr 3/1961
6		0-4-0ST	OC	HL	3343	1918	New		Scr c7/1962
7		0-4-0ST	OC	HL	3344	1918	New		Scr c7/1962
14		0-4-0ST	OC	HL	3346	1918	New		s/s c/1963
15		0-4-0ST	OC	HL	3366	1918	New		Scr 3/1961
8		0-4-0ST	OC	HL	3385	1919	New		Scr c7/1962
9		0-4-0ST	OC	HL	3429	1919	New		(8)
10		0-4-0WTST	OC	K	4382	1905	(e)		(10)
18	(KING EDWARD)	0-4-0ST	OC	HL	3915	1937	(f)		Scr /1962
(22)	(MAY)	0-4-0ST	OC	MW	756	1880	(g)		(7)
19		0-4-0ST	OC	HL	2905	1911	(h)		Scr c7/1962
20		0-4-0ST	OC	RSHN	7072	1942	New		(9)
21		0-4-0ST	OC	RSHN	7075	1943	New		(9)
23		0-4-0ST	OC	RSHN	7341	1947	New		(9)
24		0-4-0ST	OC	RSHN	7342	1947	New		(9)
25		0-4-0ST	OC	RSHN	7345	1947	New		(9)
26		0-4-0ST	OC	HL	3248	1917	(i)		Scr /1962

22		0-4-0ST	OC	HL	3348	1918	(j)	Scr /1961
No.1		0-4-0ST	OC	DL		1959	(k)	Scr c/1966
42		0-4-0ST	OC	AB	1620	1919	(k)	Scr c/1966
130	JUBILEE	0-4-0ST	OC	AB	1364	1914	(l)	Scr c/1966
159	CALEDONIA	0-4-0ST	OC	AB	1736	1921	(l)	Scr c/1966

(a) ex The Kirkleatham Ironstone Co, Kirkleatham Ironstone Mine, Dunsdale, prior to 30/9/1881, possibly in 1876.
(b) ex Kirkleatham Ironstone Mine, Dunsdale, /1887.
(c) ex HC, 4/1904; previously owned by Carnforth Hematite Iron Co Ltd, Carnforth, Lancashire. The latter had purchased three new 12in locomotives from Hopkins Gilkes, maker's numbers 246/1866, 253/1869 and 298/1873.
(d) ex HC after repair and receiving new boiler; it was ordered by Walker, Maynard on 7/3/1903. The locomotive was previously owned by Thos Wragg & Sons Ltd, Swadlincote, Derbyshire.
(e) ex Newport Works, /1922.
(f) ex Cleveland Works.
(g) ex Linthorpe–Dinsdale Smelting Co Ltd, Middleton St George, Co. Durham, /1937. It was rebuilt at some stage and fitted with a Barclay cab.
(h) ex Britannia Works, /1938.
(i) ex Dean & Chapter Colliery, Ferryhill, Co. Durham, after 22/3/1940, by 1/1/1947.
(j) ex Cleveland Works, c/1954, previously 146 FRASER.
(k) ex Acklam Works, /1962; No.1 by 29/1/1962.
(l) ex Cleveland Works, /1962.

(1) advertised for sale by J.F. Wake, 1/4/1889. Ken Fleming in *The Industrial Locomotive* 13 (ILS, 1978) suggests that it may have been used on T.D. Ridley & Sons Teesside Railway contract (which see).
(2) to Tees Scoriae Brick Co, /1898.
(3) scr c/1928; after 3/7/1927, by 1929.
(4) s/s, after 3/7/1927, by 1929.
(5) scr c/1930; after/1929, by/1931.
(6) s/s, after/1931, by 23/3/1948.
(7) to Parson Byers Quarry, near Stanhope, Co. Durham, 12/1954.
(8) to Acklam Works, c4/1962.
(9) to Acklam Works, 9/1962.
(10) to Gjers Mills & Co Ltd, Ayresome Iron Works, Middlesbrough, /1963.

DOWNEY & CO, COATHAM IRON WORKS 67

Redcar NZ 575251 – **Map H**

Alfred C. Downey joined with three other partners, including Carl Bolckow (nephew of Henry Bolckow), to form Downey & Co. A. Downey had been the Engineer at Cochrane Grove & Co's Ormesby Iron Works. The company agreed to lease 40 acres of Kirkleatham Estate land near Tod Point on the north west side of the Middlesbrough–Redcar railway on 6/4/1871 (lease not formally signed until 1/10/1874) in order to build an iron works. The intention was to erect four blast furnaces similar in design to those on the Ironmasters district in Middlesbrough but a fall off in the iron trade meant that only two were constructed (NZ 575251). The *Redcar and Saltburn News* on 12/6/1873 reported "on Tuesday, No.1 furnace at Coatham Ironworks was tapped for the first time, having been put in blast the previous morning. No.2 furnace will be blown today."

The blast furnaces were 85 feet in height, then typical of Cleveland best practice, although it appears that they were constructed of brick and therefore may have accommodated only a small number of tuyeres with a consequent loss of efficiency. The connection with the Middlesbrough-Redcar railway was immediately to the north of that for the South Gare Breakwater line. Incoming minerals in railway wagons were moved to a lift that raised them on to a prominent raised gantry carrying two railway tracks which ran along the south east side of the blast furnaces. From here, the minerals could be dropped into the calcining kilns and storage bunkers. According to the *Guisborough Exchange* "the steam hoist [was] constructed with an overhead cylinder, and the empty trucks are lowered by a balance drop". The ironstone, coke and limestone were loaded into hand barrows and six at a time were raised to the top of the furnaces by a pneumatic hoist manufactured by Hopkins, Gilkes & Co. The two vertical blowing engines were built by Cochrane, Grove & Co and there were six pipe stoves for heating the blast. Downey was allowed to tip slag on the leased land and a railway line ran from the blast furnaces north westwards over the sandy flat land. The Slag & Tarmacadam Co Ltd had a siding at the Coatham Iron Works.

With an improvement in the iron trade, Downey & Co took over Lackenby Iron Works in 1880. However, as the decade developed, the company got into increasing financial difficulties and iron production at Coatham was suspended in 1886. Downey & Co finally ceased trading on 21/3/1892 and the National Provincial Bank took control of both works on 12/4/1892, followed by attempts to sell them.

During 1896-99 William Whitwell, the Stockton Ironmaster, occupied the Coatham Iron Works but did not operate the blast furnaces. Walker, Maynard & Co purchased the works in 1900. Its subsequent history is covered under the preceding Dorman Long Steel Redcar Works entry.

Reference :   'Blast Furnaces at Redcar 1870-2019', Charles H. Morris, *The Cleveland Industrial Archaeologist* 33, Cleveland Industrial Archaeology Society, 2012, pp. 8-11

Gauge : 4ft 8½in

COATHAM No.1	0-4-0ST	OC	MW	438	1873	New	s/s
COATHAM No.2	0-4-0ST	OC	MW	456	1874	New	(1)

(1)   to Bolton Iron & Steel Co Ltd, Bolton, Lancashire.

EAGLESCLIFFE BRICKS LTD

CRADOCK, ALLISON & CO LTD   **68**
EAGLESCLIFFE WORKS, Eaglescliffe   NZ 419157 – **Map D**
Cradock, Wake & Co until /1901
COATHAM STOB ESTATES LTD   **69**
COATHAM STOB BRICK WORKS, near Eaglescliffe   NZ 412162 – **Map D**
(Subsidiaries of **Crossley Building Products Ltd**, later **Crossley & Sons Ltd**)

In 1896 Witham Hall was surrounded by fields and situated on the west side of the Northallerton-Stockton railway, ½ mile north of Eaglescliffe Station. However, this rural scene soon changed when a quarry opened in 1897. Cradock, Allison & Co Ltd appear as quarry owners and brick manufacturers in Ward's Directory 1902–03 and the 1904 *Railway Clearing House Handbook of Stations* has an entry for Witham Hall Quarry. The latter was located north of the hall and its linear orientation ran north-west to south-east. A 20 feet layer of clay, interspersed with pockets of sand, overlay the band of whinstone. Blondin aerial ropeways initially

Ruston & Hornsby 198245 of 1939 transports clay from the quarry on the 2ft 6in gauge railway to Eaglescliffe Brick Works, 5th October 1964.   *(IRS Brian Webb photograph)*

delivered the clay and stone to the works. The Eaglescliffe Works had been established on the north side of the quarry and here the stone was crushed, mixed with cement and made into concrete goods. The 1913 Ordnance Survey shows kilns located at either end of the principal building for brick manufacture. A standard gauge siding left the main line immediately north of the overbridge to Witham Hall and ran across a couple of fields to serve the works, where a large chimney marked the position of the boiler house. This was equipped with Lancashire boilers providing steam heating to the brick drying chambers. A Siding Agreement between the NER and Cradock Wake was dated 13/8/1898. Any shunting of the standard gauge sidings was presumably done by horses or the railway company.

Also by 1913, a second quarry/mine had been started further west beyond Durham Lane where a drift and air shaft were present. A 2ft 6in gauge single track railway ran from here for almost a mile to reach the Eaglescliffe Works, where the railway reversed direction at a headshunt and delivered the wagons to a discharge gantry for emptying, crossing over the standard gauge siding in the process. Cradock Allison advertised in the 19/11/1913 *Contract Journal* wanting four to six 2ft 6in gauge tip wagons. Quarrying at Witham Hall gradually extended up to Durham Lane, but increasingly activity switched to Coatham Stob Quarry, west of Durham Lane, with clay being taken from near the surface and whinstone about 180 feet below ground. Eaglescliffe Works originally had four large double-ended kilns but, by the late 1920s, these had been replaced by two Belgian kilns – a 22 chamber fully continuous kiln and a 16 chamber semi-continuous kiln.

In 1934 Crossley took over Cradock, Allison & Co and, in the following year, formed Crossley Building Products Ltd. This was made up of a number of separate companies, including Cradock, Allison (quarry owners and concrete manufacturers) and Eaglescliffe Bricks Ltd (brick makers). Crossley also established a subsidiary company, Coatham Stob Estates Ltd, to be responsible for the Coatham Stob Quarry and a new brick works opened nearby on 16/3/1938 with access off Durham Lane. A Crossley catalogue dated 10/1937 listed, amongst its products, common bricks (the about to open Coatham Stob); common bricks, hand-made sand faced bricks and "Redruf" facing bricks (Eaglescliffe Bricks) and concrete goods and whinstone aggregates (Cradock Allison). "Redruf" bricks had powdered whinstone blasted on to the faces which was claimed to make a more durable brick. The whinstone quarry closed on 2/2/1957. Cradock Allison had served notice to cease using the standard gauge connection with the main line in a letter dated 2/11/1954.

It is thought that initially Eaglescliffe Bricks Ltd worked the quarries for both works until late 1938 when Coatham Stob looked after its own requirements. It began to use dumper trucks in 1962. The Eaglescliffe Works still relied on the narrow gauge railway in 2/1966 when Peter Hutchinson filmed the line. A rake of about five side tipping wagons, each containing approximately two tons of clay, was filled by a Ruston Bucyrus electric drag line digger at Coatham Stob Quarry and a Ruston & Hornsby locomotive pulled them across Durham Lane and along the narrow gauge line bordering the former Witham Hall Quarry to Eaglescliffe Works. According to Peter Hutchinson an engine shed was located next to the line on the east side of Durham Lane. However, there was also a small brick built shed at Eaglescliffe Works. (see *Industrial Railway Locomotive Sheds 4*, IRS, 2022). Eaglescliffe Brick Works closed on 19/9/1967, but the Coatham Stob Works continued until the 1990s.

References : *Crossley's Clay & Concrete Products*, Crossley & Sons Ltd, 10/1937

Brickworks of the North East, Peter J. Davison, Gateshead Libraries & Arts Service, 1986

The Brickworks of the Stockton-on-Tees Area, Alan Betteney, Tees Valley Heritage Group, 2007

There are considerable difficulties concerning the locomotives used by the two companies, the version disliked least being given below. IRS records suggest that a Lister four-wheel petrol locomotive may have worked here also.

Gauge : 2ft 6in

Name	Type	Cyl	Builder	Works No	Built	Notes	Disposal
ZURIEL	0-4-0T	OC	WB	1917	1910	(a)	@ Scr c/1946
MAGNET II	0-4-0ST	OC	WB	1877	1911	(b)	@ Scr c/1946
-	4wDM		HE	1929	1938	(c)	* Scr c/1968
-	4wDM		RH	198245	1939	New	@ Scr c/1968
-	4wDM		RH	183726	1937	(d)	* Scr /1969
	Rebuilt				1946		
-	4wDM		RH	242914	1946	New	* Scr /1969

@ Owned by Eaglescliffe Bricks Ltd. * owned by Coatham Stob Estates Ltd.

(a) ex G. Hodsman & Sons Ltd, Port Clarence. The locomotive was advertised for sale in the *Contract Journal* on 17/9/1917. It then passed to T.D. Ridley & Sons, Middlesbrough for storage before arriving at Cradock, Allison & Co Ltd in c/1922.

(b) ex Francis R. Thompstone & Sons Ltd, Bosley, Cheshire, c/1925.

(c) new. To South Bank Brick Co Ltd, South Bank (subsidiary of Crossley & Sons Ltd), /1946 and returned /1963 to Eaglescliffe Bricks Ltd.

(d) ex Edmund Nuttall Sons & Co (London) Ltd, Letterston, Pembrokeshire. It was new as a 2ft 0in gauge locomotive to Nuttall's contract at Rhosdenni Bridge. Letterston, possibly in association with the establishment of Trecwn Armament Depot. An unknown RH arrived at Coatham Stob in 1942 and this may be it. Spares came from RH 16/5/1946 for conversion to 2ft 6in gauge.

EDF ENERGY PLC

HARTLEPOOL POWER STATION, Seaton Carew   **70**
British Energy Generation Ltd until 1/2009   NZ 532270 – **Map A**
Nuclear Electric plc until /1996
National Power (a division of CEGB) until /1991
Central Electricity Generating Board until /1990

The nuclear power station is situated at the end of the 1½ mile long Seaton Snook Branch, which leaves the main line ¾ mile south of Seaton Carew Station. The power station was designed and built by the Nuclear Power Company (Whetstone) Ltd with Taylor Woodrow Ltd responsible for the civil engineering. Construction began in 10/1968 and electricity was first supplied to the National Grid in 12/1972 from four gas-turbo generating units. The two advanced gas cooled reactors linked to generators produce 1,332 MW (gross) of electricity and began providing power to the grid on 1/8/1983. The sidings within the power station comprise two tracks curving round to finish as a single line at the engine shed, a line entering the reactor building for the delivery and collection of flasks and a short siding on which wagons can be stabled.

In 1990 the government divided the CEGB into four separate companies prior to privatisation and the Hartlepool Station became part of National Power. However, when this was privatised in 1991, the nuclear element was taken out and became Nuclear Electric plc. The duties of the locomotive at the power station are not too demanding and generally it is only used once a week. The main line locomotive used to transport the flask up to the site boundary but now takes them into the power station, with the works shunter being responsible for delivering flasks into the reactor building. An open event held over the weekend of 9-10/7/1988 saw the visit of former BR 69023 (Darlington /1951) giving footplate rides.

The locomotive at Hartlepool Power Station on 18th March 1983 was Ruston & Hornsby 544996 of 1968, a member of the LLSH class, the last to emerge from the Lincoln Works, of which only four were built. It was painted in a colour close to LNER blue because, it was said, one of the station managers had lived by the East Coast Main Line!   *(Cliff Shepherd photograph)*

Gauge : 4ft 8½in

-	4wDH	RH	544996	1968	(a)	(1)
-	0-6-0DH	JF	4240015	1962	(b)	(3)
03 (D3932, 08764) FLORENCE	0-6-0DE	Hor		1961	(c)	(2)
H003 ROSEDALE	4wDH	S	10070	1961	(d)	(4)
H058	4wDH	RR	10280	1968	(e)	(5)
-	0-6-0DH	EEV-AEI	4003	1971	(f)	

(a) ex CEGB, Mexborough Power Station, Mexborough, South Yorkshire, 18/7/1981.
(b) ex CEGB, Stella South Power Station, Blaydon, Tyne & Wear, 23/2/1987.
(c) ex Sheerness Steel Co Ltd, Sheerness, Kent, (property of R.F.S. Engineering Ltd), on hire, after 29/6/1993, by 7/10/1993.
(d) ex RMS Locotec, Dewsbury, West Yorkshire, on hire, 12/1/1994.
(e) ex RMS Locotec, Wakefield, West Yorkshire, on hire, 9/1/2008, arrived 11/1/2008.
(f) ex Chasewater Light Railway and Museum Company, Brownhills, Staffordshire, (property of Ed Murray & Sons Ltd), on hire, 23/7/2018 via haulier's yard from 20/7/2018.

(1) to Barnsley Metropolitan Borough Council, Elsecar & Cortonwood Project Group, Elsecar Workshops, Wombwell, South Yorkshire, c/1993, by 3/11/1993.
(2) to R.F.S. Engineering Ltd, Doncaster, South Yorkshire, ex hire, after 3/11/1993, by 6/2/1994.
(3) to Rutland Railway Museum, Cottesmore, Leicestershire, 14/2/1995.
(4) to RMS Locotec, Wakefield, West Yorkshire, ex hire, 11/1/2008.
(5) to British American Railway Services Ltd, Wolsingham, Co. Durham, ex hire, by 2/2019.

ERIMUS IRON CO LTD

ERIMUS IRON WORKS, Thornaby 71
NZ 464188 – **Map E**

The Erimus Iron Co Ltd was formed in 1872 and proceeded to erect its works with six puddling furnaces on the Danks principle, on the marshes between the Darlington-Middlesbrough railway and the south bank of the River Tees immediately to the west of the old course of the river. The directors were C.E. Muller, John Jones, W.S. Ayrton and J.A. Jones; the latter being one of the three UK representatives sent to America to inspect Danks' invention. The Jones brothers had also recently established the Ayrton Sheet Works in Middlesbrough. The company erected three rows of terrace houses for its workers and the unusually grand Erimus Hotel (date stone 1875) next to the Middlesbrough Road, known locally as the "Wilderness Road" (the road had opened on 5/11/1858). Unfortunately the depression in the iron trade resulted in the works closing in 12/1875 putting 300 people out of work.

Erimus Iron Works was remodelled in 1880 to produce steel using Bessemer converters but again with little success. The Stockton Extension & Improvement Plan prepared for the 1889 Parliamentary session has it as "not working" and the works was offered for sale in *The Engineer* on 5/9/1890. The layout of the railway system was shown on the 1895 Ordnance Survey. The NER had a set of four sidings on the north side of the main line to handle wagons for the Thornaby and Erimus Iron Works. From here, a railway ran into the Erimus site to a fan of five sidings, from which individual lines served the remainder of the premises. Although situated next to the river, there does not appear to have been a wharf. Wm Whitwell & Co purchased the Erimus Iron Works in 1897 and dismantled it. Although this increased its available area for slag tipping, the southern half of the site was subsequently used by the NER to lay out part of its Erimus marshalling yard.

Gauge : 4ft 8½in

-	0-4-0ST	OC	JF	1575	1873	New	s/s
-	0-4-0ST	OC	RWH	1847	1881	New	(1)

(1) to Sir Bernhard Samuelson & Co Ltd, Newport Iron Works, Middlesbrough.

Gauge : 2ft 6in

1	0-4-0ST	OC	BH	562	1880	New	(1)
2	0-4-0ST	OC	BH	563	1880	New	(1)

(1) to Robert Pitt, contractor, Liverpool. He advertised in *The Engineer* on 3/8/1883 to hire or buy cheaply a second hand 2ft 6in gauge locomotive but it is not known for which contract, nor whether it was this which promoted the move of one or both locomotives to Liverpool.

ESTON GRANGE IRON CO/W. BACON & CO

ESTON GRANGE IRON WORKS, Grangetown

approximately NZ 553219 – **Map H**

On 19/3/1869 John Thomas of Middlesbrough, William Bacon of Newcastle upon Tyne and Harrison Groves of Redcar filed a patent for an invention of "improvements in the manufacture of iron and steel, and in furnaces and apparatus employed there in". A partnership was formed comprising Bacon, Groves, Thomas and Hugh Chaytor under the style of W. Bacon & Co to act as iron refiners, although Chaytor left the company on 2/3/1872. His place was taken by Thomas Ingledew who was a brick and tile maker at Linthorpe. The company initially operated from the Acklam Iron Refinery which, according to the 17/3/1875 edition of the *Northern Echo* was near the Tees Side Iron Works and had been there "for some years". In the 1871 census, William Bacon (58) was living in East Coatham and was described as head of the firm; he died on 7/3/1873. William Bacon & Co won an action against Gjers Mills & Co for breach of contract for not delivering the proper quality of "pig bed scrap".

According to the above *Northern Echo* article, the "Acklam Iron Refinery… is shortly to be removed to Eston". This referred to the Eston Grange Iron Works and a subsequent sale notice dated 19/4/1882 for this site gives a clue as to its location. It was south of the Middlesbrough-Redcar railway near Bolckow Vaughan's new steel works and the NER was aiming to establish a station between the steel works and the Eston Grange Works. This suggests that it was probably where the Grangetown Power Station was later built. However, no sooner had the Eston Grange Iron Works been erected, the worsening depression in the iron trade forced the partnership between Thomas, Groves and Ingledew to be dissolved on 31/8/1875. (There were two notices with the same date and signatories, one for W. Bacon & Co and the other for the Eston Grange Iron Co.) Both Thomas and Ingledew called meetings of creditors in 1/1876 as part of the process of liquidation. An advertisement in the *Northern Echo* announced the sale of plant and materials at the Eston Grange Works on 1-2/3/1876, including a "nearly new locomotive". This is likely to be that shown below which had been sold to W. Bacon & Co. An advertisement for the approximately five acre site, which was connected to the NER by sidings, appeared in the *North Eastern Daily Gazette* on 19/4/1882. A subsequent notice for a sale on 27/8/1883 indicates the dismantling of the premises. Amongst the items were "2,079 yards of F.B and D.H. rails, 72 and 80lb", a Pooley platform weighing machine, about 68 tons of cast iron 10-19ft long columns, 31½ft wrought iron girders and 50,000 cleaned fireclay and red bricks.

Meanwhile, J. Thomas had earlier put up the plant at the Acklam Iron Refining Works for sale by auction on 9/8/1876. This included the contents of the refining foundry (including one egg-ended boiler), smith's shop, pattern shop and offices.

Gauge : 4ft 8½in

-	0-4-0ST	OC	YE	236	1874	New	(1)

(1) later at Bell Bros Ltd, Clarence Iron Works, Port Clarence; spares were sent there in 12/1888.

ESTON SHEET & GALVANIZING CO LTD

ESTON SHEET WORKS, South Bank.
(Subsidiary of **Bolckow Vaughan & Co Ltd** from /1916)

NZ 531219 – **Map H**

The works was situated immediately south west of South Bank Wharf on the River Tees. The Siding Agreement between the NER and Eston Sheet & Galvanizing Company was dated 28/1/1913. In 1926 it was manufacturing black and galvanised plain and corrugated steel sheets by "cold rolling, close annealing, patent flattening and pickling".

The premises had a rail connection to Bolckow Vaughan's line that ran from the iron works at South Bank and Clay Lane to the wharf. A single siding entered the Eston Sheet Works where it divided into two pairs of tracks on the east and south sides of the main building. Presumably Bolckow Vaughan's locomotives initially delivered and collected traffic. Under the 1913 Act, two additional sidings for the company's traffic, to be exchanged with the NER, were approved on the north side of the main line and Bolckow Vaughan's sidings. Revised arrangements were agreed on 23/3/1928. There were now to be seven sidings, instead of two and these were for the company's inward traffic and despatches that could not go through Bolckow Vaughan's "subway". Three new sidings for outward traffic were to be installed on the south side of the main line to be reached by the "subway". Following Eston Sheet & Galvanizing Co's purchase of its own locomotive, the LNER approved on 30/10/1929 its use over the railway company's tracks. It may have been a coincidence that Dorman Long had taken over Bolckow Vaughan's responsibilities as from 1/11/1929.

By 1927, a separate siding entered the premises from the north leaving Bolckow Vaughan's line as the latter curved round to South Bank Wharf. During the 1930s, as part of reducing surplus capacity, the National Sheet

Conference proposed to buy the Eston Sheet Works and close it down. According to the *North-Eastern Gazette* on 11/8/1939, Mrs Elsie Foster, who ran E. Hind (South Bank) Ltd, had attempted to take over the works and keep it running. Although unsuccessful, she nevertheless purchased the premises and started a refined pig iron business there producing high-grade castings for the motor industry. The implication in the article was that this had been operating for about three years and that a new cupola furnace had just been installed. The locomotive was acquired by Thos W. Ward Ltd in 10/1936 and used in dismantling the works before moving to Sheffield.

Gauge : 4ft 8½in

-		0-4-0ST	OC	HL	3684	1929	New	(1)

(1) to Thos W. Ward Ltd, Sheffield, Yorkshire (WR) which sold it on 23/1/1937 to Sir Lindsay Parkinson & Co Ltd on a contract for the Chorley Royal Ordnance Factory.

J. GARRITY & CO

SCRAPYARD, Middlesbrough

74
NZ? – **Map G**

According to a 1963 Trades Telephone Directory, Garrity's scrapyard was based at the "B.R. Dock Entrance" in Middlesbrough; presumably this was near the Dock stockyard but a specific site has not been identified. Garrity comes to our attention because of his dealings with the Tyne-Tees Steam Shipping Co Ltd. He scrapped two of its locomotives on their respective sites: SUTHERLAND (HL 2439 of 1899) at Tyne-Tees Wharf, Middlesbrough on 17/4/1961 and SWILLINGTON (HCR 150 of 1874) at Stockton Wharf about 1/1964.

Gauge : 4ft 8½in

No.1	0-4-0ST	OC	RSHN	7084	1943	(a)	Scr
No.2	0-4-0ST	OC	HL	3006	1913	(b)	(1)

(a) ex Smith's Dock Co Ltd, South Bank, after 4/1960.
(b) ex Smith's Dock Co Ltd, South Bank, after c5/1961. Bob Payne had noted it as belonging to Garrity at South Bank on 20/4/1961.

(1) to Tyne-Tees Steam Shipping Co Ltd, Tyne-Tees Wharf, Middlesbrough, c2/1961.

GJERS MILLS & CO LTD

AYRESOME IRON WORKS, Middlesbrough
Gjers Mills & Co until 29/11/1901

75
NZ 484216 – **Map F**

After his significant involvement in both the Tees Side and Linthorpe Iron Works, John Gjers decided to establish his own iron works. Gjers Mills & Co was formed during 1869 with three partners – John Mills (a banker from Stockton), Thomas Dodson and Eliezor Emmerson. A 32 acre site was purchased in the Ironmasters District, with a river frontage of 330 yards, to the north of the Britannia Works, then under construction. Work commenced on the erection of two blast furnaces at the beginning of 1870. John Gjers gave up his contact with Tees Side and Linthorpe Works in 1/1871 to concentrate on his own business. The blast furnaces were blown in on 24/3/1871 and work began immediately on two more which were operational by mid-1872.

The blast furnaces then represented the best in Cleveland iron making practice and stood dramatically in a row surrounded by open ground and marsh which would rapidly be covered by mounds of slag and railways. The furnaces were 85 feet high with brick-built iron banded stacks supported on massive brick piers. Two vertical blowing engines, eight calcining kilns, pneumatic hoists and pipe stoves were all provided to Gjers' own designs. A rail connection was installed from the end of the NER's Marsh Branch. This crossed over a strip of OME land making rail traffic passing over the 3 chains 8 yards section of line subject to MOR tolls. In 1914 it seems the tolls applied on inward access to the calcining kilns but not the outward track for its pig iron to the NER. Gjers' railway then ran either side of the blast furnaces, with the kilns and stoves to the north and the pig beds and slag discharge points to the south. Also on the south side of the site, a line continued on to serve the company's river wharf next to the Britannia Works' northern boundary by 1875. Other sidings allowed slag to be dumped on the surrounding land.

In a paper published in 1871, John Gjers described how the minerals arrived in train loads of hopper wagons, each carrying either 10 tons of ironstone or 6 tons of coke. These were placed in the storage sidings from where the works locomotive, a four-coupled 12 inch tank engine, took them as required over the weighbridge to the incline road. Here they descended by gravity to the pneumatic lift where they were raised, one by one, to the 36 feet high gantry at the top of the calcining kilns and stoves. After the minerals were tipped through the bottom boards, the trucks ran by gravity to the far end where they were let down by a pneumatic 'drop' and sent via a "self-acting return switch" to the standage sidings for empties. At the blast furnaces the slag was run into fixed boxes, five to each furnace, with the slag bogies being taken underneath by a small locomotive. Construction of the 210 feet long river wharf was underway in 1871 and this was to have two 4-ton travelling steam cranes. The wharf would be served by three railway tracks; the outer two falling towards it so that full wagons could reach the cranes and the middle road for empties descending away. The repair shops were stated to be located in one range of buildings that also contained the engine shed.

Mr Hopwood's photograph of 1873 showing the four blast furnaces (a copy is in Dorman Museum) has in front "Two horizontal boiler and one vertical boiler locomotives" with several pig carrying wagons. The early involvement of vertical boiler locomotives is confirmed by a letter from Head Wrightson & Co to Gjers Mills dated 17/4/1871: "Also, for a 6" vertical boiler geared locomotive exactly the same as that previously supplied but with duplicate cast iron pinion for the sum of £425 delivered to your works". Hjerleid & Spence from the Marsh Road Engine Works offered, on 30/6/1871, to supply a steam crane for £340. Gjers Mills then proceeded to purchase ten new Manning Wardle four-coupled saddle tanks between 1871 and 1916. These were either Class H 12 inch or latterly Class P 14 inch and most had an increased wheel diameter of 3ft 6in.

By 1893 the number of sidings at the works had increased, although the overall layout was not significantly different. The link with the Marsh Branch entered the premises to connect with a set of exchange sidings. A building (NZ 485214) with two tracks entering it and situated in the middle of these sidings was the engine shed. As before, lines continued on to the blast furnaces and river wharf. Much of the remaining land was subject to slag tipping with tracks laid to carry the slag bogies, however, by 1913 this was less evident and the area adjoining the boundary with the Britannia Works appears to have been laid out as a stockyard with four sets of tracks, three of which were served by "travelling cranes".

The weekly output from the four blast furnaces was 2,900 tons of iron in 1893. Initially the works had smelted Cleveland ironstone from Spa Mine, coke from Co. Durham and limestone from its Aycliffe Quarry but, over the years, it turned to producing haematite pig iron and special foundry iron, with most of the iron ore coming from Spain. The calcining kilns only functioned as such when foundry iron was being made, otherwise they were used as stores. In 1901 the business was reconstituted as a private limited company. During World War 1, a ¼ million shell cases were manufactured at the Ayresome Works and another blast furnace was blown in on 17/3/1917 to meet the demand for iron.

With difficult trading conditions after the war, little development took place apart from the installation of eleven Cowper stoves to replace the outdated pipe stoves. Ayresome Wharf was recorded as 750 feet long with six 5 ton electric cranes in 1926; iron ore was imported and pig iron and crushed slag shipped out. A letter from a Mr Walker, presumably from the LNER, dated 7/2/1942 agreed that Dorman Long locomotives and ladles could

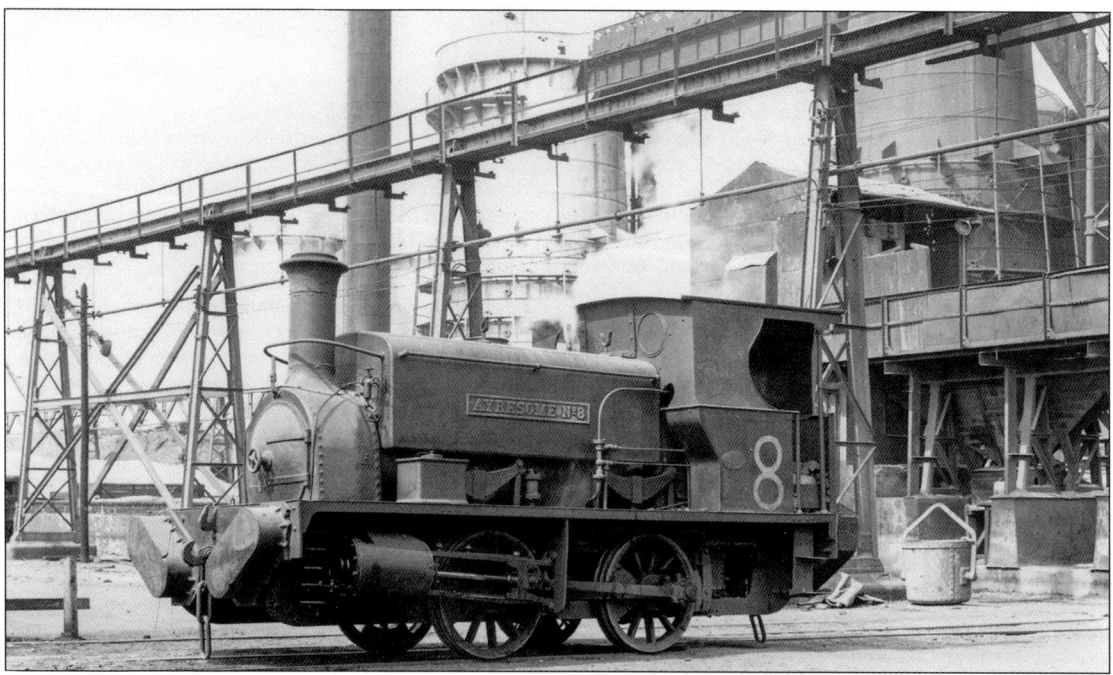

Gjers Mills operated a delightful fleet of elderly Manning Wardle locomotives at its iron works for many years. This is AYRESOME No.8 (1390 of 1898) standing in front of the blast furnaces on 20th June 1951.

(IRS K.J. Cooper photograph)

work from Gjers Mills to the Britannia Steel Works and Acklam Iron Works. This was in lieu of traffic from the Acklam to Britannia Works and the amount was limited to 600 tons per week. In 1943 the business was acquired by the Whitehead Industrial Trust Ltd and it became a public company a year later. Only two blast furnaces were working by 1953 and Nos.1 and 3 were retained just to support the furnace gantry. The plant had changed little over its 80 years existence and presented a distinctly archaic appearance. There was a belated attempt to modernise the locomotive fleet possibly stimulated by the visit of a Sentinel demonstration locomotive (maker's number 9561) to Teesside and five Sentinel 100hp vertical boiler steam engines were purchased from the manufacturer. In 1960 No.1 Blast Furnace was demolished. After the abortive nationalisation of the iron and steel industry in the early 1950s, Ayresome Iron Works was eventually sold by the Iron & Steel Holding and Realisation Agency to The Millom Hematite Ore & Iron Co Ltd in 1963. As this was a direct competitor and with demand for haematite pig iron declining, it was no surprise on 30/6/1965 that the final blast furnace (No.2) at Ayresome was taken out of blast and the remaining 450 employees paid off. The works closed on 3/7/1965 and was demolished by Thos W. Ward Ltd in 1966-67.

References :   'John Gjers: Ironmaster, Ayresome Ironworks, Middlesbrough', John K. Harrison, *De Archaeologische Pers*, Nederland, 1982

'The Gjers Mills' Sentinels', Cliff Shepherd, *Industrial Railway Record* 193, IRS, 2008

Gjers Mills and Company Limited – A case study of a Victorian Ironworks, B.E.M. Macklin, *Transactions of the Teesside Industrial Archaeology Group*, c/1969

Gauge : 4ft 8½in

-		0-4-0VBT	VCG	HW		c1870/71		Scr
-		0-4-0VBT	VCG	HW		c1870/71		Scr
	AYRESOME No.2	0-4-0ST	OC	MW	333	1871	New	(1)
	AYRESOME No.4	0-4-0ST	OC	MW	399	1872	New	(2)
	AYRESOME No.5	0-4-0ST	OC	MW	777	1881	New	(3)
6	AYRESOME No.6	0-4-0ST	OC	MW	1022	1887	New	Scr /1956
7	AYRESOME No.7	0-4-0ST	OC	MW	1161	1891	New	(6)
8	AYRESOME No.8	0-4-0ST	OC	MW	1390	1898	New	(4)
9	AYRESOME No.9	0-4-0ST	OC	MW	1457	1899	New	Scr c/1963
	AYRESOME No.10	0-4-0ST	OC	MW	1529	1901	New	Scr c/1955
11	AYRESOME No.11	0-4-0ST	OC	MW	1714	1907	New	Scr c/1963
12	AYRESOME No.12	0-4-0ST	OC	MW	1903	1916	New	(5)
1	AYRESOME	0-4-0ST	OC	P	845	1900	(a)	Scr /1967
3	(DINSDALE No.3)	0-4-0ST	OC	P	1058	1906	(b)	Scr c/1963
4	AYRESOME	0-4-0ST	OC	P	956	1903	(c)	Scr c/1963
1	AYRESOME	4wVBT	VCG	S	9566	1954	New	(7)
2	AYRESOME	4wVBT	VCG	S	9594	1955	New	(10)
3	AYRESOME	4wVBT	VCG	S	9598	1955	New	(8)
4		4wVBT	VCG	S	9600	1955	New	(11)
5	AYRESOME #	4wVBT	VCG	S	9613	1956	New	(9)
10		0-4-0ST	OC	K	4382	1905	(d)	Scr /1967
21		0-4-0ST	OC	RSHN	7075	1943	(e)	Scr /1967
179	CHURCHILL	0-4-0ST	OC	RSHD	7041	1941	(e)	Scr /1967
184		0-4-0ST	OC	RSHN	7348	1947	(f)	Scr /1967

\#   AYRESOME nameplate supplied loose by Sentinel but may not have been carried.

(a)   ex Linthorpe Dinsdale Smelting Co Ltd, Dinsdale, Co. Durham, after 23/8/1946, by 14/4/1949.

(b)   ex Linthorpe Dinsdale Smelting Co Ltd, Dinsdale, Co. Durham, after 23/8/1946, by14/4/1949. To Aycliffe Lime & Limestone Co Ltd, Aycliffe, Co. Durham, after 3/1957 and returned /1962.

(c)   ex Tarslag Ltd, Stockton after 24/3/1947. It was noted at Ayresome on 30/10/1949.

(d)   ex Dorman Long (Steel) Ltd, Redcar Works, /1963.

(e)   ex Dorman Long (Steel) Ltd, Acklam Works, 13/2/1964.

(f)   ex Dorman Long (Steel) Ltd, Acklam Works, 11/1964.

(1)   to Topham, Jones & Railton Ltd, contractors, Swansea, via MW.

(2)   to T.J. Thomson & Son Ltd, Stockton, /1927.

(3)   to Aycliffe Lime & Limestone Co Ltd, Aycliffe, Co. Durham, 6/1950.

(4)   scr c/1955, after 4/1955.

No.4 AYRESOME (P 956/1903) was in Gjers Mills' exchange sidings on the Ironmasters with one of British Railways 350hp diesels shunters, 19th May 1960. (IRS J.P. Mullett photograph)

Following a visit from the Sentinel salesman in the mid-1950s, Gjers Mills purchased five new vertical boiler locomotives and 4 (9600/1955) is operating by the engine shed. (IRS J.P. Mullett photograph)

(5) to Aycliffe Lime & Limestone Co Ltd, Aycliffe, Co. Durham, 3/1956.
(6) scr c/1960, after 21/5/1960.
(7) scr c/1967, after c/1963.
(8) scr c/1967, after 9/4/1964.
(9) scr c/1967, after 10/1964.
(10) scr c.1967, after 14/3/1966.
(11) scr c/1967, after 22/7/1966.

WILLIAM GRAY & CO LTD

SHIPYARDS AND ENGINEERING WORKS, Hartlepools
William Gray & Co (1918) Ltd until 31/12/1922
William Gray & Co Ltd until 31/12/1918
William Gray & Co until 1/1/1889
Denton, Gray & Co until /1874

J.P. Denton joined with William Gray in 1863 to build iron ships at its Middleton Shipyard. Increasing business meant that larger facilities were required and, in 6/1868, the company leased the former Pile, Spence yards next to Jackson and Swainson Docks, the first ship being launched from here in 4/1869. John Denton died in 1871 and there was a lengthy dispute about the composition of the company until William Gray gained overall control in 1874. Gray achieved the maximum output of any British shipyard for 1878.

In 1883 a 10 acre site alongside the North Basin was acquired by William Gray to establish its Central Marine Engineering Works comprising an engine shop, boiler shop, foundry and machine shops. Although nominally a separate company, it remained a wholly owned subsidiary of William Gray throughout its existence. Engines, boilers and auxiliary equipment were manufactured to power ships. Due to continuing demand for larger ships, William Gray leased a 5½ acre site at the north end of the Central Dock in 1887. It had three building births up to 450 feet in length and was separated from the Central Marine Engineering Works by the NER's graving dock. Sir William Gray died on 12/9/1898.

By 1900, the company possessed eleven building berths and employed about 3,000 people. It launched its 1,000th ship, the 11,800dwt CITY OF DIEPPE in 1929 and managed to keep going during the recession but no ships were built in 1931 and 1933. A year later, two paddle steamers were built for the LNER's service across the Humber and one, WINGFIELD CASTLE, is preserved at Hartlepool. During World War 2, Gray built 90 ships totalling 408,030 gross tons and handled the repair and overhaul of 1,750 vessels. The demand for new ships continued after the conflict but, with the 1950s, came increasing competition from the reconstructed yards of Germany and Japan. One of the last ships built was the JOYA McCANCE, an ore carrier which began operating in 1960 and, at 16,830dwt, was the largest constructed in Hartlepool. The final ship, BLANCHLAND, was completed in 1961, to be followed by the conversion of four tankers into bulk cargo carriers in 1961–62. Gray announced in 12/1962 that it was going into voluntary liquidation and the final two ships left the Old Shipyard dry docks in 3/1963. The demise of the shipyards also meant the end of the Central Marine Engine Works and Cliff House Foundry. The contents of the premises were auctioned on 7/5/1963.

Reference :   *Shipbuilders of the Hartlepools*, Bert Spaldin, Hartlepool Borough Council, 1986

OLD SHIPYARD, West Hartlepool   76
NZ 512331 – **Map B**

Gray's Old Shipyard possessed four shipbuilding berths in 1929, two facing on to Jackson Dock and two on to Swainson Dock. The LNER's dock line followed the route of the early railway from Church Street Junction passing to the south of Swainson Dock and running parallel to the main line and Timber Dock to reach the tall Greenland Signal Box, which controlled tracks in the vicinity of the Old Shipyard and Hartlepool Gas Works, before terminating in a fan of sidings between the Jackson and Union Docks. The connection into the Old Shipyard diverged off the outermost of these sidings and ran along the length of the yard next to the Timber Dock with other tracks serving both sides of the two graving docks.

Gauge : 4ft 8½in

-		0-4-0ST	OC	BH	398	1882	New*	s/s
	WILLIAM GRAY No.2	0-4-0ST	OC	BH	908	1887	New	(1)
	NUMBER ONE	0-4-0ST	OC	HL	3418	1919	New	(2)
-		0-4-0ST	OC	KS	3126	1918	(a)	(3)
-		4wDM		FH	3572	1952	(b)	(4)

* Built for stock in 1876 and sold by BH as a secondhand engine to Gray in 1882.

(a) ex William Gray & Co Ltd, EGIS Shipyard, Pallion, Sunderland, Co. Durham, 3/1932. This yard had closed in 1930.
(b) ex Central Shipyard, Hartlepool, c/1958.
(1) sold for scrap, c/1955.
(2) to ? , c/1958. It was advertised for sale by Gray on 30/10/1958.
(3) to Central Marine Engineering Works, Hartlepool.
(4) to J.D. White Ltd, Thornaby, 25/7/1963.

CENTRAL MARINE ENGINE WORKS
Hartlepool

77
NZ 518339 – **Map B**

The works was constructed in 1883–84; the foundry being soon completed so that it could cast iron columns for the roof supports of the remaining buildings. The first marine engine built was a triple expansion machine installed in the ENFIELD in 1885. The site was a triangular shape with the apex at the north west end where Gray's railway entered the premises to serve foundries, engine and boiler shops. The southern broader end of the site was formed by the North Basin and railway lines led directly from the engine and boiler shops to the quay. Sheerlegs, 110 feet high to lift 80 tons, were installed on the quayside but these were replaced later by a hammer-head crane capable of raising 100 tons. It formed a major landmark until demolished in 1964. NER dock lines ran by the other two sides of the premises defining the boundaries. Gray had running powers over the railway company's double track line from Central Marine Signal Box in the north along the south east boundary of the works and then as a single line connecting with Gray's railways at the North Basin. This was the position in 1951 when Gray was allowed to use its own locomotives and steam or diesel shunting cranes hauling empty and loaded wagons over these tracks. Archibald Henderson became the first works manager in 1887 and Peter Hogg describes how he "made a stately arrival at work on board a CMEW rail locomotive, which picked him up each morning at the Greenland crossing near his home." The new boiler shop was taken over by the Ministry of Munitions in 1916 for the production of 8 inch shells and was known as "The Hartlepools National Shell Factory". Gray installed its first direct reduction geared turbine in a merchant vessel in 1920 and had licences to build Polar and Oxford diesel engines after World War 2.

Reference : *A History of the Central Marine Engine Works 1884-1961*, Peter L. Hogg, Hartlepool Borough Council, 1995

WEST HARTLEPOOL No.2 arrived secondhand at William Gray's Central Marine Engine Works at Hartlepool Docks about 1935. It is considered to have been built by Barclays & Co (296 of 1882) although it carried a Lennox Lange plate. *(IRS Trevor Lodge collection)*

Gauge : 4ft 8½in

SPEEDY	0-4-0CT	OC	HL	2334	1896	New	Scr
WEST HARTLEPOOL No.2	0-4-0ST	OC	B	296*	1882		
	Rebuilt		Sir J. Jackson, Keyham		1903	(a)	(1)
-	0-4-0ST	OC	KS	3126	1918	(b)	(2)

* plate read "Lennox, Lange & Co, 1882". This company seems to have acted as marketing agents for Barclays & Co.

(a) ex Admiralty, Devonport Dockyard, Devon, c/1935, after 10/1935.
(b) ex Old Shipyard, West Hartlepool.

(1) to Cox & Danks Ltd, contractors, /1948.
(2) scrapped /1963. A photograph was taken on 17/6/1963 at Gray's with "LOT 874" painted on the locomotive's saddle tank; there had been a sale of plant in 5/1963.

CENTRAL SHIPYARD 78
Hartlepool NZ 516340 – Map B

The Central Shipyard launched its first ship in 12/1888 and, in 1900, two 500 feet long building berths were added. The yard was located between the Timber Ponds and the NER's central graving dock; the latter could be leased for new building and repair projects. One of the NER's dock lines ran from near the Greenland creosote works * by the side of the central graving dock to reach Queen Street Signal Box at the end of Middleton Road. Sidings left this line at Central Marine Signal Box and entered the shipyard running by and through the buildings to terminate alongside the various building berths facing Central Dock. Sidings also ran into the nearby Central Marine Engine Works. The signal boxes were operated by the NER. The central graving dock closed in the mid-1950s due to a structural fault.

Gauge : 4ft 8½in

No.2	0-4-0ST	OC	MW	1020	1887	(a)	(1)
-	4wDM		FH	3572	1952	New	(2)
No.25	4wDM		RH	210479	1942	(b)	(3)

(a) ex Price, Wills & Reeves, contractors, Immingham, Lincolnshire. The locomotive was used on either (or both) the Immingham Dock and Grimsby District Electric Railway contracts before moving to Hartlepool.
(b) ex Charles Jones of Aldridge Ltd, dealer, Aldridge, Staffordshire, c/1958; formerly Air Ministry, RAF Maintenance Unit, Sealand, Flintshire.

(1) to Michael Baum & Co Ltd, Middlesbrough, for scrap, /1952.
(2) to Old Shipyard, West Hartlepool, c/1958.
(3) to Thos W. Ward Ltd, Templeborough Works, Sheffield, Yorkshire (WR), c10/5/1963; resold to South Western Gas Board, Bath Gas Works, Somerset, c2/1964.

* *From 8/1925, the LNER's Greenland creosote works employed a 40hp eight ton standard gauge Motor Rail (3783 of 1925 although it was a refurbished 2126 of 1922). It spent its entire life at the creosote works, being taken into departmental stock by BR in 5/1949 and allocated number 15097. Shortly after, it was replaced by 11104 (FH 3466/1950) later BR Departmental 52.*

TEES SHIPYARD 79
Graythorp NZ 522270 – Map A

In 1913 William Gray & Co leased land on the north side of Greatham Creek with the intention of establishing a new yard that could construct ships of up to 20,000 tons. Much reclamation of the marshes was required before work on laying out the shipyard could begin, but the onset of World War 1 meant that the project had to be held in abeyance. After the war work resumed, the civil engineering contractor being Holloway Bros (which see). The 80 acre yard comprising ground works for four shipbuilding berths (not completed), a 500 feet long dry dock and a 1,200 feet long fitting out quay with 31 feet depth of water at high tide opened in 1924; new housing for the workers being erected on the other side of the A178 road. Gray signed an agreement with the NER on 18/4/1917 for a connection with the Seaton-on-Tees Branch on the south side of the A178 overbridge. From here, Gray's sidings ran alongside the dry dock and served the buildings at the fitting out quay. Spaldin states that the whole establishment was electrically powered, "steam only being used to power mobile cranes and locomotives". Unfortunately no details have survived of these vehicles.

In view of the difficult trading conditions in the interwar years, the yard concentrated on repair work. Following the liquidation of William Gray & Co Ltd in 1962, Smith's Dock Co Ltd took over the yard and formed Gray (Tees) Ltd to carry out ship repairs but this ceased in 2/1968. Laing Pipelines Offshore Ltd then occupied the yard and signed an agreement with BR for construction and use of the siding in connection with the Seaton-on-Tees Branch. Development of the yard began in 2/1972 with the enlargement of the basin, provision of permanent lock gates and a 300 feet long wharf. The yard went on to build oil rigs; the first being despatched in 6/1974. The business had a short life and, in 2/1981, demolition of the remaining buildings commenced. The site is used today by Able UK to dismantle ships and offshore modules, but with no rail connection. However, in 2011, some scrap was loaded over the fence on to main line trains standing on the Seaton-on-Tees Branch.

Reference : *Graythorp Works*, Laing Pipelines Offshore Ltd, 1974

GRINKLE PARK MINING CO LTD

Grinkle Ironstone Mine, near Staithes 80
Palmers Shipbuilding & Iron Co Ltd until 26/5/1899 Map I
Palmer Bros & Co until 21/7/1865 NZ 760178 NZ 798177 NZ 782176

Grinkle Mine was situated on the boundary between the present-day Teesside and North Yorkshire but the full entry is included here for convenience. Palmers' shipyard on the River Tyne was established by George and Charles Mark Palmer in 1851 and developed rapidly with blast furnaces being added from 1857. Meanwhile mining of the Main Seam ironstone had begun on the coast at Skinningrove. About 1854 Palmers set up the Staithes Ironstone Company with the intention of transporting ironstone in their ships to the Tyne. At Rosedale Wyke, the Main Seam lay 60 feet below the foreshore, with the Dogger or Top Seam in the cliffs 200 feet above and it was here that Palmers developed their Port Mulgrave harbour (NZ 798177). Two jetties were constructed to enclose an area of water and a timber gantry was erected about 30 feet above the South Jetty. Shafts linked the two seams to a horizontal gallery enabling tubs of ironstone to emerge from the cliff face and run directly on rails along the gantry where they could be emptied into storage boxes, prior to loading into small coasters and barges. Initially mining took place in the Seaton royalty close to Port Mulgrave.

Charles Mark Palmer began to acquire property inland, both as a source of additional ironstone and to occupy the hall, Grinkle Park, as a country residence. With increasing demand for ironstone, Palmers opened Grinkle Mine (NZ 760178) in 1875. This was located in the secluded valley of Easington Beck, near where it was joined by a tributary called Twizzie Gill and had the advantage that access to the Main Seam could be gained

One of the John Fowler 3ft gauge locomotives at the entrance to Seaton Drift having brought ironstone from Grinkle Mine; the engine shed is on the right. *(D. Linton collection)*

by a drift. The confined nature of the valley meant that the stream had to be put in a tunnel in order to provide space for the workshops and engine house. A rope haulage system was installed for pulling the tubs out of the workings; horses were responsible for moving them from the individual faces underground.

Once on the surface, the tubs were weighed and any shale was picked out to be added to the heap in the valley. The ironstone was then loaded into wagons for the journey to Port Mulgrave. The provision of the 3ft 0in gauge railway had required the construction of a stone and brick lined tunnel taking the railway under Ridge Lane and into a parallel valley to the south. A substantial bridge was erected at Dalehouse with the track carried on wrought iron girders supported by timber columns. Two hundred yards beyond, the railway reached Seaton Drift where there was a loop and a siding to the wooden engine shed (NZ 782176); this was the limit of locomotive working. The company introduced larger steel hopper wagons for the journey, each of which could carry eight tons of ironstone and equipped with two bottom doors. There were also special bogies known locally as "buses" to transport the miners between Dalehouse and Grinkle. The Seaton Drift comprised a brick lined tunnel which joined the existing mine workings to emerge from the cliff face at Port Mulgrave. A main and tail rope haulage powered by an engine house on the quayside moved wagons on the single track through the 1,500 yard long drift. An embankment was constructed at Port Mulgrave connecting with the seaward end of the southern jetty to enclose a triangular area which contained wooden staging, at the same height as the gantry, with two 3ft 0in gauge sidings enabling additional stocks of ironstone to be stored.

The mine and harbour were leased to the Grinkle Park Mining Company in 1899, although Palmers still played a prominent and increasing role in the new company, in addition to owning the ironstone reserves. Grinkle Mine's Easington royalty backed on to that of Pease & Partners' Loftus Mine and, beginning in March 1900, some ironstone from the Grinkle workings was transported through Loftus Mine to the Skinningrove Iron Co's blast furnaces. Rope haulages were used to move the tubs over a distance of about three miles which was mostly underground. Only Loftus tubs were used for this traffic; although both were of the same 3ft gauge, the fact that the former had sloping sides and Grinkle vertical sides may have had some bearing. About 1,500 tons of ironstone per week were going out this way in early 1914.

Palmers ordered three nine inch four-coupled tank locomotives for the 3ft 0in gauge on 25/4/1871. According to Black, Hawthorn's records, maker's numbers 206–08 were for the "Saltburn Ironstone Mines". However, there is doubt that they were delivered to Grinkle Mine. The latter did not open until 1875 and Palmers purchased three saddle tank locomotives with 9in x 14in cylinders from John Fowler in 1875–76 to operate the 3ft 0in gauge railway. The mine appears to have still relied on these three locomotives up to 1909 when a new locomotive with 9in x 15in cylinders was purchased from Hudswell Clarke. It was painted in Midland Red livery and subsequently made No.3 which suggests that one of the Fowler locomotives may have gone, although an undated photograph appears to show a locomotive on the quay at Port Mulgrave. A similar locomotive, No.4, came from Hudswell Clarke in 1914. Although the Fowler and Hudswell Clarke products would suggest just two locomotive designs, existing photographs show three types; one explanation may be that some rebuilding had taken place. There was a substantial need for coal to power the locomotives and stationary engines and this was brought in by ship from Blyth, offloaded at the coal berth by crane and taken on the railway to Grinkle Mine. A siding near the Dalehouse bridge served a coal depot for domestic sales.

Grinkle Mine was expensive to work because of its isolation and the low iron content and high level of impurities of its Main Seam. Skinningrove Iron Company's dissatisfaction with the quality of the ironstone and the lack of profit for the Grinkle Park Co resulted in the cessation of moving stone through Loftus Mine in 1914. Grinkle was now totally reliant on Port Mulgrave but the threat to shipping from German submarines meant that an alternative had to be found. Fortunately, the NER Whitby–Loftus railway had finally opened to the north of the mine in 1883. A double track inclined railway was built up the side of the valley to the NER line during World War 1. Here an electric motor hauled hopper wagons up on to a gantry so that the ironstone could be tipped into standard gauge trucks. These sidings were located opposite those for the Boulby ironstone mine. The agreement with the NER was dated 15/5/1916 with the railway company constructing the main line connection and Grinkle Park laying three sidings either side of the tippler for the NER to deliver empty wagons and collect loaded wagons respectively. An aerial ropeway was also erected passing over the NER branch and terminating at Bias Cliff on the coast for tipping spoil.

With the economic downturn after the war, Grinkle Mine closed in 2/1921 but was kept on a 'care and maintenance' basis. It reopened in 4/1927 and, in the following June, was sending out about 2,300 tons a week via the incline and LNER to the Jarrow furnaces. The tramway to Dalehouse was reported to be in a "deplorable state" and Seaton Drift was blocked. The mine closed again for over a year due to flooding caused by the collapse of the tunnel taking Easington Beck under the shale heap. Mining re-commenced but then Palmers damped down its remaining blast furnace at Jarrow and the mine shut on 22/5/1930. Palmers Shipbuilding & Iron Company finished in 1932 and the Grinkle Park Mining Co went into receivership. On 4/11/1936 the Receiver stated that the value of the mine as a going concern was nil and he intended to sell the loose plant and equipment. George Cohen, Sons & Co Ltd was responsible for dismantling salvageable material from the mine and harbour.

References :   *Grinkle Ironstone Mine*, Parts I and II, Simon Chapman, published by Peter Tuffs, 2012
Staithes and Port Mulgrave Ironstone, J.S. Owen, CIAS, 1985
'Some thoughts on Grinkle Mine and Port Musgrave' E. Jopling, CIAS Newsletter 71, 10/1998

Gauge : 3ft 0in

-	0-4-0ST	OC	JF	2376	1875	New	(1)
-	0-4-0ST	OC	JF	2377	1875	New	(1)
-	0-4-0ST	OC	JF	2378	1876	New	(1)
No.3	0-4-0ST	OC	HC	898	1909	New	(1)
No.4	0-4-0ST	OC	HC	1096	1914	New	(2)

(1) scrapped c/1936, although at least one may have been disposed of earlier.
(2) to George Cohen, Sons & Co Ltd, Machinery Depot, Stanningley, Yorkshire (WR), c/1936. Resold by Cohen to Lehane, Mackenzie & Shand Ltd, Fernilee Reservoir Contract, Whaley Bridge, Derbyshire.

HEAD WRIGHTSON & CO LTD

Head Wrightson's main factory was the Teesdale Iron Works at Thornaby but, over the years, it acquired various other premises on Teesside. Those with locomotives are described below, however, there were four sites with only limited railway involvement. These were:

Head Wrightson Machine Co Ltd. This was formed in 1939 and moved to Middlesbrough in 1945. It occupied the former Richardsons, Westgarth engine works on Commercial Street; the premises were modernised and equipped with new machine tools. A siding ran into the premises from the Vulcan Street Branch but much of the site was covered with buildings of some age. The works closed in 1979.

Head Wrightson Stampings Ltd. The factory was located on the south side of the Stockton-Hartlepool railway, just north of Seaton Snook Junction. Die stamping transferred from Teesdale to the Hartlepool works after its acquisition in 1938. An agreement with the LNER dated 15/12/1941 concerned the siding which left the Seaton Snook Branch immediately before the junction with the main line and entered the premises to serve the materials stockyard and stamp shop. The works passed to the Davy Corporation about 1976–78 and was purchased by Caparo Forgings in 1989.

Head Wrightson (Steelcast) Ltd. This was acquired by Head Wrightson in 1971 and was the former Davy United Roll Foundry (which see).

Head Wrightson Steel Foundries Ltd. The works was located on the west side of the Yarm–Norton railway, south of Stockton Station. It was purchased in 1927, modernised and produced carbon, manganese and alloy steel casting. Later, the site was enlarged to take in the former Riley Bros boiler works (which see).

At its peak, the company was said to employ 6,000 people and had become a major engineering business with a world-wide reputation. A summary of its larger products over the years included the following:

    Bridges e.g. Putney Bridge, Walnut Tree viaducts, including several in Japan and South Africa
    Piers e.g. Dover East, Port Elizabeth, Redcar
    Structural steel and iron work for 55 companies with blast furnaces (by 1903).
    Colliery head gears and skip haulages
    Ball mills and crushers
    Cast iron tunnel segments for London Underground
    Caissons, dock gates and turntables, including for India, Devonport Dockyard
    Heat Exchangers for petro-chemical plants
    Boilers for Bradwell Nuclear Power Station

TEESDALE IRON WORKS, Thornaby   81

Head, Wrightson & Co until 21/6/1890   NZ 451188 – **Map E**
Head, Ashby & Co until /1866
Head & Wright until /1860

The Teesdale Iron Works was established by a Mr Skinner in 1840 on a small site at South Stockton (later called Thornaby), north of the SDR extension to Middlesbrough, in order to manufacture cast iron sashes for windows. In 1859 Howard Head and Joseph Wright purchased the works for £1,500. Later that year, Joseph Ashby joined the company. Wright retired in 1860 and the business became Head Ashby & Co, with the adjacent vacant cotton mill being taken over and used as machine shops. Small iron castings and girders for railway bridges were then being produced. Head Ashby was responsible to the contractor, Thomas Nelson, for the design and provision of the iron bridges on the NER Deviation Line at Goathland in 1863–65, although the specification called for Staffordshire, Weardale or Scottish iron to be used. Charles Arthur Head joined his brother during 1863 and, in 1866, Ashby retired to be replaced by Thomas Wrightson. Management and ownership now belonged to C. Arthur Head and Thomas Wrightson and they were responsible for the development of the company.

Amongst broadening of the project range, Head Wrightson appears to have contemplated becoming a locomotive manufacturer. In company advertisements dated 1866 and 1869, an "IMPROVED TANK LOCOMOTIVE" was offered for sale. The drawing depicted a four-coupled inside cylinder saddle tank. It is not known whether any were built, although the 1866 advertisement claimed that three of these locomotives were ready for delivery and others to be finished shortly. Even more unusual, *The Engineer* for 15/12/1871 contained an engraving of a traditional looking outside cylinder saddle tank with the claim "Colliery Locomotive constructed by Messrs Head, Wrightson and Co, Engineers, Stockton-on-Tees". However, the illustration was based on a known photograph of a Black, Hawthorn locomotive, the plate of which dates it to about 1870. Significantly the text with the engraving states that the locomotive was "built to design sometime since prepared by Messrs Head, Wrightson," so could it be that a draughtsman had moved from Head Wrightson to Black, Hawthorn with the plans?

Possibly influenced by Cochrane & Co and Gilkes, Wilson & Co, which constructed some vertical boiler industrial locomotives about this time, Head, Wrightson also built a number of this type. Although not particularly powerful, they had the advantages of being relatively inexpensive, easy to raise steam quickly and light on the track. The number built is not known, but some have been identified, of which three have been preserved. Raine & Co, with a factory at Winlaton Mill, Co. Durham, advertised for sale between January 1890 and January 1891 a four-wheel standard gauge vertical boiler locomotive built by Head Wrightson & Co.

TEESDALE No.2 (Black, Hawthorn 905 of 1887) at Head Wrightson's Teesdale Works. The ship in the background is stated to be the KING DAVID which was constructed at Craig Taylor's nearby yard in Thornaby during 1906. *(Cultural Services, Hartlepool Borough Council)*

Maker's Number	Date Built	Customer
?	c/1864-69	Lancashire Union Railway (or Abraham Pilling, contractor), Wigan @
?	2/1869	East London Waterworks Co, Tottenham. x
21	1870	Londonderry Estates Co, Seaham +
?	c/1870-71	Gjers Mills & Co, Middlesbrough
?	c/1870-71	Gjers Mills & Co, Middlesbrough
?	1871	Dorking Grey-stone Lime Co Ltd, Betchworth +
32*	1872	Hannoversche Maschienen Fabrik Co, Hanomag
33	1873	Seaham Harbour Dock Co, Seaham +
?	c/1870-1875	At some stage with Weardale Iron & Coal Co
35 or 36	1876	Chell Colliery Co, Stoke on Trent

*   Vertical boiler?
\+   Preserved
x   See 'Railways of the East London Waterworks Company Part 1' Frank Jux, *The Industrial Locomotive*, ILS, 2016
@   Advertised for sale in the Bradford Observer at Wigan on 13-16/9/1869

The locomotives were mostly four-coupled standard gauge machines and, where photographs survive, with a conical topped vertical boiler. They came in two distinct types, either with vertical cylinders and geared drive or more conventional outside cylinders. The Dorking Grey-stone locomotive carried plates reading "T.H. Head, Engineer, 90 Cannon St, London, 1871", which was presumably the London office of the company.

Head Wrightson built a vast range of industrial plant and equipment, including numerous scale cars from the 1920s to the 1950s. This scale car bearing a Head Wrightson plate travelled under the blast furnace iron ore hoppers collecting ores from various bins and weighing them before dropping them into a skip, from which they were raised by conveyor to the top of the blast furnace.   *(IRS collection)*

For its own shunting, Head Wrightson mainly relied on four standard four-coupled saddle tanks with 12in x 19in cylinders built by Black, Hawthorn and its successor, Chapman & Furneaux. The older pair, Nos.1 and 2, were probably later re-boilered because they were equipped with Ross pop valves on the firebox with a pipe taking the steam through a modified cab roof. Nos.3 and 4 retained their Ramsbottom safety valves on the centre of the boiler. Nos.1 and 4 also had a linkage from the crosshead operating a bell which would ring continuously when the locomotive moved.

In the 1890s the works occupied a rectangular shaped site of 16 acres principally on the east side of Trafalgar Street when it employed over 1200 people. It was a low-lying area near the river and the street was prone to flooding. Here was the Bridge Yard with an iron foundry, pattern shops, forges and machine shops, with space open to the elements to lay out and erect some of the massive structures. There were six foundries alongside the western boundary – Nos.2 and 3 were provided with a light railway for handling moulds, No.5 produced rail chairs and No.6 used mainly haematite iron to manufacture ingot moulds and heavy castings assisted by a steam travelling crane. A line left the NER immediately east of Thornaby Station and formed a set of four exchange sidings before crossing Hanover Street as a single line. It then divided to serve most parts of the Bridge Yard before heading north past the South Stockton shipyard to Head Wrightson's wharf on the river. This was described in 1926 as 225 feet long with one 5 ton steam crane. A separate line headed east from the Bridge Yard over Long Row to connect with Whitwell's Iron Works rail system.

There had been only minor additions to Head Wrightson's railway by 1915 but, two years later, electric arc furnaces were installed. A second line diverged from the railway as it approached Hanover Street and ran over Nile Street to serve the properties west of Trafalgar Street, including the Thornaby shipbuilding yard. After Craig, Taylor ceased shipbuilding in 1931, Head Wrightson acquired its yard for additional erection space; the attractive offices becoming the steel foundry headquarters and laboratories. Head Wrightson also occupied much of Richardson Duck's former shipyard, apart from the fitting out wharf which was used by a ship dismantling company. During World War 2, both yards assembled some of the 238 tank landing craft prefabricated by Head Wrightson, South Durham Steel & Iron Co, Cleveland Bridge & Engineering and Whessoe of Darlington under the 'Stockton Construction Company' title. Loads of up to 60 tons could be handled at the ex-Richardson Duck slipways, such as dock gates and caissons and these were then floated out and taken away by a tug.

Head Wrightson also became an important manufacturer of rolling stock for both the railway companies and within iron and steel works. When the Iron & Steel Institute visited Teesdale in 1937, the company was producing 20 ton hopper wagons for the LNER, part of an order for 900. The Wagon Shop in the Bridge Yard became an important part of its activities after World War 2. The first batch of wagons built for the newly nationalised BR was 500 21 ton coal hoppers; another 500 were supplied in 1958. Small numbers of specialist wagons were built for BR's civil engineers department. Head Wrightson also supplied the needs of iron and steel works with hot metal and slag ladles, all-welded platform wagons for carrying heavier ingots, ingot cars that were subject to arduous working conditions and electric scale and transfer cars.

In the early 1950s, Head Wrightson purchased three new Ruston & Hornsby 88DS diesel mechanical shunters to replace its elderly steam locomotives. The amount of rail traffic for the diesels began to reduce; they brought in wagons of steel plate and coal with manufactured goods going out. Head Wrightson had also used steam rail cranes to lift and move heavy items. Albert Roxburgh thought that the locomotives were then replaced by road tractors for what little rail traffic remained. Use of the railway appears to have ceased about 1976.

From the late 1960s, Head Wrightson faced increasing international competition and falling demand at home. It merged with the Davy Corporation in 1976 but this was, in effect, a takeover. Davy saw its future in design and project management rather than manufacturing and the Teesdale Works closed in 1984. ITM (Offshore) Ltd took part of the site until 1986 when it was declared bankrupt. The site was then redeveloped by the Teesside Development Corporation.

References :   *Life at Head's*, Editor Margaret Williamson, Teesside Industrial Memories Project, 2013

    Head, Wrightson & Co Ltd Engineers, Album of Manufactures, 1903

    Vertical Boiler Locomotives and Railmotors built in Great Britain Volume Two, Philip J. Ashforth and Vic Bradley, Industrial Locomotive Society, 2016

Gauge : 4ft 8½in

No.	Name	Type		Builder	Works No.	Date		Notes
No.1	TEESDALE	0-4-0ST	OC	BH	287	1874	New	(1)
	TEESDALE No.2	0-4-0ST	OC	BH	905	1887	(a)	(2)
	TEESDALE No.3	0-4-0ST	OC	CF	1145	1897	(b)	(2)
	TEESDALE No.4	0-4-0ST	OC	CF	1179	1899	(c)	(1)
	(form. EAGLESCLIFFE No.1)							
-		4wDM		RH	312429	1951	(d)	s/s c/1967
1		4wDM		RH	312434	1951	New	(4)
2		4wDM		RH	321735	1952	(e)	(3)

(a)	new. to Eaglescliffe Foundry, /1937 and returned, /1951, by 8/9/1951.	
(b)	new. to Stockton Forge (it was there 11/6/1934) and then at Egglescliffe Foundry 18/9/1936. It probably returned to Teesdale c/1937.	
(c)	ex Egglescliffe Foundry, by 28/5/1934.	
(d)	ex Stockton Forge.	
(e)	new. to M. Henderson Clark Ltd, Cargo Fleet, c/1970 and returned c/1973. Although, it should be noted that Henderson Clark also had a yard next to the Teesdale Works.	
(1)	to J.T. Atkinsons (Metals) Ltd, Stockton, for scrap, 6/1967.	
(2)	to J.T. Atkinsons (Metals) Ltd, Stockton, for scrap, BH 905 was reported to be in a scrapyard near Thornaby Station 5/1967 and CF 1145 in 8/1967.	
(3)	to T.J. Thomson & Son Ltd, Stockton, /1975.	
(4)	to Thomas Turnbull & Sons Ltd, Thornaby, /1976.	

EGGLESCLIFFE FOUNDRY, Stockton 82
Egglescliffe Iron Co Ltd until /1894   NZ 434175 – Map D

The foundry was located on Yarm Road in the angle between the Darlington-Middlesbrough and the Northallerton-Stockton railway lines. The Egglescliffe Iron Company was formed in 1/1893 to acquire Henry Smith's and Frederick William Stoker's plant but went into liquidation in late 1894 and was finally wound up on 31/10/1895. The foundry was then leased by Head Wrightson in 1896. A siding left the Darlington-Middlesbrough railway, passed under Yarm Road and ran alongside the main line to a headshunt. From here, three sidings ran into the works; two entering the principal foundry building and the other running on to a staith acting as a coal depot next to Northall Street. Additional sidings had been installed by 1915, including two reception tracks for the delivery of pig iron and sand. On the opposite side of Yarm Road was the Eaglescliffe Brick and Tile Works operated by the Eaglescliffe Brick Co. This had a short railway, presumably narrow gauge, connecting its clay pit to the brick works but, on 20/9/1897, 5½ acres including the works was sold to Head Wrightson; the balance of the site following in 1907 (3 acres) and 1915 (2 acres). It is probable that some worked out clay pits were used for the disposal of spoil.

Until the end of World War 2, Egglescliffe Foundry acted as an overflow for the iron foundries at Teesdale and Stockton Forge, although it may have been closed in 1924-36. It comprised heavy and light moulding shops, furnaces, a fettling shop and a workshop for assembly and testing, together with pig iron and scrap storage yards. In the early 1950s it became the main iron foundry for Head Wrightson and equipment was moved from Stockton Forge to here. The foundry concentrated on producing ingot moulds, cast iron tunnel segments and rail chairs and base plates. The diesel locomotive was sold in 1967 and it is presumed that rail traffic had ceased by about this time.

Gauge : 4ft 8½in

EAGLESCLIFFE No.1	0-4-0ST	OC	CF	1179	1899	New	(1)
TEESDALE No.3	0-4-0ST	OC	CF	1145	1897	(a)	(2)
TEESDALE No.2	0-4-0ST	OC	BH	905	1887	(b)	(3)
-	4wDM		RH	312427	1951	New	(4)

(a)	ex Stockton Forge, by 18/9/1936.
(b)	ex Teesdale Iron Works, Thornaby, /1937.
(1)	to Teesdale Iron Works, Thornaby, by 28/5/1934.
(2)	to Teesdale Iron Works, Thornaby, c/1937.
(3)	to Teesdale Iron Works, Thornaby, /1951, by 8/9/1951.
(4)	to North Eastern Iron Refining Co Ltd, Stillington, 9/1967.

STOCKTON FORGE, North Shore Branch, Stockton 83
Stockton Forge Co until /1897   NZ 449199 – Map D

The Stockton Forge Works was located on the east side of the North Shore Branch, south of Norton Road and opposite the former Stockton & Hartlepool Railway's Stockton passenger station (opened in 1841, but redundant in 1852 when all passenger services were transferred to the new Leeds Northern Station; the building continuing as a goods depot). Initially known as the Lustrom Ironworks, it was established in 1865-66 but struggled financially. Joseph Dodds purchased the works in 1871 and changed its name to the Stockton Forge Co. It supplied metal work for railway stations and bridges but the company had to be restructured in 1889. It went into receivership in 1896. Head Wrightson leased the works in 1897, purchasing it in 1900. The premises were extended to take over a former brick and tile works. A Head Wrightson brochure for 1903 states that the works was capable of turning out light and heavy forgings, both from scrap and ingots. At this date, it was specialising on products for ships – stern and rudder frames, keel and stern bars and engine forgings, such as crank shafts. Later it concentrated on products for coal and metalliferous mines. The company's railway had two connections with the North Shore Branch, from which lines ran northwards to serve the main buildings including the "Bridge Yard", which was actually a large covered building next to the Branch, template shops, machine and fitting shops and stores. The largest building occupying the northern part of the site was the fabrication, assembly and welding shop. The works closed in 1969.

Reference :   *Life at Head's*, Editor Margaret Williamson, Teesside Industrial Memories Project, 2013

'*The Ironworks of Stockton-on-Tees*' A. Betteney, CIAS, 2021

Gauge: 4ft 8½in

STOCKTON FORGE No.1	0-4-0ST	OC	CF	1147	1898	New	(1)
TEESDALE No.3	0-4-0ST	OC	CF	1145	1897	(a)	(2)
-	4wDM		RH	312429	1951	New	(3)

(a)   ex Teesdale Iron Works, Thornaby; it was here 11/6/1934.

(1)   scrapped by T.J.Thomson & Son Ltd, /1952.
(2)   to Egglescliffe Foundry where it was seen, 18/9/1936.
(3)   to Teesdale Iron Works, Thornaby.

C. HERRING & SON LTD

SCRAPYARD, Longhill, West Hartlepool   84
NZ 515307 – **Map B**

Mr Christopher Herring reached agreement with BR for a siding off the Cliff House Branch on 3/12/1957. A 1958 Trade Directory gives the address as Windermere Road, Longhill. Locomotives were present in the yard occasionally for scrap, including AB1768/1922 from Cerebos Foods Ltd, Greatham in 3/1968, but there are no details of any used for shunting. The Siding Agreement was terminated on 31/5/1979.

HIGHWAYS CONSTRUCTION LTD

SLAG AND TARMACADAM WORKS, Port Clarence   85
NZ 514212 – **Map C**

In common with many of the iron works on Teesside, the slag tips at the Clarence Iron Works came to be seen as a potential source of income. Extensive tips had been created to the north and east of Bell Brothers' plant on the mud flats bordering the River Tees. Slag had a number of uses including as a fill material; it was transported to assist with the construction of the North Gare breakwater and, on 12/2/1917, Bell Bros agreed that slag could be taken from the "high tip at Clarences" for the Furness Shipbuilding Co's site at Haverton Hill. Slag could also be made into scoria blocks or bricks and, on 7/9/1904, Bell Bros Board approved an annual tenancy on about three acres of land at the Clarence Iron Works for the Erimus Slag Paving Brick Company. The latter's siding was still listed by the Railway Clearing House in 1938, although Dorman Long's locomotive (as the successor to Bell Bros) would have probably carried out the shunting.

A major use of blast furnace slag was either as crushed material or tarmacadam for roadbuilding. Two companies were involved on the tips. The first was George Hodsman, to be joined later by J.F. Wake. It appears that the latter became associated, with Highways Construction Ltd which had been registered as a company on 20/5/1913 and had sites elsewhere in the UK. The Ruston & Hornsby diesel 172908 (see below) was purchased by Highways Construction Ltd of London SW1. Also in the mid 1930s, they had the same Darlington office address, the nature of the relationship between the two companies is uncertain.

George Hodsman until 21/7/1916, **George Hodsman & Sons Ltd** until 2/11/1928, then **George Hodsman (1928) Ltd**

George Hodsman lived in York and, in 10/1885 was trading there as a merchant. He went on to own Greengates whinstone quarry in Teesdale and was also a coal, coke, lime, whinstone and slag merchant in 1899. Hodsman made the first move to establish a slag works at Port Clarence; Bell Brothers approving a lease for land and permission to erect the plant on 6/10/1914. The NER agreed on 27/3/1916 to lay a siding from its Port Clarence goods yard to what were described as the "mixer bunkers" for Hodsman; specific sidings were identified at Hodsman's Works for the NER to deliver and collect traffic. There were numerous railway lines near the Clarence Old Cottages so it would not be obvious which belonged to Bell Bros and those used by Hodsman. George Hodsman & Sons was registered 21/7/1916 as a private company to take over the business shortly

Wake's screening plant at Port Clarence about 1930 where slag from the tips behind was crushed and loaded into wagons. As a dealer, Wake also handled a number of Baldwin four-coupled saddle tanks at the end of the First World War and one has found its way to Port Clarence. The photograph has been modified to introduce advertising on the two nearest trucks but the fourth is genuinely inscribed "Messrs Wake & Co Ltd" and "Slag Works".
(Courtesy Parkhouse/Pope Archive)

A slightly spoilt negative but interesting in showing the former Central London Railway No.1 (Hunslet 695 of 1899, rebuilt by Wake at Darlington about 1925) at Highways Construction/Wake's slag works, Port Clarence on 28th July 1948. (RCTS J. Faithfull photograph)

before his death. It became a public company in 1928. The NER *Private Owner Wagons Register* reveals that George Hodsman, Middlesbrough, registered thirty five 20 ton tarmac wagons (numbers 66–100) in 9/1923 to run over the NER. These had been converted by the Rolling Stock Co from 20 ton goods vans. In addition to the works at Port Clarence, Hodsman had also acquired 13 acres of old slag tips at Grosmont in 1925. Hodsman's siding at Port Clarence continues to appear under his name in the *Railway Clearing House Handbook of Sidings* to beyond 1956, but the company probably withdrew from Port Clarence years before. The capital of Hodsman was significantly reduced in 1937. The company was wound up on 16/2/1967.

Gauge : 4ft 8½in

13		0-6-0ST	IC	HE	464	1888	(a)	s/s

(a) ex Furness Shipbuilding Co Ltd, Haverton Hill, c5/1925. It was noted at Port Clarence on 3/7/1927.

Gauge : 2ft 6in:

(EBENEZER)	0-4-0IST	OC	WB	1002	1888	(a)	s/s
ZURIEL	0-4-0T	OC	WB	1917	1910	(a)	(1)

(a) ex George Hodsman & Sons Ltd, Greengates Quarry, near Lathkirk, Yorkshire (NR). They were offered for sale by Hodsman in the *Contract Journal* on 15/1/1919 and 17/9/1919 when they were said to be at Port Clarence, together with 2½ miles of 2ft 6in gauge railway. This amount of track could indicate that the stock may have still been at Greengates with Port Clarence used as a contact address.

(1) to T.D. Ridley & Sons, Middlesbrough for storage, /1921 and then to Cradock, Allison & Co Ltd, Eaglescliffe Works, Eaglescliffe, c/1922.

Wake & Co Ltd

Following his move to the Skerne Iron Works in Darlington, John F. Wake became involved in selling "machine and hand-broken slag"; the earliest advertisement traced being dated 10/8/1898 when it could be supplied from his crushing mills at four iron works, including Skerne but not mentioning Port Clarence. Wake was later involved at Port Clarence, his first locomotive arriving in 1919, but continued operating there longer. Hodsman had two siding agreements, one with the NER dated 24/3/1921 and the other with the LNER on 8/11/1929. It is interesting that Wake's name was subsequently added to both these documents. In 1922 Wake & Co, Port Clarence, registered fifteen 12 ton open goods wagons (numbers 250–64) to run over the NER. Erimus Slag Works was located as an adjunct to Bell's extensive railway system, east of Clarence Old Cottages and the By-products plant. A series of sidings served the main crushing and screening building. The accompanying 1929

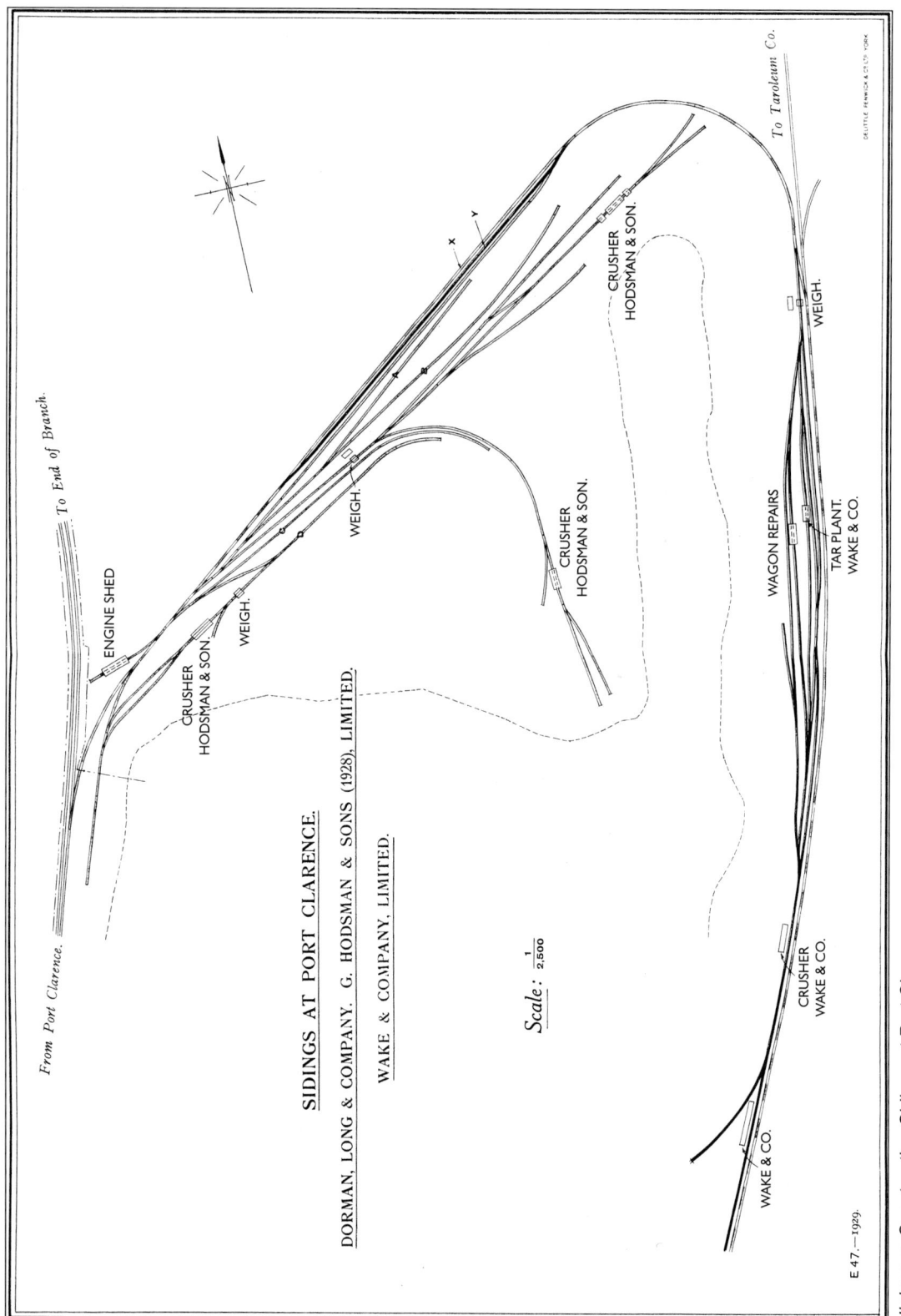

Highways Construction Sidings at Port Clarence

Sidings Agreement map shows the arrangement of the two companies' standard gauge lines with Hodsman located on the tips by the Greatham Branch, to which it had a connection. This suggests that Wake took over Hodsman's first crusher nearer the Iron Works. Eric Hannan visited Wake on 13/9/1946 and saw HE 695; it was in a derelict ruin of a shed with its wheels half buried in debris. The Hopkins Gilkes locomotive was working with "WAKE & Co No 5" painted on its saddle tank.

The 2ft 0in gauge railway is likely to have been used to transport slag from different parts of the tip to the processing plant, with locomotives being introduced as distances from the "quarrying" increased. On his visit, Eric Hannan noted two narrow gauge internal combustion locomotives – a Hudson and an unidentified machine, heavily sheeted over and only recently delivered. The works closed in 1948.

Gauge : 4ft 8½in

-		0-4-0ST	OC	BLW			(a)	Scr/1934
-		0-6-0T	IC	HE	695	1899		
		Rebuilt		Wake	2805	c/1925	(b)	(2)
No.5		0-4-0ST	OC	HG	355	1874	(c)	(1)

(a) presumed ex J.F. Wake, Darlington, c/1919.
(b) previously City & South London Railway. HE 694 and 695 were specially cut down locomotives used for equipping this railway. They were sold to Wake in 5/1922. About 1927 a locomotive was required at the slag works and it is thought that the two Hunslets were cannibalised to make one serviceable engine which bore Wake's 2805 plate.
(c) ex William Whitwell & Co Ltd, Thornaby Iron Works, Thornaby, after 25/8/1928, by /1931.
(1) to Athole G. Allen Ltd, Stockton for use as a stationary boiler, via J. Shaw (West Hartlepool) Ltd, West Hartlepool, c/1948, after 28/7/1948.
(2) to J. Shaw (West Hartlepool) Ltd, Longhill Iron Yard, West Hartlepool for scrap, 9/1948.

Gauge : 2ft 0in

CLARO	0-4-2ST	OC	TG	312	1903	(a)	s/s
-	4wPM		HU	42701	1931	New	s/s
	4wDM		RH	175135	1935	(b)	s/s
-	4wDM		RH	172908	1934	(c)	s/s

(a) ex H. Arnold & Sons Ltd, contractor c/1922 following housing estate construction at Kirk Sandall and a contract at Pilkington's Works, Doncaster, Yorkshire (WR). It came via Wake's Darlington dealership.
(b) new, ordered by Wake for the Port Clarence Slag Works and spares sent there by RH 25/11/1935 to 17/3/1943.
(c) it was new to Highways Construction Ltd, but the location was not stated. It received a replacement engine by 8/3/1943. Spares were sent to Wake at Port Clarence 14/4/1943 to 23/2/1944.

RICHARD HILL & CO LTD

NEWPORT WIRE WORKS and MARSH WIRE WORKS, Middlesbrough.   86
Richard Hill & Co (1899) Ltd until 3/1921   NZ 487209 NZ 485207 – **Map F**
Richard Hill & Co Ltd until 21/3/1899
Richard Hill & Co until /1891

Richard Hill came from Birmingham, opened his Newport Wire Works with Albert Ward in 9/1869 and eventually operated from two sites in the Ironmasters District. An 1874 Middlesbrough Directory has "Hill & Ward Newport Wire and Rolling Mills". The Newport Wire Works (NZ 487209) was at the south end of the OME's Forty Feet Road (later Forty Foot Road – the name came from it being built as a forty foot wide road) with sidings entering the premises off the NER Marsh Branch in 1875; these curved into the works from both the north and south directions in 1893. Rail traffic for the works had to pass over sections of track belonging to the MOR and so was subject to tolls.

The other site (NZ 485207) was located north of the Newport Rolling Mills. In 1874–75 it was identified as "The Middlesbrough Wrought Nail Works" and did not have a siding at that date. The premises comprised 3½ acres and the directors included J. Head and J. Fox so, at this date, it appears to have had a close relationship with the adjoining Fox, Head & Co. However, Slater's 1876–77 directory gives the owners as Hill and Ward. The NER and Ordnance Survey refer to it as the Marsh Wire Works with a road entrance close to Metz Bridge. The siding off the Marsh Branch passed under this bridge and curved round to run through to the end of the site. The principal siding agreement with the NER was dated 24/4/1883.

The two sites included a wire rod rolling mill (about 10,000 tons pa capacity), a wire drawing and galvanising plant (about 6,000 tons), a gun wire plant and a hoop, strip and bar rolling mill (10,000 tons). The company came to specialise in high quality bright and galvanised steel wire for mining ropes, ships' hawsers and engineering ropes. In 1923–24 it also began to produce electrically welded wire mesh for reinforced concrete. The strip and hoop mills were replaced by another rod mill in 1938. After denationalisation following the 1951 Iron & Steel Act, it became one of the companies in the Firth Cleveland Ltd group from 27/7/1953.

It was necessary for Hill's locomotives to work over the NER's Marsh Branch between the two works and the annual charge was fixed at ten shillings in 1898. BR realised in 1948 that a considerable quantity of rail traffic was passing between the "Newport Wire Mills" and Hill's storage ground leased from the railway company, but concluded that it was covered by the existing £10 fee and £7 tolls. Under an agreement dated 8/2/1951, Hill was given permission to operate one diesel locomotive, three steam locomotives and one diesel rail crane over the branch, the fee being £10.10s. However, Hill had to give notice of movements by the rail crane and a BR inspector had to accompany it with his wages being paid by the company. The siding layout at both works in 1951 had not substantially changed, although a narrow gauge tramway (presumably hand-worked) linked some of the buildings at the Marsh Works. Here, the standard gauge had been reduced and the narrow gauge dispensed with by 1967. A map attached to a supplementary agreement with BR dated 15/8/1956 shows that Hill had running powers to work its own traffic over the Marsh Branch and sidings, not just between its own two works, but also with the Newport, Britannia and Acklam iron and steel works. The approval covered "two diesel locomotives, one steam locomotive and one diesel locomotive crane"; the annual sum was now £12.12s.

When Ted Haigh visited the works on 10/6/1950, No.1 was painted black; No.2 was at Ridley Shaw for repairs; No.3 was black, lined red and yellow, with red motion and the diesel was green with red lining. The company had 319 employees in 1961.

Parson & Crosland, steel stockholders, took over the Newport Wire Works site according to BR in 6/1966. The same record suggests that Richard Hill & Co at the Marsh Works had become Tirhm Cleveland Reinforcements Ltd, according to a note dated 19/1/1962. External rail traffic ceased and, as a result, the payment of MOR tolls was terminated week ending from 17/8/1970 (Newport Works) and 2/10/1970 (Marsh Works).

Gauge : 4ft 8½in

-		VBT		?	?		(a)	(1)
No.1		0-4-0ST	OC	AB	868	1900	New	(3)
No.2	(form. No.3)	0-4-0ST	OC	HL	3249	1917	New	(4)
No.3		0-4-0ST	OC	AB				
			Rebuilt	Ridley Shaw		1942	(b)	(2)
No.4		0-4-0DM		JF	4210016	1949	New	(6)
No.5		0-4-0DM		RH	319285	1953	New	(5)

Inness suggests a Henry Hughes 0-4-0 saddle tank was here and moved to the Newcastle on Tyne area in 1918 but no other information has been located to confirm this.

(a) origin and identity unknown.
(b) origin and identity not known. Ted Haigh noted the number 893 on the right hand crosshead and 6653 on the right and left hand crossheads on 10/6/1950.
(1) following alterations at the Marsh Wire Works, Richard Hill advertised in the *North Eastern Daily Gazette* on 26/12/1891 that some items of plant, including a "coffee pot locomotive", were for sale and would be auctioned at the works on 6/1/1892.
(2) s/s c/1950; after 10/6/1950.
(3) to E. Hind (South Bank) Ltd for scrap, c/1954. A visitor to Richard Hill had reported No.1 repaired and for sale on arrival of the new diesel. No.1 was advertised for sale on 14/8/1953.
(4) sold for scrap, c/1958.
(5) to Lunt, Comley & Pitt Ltd, Shut End, Staffordshire, c/1963; by 10/1963.
(6) to George Cohen, Sons & Co Ltd, Cargo Fleet, /1974.

F. HILLS & SONS LTD

North Shore Branch, Stockton on Tees.

87
NZ 446203 – **Map D**

Francis Hills started his one man joinery business in Yarm during the 1840s. It developed into an important manufacturer of wood products, especially for the Durham mining industry in the nineteenth century. Large quantities of shell boxes were produced in World War 1. In 1920 F. Hills & Sons was incorporated as a limited liability company employing 25 people and it began importing softwoods to supply goods, such as doors and windows to the building industry. Following the closure of Blair's marine engine works in 1932, F. Hills purchased these premises (sale announced 4/8/1933) and moved its joinery business from Yarm, including taking over one of Blair's elderly locomotives. The buildings were totally re-equipped with woodworking machinery and a "flush" door production line introduced. A new Hibberd diesel shunter, with a 40 horse power Dorman engine, was acquired in 1935. During World War 2, the works undertook the large-scale manufacture of wooden aircraft bodies, wings and propellers. After 1945 it concentrated on producing doors, windows and general building joinery. By 6/1968 rail traffic had ceased; there was no connection to the North Shore Branch and only an isolated section of track emerging from one building.

Gauge : 4ft 8½in

-		0-4-0ST	OC	AB	239	1881	(a)	s/s
L8	MARY LOUISA	4wDM		FH	1943	1935	New	(1)

(a)   ex Blair's (1926) Ltd, Stockton, /1933 with works.

(1)   to J.D. White Ltd, Thornaby, c/1970.

E. HIND (SOUTH BANK) LTD

IMPERIAL WORKS, South Bank   88
NZ 543212 – Map G

According to the *North-Eastern Gazette*, 11/11/1939, the business E. Hind (South Bank) Ltd was founded by Mrs Elsie Foster's father some 30 years ago. It functioned as a marine store dealer and also as metal merchants at the Imperial Works in South Bank. The latter was situated on the north east corner of the long defunct Imperial Iron Works site where Normanby Road terminates at the Middlesbrough–Redcar railway. A short siding left the main line and crossed the 'Black Path' to enter the yard. The NER Way and Works Committee had agreed on 10/1/1918 that the siding ("as far as it had not been constructed") should be provided for Mrs Emily Hind to operate as a scrap metal merchant. Presumably the former Imperial Iron Works connection might have still been available. In the premises, the siding divided into two running either side of a warehouse; one line of which had a weighing machine.

The business was advertising in 1926 as a "buyer of every description of works residues and scrap materials". The above newspaper article tells us that Mrs Foster (Elsie Hinds) was running the company and three years ago had installed a large rotary aluminium furnace, in addition to five blending furnaces, to process scrap aluminium. According to an advertisement in the *Eston Urban District Council Handbook* 1946–48, the company purchased all classes of scrap, iron, brass, rags, baled tins, jars, bottles and sold sheets, plates, girders, ingots, pig iron and all building materials. It is unlikely that Hind would require a locomotive to shunt its yard because of the small amount of track. It did, however, acquire some locomotives either for resale or scrap. Hind advertised the following for sale in the 1930s – a locomotive (*Machinery Market* 22/8/1930), two secondhand 0-4-0 locomotives with 12in x 18in cylinders (*Machinery Market* 7/9/1934) and a 13 inch Manning Wardle saddle tank for £120 (*Colliery Guardian* 22/5/1936). It was announced in the *London Gazette*, 2/2/1959, that E. Hind was being wound up as a business.

Gauge : 4ft 8½in

-	0-4-0ST	OC	BH			(a)	s/s
No.1	0-4-0ST	OC	AB	868	1900	(b)	s/s

(a)   ex Lingford, Gardiner & Co Ltd, Auckland Engine Works, Bishop Auckland, Co. Durham probably late 1930/early1931 when this company closed down.

(b)   ex Richard Hill Ltd, Newport Wire Works, Middlesbrough, c/1954.

HOGG & HENDERSON, Middlesbrough   89
Map G

Hogg & Henderson appear to have been involved in a number of commercial activities. An 1880 directory gives their offices in the Zetland Buildings, Middlesbrough and describes them as "iron merchants, brokers, steamship owners and wharfingers." They certainly owned the SS HARIET in 1887 according to the *Daily Exchange*. In 1873–74 Hogg & Henderson purchased two new 8 and 10 inch four-coupled saddle tanks from Hudswell Clarke. It is unlikely that this was in its role as an agent because the locomotives carried the inscription "H & H". Hogg & Henderson owned a wharf, possibly in Middlesbrough, during the 1880s but its location has not been established.

Gauge : 4ft 8½in

H & H	0-4-0ST	OC	HCR	125	1873	New	(2)
H & H	0-4-0ST	OC	HCR	139	1874	New	(1)

(1) to Egglescliffe Chemical Co Ltd, Urlay Nook.
(2) to The Owners of the Middlesbrough Estate Ltd, Cargo Fleet Timber Yard, Cargo Fleet, /1892.

IMPERIAL CHEMICAL INDUSTRIES LTD

The business began as Synthetic Ammonia & Nitrates Ltd to operate the Billingham Works as a wholly owned subsidiary of initially Brunner Mond & Co Ltd and then Imperial Chemical Industries Ltd from 1/1/1927. As from 29/12/1931, the latter divided responsibility for its plant at Billingham between ICI Fertilizers & Synthetic Products Ltd to run the ammonia based processes, while the newer Billingham South Works (later renamed Cassel Works) came under ICI (General Chemicals) Ltd based at Runcorn. ICI Salt Ltd operated some of the salt works at the beginning of the 1940s. The various companies were brought together in 1/1/1944 under Imperial Chemical Industries Ltd but several quasi-independent divisions were created, although these were no longer separate legal entities and some later merged to form the Mond Division. After World War 2, there was an increasing development in the use of plastics and fibres based on oil. ICI lacked sufficient space to produce them at Billingham and it established a second major chemical complex at Wilton on the south side of the River Tees where an olefines cracker plant was built to break down oil into its constituents. At one stage, five ICI divisions operated at Wilton – Fibres, Mond, Organics, Petrochemicals and Plastics, with the railway being the responsibility of the Petrochemicals Division. From the late 1960s, there was a steady decline in the numbers of people employed by ICI partly due to changes in technology resulting in older plant closures and also because of increasing competition from overseas. In 1986, ICI Chemicals & Polymers Ltd was formed with responsibility for both Billingham and Wilton, but ICI decided to withdraw from the bulk commodity market and concentrate on specialist chemicals. As a result, some plants closed and others were sold to different companies. The integrated nature of both sites required the disposal of their services, such as utilities, pipelines and internal railways, to one company which would operate them on behalf of the occupiers; these were sold to the **Enron Corporation** as from 1/1/1999 and subsequently passed to **Sembcorp Utilities (UK) Ltd**. As this last phase of the railway systems at Billingham and Wilton was of relatively lesser significance, it has been listed separately under Sembcorp. ICI ceased to exist as an independent company on 2/1/2008.

BILLINGHAM WORKS, Billingham   90
NZ 470225 – **Map C**

The establishment of a major chemical works at Billingham began as a government response to the shortage of explosives during World War 1. The nitrogen required for TNT came from natural nitrates imported from Chile, but these supplies were being obstructed by the German U-boat campaign. The Ministry of Munitions appointed Brunner Mond & Co Ltd of Cheshire to develop the technology to produce synthetic ammonia – a source of nitrogen and hydrogen – and purchased the site at Billingham. Only several contractors' huts had been erected when the war ended. Brunner Mond purchased the site at a 'knockdown' price and established Synthetic Ammonia & Nitrates Ltd as a subsidiary company (registered 3/6/1920) to manufacture ammonia, not for use in explosives but as the basis of nitrogenous fertilisers. Enough railway track had been laid by 1/1/1922 to enable construction of the No.2 ammonia unit and a scheme for connection to the railway company was ready in 2/1922. The first ammonia was produced at Billingham in 12/1923. Brunner Mond & Co Ltd and Nobel Industries Ltd combined to form Imperial Chemical Industries Ltd (company registered 7/12/1926 and began trading 1/1/1927). United Alkali Ltd was one of the other companies to join the enlarged business. ICI developed into a worldwide supplier of bulk chemicals made at the Billingham site which, at its peak, employed 15,000 people.

The Billingham site occupied a huge area on the west side of the River Tees stretching 2½km in one direction and 2km in the other. Its boundaries were defined approximately by the Port Clarence Branch railway and the B1275 road to the north and west, by Billingham Beck to the south and by the Billingham Beck Branch railway and River Tees to the east. The latter railway from North Shore Junction at Stockton had been authorised in 1913 but construction was delayed by World War 1 and it opened in 1920. It joined with a line from Haverton Hill (opened 1901) to form a new alternative route from the south to the expanding industries north of the Tees.

The largest single business was the manufacture of fertilisers based on the production of ammonia. The company preferred to use anhydrite (calcium sulphate) to make ammonia rather than sulphuric acid and the Gotham Company supplied it with 25,000 tons of anhydrite mainly from Cumwhinton, near Carlisle. It then agreed to take anhydrite from the Warren Cement Works at West Hartlepool, the mineral to be transported in sheeted railway wagons and this traffic commenced in 3/1925. Meanwhile a borehole had been sunk on the Billingham site which located a 20 feet thick bed of anhydrite at a depth of approximately 900 feet in 11/1925. The mine, with its two 13 feet diameter shafts and headgear manufactured by Head Wrightson, opened on the north east corner of the Billingham site in 1927.

No.1 shaft had an electric skip winder and, with three shifts working, could raise 18,000 tons per week. After crushing, the anhydrite was taken to the ammonium sulphate plant by aerial ropeway and to the sulphuric acid plant by band conveyor. The arrangements underground were described by G. Eland Stewart in his article 'Billingham Mine' dated 12/9/1946. The locomotive haulage road, 18 feet wide and called the West Level, bisected the workings. A 2ft 6in gauge railway was laid with 40lb per yard rail for the two English Electric trolley locomotives. The ruling gradient was 1 in 80 in favour of loaded tubs, although experience had shown that 1 in 200 would have been preferable in terms of loads that could be hauled. From the West Level, rope haulage inclines were established every 120 yards. Loading up to 1931 was hand shovelling into the 25cwt tubs but then scraper loaders were introduced. The electric locomotives were eventually replaced by caterpillar tractors using wider headings.

In 3/1928, ICI had taken over Casebourne & Co (1926) Ltd (which see) as a subsidiary following successful trials of the byproduct chalk in the manufacture of cement at the Pioneer Works. Initially the chalk was transferred from the ammonium sulphate plant in railway wagons hauled by ICI locomotives along the LNER's Billingham Beck Branch. There was also an agreement dated 12/4/1929 for an ICI steam rail crane to travel over the branch from the East Grid to the cement works provided it did not haul any wagons. As production of cement reached 1,700 tons a week, ICI opened an aerial ropeway to move the chalk by 12/1929 to the cement works and for storage behind Bamletts Wharf on the river. The sulphuric acid plant was located near the mine shafts and the cement clinker byproduct was taken by another aerial ropeway to the cement works. These aerial ropeways operated until 2/1962 when ICI changed to using lorries.

Cement manufacture required clay and access to more supplies were obtained when ICI established a new clay field at Saltholme Farm. Digging commenced in the fields east of Holme Fleet in 5/1929. The original South Durham private railway was extended over Holme Fleet and clay was hauled in side-tipping wagons by ICI's steam and later diesel locomotives to the cement works. Approximately 1,400,000 tons of clay were taken from an area of 47 acres between 1929 and 11/1970. Cottrell's dust from the kilns was transported back in railway tip wagons to reclaim the clay pits. The four original kilns were replaced by a new No.2 kiln (lit 24/8/1929); the Casebourne subsidiary was formally incorporated into ICI from 1/1/1944 and peak annual production of cement reached 389,350 tons in 1965. After World War 2, cement was also despatched in bulk, initially in tanks on lorries and later by BR using Prestoflo wagons.

Early in 1929 ICI started producing a new fertiliser, Nitro-chalk, which was a combination of ammonium nitrate and chalk. When combined with other essential phosphates and nutrients, this was marketed as Concentrated Complete Fertiliser. The CCF plant was located between the Billingham Beck Branch and Haverton Hill Road

Many of the ICI steam locomotives at Billingham were named after rivers; 39 TAW (HL 3671/1927 waits to take a RCTS party on a tour of the works' railway in open wagons on 19th May 1957.
(IRS Bernard Mettam photograph)

English Electric 745 of 1929 on the 2ft 6in gauge railway in the anhydrite mine that lay underneath ICI's Billingham Works. (IRS Brian Webb collection)

complete with its own set of sidings and there is plant here still producing fertilisers although the sidings have long gone.

The nitrogen and hydrogen were made in a gas plant utilising synthesis gas from coke ovens on the site. Sixty four Simon-Carves Ltd Otto regenerative coke ovens began operating in 1929 (contract awarded 1927) and, were supplemented by 36 more in 1941 (12) and 1944 (24). The original 64 were replaced by 24 more ovens in 1946-47 and 12 in 1954–55. In 1930, what became the Cassel Works began to produce caustic soda and chlorine by the electrolysis of brine. It was located at the southern end of the site next to New Road which had been constructed between Billingham and Haverton Hill. A cyanide plant producing raw materials for plastics and nylon was established here in 1934 and it was next to the Cassel Works that perspex was developed and manufactured in 1936. With a deteriorating international situation, the Oil Works opened in 2/1935 to make a range of motor spirits by the hydrogeneration of coal. It was officially opened by Ramsay MacDonald on 15/10/1935; he had arrived at Middlesbrough on a train hauled by the LNER A4 locomotive, QUICKSILVER. The plant was consuming 1,000 tons of coal a day but, in 1939, it changed to creosote and tar as a feedstock to minimise the risk of blockages caused by bomb damage. The plant provided vital aviation fuel during World War 2. The motor spirit was sold in bulk and transported away initially in customers' rail tank wagons and later in ICI's own or hired wagons. Fairly early in the development of the factory, two riverside wharves had been built. Bamletts Wharf was intended for the disposal of waste chalk and Billingham Reach Wharf for incoming rock phosphate. A tank farm was established on the west bank of the Tees with sidings off the Billingham Beck Branch to a rail filling station. An oil jetty was provided at Billingham Reach Wharf and the first despatch of petrol by ship took place on 9/4/1935. The Oil Works also went on to produce a wide range of organic chemicals. Centralised boiler and power plants provided steam and electricity for the whole Billingham site.

Information on the early construction of Billingham Works is limited, but Watts, Hardy (1920) Ltd, based on Tyneside, made all of the side tipping wagons for the removal of spoil. A 'temporary' halt was erected on the Port Clarence Branch immediately west of the Belasis Lane level crossing (later overbridge) and signal box to bring in workers. This belonged to Synthetic Ammonia & Nitrates Ltd until taken over by the LNER under an agreement dated 12/5/1924. The Billingham Works made extensive use of aerial ropeways which were designed and constructed by R. White & Sons of Widnes. These comprised No.1 (5,300 feet long), No.2 (2,613 feet), No.3 (3,790 feet) and No.6 (1,460 feet). The first two were interconnected at a transfer station. At a point close to where Nos.1 and 3 ran together, the LNER railway line was crossed and protective netting was stretched across beneath the ropeways.

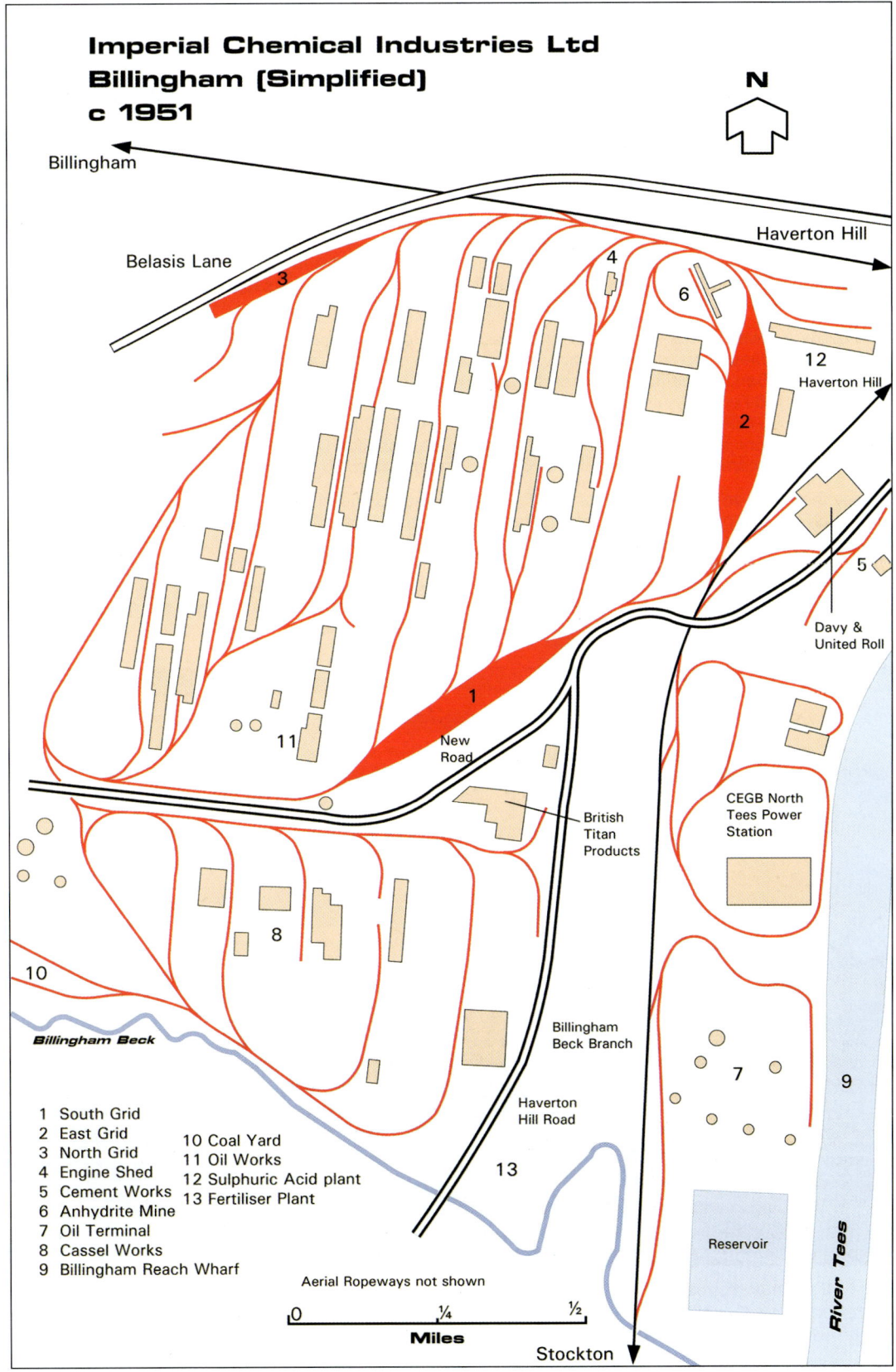

A substantial internal railway system was developed by ICI requiring the services of a sizeable fleet of locomotives. The growth of the works over a relatively short period of time resulted in a well-ordered layout. A railway line ran around the perimeter of the main site, with a branch under New Road to also envelope the Cassel/nylon works. A line diverged off this latter circuit to pass along the north side of Billingham Beck to reach the coal yard. A tip was established at Norton Bottoms. Within the main two peripheral circuits, a series of parallel sidings ran virtually north to south to serve the individual plants. Connections to the main line were made with the Port Clarence Branch at Belasis Lane on the north side of the works and with the Billingham Beck Branch near Haverton Hill South Signal Box. Separate sidings served the plants and wharves between the Billingham Beck Branch and the River Tees. The main reception sidings were located next to the Port Clarence Branch and these joined with the East Grid inward sidings (20 lines in 1930) which occupied space between the railway company's two branches. Another group of sidings on the north side of New Road was called the South Grid outward yard (23 lines in 1930). During the first years, a miscellaneous collection of locomotives was assembled to assist with construction and handle early traffic. Some of these were soon disposed of and ICI purchased a large fleet of mainly Hawthorn Leslie 0-4-0 saddle tanks; many of which were named after rivers.

After experience with a new English Electric diesel shunter purchased in 1948 and, following trials of the Yorkshire Janus diesel electric locomotive, ICI purchased 15 Janus six-coupled diesels in 1956–59, to be followed by three Sentinel four-wheel diesel hydraulics in 1960, resulting in the disposal of the steam locomotives. According to a 1959 article by J.V. Merritt of ICI Billingham in BR's North Eastern Region magazine, the internal railway totalled 90 miles of track and transport staff handled 11,000 wagons a week. All locomotives were fitted with radios communicating with a central control room. The Transport Division had its own wagon shops where maintenance was carried out on 700 internal wagons and another 700 main line tank wagons.

However, there was an increasing use of road transport; it had only been 4% in 1939 but was 40% of despatched goods in 1959. Even more significant was a change from coal to naphtha as a feedstock for some processes. Billingham Works had used 1,500,000 tons of coal each year, all brought by rail from collieries. As a result, some of the recently purchased diesels were sent to other ICI works and most of the remaining Yorkshires had left by the early 1970s. From then on, there were two or three locomotives to deal with the remaining rail traffic. Gradually most of the north-south sidings through the main works were removed leaving the reception sidings, the North Grid where wagons were often stored and the line to the Cassel Works.

One of ICI's Janus diesels, No.6 FARNDALE Yorkshire 2719 of 1958, stands on BR's metals at Bowesfield near the junction with the Stockton Goods Branch on 22nd March1967. The coupling rods have been removed from the wheels and the locomotive is probably in the course of a temporary transfer from or to another works.   *(ARPT J.M. Boyes photograph)*

By 1960, the production of ammonia based on coal had become too expensive and new plant was introduced using naphtha which doubled capacity and halved the number of employees required. This resulted in the closure of the coke ovens in 1/1962 and a reduction in coal traffic. Ammonium sulphate was no longer required to produce nitrogenous fertilisers and the anhydrite mine closed in 1971 (headgear demolished 1978). This also resulted in the finish of the anhydrite raw meal kilns which produced sulphur dioxide for the sulphuric acid plant. The loss of chalk and cement clinker meant the end of the cement works; its kilns shut down in 1970 and the last load of cement was despatched on 3/3/1972. The nylon plant was also closed in 1972 following the opening of a new plant at Ardeer. Faced with increasing competition from overseas, ICI decided to concentrate on specialist chemicals and the remaining plants at Billingham were sold off to other companies. By the end of the twentieth century, rail traffic at the Billingham site was very small and is covered under the Sembcorp Utilities UK Ltd entry.

References :   *Life at the ICI Memories of Working at ICI Billingham*, edited Margaret Williamson, TIMP, 2008

1862-1972 Casebourne's Cement Plant, V. Turley, courtesy Teesside Archives

'The Nineteenth Century Salt and Alkali Industry of Cleveland', C. McNab, *The Cleveland Industrial Archaeologist* 4, CIAS, 1975

The Salt Industry of the River Tees, David M. Tomlin, De Archaeologische Pers, 1982

'160 Years of Cement Manufacture in Cleveland', C.M. Morris, *The Cleveland Industrial Archaeologist* 16, CIAS, 1984

R. White & Sons, Widnes, Catalogue, 1934

'The Billingham Chemical Factory in the Second World War', Alf Rout, *Cleveland Industrial Archaeology Newsletter* 56, 12/1992

'Saltholme A Story of Salt, Cement and Birds', Cliff Shepherd, *Industrial Railway Record* 197, IRS, 2009 pp. 345-65

'Billingham Anhydrite Mine', *Cleveland Industrial Heritage* 34, pub. Peter Tuffs, 2014, pp. 32-33

Gauge : 4ft 8½in

No.	Name	Type	Cyl	Builder	Works No.	Year	Ref	Disposal
-		0-4-0ST	OC	AE	1055	1874		(3)
		Rebuilt		Sdn		1892	(a)	
50	WEAR	0-6-0ST	IC	MW	1290	1895	(b)	s/s after 14/4/1929
52	WYE	0-6-0ST	IC	MW	1634	1905	(c)	s/s after 14/4/1929
-		0-6-0ST	IC	MW	1772	1911	(d)	s/s
51	WEAVER	0-4-0ST	OC	MW	1072	1888	(e)	(8)
46	DON	0-6-0ST	OC	AB	1408	1915	(f)	(7)
47	ESK	0-4-0ST	OC	P	921	1903	(g)	(2)
48	NIDD	0-4-0ST	OC	KS	3112	1918	(h)	(6)
49	TAY	0-4-0ST	OC	HL	2295	1895	(i)	(14)
45	TILL	0-4-0ST	OC	HL	3638	1925	New	Scr 5/1957
44	TWEED	0-4-0ST	OC	HL	3639	1926	New	(27)
43	DEE (DAISY)	0-4-0ST	OC	AB	1179	1910	#	
		Rebuilt		KS		1924	(j)	(4)
54	CALCIUM	0-4-0ST	OC	RS	2124	1873	(k)	s/s
41	EDEN (TYNE)	0-4-0ST	OC	HL	3649	1927	(l)	(24)
42	SPEY (TEES)	0-4-0ST	OC	HL	3652	1927	New	(12)
40	ALN	0-6-0ST	OC	RS	2554	1885	(m)	(15)
39	TAW	0-4-0ST	OC	HL	3671	1927	New	(30)
38	TAFF	0-4-0ST	OC	HL	3672	1927	New	(16)
	HOWDEN DENE	0-6-0T	OC	HL	2880	1911	(n)	(1)
55	ATLAS	0-4-0ST	OC	B	236	1877	(o)	s/s after 14/4/1929
58		0-4-0ST	OC	AB	261	1883	(p)	s/s
37	AVON	0-4-0ST	OC	HL	3567	1924	(q)	(27)
36	TAME	0-4-0ST	OC	HL	3720	1928	New	s/s /1959
33	ETTRICK	0-4-0ST	OC	HL	3721	1928	New	(11)

30	TEVIOT	0-4-0ST	OC	HL	3723	1928	New	(17)
29	SANNOX	0-4-0ST	OC	HL	3730	1928	(r)	(30)
28	ANNAN	0-4-0ST	OC	HL	3731	1928	New	(26)
27	NITH	0-4-0ST	OC	HL	3734	1928	New	(28)
26	FORTH	0-4-0ST	OC	HL	3735	1928	New	(30)
24	TYNE	0-6-0T	OC	HL	3737	1928	New	(27)
25	TEES	0-6-0T	OC	HL	3738	1928	New	(27)
31	LEVEN	0-4-0ST	OC	HC	1623	1928	New	(29)
32	THAMES	0-4-0ST	OC	HC	1624	1928	New	(30)
34	SEVERN	0-4-0ST	OC	P	1589	1928	New	(5)
35	CLYDE	0-4-0ST	OC	AB	1945	1928	New	(27)
60	STELLA	0-4-0ST	OC	HL	2583	1904	(s)	
		Rebuilt		HL		1931		(9)
55	PIONEER No.1	0-4-0ST	OC	BH	298	1875		
		Rebuilt		LG		1925	(t)	(10)
56	PIONEER No.4	0-4-0ST	OC	BH				
		Rebuilt		Ridley Shaw		1927	(u)	s/s after 14/4/1929
57	HUMBER (PIONEER No.5)	0-4-0ST	OC	AB	1800	1923	(v)	(22)
	THE T.H.BROWN	4wWE					(w)	s/s c/1962
23		0-4-4-0T	4C	BP	6172	1924	(x)	Scr 4/1949
53	DANE	0-4-0ST	OC	HL			(y)	s/s here 19/9/1934
40	FIRELESS	0-4-0F	OC	HL	3765	1930	New	(20)
34	MERSEY	0-4-0ST	OC	HC	1422	1921	(z)	Scr 13/2/1957
59		0-4-0ST	OC	HL	2684	1907	(aa)	(13)
61		0-4-0ST	OC	HL	3491	1921	(aa)	(18)
62		0-4-0ST	OC	HL	3175	1916	(aa)	(19)
38	DERWENT	0-4-0ST	OC	RSHD	6973	1939	New	(28)
30	WEAR	0-4-0ST	OC	RSHD	7024	1940	New	(28)
33	SWALE	0-4-0ST	OC	RSHN	7332	1946	New	(30)
40	CAM	0-6-0T	OC	RSHN	7337	1947	New	(28)
42	ISIS	0-6-0T	OC	RSHN	7338	1947	New	(28)
	GUISBOROUGH	0-6-0DE		EE	1554	1948	New	(21)
No.11		0-4-0ST	OC	AE	1830	1919	(ab)	(23)
-		0-6-0DE		YE	2595	1956	(ac)	(25)
1	ALLENDALE	0-6-0DE		YE	2629	1956	(ad)	(39)
2	BILSDALE	0-6-0DE		YE	2665	1957	New	(41)
3	COMMONDALE	0-6-0DE		YE	2666	1957	New	s/s 2/1985
4	DANBYDALE	0-6-0DE		YE	2714	1958	New	(34)
5	ESKDALE	0-6-0DE		YE	2718	1958	New	(34)
6	FARNDALE	0-6-0DE		YE	2719	1958	(ae)	(43)
7	GLAISDALE	0-6-0DE		YE	2723	1958	(af)	(38)
8	HARWOODDALE (KILDALE)	0-6-0DE		YE	2724	1958	(ag)	(39)
9	IBURNDALE (LINGDALE)	0-6-0DE		YE	2725	1958	New	(32)
10	KILDALE	0-6-0DE		YE	2741	1959	New	(31)
11	LINGDALE	0-6-0DE		YE	2742	1959	(ah)	s/s /1985
12	MARDALE	0-6-0DE		YE	2743	1959	(ai)	(35)
13	NIDDERDALE	4wDH		S	10051	1960	New	(36)
14	PATTERDALE	4wDH		S	10052	1960	New	(33)
15	ROSEDALE	4wDH		S	10070	1961	New	(37)
07011 (D2995)		0-6-0DE		RH	480696	1962	(aj)	(40)
D1		6wDH		TH	296V	1981	(ak)	(44)

D2		6wDH		TH	297V	1981	(al)	(46)
-		0-6-0DH		RR	10220	1965	(am)	(42)
T2		4wDH	R/R	NNM	83501	1983	(an)	(50)
T3		4wDH	R/R	NNM	83502	1984	(ao)	(50)
D3		0-6-0DH		TH	167V	1966	(ap)	(45)
T1		4wDH	R/R	NNM	82503	1983	(aq)	(50)
T4	05/273	4wDH	R/R	NNM	83503	1984	(ar)	(50)
T5		4wDH	R/R	NNM	83504	1984	(as)	(50)
(T6)	(3-161)	4wDH	R/R	NNM	83505	1984	(at)	(50)
-		0-6-0DH		HE	8977	1980		
		Rebuilt		HE	9307	1992	(au)	(49)
-		0-6-0DH		TH	285V	1979		
		Rebuilt		HAB		1992	(av)	(48)
08903 (D4133) JOHN W. ANTILL		0-6-0DE		Hor		1962	(aw)	(47)
08743 (D3911) BRYAN TURNER (ANGIE)		0-6-0DE		Crewe		1960	(ax)	(50)
"WILTONIA"		2w-2DMR		Wkm	7591	1957		
Rebuilt		2w-2DHR		YEC	L112	1992	(ay)	(50)
-		2w-2PMR		Wkm	7603	1957*	(ay)	(50)

\#   Built for stock and completed 21/10/1909 but sold 18/4/1910 as new with a 1910 plate.

*   Converted to flat truck.

IRS records indicate another locomotive, 38 VALERIE built by KS and s/s, but no other information has been found on it. They also state that YE 2723 and YE 2724 went to RSH for repairs in the 1960s.

(a)   ex Ministry of Munitions, Birtley, Co. Durham, after 6/1919, by 12/1919.
(b)   ex Ministry of Munitions, previously H. Lovatt & Co Ltd, contractor. It was noted at Billingham Works, 14/4/1929.
(c)   ex Ministry of Munitions, previously Sir John Jackson, contractor. It was noted at Billingham Works, 14/4/1929.
(d)   ex Logan & Hemingway, contractor.
(e)   ex Brunner Mond & Co Ltd, Winnington Works, Northwich, Cheshire, /1922.
(f)   ex Ministry of Munitions, Houston, Glasgow; via dealer Geo Cohen, /1925.
(g)   ex Ministry of Munitions, Litherland, Lancashire, via dealer Geo Cohen, /1925.
(h)   ex Ministry of Munitions, Rainhill, Lancashire, via dealer Geo Cohen, /1925.
(i)   ex Ind Coope & Co Ltd, Burton on Trent, Staffordshire, via dealer T.F. Young, Darlington, 9/1925.
(j)   ex Brunner Mond & Co Ltd, Winnington Works, Northwich, Cheshire, /1926.
(k)   ex ICI Ltd, Allhusen Works, Gateshead, Co. Durham.
(l)   new. To ICI Ltd, 'Tennant's' Salt Works, Haverton Hill, by 9/7/1950 and returned, by 6/1952.
(m)   ex ICI Ltd, Allhusen Works, Gateshead, Co. Durham, 5/1927.
(n)   ex Strakers & Love Ltd, Brandon Colliery, Brandon, Co. Durham, on hire, /1928.
(o)   ex? /1928
(p)   ex Tees Salt Co Ltd, Tees Salt Works, Haverton Hill, c/1928. It was noted at Billingham Works, 14/4/1929.
(q)   new, ex Hawthorn Leslie, /1928.
(r)   new. to ICI Ltd, Middleton Works, Heysham, Lancashire, /1940 and returned /1942.
(s)   ex Stella Coal Co Ltd, Addison Colliery, Co. Durham, on hire, /1928.
(t)   ex Casebourne & Co (1926) Ltd, Haverton Hill, /1929.
(u)   ex Casebourne & Co (1926) Ltd, Haverton Hill, /1929.
(v)   ex Casebourne & Co (1926) Ltd, Haverton Hill, /1929.
(w)   origin? IRS records suggest that it was 1392 of 1928.
(x)   ex ICI Metals Ltd, Landore Works, Landore, Glamorgan, /1929, by 14/4/1929.
(y)   ex Brunner Mond & Co Ltd, Winnington Works, Northwich, Cheshire, /1929.
(z)   ex ICI Ltd, Dalton Works and Turnbridge Works, Huddersfield, Yorkshire (WR), 20/12/1933.
(aa)   ex William Whitwell & Co Ltd, Thornaby.

(ab)   ex Thos W. Ward Ltd, Sheffield, on hire, 14/3/1953. Earlier at Birmingham Railway Carriage & Wagon Co Ltd, Smethwick, Birmingham.

(ac)   this was a new locomotive destined for the Appleby-Frodingham Steel Co, Appleby-Frodingham Works, Scunthorpe but was initially sent for trials at ICI's Billingham Works. Other interested parties were invited to a demonstration at Billingham on 18/6/1956.

(ad)   new. To ICI Ltd, Prudhoe Works, Prudhoe-on-Tyne, Northumberland, 6/1965 and returned by 6/1967. To British Steel Corporation, Hartlepool South Works, Hartlepool, on loan, 9/1970. Returned to Billingham Works, 4/1971.

(ae)   new. To ICI Ltd, Severnside Works, Severn Beach, Gloucestershire, 11/9/1963 and returned /1964. It had another transfer to or from another works on 22/3/1967. To British Steel Corporation, Hartlepool South Works, Hartlepool, on hire, 3/1970 and returned 5/1970. To Thos. Hill (Rotherham) Ltd, Kilnhurst, South Yorkshire, 22/10/1982 and returned 27/10/1983.

(af)   new. To ICI Ltd, Wilton Works, Redcar, c/1964 and returned c/1969.

(ag)   new. To Cleveland Potash Ltd, Boulby Mine, Easington, /1974 and returned by 2/1975.

(ah)   new. To British Steel Corporation, Hartlepool South Works, Hartlepool, on hire, 5/1970 and returned 9/1970.

(ai)   new. To British Steel Corporation, Hartlepool South Works, Hartlepool, on hire 1/1970 and returned, 4/1971.

(aj)   ex Resco (Railways) Ltd, Woolwich, London, on hire, 3/1979.

(ak)   new. To Thos. Hill (Rotherham) Ltd, Kilnhurst, South Yorkshire, 4/10/1984 and returned 19/12/1984.

(al)   new. To Thos. Hill (Rotherham) Ltd, Kilnhurst, South Yorkshire, 27/6/1984 and returned 4/10/1984.

(am)   ex Thos. Hill (Rotherham) Ltd, Kilnhurst, South Yorkshire, on hire, 25/6/1984.

(an)   new. To ICI Chemicals & Polymers Ltd, Wilton Works, Redcar, by 9/4/1991 and returned c/1996; after 19/8/1994, by 12/3/1997.

(ao)   new. To ICI Chemicals & Polymers Ltd, Wilton Works, Redcar, c/1992 and returned c/1996; after 19/8/1994, by 12/3/1997.

(ap)   ex ICI Ltd, Heysham Works, Lancashire, by 5/8/1986.

(aq)   ex ICI Ltd, Stevenston Works, Stevenston, Strathclyde. To Scottish Agricultural Industries plc, Leith, Lothian, by 9/4/1991 and returned by 19/8/1991. To ICI Chemicals & Polymers Ltd, Wilton Works, Redcar, c/1992 and returned c/1994.

(ar)   ex ICI Ltd, Ardeer Works, Strathclyde, by 5/8/1986. To ICI Chemicals & Polymers Ltd, Wilton Works, Redcar, c/1992 and returned c/1996; after 19/8/1994, by 12/3/1997.

(as)   ex ICI Ltd, Heysham Works, Lancashire, by 5/8/1986. To ICI Chemicals & Polymers Ltd, Wilton Works, Redcar, c/1992 and returned c/1996; after 19/8/1994, by 12/3/1997.

(at)   ex Scottish Agricultural Industries plc, Leith, Lothian, by 19/8/1991. To ICI Chemicals & Polymers Ltd, Wilton Works, Redcar, c/1992 and returned c/1996; after 19/8/1994, by12/3/1997.

(au)   ex Hunslet Engine Co Ltd, Leeds, on hire, c/1993.

(av)   ex Hunslet-Barclay Ltd, Kilmarnock, Strathclyde, on hire, c/1993.

(aw)   ex British Rail, Doncaster, South Yorkshire, 9/5/1996.

(ax)   ex R.F.S. Engineering Ltd, Doncaster, South Yorkshire, w/c 27/1/1997.

(ay)   ex ICI Chemicals & Polymers Ltd, Wilton Works, Redcar, after 19/5/1997, by 29/1/1998.

(1)   to Strakers & Love Ltd, Brandon Colliery, Brandon, Co. Durham, ex hire, by 2/1929.

(2)   to Brunner Mond & Co Ltd, Winnington Works, Northwich, Cheshire, 12/1929.

(3)   to James W. Ellis & Co Ltd, Swalwell, Co. Durham, c/1930, by 25/6/1932.

(4)   to Mold Gas & Water Co, Mold, Flintshire. It was at the Billingham Works until at least 3/1930.

(5)   to River Wear Commissioners, Sunderland, Co. Durham, per C.J.M. Lowe, dealer. It was purchased by the Commissioners 18/8/1931 and received 1/9/1931.

(6)   to Mells Quarry Co, Bilboa Quarry, Mells, Somerset, after 6/5/1930, by 4/1932.

(7)   to S. Pearson & Co Ltd, Silent Valley Reservoir contract, Belfast, after 14/4/1929.

(8)   to T.J. Thomson & Son Ltd, Stockton, /1934.

(9)   to Stella Coal Co Ltd, Addison Colliery, Co. Durham, ex hire, /1934.

(10)   to ?, /1934.

(11)   to ICI Ltd, Winnington Works, Northwich, Cheshire, /1935.

(12)   to ICI Metals Ltd, Witton Works, Birmingham, Warwickshire, /1936.

(13)   to Ridley Shaw & Co Ltd, Cargo Fleet, rebuilt /1937; then to Hawthorn Limestone Co Ltd, Hawthorn Quarry, Co. Durham.

(14)   to ICI Ltd, Muspratt No.2 Works, Widnes, Lancashire, /1937.

(15)   to W. Shaw & Co Ltd, Middlesbrough, per W. Blenkinsop & Co, Middlesbrough.

(16)	to ICI Ltd, Gaskell-Marsh Works, Widnes, Lancashire, /1938.					
(17)	to ICI Ltd, Pilkington-Sullivan Works, Widnes, Lancashire, probably before /1940; it was noted there 16/6/1942.					
(18)	to Southern Cotton Oil Co Ltd, Trafford Park, Lancashire.					
(19)	to Kinneil Cannel & Coking Coal Co Ltd, Bo'ness, West Lothian.					
(20)	to ICI Ltd, Dalton Works and Turnbridge Works, Huddersfield, Yorkshire (WR).					
(21)	to ICI Ltd, Wilton Works, Redcar, c/1950.					
(22)	to ICI Ltd, Clarence Salt Works, Port Clarence, 6/1951.					
(23)	returned to Thos W. Ward Ltd, Sheffield, ex hire, 1953. Ward then hired it to the Nassington Barrowden Mining Co Ltd on 23/10/1953.					
(24)	to ICI Ltd, Dowlais Works, Dowlais, Glamorgan, 7/1956.					
(25)	to Appleby-Frodingham Steel Co, Appleby-Frodingham Works, Scunthorpe, Lincolnshire after trials at Billingham. It was at Appleby-Frodingham Works by 30/8/1956.					
(26)	to ICI Ltd, Dowlais Works, Dowlais, Glamorgan, 6/1958; seen en route 23/6/1958.					
(27)	sold for scrap, /1958.					
(28)	to J.N. Connell Ltd, Coatbridge, Lanarkshire, for scrap, 7/1959.					
(29)	to ICI Ltd, Dowlais Works, Dowlais, Glamorgan, 12/1959.					
(30)	to Thos W. Ward Ltd, 2/3/1961.					
(31)	to ICI Ltd, Severnside Works, Severn Beach, Gloucestershire, 10/1962.					
(32)	to ICI Ltd, Severnside Works, Severn Beach, Gloucestershire, 9/1963.					
(33)	to ICI Ltd, Dalton Works and Turnbridge Works, Huddersfield, Yorkshire (WR), c6/1964.					
(34)	to ICI Ltd, Castner-Kellner Works, Runcorn, Cheshire, 9/1972.					
(35)	to ICI Ltd, Castner-Kellner Works, Runcorn, Cheshire, c9/1972.					
(36)	to ICI Ltd South Central Workshops, Tunstead, Derbyshire, c1/1973.					
(37)	to Bell & Son (Doncaster) Ltd, Port Clarence, /1974.					
(38)	to Cleveland Potash Ltd, Tees Dock, Grangetown c12/1974.					
(39)	to Cleveland Potash Ltd, Boulby Mine, Easington, by 5/1975.					
(40)	returned to Resco (Railways) Ltd, Woolwich, London, ex hire, 12/11/1979.					
(41)	to Thos. Hill (Rotherham) Ltd, Kilnhurst, South Yorkshire, /1981.					
(42)	to National Smokeless Fuels Ltd, Smithywood, South Yorkshire, (property of Thos Hill (Rotherham) Ltd), ex hire, 12/1984.					
(43)	to Thos. Hill (Rotherham) Ltd, Kilnhurst, South Yorkshire, 3/1/1985; then to Marple & Gillott, Sheffield, South Yorkshire.					
(44)	to Wilmott Bros (Plant Services) Ltd, Ilkeston, Derbyshire, 21/6/1994.					
(45)	to T.J. Thomson & Son Ltd, Stockton, c6/1994.					
(46)	to Wilmott Bros (Plant Services) Ltd, Ilkeston, Derbyshire, 12/8/1994.					
(47)	to R.F.S.(Engineering) Ltd, Doncaster, South Yorkshire for repairs, 3/1997; then to ICI Ltd, Wilton Works, Redcar, 4/8/1997.					
(48)	to Hunslet-Barclay Ltd, Kilmarnock, Ayrshire, ex hire, after 13/8/1997, by 1/1998.					
(49)	to Hunslet-Barclay Ltd, Kilmarnock, Ayrshire, ex hire, 2/7/1997.					
(50)	to Enron Teesside Operations Ltd, 1/1/1999.					

Billingham Coking Plant

Gauge : ?

-		WSO	1292	1928	New	Scr/1962

Anhydrite Mine

Gauge : 2ft 6in

-	4wWE	EE	745	1929	New	s/s
-	4wWE	EE	1227	1943	New	s/s
-	4wWE	YE	2414	1944	New	s/s

CLARENCE SALT WORKS, Port Clarence
ICI (Salt) Ltd until 31/12/1942
The Salt Union Ltd until 28/3/1940 (subsidiary of **ICI Ltd** from 3/1937)
Bell Brothers Ltd until 2/11/1888

91
NZ 505214 – **Map C**

Bell Bros began pumping brine in 1882 from the 100 feet thick salt bed approximately 1,200 feet below the surface at Salt Holme Farm and Port Clarence. The brine was evaporated in both coal fired pans and some using waste heat from Bell's blast furnaces to produce salt. This was in demand for making synthetic alkali (soda) for the soap and glass industries, together with the production of foodstuffs for the expanding urban population. Bell Bros' initiative encouraged other companies to also establish salt works and the output from Teesside reached a peak of 318,000 tons in 1894 from 55 wells with much of the salt going to alkali works on the Tyne. Production then declined to a little over 100,000 tons in the 1920s. Also brine, rather than salt, was increasingly used as a feedstock for the production of caustic soda and chlorine which resulted in the salt works on the north side of the Tees being acquired by ICI.

The Clarence Salt Works was situated on the south west portion of Bell's Iron Works site near to its blast furnaces in order to benefit from the waste heat. An inventory of Salt Union property in 1900 states that the Clarence Salt Works comprised, in addition to the brine wells, two brine reservoirs, 25 coal fired pans, 11 gas fired pans, 5 storehouses, workshops and offices. These were served by the Salt Union's railway system which connected with the end of the Port Clarence Branch, ¼ mile beyond the passenger station. A 2 mile 5 furlong railway was authorised by the 1897 North Eastern Railway Act commencing at a junction with the Port Clarence Branch and terminating on the south side of Greatham Creek. The line (opened in 1901) ran northwards initially alongside the Clarence Salt Works. In 1908–13, Salt Union erected an extension to its works on leasehold land to the west of the Greatham Creek Branch and, by 1926, its railway included a line that climbed over the Port Clarence Branch and the iron works sidings to a jetty on the river for shipping out salt; the bridge consisted of steel lattice girders, a timber deck and brick piers.

It is likely that Bell's locomotives shunted the salt works until Salt Union took over and used its own engines. Industrial salt went out in sacks or sheeted open wagons, while packets of kitchen salt were loaded into railway vans. Ash was shovelled into railway wagons for dumping on the tip. The Salt Works single road engine shed was located near the salt pan house at the southern end of Bell's Old Side blast furnaces. The salt pans had to rely solely on coal after the blast furnaces finished in 1930. Clarence Salt Works closed at the beginning of the 1950s and SALTONIA assisted with the demolition of the premises during 1952; it was still there in 4/1955, having been sold for scrap.

Stephen Lewin at the Poole Foundry did not build many locomotives but this one (606 of 1875) was working at one of the Teesside salt works, possibly Tennant's at Haverton Hill on 2nd July 1936.

(R.G. Jarvis photograph)

References :   See Billingham Works entry

Gauge : 4ft 8½in

SALT UNION No.2	0-4-0ST	OC	AB	668	1891	(a)	(2)
BOSTOCK	0-4-0ST	OC	WkB		1881	(b)	(1)
WINSFORD	0-4-0ST	OC	FE	117	1889	(c)	(3)
SALTUNIA	0-4-0ST	OC	AB	1018	1905	(d)	(6)
HUMBER	0-4-0ST	OC	AB	1800	1923	(e)	(5)
NEWBRIDGE	0-4-0ST	OC	FE			(f)	(4)

The Lewin locomotive, WHARTON, may have been here before moving to 'Tennant's' Works.

(a) ex South Durham Salt Works, Haverton Hill, after 6/1898, by 7/1898, although Inness has /1894.
(b) ex The Salt Union Ltd, West Works, Winsford, Cheshire, after /1922. It was at Haverton Hill 8/5/1928 so may have gone to the South Durham Salt Works first.
(c) ex The Salt Union Ltd, West Works, Winsford, Cheshire, after 6/1934, by /1943.
(d) ex 'Tennant's' Salt Works, Haverton Hill, by 7/1950.
(e) ex ICI Ltd, Billingham Works, Billingham, 6/1951.
(f) ex 'Tennant's' Salt Works, Haverton Hill, after 4/1952.

(1) to 'Tennant's' Salt Works, Haverton Hill, before /1945.
(2) s/s; after 17/4/1949, by /1950.
(3) to 'Tennant's' Salt Works, Haverton Hill, 6/1952.
(4) to T.J. Thomson & Son Ltd, Stockton, after 5/1953.
(5) to ICI Ltd, West Works, Winsford, Cheshire, via Ridley Shaw & Co Ltd, Cargo Fleet, after 10/5/1953, by 21/4/1954.
(6) to T.J. Thomson & Son Ltd, Stockton, for scrap, after 4/1955.

'TENNANT'S' SALT WORKS, Haverton Hill. 92
ICI (General Chemicals) Ltd until ?   NZ 495227 – **Map C**
United Alkali Co Ltd until 3/6/1931 (subsidiary of **ICI Ltd** from 7/12/1926)
Charles Tennant & Partners Ltd until 1/11/1890

'Tennant's' was the local name for the works, although officially it became the Tees Salt Works in the ICI era. The local version has been retained to help distinguish it from the Tees Salt Works (which see). Tennant's began drilling for salt in 1885 and erected a works at Haverton Hill on the south side of the South Durham Salt Co's railway. Sidings from this railway ran to serve two large buildings containing the panhouse and stores, together with an engine shed and ash tip. One of the sidings went to a cluster of brine wells between the works and Holme Fleet. Another short line off South Durham's railway ran to some brine wells at the end of Oak and Elm Streets. Salt was despatched to Tennant's works on Tyneside which manufactured alkali by the Leblanc process. On 1/11/1890 the Leblanc Operators formed the United Alkali Co Ltd with Mr Tennant as chairman to compete against the cheaper Solvay process. When it became a subsidiary of ICI, the works at Haverton Hill had 20 coal fired pans capable of producing 40,000 tons of coarse salt and four pans giving 6,000 tons of fine salt each year.

The locomotives known to have operated at Tennant's Works are shown below, but it is likely that the list is incomplete before 1926 and one or more Tennant's/United Alkali's engines at their Hebburn Works on Tyneside may have spent some time at one of that company's Haverton Hill works. When Ted Haigh visited the works on 9/7/1950, EDEN was present on loan from Billingham Works, SALTONIA had gone to the Clarence Works and BOSTOCK was away for repairs. Coal arrived in railway wagons and was shunted up a ramp for the furnaces, while ash was being dumped on the former South Durham Works site. Tennant's Works closed in the early 1950s but one of the pan houses was not demolished until the 1970s.

References :   See Billingham Works entry.

Gauge : 4ft 8½in

	No.1 WHARTON	0-4-0ST	OC	Lewin	606	1875		
		Rebuilt				1881	(a)	(3)
	BOSTOCK	0-4-0ST	OC	WkB		1881	(b)	(2)
	TENNANT'S No.3	0-4-0ST	OC	BH	881	1886		
		Rebuilt	MW			1911	(c)	(1)
	SALTUNIA	0-4-0ST	OC	AB	1018	1905	(d)	(4)
41	EDEN	0-4-0ST	OC	HL	3649	1927	(e)	(6)

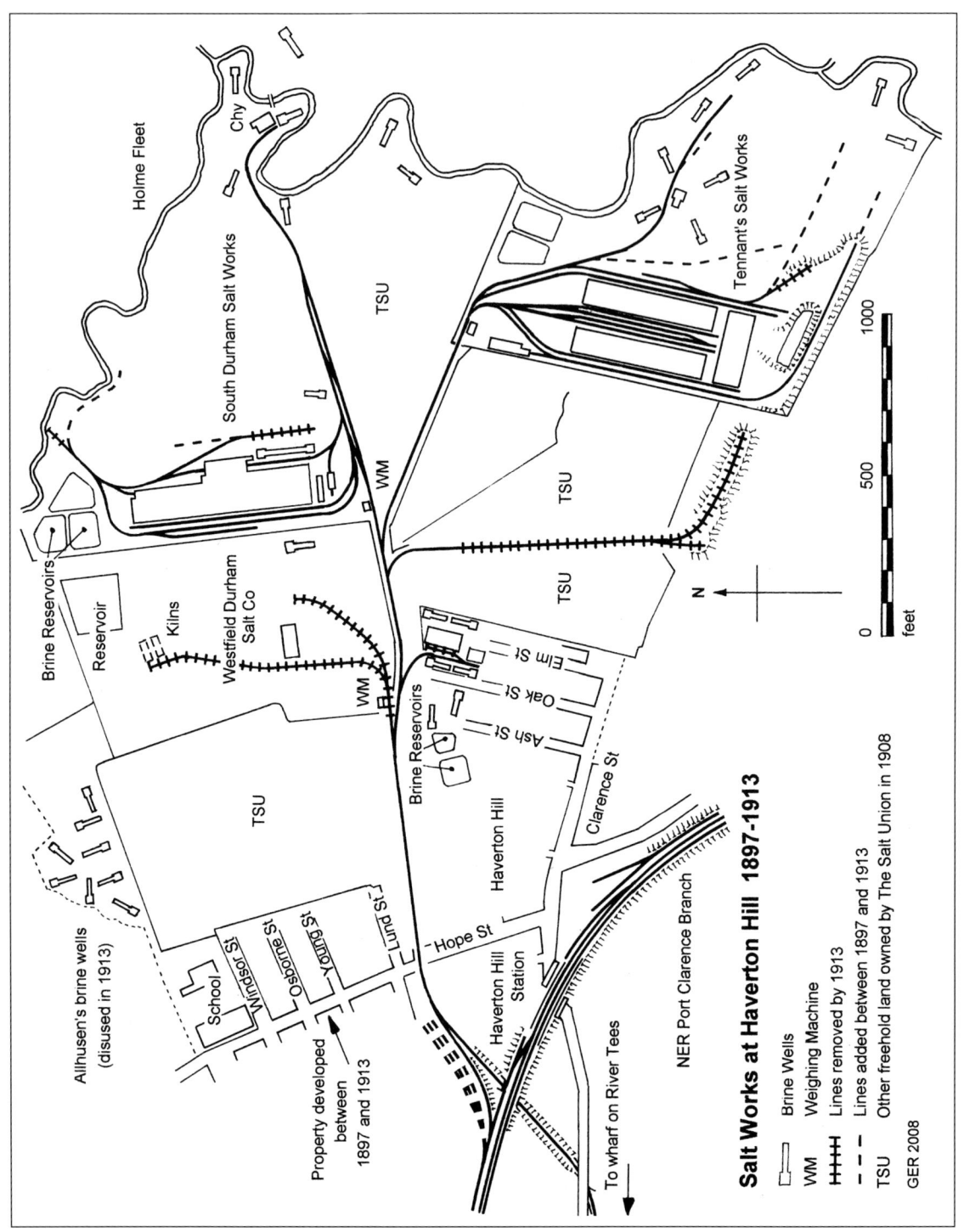

NEWBRIDGE		0-4-0ST	OC	FE			(f)	(5)
WINSFORD		0-4-0ST	OC	FE	117	1889	(g)	s/s after 6/1952

(a) ex ICI (Salt) Ltd, South Durham Salt Works, Haverton Hill. It may have also worked at the Clarence Salt Works, Port Clarence. To ICI Ltd store at Marske Airfield and returned 1/1944.
(b) ex Clarence Salt Works, Port Clarence, before /1945.
(c) ex The United Alkali Co Ltd, Tennant's Works, Hebburn, Co. Durham. A photograph of BH 881 appeared in the IRS Durham Handbook Part One (page 205) but it was almost certainly taken at a Teesside Salt Works. either Tennant's or South Durham, because drilling rigs can be seen and the same design of fence is visible in the view of Lewin 606.
(d) ex G. Shellabear & Son, Park Royal Plant Depot, London; it was advertised for sale in 4/1931. It was noted at Tennant's Salt Works in 4/1949.
(e) ex ICI Ltd, Billingham Works, Billingham, by 9/7/1950.
(f) ex ICI Ltd, Stoke Works, near Bromsgrove, Worcestershire, after 7/1950, by 7/1951.
(g) ex Clarence Works, Port Clarence, 6/1952.

(1) returned to The United Alkali Co Ltd, Tennant's Works, Hebburn, Co. Durham.
(2) to G.P. Trentham Ltd, contractor, for use on construction of Meaford 'A' Power Station, Staffordshire, /1945. It was there by 2/1946.
(3) to Thos W. Ward Ltd, South Bank, for scrap, /1948.
(4) to Clarence Salt Works, Port Clarence, by 7/1950.
(5) to Clarence Salt Works, Port Clarence, after 4/1952.
(6) returned to ICI Ltd, Billingham Works, Billingham, by 6/1952.

WILTON WORKS, Redcar 93
NZ 564216, NZ 568223 – **Map H**

As early as 1943, ICI was looking to widen its production of chemicals based on the importation of oil. In 12/1945 it purchased the Wilton estate of 3,500 acres from Colonel Lowther; 1,841 acres of which were to be laid out as a chemical industrial estate with several distinct but related plants. The works was situated between the A1085 Trunk Road and the A174, with a branch railway leaving the Middlesbrough-Redcar line east of Grangetown Signal Box and passing under the main line to run south to Wilton. Construction on the site started in 2/1946 with J.B. Edwards of London beginning to lay out the roads and drains two months later. The first plant to come on stream in 7-8/1949 produced phenol, formaldehyde and perspex; the site was officially opened on 14/9/1949. By far the most important plant was the olefine cracker which split naphtha into several simpler hydrocarbons, the main one of which was ethylene and started production in 1951. This, together with the nearby interconnected aromatics plant, provided a range of chemical building blocks for a number of major products. Amongst the plants on the site were nylon, chlorine and caustic soda, tereylene, aniline, titanium and paraxylene. The power station opened in 3/1952 and by 1958 10,000 people were employed on the site. Pipelines not only connected the various plants, but also to facilities on the north side of the River Tees via a 9 feet diameter tunnel under the river constructed of cast iron segments similar to the London Underground. Later pipelines were provided to other parts of the country e.g. Runcorn, Hillhouse.

The railway tracks were laid out by Thomas Summerson & Sons Ltd of Darlington. A double line entered the site allowing arrival and departure traffic to be separated. This was controlled by Wilton Signal Box which belonged to ICI, but was operated by a BR signalman paid for by ICI. Unfortunately this box was demolished on 28/12/1967 when a tank wagon on a departing train derailed and hit the box killing the signalman. The signal box was not replaced. According to Terry Bye, a passenger service ran for contractors and ICI personnel until the mid-1950s. Unlike the rail system at the Billingham Works, the layout at Wilton was mainly centred on the sidings along the north west side of the site, with only a couple of separate branches, one of which ran to the nylon works. Conveyors and pipelines connected the relevant plants to the sidings, thus obviating the need for a large fleet of shunting locomotives. In 1957 about 775,000 tons of raw materials for processing arrived at Wilton made up of 321,000 tons by sea and pipeline, 94,500 tons by road, 334,000 tons via the pipeline from Billingham and only 26,000 tons by rail. However, 330,800 tons of coal were purchased for the boilers. The power station was also located reasonably close to the sidings for its coal supplies until, in 1971, it was converted to burn byproduct gases and oils. In 1979 about one hundred 13 ton wagons of lime were received from Buxton each week in order to produce a slurry used in the propylene oxide plant. There were three or four trains on alternative weeks taking chlorine tank wagons to Ardeer in Scotland. The sidings serving the terminal for incoming oil tank wagons were only used infrequently by 1979. The construction of a new part of the nylon plant during the same year resulted in the loss of the old engine shed (NZ 564216) and a replacement (NZ 568223) was erected further north alongside the sidings.

The new English Electric diesel shunters purchased by ICI in 1948 were similar to those being built for the LMS. They proved useful locomotives, but obtaining spares eventually became a problem and English Electric 1554 was finally cannibalised, which explains the change of name on 1553 when it acquired the former's bonnet covers. The works railway had once been very busy but, by the mid-1980s, a lot of the traffic had gone, including coal, lime, cyclohexane, caustic soda, chlorine, glycol and diluent.

Two significant developments resulted in new rail traffic in the late 1980s, but not necessarily for ICI's shunting locomotives. In 1988 a ceremony was held to mark the completion of the power station's coal reversion project. This resulted in the arrival of two trains of 36 HAA wagons each day, Monday to Friday, from Butterwell opencast coal site. At the north west end of Wilton, a new Freightliner Terminal opened in 2/1989, replacing the depot at Stockton and this was served by container trains from Felixstowe and Leeds. With the earlier decline in traffic, the connection to BR had been reduced to one line and only a single train was allowed on the Wilton site at a time. As a result of the developments at the power station and Freightliner Depot, a new signalling system was introduced on 13/2/1989 that meant three trains could be on the Wilton site simultaneously. Workshops at Wilton were also used as a base for the restoration of main line locomotives by the North Eastern Locomotive Preservation Group and the Deltic Preservation Society.

In response to major changes in the world chemical industry, ICI split its activities in 1993 between Zeneca and ICI bulk commodities and decided to also concentrate on lighter specialist products. It had already begun to divest itself of the plants at Wilton, selling them to other world class companies. As from 1/1/1999, ICI sold its Teesside utilities and services to the Enron Corporation, which included the railway, pipelines, power station and undeveloped land at Wilton. In view of the completely changed ownership profile, the site was renamed Wilton International in 1999. (See the Sembcorp Utilities UK Ltd entry for developments after this date).

References : *Wilton The First Fifty Years*, Colin Hurworth, Falcon Press, 1999

Wilton Rail Group, T. Bye, 1989. Unpublished notes on Wilton Rail Section History.

Gauge : 4ft 8½in

	ESTON	0-6-0DE		EE	1552	1948	New	(1)
	GUISBOROUGH (REDCAR)	0-6-0DE		EE	1553	1948	New	(4)
	(GUISBOROUGH)	0-6-0DE		EE	1554	1948	(a)	(5)
No.7	GLAISDALE	0-6-0DE		YE	2723	1958	(b)	(2)
	TAME	4wDM		FH	3685	1954	(c)	(3)
(DB965076)	WILTONIA	2w-2DMR		Wkm	7591	1957	(d)	
	Rebuilt	2w-2DHR		YEC	L112	1992		(9)
(DB965088)		2w-2PMR		Wkm	7603	1957*	(d)	(9)
(07005) (D2989) (LANGBAURGH)		0-6-0DE		RH	480690	1962		
		Rebuilt	Resco		L106	1978	(e)	(10)
07011 (D2995) CLEVELAND		0-6-0DE		RH	480696	1962		
		Rebuilt	Resco		L105	1978	(f)	(8)
08502 (D3657) ANGIE		0-6-0DE		Don		1958	(g)	(10)
08503 (D3658)		0-6-0DE		Don		1958	(h)	(10)
T2 (2)		4wDH	R/R	NNM	83501	1983	(i)	(7)
1 T1		4wDH	R/R	NNM	82503	1983	(j)	(6)
T3		4wDH	R/R	NNM	83502	1984	(j)	(7)
T4 05/273		4wDH	R/R	NNM	83503	1984	(j)	(7)
T5		4wDH	R/R	NNM	83504	1984	(j)	(7)
(T6) (3-161)		4wDH	R/R	NNM	83505	1984	(j)	(7)
08903 (D4133) 903 (7)								
	JOHN W. ANTILL	0-6-0DE		Hor		1962	(k)	(10)

* converted to flat truck, c/1968

Occasionally a Class 08 shunter was hired from BR; 08212 was present on 26/7/1979.

(a) ex Billingham Works, c/1950.
(b) ex Billingham Works, c/1964.
(c) ex Imperial Metal Industries (Kynoch) Ltd, Witton, Birmingham, c/1967.
(d) ex British Rail, Gateshead Depot, /1968.
(e) ex Resco (Railways) Ltd, Woolwich, London, 18/7/1979.
(f) ex Resco (Railways) Ltd, Woolwich, London, 4/9/1980.
(g) ex British Rail, Thornaby Depot, 6/9/1988.
(h) ex British Rail, Doncaster Works, Doncaster, South Yorkshire, 26/9/1988.

ICI purchased two ex-BR diesels made redundant from shunting at Southampton Docks. The first, 07005 (RH 480690/1962) was named LANGBAURGH at Wilton by the mayor, Ivy Cole, on 12th November 1979.
(Courtesy Evening Gazette, Middlesbrough)

(i) ex Billingham Works, by 9/4/1991.
(j) ex Billingham Works, c/1992.
(k) ex R.F.S. (Engineering) Ltd, Doncaster, South Yorkshire, after repairs for ICI, 4/8/1997.

(1) to Winnington Works, Cheshire, /1949.
(2) to Billingham Works, c/1969.
(3) to Resco (Railways) Ltd, Woolwich, London, 18/7/1979.
(4) to Resco (Railways) Ltd, Woolwich, London, 9/9/1980; then to Mid Hants Railway Preservation Society, Ropley, Hampshire, 8/1982 and finally to North Yorkshire Moors Railway, Grosmont, North Yorkshire, 7/9/1983 where erroneously preserved as 12139.
(5) to T.J. Thomson & Son Ltd, Stockton for scrap, 7/9/1983.
(6) to Billingham Works, c/1994.
(7) to Billingham Works, c/1996; after 19/8/1994, by 12/3/1997.
(8) to Hastings Diesels Ltd, St Leonards, East Sussex, 17/5/1996.
(9) to Billingham Works, after 19/5/1997, by 29/1/1998.
(10) to Enron Teesside Operations Ltd with site, 1/1/1999.

ICI (SALT) LTD

SOUTH DURHAM SALT WORKS, Haverton Hill. 94
The Salt Union Ltd until 28/3/1940 (subsidiary of **ICI Ltd** from 3/1937)   NZ 493229 – **Map C**
South Durham Salt Co until 6/10/1888
Haverton Hill Salt Co until 1/12/1886

The Haverton Hill Salt Co began sinking brine wells and erecting a salt works north east of the small settlement of Haverton Hill in 1884. It appears to have also built the private railway from the works passing under the Port

Clarence Branch and Chilton's Lane to reach a wharf on the west bank of the River Tees. A curve connected the railway with the Port Clarence Branch, west of Haverton Hill Station. The agreement for this connection was dated 4/9/1885 but construction was not completed until 2/1886. The business changed its name to the South Durham Salt Co in 1886 when its premises comprised five brine wells, a panhouse with nine pans, an engine house and boilers, brine reservoirs, a salt store and workshops.

By 1897, the private railway not only served the panhouse, boilers and stores but also extended eastwards to Holme Fleet where there were brine wells and a pumping station. At the works, sidings ran either side of the principal building before terminating at a headshunt next to Holme Fleet. A short siding connected with the company's wooden engine shed, while a weighing machine was situated on the private railway to measure South Durham's traffic. The Westfield & Durham Salt Co (registered 27/9/1889) had plans in 1890 for a combined panhouse and store on land next to the South Durham Works' western boundary. A siding was laid to the new project from the South Durham's private line but the Westfield & Durham Salt Co's premises were offered for sale by auction in 1892 when the plant still appeared to be incomplete. The NER commented, at a date unknown, "No salt has been sent away but a wagon of bricks has occasionally been forwarded". The siding had been removed by 1913.

The South Durham Works did not have many locomotives but made up for it by some interesting engines, including a rare example built by Robey and one by Stephen Lewin of Poole. According to Inness, LIVERTON was rebuilt at the NER Darlington Works after a collision. In 1926, the South Durham Works had the capacity to produce 80,000 tons of coarse and 9,000 tons of fine salt annually. Following the takeover of Salt Union by ICI, the latter concentrated production on the former Tennant's Works and South Durham closed about 1940.

References :   See the ICI Billingham Works entry.

Gauge : 4ft 8½in

	SALT UNION No.1							
	LIVERTON	0-4-0ST	IC	Robey			(a)	(2)
	SALT UNION No.2	0-4-0ST	OC	AB	668	1891*	New	(1)
No.1	WHARTON	0-4-0ST	OC	Lewin	606	1875		
		Rebuilt				1881	(b)	(3)
-		0-4-0ST	OC	BH	764	1885	(c)	s/s

The much travelled and rare Robey locomotive, LIVERTON, had been used at Liverton Ironstone Mine, Cargo Fleet Iron Works and the South Durham Salt Works at Haverton Hill; however, this photograph may have been taken after it had moved to Cheshire in 1907.   (IRS collection)

* Despatched by maker 24/10/1890 but later noted with 1891 plates.

(a) ex The Cargo Fleet Iron Co, Cargo Fleet Iron Works, Cargo Fleet, by /1886.
(b) ex The Salt Union Ltd, Weston Point Works, Runcorn, Cheshire, /1907.
(c) ex Dorman Long & Co Ltd, Acklam Works, Middlesbrough, (information from Inness).

(1) to The Salt Union Ltd, Clarence Salt Works, Port Clarence, after 6/1898, by 7/1898, although Inness has /1894.
(2) to The Salt Union Ltd, Meadow Works, Winsford, Cheshire, 8/1907.
(3) to 'Tennants' Salt Works, Haverton Hill.

There is a possibility that the Walker Bros locomotive, BOSTOCK, was here in 8/5/1928 before going to the Clarence Salt Works.

ICL UK (CLEVELAND) LTD

BOULBY MINE, Easington   **95**
Cleveland Potash Ltd until 29/9/2015   NZ 763183 NZ 765189 – **Map I**

The bed of potassium salts was discovered shortly before World War 2 by a deep borehole searching for oil in the Esk Valley. ICI became interested because world demand for fertiliser was increasing and the UK depended on imported potash. It was not until 1967 that four boreholes were drilled and ICI decided to establish a joint venture with a company possessing deep mining experience. Cleveland Potash Ltd (CPL), jointly owned by ICI Ltd (50%) and the Anglo American Corporation (50%), was formed in 1969 to sink and operate Boulby Mine. Two shafts were sunk 3,774 feet (1,150 metres) to the potash deposits, which have a variable but average thickness of 7 metres. It was the second deepest mine in Europe and production commenced in 1973 with full output in 1976. ICI subsequently sold its share to Anglo American in 1979 but, in 2002, CPL was purchased by ICL Fertilizers Ltd, a division of Israel Chemicals Ltd. Approximately 2.4 million tonnes of potash (used for fertilisers, glassmaking and in the chemical industry) and 0.8 million tonnes of rock salt (used on roads, for sugar beet cultivation and in animal feed) were being produced each year. Workings extended 16km under the North Sea and went down to 1,350 metres. The site for the potash mine was established near the location of the former Boulby Ironstone Mine astride the track bed of the Loftus-Whitby railway line. Mined material is raised in skips up the shaft and conveyed to the top of the main processing building. Here it passes over mills to reduce its particle size, is cleaned and then washed in brine to separate the potash and the salt before being dried. Two conveyors carry the materials to the stores or rail load-out hoppers. Waste and sea water slurry is pumped into worked out areas underground.

In view of the heavy tonnages of potash and salt to be transported from the mine, it was a planning condition that most should leave by rail. Fortunately much of the trackbed was still available, but five miles of single line railway had to be laid to reconnect with BR at Crag Hall Signal Box, Carlin How. It also required a replacement viaduct to be constructed over the A174 road and a number of bridges, including Grinkle Tunnel, had to be strengthened. A report dated 4/8/1972 stated that most of the work on the branch and mine sidings had been completed. Although the main line railway companies are responsible for working trains to and from the mine, the track to Carlin How is a CPL responsibility; it was relaid in 2002. The pattern of working consisted of potash being taken to Tees Dock where CPL had leased a new terminal (which see) from the Tees & Hartlepool Port Authority for the material to be exported or distributed by road. Deliveries of salt and some potash were taken by train to Middlesbrough Town Goods where various companies arranged for its onward movement by road. In the early days, there was an occasional train to Severnside and a once a week delivery of oil to the mine.

The standard gauge sidings at the mine are confined to the west side of the premises and these are shunted by CPL's locomotives. The line from Carlin How enters the site and divides into two. One connects with tracks on the western boundary that are used for stabling wagons and sometimes out of use locomotives. The other divides into two tracks which pass under, or by, the main discharge hopper before joining to reach the headshunt adjacent to the A174 road. A CPL locomotive pulls each rake of potash wagons under this hopper to the headshunt and then propels the loaded wagons past the hopper to rejoin the line leading out of the site. A separate siding off the hopper bypass line serves the salt discharge hopper. There is also another siding near the connection with the branch ending in a stopblock and a winch; the latter was intended to move rakes of wagons but was never used for this purpose.

It was not surprising that the first locomotive at CPL, YE 2724, should come from ICI Billingham Works and was a stalwart for a number of years. However, potash and salt have a corrosive impact on the locomotives and wagons, which require regular attention and are prone to rusty bodywork. Occasionally, locomotives had to be hired on a short term basis to cover for failures and maintenance of CPL diesels, together with upgrading of the branch to Carlin How. This especially occurred in the summer of 1996; indeed on 25/5/1996 one of EWS's Class 37s was used to shunt trains! On 28–29/5/1996 09005 (Darlington, 1959) was hired from EWS and worked permanent way trains on the branch, before moving on to loading duties 30/5/1996 to 11/6/1996 due to

Rolls-Royce 10234 of 1965 propels potash wagons under the loading hopper at Boulby Mine on 23rd October 1996. *(Cliff Shepherd photograph)*

Wickham 10842 of 1975 was on hire at Boulby Mine from RMS Locotec Ltd in 1999-2000. It made occasional sorties along the branch to Crag Hall in connection with permanent way works. *(IRS collection)*

the non-availability of RR10225 (defective wheelsets). Later 09106 (Horwich, 1961) was used on permanent way trains in 4/8/1996 to 19/8/1996 for the duration of the summer shutdown. A similar situation occurred in 1998 when 09106 was again hired from EWS, but after derailing on 9/10/1998, was replaced by 08995 KIDWELLY (Horwich, 1959) on 18/10/1998 and this locomotive continued on hire for at least six months (after 21/4/1999). It was interesting because it had been previously modified with a reduced height cab for working the Cwmmawr Branch in South Wales. The problems of securing cover for locomotive failures were generally resolved when the shunting duties were put out to contract. Wickham 10842 was hired for track maintenance on the branch. A 1ft 6in gauge locomotive is based 500 feet underground at the charging station (NZ 765189) and is used to maintain the sea water pipeline which runs about a mile under the sea from No.3 shaft near the top of the cliffs.

On 12/11/2015 ICL UK Ltd announced that "economically feasible" reserves of potash at the mine would run out in 2018 and that the mine would change to extracting polyhalite (sold as polysulphate) containing four vital plant nutrients. The last load of potash left the mine by rail on 5/7/2018. It has now become the first and only polyhalite mine in the world with an annual production of 700,000 tonnes by the end of 2019.

Gauge : 4ft 8½in

(HARWOODDALE)	0-6-0DE	YE	2724	1958	(a)	(2)
7	0-4-0DH	S	10131	1963	(b)	(1)
No.1	0-6-0DE	YE	2723	1958	(c)	(5)
3	0-6-0DH	JF	4240014	1962	(c)	(3)
No.II (No.2)	0-6-0DE	YE	2713	1958	(d)	(5)
No.7	0-6-0DH	S	10095	1962		
Rebuilt		AB	6007	1982+		
Rebuilt		HE	9069	1982	(e)	(4)
2	0-6-0DH	RR	10234	1965	(f)	(10)
CLEVELAND POTASH No.1						
LOCO(1)	0-6-0DH	RR	10225	1965	(g)	(12)
HL 1006	0-6-0DH	HE	6294	1965	(h)	(6)
SM01	2w-2DMR	Wkm	10842	1975	(i)	(9)
No.48	0-6-0DH	HE	7279	1972	(j)	(8)
H016 08598 (D3765)	0-6-0DE	Derby		1959	(k)	(7)
124 KEN	0-6-0DH	TH	261V	1976		
Rebuilt		YEC	L124	1996	(l)	(13)
147 TAMMIE	0-6-0DH	S	10148	1963		
Rebuilt		YEC	L147	1999	(m)	(14)
MPP 0010						
DX 68010 DB 965987	2w-2DMR	Wkm	10731	1974	(n)	(11)
KAREN	0-6-0DH	S	10147	1963	(o)	(15)
MADDIE	0-6-0DH	HE	6662	1966	(p)	(18)
08913 (D4143) HYWELL	0-6-0DE	Hor		1962	(q)	(16)
(309)	0-6-0DH	YE	2825	1961	(r)	(17)
VALIANT	0-6-0DH	S	10108	1963		
Rebuilt		TH		1988	(s)	(19)
08401 (D3516)	0-6-0DE	Derby		1958	(t)	(20)
LADY POTTER	0-6-0DH	RR	10214	1964	(u)	(21)
(YOGI)	0-6-0DH	S	10166	1963	(v)	(22)
CHARLIE	0-6-0DH	S	10107	1963	(w)	
CD40 B4618/7	4wDH	CE	B4618.7	2016	(x)	
CD40 B4618/6	4wDH	CE	B4618.6	2016	(y)	(23)
CD40 B4618/2	4wDH	CE	B4618.2	2016	(z)	

+ Comprised Sentinel frame and chassis with an Andrew Barclay superstructure.

Note: Geoff Allen, who worked at Cleveland Potash, considered that 5 (S10129 of 1963) was also hired temporarily from BSC Skinningrove to cover for a locomotive failure.

(a)   ex ICI Ltd, Billingham Works, /1974 (although Geoff Allen suggested 2/1972); returned to ICI Ltd, Billingham Works by 2/1975 and back to Cleveland Potash Ltd, by 25/5/1975.
(b)   ex British Steel Corporation, Skinningrove Works, Carlin How, on hire, 9-10/1979.
(c)   ex Tees Bulk Handling Ltd, Tees Dock, Grangetown, before 11/5/1986.

(d) ex C.F. Booth (Steel) Ltd, Rotherham, South Yorkshire, 21/5/1986; previously NCB Western Area, Sutton Manor Colliery, St Helens, Merseyside.
(e) ex Tees & Hartlepool Port Authority, Tees Dock, Grangetown, on hire, 30/1/1992.
(f) ex British Steel plc, Teesside Works, Grangetown, 27/4/1992. On hire from Wilmott Bros (Plant Services) Ltd, Ilkeston, Derbyshire, from /2001.
(g) ex British Steel plc Teesside Works, Grangetown, 28/4/1992. On hire from Wilmott Bros (Plant Services) Ltd, Ilkeston, Derbyshire, from /2001.
(h) ex RMS Locotec, Dewsbury, West Yorkshire, on hire, 10/6/1996.
(i) ex RMS Locotec, Dewsbury, West Yorkshire, on hire, c7/1998, by 21/4/1999.
(j) ex RMS Locotec, Dewsbury, West Yorkshire, on hire, 12/1/1999. Returned to RMS Locotec, ex hire, 22/6/1999. Ex RMS Locotec, on hire, 21/10/1999.
(k) ex RMS Locotec, Dewsbury, West Yorkshire, on hire, 21/6/1999.
(l) ex Aggregate Industries UK Ltd, Croft, Leicestershire, (property of Wilmott Bros (Plant Services) Ltd), on hire, 31/5/2000. To Wilmott Bros (Plant Services) Ltd, Ilkeston, Derbyshire, ex hire, 3/2001. Ex Wilmott Bros (Plant Services) Ltd, on hire, by 1/4/2003.
(m) ex Wilmott Bros (Plant Services) Ltd, Ilkeston, Derbyshire, on hire, 23/3/2001.
(n) ex Jarvis Ltd, Darlington Park Lane Depot, Co. Durham, on hire, /2004; after 1/2004, by 4/2004.
(o) ex LH Group Services Ltd, Barton-under-Needwood, Staffordshire, on hire, 4/10/2004.
(p) ex LH Group Services Ltd, Barton-under-Needwood, Staffordshire, on hire, 12/10/2004. Returned to LH Group Services Ltd, ex hire, 16/10/2009. Ex LH Group Services Ltd, on hire, 19/7/2011.
(q) ex Manchester Ship Canal Co Ltd, Trafford Park, Greater Manchester, (property of LH Group Services Ltd, Barton-under-Needwood, Staffordshire), on hire, c5/2009, after 15/11/2008, by 21/5/2009.
(r) ex Corus plc, Skinningrove Works, Carlin How, on hire, 16/10/2009.
(s) ex LH Group Services Ltd, Barton-under-Needwood, Staffordshire, on hire, c29/10/2009 and returned, ex-hire, 31/1/2013. Ex LH Group Services Ltd, on hire, 23/10/2013.
(t) ex LH Group Services Ltd, Barton-under-Needwood, Staffordshire, on hire, 31/1/2013.
(u) ex Tata Steel Europe, Skinningrove Works, Carlin How (property of Ed Murray & Sons Ltd), on hire, 25/9/2014.
(v) ex Ed Murray & Sons Ltd, Casebourne Road, Hartlepool, on hire, 13/10/2014.
(w) ex Chasewater Light Railway & Museum Company, Brownhills, Staffordshire, on hire from Ed Murray & Sons Ltd, Hartlepool, 15/4/2015 via haulier's yard.
(x) ex Railway Support Services Ltd, Wishaw, Warwickshire, on hire, 20/9/2021.
(y) ex Chasewater Light Railway & Museum Company, Brownhills, Staffordshire, (property of Railway Support Services Ltd), on hire, 28/10/2021.
(z) ex Railway Support Services Ltd, Wishaw, Warwickshire, on hire, 4/3/2022.

(1) returned to British Steel Corporation, Skinningrove Works, Carlin How, ex hire, by 8/1980.
(2) to C.F. Booth (Steel) Ltd, Rotherham, South Yorkshire, 21/5/1986 for scrap.
(3) to C.F. Booth (Steel) Ltd, Rotherham, South Yorkshire, by 10/11/1988.
(4) returned to Tees & Hartlepool Port Authority, Tees Dock, Grangetown, ex hire, 28/4/1992.
(5) to Booth Roe Metals Ltd, Rotherham, South Yorkshire, between 24/9/1992 and 3/10/1992; last reported at Boulby 18/8/1992.
(6) to Faber Prest Ports Ltd, Flixborough, Humberside, (property of RMS Locotec), ex hire, 9/1/1997.
(7) to RMS Locotec, Dewsbury, West Yorkshire, ex hire, after 5/9/1999, by 14/7/2000.
(8) to South Yorkshire Railway Preservation Society, Wincobank, Sheffield, South Yorkshire, ex hire, w/e 23/6/2000.
(9) to RMS Locotec, Dewsbury, West Yorkshire, ex hire, after 18/3/2000, by 14/7/2000.
(10) to Wilmott Bros (Plant Services) Ltd, Ilkeston, Derbyshire, ex hire, c23/3/2001.
(11) to Wensleydale Railway, Leeming Bar, North Yorkshire, ex hire, 8/2004; after 7/8/2004, by 31/8/2004.
(12) to T.J. Thomson & Son Ltd, Stockton, (property of W B Power Services Ltd) ex hire, 13/10/2004.
(13) to W B Power Services Ltd, Ilkeston, Derbyshire, ex hire, 14/10/2004.
(14) to W B Power Services Ltd, Ilkeston, Derbyshire, ex hire, after 15/6/2004, by 20/10/2004. It may have moved direct to the former WD depot at West Hallam where some of W B Power Services' locomotives were stored.
(15) to LH Group Services Ltd, Barton-under-Needwood, Staffordshire, ex hire, c5/2009.
(16) to Manchester Ship Canal Co Ltd, Trafford Park, Greater Manchester, (property of LH Group Services Ltd, Barton-under-Needwood, Staffordshire), ex hire, 10/12/2009.
(17) to Tata Steel Europe, Skinningrove Works, Carlin How, ex hire, 1/12/2011.
(18) to LH Group Services Ltd, Barton-under-Needwood, Staffordshire, ex hire 11/9/2014.

(19) to Tees Dock Terminal, Tees Dock, Grangetown, (property of LH Group Services Ltd), ex hire, w/c 8/11/2014.
(20) to Celsa Manufacturing (UK) Ltd, Cardiff, (property of LH Group Services Ltd) ex hire, 24/10/2013.
(21) to Tata Steel Europe, Hartlepool Works, Hartlepool (property of Ed Murray & Sons Ltd), ex hire, 6/5/2019.
(22) to British Steel Ltd, Teesside Beam Mill, Lackenby, ex hire, 1/2/2022.
(23) to Railway Support Services Ltd, Wishaw, Warwickshire, ex hire, 4/3/2022.

Gauge : 1ft 6in

-		0-4-0BE	WR	7654	1974	New	
-		4wBE	CE	B0922B	1975	(a)	(1)

(a) ex RMS Locotec, Dewsbury, West Yorkshire, on hire, from 29/6/1999.

(1) taken off hire and moved to Loftus Goods Shed, /2000. It was with Hunslet Locomotive Hire Ltd, Killamarsh, Derbyshire in Industrial Locomotives 13EL (IRS, 2003).

TEES DOCK TERMINAL, Tees Dock, Grangetown   96
Cleveland Potash Ltd until 29/9/2015   NZ 549235 – **Map H**
Teesbulk Handling Ltd until c/1991 (a subsidiary of **Cleveland Potash Ltd**)
Cleveland Potash Ltd until c/1980

The bulk terminal is situated on the north east side of Tees Dock and has the benefit of a 365m long berth. The THPA built the terminal as part of its Tees Dock development and leased it to Cleveland Potash Ltd (CPL) in 1973. Although its main purpose was to deal with potash exports from Boulby Mine, the first cargoes in 1974 were imported phosphate rock for ICI at Billingham. By 1984, well over 500 ships had been loaded with potash delivered in trains from Boulby Mine. Trains arrived via the Tees Dock Branch, which left the Middlesbrough-Redcar line ¼ mile north east of Grangetown Station. On reaching the exchange sidings, a CPL locomotive coupled on to the rear of the wagons and hauled them to near the terminal. Here the locomotive ran round and propelled the wagons over the discharge points, before returning empty wagons to the exchange sidings. There were two discharge points, one in the open and the other below a bay extending from the silo. Usually

Sentinel 10102/1962 propels potash wagons into the unloading bay on 19th February 1987 during a period when Cleveland Potash's Tees Dock Terminal was marketed as Teesbulk Handling.

(Cliff Shepherd photograph)

the working locomotive was stabled in this bay when not in use. In 1999 40,000 tonnes of potash powder and 30,000 tonnes of granulated salt would be stored in the silo. Another siding terminated in a building used for wagon repairs and maintenance.

In the early 1980s, as there was spare capacity at the terminal, a greater emphasis was placed on handling a range of bulk commodities beginning with soda ash which arrived in lorries. Other goods to be despatched during this period included coal from Easington Colliery for the London power stations. Strikes in Poland had enabled the NCB to pick up extra business resulting in less capacity at the usual ports. Grain arrived in polybulk wagons via the Speedlink service. It was during this period that the terminal operated under the Teesbulk Handling title.

In view of ICI Ltd's initial involvement, it sent one of its surplus Janus Yorkshire diesels to shunt wagons at the terminal. Potash and salt are very corrosive and eventually the locomotive had to be sent away for overhaul. Short-term loans were arranged locally with BSC, such as RR 10292/1971 and S 10102/1962 in 8/1982. A more surprising hiring was the four-coupled S 10127/1963 from BSC Skinningrove. Subsequently CPL has relied on specialist locomotive hirers to meet its shunting requirements. In 1997, the Tees & Hartlepool Port Authority Ltd expanded the container provision, involving the relocation of the exchange sidings to the north east side of, and parallel to, the dock, with a revised connection to the bulk terminal.

According to the 2000/01 THPA Annual Report, six trains arrived at the terminal daily carrying 3,500 tonnes of potash and rock salt. The terminal also acts as a distribution centre for despatches by road. There was a reduction in rail traffic with the change from potash to polyhalite production but this is increasing again.

Gauge : 4ft 8½in

7 (GLAISDALE)	0-6-0DE	YE	2723	1958	(a)	(2)	
3	0-4-0DH	S	10127	1963	(b)	(1)	
1 (102)	0-6-0DH	S	10102	1962	(c)	(4)	
2	0-6-0DH	AB	487	1964	(d)	(3)	
(No.31)	0-6-0DH	JF	4240014	1962	(d)	(2)	
-	0-6-0DH	RR	10257	1966	(e)	(6)	
20/110/707	0-6-0DH	AB	608	1976	(f)	(7)	
023 (08874) (D4042)	0-6-0DE	Dar		1960	(g)	(5)	
007 (08077) (D3102) JAMES	0-6-0DE	Derby		1955	(g)	(5)	
H012	0-6-0DH	RR	10289	1970	(h)	(12)	
H029 DUNCAN	0-6-0DH	RR	10286	1969	(i)	(11)	
08743 (D3911) BRYAN TURNER	0-6-0DE	Crewe		1960	(j)	(8)	
11	0-6-0DH	EEV	3994	1970			
	Rebuilt	YEC	L180	2000	(k)	(9)	
H006 15	0-6-0DH	HE	6294	1965	(l)	(10)	
EMILY	0-6-0DH	AB	660	1982			
	Rebuilt	HAB	6769	1990			
	Rebuilt	HE		2004	(m)		
SAM	0-6-0DH	AB	659	1982			
	Rebuilt	HAB	6768	1990			
	Rebuilt	HE		2004	(n)	(17)	
VALIANT	0-6-0DH	S	10108	1963			
	Rebuilt	TH		1988	(o)	(13)	
277 WATERFALL	6wDE	GECT	5475	1978	(p)	(14)	
268	6wDE	GECT	5465	1977	(q)	(16)	
261	6wDE	GECT	5430	1977	(r)	(15)	
BLUEBIRD	0-6-0DH	HE	8998	1981	(s)		

(a) ex ICI Ltd, Billingham Works, c12/1974. to Shell (UK) Ltd, Grangetown for overhaul, 3/1982 and returned by 1/11/1982.
(b) ex British Steel Corporation, Skinningrove Works, Carlin How, on hire, 2/1982.
(c) ex British Steel Corporation, South Teesside Works, Lackenby, initially on hire c5/1982. It was later purchased by Teesbulk Handling Ltd.
(d) ex Central Electricity Generating Board, Blyth Power Station, North Blyth, Northumberland, 10/1982. Although AB 487 was wanted for the terminal, JF 4240014 was for resale and stored for Peter Manuel Plant Hire, Haverton Hill.

(e) ex R.F.S. Engineering Ltd, Kilnhurst, South Yorkshire, /1990.
(f) ex Douglas Engineering Services Ltd, Grangetown, c1/1994, after 15/1/1994.
(g) ex R.F.S. Engineering Ltd, Doncaster, South Yorkshire, on hire, c5/1994; after 12/4/1994, by 16/6/1994.
(h) ex RMS Locotec, Dewsbury, West Yorkshire, on hire, 16/6/1997. Returned to RMS Locotec, ex hire, after 7/5/2000, by 14/7/2000. Ex RMS Locotec, on hire, c2/2001. Became property of Harry Needle Railroad Co Ltd, from c/2008.
(i) ex RMS Locotec, Dewsbury, West Yorkshire, on hire, 16/11/1999. Became property of Harry Needle Railroad Co Ltd, from c/2008.
(j) ex Sembcorp Utilities (UK) Ltd, Wilton International, Redcar, on hire, 7/2005.
(k) ex P.D. Ports plc, Teesport, Grangetown, c3/2007, by 14/3/2007; on hire from RMS Locotec, Wakefield, West Yorkshire.
(l) ex P.D. Ports plc, Teesport, Grangetown, /2007, by 10/2007; on hire from RMS Locotec, Wakefield, West Yorkshire.
(m) ex LH Group Services Ltd, Barton-under-Needwood, Staffordshire, on hire, 5/2008. To LH Group Services Ltd for repairs, w/e 8/11/2014 and returned 11/2/2015. To G.C.S. Johnson Ltd, Barton Park, Barton, North Yorkshire, ex hire, 6/7/2017. Ex Tata Steel Europe, Trostre Works, Llanelli, (property of LH Group Services Ltd), on hire, 18/10/2018. Became property of Hunslet Ltd, Barton-under-Needwood, Staffordshire but still on hire.
(n) ex LH Group Services Ltd, Barton-under-Needwood, Staffordshire, on hire, 19/6/2008. To G.C.S. Johnson Ltd, Barton Park, Barton, North Yorkshire, property of LH Group Services Ltd, ex hire, 6/7/2017. Ex LH Group Services Ltd, Barton-under-Needwood, Staffordshire, on hire, c28/8/2021.
(o) ex Cleveland Potash Ltd, Boulby Mine, Boulby, (property of LH Group Services Ltd), on hire, w/e 8/11/2014.
(p) ex British Steel Ltd, Teesside Beam Mill, Lackenby, (property of Ed Murray & Sons Ltd), on hire, 11/8/2016.
(q) ex Aggregate Industries UK Ltd, Merehead Stone Terminal, Somerset, (property of Ed Murray & Sons Ltd) on hire, 19/12/2016, via Chasewater Railway, Brownhills, Staffordshire from 10/2016.
(r) ex Chasewater Light Railway & Museum Company, Brownhills, Staffordshire, (property of Ed Murray & Sons Ltd), on hire, 14/10/2016.
(s) ex LH Group Services Ltd, Barton-under-Needwood, Staffordshire, on hire, 5/2/2019.

(1) to British Steel Corporation, Skinningrove Works, Carlin How, ex hire, c12/1982, after 23/12/1982.
(2) to Cleveland Potash Ltd, Boulby Mine, Easington, before 11/5/1986.
(3) scr w/c 31/3/1991.
(4) scr on site c3/1994; after 16/3/1993, by 16/6/1994.
(5) to R.F.S. Engineering Ltd, Doncaster, South Yorkshire, ex hire, 9/1994; after 4/9/1994, by 27/9/1994.
(6) to RMS Locotec, Dewsbury, West Yorkshire, ex hire, c2/2001.
(7) to T.J. Thomson & Son Ltd, Stockton, 23/4/2002.
(8) to Sembcorp Utilities (UK) Ltd, Wilton International, Redcar, ex hire, 10/2005.
(9) to Cobra Railfreight Ltd, Middlesbrough, (property of RMS Locotec, Wakefield, West Yorkshire), ex hire, 13/12/2007.
(10) to RMS Locotec, Wakefield, West Yorkshire, ex hire, /2008, by 6/2008.
(11) to T.J. Thomson & Son Ltd, Stockton, 10/7/2009.
(12) to T.J. Thomson & Son Ltd, Stockton, 7/2009; after 10/7/2009, by 18/7/2009.
(13) to LH Group Services Ltd, Barton-under-Needwood, Staffordshire, ex hire, 12/2/2015.
(14) to Chasewater Light Railway & Museum Company, Brownhills, Staffordshire, ex hire, 9/4/2018.
(15) to British Steel Ltd, Teesside Beam Mill, Lackenby, ex hire, 7/11/2018.
(16) to British Steel Ltd, Teesside Beam Mill, Lackenby, ex hire, 1/2019; after 18/1/2019, by 28/1/2019.
(17) to LH Group Services Ltd, Barton-under-Needwood, Staffordshire, ex hire 22/4/2021, following a fire on 19/4/2021.

JACKSON, GILL & CO LTD

IMPERIAL IRON WORKS, South Bank   **97**
Jackson Gill & Co until c/1872   NZ 532212 – **Map G**

Construction of the Imperial Iron Works began in 1870 comprising 20 puddling furnaces, one four-ton hammer and a forge train. It opened in Spring 1871 and was located on the south side of the Middlesbrough-Redcar railway between it and North Street, South Bank. The company, managed by Thomas Gill of Middlesbrough, purchased pig iron and manufactured semi-finished puddled iron bars and billets. By 9/1872, it had become a limited liability company and installed 20 furnaces. The works was served by a siding off the main line which curved round by Normanby Road. Unfortunately, investment in the works was ill-advised as there was an economic depression in the iron trade during the late 1870s and steel was beginning to supersede wrought iron for some uses. The owners went into liquidation in 1879 and The *Colliery Guardian* (3/6/1881) reported that the works had been auctioned at the Royal Exchange, Middlesbrough but no bids had been received. According to the *North-Eastern Daily Gazette* (24th November 1881), new owners were to take over the works in December 1881. However, on 6/10/1887, The *Colliery Guardian* announced that Bradshaw Brown was to auction the plant although its reference to "eight locomotives" was probably in error; *The Engineer* quoted the same figure. The *North-Eastern Daily Gazette* on 17/9/1887 gave more precise details of the plant material to be auctioned, it included "the forge, 51 puddling and mill furnaces, and 3 iron mills", "6 steam engines, 13 horizontal and stack boilers, 4 steam hammers, locomotive steam engine," etc; so there was just one locomotive left by 1887.

The land at the west end of the site behind the NER South Bank Goods Shed was occupied by the Tees Tilery Brick Works (later Tees Brick and Tile Works) with a tramway running from its buildings southwards into an adjacent clay pit. The owners were stated to be Johnson & Maw in 1880 (Davidson) and Maw & Son in 1898 (NER). By the 1892–93 Ordnance Survey, sidings still entered the iron works site, although its building was labelled "Imperial Brick Works (Disused)". Also shown was a coal depot on North Street connected to the former iron works siding and to a railway that ran over Normanby Road to the Clay Lane Iron Works.

On 28/2/1907, the NER Way and Works Committee agreed that Thos. D. Ridley & Sons could use the reinstated connection with the main line. However, the more significant agreement between the NER and Ridley was dated 22/3/1909. Bolckow Vaughan, owners of the Clay Lane Iron Works, had also acquired the disused Tees Brick and Tile Works and Ridley was given approval to transport "molten slag and other materials" from the Clay Lane Iron Works to fill in the clay pits at the brickworks. The railway line used the existing crossing of Normanby Road and ran to the former Imperial and Tees Brick and Tile Works sites; Ridley intended to use his own locomotives and wagons. This was regarded as a short-term project as the NER charged £100 pa for two years, with provision for extension. The contractor was not to interfere with the new road to be constructed from North Street, across the middle of the iron works site and over the main line. By 1913, both sites had been levelled, the road had been constructed and Ridley had terminated his agreement for use of the NER connection.

Reference : 'A Hunslet Standard Gauge 0-4-0ST in Australia – an update', John Browning, *The Industrial Locomotive* 162, ILS, 2017, pp. 90-94

Gauge : 4ft 8½in

ELFIN		0-4-0ST	OC	HE	41	1870	New	(1)
IMPERIAL		0-4-0ST	OC	YE	287	1876	New	s/s

(1) an entrepreneur from Australia, Joseph Lee, visited the UK in 1888-89 and subsequently, in 6/1889, advertised in the *Sydney Morning Herald* a Hunslet 8 inch locomotive for sale, which the reference suggests was ELFIN. If so, it moved in 1895 to C.E. Jeanneret's Parramatta River Steamers & Tramway Co, New South Wales, Australia where it was photographed in about the following year.

W. JONES & CO

MIDDLESBROUGH CHEMICAL WORKS, Middlesbrough 98
NZ 503203 – **Map G**

William Jones and his partner, Isaac Sharp, were both involved in the Guisborough Alum Company. They were Quakers and enjoyed a close association with the Pease family; the site of the Middlesbrough Chemical Works lay on the east side of the Middlesbrough-Guisborough line shortly after it diverged from the railway to Redcar and occupied part of the OME's New Brick Yard. The Pease family was keen to make use of some pyrites being extracted with the ironstone from its Belmont Mine at Guisborough. In 1860 Jones applied for an early form of development approval, its purpose being "the manufacture of superphosphate of lime or other artificial manures". This required sulphuric acid which could be made using the pyrites (iron sulphide). According to Lillie, the business began on 11/12/1860 although there were soon complaints about pollution and Jones and Sharp were fined.

However, the business prospered especially after alkali production began and William Jones introduced technical improvements to the Leblanc process furnaces. The equipment used for the Leblanc process could be employed to produce other chemicals when different raw materials were introduced. In 1877 he employed 200 men, manufactured a wide range of chemicals and was producing about 320 tons per week. The *Daily Exchange* in 1880 referred to the chemicals being sold and claimed that the company was "now the largest oxalic acid makers in the world." Manures were also mentioned by the newspaper suggesting that artificial fertiliser production was continuing. Three sets of sidings entered the premises from the Guisborough Branch and the company owned a locomotive to carry out its shunting. In 1880 William Jones, at the age of 55, decided to sell his business to Samuel Sadler (which see), who occupied the adjoining site. Under Sadler, the works continued as a separate operating organisation but gradually declined as the emphasis switched to Sadler's manufacture of chemicals from coal tar.

Reference : 'William Jones and the First Chemical Factory in Middlesbrough', Ian Pearce, *The Cleveland Industrial Archaeologist* 32, CIAS, 2008

Gauge : 4ft 8½in

-		0-4-0ST	OC	AB	148	1874	(a)	(1)

(a) the limited surviving Andrew Barclay records have "W.Jones & Co Chemical Manufacturers" in pencil; against this maker's number so possibly Jones was not the original owner. The name of "C.G. Johnson & Co Middlesbrough" also appears against the entry. Johnson was an iron merchant and engineer; it is more likely that this company was involved as an agent.

(1) probably to Samuel A. Sadler with the works, /1880.

CHARLES P. KINNELL & CO LTD

VULCAN IRON WORKS, Thornaby

99
NZ 449182 – **Map E**

The site occupied a narrow piece of land on the east bank of the River Tees, with the Darlington-Middlesbrough railway forming its eastern boundary and the Tees Bridge just to the south. Initially the Tees Bottle Works and a sawmill were situated on the site, the foundation stone of the former was laid on 2/4/1839 and Richmond comments "This was the first work of the kind undertaken on the banks of the Tees". In 1847 the Tees Crown Glass Co was producing glass for windows but, by 1857, it was called the Tees Bottle Works. Peak production occurred in 1865 when 600,000 bottles were produced by 220 employees. A single short siding off the main line enabled coal to be brought in for the furnaces. The works seem to have closed about 1890.

The site was then taken over by Charles P. Kinnell & Co Ltd which established the Vulcan Iron Works in 1899 to manufacture cast iron piping, particularly for heating systems. In an agreement dated 27/9/1899, Kinnell requested the NER to reconnect and complete the sidings at the former Tees Bottle Works. By 1915, the rail connection with the main line had been reversed to enter the premises from the south, running to a headshunt and then coming back as a double siding alongside the main line. This was still the arrangement in the 1950s. The works and plant, including AB 1014, were for sale by auction on 24/6/1958. The premises were then taken over by Thomas Turnbull & Son Ltd (which see).

Gauge : 4ft 8½in

-		0-4-0ST	OC	FJ	101	1872	(a)	s/s
No.1		0-4-0ST	OC	AB	1014	1906	(b)	scr 8/1961
No.2	PULL	0-4-0ST	OC	FJ	76	1867	(c)	s/s c/1958

(a) ex William Whitwell & Co Ltd, Thornaby Iron Works, Thornaby.
(b) ex J. & R. Howie Ltd, Muirside Colliery, Crosshouse, Ayrshire, /1920.
(c) ex Tyne-Tees Steam Shipping Co Ltd, Stockton Wharf, Stockton, c/1949.

LIBERTY PIPES (HARTLEPOOL) LTD

42 and 84 INCH PIPE MILLS, HARTLEPOOL

Tata Steel Europe until 1/8/2017 North Works Steel Plant NZ 517313
Corus plc until 27/9/2010 North Works Blast Furnaces NZ 517309
(Became a subsidiary of **Tata Steel** 2/4/2007 South Works 20 inch Mill NZ 508289
and rebranded **Tata Steel Europe**, 27/9/2010) South Works 44/42 inch Mill NZ 504277
British Steel plc until 6/10/1999
British Steel Corporation until 28/7/1988
South Durham Steel & Iron Co Ltd until 28/7/1967 (operation of works not taken over until 1/7/1968)
See following text for earlier companies.

The South Durham Steel & Iron Co's works at West Hartlepool was formed from a number of separate plants and these are described below. Later, as a result of closures, only the Pipe Mill on the former South Works site remained.

Pile, Spence & Co

John Pile began to lay out his shipyard at the Jackson Dock, West Hartlepool in 1853 and, at the end of the following year, launched his first ship from there, with an iron ship being completed in 1856. In order to obtain the necessary iron plates and sections, Pile had opened the West Hartlepool Rolling Mills in 1855. The business became Pile, Spence & Co in 1859 when the operation of a second shipyard was included. There is some doubt about the location of the iron works as the 1861 Ordnance Survey map shows the West Hartlepool Iron Works on the west side of the Norton-Hartlepool railway immediately to the **north** of the Cliff House Iron Works. A siding ran into the premises off the main line. To the south of the Cliff House Iron Works was a larger site identified on the map as "Iron Works" and this could be the blast furnaces if constructed by Pile Spence. Although also situated on the west side of the main line, this iron works was served by a railway constructed by the West Hartlepool Harbour & Railway Company south from Newburn Junction to the west of the premises. This railway was later extended southwards becoming the NER Cliff House Branch to Cliff House South Junction on the main line. Interestingly, Head & Ashby erected rolling mills for the Hartlepool Rolling Mill Co. More research is required to clarify the relationship between the various businesses and plants. In 1866 Pile, Spence & Co was ruined by the failure of the bankers, Overend Gurney & Co, being formally wound up in December. The only locomotive that we know about appears to have been associated with the shipyard.

West Hartlepool Iron Co Ltd, West Hartlepool Iron Works   100
NZ 516311 – **Map B**

The Pile, Spence & Co iron works then stood idle until it was purchased by Thomas Richardson & Sons in 7/1868. According to Robert Wood, by 1874 "the Mills had extended from six acres to twelve and three new blast furnaces took in twenty more acres". He reckoned that the mills could "turn out 1,000 tons of rails and 250 tons of iron plates in a week". This expansion took place at the "Iron Works" mentioned above and involved acquiring land to the south. The site of the original West Hartlepool Iron Works had mostly become the Albion Saw Mills by 1896 with the Cliff House Foundry at its southern end. The foundry was purchased by William Gray & Co Ltd in 1898 from John Muir & Co and it concentrated on producing haematite iron castings until Gray's liquidation in 1962.

Also in the mix of ownerships was the Hartlepool Malleable Iron Co (HMI). It had 32 puddling furnaces and two mills/forges in 1872 while T. Richardson & Sons operated 109 puddling furnaces and three mills/forges. This was generally the position in 1877. The HMI had a tank locomotive repaired by Black, Hawthorn in 1874. In the *NER Collieries, Works and Sidings Book* Volumes I – II, 1895, the HMI works was stated to be dismantled and its "ballast" siding was used by the West Hartlepool Steel & Iron Co. Confusingly, the "ballast" siding was stated to be 30 chains from "Cemetery Junction South".

In 5/1874 it was reported that Thomas Richardson 2 was forming the West Hartlepool Iron Co Ltd to take over the Richardson iron works and operate the blast furnaces and mills. The local newspaper gave details of the premises – the works stand on about 30 acres of land; there were three blast furnaces, 116 puddling furnaces, a rail mill making 42,000 tons pa and a plate mill making 11,000 tons pa. Given the downturn in the iron trade, the West Hartlepool Iron Co suffered a heavy financial loss in its first year and went into liquidation in 1875. The works stood vacant for the next five years and it is likely that most of the locomotives were sold off to meet the demands of the creditors. Although those below were ordered from Hartlepool, it is probable that they operated at the iron works.

Gauge : 4ft 8½in

No.1					(a)	s/s
No.2	0-4-0ST	OC	HE	30	1869 New	(2)
No.3	0-4-0ST	OC	HE	60	1871 New	(4)
No.4	0-4-0ST	OC	HE	78	1872 New	(3)
No.5	0-4-0ST	OC	HE	80	1873 New	(1)

(a)   existence assumed.
(1)   to Tees Conservancy Commissioners, South Gare Breakwater, Redcar, by 10/1875.
(2)   to Hull Dock Co, Hull, Yorkshire (ER), via Hunslet Engine Co Ltd, Leeds, 12/1875, although the possibility remains that the locomotive may have been exported via Hull rather than worked there.
(3)   to Terry, Greaves & Co, Old Roundwood Colliery, Ossett, Yorkshire (WR), by 1/1876.
(4)   to Hutchinson's Trustees, Widnes, Lancashire, by 10/1897 but probably much earlier.

West Hartlepool Steel & Iron Co Ltd, West Hartlepool Iron Works   101
NZ 516313 – **Map B**

After standing idle, the northern part of Richardson's iron works comprising the puddling furnaces and rolling mills were taken over by Matthew Gray, son of William Gray, and Arthur Gladstone in 1881 with the title of West

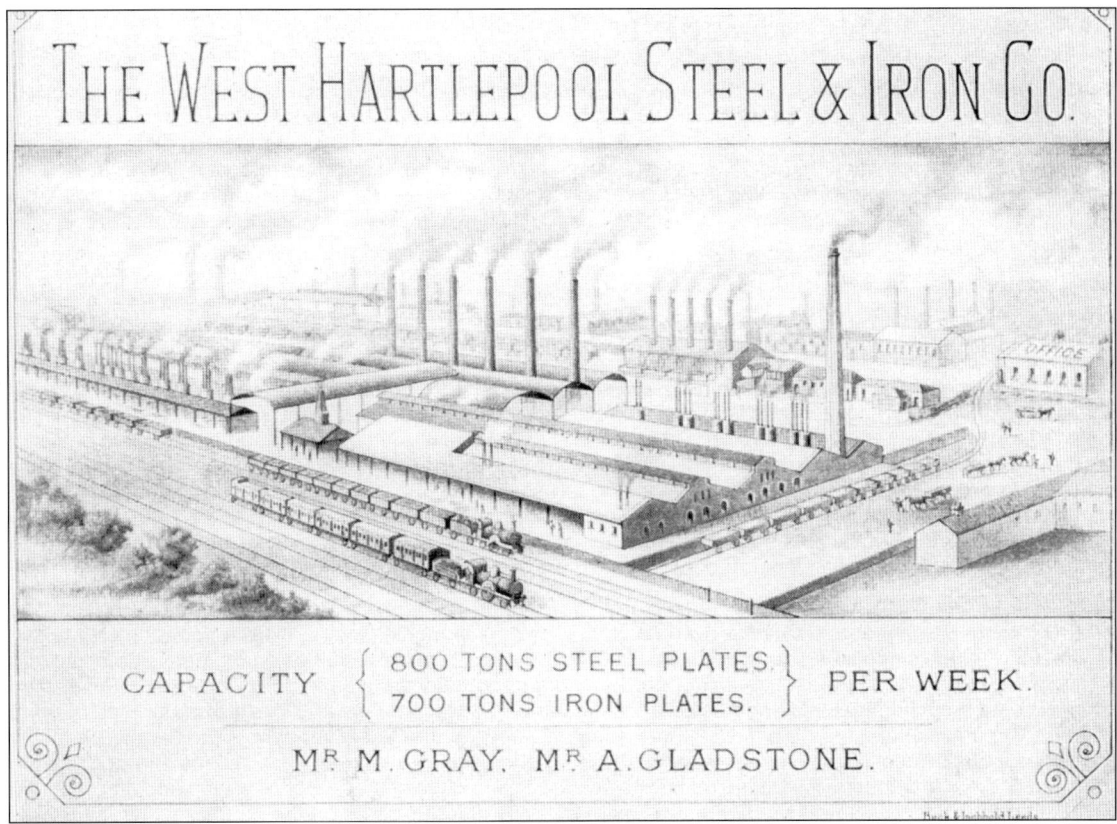

An engraving showing the furnaces and rolling mills at the West Hartlepool works alongside the NER Norton-Hartlepool line.

Hartlepool Steel & Iron Co Ltd. The works concentrated on the manufacture of iron plates until, in 1887, some of the puddling furnaces were demolished to make way for a new steel plant containing six 20 ton open hearth furnaces; the first steel plate was rolled in 8/1888. The remaining puddling furnaces were located at the south end of the works, with the melting shop alongside to the west, and the cogging mill and Nos.1 and 2 plate mills on the northern part of the site. A third mill was under construction in 1888 capable of rolling the largest boiler plates. In 1893 these plants occupied much of the 12 acre site next to the NER's Norton-Hartlepool railway. The casting pit for the open hearth furnaces was serviced by three "locomotive cranes" that delivered steel ingots to a hydraulic charging crane. After rolling into slabs, the steel was lifted by a 4 ton "locomotive crane" and transferred to bogies for conveyance to the plate mills. A double track line connected with the NER Cliff House Branch and entered the works from the north on a similar alignment to that shown on the 1861 plan. At the southern end of the site, sidings emerged from the mills building to converge and join the main line. In addition, another siding headed south into the Seaton Carew Iron Works (see below) and this was, no doubt, much of the source of its iron. When Matthew Gray died in 1896, his executors led by Sir William Gray and then William Creswell Gray, continued to run the company until the formation of the South Durham Steel & Iron Co Ltd in 1898.

Gauge : 4ft 8½in

WEST HARTLEPOOL No.1	0-4-0ST	OC	BH	613	1881	New	(1)
No.5	0-4-0ST	OC	Grange Iron Works		1873	(a)	(2)
-	0-4-0ST	OC	AB	299	1888	New	(2)
No.3	0-4-0ST	OC	HL	2134	1889	New	(2)

(a)   origin unknown

(1)   to Castle Eden Coal Co Ltd, Castle Eden, Co. Durham. It was auctioned there 17-20/4/1894.

(2)   to South Durham Steel & Iron Co Ltd, with works, 28/12/1898.

Seaton Carew Iron Co Ltd, Seaton Carew Iron Works   102
NZ 516308 – **Map B**

The company, with a prominent Whitehaven presence in the leading shareholders, was registered on 25/4/1882 to take over the blast furnaces on a 41 acre site formerly owned by the West Hartlepool Iron Co Ltd. There were three blast furnaces at the Seaton Carew Iron Works, of which two were operational in 1883. At the end of 1886, the Board instructed that the No.3 blast furnace be blown in "with all possible speed". Another blowing engine was bought from the Tees Side Iron & Engine Works Co Ltd, through the Carlton Iron Co, in 1890. On the night of 6/1/1891, there was a major boiler explosion resulting in the death of one man and the wrecking of two boilers. The company bought six new boilers from Thomas Beeley of Hyde in 1893.

The works occupied all of the land between the Norton-Hartlepool railway and the Cliff House Branch, south of the West Hartlepool Steel & Iron Co's works, by 1896. The Seaton Carew Iron Co's (SCICo) rail connections with the NER joined the Cliff House Branch and it possessed a set of sidings next to the branch. The blast furnaces stood in a row, north to south, with the calcining kilns and mineral stores served by a single railway on their west side and the discharges for iron and slag to the east. To begin with, the iron was cast in pig beds but later the molten iron could be taken along a line directly to the West Hartlepool Steel & Iron Co's furnaces in 20–25 ton capacity ladles mounted on carriages. The slag was initially tipped on land in the angle between the branch and main line, but an agreement dated 10/7/1884 was signed with the NER for a "slag bridge" to be constructed across the Cliff House Branch. A siding ran south from the works over the bridge before dividing into three tracks on an already extensive area of tipping south of Seaton High Lighthouse by 1896. In the early twentieth century, the brick lined slag bogies were known as "Jumbos". Later the "Dewhurst" and then the much larger capacity "Pollock" ladles, operated by compressed air, were introduced for this traffic.

On 12/11/1897 the SCICo's Board approved the erection of new coke ovens and by-product plants. By 25/10/1900, the 50 Semet-Solvay coke ovens and "sulphate of ammonia factory" were reported to be working satisfactorily. At the same time, the SCICo paid £1,300 to the NER for extra sidings on the east side of the iron works giving access to the main line for outward traffic and on the west side to the Cliff House Branch. The latter were for incoming materials and could accommodate 450 wagons, each with a maximum load of up to 12 tons. The erection of a first Talbot tilting open hearth furnace at the West Hartlepool Works, now owned by the South Durham Steel & Iron Co, encouraged the SCICo to begin planning for a fourth blast furnace and the transport of greater amounts of molten iron north. Also the "slag bridge" was in poor condition and a new bridge was constructed alongside it to reach the now extensive network of sidings on the tip; the old structure was later demolished. No.4 blast furnace was built in line with the other three and it was ordered on 20/11/1911 to be blown in.

The SCICo relied on a fleet of four-coupled saddle tank locomotives, all of which were purchased new and carried names associated with the chairman and family of William Thomlinson. In the latter half of 1906, a note in the company's Cast Book lists expenditure on a new locomotive shed and alterations to the railway sidings. The work was carried out by the Darlington Construction Co and the old shed was removed. Pencil notes were added that the cost was "Refunded to us by South Durham S & I Co" and the track changes involved the rolling mill which suggests that the new shed **may** have been on South Durham Steel & Iron Co's site.

In 1916 the SCICo purchased the 16 acre Longhill Brick Works site on the west side of the slag tip and limited production of purple ore bricks began in the following year; the same year that it acquired Casebourne's former cement works as an ore stocking ground. This was known as "Diamond Sidings" and unloading was carried out by two Priestman grab cranes. Previously, ore stocking had taken place on the slag bank. The Government was pressing the South Durham Steel & Iron Co to construct a new integrated iron and steel works capable of rolling 1,000 tons of plate a day. The SDSI helped form the East Coast Steel Corporation in 1917 to build the new plant. It did not proceed because of Government conditions but Benjamin Talbot used the East Coast Steel Corporation to strengthen the associated companies. On 9/1/1919, the majority of SCICo's shares were transferred to the East Coast Steel Corporation, although the SCICo kept its separate identity. In 1924, the entire share capital of the SCICo was purchased by SDSI from the East Coast Steel Corporation and new Articles of Association were drawn up, although it may not have been formally absorbed until 1928.

Gauge : 4ft 8½

No.1	GLADYS	0-4-0ST	OC	B	292	1882	New	(1)
No.4	DAISY	0-4-0ST	OC	P	467	1888	New	(1)
No.5	WALTER MORRISON	0-4-0ST	OC	P	657	1897	New	(1)
No.2	FRANCIS	0-4-0ST	OC	HL	2378	1898	New	(1)
No.3	AILEEN	0-4-0ST	OC	HL	2445	1899	New	(1)
No.6	MAJOR	0-4-0ST	OC	AB	1363	1914	New	(1)
	COLONEL	0-4-0ST	OC	AB	1501	1917	New	(1)
"1918"		0-4-0ST	OC	AB	1609	1918	New	(1)

(1) to South Durham Steel & Iron Co Ltd, with works, probably /1924.

An Andrew Barclay list, dated 1917, records the above numbers. It is considered that these were carried by the locomotives as the SCICo's Cost Book records repairs to "(FRANCIS) No.2" from 11/1909 to 5/1910 costing £324. John Henry Proud's reminiscences list the above, but also a locomotive named KATHLEEN after one of William Thomlinson's daughters.

South Durham Steel & Iron Co Ltd, West Hartlepool Iron & Steel Works (later North Works) 103
NZ 516311 – **Map B**

In order to reduce wasteful competition and achieve better financial returns, the South Durham Steel & Iron Co Ltd (SDSI) was formed on 28/12/1898 by the amalgamation of the West Hartlepool Steel & Iron Co Ltd, the Moor Steel & Iron Co Ltd and the Stockton Malleable Iron Co Ltd. William Cresswell Gray was the chairman until 3/1900 when it was converted into a public company chaired by Sir Christopher Furness. Output from SDSI's three constituent works in its first year totalled 284,520 tons comprising steel plates (73%), iron plates (20%) with the balance made up of angles, bars and sheets.

Prior to 1907, the West Hartlepool Works consisted of six 65 ton open hearth furnaces. SDSI modernised the plant by installing a 180 ton Talbot tilting open hearth furnace in 1907, to be followed by another in June 1913. Additional land had been obtained by the diversion of the southern length of Mainsforth Terrace to run alongside the Cliff House Branch in 1906 and the siding capacity was increased. More molten iron was transported in ladles from the Seaton Carew blast furnaces over the railway tracks that connected them to the West Hartlepool steel plant. SDSI had taken over the former Cliff House Pottery site, a siding agreement with the NER was dated 29/11/1904 and, by 1914, a track now crossed the road to serve those buildings. During 1920 some of the original open hearth furnaces were dismantled and a third tilting furnace added. In 8/1923 SDSI had running powers over the Cliff House Branch for its locomotives from the West Hartlepool Works to shunt the Expanded Metal Co Ltd's premises; SDSI provided the thin sheets of steel from which trellis and reinforcements were made. Expanded Metal's Stranton Works was located on the north side of Greatham Street and next to the Cliff House Branch. The combined West Hartlepool Works (NZ 516311) now covered an area of 303 acres and employed about 3,000 people. However, in 1925, the blast furnaces were taken out of

SDSI's No.10 (Hunslet 1405 of 1920) at what became the North Iron and Steel Works in West Hartlepool.
(NERA R. Inness photograph)

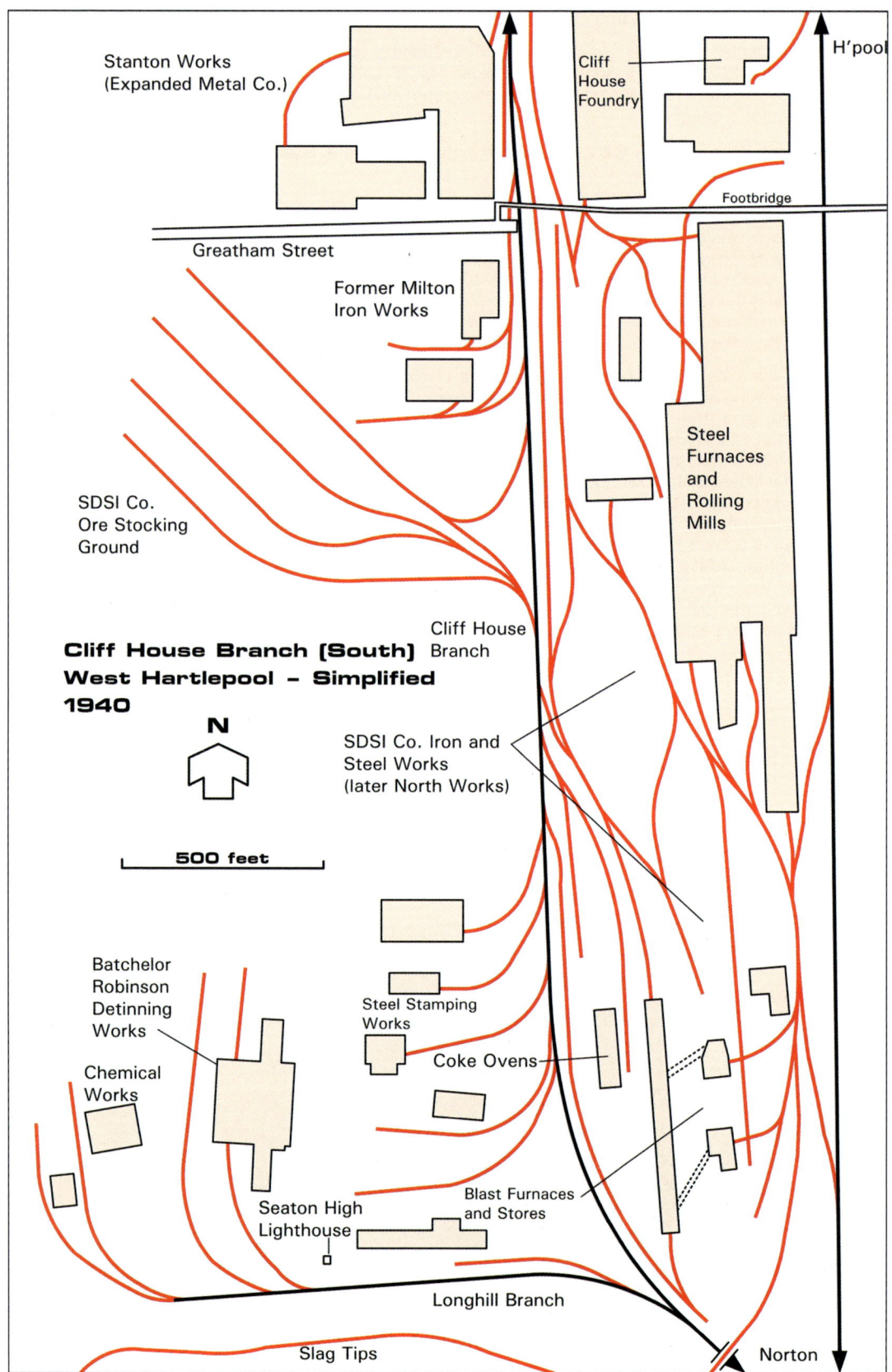

blast because of the economic depression and iron was supplied from the SDSI's Cargo Fleet Works; auxiliary hearths heated by exhaust gases from the open hearth furnaces being installed in 1927–28 to melt the pig iron. The coking plant also shut at the same time.

The West Hartlepool steel plant in 1937 comprised three basic 180 ton tilting open hearth furnaces and a 50 ton acid lined furnace. All of the cold materials, except dolomite, were brought into the 625 feet long Loading Bay in railway wagons prior to being put into charging boxes. At this time, improving economic conditions and the prospect of war encouraged SDSI to erect two new blast furnaces on the site of the old Seaton Carew structures; 30 Gibbons Kogag coke ovens next to the blast furnace bunkers were started up on 7/7/1937 and a new light plate mill installed. By 1945 the disposition of the plant was as follows (south to north): the steadily expanding slag tip south west of the Cliff House Branch; the two blast furnaces with the calcining kilns and bunkers served by an overhead railway track together with the coke ovens next to the Cliff House Branch; what was described on a SDSI plan as the "New Loco Shed" was immediately north of the blast furnaces; then came the steel plant sidings and the open hearth furnaces with a row of soaking pits be followed by the mills.

Meanwhile the layout of the ore stocking ground sidings reached by SDSI locomotives travelling along a short section of the Cliff House Branch was altered during the war to incorporate ore stocking and scrap stocking gantries. An agreement dated 11/7/1951 between the British Transport Commission and SDSI specified that the latter would construct slag pits on the ex Casebourne site and that its locomotives would tip 500–800 tons of slag into the pits per week. BR would load the slag into its railway wagons and pay nine pence per ton for all slag taken.

As a result of the first nationalisation of the steel industry, West Hartlepool Works passed to the Iron & Steel Corporation of Great Britain on 15/2/1951 but, with a change of government in 10/1951, the nationalisation was halted. The works transferred to the Iron & Steel Holding & Realisation Agency and was offered back to the private sector in 1/1956, although SDSI had continued to be responsible for its day to day operation and secured agreement to a programme of modernisation. Iron ore was delivered in railway wagons to a new ore preparation and sinter plant (1953–58). Coke production was increased by the installation of 78 Gibbons–Wilputte ovens between 12/1953 and 4/1957 replacing the 1937 plant. The existing blast furnaces were superseded by two new structures; No.1 commencing in 1956 and No.2 in 10/1959. The "New Loco Shed" was replaced by a larger building next to the main line. The improvements in the 1950s to what became the North Works were matched by the replacement of steam by diesel shunters involving the acquisition from John Fowler of eleven Class 421 0–4–0 diesel mechanicals equipped with McLaren 150hp engines.

South Durham Steel & Iron Co Ltd, Hartlepool South Works   104
NZ 508282 – **Map A**

SDSI did not halt its expansion there because it was already considering constructing a new iron and steel works on the 110 acres of slag tip to the south of the existing plant; indeed, a plan exists in Teesside Archives showing the layout of a new steel plant. This proved too expensive because it would have required removing 3 million tons of spoil. Instead SDSI acquired the 170 acres of the former Greatham Airfield as part of a 537 acre site between Seaton Carew and Greatham. This was to become the South Works (NZ 508282), with the West Hartlepool plant being renamed North Works. Construction took place between 1957 and 1961 and included a new blast furnace said to be then the largest on the north east coast (commissioned 2/1964) and an ore handling plant; 78 Gibbons-Wilputte coke ovens (started up 8/1960–9/1961); a melting shop containing five 360 ton open hearth furnaces (first steel tapped 16/8/1960); a slabbing and plate mill that could roll plate up to 136 inches wide (first day 3/9/1960). The two works were connected by a two-track private railway that ran alongside the BR line enabling the transfer of hot metal, ingots and slabs.

There had been a railway bridge over the Cliff House Branch to reach the slag tips for many years, but the new hot metal double track between the two works crossed the branch at grade and right angles south of the existing bridge. According to an agreement dated 15/3/1968, this was known as the "Manganese Crossing". It was anticipated when both works were operating at full capacity, that over one million tons of foreign iron ore would be required each year and mainly to be imported via Hartlepool Docks. After the opening of South Works, most of the SDSI locomotives were transferred to its engine shed (NZ 510288) about 1963, although a few steam engines lay around parts of North Works until scrapped. The construction of the South Works led to SDSI also purchasing a batch of new diesels from John Fowler. It chose the Class 424 six-coupled diesel hydraulic design incorporating Leyland 230hp engines.

In 1966 SDSI agreed to merge with Dorman Long & Co Ltd and Stewarts & Lloyds Ltd as British Steel & Tube Ltd but this came into being only a month before nationalisation. Vesting day for the British Steel Corporation was 28/7/1967, although responsibility for the operation of SDSI was not taken over until 1/7/1968 and SDSI did not cease to be a legal entity until 29/3/1970. Plans had been drawn up in 1966 for the construction of welded pipe mills at South Works and separate 44 inch and 20 inch mills commenced operations in 1968.

Most of the buildings at South Works were constructed parallel to the Norton-Hartlepool railway, with the exception of the 20 inch pipe mill (NZ 508290) at the northern end of the site allegedly at the insistence of

The steel plant at the British Steel Corporation's South Works, Hartlepool on 20th June 1977 with 8 (John Fowler 42400007 of 1960) shunting. (IRS Martin Shill photograph)

Stewards & Lloyds. Next to the main line were the South Works exchange sidings, which had their own signal box and connections with BR at both the north and south ends. *Steel News* on 4/3/1971 described the iron making operations at Hartlepool immediately prior to the rundown of both works. At North Works, No.1 blast furnace was being relined as a spare and No.2 was producing 5,000 tons of iron per week. Iron ore was delivered by rail, tipped into a roll crusher to be followed by processing in the sinter plant. Molten iron was transported to the open hearth plant at South Works. At the latter, No.4 blast furnace was producing about 8,000 tons of iron per week.

BSC took the strategic decision to concentrate its integrated iron and steel works on a limited number of sites including Lackenby south of the Tees. Faced with increasing financial losses, the rundown of North Works began. The light plate mill closed 30/7/1971 and the six acre site was sold to The Expanded Metal Co. No.2 blast furnace stopped in 7/1971, with No.1 being dismantled in 1977. No.2 was given a temporary lease of life when it was relit on 5/4/1975 to operate while No.4 blast furnace at South Works was relined. Fifty two of the coke ovens at North Works finished in 1972/73, but the remainder kept going until No.2 blast furnace was blown out in 4/1976. North Works was demolished in 1976-78.

At the South Works, No.4 blast furnace finished in 9/1977 and the last steel was produced in 12/1977. Much was demolished, leaving the coking plant, plate mill and the 20 inch and 44 inch pipe mills. The coking plant continued because of problems with the new Redcar coke ovens but it eventually closed in 9/1982. The plate mill had been supplying the 44 inch pipe mill but, with fluctuating demand for pipes, the plate mill finished in 2/1985.

Hartlepool Pipe Mills

The remnants of South Works had, in effect, become the North East Pipe Mills with the 20 inch and 44 inch plants. Only the southern junction with the BR main line was retained. From the exchange sidings, a line curved round with one track entering the 44 inch mill and another running as the ex-shear line into the former plate mill. Part of the latter housed the workshops where locomotives were repaired; the modern engine shed having been abandoned. Another line headed across the site to the 20 inch mill. Steel plate was delivered by BR from Lackenby and later Scunthorpe and Dalzell Works. The finished pipes were despatched by either road or rail. Only one BSC locomotive was now required for shunting, taking plate to the mills and collecting pipes from the 'Goliath' gantry line for haulage to the exchange sidings. This was either a GEC Traction or Thos. Hill locomotive in the 1990s, although it was reported that the 20 inch mill was under separate management at this time and only BR locomotives worked through the 44 inch mill exchange sidings to the 20 inch mill. The latter was modernised in 1990 and the 44 inch mill replaced by a secondhand Japanese 42 inch mill which started operating in 10/1993. In 2001, 84 inch pipes were being received from the Stockton Works, welded into 27 metre lengths and despatched by rail through Hartlepool or Grangemouth Docks. The 84 inch mill was then transferred from Stockton and commenced work at Hartlepool next to the 42 inch mill on 14/2/2006. Meanwhile, in 2003, Ed Murray & Sons Ltd had taken over responsibility for shunting at Hartlepool Pipe Mills using its own locomotives. It should be noted that some of these locomotives were used on the Pipe Mills contract while others were stored there until required.

On 1/8/2017, the 42 and 84 inch pipe mills were purchased by Liberty Pipes (Hartlepool) Ltd, with Tata retaining the 20 inch mill linked to steel coming from its Port Talbot Works. While Tata relied on the main line locomotives to shunt its 20 inch mill, the wagons for Liberty Pipes were collected from the exchange sidings and taken to the 42 inch and 84 inch mills by the contractor's industrial locomotive. When loaded, they were moved back to the exchange siding to be formed into a train for eventual collection by the main line locomotive.

References :   *History of the South Durham Steel & Iron Co Ltd*, W.G. Willis, SDSI, 1969

Institute of Mechanical Engineers visits to works, 1893

The Engineer magazine, 26/10/1888

Iron and Steel, South Durham Cargo Fleet Steel, SDSI, 1936

Seaton Carew Iron Co Ltd Minutes of Board of Directors 1882–1919

'Practice at the Works of the South Durham Steel & Iron Co Ltd', James Winter, *Symposium on Steel Making in North East District*, 1938

'Continuing Improvements for Hartlepool's Plant', David Robinson, *Steel News*, British Steel Corporation, 3/4/1971

'Seaton Carew Ironworks John Henry Proud's Reminiscences of the Early 20th Century with Commentaries and Notes', J.H. Proud and J.K. Almond, *The Cleveland Industrial Archaeologist* 31, CIAS, 2006 pp.3–48

SDSI obviously high hopes for its South Iron and Steel plant at Hartlepool which included extensive exchange sidings and a new signal box containing this lever frame. However, it was no longer in use when the interior was photographed in September 1986. (Kevin Lane photograph)

West Hartlepool / North Works

Gauge : 4ft 8½in

No.	Name	Type	Cyl	Builder	Works No.	Year	Origin	Disposal
No.5		0-4-0ST	OC	Grange Iron Works		1873	(a)	Scr c/1918
-		0-4-0ST	OC	AB	299	1888	(a)	s/s
No.3		0-4-0ST	OC	HL	2134	1889	(a)	Scr c/1918 #
No.1		0-4-0ST	OC	HL	2412	1899	New	(1)
No.4		0-4-0ST	OC	HE	608	1895	(b)	Scr /1958
No.6		0-4-0ST	OC	HE	951	1907	(b)	Scr c/1957 after 27/4/1957
No.7		0-4-0ST	OC	HE	1086	1911	(b)	Scr /1958
No.8		0-4-0ST	OC	HE	1108	1912	(b)	Scr c/1957 after 27/4/1957
No.1		0-4-0ST	OC	HL	3354	1918	New	(9)
SDS & I Co No.3		0-4-0ST	OC	HL	3355	1918	New	(9)
No.5		0-4-0ST	OC	KS	3095	1918	New	Scr /1969
No.12		0-4-0ST	OC	MW	1967	1918	New	Scr /1969
No.10		0-4-0ST	OC	HE	1405	1920	(b)	Scr c/1960
	GLADYS	0-4-0ST	OC	B	292	1882	(c)	(2)
	DAISY	0-4-0ST	OC	P	467	1888	(c)	Scr c/1936
	WALTER MORRISON	0-4-0ST	OC	P	657	1897	(c)	Scr c/1936
	FRANCIS	0-4-0ST	OC	HL	2378	1898	(c)	(3)
	AILEEN	0-4-0ST	OC	HL	2445	1899	(c)	
	Rebuilt			SDSI		c/1940		Scr after /1945
	MAJOR	0-4-0ST	OC	AB	1363	1914	(c)	(6)
	COLONEL	0-4-0ST	OC	AB	1501	1917	(c)	(7)
"1918"		0-4-0ST	OC	AB	1609	1918	(c)	(8)
No.5		0-4-0ST	OC	HE	894	1905	(d)	Scr
No.13		0-4-0ST	OC	HL	3935	1937	New	(5)
No.9		0-4-0ST	OC	RS	3057	1904	(e)	Scr c/1956
No.11	STANGHOW	0-4-0CT	OC	HL	2516	1902	(f)	
	Rebuilt	0-4-0T		SDSI		1941		Scr c/1953
No.14		0-4-0ST	OC	AB	2105	1940	New	Scr c6/1960
No.15		0-4-0ST	OC	RSHN	7045	1942	New	(4)
-		0-4-0DM		HE	2652	1942	New	(14)
No.16		0-4-0ST	OC	HE	1087	1911	(g)	Scr c/1956
No.17		0-4-0ST	OC	AE	1801	1918	(h)	Scr c/1956
No.18		0-4-0DM		JF	4210086	1953	New	(13)
No.19		0-4-0DM		JF	4210089	1953	New	(14)
No.20		0-4-0DM		JF	4210091	1954	New	(14)
No.21		0-4-0DM		JF	4210094	1954	New	(14)
No.22		0-4-0DM		JF	4210099	1955	New	(14)
No.23		0-4-0DM		JF	4210102	1955	New	(14)
No.24		0-4-0DM		JF	4210107	1955	New	(14)
No.25		0-4-0DM		JF	4210110	1956	New	(14)
No.26		0-4-0DM		JF	4160009	1953	(i)	(10)
No.27		0-4-0DM		JF	4210128	1957	New	(14)
No.28		0-4-0DM		HC	D978	1957	New	(14)
No.1		0-4-0DM		JF	4210146	1958	(j)	(11)
No.29		0-4-0DM		HC	D1052	1958	New	(14)
No.30		0-4-0DM		HC	D1141	1959	New	(14)
31		0-4-0DM		JF	4210148	1958	(k)	(14)
32		0-4-0DM		JF	4210147	1958	(k)	(12)

\# see SDSI Moor Works for an alternative possibility.

The Hartlepool 42 inch Pipe Mill relied on hired shunters from Ed Murray & Sons Ltd on 29th July 2008. This is the French built Moyse 1464 of 1979; Astontrack was a marketing name adopted by Murray for a brief period.
(Cliff Shepherd photograph)

(a) ex West Hartlepool Steel & Iron Co Ltd, with works, 28/12/1898.
(b) ex SDSI, Moor Steel and Iron Works, Stockton. No dates for transfers are known but they would have been in the period 1898 to 1922.
(c) ex Seaton Carew Iron Co Ltd, with works, probably /1924.
(d) ex SDSI, Malleable Works, Stockton, /1936.
(e) ex Ritsons (Burnhope Collieries) Ltd, Burnhope, Co. Durham, /1939.
(f) ex Cochrane's (Middlesbrough) Foundry Ltd, Cargo Fleet, c/1940.
(g) ex SDSI, Malleable Works, Stockton, /1949.
(h) ex T.J. Thomson & Son Ltd, Stockton, /1949.
(i) ex SDSI, Malleable Works, Stockton, /1957, although JF records show as new to West Hartlepool Works.
(j) ex SDSI, South Works, by 8/1958. To South Works, 9/1958 and returned by 12/1958. To South Works, 3/1959 and returned c3/1962.
(k) ex SDSI, South Works, by 1/1960.

(1) to Weardale Steel, Coal & Coke Co Ltd, Tudhoe, Co. Durham. /1917.
(2) to John F. Wake, Darlington, Co. Durham, /1935, although Inness suggests that it may have gone to the Malleable Works first.
(3) boiler fitted to AILEEN c/1940 and remainder scrapped.
(4) to SDSI, Cargo Fleet Works, 6/1957.
(5) to SDSI, Cargo Fleet Works, c/1957.
(6) to SDSI, Irchester Ironstone Quarries, Northamptonshire, 11/1957.
(7) to SDSI, Cargo Fleet Works, 11/1957.
(8) to SDSI, Irchester Ironstone Quarries, Northamptonshire, 2/1959.
(9) to Seaham Harbour Dock Co Ltd, Seaham, Co. Durham, 11/1961.
(10) to SDSI, Malleable Works, Stockton, 1/1962.
(11) to SDSI, South Works, c9/1962.
(12) to SDSI, Cargo Fleet Works, 12/1962.
(13) to SDSI, Malleable Works, Stockton, 9/1963.
(14) to SDSI, South Works Engine Shed, c/1963.

North Works Coke Ovens

Gauge : 4ft 8½in

-		4wWE	GB	1448	1936	New	(1)

(1) to British Steel Corporation, South Works, for repairs after closure of the coking plant, 4/1977.

North Works Cogging Shop

Locomotives used for working furnace ladles.

Gauge : 2ft 6in

-		4wDM	HE	3308	1946	New	s/s c/1954
-		4wRE	SDSI			New	s/s

South Works/Hartlepool Pipe Mill

Gauge: 4ft 8½in

1	(No.1)	0-4-0DM	JF	4210146	1958	(a)	Scr 1/1973
No.2		0-4-0DM	JF	4210147	1958	New	(1)
31	(No.3)	0-4-0DM	JF	4210148	1958	(b)	Scr 1/1973
2	(No.2)	0-6-0DH	JF	4240001	1959	New	Scr /1979
3	(No.3)	0-6-0DH	JF	4240002	1959	New	Scr /1979
4	(No.4)	0-6-0DH	JF	4240003	1959	New	Scr c/1976
5	(No.5)	0-6-0DH	JF	4240004	1959	New	Scr c12/1979
6	(No.6)	0-6-0DH	JF	4240005	1960	New	Scr c3/1979
7	(No.7)	0-6-0DH	JF	4240006	1960	New	Scr c12/1979
8	(No.8)	0-6-0DH	JF	4240007	1960	New	Scr c12/1979
9	(No.9)	0-6-0DH	JF	4240008	1960	New	Scr c3/1979
10	(No.10)	0-6-0DH	JF	4240009	1960	New	Scr c12/1979
11	(No.11)	0-6-0DH	JF	4240011	1961	New	Scr c3/1979
-		0-4-0DM	HE	2652	1942	(c)	(2)
19	(No.19)	0-4-0DM	JF	4210089	1953	(d)	
		Rebuilt	JF		1966		Scr c 3/1979
20	(No.20)	0-4-0DM	JF	4210091	1954	(c)	Scr 1/1973
21	(No.21)	0-4-0DM	JF	4210094	1954	(e)	
		Rebuilt	JF		1966		(6)
22	(No.22)	0-4-0DM	JF	4210099	1955	(c)	Scr 1/1973
23	(No.23)	0-4-0DM	JF	4210102	1955	(c)	Scr 1/1973
24	(No.24)	0-4-0DM	JF	4210107	1955	(c)	Scr 7/1972
25	(No.25)	0-4-0DM	JF	4210110	1955	(c)	Scr 1/1973
27	(No.27)	0-4-0DM	JF	4210128	1957	(c)	Scr 7/1972
28	(No.28)	0-4-0DM	HC	D978	1957	(c)	Scr/1976
29	(No.29)	0-4-0DM	HC	D1052	1958	(c)	Scr /1976
No.30		0-4-0DM	HC	D1141	1959	(c)	Scr /1976
15		0-4-0DH	JF	4220027	1964	(f)	(9)
16		0-4-0DH	JF	4220028	1964	New	Scr c12/1979
33	BOYLE	0-4-0DE	RH	408309	1957	(g)	(7)
34	JAMES WATT	0-4-0DE	RH	381757	1955	(g)	(7)
12	MARDALE	0-6-0DE	YE	2743	1959	(h)	(5)
32		0-4-0DM	HC	D1013	1957	(i)	Scr 9/1971
33		0-4-0DM	HC	D1081	1958	(i)	Scr 9/1971
36		0-4-0DM	HC	D1032	1958	(i)	Scr 7/1972
6	FARNDALE	0-6-0DE	YE	2719	1958	(j)	(3)
11	LINGDALE	0-6-0DE	YE	2742	1959	(k)	(4)
1	ALLENDALE	0-6-0DE	YE	2629	1956	(l)	(5)
450		4wDH	TH	231V	1971	New	(12)
451		4wDH	TH	232V	1971	New	(14)

452		4wDH	TH	233V	1971	New	(11)	
453	(CHURCHILL)	4wDH	S	10025	1960	(m)	(8)	
454		4wDH	S	10018	1960	(n)	(8)	
455		4wDH	TH	258C	1975	(o)	(10)	
456		4wDH	TH	259C	1975	(p)	(10)	
457		4wDH	TH	260C	1975	(q)	(10)	
10		4wDH	S	10010	1959	(r)	Scr c3/1979	
15		4wDH	S	10015	1959	(r)	Scr c3/1979	
28		4wDH	S	10028	1960	(r)	Scr c3/1979	
38		4wDH	S	10038	1960	(r)	Scr /1980	
42		4wDH	S	10042	1960	(r)	Scr /1980	
453		4wDH	TH	221V	1970	(s)	(11)	
36		4wDH	TH	222V	1970	(s)	Scr c10/1985	
37	(454)	4wDH	TH	223V	1970	(s)	Scr c8/1989	
38		4wDH	TH	224V	1970	(s)	Scr c10/1985	
264	PORT MULGRAVE	6wDE	GECT	5461	1977	(t)		
279		6wDE	GECT	5478	1979	(u)	(13)	
425		6wDE	GECT	5425	1977	(v)	(17)	
MURR1	(2)	4wDE	Moyse	1364	1976	(w)		
MURR2	(1)	4wDE	Moyse	1464	1976	(x)	(21)	
-		0-4-0DH	HE	7425	1981	(y)	(15)	
-	(01560)	4wDH	RR	10229	1965	(z)	(19)	
Z.Z.267		0-6-0DH	EEV-AEI	4003	1971	(aa)	(16)	
	LADY POTTER	0-6-0DH	RR	10214	1964	(ab)	(20)	
306		0-6-0DH	GECT	5383	1973	(ac)	(18)	

(a) new. To SDSI North Works, by 8/1958 and returned, 9/1958. To North Works by 12/1958 and returned, 3/1959. To North Works, c3/1962 and returned, c9/1962.
(b) new. To SDSI North Works by 1/1960.
(c) ex SDSI North Works, c/1963.
(d) ex SDSI North Works, c/1963. Rebuilt by JF 7/1965 to 3/1966 with 203hp engine.
(e) ex SDSI, North Works, c/1963. Rebuilt by JF 3/1966 to 8/1966 with 203hp engine.
(f) new, ex demonstration locomotive.
(g) ex ICI Ltd, Burn Naze Works, Fleetwood, Lancashire, 3/1965.
(h) ex ICI Ltd, Billingham Works, on hire, 1/1970.
(i) ex BSC, Cargo Fleet Works, 2/1970, for scrap.
(j) ex ICI Ltd, Billingham Works, on hire, 3/1970.
(k) ex ICI Ltd, Billingham Works, on hire, 5/1970.
(l) ex ICI Ltd, Billingham Works, on hire, 9/1970.
(m) ex BSC, South Teesside Works, 15/10/1971.
(n) ex BSC, Britannia Works, 19/7/1972.
(o) ex TH, 5/1975; rebuild of S 10001 of 1960 formerly at BSC South Teesside Works.
(p) ex TH, 8/1975; rebuild of S 10041 of 1960 formerly at BSC South Teesside Works.
(q) ex TH, 9/1975, rebuild of S 10011 of 1961 formerly at BSC South Teesside Works.
(r) ex BSC, South Teesside Works, 11/1976.
(s) ex BSC, Consett Works, 5/1981.
(t) ex British Steel, Teesside Works, 3/1992. Became property of Ed Murray & Sons Ltd after many years out of use, 2019.
(u) ex Corus, Teesside Works, c2000–01, by 4/3/2000.
(v) ex Corus, Teesside Works, after 18/7/2000, by 2/4/2001. Became property of Ed Murray & Sons Ltd after many years out of use, 2019.
(w) ex Ed Murray & Sons Ltd, Casebourne Road, Hartlepool, on hire, 24/9/2003.
(x) ex Ed Murray & Sons Ltd, Casebourne Road, Hartlepool, on hire, c10/2003. To Chasewater Light Railway & Museum Company, Brownhills, Staffordshire for repair, 5/8/2013 and returned, 15/9/2015. To Longs Steel UK Ltd, (subsidiary of Tata Steel Europe), Teesside Beam Mill, (property of Ed Murray & Sons Ltd), on hire 10/2015. Returned ex hire w/c 5/8/2019.
(y) ex Ed Murray & Sons Ltd, Hartlepool, after c5/2005, by 10/1/2007 for storage.

(z)		ex Tata Steel Europe, Skinningrove Works, property of Ed Murray & Sons Ltd, Hartlepool, 21/12/2011. Locomotive acts as a spare for Hartlepool or Skinningrove Works shunting contracts. To Tata Steel Europe, Skinningrove Works, on hire, c25/9/2014; returned 9/12/2019.				
(aa)		ex T.J. Thomson & Son Ltd, Stockton, property of Ed Murray & Sons Ltd, 23/2/2017.				
(ab)		ex ICL UK (Cleveland) Ltd, Boulby Mine, Loftus, property of Ed Murray & Sons Ltd, 6/5/2019.				
(ac)		ex Chasewater Light Railway & Museum Company, Brownhills, Staffordshire, (property of Ed Murray & Sons Ltd), 30/10/2019, formerly Tata Steel Europe, Llanwern Works, Newport, South Wales.				

(1) to SDSI, North Works, by 1/1960.
(2) to Bell & Son (Doncaster) Ltd, Port Clarence, 3/1968, by 23/3/1968.
(3) returned to ICI Ltd, Billingham Works, ex hire, 5/1970.
(4) returned to ICI Ltd, Billingham Works, ex hire, 9/1970.
(5) returned to ICI Ltd, Billingham Works, ex hire, 4/1971.
(6) to North Yorkshire Moors Railway, Grosmont, North Yorkshire, 5/1972.
(7) to T.J. Thomson & Son Ltd, Stockton, for scrap /1976.
(8) to BSC, River Don Works, Sheffield, South Yorkshire, 7/1978.
(9) to North East Iron Refining Co Ltd, Stillington, 4/1979.
(10) dismantled as source of spares for other locomotives from /1980; scrapped c10/1985.
(11) scrapped on site by Stephenson Demolition, Billingham, 4/1991.
(12) to T.J. Thomson & Son Ltd, Stockton, 21/6/1999.
(13) to Corus, Teesside Works, after 18/7/2000, by 2/4/2001.
(14) to Corus, Skinningrove Works, /2002.
(15) to Bonlea Fabrications, Hartlepool, for repairs, c7/2007 and then to Ed Murray & Sons Ltd, Brenda Road, Hartlepool.
(16) to Chasewater Light Railway & Museum Co, Brownhills, Staffordshire, 3/5/2017.
(17) to Chasewater Light Railway & Museum Co, Brownhills, Staffordshire, 29/10/2019.
(18) to Ed Murray & Sons Ltd, Casebourne Road, Hartlepool, c1/2020; after 9/12/2019, by 8/2/2020.
(19) to Ed Murray & Sons Ltd, Casebourne Road, Hartlepool, w/e 4/4/2020.
(20) to Ed Murray & Sons Ltd, Casebourne Road, Hartlepool, w/c 27/7/2020.
(21) to British Steel Ltd, Teesside Beam Mill, Lackenby, 1/5/2021.

South Works Coke Ovens

Gauge : 4ft 8½in

-		0-4-0WE	GB	2937	1960	New	Scr c10/1985
-		4wWE	GB	1448	1936	(a)	Scr c10/1985
-		4wWE	GB	420306	1972	(b)	Scr c10/1985

(a) ex BSC, North Works, for repairs, 4/1977.
(b) ex BSC, Consett Works, Consett, 8/1981.

LINTHORPE-DINSDALE SMELTING CO LTD

LINTHORPE IRON WORKS, Middlesbrough.   105
Edward Williams until 3/4/1903*   NZ 489215 – **Map F**
Lloyd & Co until /1879

* Although the business was sometimes referred to as "Executors of the late Edward Williams" c/1886–1891.

The two blast furnaces at the Tees Side Iron Works appear to have had insufficient capacity to meet demand, because Hopkins & Co was purchasing pig iron to supplement its output. By late 1863 the proprietors of the iron works decided to establish a new company under the name of one of the partners, Robert Lloyd, who was the son-in-law of Thomas Snowdon, a founder of the Tees Side Iron Works. Also Lloyd & Co employed John Gjers, manager at the Tees Side Iron Works, to design and supervise construction of the four Linthorpe blast furnaces and ancillary plant. The blast furnaces were blown in on 18/8/1865 and supplemented by two more which were lit on 1/7/1870. All six were operating in 1873 producing Nos.1 and 3 pig iron.

The Linthorpe Iron Works was located in the Ironmasters district immediately to the west of the Tees Side Iron Works. The latter had its row of blast furnaces at right angles to the river, but Linthorpe's were located more parallel to the river. Lloyd's railway diverged from the NER Old Town Branch and curved round to serve a fan of sidings to the south west of the blast furnaces. It was also on this side that the calcining kilns and mineral stores were located. A pneumatic lift raised single wagons on to the 50 feet high gantry where three tracks extended

600 feet over the kilns and stores. The pig beds and slag discharge points were on the north east side of the blast furnaces. A railway from the fan of sidings went past the end of the blast furnaces before turning north to reach the river bank. Here it extended in a downstream direction to serve Linthorpe Wharf and then continued to enter the Tees Side Iron Works.

As a result of the depression in the mid to late 1870s, the Linthorpe Iron Works had financial difficulties and suspended payments. The *London Gazette* on 20/5/1879 gave details of meetings with creditors under the proceedings for liquidation by William Randolph Hopkins, Isaac Wilson and Edgar Gilkes trading together at the Linthorpe Iron Works and Lackenby Iron Works under the style of Lloyd & Company. An advertisement in *The Engineer* on 5/9/1879 stated that an auction was to take place at the Linthorpe Iron Works, including seven locomotives; not all of which have been identified. Edward Williams had been General Manager of Bolckow Vaughan & Co from 1865 and had played a major role in the development of that business. In 1879 he became Chief Proprietor of the Linthorpe Iron Works which he ran until his death in 1886. His successors continued to operate the works although, by the start of the twentieth century, only three or four blast furnaces were operating.

The Iron Works produced two brands of pig iron – Cleveland quality "Linthorpe" and Bessemer quality "E.W. Hematite". In 1889 it had received a very large single consignment of 3,300 tons of Spanish ore. Warner's purchased pig iron from Linthorpe in 1901–05 and also paid it for wharfage. The internal railway system had not significantly changed, but the area between the blast furnaces and the river had been used for slag tipping, although slag was also dumped at sea using hopper barges.

The Dinsdale Smelting Co Ltd operated an iron works at Middleton St George on the railway to Darlington. In April 1903, this company amalgamated with 'Edward Williams', owner of Linthorpe Iron Works, to create Linthorpe–Dinsdale Smelting Co Ltd. The latter business was later restructured and re-registered with the same name on 25/5/1920. Linthorpe Dinsdale increased the size of its holding at the Linthorpe Iron Works by purchasing eight acres of the adjoining former Tees Side Iron Works, although it had to maintain the old siding across this land connecting with Richardsons Westgarth's premises. The latter could use the siding, without payment, to transport boilers, machinery and goods up to 20 feet wide, 25 feet high and with a net weight of 100 tons.

In 1937 Dorman Long inspected both of the Linthorpe-Dinsdale works and its findings make sorry reading. The Middleton Iron Works comprised four very old blast furnaces, two of which had been relined in 1931 but had not worked since because of the closure of the plant. There were three locomotives in working order and sufficient rolling stock to work two blast furnaces but Dorman Long concluded that the works was not worth purchasing.

According to a 22/4/1937 report, the Linthorpe plant had hardly worked since 1920. There were still six blast furnaces, all in varying stages of disrepair, served by old reciprocating blowing engines supplied with steam from 24 now derelict Lancashire boilers. The furnace hoist "was bad, both in design and condition and the general mechanical equipment is little better than scrap". The report went on to state that the "wharf and wharf cranes are useless". Dorman Long decided to purchase the plant and site for its scrap value and strategic location next to the Acklam Iron and Steel Works. Its offer of £65,000 was agreed probably in 7/1937 and a detailed valuation of the loose plant (£9,635) and fixed plant (£5,100) took place in the following month. Included were six locomotives of which only three were serviceable, 18 flat wagons, 2 pig bed cranes, 2 breaker cranes, a Priestman grab crane, 3 sea-going slag hopper barges and 2 wharf cranes. The engine shed was valued at £300. All of the plant had been removed from the site by 9/1940.

Reference :   Correspondence concerning Dorman Long's purchase of the Linthorpe Iron Works, 1937, Teesside Archives (BS.DL/3/2/1/3/70)

Gauge : 4ft 8½in

	Name	Type	Cyl	Builder	Works No	Date		
	ROBERT	0-4-0tank	VC	HG	211	1865	(a)	s/s
	LINTHORPE	0-4-0ST	OC	MW	191	1866	New	s/s
	ARTHUR	0-4-0tank		HG	276	1870	New	(1)
	MAY	0-4-0ST	OC	MW	756	1880	New	(2)
5		0-4-0ST	OC	AB	300	1888	(b)	(3)
6		0-4-0ST	OC	AB	671	1891	(c)	(3)
7		0-4-0ST	OC	AB	897	1900	New	(3)
8		0-4-0ST	OC	AB	1102	1907	New	(3)

Inness suggests a four-coupled saddle tank with 14in x 21in cylinders by Dick & Stevenson was also here but no more information is known.

(a)   new. Purchaser recorded as "Hopkins, Lloyd & Co".
(b)   new. Acquired via the agent, J. Torbock.
(c)   new. Acquired via the agent, Executors of J. Torbock.

(1) to The Hutton Henry Coal Co Ltd, Hutton Henry Colliery, Station Town, Co. Durham.
(2) to Linthorpe-Dinsdale Smelting Co Ltd, Middleton Iron Works, Middleton St George, Co. Durham; later sent from there to Dorman Long's Redcar Works in /1937.
(3) to Dorman Long & Co Ltd, Acklam, Britannia, Newport Combined Fleet, Middlesbrough, c/1937.

LONDON & NORTH EASTERN RAILWAY

LACKENBY SLAG PLANT, Grangetown   106
North Eastern Railway to 31/12/1922   NZ 555228 – **Map H**

The owners of the Lackenby Iron Works had constructed a bridge over the Middlesbrough-Redcar railway in order to tip slag on the north side of the line. The slag was deposited on the marshes bordering the River Tees and these tips grew to an enormous size. The NER laid its own sidings alongside the tips, possibly by 1910 and certainly by 1913, with a connection to its railway controlled by Lazenby Signal Box to the east of Iron Works overbridge. The siding ran parallel to the main line before curving towards the river where it divided into four tracks serving the NER's slag breaking plant and then terminating at a headshunt. The plant was provided by Heenan & Froude Ltd and comprised slag breaking machinery and three bunkers, each containing different sizes of slag, with hoppers for dropping it into the railway company's wagons. The slag was extracted from the tips and loaded into one cubic yard side tipping tubs which ran on a narrow gauge railway system to the plant where each wagon was attached to a creeper chain and hauled up to the feeding platform. The *Harrogate Advertiser* reported on 1/3/1913 that the NER was reballasting the whole main line between Shaftholme Junction and Berwick replacing ash with slag from its "slag breaking plant near Redcar". By 1925 the LNER was using a Simplex locomotive to haul the tubs between the tip faces and the plant. Pease & Partners is reported to have sold the tips containing some two million tons of slag to the LNER in 1928 and the Lackenby Iron Works closed in 1931. The LNER continued to use the slag and its magazine in 1938 reported that "during 28 years of working by this company" (sic), the plant had contributed nearly 1,660,000 tons of ballast. Approximately 100,000 tons a year were then being processed.

The LNER Siding Diagram dated 8/1924 shows a separate group of lines, called the Wilton Estate Tip Sidings, had been laid from the Lazenby Signal Box connection to the east of the Lackenby Slag Plant. It appears that these were used for wagons of spent ballast to assist reclamation of the marshes and also provide a source of empty wagons for the outgoing slag. The standard gauge lines of both sites were shunted by railway company locomotives. Wooden open wagons were used, some had their sides braced to withstand the thrust of the load. A batch of 25 ton steel ballast hoppers, built by Metropolitan Cammell Carriage & Wagon Co Ltd in 1930 and twin plough ballast brake vans were allocated to the Lackenby Slag Plant and lettered accordingly. Part of the film in the references shows slag being processed at the Lackenby plant, tipped into wagons and being taken away by a LNER Class J77 locomotive. According to the LNER Sectional Appendix, dated 1/11/1947, its trains from the Lackenby slag crusher and Wilton Estate tip could travel in the wrong direction on the down goods 'independent' line to Grangetown Signal Box provided the signalman gave approval. It is not known when the plant closed, but as late as 1951, a Metropolitan Cammell manufacturer's photograph shows a new 25 ton self-discharging hopper ballast wagon bearing the inscription "ED EMPTY TO LACKENBY SLAG PLANT GRANGETOWN".

References :   'The Limestone and Slag Quarries of the North Eastern Area', F.L. Pawley, *LNER Magazine*, 9/1938

'LNER On the Narrow Gauge Part Two', Andrew Dow, *Steam World*, 2/2010

'Blast Furnace Slag for Ballasting Purposes', J. Kearney, R. Tidswell, D.J. Williamson and E. Scarlett, *North Eastern Express* 222, NERA, 5/2016

www.yorkshirefilmarchive.com/film/stone-ballasting-hulands-quarry

Gauge : Narrow

-	4wPM/DM	MR?	(a)	s/s

(a) here by 10/1925.

E. MARSON & CO LTD

FERTILISER WORKS, Port Clarence   107
NZ 501214 – **Map C**

Following closure of the Anderston Foundry in Port Clarence, the premises were occupied by Bell & Son (Doncaster) Ltd (which see) and E. Marson, although the arrangements within the site are not known. A report in the IRS Bulletin, issued in October 1974, records that HE 5306 was in the yard with a number of fitted vans.

The small 71hp Hunslet Yardmaster locomotive only weighed 13 tons and was designed to operate in small good yards. Maker's number 5306 of 1958 was at the Marson Fertiliser Works at Port Clarence on 23rd August 1974. (IRS Brian Webb photograph)

In a small shed was S10070 under repair and still in ICI livery. It is thought the Sentinel actually belonged to Bell. Rail traffic ceased about 1974. E. Marson & Co Ltd's name appears in the 1972 telephone directory at the Anderston Foundry but is absent from the 1975 issue.

Gauge : 4ft 8½in

-		4wDM	HE	5306	1958	(a)	(1)

(a)   ex J. Walker, dealer, Aston, Birmingham; previously George Richards & Co (Engineers) Ltd, Broadheath, Altrincham, Cheshire, c/1970.

(1)   to Andrew Barclay, Sons & Co Ltd, Kilmarnock, Ayrshire, c10/1974.

METROPOLITAN-VICKERS-BEYER PEACOCK LTD

BOWESFIELD WORKS, Stockton

108
NZ 435174 – **Map D**

The company was formed combining the electrification experience of Metropolitan-Vickers Electrical Co Ltd and the locomotive construction skills of Beyer Peacock & Co Ltd in 11/1949 to build "locomotives other than steam". The premises already existed comprising 95,000sq ft of factory floor space and offices. It was erected by the Government, possibly as a 'shadow factory' during World War 2 and was initially to be leased by the Board of Trade in 1946 to Le Tourneau (Great Britain) Ltd for the manufacture of heavy earth moving equipment, but this fell through and Metropolitan Vickers took over the lease in 1947. To begin with, it was used as an extension factory for its Trafford Park Works, but the original building was increased by 33,500sq ft of manufacturing floor space to enable locomotive construction by the new joint company.

A single branch line was laid connecting with the Darlington-Middlesbrough railway at Bowesfield Junction. This headed south past the clay pits and tips of the Bowesfield Brick Works. Approximately halfway along the branch, the line divided into three tracks forming exchange sidings before continuing as a single line to the factory on the east side of Yarm Lane. Although the company was responsible for the upkeep of most of the branch, BR locomotives worked as far as the exchange sidings. At the works, two sidings served the factory, one of which ran to a turntable from where three tracks entered the main building. Subsequently the branch also gave access to the Power-Gas Corporation Ltd's South Works (which see).

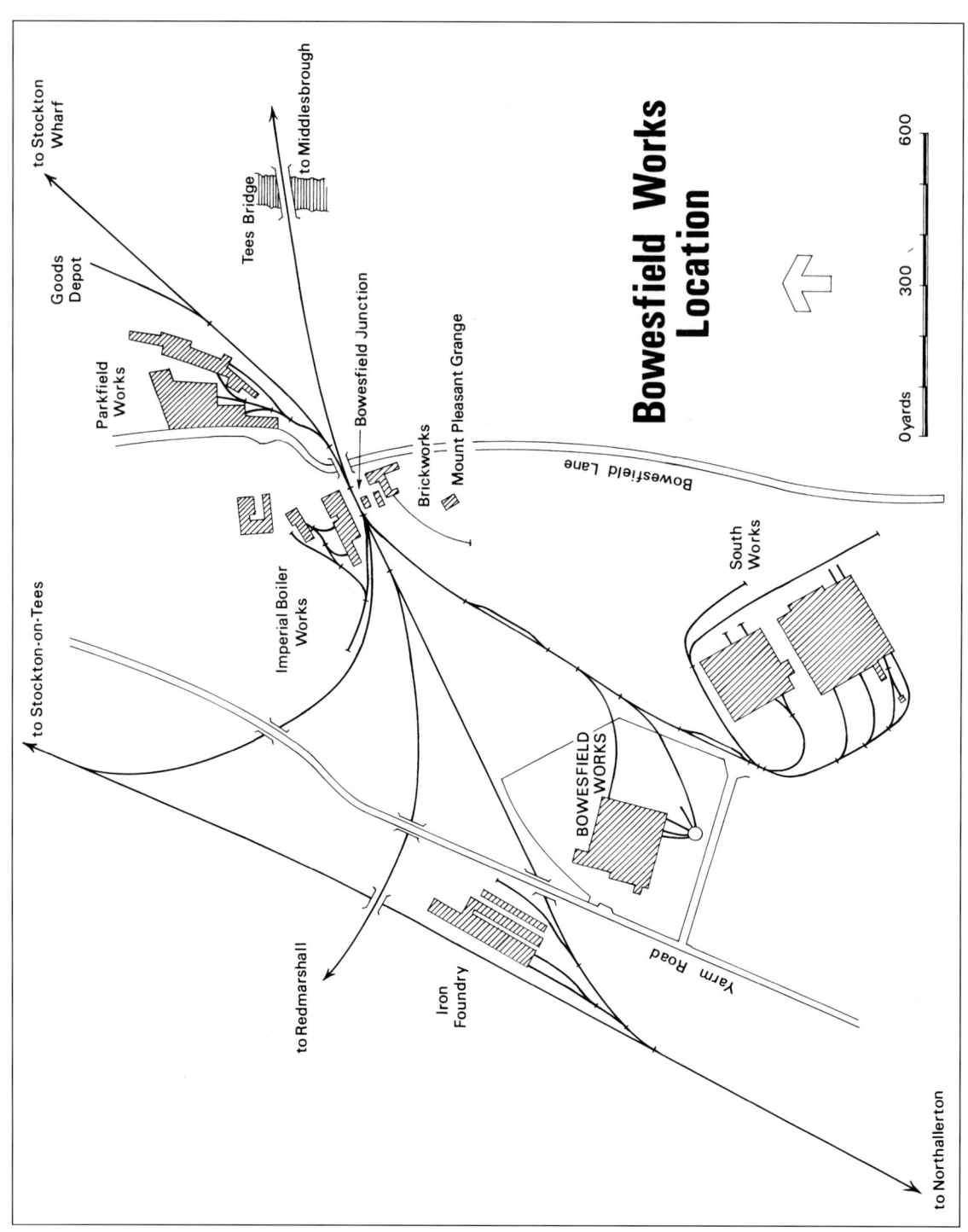

After conversion from gas turbine to electric power, number 18100 was being towed along the Bowesfield Works branch by Metropolitan-Vickers-Beyer Peacock's Electromobile battery shunter (W247/1927) for delivery to BR on 31st August 1958. *(ARPT J.W. Armstrong photograph)*

New locomotives under construction at the Bowesfield Works of Metropolitan–Vickers-Beyer Peacock, including BILSTON GLEN No.2 (maker's number 1016 of 1959) and one of the D57xx diesels for BR in the background. *(IRS collection)*

Operations at the Bowesfield Works began with a healthy order book of 24 electric locomotives for Brazil, 40 electric locomotives for New South Wales and 48 diesel electrics for Western Australia. Although the electrical equipment came from other Metropolitan-Vickers factories, the mechanical parts and assembly were dealt with at Bowesfield. The company also owned a former War Department four-wheel battery electric locomotive built by Electromobile for shunting at the works. It went on to construct 20 not very successful 1,200hp diesel electrics for BR (D5700–19) and a number of 93hp underground battery locomotives for the NCB. One other notable product was a gas turbine powered locomotive, 18100, which had been delivered to BR's Western Region, but was later converted at Bowesfield to a conventional electric locomotive.

With the approaching completion of 100 electric locomotives for South African Railways, there was little prospect of future orders and the closure of the Bowesfield Works was announced to its 324 employees. Metropolitan-Vickers-Beyers Peacock Ltd went into voluntary liquidation on 2/2/1961 and this marked the closure of the factory.

Locomotives built at Bowesfield Works

Maker's No.	Date built	Type	Power	Gauge	Customer	Numbers
732-45	1949	BoBoWE	1072hp	Metre	Rede Mineira de Viacao, Brazil	901-14
772-81	1950-51	BoBoWE	1072hp	Metre	Rede de Viacao, Parana-Santa, Catarina, Brazil	2000-09
786-825	1954-55	CoCoWE	3820hp	4ft 8½in	N.S.W.R.G.	4601-40
830-35	1953	2Do2DE	1105hp	3ft 6in	W.A.G.R.	X1001-06
836-59	1954	2Do2DE	1105hp	3ft 6in	W.A.G.R.	X1007-30
860-73	1954	2Do2DE	1105hp	3ft 6in	W.A.G.R.	Xa1401-14
874-75	1956	2Do2DE	1105hp	3ft 6in	W.A.G.R.	Xa1415-16
876-77	1956	2Do2DE	1105hp	3ft 6in	W.A.G.R.	X1031-32
993-1000	1958	CoBoDE	1200hp	4ft 8½in	British Railways	D5700-07
1001-12	1959	CoBoDE	1200hp	4ft 8½in	British Railways	D5708-19
1013-14	1959	4wBE	90hp	3ft 0in	NCB Rothes Colliery	No.10, No.11*
1015-17	1959	4wBE	90hp	3ft 0in	NCB Bilston Glen Colliery	No.1, No.2, No.4
1018-19	1959	4wBE	90hp	3ft 0in	NCB Rothes Colliery	No.9, No.13*
1020	1959	4wBE	90hp	3ft 0in	NCB Bilston Glen Colliery	No.3
1032-1131	1958-60	BoBoWE	2280hp	3ft 6in	S.A.R & H.	E364-E463
1167-68	1959	4wBE	90hp	3ft 0in	NCB Bilston Glen Colliery	No.5, No.6

Notes

* It is possible that these carried Nos.1 to 4 initially.
N.S.W.G.R. New South Wales Government Railways, Australia.
W.A.G.R. West Australia Government Railways.
S.A.R & H. South African Railways and Harbours.
The locomotives built for the NCB were classified as Type DBF12 and, although shown in the listings as 90hp, they were in fact fitted with a pair of 46½hp motors, equivalent to 93hp. All of these NCB locomotives were for use underground.

Reference : 'The Bowesfield Works, Stockton-on-Tees', R.D. Darvill, *Industrial Railway Record* 148, IRS, 1997

Gauge : 4ft 8½in

-		4wBE	Electromobile	W247	1927	(a) (1)

(a) ex Thos W. Ward Ltd, Grays, Essex; previously War Department, Shoeburyness, Essex. It was advertised for sale by Ward in the *Contract Journal*, 3/1949, 9/1949 and 12/1949.

(1) to Richard Garrett Engineering Works Ltd, Leiston, Suffolk, 4/1962. This company was a subsidiary of Beyer Peacock.

MIDDLESBROUGH STEEL, STRIP & HOOP CO LTD

Middlesbrough   **109**
J.F. Pease & Co Ltd until /1904   NZ 498211 – **Map G**
Steel Strip & Nail Co Ltd until c/1899
Atlas Steel Hoop & Wire Rod Works Ltd until c/1895
Isaac Wilson & Co until /1887
Middlesbrough Earthenware Co until /1852
Middlesbrough Pottery Co until /1844

One of the earliest industries in the new town was the Middlesbrough Pottery. Richard Otley, first secretary of the SDR, reached agreement with the OME on 27/1/1834 to purchase 1½ acres of its land on the east side of Holmes' Shipyard (see Northern Gas Board entry). The Middlesbrough Pottery Company was established 1/3/1834 and the first kiln fired in the following month. As part of the deal, the OME agreed to construct a railway (the start of the Vulcan Street line) from the SDR near the Port Darlington coal drops to the pottery. It also said that it would provide a 30 feet wide right of way from the pottery to the river. An 1840 plan shows the railway running from the packet station approach lines, crossing over the shipyard and terminating within the eastern boundary of the pottery. Further east, land was still being reclaimed from the marshes. A separate single line left the centre of the pottery, crossed the above railway at right angles and continued to the end of the pottery jetty on the River Tees. The jetty enabled china clay and flints to be brought in with pottery shipped out; the first order being sent to Gibraltar. By John Dunning's 1856 map, a siding also curved round to enter the eastern side of the pottery premises serving the boiler house. There was a separate narrow gauge line within the pottery for hand-propelled tubs. The most prosperous period was in the early 1850s when 164 adults and children were employed. By 1875, the line from the jetty had been revised to join the curve from Vulcan Street into the east side of the works. The pottery closed in 1887 and was advertised for sale on 17/9/1887.

According to Postgate (1889), the premises had been converted into a "steel strip and mill works". This was the Atlas Steel Hoop & Wire Rod Works Ltd. An agreement between the NER and this company for a water pipe under the Vulcan Street railway was dated 26/4/1892 and its name appears on the 1893 Ordnance Survey. Ward's directories show the factory as occupied by the Steel Strip & Nail Co Ltd in 1896–99 and J.F. Pease & Co Ltd in 1900–03. However, the latter was in financial difficulties in 1903 and the Middlesbrough Steel, Strip & Hoop Co Ltd was registered to acquire the factory in Lower Commercial Street according to the 2/1904 issue of *Machinery Market*.

Under the title "Steel, Strip & Hoop Co Ltd, Middlesbrough", it placed various advertisements in *Machinery Market* concerning locomotives. Possibly the 11 inch side tank locomotive had come with its acquisition of the works and, given the limited extent of its railway system, it was looking to hire it out. Nevertheless, the company still appears to have owned the locomotive in 1916. Later, on 18/6/1920, it was wanting to purchase an 11 or 12 inch standard gauge locomotive because, on 2/7/1920, it had a standard gauge vertical boiler locomotive with 7½in x 14½in cylinders for sale.

The days of the company were numbered as the LNER signed a third party agreement on 28/12/1923 with a Mr T. Oldham in order that he could send out scrap from these premises on to the MOR. He had already forwarded several wagons in the previous month. By 1927, the railway company noted that it had not dealt with any traffic for some considerable time and Oldham had ceased business in the previous year.

References :   *The Pottery that began Middlesbrough*, Mary Williams, C. Books, 1985
                 The Industrial Heart of Old Middlesbrough, J.K. Harrison, CIAS, 2010

Gauge : 4ft 8½in

No.1		0-4-0T	OC	FJ	165	1880	(a)	(1)
	ALLIANCE	0-4-0ST	IC	JF		1871	(b)	(2)
		VB		?	?	?	(c)	(3)

(a)   Inness records this locomotive at J.F. Pease in his lists started in 1911. He describes it as a 12 inch locomotive and was originally intended for delivery to France. It is not known whether this was the 11 inch locomotive offered for hire by Middlesbrough Steel, Strip & Hoop Co on 6/7/1906 in *Machinery Market*.
(b)   origin unknown.
(c)   origin and identity unknown.

(1)   s/s. An 11 inch FJ four-wheel side tank was offered for sale in *Machinery Market* on 24/12/1915 and was still for sale according to a 17/3/1916 advertisement. It may have originally been a 10 inch locomotive but with the cylinders bored out.
(2)   s/s. Inness has a note "This engine at Dton".
(3)   s/s. A vertical boiler locomotive was offered for sale on 2/7/1920 (*Machinery Market*).

MINISTRY OF DEFENCE, NAVY DEPARTMENT

EAGLESCLIFFE SPARE PARTS DISTRIBUTION DEPOT, Eaglescliffe — 110
Admiralty until 1/4/1964
NZ 410148 – **Map D**

This was the former Morris Motors Ltd Metal Produce Recovery Depot (which see). Although the open storage areas to the north returned to nature, the warehouses and railway were used to store and distribute parts for the Royal Navy. The premises generally relied on a couple of diesel locomotives, with one shunting and the other spare. According to a 7/12/1976 BR plan attached to the Private Siding Agreement dated 30/11/1954, the revised arrangements from 1/1/1977 were for the connections between the main line and a parallel siding to be controlled by Urlay Nook Signal Box at the west end and a two lever ground frame at the east end. A line then left this siding, passed through a gate in the security fence before dividing to serve two "'A' sidings, transit shed, engine shed and warehouse, together with the RNSPD distribution sidings. With declining rail traffic, a new road/rail Unimog vehicle was purchased in 1978 to deal with the remaining wagons being exchanged with BR while, on 1/4/1980, RH 375717 (green) and DC 2167 (yellow) stood on a siding by the main line awaiting disposal. The depot eventually closed in 1/1997.

Gauge : 4ft 8½in

-		0-4-0ST	OC	P	2048	1944	(a)	(1)
-		4wDM		RH	224352	1945	(b)	(2)
No.1	YARD No.115	0-4-0DM		JF	22945	1941	(c)	(3)
No.2	YARD No.114	0-4-0DM		JF	22938	1941	(c)	(3)
	YARD No.WD 6692/ YARD No.736 EARL LEOFRIC OF MERCIA	0-4-0DM		RH	375717	1955	(d)	Scr 10/1980
	RISLEY YARD No.106 MED	0-4-0DM		DC	2167	1942		
				VF	4859	1942	(e)	Scr 10/1980
-		4wDM	R/R	Unimog	031047	1978	(f)	(4)

(a) ex Morris Motors Ltd, Eaglescliffe with depot, c/1948-49.
(b) ex Admiralty, Royal Naval Dockyard, Rosyth, Fife, /1951.
(c) ex Royal Ordnance Factory, Risley, Lancashire.*
(d) ex Admiralty, Ditton Priors Munitions Depot, Ditton Priors, Shropshire, c/1965; via Ruston & Hornsby Ltd, Lincoln, c6/1966.
(e) ex Royal Ordnance Factory, Risley, Lancashire, c/1970.*
(f) new, arrived c4/1979.

(1) to Royal Ordnance Factory, Risley, Lancashire, c/1958.*
(2) returned to Admiralty, Royal Naval Dockyard, Rosyth, Fife.
(3) to T. Ottewel & Co, Dewsbury, Yorkshire, (WR), for scrap, c5/1967.
(4) to MODND, Beith, Strathclyde, c12/1992, (possibly went first to MODND, Bedenham, Hampshire), by 3/1/1992.

* This might be the RN Stores Depot at Risley.

MINISTRY OF FUEL AND POWER

The responsibility for these sites passed from the Board of Trade Mines Department to the Ministry of Fuel and Power on 11/6/1942.

COAL DISPOSAL POINT, SEATON CAREW — 111
approx. NZ 520273 – **Map A**

The information about this site is taken from the former Durham Handbook (IRS, 1977) but it has not been possible to find out any more information about it. The depot was established and operated by Sir Lindsay Parkinson & Co Ltd on behalf of the Government during World War 2 and was served by sidings off the LNER Seaton Snook Branch.

Gauge : 4ft 8½in

-	0-6-0ST	IC	MW	1379	1898	(a)	(2)
RUBY	0-6-0ST	IC	MW	1418	1898		
	Rebuilt		HE		1931	(a)	(1)

200	JEANETTE	0-6-0ST	IC	HC	1699	1938	(b)	(3)
	NILE	0-6-0ST	OC	AB	1770	1921	(c)	(4)

(a) ex SLP, ROF Risley Contract, 12/1941.
(b) ex SLP, Dale Airfield, 1941-43 contract, Haverford West, Pembrokeshire, by 1/1943.
(c) ex Pauling & Co Ltd, Park Royal Plant Depot, Middlesex.

(1) to SLP Earlestown Sidings, Lancashire, by 2/1942.
(2) to SLP, Woodkirk, near Dewsbury, Yorkshire (WR), by 8/1943.
(3) to Coal Storage Site, West Hartlepool, by 5/1944.
(4) returned to Pauling & Co Ltd, Park Royal Plant Depot, Middlesex, by /1947.

COAL STORAGE SITE, WEST HARTLEPOOL   112
NZ 510337 – **Map B**

The Greenland area of the docks at the Hartlepools was located between the Timber Ponds at the former Slake and the main line running parallel to Clarence Road. A series of LNER sidings served the sawn wood and pit prop yard and drivers of locomotives on the main line had instructions to avoid the risk of sparks causing fires. The LNER used its class J77 tank locomotives to shunt the area, with J26 and J27 engines hauling timber trains to the Cliff House storage site. During World War 2, the lines in the Greenland area were reorganised to enable the storage of coal and this was operated by Sir Lindsay Parkinson & Co Ltd on behalf of the Ministry of Fuel and Power. According to Stan Wolfe "two quite large industrial tank locomotives were based there". After the war, the storage of pit props was resumed.

Gauge : 4ft 8½in

	JEANETTE	0-6-0ST	IC	HC	1699	1938	(a)	(1)
No.1	RISLEY	0-6-0ST	IC	HC	1606	1929	(b)	(2)

(a) ex Coal Disposal Point, Seaton Carew. It was noted at West Hartlepool in 5, 8 and 9/1944.
(b) stored at SLP Plant Yard, Winwick, Lancashire, 2/1945. At timber and pit prop yard, West Hartlepool, 29/7/1945.

(1) at Hudswell Clarke being converted to oil firing 6/1945.
(2) at SLP Temple Newsham Store, Leeds, by 8/6/1947.

MORRIS MOTORS LTD

METAL PRODUCE AND RECOVERY DEPOT No.2, Eaglescliffe   113
(Operated on behalf of the **Ministry of Supply**)   NZ 410148 – **Map D**

The depot was established by the Government during World War 2 to recover aluminium from crashed aircraft in the north of England, the remains of which were brought by rail and road. It comprised a series of large warehouses with an extensive area of open land to the north west of the buildings divided into rectangular storage areas separated by roads and presumably used for storage of the wreckage. A rail connection was provided off the Darlington-Middlesbrough line, east of Urlay Nook Signal Box. From here, some reception sidings were situated parallel to the main line and two railway tracks curved round to run either side of the block of warehouses. The engine shed was located on the southern group of sidings. According to a Parliamentary reply in 7/1947, agreement in principle had been reached in 9/1946 for Morris Motors to take over the factory but Ministry of Supply work was not expected to decline before the early part of 1948. In the event, it closed and the premises were taken over by the Admiralty (see Ministry of Defence, Navy Department).

Gauge : 4ft 8½in

-	0-4-0ST	OC	P	2047	1943	New	(1)
-	0-4-0ST	OC	P	2048	1944	New	(3)
-	0-4-0ST	OC	P	2049	1944	New	(1)
-	0-6-0T	OC	Dav	2505	1943	(a)	(2)

(a) ex War Department, number 1940, previously U.S. Army Transportation Corps, 6/1947.

(1) to Morris Motors Ltd, Metal Produce Recovery Depot No.1, Cowley, Oxford, /1948.
(2) to Austin Motor Co Ltd, Longbridge Works, Birmingham, per Abelson, 10/1949.
(3) to Admiralty, with depot, c/1948-49.

NEWPORT ROLLING MILLS LTD

NEWPORT ROLLING MILLS, Middlesbrough
John Hill & Co until /1909
Fox, Head & Co until 5/1888

114
NZ 484206 – Map F

The works, erected by Theodore Fox and Jeremiah Head, was reported completed and nearly in full operation in October 1864; Fox was the brother-in-law of J.W. Pease. Jeremiah Head, a Quaker, had served an apprenticeship at Robert Stephenson & Co's Newcastle upon Tyne works and collaborated with John Fowler on the development of the steam plough. In 1863 Fox and Head concluded that the manufacture of boiler and ship iron plates could be a lucrative enterprise. The works was located immediately north of the Newport Iron Works in the Ironmasters district. Road access was initially via a level crossing over the Old Town and Marsh Branches from North Road, but Metz Bridge was subsequently provided over these railway lines. The premises comprised 30 puddling furnaces – the number varied between 40 and 46 up to 1880 – for producing wrought iron, together with rolling mills said to total four in 1880.

It soon faced difficulties due to a puddlers' strike in 1866 and, later that year, it was transformed into a co-operative with any financial surplus being divided in two equal parts – one for the capitalists as their profit and the other to those receiving wages. Although the 1870s witnessed a significant decline in the use of iron for rails, there remained a strong demand for iron plates. A Middlesbrough Directory for 1874 stated that the works occupied 22 acres, employed about 550 people and manufactured boiler, bridge and ship iron plates. The company continued to rely predominantly on manufacturing iron plate but, in 10/1885, it was operating at a loss and closed. In 5/1888 the works was leased by John Hill & Co, which subsequently purchased and remodelled it with production restarting on 16/7/1888. In 1893 the works was stated to have 50 puddling furnaces with an output of 800 tons per week of both iron and steel plates for ships and bridges.

A series of railway sidings ran into the east side of the works from the Marsh Branch next to the numerous lines to the Newport Iron Works. All rail traffic for the NER would have entered and left by this route. Sidings also came into the west side of the works and these connected with the Newport Iron Works; presumably this was a source for some of its iron. Fox, Head & Co appear to have had two locomotives in 1883 and advertised a 5½ inch geared "coffee pot" locomotive for sale in *Machinery Market* on 2/4/1883 which had recently received a new vertical boiler. John Hill & Co is in a list, dated 5/3/1897, of owners possessing private locomotives working over the NER and MOR lines. The types of charges levied for this activity were;

By 1908, the Newport Rolling Mills belonging to John Hill & Co were said to have the only iron plate rollers remaining in the Middlesbrough district. (IRS J.A. Peden collection)

Marsh Branch *MOR, Wayleave*
Vulcan St and Middlesbrough Dock *NER/MOR Carriage, Wayleave*
Traffic over "Independent" to Council's Wharf *NER Carriage, Wayleave*

No charge was made for working over NER lines from the Newport Rolling Mills to the Marsh Branch sidings with traffic for the main line and ballast for the permanent way department. An agreement with the NER dated 31/1/1902 allowed John Hill & Co to run its "locomotive steam crane" on a NER's siding adjacent to the Newport Iron Works, provided that the jib of the crane was hoisted and not used for any purpose. The agreement was cancelled 10/1910.

The market for iron plates declined as steel was increasingly used. In 1908, during a visit by the Iron & Steel Institute, John Hill & Co was said to possess "the only iron plate rollers in the Middlesbrough district". The plant then consisted of 35 puddling furnaces, five heating furnaces, one plate mill and one forge train. According to the prospectus for the Newport Rolling Mills Ltd (issued 15/12/1909), the company had been formed to take over the rolling mills. The plant was stated to then include three steam jib travelling cranes (7, 5, and 3 tons), two 12 inch locomotives and about two miles of railway sidings. A new agreement with the NER on 4/8/1910 allowed Newport Rolling Mills Ltd's locomotives and "steam locomotive crane" to traverse some of the NER's lines but this was cancelled on 20/11/1914. The rolling mills were being dismantled in 1914 and plant, including the two 12 inch locomotives, was offered for sale.

References : The Co-operative Scheme of Messrs Fox, Head & Co, Middlesbrough, 24/2/1872 (Meeting to report on 1871 accounts)

'The Lost Village of Fox Heads', Jenny and Geoff Braddy, *Cleveland Industrial Heritage* 26, pub. Peter Tuffs, 2010 pp 30-34

Gauge : 4ft 8½in

No.1	0-4-0ST		?			(a)	s/s
F.H. & Co. No.2	0-4-0ST	OC	BH	236	1873	New	s/s
	VB		?			(a)	(1)
No.1	0-4-0ST	OC	HL	2431	1899	New	(2)

(a) origin & identity unknown.
(1) advertised for sale in *Machinery Market*, 2/4/1883.
(2) to Sir Hedworth Williamson's Fulwell Limeworks Ltd, Fulwell and Southwick Quarries and Limeworks, Fulwell, Sunderland, Co. Durham, by 12/1914.

NORMANBY IRON WORKS CO LTD

NORMANBY IRON WORKS, Cargo Fleet   **115**
Pease & Partners, Normanby Iron Works Co Ltd until 7/10/1953 *   NZ 518208 – **Map G**
Normanby Iron Works Co Ltd until 15/7/1947 (subsidiary of **Pease & Partners Ltd**)
Normanby Ironworks Co Ltd until 1/7/1911
Jones, Dunning & Co Ltd until /1895
Jones, Dunning & Co until /1888

* This was the title adopted when Pease & Partners Ltd became a holding company but it was not a registered title.

The Normanby Iron Works was situated on a rectangular site north of the Middlesbrough-Redcar railway to the east of Normanby Beck. Prior to its construction, this area had been mainly mudflats adjoining the River Tees. To the west, land had been reclaimed for the Ormesby Iron Works. On the eastern boundary was Cargo Fleet Point with its corn mill and the SDR "Cleveland Port" railway station; the original location of Cargo Fleet Station before its move west. Edwin Jones (latterly blast furnace manager at the Ormesby Iron Works) and John Dunning (OME's land surveyor) began constructing three blast furnaces in 1859, but there were delays and they were not blown in until 1861. Jones only designed the furnaces to be 60 feet high, rather than his preferred 65 feet, because the patterns for the 15 feet long columns used in Belah railway viaduct were still available for the castings to make the furnace hoist supports. By 1887, the three blast furnaces were producing pig iron which was mostly shipped to Grangemouth. The works had a capacity of 1,150 tons per week although, in 1893, it was also making 600 tons of haematite iron each week using Spanish ore, by which date the blast furnaces had been increased to 75 feet in height.

The blast furnaces stood in a row approximately parallel to the NER's line with the calcining kilns and mineral stores, together with the exchange sidings, in the intervening space. Incoming wagons were shunted past a "weigh office" to a hoist for the gantry over the kilns and stores. A steam ramp operated from a jacketed cylinder in the ground to lift the wagons (each weighing about 15 tons) to the gantry. The pig beds were north of the blast furnaces and two fans of sidings facilitated the removal of the iron and slag. There was space on the remainder of the site initially for dumping slag and railway lines ran across this area serving a "slag breaking works", although later slag was tipped into the sea using hopper barges. The pig iron could be hauled in wagons by locomotives to the exchange sidings or pulled along to the company's river wharf. In 1893 the wharf was equipped with three steam cranes of 7, 5, and 3 tons respectively.

By T.M. Smith's map of 1882, Cargo Fleet Station had been relocated to the west of the Normanby Iron Works, opposite the Deepwater Wharf site, but was to move west again to its final location next to the Ormesby Iron Works in 1885.

As some of its land for the iron works had been sold to Jones, Dunning & Co, OME had retained the wayleave for all rail traffic entering and leaving the works; hence MOR tolls had to be paid. OME also extended its line from Dock Lock Bridge past the Ormesby Iron Works and over Miss Brown's land (later Deepwater Wharf site) to reach the Normanby Iron Works. Here it was laid in a broad sweep across the northern part of the iron works site, presumably because lack of space or land ownership reasons blocked a route alongside the main line. MOR tolls were paid for traffic on this line. According to an 1897 list, Normanby Iron Works locomotives were hauling wagons over the MOR line to the Ironmasters district. A subsequent LNER memorandum (W/T1443) stated "The whole of the M.O. line passing through Normanby Ironworks was single track but we have the rights to lay double track if we wish. Certain curves on this line prohibited the use of our engines".

In 1895, the Normanby Ironworks Co Ltd was formed as a private company to acquire the business. However, in 1900, a public company was established with the same name to operate the iron works. The Engineers report in the prospectus dated 20/7/1900 for this company included reference to the following being taken over: five locomotives, four steam cranes, 35 railway wagons and two hoppered slag barges each capable of carrying 250 tons. Information on two of the early locomotives is lacking, although Inness' records contain a pencil note "2 Coffee Pots" so it is possible that these were locally built vertical boiler engines. Another option concerns the contractor, S. Pearson & Son, which used a locomotive named NORMANBY, built by Hopkins Gilkes with 12in x 20in cylinders, on the construction of the Bentinck Dock at King's Lynn in 1883.

Although officially identified as the Normanby Iron Works Co Ltd, the plant was a subsidiary of Pease & Partners and hence the inscription on Robert Stephenson 2748 of 1892. (NERA R. Inness photograph)

WASP, with its unusual combined vertical and horizontal boiler, was at the Normanby Iron Works on 5th July 1923. *(NERA R. Inness photograph)*

An additional blast furnace to the original three was in course of construction in 1906. The number of furnaces was reduced to three in 1921 and this remained the total up until closure, although only one was operational in 1925. The works concentrated on producing refined and malleable haematite iron; Warner & Co Ltd being one of its customers in 1903–05. The Normanby Iron Works had the reputation for many years of being the oldest fashioned on the River Tees. The works continued to rely on its sales of pig iron and, with one or two exceptions, there was little development of ancillary plant. More sidings had been laid over the northern 'open' part of the site by 1913 and the original engine shed had been replaced. Following Pease & Partners Ltd takeover of the works, discussions took place with the NER on 30/10/1913 for Pease to acquire the MOR line through the northern part of the site so that it could extend the river wharf and build coke ovens. It was intended to divert the MOR line alongside the NER main line but the idea was abandoned because of lack of space and objections from Lord Furness' interests. Nevertheless, improvements took place at the wharf; the new berth was stated in 1926 to be 450 feet long and served by three 5 ton electric cranes. At some date it was also connected to the stocking grounds and blast furnaces by overhead conveyor belts. There were still ideas of moving the MOR's alignment a little further south, with which the NER agreed as some of the curves were "rather sharp". Parliamentary approval for the diversion was obtained in 1921 and the agreement between the two parties signed on 22/9/1924. According to the South Bank Yardmaster, Pease & Partners Ltd had already made a start on the diversion, but it was not completed due to the "state of trade". Mr A.S. Dewhirst, Terminal Rate Clerk at Middlesbrough, walked the line in 1934 and commented – I found the "old slag line" blocked with private wagons loaded with slag, two of the trucks in fact being off the rails. There is also a portion of line adjacent to the defunct Cargo Fleet Salt Works which is disconnected and I understand this was done several years ago in connection with the erection of a bridge nearby. The salt works was later taken over by Foam Slag (Tees-side Production) Ltd about 1946 which produced lightweight building and insulating materials by mixing slag with sand and cement. Rail traffic for Foam Slag travelled via the LNER, MOR and iron works lines and the address of the company was given as Normanby Iron Works.

The LNER identified on 17/10/1946 the annual receipts from MOR tolls associated with Normanby Iron Works; –

- Traffic from the ironworks westwards as far as Dent's Wharf over the MOR. Annual average over the last seven years £36
- Traffic to and from the iron works via the MOR/LNER £405

- Iron ore Normanby Wharf to Skinningrove Iron & Steel Works £5.13s 0d
- "Foam Slag (Tees-side) Production Ltd" [sic] (18,000 tons pa) £18.15s 0d
- Deepwater Wharf to and from the iron works. There had been no traffic for several years but since Pease & Partners had acquired the wharf, 10,413 tons had passed in the first half of 1946. £65
- Cleveland Slag Roads Ltd. A nominal amount because it had only recently taken possession of a siding at the iron works. £5

it was not possible to substantiate any tolls due from internal traffic between Normanby Wharf and the iron works. From the total receipts, a sum of £61 had to be taken off for wayleave payments to Deepwater Wharf.

Normanby Iron Works employed a fleet of four-coupled saddle tank locomotives from various makers. Members of The Industrial Locomotive Society had been taken round the works on 16/4/1954, when they were informed that the stock in 1915 comprised numbers 1 to 6 and JOHN. They were informed that previously CLARENCE had been loaned from Ridley Shaw and SODIUM from OME.

After the brief nationalisation of the iron and steel industry was reversed in 1951, the Government established the Iron and Steel Holding and Realisation Agency to dispose of the works back to the private sector. While the major works soon moved back, smaller plants, such as the Normanby Iron Works, did not. Therefore, although a nominally independent company, its share capital was owned by the state. On 8/7/1959 the Agency announced that the works was to close and the blast furnaces were taken out of blast on Friday 24/7/1959. In answer to criticism from the local MP, the Government stated that the plant was obsolete; the three blast furnaces had been last rebuilt in 1900, 1906 and 1923 respectively and production was running at only 60% of capacity; there being an excess in the haematite iron industry. The works was for sale by order of the liquidator in 3/1960 and George Cohen, Sons & Co Ltd proceeded to demolish the works in 1961–62. Scrap was forwarded to Cleveland and Lackenby Steel Works with the remaining stocks of iron ore and limestone being sent to the Acklam Iron Works; MOR tolls still having to be paid!

Gauge : 4ft 8½in

	NORMANBY	0-4-0ST	OC	MW	123	1864	New	s/s
	WASP	0-4-0VBT	OC	GW	169	1863	(a)	s/s
	BEE	0-4-0VBT	OC	HW		1871	(a)	s/s after 5/7/1923
	NORMANBY No.4	0-4-0T	OC	K	1790	1872	New	(1)
	NORMANBY No,5	0-4-0ST	OC	K	2237	1878	(b)	s/s
No.5	(6)	0-4-0ST	OC	RS	2748	1892	New	(4)
No.3		0-4-0ST	OC	AE	1352	1896	New	(7)
(1)		0-4-0ST	OC	RS	3045	1901	New	(6)
No.5	ERIMUS	0-4-0ST	OC	YE	236	1874	(c)	(2)
No.6	(NORMANBY)	0-4-0ST	OC	HL	2870	1911	New	(8)
	JOHN	0-4-0ST	OC	AB				
		Rebuilt				1902	(d)	(3)
7		0-4-0ST	OC	AB	1597	1918	(e)	(8)
(8)		0-4-0ST	OC	KS	4144	1919	(f)	(8)
	VICTORIA	0-4-0ST	OC	LE	240	1900	(g)	(8)
(No.2)	KILMARNOCK	0-4-0ST	OC	B	+			
		Rebuilt	AB		7571	1907	(h)	(5)
	QUEEN MARY	0-4-0ST	OC	AB	1606	1918	(i)	(8)
	QUEEN ELIZABETH	0-4-0ST	OC	AB	1602	1918	(j)	(9)
	KING GEORGE V (MERCURY)	0-4-0ST	OC	AB	1317	1914	(k)	(8)

+ Russell Wear suggests that this may be maker's number 231 of c/1876.

Inness suggests that a 0-4-0 saddle tank named CASTELL constructed by John Harris of Darlington was here. He stated that it was ex Tees Iron Works but it does not appear in his list for the latter place.

(a) ex Pease & Partners Ltd, Tees Iron Works, Cargo Fleet; WASP prior to 5/7/1923 and BEE after 5/7/1923. Both were fitted with a combined horizontal and vertical boiler.
(b) new, delivered 12/4/1879.
(c) ex T.D. Ridley & Sons, Middlesbrough, possibly on completion of contracts on Tyneside.
(d) origin unknown.
(e) new. On ILS visit in 1954, it carried an AB 1597 plate on one side and AB 1606 on the other!

Iron making ceased at the Normanby Iron Works in 1959 and Lowca Engineering 240 of 1900, seen here on 19th June 1960, was scrapped soon after. (IRS J.P. Mullett photograph)

(f) new, a locomotive built for stock by Kerr, Stuart.
(g) ex Pease & Partners Ltd, Tees Iron Works, Cargo Fleet, c/1929.
(h) ex Pease & Partners Ltd, Lackenby Iron Works, Grangetown, c/1930. It was noted at Lackenby Iron Works in 7/1927 as 3 KILMARNOCK.
(i) ex Pease & Partners Ltd, Tees Iron Works, Cargo Fleet.
(j) ex Air Ministry, Farnborough, Hampshire, late 1948. On arrival, carried plates "R.A.F. LOCO. No.4" and "A.1873". Name fitted c1949/50.
(k) ex Glasgow Iron & Steel Co Ltd, Wishaw, Lanarkshire, 9/1948. It had been noted at Wishaw in 10/1947. It came as MERCURY and received its new name by 3/1950.

(1) to Pease & Partners Ltd, Fine Burn Quarry, near Frosterley, Co. Durham.
(2) to T.D. Ridley & Sons, Middlesbrough.
(3) to Pease & Partners Ltd, Tees Iron Works, Cargo Fleet, /1921.
(4) scrapped c/1933.
(5) scrapped c/1951; after 27/5/1950.
(6) scrapped by 11/1951.
(7) scrapped c4/1956. It had been advertised for sale in the *Contract Journal* on 15/7/1953 but there was no interest and it stood derelict by /1955.
(8) scrapped c/1960; AB 1597, LE 240, KS 4144 after 1/5/1960.
(9) scrapped c/1962.

NORTH EASTERN IRON REFINING CO LTD

STILLINGTON WORKS, Stillington

116
NZ 372237 – Map D

The company was established in 1926 and erected two cupula furnaces at the former briquetting building on the Carlton Iron Works site, then owned by Dorman, Long & Co Ltd (which see). The North Eastern Iron Refining Co Ltd took over the production and marketing of a specialised refined pig iron that had been made previously by Seaton Carew Iron Works. Pig iron and scrap were melted in the furnaces to produce 'refined'

North Eastern Iron Refining's AB 352/1941 stands among the degenerating buildings at the former Carlton Iron Works on 25th July 1979. (Cliff Shepherd photograph)

iron that was suitable for particular types of castings carried out by small foundries, especially in the West Midlands. Annual output varied between 2,200 and 4,200 tons in the 1930s. It is likely that initially Dorman Long's Carlton Iron Works locomotives hauled rail traffic between the exchange sidings, the works and the slag tip until the first diesel was obtained in 1937. During World War 2, the furnaces were also used for reheating imported ferro-silicon and steel scrap to produce an alloy used in steel production. After the war, the original cupulas were replaced by three new, but similar, furnaces, together with a casting machine. Electric furnaces were installed in 1957–58.

In 1948 the company established Stillington Estates Ltd, with the same directors, to manage the former iron works site and the remaining private railway lines, but not the refinery. By the 1970s, there had been a significant reduction in the demand for 'refined' iron and, by then, the company only employed about 30 people. The original iron works rail connection with the Ferryhill–Stockton line had been retained but, by this date, the system comprised only a single track running round to a headshunt. From here, one siding entered the corrugated iron buildings housing the furnaces, whilst another disappeared into the long grass near the remnants of the iron works. During a visit on 25/7/1979, the Fowler was the working locomotive, RH 312427 stood out of use near the slag heaps and the diminutive Barclay was off the rails in a gloomy cavernous empty building. Rail traffic ceased in 12/1982.

References : *Stillington in the Borough of Stockton-on-Tees, Industry, Church and School 1860 to 1991*, Parish Map Community Programme Project, 1991

There was a Green Hill, The History of Stillington from its Beginnings until 1950, J.D. Tuffs, pub. by the author, 1999

Gauge : 4ft 8½in

-	4wDM	RH	187071	1937	New	(1)
2	4wDM	RH	279593	1949	New	(1)
No.3	0-4-0DM	AB	352	1941	(a)	(2)
-	4wDM	RH	312427	1951	(b)	Scr 6/1987
-	0-4-0DH	JF	4220027	1964	(c)	Scr 1/1988

(a) ex War Department, Royal Ordnance Factory, Birtley, Co. Durham, 2/1962.
(b) ex Head Wrightson & Co Ltd, Eaglescliffe, 9/1967.
(c) ex British Steel Corporation, Hartlepool South Works, Hartlepool, 4/1979.

(1) to M. Henderson Clark Ltd, Cargo Fleet, 11/1968.
(2) to Market Overton Industrial Railway Association, Cottesmore, Rutland, 29/10/1980, for preservation.

NORTH-EASTERN ELECTRIC SUPPLY CO LTD

GRANGETOWN POWER STATION, Grangetown 117
Cleveland and Durham County Electric Power Co Ltd until 30/9/1932   NZ 553219 – **Map H**

The Cleveland and Durham County Electric Power Co was incorporated by Act of Parliament in 1901. By 1906, it had erected power stations at Bishop Auckland and Consett, with a head office situated at Queens Square, Middlesbrough. Plans were also underway to construct a power station at Grangetown served by a siding from the Middlesbrough-Redcar railway line, a short distance north east of Grangetown Station. The agreement between the NER and Power Company was dated 9/9/1905, following approval by the NER Way and Works Committee on 18/7/1905. This siding curved round past the main building to a headshunt, from which two railway lines led to the power station. A line alongside the boiler house served a "coal receiver and elevator". According to *Tramway & Railway World* 14/9/1905, the order for the locomotive had been placed by the company's consulting engineers, Bramwell & Harris. Electricity generation started in 12/1906 and it was one of the earliest power stations in the world to generate at 11,000 volts, which became the standard for many years. By late 1907, it had between four and six times the installed capacity of the municipal power stations at Middlesbrough and Stockton. It tended to supply large industrial users, including some of the ironstone mines. The Newcastle upon Tyne Electric Supply Co Ltd was looking to extend its sphere of operations to Teesside and, from 1917, owned all of the Cleveland company's shares. With demand increasing, a new 'baseload' power station was commissioned at North Tees in 3/1921 (which see). The North-Eastern Electric Supply Co Ltd took over both the Cleveland and Newcastle companies on 30/9/1932. Grangetown Power Station closed in 1937, but the siding agreement was not taken over by Dorman Long until 1/4/1958 and the generating building was not demolished until 1969 to make way for the British Steel Corporation's Continuous Casting Plant.

Reference :   'Early Electricity Supplies on Teesside', D.W. Pattenden, *The Cleveland Industrial Archaeologist* 17, 1995, pp.15-20

Gauge : 4ft 8½in

| | | 4wBE | BT | | 1905-06 | New | (1) |

(1) s/s. One possibility is that the locomotive was transferred to the North Tees Generating Station but no evidence has yet been found.

NORTHERN GAS BOARD

MIDDLETON ROAD GAS WORKS, West Hartlepool 118
Hartlepool Gas & Water Co Ltd until 1/5/1949   No.1 Works NZ 510334 – **Map B**
No.2 Works NZ 508342

The first gas works in Hartlepool was erected as a private venture by James A. West between Commercial Street and Northgate in 1836 near the recently opened Hartlepool Dock & Railway Co's terminus. Under an Act of Parliament dated 26/6/1846, the Hartlepool Gas & Water Company (HGW Co) was established and purchased West's gas works which continued to operate until 1880. Meanwhile a small gas works had been erected south of Middleton Road to supply the dock premises and this was taken over by the HGW Co in 1848. Between this date and 1856, the HGW Co constructed what became known as the No.1 Gas Works on the north side of Middleton Road and the dock facility was demolished. In 1866 the Middleton Road Gas Works had 190 circular retorts and three gas holders. Initially, the works was connected to the NER dock line that passed under Middleton Road to reach the subsequent Greenland Timber Storage Yard. From this line, a siding reversed into the gas works. By 1896 another NER dock line ran alongside the main line and also passed under Middleton Road to enter the other end of the gas works premises. The Middleton Road Gas Works occupied a constricted site relying on several wagon turntables but the two sets of sidings do not seem to have been connected within the premises. As its name suggests, the company also supplied the Hartlepools with water and it had a pumping station on the opposite side of the main line to the Middleton Road Gas Works, although it was not rail served.

In view of the constricted site, the HGW Co erected another works on 20 acres (later 25 acres) of land at Dyke House Farm in 1900 comprising a carburetted water gas plant with a relief gas holder. Known as No.2 Gas

Works, it was served by a lengthy siding that ran from the north end of the NER's Greenland Timber Storage Yard, passing under the main line, to travel by Montague Road to the works. An agreement had been signed with the NER for the new railway involving removal of part of an embankment as early as 6/2/1891.

Reference : *History of the Hartlepool Gas & Water Company 1846-1930*, Hartlepool Gas & Water Co Ltd

Gauge : 4ft 8½in

HURWORTH	0-4-0ST OC	BH	306	1875	(a)	
	Rebuilt	Ridley				(1)
-	4wDM	RH	235512	1945	New	(2)

(a) new. Locomotive was actually ordered for stock on 17/10/1873 but was not sold to the Hartlepool Gas & Water Co Ltd until 19/2/1875.

(1) to Anderston Foundry Co Ltd, Port Clarence, c/1947.

(2) to Stockton Gas Works, Stockton, /1962.

MIDDLESBROUGH GAS WORKS, Middlesbrough 119
County Borough of Middlesbrough Gas Department until 1/5/1949 NZ 498210 – **Map G**
The Owners of the Middlesbrough Estate until /1856

Gas lighting began to be used in the UK during the first two decades of the nineteenth century. Gas was being produced by the SDR in 1833 to illuminate the Port Darlington coal drops, but a gas company was not established by the OME in Middlesbrough until 1838. The gas works was located on the south side of Lower Commercial Street opposite the Middlesbrough Pottery. In 1856 the premises contained a retort and purifying house, a "coal shed" and two gas holders. Under the Middlesbrough Improvement Act (received royal assent 7/7/1856), the Corporation was empowered to purchase the gas company, the OME having agreed to sell the 3.63 acre site on 9/11/1855. Various improvements continued to be made as gas usage increased. Another gas holder was ordered from Gilkes, Wilson & Co in 9/1857 and, by 1866, a condenser retort house, purifier and 20 ton weighbridge had been added. The site was extended to Bath and Gas Streets. In 1885, tenders were accepted for the construction of a retort house and "coal shed" in the yard next to the No.3 gas holder representing a 30–40% increase in retort capacity. A coke handling plant was also erected in 1905.

Middlesbrough Gas Works' Ruston & Hornsby 432480 of 1959 had come out on to the Vulcan Street railway on 25th July 1962. The slag wall marked the southern boundary of the Tyne-Tees Wharf.

(Courtesy David Webb)

The railway connection to the gas works does not appear on a plan until 1875. This left the Vulcan Street railway near the latter's crossing of Ferry Road and curved round past the former Holmes' sail loft and over Lower Commercial Street to enter the premises. A number of tracks within the gas works are shown on the 1893 survey for the Ordnance Survey plan, but the site was quite constricted and initially horses may have been used for shunting. In 1902 the Tees Union Shipping Company complained that the wagons of coal for the Corporation Gas Works were being placed (and had been for some time) on sidings at Vulcan Street that had been used exclusively for Tees Union traffic for many years. According to the NER on 8/5/1906, it delivered wagons for the gas works to a point in Vulcan Street, 67 chains from Old Town Junction, from where it was dealt with by a private locomotive.

There was another factor that limited both the development of the works and amount of rail traffic. Middlesbrough Corporation signed an agreement on 14/12/1912 with Sir B. Samuelson & Co Ltd under which the latter would extend its coke oven plant at Newport Iron Works and supply 1.6 million cu ft of spare gas a day beyond its own needs. A 20 inch gas main was laid from Newport to the gas works and the expanded coke ovens began production in 1914. A new agreement was signed with Dorman Long, now owners of Newport Iron Works, in 10/1925. A total of 17,257,344,000 cu ft of coke oven gas was supplied between 1912 and 1934 to the Corporation saving 1½ million tons of coal per year. Additional gas holders were erected on separate sites at Snowdon Road and Cannon Street (gas holders commissioned 28/10/1926 and 11/8/1936). By 1930 the Middlesbrough Sand & Gravel Company also had use of a siding at the gas works and it, as well as the Corporation, paid MOR tolls for traffic on the Vulcan Street lines. On 30/7/1938 Dorman Long announced the impending closure of Newport coke ovens but, in 2/1939, agreed to supply gas from its Cleveland Works via a 5 mile long main to the Lower Commercial Street works.

By the end of World War 2, the continuing demand for gas threatened to outstrip the supply, despite the substantial but fixed amount coming from Dorman Long. On 26/11/1946 Messrs West's Gas Improvement Co Ltd's tender of £307,024 for a new carbonising plant at the gas works was accepted by the Corporation's Gas and Electricity Committee. It also agreed that an "oil engine locomotive" be purchased at an estimated cost of £3,600 and tenders were sought. At the committee meeting on 29/4/1947 approval was given to Ruston & Hornsby's tender to supply a new diesel locomotive at a price of £6,450 less 2½%. The committee had been told earlier that "At present this work is carried out for us by a nearby firm at a charge for each truck handled". This was T. Roddam Dent and the Corporation had paid its shunting charges since at least 1933. However Dent had given notice that it wished to terminate its contract by the end of 5/1947. At a meeting of the businesses affected by Dent's decision, it had been agreed to arrange for a joint hiring or purchase of a locomotive. As this had not been successful, the 29/4/1947 committee meeting also agreed to the purchase of a secondhand steam locomotive for £375 from the Grantham Boiler & Crank Co Ltd, subject to a satisfactory inspection, as an interim measure. Nothing more was heard of this proposal and T. Roddam Dent continued to carry out shunting for the gas works presumably until the new diesel locomotive was delivered on 11/1/1949.

The gasworks was nationalised in 1949 and became the responsibility of the Northern Gas Board. According to a plan dated 7/9/1953 and agreement of 8/7/1954, BR was to provide a siding off the Vulcan Street branch into the salt works site opposite for traffic to be conveyed between the gas works and the "Trader's premises at Cleveland Salt Works. This was terminated in 6/1966. With the changeover to natural gas, rail traffic to the gas works ceased about 1970, the agreement with BR being terminated on 1/7/1970 and the works closed in 1973.

References :   *A History of the Middlesbrough Gas Works 1838-1926*, Preston Kitchen, County Borough of Middlesbrough, 1926

The History of Middlesbrough, William Lillie, Middlesbrough CB, 1968

Proceedings of Middlesbrough Town Council, Minutes of Gas and Electricity Committee, various dates

Gauge : 4ft 8½in

-		0-4-0DM	RH	252685	1949	New	(1)
-		4wDM	RH	432480	1959	New	(3)
-		4wDM	RH	235512	1945	(a)	(2)

(a)   ex Stockton Gas Works, /1965.

(1)   to Howden Lane Gas Works, Willington Quay, Northumberland, 20/3/1959.
(2)   s/s c/1966.
(3)   to Golightly (Developments) Ltd, (acting as dealers), Ferryhill, Co. Durham, c10/1970; which resold it to Doxford & Sunderland Ltd, Wolsingham Works, Co. Durham, 11/1970.

STOCKTON GAS WORKS, Stockton — 120
Stockton-on-Tees Gas Co until 1/5/1949 — NZ 443198 – **Map D**

The original gas works in Stockton opened on the corner of Yarm Road and High Street in 1822. The Corporation took over this works in 1857 and it operated as the Stockton-on-Tees Gas Company. A larger gas works was opened at Thompson Street in c1860 and the original site was vacated to be replaced by a courthouse. The gas works was situated to the east of the Yarm-Norton railway, just to the north of Stockton Station. A siding left the North Shore Branch, immediately south of North Shore Junction and ran parallel to the main line before dividing into two; one reversed to enter the Corporation's store yard while the other continued on to the gas works. Here it passed two gas holders next to Kirby Street before running past the end of Oxford Street to reach the gas product plant.

By the 1920s, the works relied on its two locomotives, BULLER and EVELYN JEANIE. The latter had been overhauled in 1924 but, by 1930, it required extensive repairs. The locomotive was 55 years old and practically all of the coal was now received in 20 ton wagons so a more powerful engine was required. This took the form of an ex NER class H 0-4-0 side tank, number 945, which required alterations to the chimney, safety valve, whistle and cab to enable it to run along the coal gantry. The cost was £250 (purchase), plus £37 for the alterations and £25 for painting. In 1945, a Council committee agreed that this locomotive should be sent for overhaul at the LNER's Darlington Works. It was also to be named WILLIAM BROWN in view of the valued assistance given by Councillor Brown. The Corporation was looking to hire a locomotive from the railway company while the overhaul was taking place.

The gas works was nationalised in 1949 and became the Northern Gas Board's responsibility. The increasing use of natural gas resulted in the threatened closure of the Stockton Works in the 1960s, but it was reprieved for a short while to supply coke as a smokeless fuel to coal merchants on Teesside. In 10/1970 production had been doubled to 800 tons of coke per week, but the works closed in 3/1971. Much of the premises were then demolished, although two gas holders off Albert Street were not dismantled until 1985.

References : *Proceedings of Stockton Borough Council*, Minutes of Gas Committee, various dates

'A Shunter for Stockton Gas Works', Geoffrey Horsman and John G. Teasdale, *North Eastern Express* 180, NERA, 11/2005

Gauge : 4ft 8½in

EVELYN JEANIE	0-4-0ST	OC	MW	539	1875	New	(1)
BULLER	0-4-0ST	OC	P	904	1901	(a)	(2)
WILLIAM BROWN	0-4-0T	IC	Ghd		1888	(b)	Scr 8/1953

The former NER Class H locomotive, built at Gateshead in 1888 and now named WILLIAM BROWN, poses in front of Stockton Gas Works on 30th August 1947.   (L.W. Perkins photograph, courtesy F.A. Wycherley)

	0-4-0ST	OC	RSHN	7308	1946		(3)
		Rebuilt	Ridley Shaw		1953	(c)	(3)
NORTHERN GAS BOARD No.1	0-4-0ST	OC	P	2142	1953	New	(4)
-	4wDM		RH	235512	1953	(d)	(5)
-	0-4-0DE		EES	8453	1964	New	(6)
-	4wDM		RH	305320	1951	(e)	(6)

(a) ex Derwent Valley Water Board, Howden Reservoir, Derbyshire, 11/1913 (date of sale).
(b) ex LNER, 945, 7/1930.
(c) ex Hawthorn Limestone Co Ltd, near Seaham, Co. Durham, via Ridley Shaw, /1953.
(d) ex Hartlepool Gas Works, Hartlepool, /1962.
(e) ex Darlington Gas Works, Darlington, Co. Durham, /1964.

(1) advertised for sale, presumably as scrap, 11/1930. Tender of Mr M. Hay of Stockton for £48 approved but then he withdrew and the next tender was accepted, but it was not scrapped until /1934.
(2) derelict in /1954 and s/s c/1956.
(3) to T.J. Thomson & Son Ltd, Stockton, for scrap, /1964.
(4) to Darlington Gas Works, Darlington, Co. Durham, c9/1964, after 19/8/1964.
(5) to Middlesbrough Gas Works, Middlesbrough, /1965.
(6) scrapped on site by Golightly (Developments) Ltd, Ferryhill, Co. Durham, 4/1972.

NORTON IRON CO LTD/THE STOCKTON STONE & CONCRETE CO LTD

Norton Iron Works, Norton   **121**
Warner's, Lucas & Barrett until 1865   NZ 440229 – **Map D**

The iron works was established on an approximately 20 acre site on the north side of the West Hartlepool Harbour & Railway line (former Clarence Railway), immediately east of the later Norton-on-Tees Station. A halt was provided here for the iron works which later 'evolved' into the station. According to a contemporary newspaper account, the West Hartlepool Ironworks Company began constructing the blast furnaces in 1855, but Warner's, Lucas & Barrett were the owners by about the end of that year. The Warner family, based in London, were well known for bell making. Three 50 feet high blast furnaces standing in a row at right angles to the main line were completed in 1856–57. Two sidings ran into the site allowing rail traffic to enter and leave from either direction. Shunting was probably carried out by horses as there was no reference to locomotives in an 1865 inventory. The main products were pig iron and rail chairs, but the works' main claim to fame was that, in 1856, it cast the first great bell weighing 15 tons 18 cwt for Big Ben at the Houses of Parliament. This was taken on a railway wagon to Hartlepool for shipping to London. Unfortunately, after rigorous testing, the bell cracked and had to be replaced by that from another foundry.

In 1865, the works was sold to the Norton Iron Co Ltd (registered 21/8/1865) and the latter erected three more blast furnaces in 1866–67 in line with the earlier furnaces but further away from the main line. Meanwhile, back in 1863, the Norwegian Titanic Iron Co Ltd had been established and it purchased collieries at Neville Hill, near Leeds and iron ore mines near Saggendal in Norway. The first shipment of ore from Norway smelted in one of the Norton Iron Co's furnaces appears to have been in 8/1867. The Titanic Iron Co leased the three old and less efficient blast furnaces at the Norton Iron Works in 1869; two were operating in 7/1870 but struggled to make a profit with the losses being subsidised by the surplus from Neville Hill collieries. The old blast furnaces were at a standstill in 1/1878 and the Titanic Iron Co went into liquidation in 1886 but had been dormant for years before.

Trade was depressed by the late 1870s and the Norton Iron Co announced in 7/1877 that it was suspending operations. It was wound up on 26/11/1877. An attempt to sell the assets of the company in 1880 stated that the site measured 43 acres, of which 19 acres were available for slag tipping. Amongst the equipment were two railway locomotives valued at more than £1,000, although only the identity of one is known. The blast furnaces were demolished in the 1880-90s but the loose plant was auctioned on 27-28/1/1881 including "6 locomotives [sic], 18 side and end tip waggons. 3 W.I. Chaldron waggons, 50 slag bogies".

The Stockton Stone & Concrete Co Ltd

The iron works site remained derelict until it became the Stockton Concrete Works operated by John Burn trading as Norton Stone & Concrete Company from 5/1899, which made use of the slag to produce "road metal", artificial stone and concrete goods on the 27 acre site. The agreement with the NER dated 13/6/1900 stated that inward traffic was placed in the standage siding by a NER locomotive and taken into the works by a private engine. The Norton Stone & Concrete Co Ltd was incorporated in 1902 but sold the works to The Stockton Stone & Concrete Co Ltd on 11/3/1905. The sale particulars in 1904 show that, on entering the premises, the standard gauge railway divided into five sidings; the main one running past the 350ft long "Architectural Artificial Stone Department" building to the crusher. Another served the single road 75 feet long engine shed which had an adjoining cart shed. A nine inch cylinder locomotive was included in the sale. From the crusher, an endless haulage inclined narrow gauge railway climbed over the 7-8 acres of slag tip on a 250ft long timber gantry. At the north end, a narrow gauge railway ran by the edge of the tip to collect the slag in side tipping wagons. Stockton Stone & Concrete Co Ltd gave notice to terminate the siding agreement on 28/4/1953. The company continued in business until it was taken over by Marshalls in 1967 and the works did not close until 1996.

Reference :   *A History of the Ironworks at Norton and the Story of Big Ben*, Alan Betteney, Tees Valley Heritage Group, 2011

Gauge : 4ft 8½in

-	0-4-0T	OC	FJ	108	1872	New	(1)
-	0-4-0ST	OC	HH	?	?	(a)	(1)
-	0-4-0VB		HW	?	?	(a)	s/s

(a)   origin and identity unknown.

(1)   The details are based on Inness records which suggest that these locomotives moved to Northumberland, becoming Killingworth No.3 and No.4 respectively, (presumably the Seaton Burn waggonway then operated by John Bowes & Partners Ltd).

NORTON JUNCTION SAND & GRAVEL CO LTD

Stockton   **122**
Norton Junction Sand & Gravel Co until 2/1935   NZ 428228 – **Map D**
Norton Junction Sand Co until 9/1933

The first record of an advertisement by the company appeared in the *Hartlepool Northern Daily Mail* in December 1930. It was supplying sand and gravel to Stockton Borough Council in the 1930s. The Norton Junction quarry was situated to the north of the LNER's triangular Junction and was connected by its own short single line to a siding off the north curve between Norton West and East Signal Boxes. The siding is shown on the 1939 Ordnance Survey but the workings are marked as "Old Quarry"; the line was certainly out of use in 1940. Another quarry existed to the west, possibly associated with a tile works and was served by a road from Blakestone Lane. The company also owned a large quarry (NZ 359225) about a half mile north of Bishopton village which was not rail-connected. In the 1965/66 *Directory of Quarries*, the Norton Junction site was described as a sand quarry and Bishopton as a gravel quarry. There is no information available on the use of the locomotive but it was an example of a small Lister "Auto-Truck" design that had been adapted for light rail use. The possibility remains that the locomotive may have worked at the Bishopton quarry. The company remained in existence up until the 1970s and went into voluntary liquidation on 24/12/1975.

Gauge : narrow

-	4wPM	L	960	1929	(a)	s/s

(a)   ex R.A. Lister & Co Ltd, Dursley, Gloucestershire, after overhaul, 3/9/1935.

THE OWNERS OF THE MIDDLESBROUGH ESTATE LTD

The Owners of the Middlesbrough Estate until 14/7/1886
CARGO FLEET TIMBER YARD   **123**
Cargo Fleet   NZ 509206 – **Map G**

According to an OME minute dated 24/11/1869, the new sawmill near the Tees Iron Works was almost complete and the adjacent timber pond, with a link for water from the Middlesbrough Dock entrance channel, was being formed by slag tipping from Gilkes, Wilson, Pease and Co. Timber was floated to the pond from the OME's wharf or the deep water berths opposite. The sawmills and workshops were aligned north–south, with the pond on the north west side. By 1875 a siding from the Middlesbrough-Redcar line entered the timber yard immediately after crossing over the rail access to the Tees Iron Works and headed north through the site to the OME wharf on the river. A second siding ran from the timber yard past the pond to the MOR line leading to Dock Lock Bridge. Sidings within the yard served the large drying sheds and saw, planing, turning and moulding mills and, by 1892, additional lines had been laid south of the timber pond. Large stocks of long pitch pine logs suitable for ship masts, spars and piles for wharves were available in 1887.

It is presumed that horses or travelling cranes shunted about the premises initially. The first locomotive known to have come to the yard arrived in 1892, to be followed in 1898 by a new four-coupled saddle tank with 11in x

17in cylinders purchased from Chapman & Furneaux. The NER had included the OME's Timber Department in a list dated 5/3/1897 of private locomotives that could travel via Vulcan Street to the Ironmasters district. The main change to the internal railway took place as a result of the final expansion of Middlesbrough Dock at the turn of the century. The timber pond was filled in and laid out with sidings to become part of the NER's Dock storeyard. The OME's wharf had to be replaced and it was reported on 16/12/1904 that this contract had been let to Ridley & Co. The new wharf was built adjacent to Wilsons, Pease & Co Ltd's property with a river frontage of 312 feet. The OME complained to the NER on 6/11/1905 that steamers had not been able to discharge at the new wharf because dredging had not been completed. A 1951 plan reveals the timber yard to be long and narrow, aligned north to south, with the sidings entering the sawmills from the south rather than the north and with timber stored on the western part of the site.

When Ted Haigh visited on 6/10/1951, CF 1167 was in the shed with CF 1156 painted green, lined red and black, out of use. The latter was replaced by a small Ruston & Hornsby 48DS diesel shunter in 1957. By the late 1960s, the timber yard contained a modern sawmill producing packing cases and pallets. MOR tolls on OME's traffic were increased from 1/2/1968 but rail transport at the timber yard appears to have ceased about 1969, with the private siding connection being officially closed from 22/7/1975.

Reference : *The Owners of the Middlesbrough Estate Business Book of Directors* 1886-1901, 1902-11, 1911-21 (Teesside Archives)

Gauge : 4ft 8½in

-	0-4-0ST	OC	HCR	125	1873	(a)	s/s
ESTATE No.1	0-4-0ST	OC	CF	1156	1898	New	
	Rebuilt		Ridley Shaw		1946		(1)
No.2	0-4-0ST	OC	CF	1167	1899	(b)	s/s c/1957
RICHARD PEASE	4wDM		RH	411318	1957	New	(2)

(a) ex Hogg & Henderson, Middlesbrough, /1892.
(b) ex Toll Bar Sidings, c/1951, by 6/10/1951.

(1) some OME plant, including a Chapman & Furneaux locomotive, was to be auctioned on 27/9/1957. To M. Henderson Clark Ltd, Cargo Fleet, for scrap, c/1957.
(2) to M. Henderson Clark Ltd, Cargo Fleet, c/1969.

ESTATE No.1 (CF 1156/1898) was at the OME's Cargo Fleet timber yard on 20th June 1951. In the background, across the main line, are the Tees Side Bridge & Engineering Works offices and, on the left, the roof of Whitehouse Signal Box; the latter was demolished in 2021. (IRS Ken Cooper photograph)

TOLL BAR SIDINGS AND BALLAST LINE, Middlesbrough 124
NZ 502204 NZ 492186 – **Map G**

The OME's Toll Bar Sidings (NZ 502204) had two connections with the NER's Middlesbrough-Nunthorpe branch, 16 and 28 chains from Guisborough Junction, where the branch joined the Middlesbrough-Redcar line. These sidings served a number of small businesses mostly on the north east side of North Ormesby Road. A charge was made for people using the road and the Toll Bar cabin, jointly owned by OME and J.W. Pennyman of Ormesby Hall, was situated next to the branch level crossing. The Toll Bar operated from 1854 to 1916 when Middlesbrough Corporation took it over and stopped the charges. The sidings and later ballast line were excluded from the 1884 Act when the NER took over a number of Middlesbrough Owners' lines. In the mid-1860s the OME constructed a line from Toll Bar Sidings over North Ormesby Road southwards to facilitate the laying out of Albert Park. The purchase of 70 acres of land for the park and associated works was paid for by Henry Bolckow. The *Middlesbrough Weekly News* dated 26/7/1867 describes an accident involving "the engine that is employed in the park". The incident had taken place on the previous Monday when the locomotive had been shunting some wagons into the park and was "about 300 yards between Marton Road and the Park"; the firemen had jumped off the engine and was killed. The park was formally opened on 11/8/1868 and the OME's line was then removed to the east side of Marton Road.

The businesses served by the OME's Toll Bar sidings are listed below (from north to south). Each property generally had a single siding.

> Charles Street Coal Depot. Established 1877 and remained to end of railway.
> Craggs Street Marine Store Dealer.
> Britannia Iron Foundry. Began 1874 and continued for life of railway.
> Atlas Coal Depot. Established c/1896 and survived until about 1966.
> Atlas Foundry. Founded 1867 and operated until 1907.
> Steam Flour Mill. Opened 1887; later belonged to hay and straw dealer.
> Toll Bar Coal Depot. Located on the south side of North Ormesby Road crossing and operated 1869–c/1972.
> Smiles' Cement Works. Located on north side of OME line adjoining Borough Road.
> Saltwells Road site. Contained bore holes for OME's Middlesbrough Salt Works. Coal was delivered to the boiler house to provide steam for the machinery.
> Dean Bros (builders). Occupied site on Marton Road served by a siding from OME line.

SODIUM (BH 935/1889) was working on the OME's Ballast Line in Middlesbrough. (ILS F. Jones collection)

Various works were established at Linthorpe to the south of Middlesbrough to provide bricks for the rapidly expanding town. However, as housing developments spread southwards, two of the brickworks closed in the 1890s. The OME Star Brick Works continued to about 1914 and the Ordnance Survey maps show a single line with a passing loop – possibly a rope-worked narrow gauge tramway – connecting the clay pit to the works. The OME decided to reclaim the former brick works land at Linthorpe for development and purchased it between 10/12/1897 and 20/10/1898. It extended its existing line from Toll Bar Sidings over Marton Road and alongside Park Vale Road past Albert Park and the Clairville Recreation Ground, where there was a passing loop, to reach the Linthorpe brickfields. This was the Ballast Line which opened to allow filling to commence in 1898 and it continued until the mid-1920's. Spoil came from local iron and steel works and, Raymond Kitching states, about 600 tons of ash per week from the engine sheds at Middlesbrough, Newport, Haverton Hill and Stockton. Fires were not uncommon if the slag was still hot! The Ballast Line ran as a single track across the brickfields area with a loop and temporary sidings to where tipping was taking place. It terminated at Cumberland Road (NZ 492186) where there were four sidings – two serving a coal depot, one a yard and the last had a common user function. The coal depot was opened by T. Robinson in 1898 but had been taken over by Middlesbrough Co-operative Society Ltd about 1908.

There were considerable fluctuations in traffic passing along the Ballast Line. On 6/11/1905 it was reported that Cochrane & Co had been sending about 200 trucks a month and Wilsons, Pease & Co about 40 to 50 trucks, but these were being discontinued for a period of "anywhere from 2 to 5 months". This would leave only 8 to 10 wagons a week. In 1906 the Ballast Line was used to deliver exhibits for the Yorkshire Agricultural Society's show which was being held nearby. On 17/3/1915, arrangements had been made for three train loads per week to be delivered from Messrs Samuelson for tipping. The Ballast Line was protected by gates where it crossed most of the roads and these had to be opened by the locomotive fireman. Unfortunately, there was an incident at Park Road South crossing in 1916 involving a locomotive, four wagons and "no look-out man". In response, OME said that its crews were given instructions about the procedure to be used and suggested that Middlesbrough Corporation should erect warning notices in the streets about the level crossings. One survived at the North Ormesby Road crossing until 1980. After World War 1, Middlesbrough Corporation began work on its Marton Grove housing estate and, on 12/4/1920, the conveyance of 16 acres of the OME's 75½ acres was noted, including the provision of a siding from the Ballast Line near Marton West Beck: "The Co's railway has been extended to convey materials and arrangements have been made for the haulage by our locos – 10% above cost to come to the coy". The siding was lifted in 1924 when deliveries began to arrive by road. After the mid-1920s OME laid a siding on the west side of Marton West Beck to raise and level the land in this area. By 1934 infilling was complete and the Ballast Line west of Marton Road closed and was removed in June 1935. The formal closure of the Cumberland Road Depot siding was 31/12/1934.

OME's one road engine shed was situated alongside the Toll Bar Sidings, south east of the Charles Street Coal Depot. The date for the arrival of SODIUM from The United Alkali Co Ltd is not known, although OME records dated 6/7/1909 state "Cost of locomotive engine. Question of proportion to be charged against ballast tipping". It appears that they were then using two locomotives because on 12/7/1915 another note states "Ballast deliveries to Brickyard. These have declined to an average of about 21 per day; given Engine Driver of the extra Loco a week's notice". Later the locomotives are said to have been painted green and face north, with SODIUM capable of hauling 10 wagons on the Ballast Line and ESTATE No.2 only five. With declining traffic and both track and the remaining locomotive in poor condition, the OME withdrew and sold the line in about 1957 to a consortium of users known as Toll Bar Sidings Ltd, except the Toll Bar Coal Depot siding which was purchased separately by that occupier. According to Raymond Kitching "two diesel locomotives were purchased to replace the steam power used by the Middlesbrough Owners and much of the track was relaid," but nothing is known to substantiate this statement. By the late 1960s there was little rail traffic and only W. Richards & Sons Ltd at the Britannia Iron Foundry made infrequent use of the sidings. The last load was delivered on 21/4/1971 and BR formally closed the connection to Toll Bar Sidings Ltd on 29/3/1972.

References : *The Middlesbrough Owners Railway North Ormesby to Linthorpe*, R Kitching, pub. R. Kitching, 1981

The Owners of the Middlesbrough Estate Business Book of Directors 1886-1901, 1902-11, 1911-21 (Teesside Archives)

Gauge : 4ft 8½in

		0-6-0					(a)	(1)
No.1		0-4-0ST	OC	MW	566	1876	New	s/s
	ESTATE No.2	0-4-0ST	OC	CF	1167	1899	New	(2)
	SODIUM	0-4-0ST	OC	BH	935	1889	(b)	
		Rebuilt	Ridley Shaw		1953			(3)

(a) origin and identity unknown.
(b) ex The United Alkali Co Ltd, Gateshead, Co. Durham.

(1)   offered for sale by OME in *Colliery Guardian* 28/1/1876 where described as an "almost new tank loco" with 3ft 6in diameter wheels and 14in x 20in cylinders.
(2)   to OME, Cargo Fleet Timber Yard, c/1951, by 6/10/1951.
(3)   to A. Ward & Son (Middlesbrough) Ltd, for scrap, c/1956; it was derelict there 6/1956.

PD PORTS plc

TEES DOCK, Grangetown   **125**
Tees & Hartlepool Port Authority Ltd until 1/4/2003   NZ 546231 – **Map H**
Tees & Hartlepool Port Authority until /1992
Tees Conservancy Commissioners until 1/1/1967

Teesport is the name used to describe PD Ports' facilities, including Tees Dock and associated estate, Hartlepool Dock and operations on the river, together with other wharves and jetties especially those owned by the petro-chemical industry.

The Tees Conservancy Commissioners (TCC) had supervised the river since 1852 and they began the construction of Tees Dock. However, it was decided that a single organisation should be responsible for the river and the Tees & Hartlepool Port Authority was formed by Act of Parliament and became operational on 1/1/1967. With privatisation, a statutory instrument came into force on 27/12/1991 allowing the THPA's assets to be transferred and in 1992 Teesside Holdings comprising Powell Duffryn, 3i and Humberside Holding Ltd each took a third interest in the THPA. Powell Duffryn plc gained a 100% of the business in 1995, although the title remained Tees & Hartlepool Port Authority Ltd. In 2000, the Japanese banking group, Nikko Principal Investments acquired Powell Duffryn. On 1/4/2003 the ports and shipping related part of the business was rebranded to become PD Ports Logistics and Shipping and, in 12/2004, was sold to PD Ports plc a new company listed on the stock exchange. Subsequently in 2/2006 PD Ports plc was bought by an Australian investment company, Babcock & Brown Infrastructure and the business was delisted from the stock exchange. Since 11/2009, PD Ports has been owned by the Canadian based Brookfield Asset Management Inc.

The concept of a dock at Lackenby had been in existence for a long time. As far back as 1864, a scheme for such a development had been prepared. Ten years later, the NER had plans for a new dock and timber basin at Lackenby and, in 1913, it purchased 1,000 acres to construct the dock but the onset of World War 1 intervened. Some work was undertaken by the LNER to level the land after the war, but the depressed economy meant that the project was not taken any further.

Tees & Hartlepool Port Authority Ltd's Andrew Barclay 614 of 1977 at Tees Dock with a ship moored opposite at the Cleveland Potash quay on 5th June 1997.   *(Cliff Shepherd photograph)*

Inside the Steel Terminal at Tees Dock on 17th August 2012 as OLD GEOFF 08648 (Horwich Works /1959), on hire from British American Railway Services Ltd, waits for steel to be unloaded from its train.
(Cliff Shepherd photograph)

With the increasing size of ships, the TCC recognised the need to establish port facilities down river where there was deeper water. The TCC's plans were sanctioned by Parliament in 8/1946 providing statutory powers to acquire land, including that owned by the LNER. These envisaged the construction of a new Tees Dock, two deep water jetties, the dock road and deepening of the river channel. The site for these developments was immediately down river of Dorman Long's slag tip and the remains of the former submarine base. Locally it was considered that the scheme would complement Dorman Long's intention to build a Universal Beam Mill at Lackenby and ICI's Wilton scheme, but Central Government approvals were slow in forthcoming for the TCC. A contract was awarded in 1949 to Geo Wimpey & Co Ltd to construct an access road to serve two new oil jetties and the later Tees Dock. The jetties were built by Cleveland Bridge & Engineering Co Ltd; No.1 (later Queen Elizabeth II) was commissioned in 10/1950, at about the same time as the 2½ mile Tees Dock Road and No.2 (later West Byng) in 1955. Two sites adjacent to the berths were leased for tank farm compounds; one to ICI for feedstock to be transferred to its plants by pipeline and the other to Shell-Mex and BP Ltd (which see).

Apart from carrying out a major part of the dredging, the contract for the construction of the first phase of the dock was not let until 1959 to Demolition & Construction Ltd. A dock water area of 400 feet wide and 3,250 feet long was planned with the first phase comprising the full length of the quay on the south west side, a transit shed behind No.1 Quay and the balance of the dredging. A crane track and four railway lines were laid along the quay. The 75 ton capacity wagon weighbridge was installed at the exchange sidings with the latter connected by rail to No.1 Quay. The dock started operating in 1962 and was officially opened on 4/10/1963, the programme for the event claiming "when the railway facilities are completed, standage for over 1,000 wagons will be available". Authority to construct the quay on the east side of the dock had been received in early 1963 to give a total of five deep water tidal births. In the late 1960s, the Cleveland Potash Ltd bulk terminal was commissioned (which see). A specialised covered steel export terminal opened in 1976 with a rail link to BSC's Lackenby Works. Roll on/roll off goods ferry services to Rotterdam and Zeebrugge had started in 1973.

The Tees Dock Branch left the Middlesbrough-Redcar line ¼ mile north east of Grangetown Station and ran to a set of exchange sidings with an accompanying headshunt. Two internal railways then reversed off these sidings; the first curved round on the north east side of the dock to serve the Cleveland Potash Ltd bulk terminal. A siding also ran to the West African berth (opened 1979) but this was later disconnected. The other railway headed round the end of the dock dividing into two to give a direct link to the south west berths or connect with a set of steel handling sidings. From the latter, lines led to the steel terminal building, the locomotive workshops and the cargo berths. There was a siding across the Tees Dock Road to the Shell-Mex and BP tank farm.

The BSC line from Lackenby left the Grids, crossed over the Middlesbrough-Redcar railway and ran down towards the tank farms before curving round to enter the steel reception sidings. This resulted in BSC diesel locomotives being a common sight at Tees Dock. Later, with increasing use of road transport to move steel slab and sections on the Lackenby site, the BSC link closed. When Teesside Works changed to become a major exporter of steel slab, it was necessary to return to rail transport. From the summer of 2003, EWS started working 24 wagon trains of slabs from the Lackenby Grids to the Tees Dock Branch exchange sidings. Here they were split into two sets of 12 wagons for PD Ports locomotives to take round to either the steel terminal or quays. The Corus subsidiary, Teesside Cast Products, then decided that it was more efficient for it to move the bulk of the steel slab by rail and the 1⅓ mile internal railway between Lackenby and Tees Dock was reinstated with a slightly eased curve to enter the steel reception sidings. The new line opened on 20/7/2006 and the reception sidings were revamped with five tracks; TCP's GECT diesels worked to the dock again.

Meanwhile the increasing importance of containers for more general cargoes had encouraged the Port Authority to provide a terminal for this traffic. It was located at the former West African berth, now identified as Tees Container Terminal 1 (TCT) and measured 294m long. After the former Shell refinery site (which see) had stood vacant for several years, PD Ports decided on the establishment of a second container berth on the river at the former Shell oil jetty. This is known as TCT 2; it opened in 9/2003, is 360 m long and used by deep sea cargo vessels. The land behind is also being developed as hardstanding for containers and distribution warehousing. Both Asda (325,000 sq ft, opened 2006) and Tesco (1,200,000 sq ft, opened 2009) have warehouses here.

This development has resulted in changes to the railway system. The main alteration was to replace the exchange sidings by a new set behind and parallel to the north east side of the dock. This gives direct access to TCT 2 and the Cleveland Potash Ltd's terminal. The curve round the end of the dock to the steel reception sidings was also eased to allow expansion of the RO-RO terminal. Trackwork Ltd (which see) had the contract for making the changes to the railway.

With the mothballing of the Redcar blast furnace on 19–20/2/2010, there was a drastic reduction in steel traffic along the line to the port. The last deep sea ship with slabs left on 2/3/2010. Two years passed with very little steel traffic until the restarting of the Redcar blast furnace by SSI (UK) and the production of steel commenced on 18/4/2012, the first vessel sailing with slabs for Thailand on 11/5/2012 and, by 10/2012, one million tonnes of steel slab had been exported through Teesport.

Following the practice of its neighbour, Dorman Long, the TCC purchased two Sentinel diesel hydraulics to shunt at Tees Dock, with a similar Rolls-Royce machine added in 1977. The fleet expanded with a couple of Hibberd locomotives made redundant from Middlesbrough Wharf and three ex BR Class 03/04 diesel mechanicals from the closed Middlesbrough Dock, but these were soon disposed of and subsequent purchases were six-coupled locomotives. Ford industrial tractors were sometimes used on the quayside for positioning wagons. In 2003 PD Ports began to 'outsource' responsibility for shunting at the dock to RMS Locotec #. Amongst the hire locomotives used at Tees Dock have been three ex Netherlands Railways units which were similar to the ex BR 08s.

When Sahaviriya Steel Industries UK Ltd restarted the Redcar blast furnace and the Lackenby steel plant, ex Norwegian State Railways Di8 locomotives began hauling 15 wagon trains of steel slab to the steel reception sidings at the dock. PD Ports hired locomotives then pulled the wagons on to the quay or into the steel terminal. Loading of the first ship began on 11/5/2012. This traffic came to an end with the liquidation of Sahaviriya Steel Industries UK Ltd on 2/10/2015. Meanwhile, PD Ports had continued to develop its container facilities. An expansion project costing £16.7 million was carried out in 2011 and Freightliner transferred its rail operations from Wilton to a new terminal at Tees Dock which opened on 15/11/2014.

\#   RMS Locotec was based at Dewsbury, West Yorkshire before moving its depot to Wakefield. From 9/2008, it was owned by British American Railway Services Ltd (a subsidiary of Iowa Pacific Holdings) and trading as RMS Locotec based at Wolsingham, County Durham. Rail Management Services Ltd (part of the Proviso Holdings Group) became the owner from 4/6/2020 but still trading as RMS Locotec.

Gauge : 4ft 8½in

No.1 (TCC No.1)	0-4-0DH	S	10137	1962	(a)	(7)
No.2 (TCC No.2)	0-4-0DH	S	10170	1964	New	(7)
-	4wDM	FH	3964	1961	(b)	(1)
3 (TYNE-TEES WHARF No.5)	4wDM	FH	3949	1960	(b)	(2)
No.3	0-4-0DH	RR	10208	1965	(c)	(3)
4 (D2024)	0-6-0DM	Sdn		1958	(d)	(6)
5 (3, D2023)	0-6-0DM	Sdn		1958	(e)	(4)
6 (1, D2205)	0-6-0DM	VF	D212	1953		
		DC	2486	1953	(e)	(5)
No.6	0-6-0DH	RR	10215	1965	(f)	(20)

No.7		0-6-0DH	S	10095	1962		
		Rebuilt	AB	6007	1982	+	
			HE	9069	1982	(g)	(8)
-		2w-2PMR	Wkm	(6607	1953?)		
		Rebuilt	DMR			(h)	(11)
5 20/110/710 (No.3)		0-6-0DH	AB	614	1977	(i)	(19)
1 (JULIAN)		0-6-0DH	S	10161	1964	(j)	(18)
2		0-6-0DH	RR	10239	1965	(k)	(20)
-		0-6-0DH	HE	6295	1965	(l)	(9)
HIPPO		0-6-0DH	EEV	5352	1971	(m)	(10)
19 H028 08622 (D3789)		0-6-0DE	Derby		1959	(n)	(21)
9 H041 08754 (D3922)		0-6-0DE	Hor		1961	(o)	(12)
16 H043 (625 690)		0-6-0DE	EE	2122	1956		
			VF	D312	1956	(p)	(23)
21 08375 (D3460)		0-6-0DE	Dar		1957	(q)	(25)
9 H046 (687)		0-6-0DE	EE	2129	1956		
			VF	D319	1956	(r)	(16)
-		0-6-0DH	EEV	3994	1970	(s)	(13)
12 29		0-6-0DE	YE	2938	1964	(t)	(14)
13 H049 (692)		0-6-0DE	EE	2146	1956		
			VF	D336	1956	(u)	(15)
14 H011 (08423) (D3538) LOCO 2		0-6-0DE	Derby		1958	(v)	
15 H006 (01578)		0-6-0DH	HE	6294	1965	(w)	(17)
17 08588 (D3755)		0-6-0DE	Crewe		1959	(x)	(22)
18 H042 08885 (D4115)		0-6-0DE	Hor		1962	(y)	(26)
20 08648 (D3815) OLD GEOFF		0-6-0DE	Hor		1959	(z)	(28)
22 08871 (D4039)		0-6-0DE	Dar		1960	(aa)	(27)
H061 08523 (D3685)		0-6-0DE	Don		1958	(ab)	(24)
23 08308 (D3378)		0-6-0DE	Derby		1957	(ac)	(30)
24 08809 (D3977)		0-6-0DE	Derby		1960	(ad)	(29)
08613 (D3780)		0-6-0DE	Derby		1959	(ae)	(31)
08847 (D4015) Loco 1		0-6-0DE	Hor		1961	(af)	
H068 08788 (D3956)		0-6-0DE	Derby		1960	(ag)	

+ Comprised Sentinel frame and chassis with a new Barclay superstructure.

(a) ex Sentinel, initially on hire, 13/12/1962.
(b) ex Middlesbrough Wharf (former Tyne-Tees Wharf), c/1970.
(c) ex Thos Hill Ltd, Kilnhurst, South Yorkshire, 15/12/1977; previously BSC Corby Works, Corby, Northamptonshire.
(d) ex Middlesbrough Dock, 25/10/1980.
(e) ex Middlesbrough Dock, 22/11/1980.
(f) ex Thomas Hill Ltd, Kilnhurst, South Yorkshire, 27/7/1981; previously Stanton & Staveley Ltd, Stanton Works, Derbyshire. To Hartlepool Docks, Hartlepool c1/1994; after 23/4/1993, by 15/1/1994 and returned after 6/7/2002, by 1/4/2003.
(g) ex Andrew Barclay, Sons & Co Ltd, Kilmarnock, Ayrshire, 12/1982; previously BSC Teesside Works, Grangetown. To Cleveland Potash Ltd, Boulby Mine, Loftus, on hire, from 30/1/1992 and returned, 28/4/1992.
(h) origin unknown. Used for weed killing duties once or twice a year when coupled to a trolley.
(i) ex Douglas Engineering Services Ltd, Tees Dock, Grangetown c3/1993, by 16/3/1993. It may have been initially hired before being acquired.
(j) ex Staffordshire Locomotives Ltd (care of Telford Horsehay Steam Trust, Telford, Shropshire), on hire from 22/1/1999 and purchased by 4/1999.
(k) ex Staffordshire Locomotives Ltd (care of Telford Horsehay Steam Trust, Telford, Shropshire), 26/4/1999.
(l) ex RMS Locotec, Dewsbury, West Yorkshire, on hire, 13/9/2003.

(m) ex Coopers (Metals) Ltd, Swindon, Wiltshire (property of RMS Locotec, Dewsbury, West Yorkshire), on hire, 7/6/2004.
(n) ex RMS Locotec, Dewsbury, West Yorkshire, on hire, 6/8/2004 and returned, ex hire, 14/2/2006. Ex Corus plc, Trostre Works, Llanelli, Carmarthenshire (property of RMS Locotec, Wakefield, West Yorkshire), on hire, 3/2008 and returned to Trostre Works, 5/2009. Ex Trostre Works (property of British American Railway Services), on hire, w/c 9/11/2009.
(o) ex Reading Permanent Way Yard, Reading, Berkshire (property of RMS Locotec, Dewsbury, West Yorkshire), on hire, after 11/5/2005, by 20/6/2005. To RMS Locotec, Dewsbury, West Yorkshire, 1/9/2005 and returned, on hire 15–16/9/2005.
(p) ex RMS Locotec, Dewsbury, West Yorkshire, on hire, 1/9/2005.
(q) ex RMS Locotec, Dewsbury, West Yorkshire, on hire, 15/10/2005. To Wabtec Rail Ltd, Doncaster, South Yorkshire for repairs, ex hire, 15/2/2006. Ex Castle Cement (Ketton) Ltd, Ketton, Rutland (property of British American Railway Services Ltd) on hire, 28/2/2011.
(r) ex RMS Locotec, Dewsbury, West Yorkshire, on hire, 24/11/2005 and returned, ex hire, /2006. Ex RMS Locotec, Wakefield, West Yorkshire, on hire, 18/4/2007.
(s) ex UK Coal Mining Ltd, Oxcroft, Derbyshire (property of RMS Locotec, Dewsbury, West Yorkshire, on hire, 14/2/2006.
(t) ex Corus plc, Appleby-Frodingham Works, Scunthorpe, on hire? 15/2/2006.
(u) ex RMS Locotec, Wakefield, West Yorkshire, on hire, by 19/7/2006.
(v) ex RMS Locotec, Wakefield, West Yorkshire, on hire, w/c 23/2/2007. To Faber Prest Ports Ltd, Flixborough Wharf, ex hire, late /2007 and returned on hire, by 4/2008. To British American Railway Services, Wolsingham, Co. Durham, ex hire, 1/8/2014 and returned 27–28/8/2014. On hire from Rail Management Services Ltd from 4/6/2020.
(w) ex RMS Locotec, Wakefield, West Yorkshire, on hire, /2007, by 5/4/2007. To Cleveland Potash Ltd, Tees Dock, Grangetown, ex hire, /2007, by 10/2007. Ex British American Railway Services Ltd, Wakefield, West Yorkshire, on hire, /2008, by 26/2/2009.
(x) ex Network Rail, Whitemoor Yard, March, Cambridgeshire (property of RMS Locotec, Wakefield, West Yorkshire, on hire, 7/2008.
(y) ex Network Rail, Whitemoor Yard, March, Cambridgeshire (property of British American Railway Services Ltd, Wolsingham, Co. Durham), on hire, 3/4/2009.
(z) ex Wabtec Rail Ltd, Doncaster, South Yorkshire after repairs (property of British American Railway Services Ltd, Wolsingham, Co. Durham), on hire, 17/1/2011.
(aa) ex Cemex Rail Products, Washwood Heath, Birmingham (property of British American Railway Services Ltd, Wolsingham, Co. Durham), on hire, 13/4/2012, but arrived 16/4/2012 after weekend storage in a haulier's yard. To British American Railway Services Ltd, Wolsingham, Co. Durham, ex hire, 22/8/2012 and returned on hire, 4/12/2012.
(ab) ex British American Railway Services Ltd, Wolsingham, Co. Durham, on hire, 22/8/2012.
(ac) ex British American Railway Services Ltd, Wolsingham, Co. Durham, on hire, 1/8/2014.
(ad) ex British American Railway Services Ltd, Wolsingham, Co. Durham, on hire, 27/8/2014.
(ae) ex British American Railway Services Ltd, Wolsingham, Co. Durham, on hire, 3/10/2018.
(af) ex Mid-Norfolk Railway Society, Dereham, Norfolk, (property of British American Railway Services Ltd, Wolsingham, Co. Durham), on hire, 14/5/2019.
(ag) ex Tata Steel Europe, Shotton Works, Hawarden, Flintshire, (property of British American Railway Services Ltd, Wolsingham, Co. Durham), on hire, 10/7/2019.

(1) to United Glass Ltd, Sherdley Works, St Helens, Lancashire, c/1972.
(2) to Thos Hill Ltd, Kilnhurst, South Yorkshire, 1/1978.
(3) scrapped 4/1982, although the frame was still present on a wagon in 12/1982.
(4) to Kent & East Sussex Railway, Tenterden, Kent, 14/8/1983.
(5) to Kent & East Sussex Railway, Tenterden, Kent, 21/8/1983.
(6) to Kent & East Sussex Railway, Tenterden, Kent, 4/9/1983.
(7) to R.F.S. (E) Ltd, Doncaster, South Yorkshire, c5/1996; after 13/3/1996, by 23/5/1996.
(8) to Staffordshire Locomotives Ltd (care of Telford Horsehay Steam Trust, Telford, Shropshire, 27/4/1999.
(9) to RMS Locotec, Dewsbury, West Yorkshire, ex hire, 7/6/2004.
(10) to RMS Locotec, Dewsbury, West Yorkshire, ex hire, 15/7/2004.
(11) to Andrew Briddon, Long Marston, Warwickshire, but stored elsewhere, 5/2005.
(12) to RMS Locotec, Dewsbury, West Yorkshire, ex hire, 25/11/2005.
(13) to Cleveland Potash Ltd, Tees Dock, Grangetown (property of RMS Locotec, Wakefield, West Yorkshire), ex hire, 3/2007, by 14/3/2007.

(14)	to Corus plc, Appleby-Frodingham Works, Scunthorpe, Lincolnshire, ex hire?, after 19/7/2006, by 27/7/2007.	
(15)	to A.C. Electrics Ltd, Motorail Logistics, Long Marston, Warwickshire, (property of RMS Locotec, Wakefield, West Yorkshire), ex hire, 8/2008.	
(16)	to Midland Railway Trust Ltd, Butterley, Derbyshire (property of British American Railway Services Ltd, Wolsingham, Co. Durham) ex hire, c25/2/2009 via Motorail Logistics, Long Marston, Warwickshire and Allely's Heavy Haulage Ltd, Studley, Warwickshire.	
(17)	to Cobra Railfreight Ltd, Middlesbrough (property of British American Railway Services Ltd, Wolsingham, Co. Durham), ex hire, 3/4/2009.	
(18)	to T.J. Thomson & Son Ltd Stockton, 7/10/2009	
(19)	to T.J. Thomson & Son Ltd, Stockton, 8/10/2009.	
(20)	to T.J. Thomson & Son Ltd, Stockton, 9/10/2009.	
(21)	to British American Railway Services Ltd, Wolsingham, Co. Durham, ex hire, 18/1/2011.	
(22)	to British American Railway Services Ltd, Wolsingham, Co. Durham, ex hire, 1/3/2011.	
(23)	to Boden Rail Engineering Ltd, Washwood Heath, Birmingham (property of British American Railway Services Ltd, Wolsingham, Co. Durham), ex hire, c17/4/2012.	
(24)	to British American Railway Services Ltd, Wolsingham, Co. Durham, ex hire, 4/12/2012.	
(25)	to Tata Steel Europe, Shotton Works, Hawarden, Flintshire (property of British American Railway Services Ltd, Wolsingham, Co. Durham), ex hire, c 27/3/2013.	
(26)	to British American Railway Services Ltd, Wolsingham, Co. Durham, ex hire, w/c 15/4/2013.	
(27)	to British American Railway Services Ltd, Wolsingham, Co. Durham, ex hire, 1/8/2014.	
(28)	to British American Railway Services Ltd, Wolsingham, Co. Durham, ex hire, 27/8/2014.	
(29)	to Tata Steel Europe, Shotton Works, Hawarden, Flintshire (property of British American Railway Services Ltd, Wolsingham, Co. Durham), ex hire, /2017; after 26/1/2017, by 17/6/2017.	
(30)	to British American Railway Services Ltd, Wolsingham, Co. Durham, ex hire, 3/10/2018.	
(31)	to British American Railway Services Ltd, Wolsingham, Co. Durham, ex hire, 11/7/2019.	

E. PEARSON (STEEL & NON FERROUS) LTD

Middlesbrough **126**
NZ 492214 – **Map F**

E. Pearson operated a scrapyard at Dent's Wharf, Depot Road, Middlesbrough. The locomotive was probably acquired with a view to its resale. Pearson vacated the site at the end of 1981.

Gauge : 4ft 8½in

	15	ROSEDALE	4wDH	S	10070	1961	(a)	(1)

(a)	ex Bell & Son (Doncaster) Ltd, Anderston Foundry, Port Clarence, c/1977, by 24/10/1978.
(1)	to Fairfield-Mabey Ltd, Chepstow, Gwent, after 14/2/1979, by 24/2/1979.

JOHN PEARSON LTD

WHINSTONE QUARRY, Stainton **127**
NZ 485139 – **Map F**

The company owned a whinstone quarry south of Middlesbrough and also operated as a contractor. Information on this latter activity appears in the Contractors' Locomotives section.

The quarry was located on the east side of Stainton, a field beyond the parish church and vicarage and connected by road to the village. The workings took place next to the Quarry Plantation and, in 1893, a single line tramway extended from the drift westwards to a weighing machine at the road entrance. The quarry was owned by William Pearson (1901), Pearson Bros (1905) and John Pearson Ltd (1913). In this latter year, the quarry had been extended eastwards and an additional siding served shafts suggesting that the workings operated on two levels. Pearson was already supplying whinstone to Middlesbrough Council and others before branching out into contracting. Eventually John Pearson Ltd ended up in liquidation and the quarry, plant and 2ft 0in gauge locomotive were put up for sale in 1/1936.

Gauge: 2ft 0in							
	0-4-0ST	OC	AE	1593	1910	(a)	(1)

(a) ex Board of Trade Timber Supply Department, Shull Timber Camp, Wiserley, near Wolsingham, Co. Durham. An Avonside was part of a sale of three locomotives and track auctioned there on 3-4/3/1920.

(1) to Thos. W. Ward Ltd, Jarrow, Co. Durham, c/1936.

PEASE & PARTNERS LTD

The Pease family of Darlington built up one of the largest industrial empires of the nineteenth century, with interests in collieries, coke ovens, quarries, iron works, brick works and railways in both Co. Durham and Yorkshire (North Riding). The family operated these businesses through **Joseph Pease & Partners**, formed in the 1850s, which was mainly concerned with the collieries and **J.W. Pease & Co** (established 1852) that owned the ironstone mines, iron works and limestone quarries. The two firms were combined on 19/8/1882 to form Pease & Partners Ltd, although the seven shareholders remained the same. This became a public company on 11/10/1898. J. & J.W. Pease was a local bank that stopped making payments in 1902. The Peases also controlled other subsidiary companies, had a substantial holding in Robert Stephenson & Co and held personal shares in the NER. After 1900, the fortunes of Pease & Partners Ltd declined for a variety of reasons. However, by the mid 1930s, the slimmed down company was profitable again. With the loss of its collieries following nationalisation, Pease & Partners Ltd withdrew from quarrying and converted most of its remaining involvement in iron works and wharves into subsidiary businesses in 1947, becoming a holding company in the process. It ceased to exist about 8/1959. For a useful insight into the various Pease holdings, see 'A Credit Crunch: The Story of J & J W Pease' by Christopher Dean in *North Eastern Express*, 195–96, NERA, 2009.

LACKENBY IRON WORKS, Nr. Grangetown — 128
Pease & Partners Ltd until /1931
NZ 556223 – **Map H**
Tees Furnace Co Ltd until 10/12/1924
Bolckow, Vaughan & Co Ltd until /1901
Downey & Co until 21/3/1892
Lloyd & Co until 5/1879
Lackenby Iron Co until 30/10/1877

This works was a product of the boom in the Cleveland iron industry during the early 1870s. It was erected by the Lackenby Iron Co on the north eastern side of the various developments at South Bank and Grangetown. Beyond, to the north east, lay the marshes bordering the Tees and the fields of West Coatham Grange; it would be another eighty years before these fields were occupied by the iron and steel industry. The co-signatures on the 99 year lease for the site to run from 1/4/1870 were Thomas Light Elwon, son-in-law of John Vaughan and James MacKean. The lease was initially for 20 acres, later extended to 69 acres. Two blast furnaces were constructed at a right angle to the Middlesbrough-Redcar railway and 'blown in' during 1871, with a third added in line in 1872. Situated on the southern side and parallel to the blast furnaces were the calcining kilns and minerals stores. Steam from the blast furnace blowing engines also operated the hoist which lifted the railway wagons up to the gantry running along the top of the kilns and stores. The pig beds and slag discharge points were located on the north side of the furnaces. The Lackenby Iron Co also developed West Hunwick Colliery in County Durham during 1872.

The iron works was connected to the Middlesbrough-Redcar railway approximately ½ mile north east of the later Eston Grange Station and lines ran into the site to connect with a fan of reception sidings and the mineral gantry before continuing southwards to a headshunt. From here, railway lines returned past the north side of the furnaces before rejoining the main line. Also on this side, other sidings enabled bogies to be filled with slag. These appear to have been initially emptied on land next to the main line but at some date, prior to 1893, a bridge was constructed over the main line and a siding extended across it so that slag could be tipped on the marshes bordering the river, thus facilitating land reclamation. An undated map, but probably circa 1910, shows Tees Slag Wool Co Ltd, Messrs Brand's slag works and the Tees Scoriae Brick Co Ltd all having sidings at the iron works premises. Another siding ran from near the blast furnaces to an "Annealed Concrete Works" close to Low Lackenby houses, although it had disappeared by 1913.

The Lackenby Iron Works soon ran into difficulties as a result of the depression in the latter 1870s and was put up for sale by auction in 30/10/1877. It was purchased by Joseph Dodds MP for £50,100 and, from mid 1878, was being operated by Lloyd & Co which also ran the Linthorpe Iron Works. The principal partners were Isaac Wilson, William Hopkins and Edgar Gilkes. Trading in Lloyd & Co was suspended on 15/5/1879 and it went into liquidation in the following month. *The Engineer* announced that C. Willman was to auction the Lackenby Iron Works, including plant and three locomotives, on 15–16/9/1879. Downey & Co, which owned the Coatham Iron Works, acquired Lackenby Iron Works in 1880. There were 48 terrace houses at Low Lackenby, near the site of the original railway station, inhabited by Messrs Downey's workmen. Unfortunately, the business had

The Lackenby Iron Company purchased a new saddle tank with 8in x 15in cylinders from Hudswell Clarke & Rogers, maker's number 112, in 1871.

increasing financial problems and it ceased trading on 21/3/1892, despite all three furnaces having been in blast for the last five years. The National Provincial Bank took control of the works on 12/4/1892 and leased it to Bolckow, Vaughan which operated it until 1901. From 1897 only two of the blast furnaces had been working. The Tees Furnace Co Ltd, owners of the Tees Side Iron Works in Middlesbrough, purchased the iron works in 1902 and operated the blast furnaces producing foundry pig iron. During World War 1, the Ministry of Munitions pressed the company to rebuild the third blast furnace and the works was shown with three after the war. In 1922 the company registered ten 12 ton open goods wagons built by the Chorley Wagon Co to run on the main line. The Tees Furnace Co went into voluntary liquidation on 10/12/1924, but the furnaces continued to operate and later Pease & Partners Ltd ran them until final closure in 1931. The premises were demolished and the site later incorporated into Dorman Long's Lackenby development.

Reference : 'Blast Furnaces at Redcar 1870-2010', Charles H. Morris, *The Cleveland Industrial Archaeologist* 33, CIAS, 2012, pp.3-7

Gauge : 4ft 8½in

	Name	Type		Maker	No.	Date		
	LACKENBY No.1	0-4-0ST	OC	K	1705	1871	New	(1)
	LACKENBY No.2	0-4-0ST	OC	HCR	112	1871	New	(2)
-		0-4-0ST	OC	D	857	1875	New	s/s
No.3	KILMARNOCK	0-4-0ST	OC	B?				
		Rebuilt		AB	7571	1907	(a)	(3)
No.2		0-4-0ST	OC	BH	531	1879	New	s/s
No.1		0-4-0ST	OC	YE	480	1891	(b)	s/s
No.4	MACDONALD	0-4-0ST	OC	P	956	1903	(c)	(4)
	ROSEBERRY	0-4-0ST	OC	HE	1294	1918	New	(4)

(a) Russell Wear suggests that this may be Barclays 231/c1876 that was new to Pease & Partners (sic), Middlesbrough and rebuilt by Andrew Barclay in 1907. He states that it was with Pease by 3/1914.
(b) ex Hudswell Clarke & Co Ltd, Leeds, Yorkshire (WR); previously Leeds Forge Co Ltd, Leeds. It was seen at the Lackenby Works on 3/7/1927.
(c) ex Bentley & Jubb, Machinery Merchants, Ince, Lancashire, c/1916; it was previously with the Derwent Valley Water Board, Howden, Derbyshire, (for sale there 6/1913).

(1) to Pease & Partners Ltd, Tees Iron Works, Cargo Fleet before 1922.
(2) to Furness Shipbuilding Co Ltd, Haverton Hill.

(3)   to Normanby Iron Works Co Ltd, Cargo Fleet, c/1930.
(4)   to Pease & Partners Ltd, Tees Bridge Iron Works, Stockton.

TEES BRIDGE IRON WORKS, Stockton   129
Tees Bridge Iron Co Ltd until /1916   NZ 443175 – **Map D**
Tees Bridge Iron Co until /1879

The iron works was established by Joseph Dodds on a 32 acre site alongside Bowesfield Lane in 1871 and it had two blast furnaces operating by 1874, with a third coming on stream in 1875. From 1890 it was usual to only have two in blast. The Tees Bridge Iron Co Ltd was taken over by Pease & Partners in early 1916. The brands of pig iron produced became named "TEES BRIDGE" and "P & P". The iron works' private sidings connected with the NER's Darlington-Middlesbrough railway, east of Bowesfield Junction. From here, its line ran south alongside Bowesfield Lane past the Bowesfield Iron Works to reach the row of three blast furnaces. The pig beds and slag discharge spouts were on the east side of the blast furnaces served by more sidings, some of which enabled slag tipping to take place on land adjoining the River Tees.

The extensive areas of slag attracted the attention of the Tees Scoriae Brick Co and this had a works on the tips making slag blocks by 1885. It was still there in 1914. The NER's *Plans Book of Collieries, Works & Sidings*, dated 1895, also shows R. Walker & Co's slag crushing works and, no doubt, both concerns made use of the Tees Bridge sidings. In 1904 the slag works was being operated by John Burns' Norton Stone & Concrete Co "under a yearly rental and a charge for the haulage of goods" by the Tees Bridge Iron Co. A narrow gauge railway with end and side tipping wagons was used by the slag works. The iron works railway continued south to serve the Tees Bridge Brick Works which had a confusing history. Alan Betteney suggests that the Tees Bridge Brick Co operated from 1896–1907 and was then replaced by the Cleveland Magnesite and Refractory Co Ltd making silica bricks from ganister brought in from Co. Durham. The NER has a note of this siding in 1908. This activity appears to have ceased by the mid 1920s. However, the 1895 plans show the location as the Tees Cement Works and the Ordnance Survey dated 1896–97 has the South Durham Chemical Works. By the 1923 edition, the cement works site was covered by slag tips and the brickworks buildings were near Bowesfield Lane.

Tom Sowler wrote that the last of the blast furnaces "was razed to the ground on January 14th 1930. The large structure weighing about 1700 tons of metal and bricks was supported upon ten hollow cast iron pillars about 12ft high and 18 inches at its base. At its destruction the furnace had been out of operation for two and a half years and was still burning when it fell, the mixture of coke and metal giving quite a glare at the bed of the furnace." Although iron making had ceased about 1927, it appears that Pease & Partners retained an interest in the site, possibly providing a locomotive (s) to assist with Tarslag's (which see) traffic. When Ted Haigh visited here on 24/4/1934, Hunslet 1294 was still present and he showed the location as Pease & Partners Tees Bridge (Tarslag).

References :   *The Brickworks of the Stockton-on-Tees Area*, Alan Betteney, Tees Valley Heritage Group, 2007

                 A History of the Town and Borough of Stockton-on-Tees, Tom Sowler, Teesside Museums and Art Galleries, 1972

Gauge : 4ft 8½in

1		0-4-0VBT						s/s
	TEES BRIDGE No.1	0-4-0tank	OC	FJ	88	1872	New	s/s
	TEES BRIDGE No.4	0-4-0ST	OC	CF	1183	1899	New	(1)
5		0-4-0ST	OC	CF	1188	1900	New	s/s
	TEES BRIDGE No.6	0-4-0ST	OC	HL	2872	1911	New	s/s
	TEES BRIDGE No.7	0-4-0ST	OC	HL	2971	1912	New	(2)
	MARY	0-4-0ST	OC	AB	1282	1912	(a)	(3)
4	MACDONALD	0-4-0ST	OC	P	956	1903	(b)	(4)
	ROSEBERRY	0-4-0ST	OC	HE	1294	1918	(b)	(5)

(a)   ex Royal Arsenal, Woolwich, London, /1921.
(b)   ex Pease & Partners Ltd, Lackenby Works, Grangetown, possibly when Lackenby closed in /1931.

(1)   to The West Hunwick Silica & Firebrick Co Ltd, Hunwick, Co. Durham, by 15/8/1931.
(2)   to Pease & Partners Ltd, Thorne Colliery, Moorends, Yorkshire (W.R.).
(3)   to Pease & Partners Ltd, Tees Iron Works, Cargo Fleet.
(4)   to Tarslag Ltd, Stockton (with part of site?).
(5)   to Skinningrove Iron Co Ltd, Skinningrove, /1937.

TEES IRON WORKS, Cargo Fleet
Tees Foundries Ltd until /1958
(Subsidiary of **Pease & Partners Ltd**)
Pease & Partners Tees Foundries Ltd until c/1953
Pease & Partners Ltd until 3/1947
Wilsons, Pease & Co until 6/1901
Gilkes, Wilson, Pease & Co until /1880
Gilkes, Wilson, Leatham & Co until c/1858

130
NZ 510207 – **Map G**

The iron works was situated north of the Middlesbrough-Redcar railway line, immediately west of the final site of Cargo Fleet Station. The SDR's Middlesbrough Dock required the construction of training walls enclosing an entrance channel to enable ships to reach the dock from the deeper water of the River Tees. As a result, land was reclaimed extending from the railway to the channel and it was here in 1852 that Gilkes, Wilson and Leather began to erect 55 feet high blast furnaces. Four were in existence by 1854 and, two years later, were producing 500 tons of pig iron each week. A fifth blast furnace had been added by 1858/59. Albert Leatham (a son-in-law of Joseph Pease) died in 1858 and his place was taken by J.B. Pease (a son-in-law of Isaac Wilson) and the business became Gilkes, Wilson, Pease & Co.

Unlike the nearby Ormesby and Normanby Iron Works, the blast furnaces were set at right angles to the main line. Therefore the company's railway left the main line, just to the west of Whitehouse Crossing, before curving eastwards to enter the works where it divided into numerous sidings which headed north to the river. Here, there was a wharf which, in 1926, was 350 feet long with two 5 ton electric cranes. En route the sidings serviced the blast furnaces and accompanying plant. The pair of 85 feet high furnaces (New Side) in the middle of the site had their calcining kilns on the west side and the pig beds on the east. Unusually, the other three 75 feet high blast furnaces (Old Side) on the eastern boundary of the site had both their kilns and pig beds on the same or western side. There were seven kilns on the 'Old Side' and six on the 'New'. Gantries ran over both sets with the railway wagons climbing an incline on the 'Old Side' and being raised by a direct acting steam ram on the other. Not surprisingly, some of the company's earliest locomotives were provided by Gilkes, Wilson & Co, although the choice of insect names was unusual and probably, even then, reflected their small size.

Immediately after leaving the main line, Tees Iron Works rail traffic was subject to MOR tolls. Its railway was also crossed by the MOR track that connected Cargo Fleet with the Ironmasters district via Vulcan Street. In 1897 Wilsons, Pease & Co's tolls over the MOR were ¼d per ton in respect of all traffic (except certain iron and slag refuse when conveyed for deposit and not for sale) entering or leaving its premises for the NER Middlesbrough-Redcar line; 1d in addition to the above toll for its traffic over Dock Lock Bridge; 3d per ton for

ALEXANDRA (Andrew Barclay 979 of 1903) and its crew pose for their photograph in 1955 at Tees Foundries Ltd, the former Tees Iron Works.   *(ILS Frank Jones photograph)*

all its traffic between the works and the Normanby Iron Works and Tees Conservancy Commissioners' graving dock and ¼d per ton on all pig iron put into its stock. It was also noted that the Tees Iron Works sent one of its locomotives to Crewdson, Hardy & Co's nearby premises to carry out shunting.

Practical benefit was obtained from some of the slag and a report data 4/8/1893 stated that "slag wool, with the invention of which Mr Charles Wood's name has always been associated, is still manufactured under his supervision at Messrs Wilsons, Pease & Co's works on a considerable scale." An excess of 1,000 tons was being produced each year. It was manufactured by blowing jets of steam across a fine stream of molten slag leaving the furnace. One of its principal uses was as an insulating material and no less a person than Samuel Johnson of the Midland Railway was quoted as having used it for several years to insulate his locomotive boilers. By about the turn of the century, there had been major changes to the iron-making plant; the blast furnaces on the eastern boundary had been demolished and another added at the north end of the central pair to give three 82 feet high blast furnaces.

Wilsons, Pease & Co was heavily in debt to the family bank of J.& J.W. Pease and when this ceased operating in 1901, the capital value of the iron works had to be written down and it became the property of Pease & Partners Ltd. The company did not develop a steel plant at Tees Iron Works but concentrated on producing iron castings. Wilsons, Pease & Co had supplied iron chairs to the Whitby, Redcar & Middlesbrough Railway prior to its opening in 1883. When the Iron & Steel Institute visited the works in 1908, it had a range of foundries including open sand, direct casting and one with a cupola furnace. About 600 tons of rail chairs, brake blocks and general castings were produced each week. Two of the blast furnaces were operating in 1925 producing haematite iron but ceased within a year or two. The works then concentrated on its foundry activities; presumably Pease & Partners could use iron from its blast furnaces elsewhere.

The *Northern Echo* on 10/11/1936 reported that E.R. Colwell Ltd, iron and steel merchants and dismantling contractors of Depot Road, had bought for demolition the whole of the Tees Iron Works blast furnace plant which had been out of operation for about eight years. The tall chimneys and mineral hoists had been a major landmark on the south side of the Tees. The work was expected to take 50 men about 12 months to complete.

In its later years, the Tees Iron Works used a fleet of four-coupled saddle tank locomotives usually built by Andrew Barclay or Hawthorn Leslie. Some had come secondhand from other premises in Pease & Partners' empire. Unlike some of the other major colliery companies in Co. Durham, Pease did not own a central workshop for locomotive repairs. There were small engineering shops at Pease's West, Crook but the Tees Iron Works also undertook a number of major overhauls of Pease & Partners' locomotives, especially in the 1930s. When Ted Haigh visited Tees Iron Works on 27/5/1950 he noted MERRYBENT (HL 3053); ALEXANDRA (AB 979) under repair; HL 2247 green, lined yellow and green; GEORGE (K 1705) for scrap and HAREHOPE (HL 2799). The premises exhibited a much simpler railway layout serving the main foundry buildings, the river wharf and a single road engine shed located in the centre of the works.

When Pease & Partners Ltd withdrew from direct involvement in individual works following nationalisation of the coal industry in 1947, it established a subsidiary called Tees Foundries Ltd to run the iron works. Having become the north's principal cast iron rail chair maker, it was one of the first casualties when that business collapsed and the contract with BR was not renewed. Pease & Partners Ltd closed the Tees Iron Works in 9/1959. H. Butcher attempted to sell the business but much of the works was demolished by George Cohen, Sons & Co Ltd in 1961, the latter being liable for any MOR tolls from w.e.f. 22/11/1960.

Gauge : 4ft 8½in

	SPIDER	0-4-0ST	OC	GW	108	1861	(a)		(1)
	WASP	0-4-0VBT	OC	GW	169	1863	(b)		(3)
	BEETLE	0-4-0T		GW		1863	(c)		Scr
	MIDGE	0-4-0VBT		GW	188	1864	New		(2)
	BEE	0-4-0VBT	OC	HW		1871	(d)		(3)
	HORNET	0-4-0T	OC	FJ	116	1873	New		
		Rebuilt as ST							Scr
	HILDA	0-4-0ST	OC	HG	356	1874	(e)		Scr
	VICTORIA	0-4-0ST	OC	LE	240	1900	New		(4)
	ALEXANDRA	0-4-0ST	OC	AB	979	1903	New		
		Rebuilt		TIW		1936			(9)
1	PATRICIA	0-4-0ST	OC	HL	2993	1913	New		(5)
No.17	GEORGE	0-4-0ST	OC	K	1705	1871			
		Rebuilt		Tait*	90	1920	(f)		Scr 9/1950
	LASCELLES	0-4-0ST	OC	AB	136	1872	(g)		
		Rebuilt				1922			(6)
	MARY	0-4-0ST	OC	AB	1282	1912	(h)		s/s

Page 268 – Locomotive Worked Sites

	QUEEN MARY		0-4-0ST	OC	AB	1606	1918	(i)	(7)
	JOHN		0-4-0ST	OC	AB				
			Rebuilt				1902	(j)	Scr /1928
9			0-4-0ST	OC	HL	2247	1892		
			Rebuilt	TIW			1935	(k)	(8)
28	HAREHOPE		0-4-0ST	OC	HL	2799	1909		
			Rebuilt	TIW			1938	(l)	Scr 9/1950
21	No.1 MERRYBENT		0-4-0ST	OC	HL	3053	1914	(m)	(9)

* Carried a plate "James Tait Jnr & Partners Ltd, Engineers, Middlesbrough, No.90, 1920".

TIW Tees Iron Works

(a) new. Although Harman has it as a saddle tank from new, Inness suggests it had a tank (but does not specify what type) and was rebuilt with a saddle tank later.
(b) new, fitted with a horizontal and vertical boiler.
(c) Inness lists this 12 inch locomotive at the Tees Iron Works.
(d) IRS had "Gorton & Co of 1887" but Inness lists it as possibly built by Head Wrightson in 1871. It was fitted with a boiler having horizontal and vertical features.
(e) ex The Tees Side Iron & Engine Works Co Ltd, Tees Side Iron Works, Middlesbrough.
(f) ex Lackenby Iron Works, Grangetown, before /1922. To Pease & Partners Ltd, Ushaw Moor Colliery and Coke Ovens, Ushaw Moor, Co. Durham, /1922 and returned. At TIW on 3/7/1927. To Pease & Partners Ltd, Chilton Quarry and Lime Works, Chilton, Co. Durham, /1932. Ex Ushaw Moor Colliery and Coke Ovens, after 16/6/1946, by 5/8/1947.
(g) ex Tees Scoriae Brick Co Ltd, Cargo Fleet.
(h) ex Tees Bridge Iron Works, Stockton
(i) new. To T.& R.W. Bower Ltd (a Pease & Partners Ltd subsidiary), Allerton Main Collieries, Swillington, Yorkshire (WR), on hire, /1922 and returned /1922.
(j) ex Normanby Iron Works, Cargo Fleet, /1921.
(k) ex Henry Stobart & Co Ltd (a Pease & Partners Ltd subsidiary), Thrislington Colliery and Coke Ovens, West Cornforth, Co. Durham, 31/12/1946. This locomotive had made an earlier visit to Tees Iron Works for rebuilding.
(l) ex Pease & Partners Ltd, Frosterley Quarry, Frosterley, Co. Durham, by 5/1947. This locomotive carried a small rebuilding plate dated 4/1938 indicating an earlier visit to the Tees Iron Works.
(m) ex Pease & Partners Ltd, Frosterley Quarry, Frosterley, Co. Durham, by 5/1947.

(1) to The Witton Park Slag Co or Auckland Rural District Council, Witton Park Slag Works, Witton Park, Co. Durham.
(2) s/s. The Skinningrove Iron Co had a four-coupled vertical boiler locomotive called MIDGE, although this may have been one of the former Cochrane fleet.
(3) to Normanby Iron Works, Cargo Fleet, named WASP prior to 5/7/1923, BEE after 5/7/1923.
(4) to Normanby Iron Works, Cargo Fleet, c/1929; after 3/7/1927.
(5) to Pease & Partners Ltd, Bowden Close Colliery and Coke Ovens, Helmington Row, Co. Durham, after 3/7/1927.
(6) to Crossley's Commondale Pipe Co Ltd, Commondale Brick and Pipe Works, Commondale, Yorkshire (NR), c/1936.
(7) to Normanby Iron Works, Cargo Fleet.
(8) scrapped w/c 8/8/1959.
(9) scrapped by George Cohen, Sons & Co Ltd, 10/1961.

UPLEATHAM IRONSTONE MINE, New Marske 131
J.W. Pease & Co to 19/8/1882 NZ 625204 – **Map I**
Derwent Iron Co to 7/1857

The Derwent Iron Company (DIC) had begun erecting blast furnaces at Conside (later Consett) in 1840. As its own local supplies of ironstone became depleted, it began to receive ore from Grosmont and, early in 1851, from Bolckow & Vaughan's Eston Mine. The DIC obtained a lease from the Earl of Zetland and established Upleatham Mine on the northern face of the hill, just below Errington Wood, where the Cleveland Main Seam ironstone outcropped. The underground workings were to stretch south to the hamlet of Upleatham and as far as Tocketts. The mine became the second largest in the Cleveland ironstone field.

In 6/1851 the DIC started to construct a 2ft 1in gauge line, about 5 miles in length, from the mine to the Middlesbrough-Redcar railway at Coatham. This led from the ironstone workings (NZ 625204) – initially consisting of open quarries – via a self-acting incline down to the flatter countryside. The next level section of the line was operated by a stationary engine at each end: Long Beck Lane (NZ 623213) to west of Redcar Lane (NZ 607231). The tubs passed under Long Beck Lane (Engine Tunnel), Kirkleatham Lane (Marske Tunnel) and Redcar Lane (Redcar Tunnel). The final two miles were worked by locomotives, although a report in the *Darlington and Stockton Times* quoted William Cockburn, Pease's Mining Engineer, as saying that the tubs were "brought down by one engine as far as the culvert [presumably Coatham Tunnel under Green Lane], through which a horse dragged it to the other side, where it was again received by a second engine". At Coatham, the tubs were weighed and the contents tipped into wagons on the standard gauge siding, which then ran westwards alongside the main line, before joining it at Upleatham Junction, immediately north of Marsh House. The narrow gauge line opened in 4/1852 and generally followed field boundaries resulting in some sharp curves. Locomotives built by Thomas Richardson & Sons, Hartlepool, were supplied to work the line.

In 7/1857 Joseph Pease and Joseph Whitwell Pease agreed with the DIC to take over the mine and supply the latter with ironstone, just three months before the DIC failed as a company. The transfer was ratified by the Earl of Zetland on 13/11/1857. There were three drifts into the hillside to reach the ironstone, the Main Winning was near to the top of the self-acting incline and this was the location of the mine's workshops, stores and offices. From here, 2ft 1in gauge railways ran along the hillside to the East Winning and the West Winning. Stationary engines hauled the ironstone tubs out of each drift and the original locomotives pulled the wagons from the drift entrances to the incline bankhead. According to George Baker Foster, Lord Zetland's mining engineer consultant, on 27/10/1857 one of the locomotives had been brought up from Coatham and was working to the East Winning and another was intended to be moved to operate on the west side.

The narrow gauge line to Coatham had already been seen as inadequate. Pease wanted to use locomotives over the whole route, but two tunnels beneath roads precluded this and the parish council objected to level crossings, so a standard gauge railway was planned to connect the mine with the Middlesbrough-Redcar line at Coatham. G.B. Foster's report on 27/10/1857 stated that Pease had already started to lay the heavier rails and that it was intended to install a new incline – 250 yards long with a gradient of 1 in 11 – alongside and replacing the narrow gauge incline. There was a slight delay in the completion of the standard gauge branch, because the SDR (in which the Peases were heavily involved) received powers on 23/7/1858 to extend its Redcar line to Saltburn and this incorporated part of Pease's branch into its route.

The 2ft 1in gauge railway system of the Upleatham Ironstone Mine probably near the Main Winning where the principal buildings were located. (Kirkleatham Museum, Redcar & Cleveland Council collection)

Each mine tub held about 31cwt of ironstone and, on reaching the bankhead, they were turned over and the contents tipped down iron spouts into standard gauge wagons. These were let down the self-acting incline to the sidings at the bankfoot, alongside the 226 workers' terrace houses built by Pease (starting in 1865) at New Marske. A fine model of the incline was made in 1862 and resides in the Science Museum. The SDR collected the wagons and took them north, passing over Long Beck by a level crossing, until it reached the main line, just after passing under Black's Bridge near Marske. Additional sidings were provided here, the junction being controlled by Upleatham Signal Box and these continued in use after the mine and branch closed in 1923. According to Ken Hoole, the Upleatham Branch was 1 mile 62 chains long, of which 39 chains were owned by the railway company and the remainder by Pease. The SDR used to run a miners' train between Redcar and New Marske.

There is a suggestion in the Ken Fleming reference that the narrow gauge at Upleatham Mine was converted to 3ft 0in in 1871 and that the three new Black, Hawthorn locomotives, maker's numbers 206–08, ordered by Palmer's Shipbuilding & Iron Co Ltd for "Saltburn Ironstone Mines" were destined for Upleatham rather than Grinkle Mine. This seems unlikely and there is little mention of a change of gauge in the extensive reports from the mine.

Although Upleatham Mine had a good record for consistent working – it produced 606,953 tons in 1871 – with time, its reserves became depleted. Small amounts of ironstone were obtained from the reworking by Pease of Hob Hill Mine near Saltburn. A 1,938 yards long aerial ropeway was erected between Hob Hill and Upleatham Mine east end. It commenced operations in the week beginning 14/6/1902 and continued until early in 1921. A limited amount of working also took place in the Pecten Seam near the East Winning starting in 1910. At Upleatham Mine, the West Winning closed in 1912, the East Winning in 1921 and the Main Winning in 1923; the date of official abandonment was 30/12/1923. In the following year, the standard gauge branch was removed and Watts, Hardy & Co (1920) Ltd dismantled the plant.

References :   *Upleatham Ironstone Mine*, Simon Chapman, pub. Peter Tuffs, 2000

'Upleatham Ironstone Mines', Ken Fleming, *The Industrial Locomotive* No.20, ILS, 1980, pp.128-34 and letters in issue No.23, 1981, pp.198-99

'A Few Notes on Cleveland Ironstone Mines', Frank Jux, *The Industrial Locomotive* No.106, ILS, 2002, pp. 256-57

New Marske Looking Back, Ena Holloway and Alan Hughes, pub. Alan Hughes, 1982

Gauge : 2ft 1in

Name	Type	Cyls	Builder	Works#	Date	Origin	Notes
			* TR	208	1852	New	s/s
			* TR	209	1852	New	s/s
			* TR	210	1852	New	s/s
			* TR	232	1853	New	s/s
AMMONITE	0-4-0tank	OC	HG	198	1865	New	(1)
No.19			?				
			Rebuilt Lingford Gardiner			(a)	(2)
No.20			TG			(a)	(2)

* According to the ILS Journals of 7/1948 and 1/1949 quoting information from Thomas Richardson records, these locomotives were built for the Consett Iron Company (although it was then the DIC). An article in Engineering, 30/11/1866 confirms that some and possibly all came to Upleatham and that they were tender locomotives. A report dated 5/9/1863 of the British Association visit states that the participants were pulled "between the extreme points of the external works" by the little engines JOHN WHITWELL and GURNEY decorated with evergreen foliage!

(a) identity unknown.

(1) in the 7/1888 issue of *Machinery Market*, A. Bainbridge (which see) offered for sale a 2ft 1in gauge locomotive built by Hopkins Gilkes; this was probably AMMONITE.

(2) included in an Upleatham Mine List of Plant for Disposal, 10/1923. Both locomotives were specifically stated to be 2ft 1in gauge. A photograph of the cab of a locomotive and its crew taken in 1919 appears in the Holloway reference.

WHITECLIFFE IRONSTONE MINE, Carlin How 132
J.W. Pease & Co Ltd until 19/8/1882 NZ 711189 – Map I
Swan, Coates & Co until 10/1876
North Cleveland Ironstone Co Ltd until 8/1872

The mine was situated a short distance up the valley from Loftus Mine on the west side of the Carlin How–Loftus Road. John Marley acquired the lease for the ironstone on 18/3/1871 and then formed the North Cleveland Ironstone Co Ltd (incorporated 29/4/1871) with John Swan, ironmaster, as one of the partners. Ironstone began to be taken out of the mine on 12/10/1871 with the consignment being despatched to Messrs Gilkes, Wilson & Pease's iron works. The ironstone company's standard gauge railway was constructed by Walker & Dickinson and ran from the shafts, passed under the Carlin How–Loftus Road and then continued alongside the eastern bank of Kilton Beck in a very tight curve on unstable ground before reaching the Loftus Mine tip yard where the wagons could be handed over to the NER. By the time of the first train, the ballasting had not been completed, more sidings needed to be constructed and a wagon weighing machine installed. A month later the NER complained that the sharp curvature on the branch was liable to damage its rolling stock. The North Cleveland Ironstone Co had also purchased an engine to work the line and this was described as a "14hp locomotive" on 31/12/1871. Earlier in the year, John Marley reported that he had been offered a cheap John Fowler locomotive. Trevor Lodge has suggested that this may have been one from a batch (makers numbers 1539–43) constructed in 1871 and left on the manufacturer's hands; most of them ended up with companies in North East England. The NCICo certainly had two Fowler stationary engines.

The North Cleveland Ironstone Co was acquired in 8/1872 by Swan, Coates & Co which had its own iron works at Cargo Fleet. On 6/11/1872 it was recorded that the tubs each carried 26 to 30cwt of ironstone and there were 15 horses used in the pit. An improved alignment for the railway to Loftus Mine was agreed with the landowner, J.T. Wharton, on 6/12/1875. The curve round Kilton Beck was abandoned and the line now headed direct to the Loftus tip yard crossing the river on two bridges. It was brought into use in 1876. A valuation of equipment at Whitecliffe Mine dated 17/7/1876 included one *"Tank Locomotive"* kept in a corrugated iron engine shed. The same report referred to *"1 Badly Constructed Branch Railway & Sidings probably 1,200 yards of road"*!

Swan, Coates & Co was having financial difficulties and went into liquidation on 31/8/1876. J.W. Pease & Co acquired the mine on 11/11/1876 but preferred to concentrate production at its other Cleveland mines. Unable to agree a new tonnage rental with the land owner, Whitecliffe Mine closed on 25/1/1879 and most of the surface buildings and railway were dismantled in 2/1885. Later the ironstone reserves were worked from Pease's Loftus Mine South Drift which was extended in a straight line underground for some two miles to the north east corner of the royalty.

Reference : *Whitecliffe Ironstone Mine*, Simon Chapman, pub P. Tuffs, 1998.

Gauge : 4ft 8½in

	SPIDER	0-4-0ST	OC	GW	108	1861	(a)	(1)
	-	0-6-0T	IC	JF	1540	1871	(a)	(1)

(a) new to North Cleveland Ironstone Co, Loftus.

(1) in view of the early closure of the mine, it is possible that the locomotives were moved to Swan, Coates' Cargo Fleet Iron Works. At some later date, JF 1540 passed through the hands of dealer, Lingford Gardiner & Co Ltd of Bishop Auckland, which sold it to The North Birchburn Coal Co Ltd, North Birchburn Colliery and Brick Works.

PILE, SPENCE & CO LTD

SHIPYARD, West Hartlepool 133
Pile, Spence & Co until 2/1865 NZ 512332 – Map B
John Pile & Co until /1859
John Pile until /1857

In the 1840s the Pile family was building ships on the River Wear at Sunderland. As part of the development of Jackson Dock, Ralph Ward Jackson had allocated land for a shipyard on the west side. The dock opened in 1862 and John Pile began to lay out his shipyard on the site in the following year, including two building berths and a 335 feet long graving dock. John Pile launched the first ship to be constructed in the new town of West Hartlepool on 20/12/1854. He also established the West Hartlepool Rolling Mills (which see) to provide iron plates and launched an iron ship in 1856. With the completion of Swainson Dock in 1856, a second shipyard and graving dock was opened adjacent to John Pile's original yard. Although leased by Wood, Spence & Co, John Pile agreed to manage it. In 1859 Pile, Spence & Co took on the management of both yards. The West Hartlepool Harbour & Railway Co's line ran around the ends of the two docks and a siding entered the west side of Pile's shipyard. The premises totalled over 8 acres by 1861 with approximately 1,500 people employed. The north quay of Jackson Dock was used as a fitting out basin and had a set of sheerlegs that could lift 80 tons. In 1865 the shipyards, rolling mills and West Hartlepool Steam Navigation Company became a limited company but, in the following year, the business failed as a result of the collapse of the bankers, Overend, Gurney & Co and it ceased trading in 7/1866. A couple of ships were completed by the liquidator and Pile, Spence & Co Ltd was formally wound up in 12/1866. Denton, Gray & Co took over the shipyards in 1868 (see William Gray).

The Engineer on 25/1/1867 advertised that R. Benson was to auction on 7-8/2/1867, re the Liquidators of Pile, Spence & Co Ltd at the dockyard, West Hartlepool, plant including one four-wheel tank locomotive, with 7½in x 14in cylinders and 2ft 3in diameter driving wheels, by Adamson & Co.

Reference: *Shipbuilders of the Hartlepools*, Bert Spaldin, Hartlepool Borough Council, 1986

C.L. PROSSER & SONS LTD
Cargo Fleet   **134**
NZ 517206 – **Map G**

The company's premises are situated at the north east end of the Works Road Bridge on what was formerly part of the Normanby Iron Works site. It is mainly a waste processing business but has occasionally scrapped a locomotive.

Gauge : 4ft 8½in

-	4wDM	RH	275881	1949	(a)	Scr 9/2006
08807 (D3975)	0-6-0DE	Derby		1960	(b)	Scr 1/2017

(a)   ex G.C.S. Johnson (Skeeby) Ltd, Barton, near Richmond, North Yorkshire, 5/2006.
(b)   ex A.V. Dawson Ltd, Middlesbrough, 7/1/2017.

W. G. READMAN LTD
REDSTEEL WORKS, Cargo Fleet   **135**
NZ 513206 – **Map G**

After the closure of the Ormesby Iron Works in 1971, some of the land was cleared. W.G. Readman Ltd then occupied the western half of the site as steel stockholders, although substantial amounts of scrap were present in some later years possibly from the demolition of the Teesside Engineering Works on the other side of the Middlesbrough-Redcar railway.

Just one siding entered Readman's premises from the main line in 1976 dividing into two to enter a large building on the southern half of the site. MOR tolls continued to be paid between 1975 and 1982, except 1980–81. No records have survived after 1982. The Dockside Spine Road was constructed along the north side of the main line to give better vehicular access to works in 1985. This smartened up the front of the premises, Cochrane's derelict office block having been demolished in the previous year. There was very little, if any, rail traffic after 1982; indeed the Sentinel locomotive was being used as a mooring post for a barge at the river wharf in 1988.

Gauge : 4ft 8½in

-	4wDM	RH	425482	1958	(a)	(1)
-	4wDH	S	10017	1960	(b)	Scr 2/1989
-	4wDM	FH	4011	1966	(b)	Scr 17/3/1983

(a)   ex Stanton & Staveley Ltd, Ormesby Iron Works, Cargo Fleet, c/1973.
(b)   ex George Cohen, Sons & Co Ltd, Cargo Fleet, 11/1981; after 31/10/1981, by 19/12/1981.

(1)   scrapped in the week previous to 8/4/1982.

REDPATH DORMAN LONG LTD
TEESSIDE ENGINEERING WORKS, Cargo Fleet   **136**
(Subsidiary of **British Steel Corporation** until 4/1982 when sold to the **Trafalgar House Group**)
NZ 508203, NZ 510208 – **Map G**

British Steel Corporation until 1/6/1972
Teesside Bridge & Engineering Ltd until 28/7/1967 (Operation of works not taken over until 1/7/1968)

TEES SIDE No.2 (Manning Wardle 1327 of 1897) was shunting alongside the road separating the Tees Side Bridge Works from Warner's iron refiners premises on 8th October 1952.   (D.G. Charlton photograph)

Tees Side Bridge & Engineering Works Ltd until 2/6/1966

The origins of the company can be traced back to the formation of Gilkes, Wilson & Co in 1845 as an engineering business at Commercial Street in Middlesbrough. This amalgamated with Hopkins & Co, Teesside Iron Works to become Hopkins, Gilkes & Co in 1865. The latter's fateful involvement in the Tay Bridge disaster led to restructuring and Wilsons, Pease & Co provided new investment, changing the name of the business to the Tees Side Iron & Engine Works Co Ltd, but it continued to struggle. In 1895–96 Sir Christopher Furness negotiated the sale and restructuring of the company. He purchased that part of the business involving bridge building and construction engineering, together with its order book. The Tees Side Bridge & Engineering Works Ltd (TSB) was incorporated on 25/7/1896 under the leadership of Sir Christopher Furness and much of the plant and machinery from Commercial Street was transferred to a new site at Cargo Fleet.

The 4½ acre site was situated in the angle between Cargo Fleet Road and Marsh Road near the Whitehouse Signal Box. It had started life as the East Yorkshire Iron Works, but had been leased in 10/1881 by Westgarth, English & Co which began to construct steam engines for blast furnace blowers, rolling mills and ships, together with boilers. By 1895 Sir Christopher Furness controlled Westgarth English and it was subsequently reorganised to become Furness, Westgarth & Co with a move to the former Tees Engine Works premises in Commercial Street, Middlesbrough. TSB took over the buildings, cranes, fixed engines and railway on the Cargo Fleet site; there was no mention of locomotives. The name on the siding agreement was amended 22/12/1897. The premises had the advantage of space for expansion with open marsh to the west as far as Ormesby Beck belonging to the NER and OME. There were two main buildings on the site with open stock yards serviced by gantry cranes and a few short sidings connecting with the NER opposite the Navigation Inn. The NER's branch ran from the Middlesbrough–Redcar railway near Cargo Fleet Station to three of its sidings alongside Marsh Road. In addition to the Victoria Foundry on the east side of the road, these also served the North Ormesby Gas Works (company in existence 5/1/1866).

TSB was an early pioneer of large steel framed buildings, including involvement in the erection of the Manchester, Sheffield & Lincolnshire Railway's St John's Wood warehouse using 2,000 tons of steel. The main building at the Tees Side Bridge and Engineering Works in 1914 was the fabrication shop parallel to Cargo Fleet Road. The remaining shops were open gantries exposed to the elements. A goliath travelling steam crane operated on a short length of track at the Bridge Yard. Although an agreement with the NER dated 15/9/1903 gave permission for TSB to run its locomotives over the NER's tracks at Marsh Road, it is thought that TSB must have relied on rail cranes for some years.

The NER constructed additional sidings (known as Whitehouse sidings) by 1905 from Marsh Road to the south of TSB's works in order to reach a coal depot near Louisa Street. An agreement between the TSB and NER on 20/10/1905 took another line of the siding's westwards through the works over the marsh as far as Ormesby Beck. This was used by TSB to provide a headshunt on which 15 wagons could stand. The NER owned land next to Ormesby Beck adjoining the TSB premises and, on 23/2/1906, it gave approval for Warner & Co to tip "ballast" there using Warner's locomotive and wagons, provided that these did not block the crossing of TSB's headshunt. Unusually the railway company continued to own land within TSB's works, despite some sales to the company in 1932, 1939 and 1943. This LNER land included some of the central sidings, the area beyond Ormesby Beck and the rail access into the works from Whitehouse. Permission for TSB locomotives to run over these lines was renewed by the LNER in 1946 and the British Transport Commission confirmed in 1959 that steam cranes could also use them.

Government orders for 4in and 6in shell casings in 1915 resulted in the erection of a well-equipped machine shop. In the following year, North Ormesby Gas Works was purchased for £13,000 in order to provide a foundry. The retort house was transformed into a production bay with two 6ft 6in cupolas heating the metal and specific mention is made of the purchase of a locomotive and cranes. A rail chair foundry was also added and in 2/1923 began producing pressed steel sleepers, railway chairs, bearing plates, keys and clips. With the rundown of war work, a wagon building shop was laid out and in 1922 TSB built 100 steel hopper wagons for Bell Bros. It began constructing "transmission towers" in 1929 and erected its own pickling and galvanising plant in the following year. It also provided the steel work in 1929 for George Bennie's monorail; this Glasgow inventor went on to launch his prototype Bennie "rail plane" in the following year.

In 1930, TSB agreed to Dorman, Long & Co Ltd purchasing most of its shares and, over the years, much work originated from Dorman Long, such as 20,000 tons of ingot moulds annually and structures for new plant at Lackenby. The Bridge Yard also contributed to Dorman Long's orders for the Runcorn-Widnes and Thelwall Viaduct Road bridges. In World War 2, TSB constructed aircraft hangars and also took over the disused Cleveland Dockyard in 11/1940 (see Michael Baum & Co Ltd entry) to build landing craft, gunboats and rocket firing craft. After the war, an attempt by TSB to construct barges at the Cleveland Dockyard proved unsuccessful so it took over approximately five acres of the former Linthorpe Iron Works site. Two railway tracks, 50 feet apart, ran across the land and into the river. Four-wheel bogies travelling on the tracks carried the 90 feet long barges transversely down to the river. A total of 612 barges were fabricated here between 1945 and 1951. The line serving this site was known as the "Linthorpe Gully Sidings".

After World War 2, the railway companies embarked on a major programme of rolling stock replacement. TSB laid out production lines in the fabrication shops and a total of 20,200 wagons were constructed over about ten years for the LNER, BR and Dorman Long, It included 24 locomotive tenders purchased from BR and converted to slab carriers for Dorman Long. The company also supplied locomotive frames to Sentinel.

Railway Wagons Constructed by TSB post 1945.

Type for LNER/BR	Total	Type for Dorman Long & Co Ltd
21 ton hopper wagon	350	60 ton pig iron wagon
25 ton hopper wagon	24	40 ton ore wagon
20 ton coke wagon	100	25/30 ton general purpose wagon
55 ton trestrol wagon	4	25 ton pan wagon
120 ton trolley wagon	1	45 ton hopper wagon
25 ton low machine wagon	6	60 ton steel framed bogie wagon
30 ton bolster 'c' wagon	100	60 ton crop wagon
22 ton machine wagon	3	60 ton bloom transfer wagon
42 ton trestle ED wagon	48	60 ton ferro-manganese wagon
50 ton bogie rail wagon (steel)	79	scrap pan bogies
16 ton mineral wagon	17,777	scrap skip transfer carriages
16 ton mineral wagon with vacuum brakes	500	24 tenders converted to slab carriers
50 ton bogie rail wagon (timber)	137	Total 191
55 ton flat arm 'ET' wagon	28	
42 ton bogie bolster 'D' wagon	850	
40 ton flat trolley	2	
	Total 20,009	

Source : *Draft history of Tees Side Bridge & Engineering Works Ltd* by L. Burns, 1966.

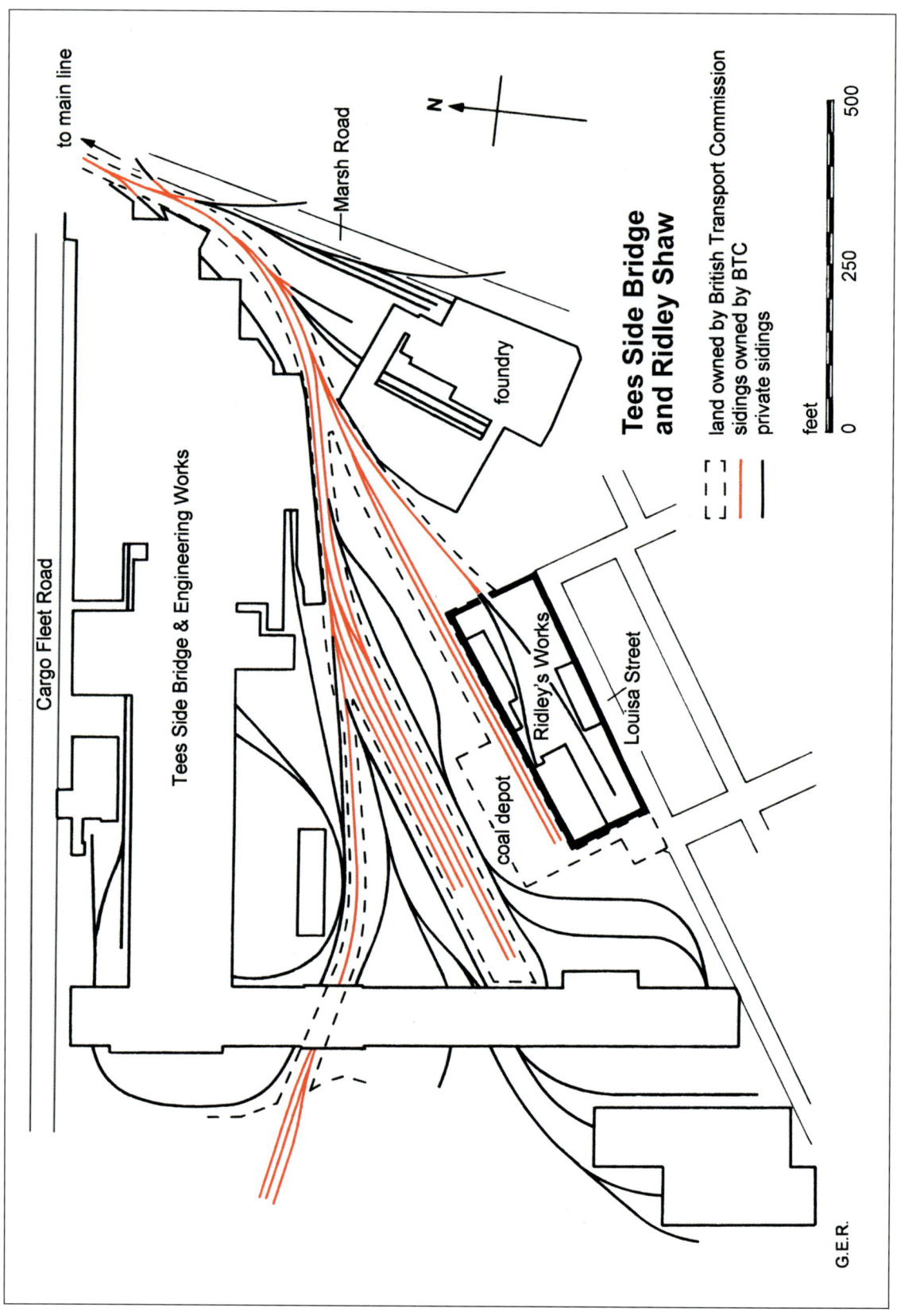

The area of the works had increased by nearly 10 acres between 1939 and 1950, with another 13½ acres added in the following decade. This included land on the east side of Marsh Road, north of Warner's, occupied by the 1938 office block, a modern galvanising plant erected in 1955 and a sports field. Much business came from the sales of electricity pylons for the 400kv national grid and steel for the construction of new power stations.

Indicative of its ad hoc assemblage, the coal depot remained within the site with road access from Louisa Street. Another unusual feature was that there were two small separate engine sheds – that near the foundry (NZ 510203) was described as a "brick building, slated roof with repair pit" in 1929 and the other was near the fabrication shops (NZ 508203). The single track BR connection to the main line near Cargo Fleet Station remained in use. TSB's engines could transfer traffic to BR's sidings alongside Marsh Road and there was also a railway which enabled them to reverse across this road to reach the galvanising plant.

Given Dorman Long's and TSB's involvement with Sentinel, it was perhaps not surprising that a new vertical boiler geared steam locomotive was bought from that manufacturer in 1957. Block type buffers of Dorman Long pattern were specified; it was painted Brunswick Green "with white diagonal line" and the nameplates were provided by TSB for fitting. Sentinel had earlier sent a 100hp demonstration locomotive, 9561 of 1953, which arrived at Teesside Bridge by 28/4/1954 and then moved to Patent Shaft Steel's works at Wednesbury on 7/5/1954. However, it was not long before TSB opted for diesel shunters and the Sentinel 9629 was eventually preserved at the National Railway Museum, although its current name FRANK GALBRAITH was not carried at TSB.

Following nationalisation, TSB became part of Redpath Dorman Long Ltd, a subsidiary of the British Steel Corporation. By the end of the 1970s, the works had become very rundown. A visit on 15/8/1979 revealed little activity – a few bogie bolster wagons were present, TH 105V was running light about the yard, a row of rail cranes stood out of use, but there were still some 21 ton wagons nearby for the coal depot. By 5/1982 FH 3933 was used about every two weeks to collect wagons from BR. It came as little surprise when the *Evening Gazette* on 3/1/1985 announced that the works was to close with the loss of the remaining 226 jobs. The decision was made by the Trafalgar House Group which had taken over the factory just under three years previously. The works was demolished by Vickers Dismantling Ltd in 3-5/1986.

References : *Tees Side Bridge The Rise, Fortunes and Dissolution of a Private Company,* Thomas R. Tighe, pub. British Steel Corporation, 1980.

'All-rounders, Battling Strongly to Double Century', *Steel News*, British Steel Corporation, 10/1968.

Gauge : 4ft 8½in

Name	Type		Maker	No.	Date		
TEES SIDE No.1	0-4-0ST	OC	HL	2646	1906	(a)	
	Rebuilt	Ridley Shaw			1944		s/s c/1961
TEES SIDE No.2	0-4-0ST	OC	MW	1327	1897		
	Rebuilt	Galloways			1922	(b)	
	Rebuilt	TSB			1936		Scr/1958
TEES SIDE No.3	0-4-0ST	OC	HL	2703	1907		
	Rebuilt	HL			9/1931	(c)	Scr c/1960
-	0-4-0ST	OC	HL	2839	1910	(d)	(1)
TEES SIDE No.4	0-4-0ST	OC	RSHN	7129	1945	New	(2)
TEES SIDE No.5	4wVBT	VCG	S	9629	1957	New	(3)
TEES SIDE No.6	4wDM		FH	3933	1960	New	(6)
TEES SIDE No.7	4wDH		TH	105V	1961	(e)	(5)
TEES SIDE No.8	4wDH		TH	115V	1962	New	(4)
TEES SIDE No.9	4wDM		FH	3935	1960	(f)	(6)

(a) ex Mold Collieries (1915) Ltd, Bromfield Colliery, Mold, Flintshire. The colliery had closed 7/1916 and the locomotive was advertised for sale there 7/1916.

(b) ex H. Widdowson & Sons Ltd, Carr Street Works, Nottingham. The locomotive had come to the dealer Widdowson from Galloways Ltd, /1934. It was sent from Tees Side Bridge to Cerebos Foods Ltd, Greatham, on hire, 1/1947 and returned 5/1947. To Smith, Patterson & Co Ltd, Pioneer Foundry, Blaydon Haugh, Co. Durham, on hire, /1947, after 5/1947 and returned /1947, after 2/9/1947.

(c) the locomotive originally belonged to William Cochrane Carr Ltd's South Benwell Colliery and Firebrick Works, Newcastle-upon-Tyne. When the colliery and firebrick works closed in /1931, the locomotive was sold to R. Frazer & Sons Ltd, dealer and, after repair by HL, was resold to Tees Side Bridge, /1940.

(d) ex The Darlington Forge Ltd, Darlington, Co. Durham (property of Thos W. Ward Ltd, Sheffield), on hire 11/3/1944.

(e) new. To NCB Lambton Railway, Philadelphia, Co. Durham for trials, 10/9/1962; then to NCB Seaham Colliery, Seaham, Co. Durham and returned to Philadelphia. To Tees Side Bridge, 27/9/1962.
(f) ex British Steel Corporation, South Teesside Works, Grangetown c11/1971.
(1) returned to Thos. W. Ward Ltd, Sheffield, ex-hire, c6/1945.
(2) to M. Henderson Clark Ltd, Middlesbrough, for scrap, 5/1967.
(3) to North Yorkshire Moors Railway, Grosmont, North Yorkshire, 7/1975.
(4) to Redpath Dorman Long Ltd, Britannia Works, Middlesbrough, c3/1979.
(5) scrapped on site by Vickers Dismantling Ltd, 3/1986.
(6) scrapped on site by Vickers Dismantling Ltd, 5/1986; FH 3935 was being cut up 1/5/1986.

RICHARDSONS WESTGARTH (HARTLEPOOL) LTD

HARTLEPOOL ENGINE WORKS, Hartlepool.   **137**
Richardsons, Westgarth & Co Ltd until 3/1938   NZ 521336 – **Map B**
Thomas Richardson & Sons Ltd until /1900
Thomas Richardson & Sons until /1871
Richardson Brothers until 1857

The Engine Works was located at Hartlepool Docks between the Central Dock and North Basin to the west and the Old Harbour to the east. However, the company's origins go back to the early days of the railways. In 1832 Thomas Richardson had leased land at Castle Eden in Co. Durham near the projected line of the Hartlepool Dock & Railway Company. The works was ideally situated to service the new railway and provide equipment for nearby collieries. It also built over 50 locomotives between 1840 and 1857.

Thomas Richardson joined with Joseph Parkin in 1836 to start building wooden ships at the Headland in Hartlepool. The first, called CASTLE EDEN, was completed in the following year but a section of the old town wall had to be taken down before it could be launched. As this was not very practical, it was decided to establish a shipyard on the west side of the Old Harbour. Parkin left the partnership in 1839 and Richardson mostly concentrated on his Castle Eden business but launched two ships in 1845–46. Thomas Richardson took over

Richardsons Westgarth's Hartlepool Engine Works mainly relied on Peckett 1510 of 1919 to carry out its shunting, 22nd April 1959.   *(D.G. Charlton photograph)*

the lease of the Middleton Iron Works in 1847, which was adjacent to his shipyard and began to develop a marine steam engine business renaming the premises Hartlepool Works. He died in 1850 and was succeeded by his two sons, the business becoming Richardson Brothers. The Castle Eden works closed in 1853 and, between 1855 and 1857, Richardson Brothers launched nine ships. Then financial difficulties caused a cessation of shipbuilding, with the yard passing to new owners and Richardsons concentrated on constructing marine engines and boilers for which there was an increasing demand. In 9/1865 Thomas Richardson & Sons joined with Denton, Gray and Richardson, Duck & Co of Stockton to form Richardson, Denton & Duck Co Ltd but this lasted barely a year because of the financial crisis in 1866 and the companies reverted to their old identities. The business was again in financial difficulties in 4/1871 and it was converted into a limited liability company, Thomas Richardson & Sons Ltd. It amalgamated with Sir Christopher Furness, Westgarth & Co Ltd of Middlesbrough and William Allan & Co Ltd of Sunderland to form Richardsons, Westgarth & Co Ltd (prospectus issued 2/11/1900).

Two of the NER's dock lines ran past Gray's Central Marine Engine Works, over the North Basin swing bridge and by the Hartlepool Engine Works on the opposite side of Central Road. The connection for the Hartlepool Engine Works sidings and the Middleton shipyard was initially near The Slake sluices, but the construction of the North Basin forced a change and, by the 1896 Ordnance Survey, the link was further south and controlled by the NER Princes Street Signal Box. Within the works, the siding ran through to a headshunt before reversing to serve the various buildings, including a line through to the Middleton Shipyard's rail system. The *NER Collieries, Works and Sidings Book* dated 1895 states "Messrs Richardson's locomotives work mineral traffic from N.E. sidings to their works, a distance over N.E. line of 33 chains and also to Messrs Furness, Withy & Co's works, a distance over N.E. line of 27 chains." Hartlepool Engine Works began by building two-cylinder simple steam engines but later developed compound and triple expansion engines. A new turbine shop was erected in 1903 on the south side of Princes Street and this resulted in one of Richardson's sidings being extended across the road to serve the new facility. The company later branched out into building steam driven alternators for factories and Doxford diesel engines for ships.

The shed where Richardsons maintained its locomotives and steam cranes was located alongside Central Road. Information on the early locomotives is scarce; it is believed to have had one or more built by John Harris of Darlington but no details have survived. A dispute arose between William Jackson at Devonport and Alexander Shanks & Son Ltd of Arbroath concerning a locomotive Shanks had advertised for sale in August/September 1864. The locomotive had been tried at "Messrs Richardson, engineers, West Hartlepool" on 21/4/1864 and a Mr Brown had reported that, with several days' work, it would be fine. Accordingly, the sale was agreed between Shanks and Jackson, but the purchaser then argued that the locomotive did not measure up to the guarantees given. The *South Durham Herald* in 12/1865 reported that Richardson, Denton & Ducks Hartlepool Works had acquired a newly patented locomotive from a Leeds builder "which carries its own water"!

The *Darlington & Stockton Times* on 3/7/1869 contained a report of an accident to two lads which took place on "a private piece of line between the North Eastern Railway and the extensive Middleton Ironworks of Messrs T. Richardson & Sons". It involved "a train of coal wagons being pushed along towards Hartlepool by an engine belonging to the firm". A new Peckett locomotive was purchased in 1919 and this became the works shunter for the next 40 years.

No more marine propulsion units were produced from 1961; the company had supplied engines to more than 500 ships since 1900. This also coincided with a greater use of road transport and rail traffic ceased about 1962. The company stopped making rotative machinery in 1967 and the southern part of the works closed. It concentrated on non-marine work such as turbo-alternators for power stations. The works became part of the Clark, Hawthorn division of British Shipbuilders on nationalisation vesting day, 1/7/1977, but closed on 18/12/1981.

References :   *"Richies" A History of Thomas Richardson & Sons and Richardsons, Westgarth & Co Ltd* 1832-1994, Peter L. Hogg, Hartlepool Borough Council, 1994

   Shipbuilders of Hartlepool, Bert Spaldin, Hartlepool Borough Council, 1986

Gauge : 4ft 8½in

-		0-4-0T(?)		HH			(a)	(1)
-		0-4-0T					(b)	s/s
-		0-4-0ST(?)					(c)	(2)
-		0-4-0ST	OC	AB	*	1883	(d)	Scr c/1922
1	R.W.C. LTD 533	0-4-0ST	OC	AB	823	1898	(e)	(3)
-		0-4-0ST	OC	P	1510	1919	New	Scr /1962
	WEAVER	0-4-0ST	OC	MW	1072	1888	(f)	(4)

*   carried Lennox Lange plate.

(a) origin doubtful; could be locomotive ex Carr House Iron Co, Carr House Iron Works, West Hartlepool.
(b) origin unknown.
(c) believed named FORREST or FORCETT; if latter, could be MW 549 of 1875, ex T.D. Ridley, West Hartlepool, on hire; originally Forcett Limestone Co Ltd, Forcett, Yorkshire (NR).
(d) ex a contractor, Hartlepool. The dealer, J. Torbock of Middlesbrough may have been involved.
(e) ex Middlesbrough Works, Middlesbrough.
(f) ex T.J. Thomson & Son Ltd, Stockton, on hire, /1947.

(1) for sale 7/1903, s/s by /1909.
(2) returned to T.D. Ridley, West Hartlepool, ex hire, (if MW 549 /1875).
(3) to Ashmore, Benson, Pease & Co Ltd, Stockton, c/1922.
(4) returned to T.J. Thomson & Son Ltd, Stockton, ex hire.

RICHARDSONS, WESTGARTH & CO LTD

Middlesbrough 138
Sir Christopher Furness, Westgarth & Co Ltd until /1900 NZ 499210 – **Map G**
Furness, Westgarth & Co Ltd until?

As part of the restructuring involving the Tees Side Iron & Engine Works Co Ltd, led by Sir Christopher Furness in 1896, some of the plant and equipment was transferred from the Tees Engine Works to the newly established Teesside Bridge & Engineering Works Ltd at Cargo Fleet (which see). In the reverse direction came the engineering business of Westgarth, English & Co, which was reformed as Furness, Westgarth & Co and occupied the Tees Engine Works premises. Shortly after, it changed its name to Sir Christopher Furness, Westgarth & Co Ltd and functioned as marine and general engineers. In addition to ships' engines and boilers, the company constructed plant for iron and steel works, together with gas engines. It also owned the strip of land on the north side of the Vulcan Street railway leading to the former Pottery Wharf, which was described as "Westgarth's Wharf and Box Store Yard". An agreement dated 7/9/1897 with the NER enabled Sir Christopher Furness, Westgarth & Co Ltd to use the Vulcan Street junctions and sidings. The NER charged £25, in lieu of a wayleave, for the use of its railway linking Westgarth's premises on either side of Vulcan Street. During 1900 Sir Christopher Furness, Westgarth & Co Ltd, Thomas Richardson & Sons Ltd of Hartlepool and William Allan & Co Ltd of Sunderland amalgamated to form Richardsons, Westgarth & Co Ltd; the formal prospectors being issued on 2/11/1900.

The internal railway layout within the works was not significantly different to that in 1875, although it still had a residue of short tight sidings. The constricted nature of Westgarth's sidings might suggest that it did not possess its own locomotive. However, a letter from the NER Middlesbrough Goods Agent dated 5/3/1897 includes "Furness, Westgarth & Co" in a list of private locomotives operating over the NER and MOR lines with rights to run over the Marsh Branch, Vulcan Street and the dock and Tees Union Wharf via Depot Road level crossing. Also, an agreement between Sir Christopher Furness, Westgarth & Co Ltd and the NER dated 18/10/1900 allowed the company's locomotive to pass over the NER lines between Westgarth's Wharf and R. Craggs & Sons' Tees Dockyard on part of the former Tees Side Iron Works site. The *Contract Journal* (24/6/1903) announced that an auction was to take place on 3/7/1903 of plant, including a 10 inch Henry Hughes locomotive, following reconstruction works at Richardsons, Westgarth & Co Ltd, Middlesbrough.

Richardsons Westgarth had purchased a site bordering the river, which previously formed part of the Tees Side Iron Works. A siding ran from the Old Town Branch across Depot Road and the remaining former Tees Side Iron Works to reach the site; the agreement for the NER to deliver traffic was dated 21/11/1910. Richardsons, Westgarth also had rights to berth vessels on the upstream section of the iron works river frontage, although there was no wharf on it. The premises functioned as an engineering works.

During World War 1, the Vulcan Street works was busy manufacturing shells, together with engines and boilers for ships. By 1923 it had built over 380 engines since the early days of Westgarth English, but the decreasing numbers of available ship's engine orders mostly passed to the Hartlepool Works. The Middlesbrough Works' last engines went on their trials in 1/1930 and the premises were then closed. The annual payment of £5 to the NER was terminated on 1/10/1930. The detached site up river, with its massive sheerlegs, then stood disused until it was acquired by T. Roddam Dent in 1947.

It appears that some salvage work was underway at the Vulcan Street premises because the LNER agreed on 31/1/1935 that the Hughes Bolckow Shipbreaking Co Ltd of Blyth could use its "locomotive steam crane" to haul wagons during daylight hours on the Vulcan Street sidings and the link into Richardson, Westgarth Engine Works.

The Vulcan Street premises were later leased by Head Wrightson in the late 1930s and specialised in the design and manufacture of finishing equipment for steel and non ferrous metals industries. Bomb casings were made here during World War 2. Head Wrightson purchased the premises after the war and, by 1951, individual sidings from Vulcan Street and Dock Street remained but much of the site was occupied by large buildings. The siding agreement was terminated 1/10/1970.

Reference : *"Richies" A History of Thomas Richardson & Sons and Richardsons, Westgarth & Co Ltd* 1832-1994, Peter L. Hogg, Hartlepool BC, 1994.

Gauge : 4ft 8½in

GEORGE LEEMAN	0-4-0ST	OC	*		1871	(a)	s/s
No.1	0-4-0ST	OC	AB	823	1898	(b)	(1)

* Inness states that it was built by "Hoppers, Britannia Iron Works, Fencehouses", presumably Hopper, Radcliffe & Co of the Britannia Iron Works. This company only produced about six locomotives. The Durham Handbook suggests it was built by Gilkes Wilson.

(a)   ex The Rosedale & Ferry Hill Iron Co Ltd, Ferry Hill Iron Works, West Cornforth, possibly via a dealer, J. Appleyard, who obtained it 12/1880.

(b)   the locomotive was ordered by the agent, Joseph Torbock, for "R. Hickman" according to Andrew Barclay records, but Inness shows it at the Richardsons, Westgarth Middlesbrough Works.

(1)   to Hartlepool Engine Works, Hartlepool.

RICHMOND IRON & STEEL CO

RICHMOND IRON WORKS, Stockton   139
Richmond Iron Works Ltd until c/1900   NZ 444176 – Map D

There is not much information on the early years of the Richmond Iron Works. In Samuel Griffith's *Guide to the Iron Trade of Great Britain* (1873), the Richmond Iron Works was operated by R. Jaques & Co in 1871, but was quite small with six puddling furnaces. It was located between the Bowesfield and Tees Bridge Iron Works, south of the Darlington-Middlesbrough railway. Robert Jaques was made bankrupt in 1875.

In 1890 the Richmond Works was offered for sale by William Henry Cowper, having been acquired from the late mortgagees. The works then occupied a site of four acres and consisted of 20 puddling furnaces, seven heating furnaces, one annealing furnace, a 20 inch forge train, 10 inch and 14 inch bar mills, together with a 20 inch sheet mill. It was said to be capable of producing each week, 250 tons of puddled bars, 350 tons of finished bars and 50 tons of iron sheets. It was proposed to establish a limited liability company – Richmond Iron Works Ltd – and shares were offered in it dated 26/2/1890. Amongst the seven subscribers to the Memorandum of Association were Henry Smith of the Eaglescliffe Foundry [sic] and John Stead. On 30/7/1890 the Richmond Iron Works Ltd signed an agreement with the NER to lease a parcel of land from the railway company next to the main line and constructed three sidings, two of 28 yards clear standage and one of 32 yards.

A connection with the main line was made at the NER sidings east of Bowesfield Junction. From here, a private railway ran south parallel to Bowesfield Lane to the Tees Bridge Iron Works. The sidings to the Richmond Iron Works diverged from this private railway near the Tees Bridge blast furnaces. The works comprised a large multi-roofed building containing the puddling furnaces producing wrought iron. One of the sidings ran through the building, while others enabled slag to be tipped on land to the east next to the River Tees.

The *Guide to the Ports on the Tees & the Hartlepools*, published in 1900, stated that the premises had just been reopened by James Riley as the Richmond Iron & Steel Co. The NER had altered the Siding Agreement to James Riley on 22/7/1897. No records of any locomotive exist, but it seems likely that there were some, unless the Tees Bridge Iron Works engines performed this role. A photograph in the 1900 Guide has a suggestion of a locomotive at the Richmond Iron Works. The NER recorded that traffic between the works and its lines was hauled by "private engines". A 1912 advertisement by the Richmond Iron & Steel Co indicated that it was a manufacturer of iron and steel in bars, strip, small angles and light rails. The company went bankrupt in 1913 and the works was purchased by Marple & Gillott Ltd in the following year probably for scrap. The site was then used by Marple & Gillott as a scrapyard until it was offered for sale in 1933 when it still possessed a railway siding. The Siding Agreement between Marple & Gillott and the LNER was terminated on 7/11/1936.

W. ROBINSON & CO LTD

CLEVELAND TIMBER YARD, North Ormesby   140
W. Robinson & Co until?   NZ 511201 – Map G

W. Robinson & Co's Cleveland Timber Yard was originally at "Dock Hill" and presumably near Middlesbrough Dock. Interestingly, at the same time in 1887/88, a W. Robinson was manager of OME's Cargo Fleet Timber

The elderly Andrew Barclay 148 of 1874 at Robinson's Cleveland Timber Yard in June 1952.

Yard. An entry for the Cleveland Timber Yard "off Smeaton Street, North Ormesby" first appears in a 1900/01 directory. The timber yard was situated on the south side of South Bank Road near its junction with Smeaton Street on the eastern edge of North Ormesby. In the early 1890s, the NER sidings alongside Marsh Road had terminated at William's coal depot at the end of Short Street but had not yet crossed South Bank Road.

The single line railway to the timber yard was constructed from the Marsh Road sidings across South Bank Road into Robinson's premises. A siding agreement was signed between the NER and Mr William Robinson trading as W. Robinson & Co on 5/5/1904 which allowed him to run with his locomotive over the NER lines adjacent to Marsh Road. This also permitted Robinson to reach the nearby premises of the Tees Side Bridge & Engineering Works Ltd and Warner & Co Ltd. The first payment had been made by Robinson under an agreement on 13/5/1902; the NER having listed the siding as an addition on 18/6/1900.

By 1913, the railway entering the timber yard split into three; two lines ran eastwards across the site to Middle Beck. The other headed south along the back of Stephenson Street to serve another site on the southern boundary of the timber yard. This site was shown as a casement window factory in 1952 but no longer served by rail. At this date, five large buildings with sidings dominated Robinson's timber yard. One of the sidings passed through the engine shed at NZ 511201.

Although Robinson had powers to operate a locomotive over the NER at Marsh Road, it did not mean that he necessarily owned one and he made payments to Warner & Co Ltd at least between 1903 and 1916 for shunting carried out on his behalf. Eventually Robinson purchased a secondhand elderly Andrew Barclay locomotive with 10in x 18in cylinders from Sadler & Co Ltd and this was belatedly replaced by a small diesel shunter in 1963. The yard closed about 1968.

Gauge : 4ft 8½in

BLUE GOWN	0-4-0VBT	VCG	HW			(a)	s/s
No.2	0-4-0ST	OC	AB	775	1896	(b)	s/s
-	0-4-0ST	OC	AB	148	1874		
	Rebuilt	#			1922	(c)	Scr/1963
-	4wDM		RH	235514	1945	(d)	(1)

\# Rebuilt by Cooke, Grundy & Stoddart, Barrow-in-Furness.

(a) ex Forcett Limestone Co Ltd, Forcett Quarry, Forcett, Yorkshire (NR). Information from Inness.
(b) ex Calders Ltd, Timber Yard and West Wharf, Middlesbrough.

(c)   ex Sadler & Co Ltd, Middlesbrough by 9/7/1949.
(d)   ex Anderston Foundry Co Ltd, Port Clarence, per Tomlinson, Hall & Co, Stockton, /1963, after 12/6/1963.

(1)   to George Cohen, Sons & Co Ltd, Cargo Fleet, c/1968.

T. RODDAM DENT & SON LTD

DENT'S WHARF, Middlesbrough   141
T. Roddam Dent & Co until /1911   NZ 494214 – **Map F**
George Watson until /1897

The SDR coal drops at Port Darlington extended along the south bank of the River Tees from a short distance west of the end of the Old Town Branch to approximately opposite the corner of Stockton Street and Commercial Street. By 1856 the coal drops had been dismantled and much of the site was vacant apart from the Patent Fuel Works at the west end, a rail served coal staith near the eastern boundary and a small building labelled "creosote works". The western half of the site was occupied by the Tees Side Iron Works and the smaller eastern portion was named creosote works by 1875. In the Agreement accompanying the 1884 NER Act, the eastern portion had been divided into two with the western half shown as belonging to the NER and the eastern half to Calder & Co's timber yard. Both sites had a series of short sidings curving off the Vulcan Street railway and terminating near the river bank.

George Watson rented the NER site and operated from it as a riverside wharf. In 10/1889 Mr T. Roddam Dent entered into partnership with George Watson. An 1892 plan describes the two sites as G. Watson (wharf) and Calder & Co timber yard, however, the 1895 OS plan surveyed in 1893 suggests that there was no physical boundary between the two and Watson may have had use of the river frontage on both sites with a continuous wharf backed by rail tracks for travelling cranes. There were nine sidings, each with space ranging from 31 to 62 yards to stable wagons. Following the death of George Watson in 1897, T. Roddam Dent took control of the wharf; the section on the land rented from the NER becoming No.2 Wharf at 238 feet long.

The complexity of the NER's attempts to charge for the use of its local lines and MOR tracks is well illustrated by the accompanying notes from the railway company's register. Both George Watson and Roddam Dent had possessed their own locomotives, according to the NER, although nothing is known of the former. "We have no powers to charge wayleaves at Watson's wharf on traffic between it and NE stations via NE line, or between it and works in the West Marshes," (5/11/1898). This did not include iron ore going to Normanby Iron Works,

Andrew Barclay 1196 of 1912 at Dent's Wharf on 22nd May 1960; the locomotive having moved from Deepwater Wharf two years previously.   *(IRS John Hill photograph)*

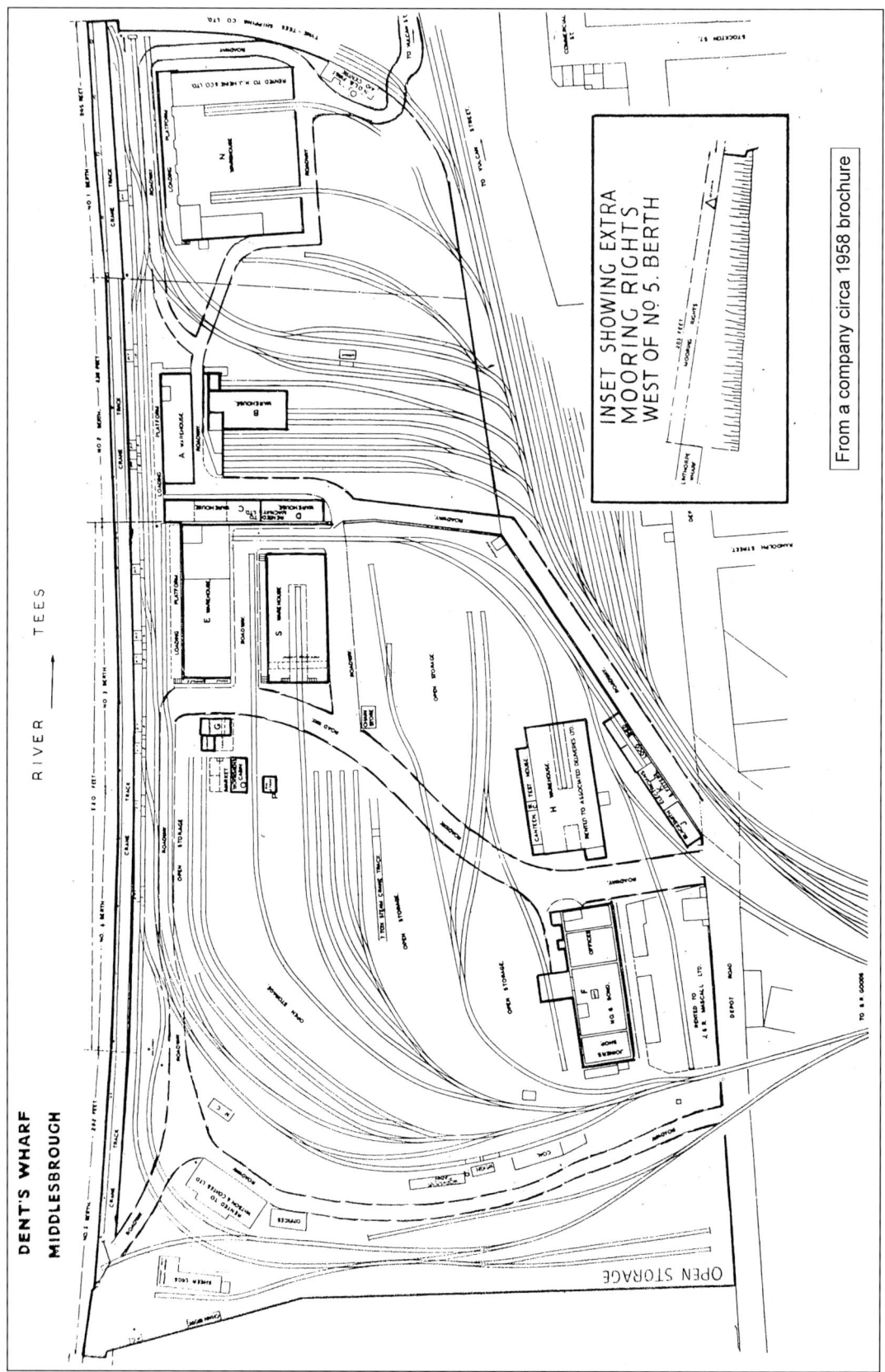

D2 (RSHN 7902/1958) shunts at Roddam Dent on 23rd April 1961. (ARPT Frank Bell photograph)

on which the latter's Dock Lock Bridge toll of 1d per ton was to be charged, plus the NER 'short distance' rate of 2d per ton for the section of NER line passed over at Watson's Wharf (27/1/1899). This latter rate, payable on traffic in owners' wagons hauled by owners' engines, freed the traffic from any wayleaves at that end (2/3/1899). If the iron ore was carried in NER wagons by a private locomotive via Dock Lock Bridge, the NER's view was that (5/4/1899), the charge would be 2d per ton 'short distance' rate, plus 1d per ton MOR's rate plus the wagon hire charge. Arrangements had to be made to collect the 'short distance' rate; the wagon hire would be picked up through the mineral accounts and MOR's rate through the wayleave returns!

In 1911 T. Roddam Dent & Son Ltd was formed as wharfingers, warehousemen and forwarding agents and purchased Messrs Craggs' shipyard immediately upriver from the existing premises. The shipyard was dismantled and a 520 feet long quay constructed with new cranes and warehouse accommodation. This became the company's Nos.3/4 berths. An agreement between the NER and Roddam Dent dated 22/1/1914 covered the new construction. The river frontage at the timber yard was repiled in 4/1926, the quay strengthened and a modern transit shed built. At 245 feet length, this became the No.1 berth and brought Roddam Dent's total wharfage up to approximately 1,000 feet with seven large warehouses. Vessels of up to 8,000 tons could be accommodated and there were twelve cranes ranging in size from 3 to 8 tons. Roddam Dent relied on a number of four-coupled saddle tank locomotives, mostly built by Hawthorn Leslie; some of the earlier engines were fitted with both spring and block buffers.

In 6/1947 Richardson Westgarth & Co Ltd's site adjoining the upriver boundary of Roddam Dent's premises was acquired and its 292 feet long wharf became the No.5 berth. The purchase included sheerlegs with a capacity to raise 80 tons and a 15 ton steam rail crane. In 10/1946 the LNER stated that this siding was "still operative but no traffic recorded as having passed during the past 20 years". Improvements were carried out to the wharf by sheet piling and new crane rail tracks were laid to connect the wharf with Nos.3 and 4 quays. New roadways were also constructed because of the increasing use of road vehicles. As part of the deal, mooring rights were secured for a further 300 feet upstream.

The land containing No.2 berth was finally purchased from BR in 1953. Sentinel's 100hp demonstration vertical boiler locomotive was seen at Roddam Dent in 3/1954, presumably for trials, between visits to Short Bros Ltd, Pallion Shipyard in County Durham and the Tees Side Bridge and Engineering Works. By 1958 Roddam Dent could handle vessels up to 500 feet in length, with a carrying capacity of 13,000 tons and had an extensive railway system serving its premises. The former Calders' timber yard was occupied by a large warehouse and the original Watson's site had another four smaller warehouses. These were served by six sidings connecting with the Vulcan Street railway and its associated lines. The remainder of the premises, with berths 3 to 5, had more and longer sidings with several areas designated for open storage. The company owned five steam rail cranes with a lifting capacity of up to eight tons for use on the open storage tracks; these formed part of a stock of twelve rail cranes. There were three separate crossings of Depot Road to enable berths 3-5 to be connected to the lines at Middlesbrough Town Goods Yard. According to a brochure published by Roddam Dent about 1958, all of the shunting in the yards and the movement of wagons to and from the exchange sidings was carried out by the company's six locomotives. The single road engine shed (NZ 494214) was located at the end of a block containing the blacksmiths, fitters and electricians behind a wall and next to the Vulcan Street railway's Depot Road crossing. At the end of the 1950s Roddam Dent began to dieselise its fleet of locomotives and purchased two new four-coupled diesels with mechanical transmission, to be followed by a couple of secondhand diesels in 1960.

On 21/10/1960 it was announced that the Tyne-Tees Steam Shipping Co Ltd had bought T. Roddam Dent to become owners of Dent's Wharf. However, both premises passed to the Tees & Hartlepool Port Authority (which see) on 1/1/1967.

Reference : *Dents Wharf*, brochure produced by T. Roddam Dent & Son Ltd, c/1958

Gauge : 4ft 8½in

Name	Type	Cyl	Builder	Works No	Year	Acquired	Disposal
-	0-4-0ST	OC	Joicey	338	1890	(a)	(1)
(No.2)	0-4-0ST	OC	CF	1212	1901	New	Scr 9/1959
T.R. DENT No.1	0-4-0ST	OC	HL	2662	1906	New	Scr/1961
T. RODDAM DENT & SON LTD No.3	0-4-0ST	OC	HL	2940	1912	New	(3)
T. RODDAM DENT & SON LTD No.4	0-4-0ST	OC	HL	2998	1913	New	(3)
-	0-4-0ST	OC	BH	421	1877	(b)	s/s
T. RODDAM DENT & SON LTD No.5	0-4-0ST	OC	HL	3911	1937	New	(3)
EAST LAYTON	0-4-0ST	OC	HL	2871	1911	(c)	(2)
T. RODDAM DENT & SON LTD No.6	0-4-0ST	OC	RSHN	7311	1948	New	Scr /1963
No.7	0-4-0ST	OC	P	2065	1945	(d)	Scr c4/1964
No.1 (No.8)	0-4-0ST	OC	HL	2730	1907	(e)	Scr /1962
No.2	0-4-0ST	OC	AB	1196	1912	(e)	(3)
D2	0-4-0DM		RSHN	7902	1958	New	(6)
D1	0-4-0DM		RSHN	7925	1959		
			DC	2592	1959	New	(6)
3 (ROF No.8, No.2)	0-4-0DM		JF	22942	1941	(f)	(4)
T. RODDAM DENT & SON LTD No.4	0-4-0DM		RSHN	6978	1940	(g)	(6)
KILMARNOCK	0-4-0ST	OC	AB	814	1898	(h)	(5)

(a) ex T.D. Ridley & Sons, Middlesbrough.
(b) ex Connal & Co Ltd, Middlesbrough. It was seen at Dent's Wharf by T. Haigh on 16/4/1934.
(c) ex Forcett Limestone Co Ltd, Forcett, Yorkshire (NR), on hire, by 1/1/1947.
(d) ex Thos W. Ward Ltd, Templeborough Works, Sheffield, Yorkshire (WR), 10/1957; previously ICI Ltd, Weston Point Works, Runcorn, Cheshire.
(e) ex The Deepwater Wharf Ltd, Cargo Fleet, /1958.
(f) ex G. Stephenson (Builders & Contractors) Ltd, Bishop Auckland, Co. Durham, 1/1960 per H. Dunn; previously ROF Thorp Arch, Yorkshire (WR).
(g) ex G. Stephenson (Builders & Constructors) Ltd, Bishop Auckland, Co. Durham, 1/1960 per H. Dunn; previously Air Ministry, Warton, Lancashire.
(h) ex Tyne-Tees Steam Shipping Co Ltd, Middlesbrough, on hire, /1961.

(1) replaced /1906, to Sir S.A. Sadler Ltd, Cleveland Chemical Works, Middlesbrough.
(2) to Forcett Limestone Co Ltd, Forcett, Yorkshire (NR), ex hire, after 1/1/1947.
(3) to A. Bainbridge Ltd, Thornaby, for scrap, c7/1960, after 22/5/1960.
(4) s/s, after 23/4/1961.
(5) returned to Tyne-Tees Steam Shipping Co Ltd, ex hire, c9/1961.
(6) taken over by Tees & Hartlepool Port Authority, with site, 1/1/1967.

SADLER & CO LTD

CLEVELAND CHEMICAL WORKS, Middlesbrough
142
Sir S.A. Sadler Ltd until?
NZ 505202 – **Map G**
Sadler & Co Ltd until /1901
Samuel A. Sadler until 29/9/1883

Samuel Sadler was employed running the laboratories of Messrs Chance, chemical and glass makers, at its Oldbury factory in the West Midlands. He experimented with coal tar products and then established a small aniline works, using benzene derived from coal tar in Marton Road, Middlesbrough during 1862. He expanded his activities by starting the Cleveland Chemical Works in 1869 on a site next to William Jones' Middlesbrough Chemical Works (which see). The new factory was located on the east side of the NER line to Guisborough shortly after that branch diverged from the Middlesbrough-Redcar railway. The premises eventually occupied an inverted triangular-shaped site of 24 acres with its base at the northern end next to Cargo Fleet Road. A railway connection was provided off the Guisborough Branch joining one of Sadler's tracks which formed a headshunt at the southern apex of the triangle. The 1913 Ordnance Survey plan shows an "engine house" at the end of the headshunt and near the Branch level crossing at North Ormesby Road. A NER siding agreement with Sadler dated 23/6/1909 gave approval for the latter to take water from the column and run its locomotives over the Railway Company's railway and sidings near the North Ormesby crossing. A charge of 6d was made for each time Sadler's locomotive used the water column.

The expansion of the iron industry with its reliance on coke and the increasing use of gas to light the streets resulted in large amounts of the by-product coal tar becoming available and Sadler used this to his benefit. An undated plan in OME records reveals that W. Jones occupied a larger site than Sadler but, in 1880, the latter acquired the Middlesbrough Chemical Works and incorporated it into his own premises. The two works continued to operate separately for some years – the former Jones site making sulphuric acid and oleum from pyrites (for 'pickling', cleaning finished steel) and producing alkali by the Leblanc process, while Sadler's original site distilled benzene from tar and used it to make aniline, the basis of a range of synthetic dyes. Increasingly the overall premises concentrated on the manufacture of chemicals from coal tar and later oil.

Sadler formed a limited company in 1883 with Sir Raylton Dixon and two directors from Samuelson's, by which time he employed 500 people. Subsidiary works were established at Ulverston, Portsmouth, Carlton and Stockton, although the Middlesbrough plant retained the central focus. At one stage it also owned seven collieries. Samuel Sadler played a major role in civic affairs, was knighted in 1901 and, when he died in 1911, a monument funded by public subscription was erected in Victoria Square, Middlesbrough.

The connection to the railway company had been rationalised by 1884 and a single track left a NER siding next to the Guisborough Branch to run behind a signal box before entering the works. However, OME had retained ownership of a narrow strip of land which allowed MOR tolls to be charged on traffic entering and leaving the works although, at some date after 1897, these no longer applied. It is likely that the locomotive information

Sadler's No.4 was interesting in that it was originally constructed by the Butterley Co in Derbyshire as a four-coupled tender locomotive. It was rebuilt by Butterley in 1895 as a saddle tank and moved to the Cleveland Chemical Works about 1942. *(Kevin Lane collection)*

for Sadler is incomplete, but a chemical works would have less need for shunting within the premises because many of the liquids can be moved through pipes. By the 1950s, Sadler relied on two steam locomotives, Nos.3 and 4, until it purchased a small Ruston & Hornsby 88DS diesel shunter in 1958.

Not surprisingly, Sadler was a major user of railway tank wagons to transport its chemicals around the country. Although the NER often insisted that its wagons were used, this did not apply to specialist vehicles and private companies had to purchase their own tank wagons. These were not even taken over by the state in 1948 and continued to be built for private businesses. The NER *Private Owner Wagons Register* gives details of Sadler's vehicles allowed to run on the railway company's network, starting with 10 ton tank wagons Nos.1 and 2 built by McLachian & Co of Darlington in 1892. By 1915 the numbers allocated to these vehicles had reached 210. They ranged in size from 8 to 20 tons; tanks were either rectangular or cylindrical in shape and were constructed by various companies, including Charles Roberts, the Rolling Stock Co, Darlington Wagon & Engineering Co and Hurst Nelson. The wagons built by Charles Roberts for Sadler were painted red and a peculiarity of some was that "Middlesborough" was misspelt. When five tank wagons for sulphuric acid were delivered by the Gloucester Railway Carriage & Wagon Co Ltd, the fleet numbers had reached 314. The final known order for Sadler was twelve tank wagons, 319–30, received in 1947. Sadler's works had a wagon shed with two through roads to maintain its rolling stock.

After Samuel Sadler's death, there was a move to concentrate the company's activities at Middlesbrough and dispose of the other works. However, to augment its supplies of crude tar, Sadler owned coke ovens at Malton Colliery and Randolph Colliery in Co. Durham. By the late 1940s, it continued to specialise in tar distillation producing refined naphthalene, disinfecting and sanitary fluids, sulphuric and hydrochloric acid, road tar and bituminous emulsions. In 1955 it created five new registered companies based on its various activities under the parent Sadler & Co Ltd. The NCB withdrew supplies of coal tar from Sadler in 1962 so that it could carry out its own processing. A 1968 Directory gives Sadler's works as Cargo Fleet and at Randolph Coke Ovens, but Sadler increasingly turned to become a storage and distribution business. It constructed a set of storage tanks at Middlesbrough Dock and in 1964 joined with Gebr. Broere NV of Holland to form Tees Storage Ltd (which see). On 22/2/1966 the Hays Wharf Group took over Sadler and, in 1971, the Middlesbrough works closed. The site is still occupied by a chemical company but no railways are involved.

References : *The Life and Times of Sir S.A. Sadler 1842-1911*, J.T. Smith, Remember Middlesbrough Society, 1994

'The Early Chemical Works of Tees-side', D.M. Tomlin, *The Cleveland Industrial Archaeologist* 4, CIAS, 1975

Private Owner Wagons A Fourteenth Collection, Keith Turton, Lightmoor Press, 2016, pp101-02

Gauge : 4ft 8½in

No.1		0-4-0ST	OC	Joicey	338	1890	(a)	Scr
No.2		0-4-0ST	OC	AB	148	1874	(b)	
		Rebuilt		#		1922	(1)	
-		0-4-0ST	OC	BH	479	1880	(c)	s/s
No.3		0-4-0ST	OC	P	1821	1931	New	(2)
No.4		0-4-0ST	OC	Butterley				
		Rebuilt		Butterley		1895		
		Rebuilt		Butterley		1917		
		Rebuilt		Ridley Shaw		1942	(d)	Scr c/1958
No.5		4wDM		RH	425482	1958	(e)	(3)

#	Rebuilt by Cooke, Grundy & Stoddart, Barrow-in-Furness.
(a)	ex T. Roddam Dent & Co, Dent's Wharf, Middlesbrough. The locomotive was replaced at Dent's Wharf in 1906 and confirmed at Sadler's 23/3/1915.
(b)	probably ex W. Jones & Co, Middlesbrough Chemical Works, Middlesbrough with premises. In view of the rebuilding, it may have been at Sadler's Sandhall Chemical Works, Ulverston, Lancashire, on loan c/1922.
(c)	ex Sadler's Sandhall Chemical Works, Ulverston, Lancashire, c/1925.
(d)	ex Butterley Co Ltd, Ripley, Derbyshire, c/1942, per dealer Steel & Co, Sunderland and via Ridley Shaw where rebuilt /1942.
(e)	new. Ordered by North Eastern Tar Distillers (Sadlers) Ltd for Cleveland Chemical Works.
(1)	to W. Robinson & Co Ltd, Cargo Fleet, by 9/7/1949.
(2)	to Tunnel Portland Cement Co Ltd, West Thurrock, Essex, c11/1958.
(3)	to Stanton & Staveley Ltd, Ormesby Iron Works, Cargo Fleet, c/1966.

SAHAVIRIYA STEEL INDUSTRIES UK LTD
TEESSIDE WORKS, South Bank, Lackenby, Redcar

143
NZ 565257 NZ 554214 – **Map H**

See the British Steel Ltd entry for the earlier history of the works.

Following the mothballing of the iron and steel plants at Teesside Works on 20/2/2010, strenuous efforts were made to sell them by Corus, which was now a subsidiary of Tata Steel. A memorandum of understanding was signed by Tata Steel and Sahaviriya Steel Industries (SSI) of Thailand for the sale. Corus was rebranded as Tata Steel Europe at the end of 9/2010. SSI formally took control of the site at midnight 24/3/2011. The GECT locomotives and torpedo ladles were removed from store, but many of the diesels were in poor condition after a hard working life. Eventually GBRF (a subsidiary of Veolia) won a 10 year contract on 25/2/2011 to take over the rail operations for SSI. GBRF leased a fleet of ex-Norwegian State Railways Di8 locomotives from Beacon Rail to haul the hot metal torpedoes and handle steel slab going to Tees Dock. Maintenance of the locomotives was subcontracted to Electro-Motive Diesel Ltd which took over the locomotive workshop at Lackenby in 12/2011. A substantial amount of work was required on the blast furnace and it was not until 15/4/2012 that it was relit with the first ship starting taking on steel slab at Tees Dock for rolling at the SSI works in Thailand on 11/5/2012. The two coke oven plants also formed part of the sale. Redcar Wharf became a jointly operated facility between SSI and Tata Steel. SSI was also responsible for shunting Tata Steel's slab for its Universal Beam Mill. By the end of 1/2013, two million tons of steel slab had been cast since the restart. To aid the long-term viability of the blast furnace, a pulverised coal injection plant was opened in 2013 in order to achieve cost savings. Various adjustments were required to the Di8 locomotives before they could shunt the concast and stocking areas and a few of the GECT locomotives carried out this work until the week ending 30/4/2014.

After a significant reduction in the world price of steel, SSI announced on 18/9/2015 that it was "pausing" iron manufacture at Redcar and mothballing the South Bank Coke Ovens. The final iron was run off from the blast furnace and the South Bank Coke Ovens closed on 19/9/2015. The financial problems were such that Sahaviriya Steel Industries UK Ltd went into liquidation on 2/10/2015. As no buyers came forward for the plant, the Official Receiver announced on 12/10/2015 that the Redcar blast furnace and coke ovens would close. The final coke from the Redcar Coke Ovens was produced on 15/10/2015.

On arrival, five of the Di8 type locomotives were unloaded at Tees Dock on 19th December 2011 from the ship EEMS CHRYSTAL and taken by GECT number 277, with 276 at the rear, to SSI's Teesside Works. The company's BOS steel plant is in the background. *(Cliff Shepherd photograph)*

Lengthy discussions were required with Thai banks before the former SSI Teesside Works site was acquired by South Tees Development Corporation, trading as South Tees Site Co Ltd, from 20/2/2020 with a view to redevelopment. This included the coke oven locomotives at South Bank and Redcar. Demolition of South Bank coke ovens took place in 2021 and is continuing on the remainder of the SSI site; the iconic former Dorman Long coal storage tower being blown up on 19/9/2021 and the Redcar blast furnace on 23/11/2022.

Gauge : 4ft 8½in

No.	Name	Type	Builder	Works No.	Year	Note	Ref
251	WALTER URWIN*	6wDE	GECT	5414	1976	(a)	(13)
252	BOULBY*	6wDE	GECT	5415	1976	(a)	(15)
253	ESTON*	6wDE	GECT	5416	1976	(a)	(17)
255	LIVERTON*	6wDE	GECT	5418	1976	(b)	(10)
257	NORTH SKELTON*	6wDE	GECT	5426	1977	(a)	(17)
258	GRINKLE	6wDE	GECT	5427	1977	(c)	(7)
260	ROSEDALE*	6wDE	GECT	5429	1977	(e)	(17)
261	STAITHES*	6wDE	GECT	5430	1977	(b)	(8)
262	LOFTUS*	6wDE	GECT	5431	1977	(a)	(9)
265	ROSEBERRY*	6wDE	GECT	5462	1977	(a)	(17)
266	SHERRIFFS*	6wDE	GECT	5463	1977	(b)	(17)
267	SLAPEWATH*	6wDE	GECT	5464	1977	(a)	(10)
268	KIRKLEATHAM*	6wDE	GECT	5465	1977	(c)	(8)
269	LONGACRES*	6wDE	GECT	5466	1977	(a)	(17)
270	CHALONER	6wDE	GECT	5467	1977	(d)	(4)
271	GLAISDALE*	6wDE	GECT	5469	1978	(c)	(16)
273	KILTON	6wDE	GECT	5471	1978	(d)	(3)
274	ESKDALESIDE	6wDE	GECT	5472	1978	(d)	(2)
276	SPAWOOD*	6wDE	GECT	5474	1978	(c)	(17)
277	WATERFALL*	6wDE	GECT	5475	1978	(c)	(8)
278		6wDE	GECT	5468	1977	(d)	(1)
-		6wDE	GECT	5421	1977	(d)	(5)
279		6wDE	GECT	5478	1978	(d)	(6)
LM20	LACKENBY 1*	4wDM	Robel 54.12-107 AD184		1980		
			Rebuilt Lackenby		2010	(c)	(8)
(LM21) 2 REDCAR 22*		4wDM	Robel 54.12-107 AD183		1980	(c)	(17)
8.712+		BoBoDE	Mak	1600.012	1996	(f)	(14)
8.716+		BoBoDE	Mak	1600.016	1996	(f)	(14)
817 (8.717)+		BoBoDE	Mak	1600.017	1996	(f)	(14)
8.719+		BoBoDE	Mak	1600.019	1996	(f)	(14)
820 (8.720) POPPY+		BoBoDE	Mak	1600.020	1996	(f)	(14)
8.701+		BoBoDE	Mak	1600.001	1996	(g)	(14)
8.703+		BoBoDE	Mak	1600.003	1996	(g)	(14)
8.704+ PAT		BoBoDE	Mak	1600.004	1996	(g)	(14)
8.711+		BoBoDE	Mak	1600.011	1996	(g)	
8.718+		BoBoDE	Mak	1600.018	1996	(g)	(14)
8.702+		BoBoDE	Mak	1600.002	1996	(h)	(12)
8.708+		BoBoDE	Mak	1600.008	1996	(h)	(11)

* After the SSI UK Ltd liquidation, these locomotives were purchased by Ed Murray & Sons Ltd on 17/11/2015 at the Teesside Works. While some were brought back into use, 253, 257, 260, 265, 266, 269, 276 and number 2 Robel remained out of use near the former Lackenby Workshops on what had become South Tees Site Co Ltd's premises.

+ Leased locomotives from Beacon Rail, operated by GBRF Ltd and maintained by Electro-Motive Diesel Inc. Following the cessation of iron making, all of the locomotives except for 8.711 were stored at PD Ports, Tees Dock from c24/9/2015 until 4/2016. 8.711 remained at Lackenby, having never run there and being partially dismantled.

(a)   ex Tata Steel Europe 25/3/2011 with part of site. Mothballed locos removed from Torpedo Shop and stored next to Lackenby Workshops.

(b)   ex Tata Steel Europe 25/3/2011 with site. Mothballed locos removed from Torpedo Shop and saw some use.

Just one day after the closure of the Redcar blast furnace was announced by SSI, 8.708 still in its former Cargonet livery was helping remove the last of the iron from the furnace for tipping into the iron ponds on 19th September 2015. *(Cliff Shepherd photograph)*

(c) ex Tata Steel Europe 25/3/2011 with part of site.
(d) ex Tata Steel Europe 25/3/2011 with part of site. Locos dumped out of use.
(e) ex Tata Steel Europe, Shapfell Quarries and Works, Shap, Cumbria, 8/8/2011.
(f) ex Norges Statsbaner (NSB), imported into Tees Dock via Uddevalla Docks, Sweden, 19/12/2011.
(g) ex Norges Statsbaner (NSB), imported into Tees Dock, 1/2/2012.
(h) ex Cargonet, Norway, 11/2012.

(1) removed from store c5/2011 and overhauled at Lackenby Workshops using the final rebuilt Dorman engine, to Tata Steel Europe, Shapfell Quarries and Works, Shap, Cumbria, 10/8/2011.
(2) out of use on a wagon next to the Torpedo Shop by 15/4/2010. Scrapped on site c5/2013.
(3) scrapped on site c6/2013.
(4) out of use by 5/2004 (?) and located near Lackenby Workshops by 25/1/2011. Scrapped on site c5/2013.
(5) out of use since arrival at Lackenby; scrapped on site, c5/2013.
(6) scrapped on site, c5/2013.
(7) scrapped on site, c10/2013; after 24/8/2013, by 1/1/2014.
(8) to Longs Steel UK Ltd, Teesside Beam Mill, Lackenby, property of Ed Murray & Sons Ltd, 17/11/2015. The locomotives remained on the Teesside Works site but changed ownership after the SSI UK Ltd liquidation. (see British Steel Ltd, Teesside Works – Teesside Beam Mill entry).
(9) to Chasewater Light Railway & Museum Company, Brownhills, Staffordshire, w/e 9/1/2016 (property of Ed Murray & Sons Ltd); via Moveright International, Wishaw, Warwickshire, from 15/12/2015.
(10) to Chasewater Light Railway & Museum Company, Brownhills, Staffordshire, (property of Ed Murray & Sons Ltd), 19–20/12/2015.
(11) to Tata Steel Europe, Appleby-Frodingham Works, Scunthorpe, Lincolnshire, w/c 21/12/2015.
(12) to Tata Steel Europe, Appleby-Frodingham Works, Scunthorpe, Lincolnshire, 5-6/1/2016.
(13) to Chasewater Light Railway & Museum Company, Brownhills, Staffordshire, (property of Ed Murray & Sons Ltd), w/c 9/1/2016.
(14) to Tata Steel Europe, Appleby-Frodingham Works, Scunthorpe, Lincolnshire, 5-15/4/2016. The individual locomotives moved on the following dates – 820 (5th), 8.719 (6th), 8.716 (7th), 817 (8th), 8.712 (11th), 8.718 (12th), 8.703 (13th), 8.701 (14th), 8.704 (15th).

(15) dismantled behind the torpedo shop and remains scrapped c/2016: after /2015, by /2019.
(16) to Ed Murray & Sons Ltd, Casebourne Road Depot, Hartlepool, 9/6/2017.
(17) to British Steel Ltd, Teesside Beam Mill, Lackenby, (property of Ed Murray & Sons Ltd), 25/3/2022.

South Bank Coke Ovens, South Bank   NZ 536214

Gauge : 4ft 8½in

1		4wWE	BSC	Hartlepool	1986	(a) (1)
2		4wWE	BSC	Hartlepool	1986	(a) (1)

(a) ex Tata Steel Europe with coke ovens, 25/3/2011.

(1) scrapped on site, probably by Thompson's of Prudhoe, w/c 7/3/2022.

Redcar Coke Ovens, Redcar   NZ 562257

Gauge : 4ft 8½in

1		4wWE	GB	420355/1	1976	(a)
2		4wWE	GB	420355/2	1976	(a)
3		4wWE	GB	420408	1977	(a)

(a) ex Tata Steel Europe with coke ovens, 25/3/2011.

SIR BERNHARD SAMUELSON & CO LTD

SLAPEWATH IRONSTONE MINE, Charltons, near Guisborough   **144**
B. Samuelson & Co until 10/1887   NZ 646148 – **Map I**
T. Charlton & Co until 2/1880

Thomas Charlton, who owned collieries in Co. Durham, leased the ironstone under Hollin Hill Farm from 1/1/1864. He had previously been Bolckow Vaughan's mining engineer at Eston. A drift was constructed near the Guisborough–Whitby Road to reach the ironstone. The Cleveland Railway's line had been extended from Slapewath to Boosbeck Lane in 1862 and a railway was constructed from it at Aysdalegate Junction Signal Box. The branch curved round as two tracks across rough ground before passing under the Guisborough-Whitby Road. This required a wayleave to be obtained (operative from 1/7/1864) in order to pass over Thomas Chaloner's land. Materials and equipment to establish the mine were carried over this line during 1865-68. The first ironstone left the mine in 7/1868 but working soon ceased when it was discovered that a washout had removed or degraded some of the ironstone and was not resumed until 3/1869.

Although Thomas Charlton died on 30/11/1872, his sons continued to operate Slapewath as a small mine with an annual output of about 40 to 50,000 tons. Most of it was sold to Gjers, Mills and Thomas Vaughan on Teesside. The ironstone seams dipped southwards while the land rose up towards the North York Moors. In 1873 the company commenced sinking a winding shaft, 286 feet deep, at the southern end of the Hollin Hill royalty and all of the ironstone was then raised up this shaft, with the original drift functioning as a 'travelling road' for the miners to gain access to the mine. The company also constructed two rows of houses for its employees between the drift and main road; the settlement was called Charlton Terrace and is still today known as Charltons. The private railway was extended southwards to handle the output from the winding shaft, where it passed under the loading chutes to terminate in a headshunt dug into the hillside. Some shale was thrown out within the mine, but any remaining was picked out as the ironstone was tipped from the tubs into the railway wagons below. According to consultant mining engineer, George Baker Forster's report on 9/6/1876 "the New Pit has been fitted up with very good Machinery and Plant and the Railway has been laid to it, which is worked by a tank locomotive belonging to the Mines." T. Charlton & Co was in financial trouble and petitioned for liquidation on 4/9/1876 but continued to operate. In the year up to 6/1879, Slapewath Mine produced 77,100 tons of ironstone, but then it was decided to put the mine up for sale. A report on the business was prepared by William Armstrong and amongst his comments on 22/12/1879 was "W. Charlton [the mine manager and son of Thomas] states the Locomotive Engines and additional rails have since the valuation been bought"; the plant had previously been valued in 1875. Armstrong also informs us that there were 160 tubs each carrying 35cwt of stone, "some of them much out of repair". Only the identity of one locomotive is known.

The sale of Slapewath Mine and the adjoining houses was held at the Royal Exchange, Middlesbrough on 20/1/1880, but the reserve price was not achieved and the mine was later sold privately (reported 11/2/1880) to B. Samuelson & Co of the Newport Iron Works. Samuelson immediately sank a ventilation shaft with a furnace at its base near the winding shaft at the mine. A new haulage engine was installed underground on 27/10/1881 to haul tubs of ironstone from both Hollin Hill and Tidkinhow royalties. Samuelson also replaced the section of locomotive worked line from near Hollin Hill Farm to the mine by an approximately 700 yards long self-acting rope hauled incline. The annual output of ironstone had increased to 172,701 tons in 1881.

In 1889 Samuelson purchased Hollin Hill Farm and he proposed to acquire the Spawood royalty of Mr Chaloner; the Weardale Iron & Coal Co had last worked Spawood Mine in 8/1886. The first ironstone from the Spawood royalty was taken on 1/7/1890 probably from drifts driven through from Slapewath Mine. On 25/2/1897 Samuelson decided to develop Spawood Mine as the main output location instead of Slapewath. The existence of drifts and the availability of sidings next to the ex Cleveland Railway meant that production could be increased and there was space for a picking belt. Spawood Mine was fully operational on 13/5/1898 and Slapewath probably stopped winding ironstone sometime during the next five years. In 1900 Samuelson was able to add the Aysdalegate royalty to the workings. The ventilation shaft at Slapewath continued to function and the incline reverted from the 'three rail principle' to a single track; presumably the occasional wagon of coal still needed to be delivered for the furnace.

Reference : *Guisborough District Mines*, Simon Chapman, pub. by Peter Tuffs, 2001

Gauge : 4ft 8½in

-	0-4-0ST	OC	YE	262	1875?	(a)	(1)

(a) new. This locomotive was purchased by T. Charlton. Thomas Charlton & Co had interests in some Co. Durham collieries, as well as Slapewath Mine, so it is not certain that the locomotive came here. However, it is quite likely as Samuelson purchased the mine shortly after the sale of the mine's locomotives.

(1) see comment above. Probably to B. Samuelson & Co, Newport Iron Works, Middlesbrough, c/1880.

SEMBCORP UTILITIES UK LTD

(Part of Sembcorp Industries Group)
Sembcorp Utilities Teesside Ltd until 10/5/2004
Enron Teesside Operations Ltd until 15/4/2003

The last remnants of the former ICI railways at its Billingham and Wilton sites were purchased by Enron Teesside Operations Ltd (ETOL) as from 1/1/1999, including the pipeline corridors, road and electrical distribution systems, water treatment plant, the undeveloped sites and the Wilton Power Station. Following the collapse of the main Enron company on 29/11/2001, ETOL continue to operate as a stand-alone business until it was placed on the market in 8/2002. Semb Utilities of Singapore purchased the assets of ETOL from Enron's receivers on 15/4/2003. It established a subsidiary company to operate services at Billingham and Wilton, Sembcorp Utilities Teesside Ltd, but this name was changed to Sembcorp Utilities UK Ltd with effect from 10/5/2004. The various manufacturing plants at the two sites were operated by a number of separate companies following their sale by ICI.

BILLINGHAM SITE, Billingham   145
NZ 470225 – **Map C**

When Enron took over responsibility for the railway system at Billingham Works, it was a shadow of its former self. The remaining connection with the main line was controlled by Belasis Lane Signal Box on the Port Clarence Branch (now renamed Seal Sands Branch). From here, the railway entered Billingham Works curving round to reach the East Grid sidings where traffic was exchanged with the railway companies. Sidings from the Grid reversed direction to serve the engine shed in the former wagon shop (NZ 475228) and ammonia loading terminal, the carbon dioxide terminal and the North Grid (former reception) sidings. Marcroft Engineering had a siding at the East Grid to carry out wagon servicing. From the south end of the East Grid, another private line ran parallel to New Road, with a siding part way along to the Amines plant, before terminating in a headshunt from where a siding passed under the road to enter the Cassell Acrylics Works. A remnant of the Billingham Beck Branch had been retained by ICI to serve the Billingham Reach Oil Terminal; this line had connected at its north end with the East Grid but had been disused since about 1992. There were some tight curves on the Billingham layout and 08743 (15/3/1999) and 08903 (4/2/2000) visited Thornaby Depot for tyre turning.

The last rail traffic for a plant on the Billingham site was the daily hydrocyanic tank train that ran from the BASF works at Seal Sands to the Ineos acrylics plant at the Cassel Works. One of the former Class 08 diesel shunters took these wagons from East Grid to the terminal until the last train ran on 25/4/2002. Subsequently a Sembcorp locomotive was used occasionally to move some redundant tank wagons before these were taken to Tees Yard and cut up in 2004. Freightliner also used the East Grid to stable rakes of empty coal wagons at weekends but this did not involve the works' shunters. The remaining locomotives were moved across to the Wilton site and the engine shed shut.

BRYAN TURNER, ex D3911 (built Crewe Works in 1960), was under repair in ETOL's Billingham shed on 2nd April 2001; this was previously ICI's wagon shop. *(Cliff Shepherd photograph)*

Gauge : 4ft 8½in

08743 (D3911) (024)							
BRYAN TURNER (ANGIE)	0-6-0DE		Crewe		1960	(a)	(5)
T1	4wDH	R/R	NNM	82503	1983	(a)	(3)
T2	4wDH	R/R	NNM	83501	1983	(a)	(3)
T3	4wDH	R/R	NNM	83502	1984	(a)	(1)
T4 05/273	4wDH	R/R	NNM	83503	1984	(a)	(4)
T5	4wDH	R/R	NNM	83504	1984	(a)	(4)
(T6) (3-161)	4wDH	R/R	NNM	83505	1984	(a)	(2)
"WILTONIA"	2w-2DMR		Wkm	7591	1957		
Rebuilt	2w-2DHR		YEC	L112	1962	(a)	(4)
-	2w-2PMR		Wkm	7603	1957*	(a)	(4)
08903 (D4133) 903 (7)							
JOHN W. ANTILL	0-6-0DE		Hor		1962	(b)	(6)

* converted to flat truck.

(a)   ex ICI plc, 1/1/1999.
(b)   ex Wilton International, Redcar, 9/1999. To Wilton International, Redcar, /2005 and returned 17/4/2007.
(1)   scrapped 1/2000; the remains were noted in a skip owned by T.J. Thomson Ltd.
(2)   scrapped after 2/4/2001, by 27/9/2003.
(3)   to Wilton International, Redcar, c/2002; after 2/4/2001, by 18/6/2002.
(4)   to Wilton International, Redcar, c/2003; after 2/4/2001, by 27/9/2003.
(5)   to Wilton International, Redcar, 4/2004.
(6)   to Wilton International, Redcar, /2008.

WILTON INTERNATIONAL, Redcar
NZ 568223 – Map H

After Enron took over responsibility for the Wilton infrastructure from ICI, the railway system was confined to a small railway yard occasionally used for storing rolling stock and sidings serving the power station and Freightliner depot. Most rail movements were handled by incoming main line locomotives. Even some of these traffic flows have ceased. The No.5 boiler at the power station was converted to burn natural gas in 2006 and a new wood burning biomass plant began operating resulting in the cessation of coal trains. Also the cost of moving containers by road from Teesport to Freightliner's depot at Wilton proved prohibitive. Freightliner transferred its container terminal to Teesport in 2014 and the last train to visit the Wilton terminal was Freightliner's 66541 which arrived from Felixstowe on 13/11/2014 and returned there with a train on the next day. Sita Sembcorp UK Holdings Ltd was then established to construct an Energy from Waste plant near the former Freightliner terminal. This processes domestic waste from Merseyside and regular trains began running in 2016, although any shunting is done by the main line locomotive.

Gauge : 4ft 8½in

(07005) (D2989)								
(LANGBAURGH)		0-6-0DE		RH	480690	1962		
	Rebuilt			Resco	L106	1978	(a)	(1)
08502 (D3657) ANGIE		0-6-0DE		Don		1958	(a)	(2)
08503 (D3658)		0-6-0DE		Don		1958	(a)	(3)
08903 903 (D4133)								
JOHN W. ANTILL		0-6-0DE		Hor		1962+	(b)	
T1		4wDH	R/R	NNM	82503	1983	(c)	(4)
T2		4wDH	R/R	NNM	83501	1983	(c)	(4)
T4	05/273	4wDH	R/R	NNM	83503	1984	(d)	(4)
T5		4wDH	R/R	NNM	83504	1984	(d)	(4)
R0.005 "WILTONIA"		2w-2DMR		Wkm	7591	1957		
	Rebuilt	2w-2DHR		YEC	L112	1992	(d)	
-		2w-2PMR		Wkm	7603	1957*	(d)	
08743 (D3911)								
BRYAN TURNER		0-6-0DE		Crewe		1960+	(e)	

+ The locomotives were placed in dry storage but had an unusual outing when transported by road to the Wensleydale Railway at Leeming Bar on 8/9/2016 to haul a special train to Redmire on 25/9/2016 before returning to their shed at Wilton on 27/9/2016.

* Converted to flat truck, c/1968. T. Bye suggests early 1970s.

(a) ex ICI plc, Wilton Works with site, 1/1/1999.
(b) ex ICI plc, Wilton Works with site, 1/1/1999, to Billingham Site, 9/1999 and returned, /2005, to Billingham Site, 17/4/2007 and returned /2008.
(c) ex Billingham Site, c/2002; after 2/4/2001, by 18/6/2002.
(d) ex Billingham Site, c/2003; after 2/4/2001, by 27/9/2003.
(e) ex Billingham Site 4/2004, to Cleveland Potash Ltd, Tees Dock on hire, 7/2005 and returned 10/2005, to LH Group Services Ltd, Barton-under-Needwood, Staffordshire for repairs, 13/10/2011 and returned, 2/5/2012.

(1) to Barrow Hill Engine Shed Society, Staveley, Derbyshire, (property of Harry Needle Railroad Co Ltd), 20/12/2000.
(2) sold to Harry Needle Railroad Co Ltd, 2/2007, to Barrow Hill Engine Shed Society, Staveley, Derbyshire, for repaint, 3/8/2007.
(3) sold to Harry Needle Railroad Co Ltd, 2/2007, to Heanor Heavy Haulage, Langley Mill, Derbyshire for storage 7/12/2007.
(4) to Bombardier Transportation Inc of Canada, Crewe Works, Cheshire, (property of Harry Needle Railroad Co Ltd), 24/9/2008.

W. SHAW & CO LTD

WELLINGTON FOUNDRY, Middlesbrough
W Shaw & Co until 2/1923
W. Shaw, Kirtley & Co until /1892

147
NZ 487210 NZ 488210 – **Map F**

William Shaw had been Managing Director of The Cast Steel Foundry Co Ltd, but joined with John Kirtley in 1885 to start his own business further south on the west side of Forty Foot Road in the Ironmasters district. There had been a foundry on this site with its own siding as early as 1875 (Mellanby & Co in 1882) but a new works was being constructed there in 1887; an agreement between W. Shaw, Kirtley & Co and the NER was dated 24/7/1890. The partnership with Kirtley was dissolved in 1892. Shaw's works was designed for the production of steel castings up to 10 tons using a charge of 10% haematite pig iron and 90% steel scrap. Sidings entered the premises off the MOR Old Town Branch Loop near its junction with the Marsh Branch and its rail traffic was subject to MOR tolls. In 1899, Shaw began work on a new building east of Forty Foot Road, some of which collapsed in strong winds injuring several workmen.

For three years following 1/7/1898, Shaw paid £2.10s.0dpa in lieu of wayleave "on the Goods, chiefly Moulding Boxes, conveyed by their own engines in their own bogies, or wagons, between W. Shaw & Co's Foundry and Stockyard…". Shaw was charged £10pa for use of the railway between its foundry and the NER. For several years after 1914, it was also paying 10s.0dpa for rail traffic between the foundry and stockyard, including that being taken up to the Ayrton Sheet Works weighbridge for weighing and return. The stockyard was located at the northern end of Shaw's premises and served by a separate siding. Shaw had permission for its locomotives and a "steam crane" to work over the railway company's lines in the vicinity of the works.

The foundry site grew to about four acres and, in the mid-1950s, had a melting plant, pattern shop, foundries, dressing and machine shops. A substantial building occupied most of the site (NZ 487210), west of Forty Foot Road, with sidings entering into it. The company also owned a smaller building (NZ 488210), immediately opposite on the east side of the road, with a set of five sidings alongside and a single line connecting with BR's Middlesbrough Town Goods Yard. Some narrow gauge track is shown on the 10/1951 Ordnance Survey next to the set of sidings and it is presumed that this is where the Motor Rails worked, possibly in a stockyard. These tracks had gone by 1967 and the works siding connection with the BR was terminated on 1/10/1970.

Gauge : 4ft 8½in

No.1		0-4-0ST	OC	HCR	151	1870	(a)	Scr /1930
W. SHAW & CO LTD No.2		0-6-0ST	OC	RS	2554	1885	(b)	
		Rebuilt	Ridley Shaw			1941		(1)
W. SHAW & CO LTD No.3		0-4-0ST	OC	RS	3056	1904	(c)	
		Rebuilt	Ridley Shaw			1917,1940		Scr c/1957
W. SHAW & CO LTD No.4		4wDM		FH	3569	1951	New	(2)

(a) ex North Eastern Steel Co Ltd, Acklam Iron and Steel Works, Middlesbrough.
(b) ex ICI Ltd, Billingham Works, Billingham per W. Blenkinsop & Co, Middlesbrough.
(c) the locomotive had moved to Sir W.G. Armstrong, Whitworth & Co Ltd's Elswick Works, Newcastle upon Tyne c/1921 and it is presumed that it went from there to W. Shaw at some time after that date. It was hired to Davy & United Roll Foundry Ltd at Haverton Hill while that company's locomotive was away being repaired, c/1947 and returned.

(1) to J. D. White, scrap merchant, Thornaby, c/1954.
(2) to J. D. White, scrap merchant, Thornaby, c3/1966.

Gauge : 2ft 2½in

-	4wPM	MR	1848	1919	New	s/s
-	4wPM	MR	7091	1940	New	s/s

SHELL-MEX AND B.P. LTD,

OIL DISTRIBUTION DEPOT, Tees Dock, Grangetown

148
NZ 543230 – **Map H**

Two oil jetties were constructed by the Tees Conservancy Commissioners on the south bank of the River Tees immediately west of the later site of Tees Dock. No.1 jetty (subsequently named Queen Elizabeth II) opened in 10/1950 was used by ICI to provide feedstock by pipeline for its sites at Wilton and Billingham. The other jetty, No.2 (later West Byng), was commissioned in 1955 to serve Shell-Mex and B.P. Ltd's (this was the registered title) 15 acre oil storage and bunkering depot, which opened in the following year with distribution taking place by rail and road. A railway line ran to the depot from off the TCC system. In 1960 there were four parallel sidings terminating in a headshunt on the east side of the depot between the storage tanks and a drainage cut. One of these sidings had a short reverse spur running into a building which **may** have been an engine shed. The building was still there in 1972 but no longer rail connected, although the other sidings remained.

Gauge : 4ft 8½in

(No.17)		0-4-0DM	DC	2165	1941		
			EE	1196	1941	(a)	(1)

(a) ex Shell-Mex and B.P. Ltd, Saltend Storage Depot, Hull, Yorkshire (ER), c/1956.

(1) it is assumed that the locomotive came to this depot initially, because it was well before work started on the Shell refinery to the east of Tees Dock. It probably then moved on to the Shell refinery c/1966.

SHELL (UK) OIL LTD

TEESPORT REFINERY, Grangetown.   149
Shell Refining Co Ltd until 1/1/1976   NZ 588233 – **Map H**

The growth in demand for Shell oil products in 1964 encouraged the company to construct a refinery adjoining the river to the north east of Tees Dock. A 140 acre site consisting of partially reclaimed land was levelled with silt dredged from the river and surfaced with blast furnace slag by Tarmac Ltd. The refinery with a processing capacity of approximately 6 million tons pa and costing £25½ million began operating in 1968. A jetty alongside provided a minimum depth of 45 feet to accommodate tankers of up to 80,000 tons dead weight fully laden from North Africa, the Middle East and Venezuela. The new plant supplemented Shell's existing UK refineries at Ardrossan, Heysham, Stanlow and Shellhaven. Products were distributed by road, rail and coastal shipping.

The railway to the refinery left the Middlesbrough-Redcar line 1¼ miles north east of Grangetown Station. It diverged off the branch leading to the ICI Wilton site and climbed over the ICI branch, ran past the end of the Tees Dock system to enter a series of sidings running parallel to Dabholm Gut and terminating in a headshunt. Another line reversed round from these sidings to reach the filling facilities, which were situated at the south east end of the site with the processing plant and storage tanks to the north west. In 7/1979, the refinery was stated to have 8½ miles of track.

The first two small locomotives already belonged to Shell Mex & B.P. Ltd and assisted with the development of the refinery. When it opened, BR won the contract to supply the locomotives for shunting. Three Class 03 204bhp locomotives, numbers D2046, D2057 and D2093, recently adapted with flameproofing installed by Hunslet, were used. No other BR locomotives were allowed to go beyond the exchange sidings. Two operated on each shift with one spare. The Thornaby shed fitters found it a struggle to keep them running due to the flameproofing modifications and the toll of shunting rakes of several 100 ton wagons. The contract ended in 1971 when Shell purchased three new Thos Hill diesel hydraulics.

However, in October 1973 a new Moyse 500hp diesel electric locomotive was demonstrated at BSC's South Teesside Works and representatives from local companies, including Shell (UK) Ltd, attended. Possibly as a result, Shell purchased two new 310hp four-wheel diesel electrics from Moyse; the use of these French built locomotives in the UK was very unusual. They did not arrive until 1976 so the refinery was dependent on one Thos Hill, the other two having left for the Stanlow Refinery. It was no surprise that three BR locomotives were reported to be on hire during this period. A third new Moyse diesel electric was purchased in 1980. According to the *Evening Gazette* on 23/1/1980, it was supplied through the dealers, Thos Ward (Railway Engineers) Ltd of Nottingham, which claimed that the locomotive was capable of hauling 1,000 tonnes and was expected to work round the clock six days a week.

The engine shed was situated near the filling sidings and contained two lines. The principal duties of the locomotives were to shunt large bogie oil tank wagons between the filling point and the sidings at Dabholm Gut. The closure of Shell's Teesport Oil Refinery was announced on 28/9/1984 with the loss of 590 jobs. The site was later cleared and incorporated into the Teesport estate.

Gauge : 4ft 8½in

(No.17)		0-4-0DM	DC	2165	1941		
			EE	1196	1941	(a)	s/s c/1971
22		0-4-0DM	JF	22998	1943	(b)	s/s c/1971
D2046		0-6-0DM	Doncaster		1958		
	Rebuilt		HE	6644	1967	(c)	(1)
D2057		0-6-0DM	Doncaster		1959		
	Rebuilt		HE	6645	1967	(c)	(1)

D2093		0-6-0DM	Doncaster		1960		
		Rebuilt	HE	6643	1967	(c)	(1)
1		4wDH	TH	234V	1971	New	(4)
2		4wDH	TH	235V	1971	New	(3)
3		4wDH	TH	236V	1971	New	(3)
-		4wDH	S	10012	1959	(d)	(2)
2	BN1364	4wDE	Moyse	1364	1976	New	(5)
3	BN1365	4wDE	Moyse	1365	1976	New	(5)
	BN1464	4wDE	Moyse	1464	1979	New	(5)

(a)   probably ex Shell-Mex & B.P. Ltd, Oil Distribution Depot, Tees Dock, c/1966.
(b)   ex Shell-Mex & B.P. Ltd, Saltend Storage Depot, Hull, Yorkshire (ER), c/1966.
(c)   ex BR, Thornaby Depot, on hire, c/1968.
(d)   ex Thos Hill, Kilnhurst, Yorkshire (WR), on hire, 27/10/1972.
(1)   returned to BR, Thornaby Depot, ex hire, /1971. The locomotives were withdrawn by BR in 10/1971 and sold to Gulf Oil Co Ltd, Waterston, Milford Haven, Pembrokeshire (D2046) and NCB Grimethorpe Colliery, Barnsley, Yorkshire (WR) (D2057, D2093).
(2)   returned to Thos Hill, Kilnhurst, Yorkshire (WR), ex hire, 11/12/1972.
(3)   to Thos Hill, Kilnhurst, South Yorkshire, /1974 (BR Train Movement Advice 26/7/1974, locomotives to be hauled dead in special freight on 30/7/1974); then to Stanlow Refinery, Stanlow, Cheshire, 8/1975.
(4)   to Thos Hill, Kilnhurst, South Yorkshire, 22/1/1980; then to Stanlow Refinery, Stanlow, Cheshire after rebuilding, 18/12/1980.
(5)   to Stockton Haulage Ltd, Middlesbrough, by 9/1985; Stockton Haulage was reported to have its own locomotive on 18/4/1985.

T. SMITH & SON

SCRAPYARD, Hartlepool   150
NZ 513313 – **Map B**

The premises were located at Casebourne Road on the Longhill Industrial Estate in Hartlepool. This had been previously part of the Seaton Carew Iron Co's ore stocking ground.

Gauge : 4ft 8½in

3	4wDM	FH	3942	1960	(a)	Scr c1/1982
-	4wDM	RH	312434	1951	(b)	s/s

(a)   ex British Steel Corporation/ Redpath Dorman Long Ltd, Britannia Works, Middlesbrough, 9/1980.
(b)   ex Thomas Turnbull & Sons Ltd, Thornaby, c7/1982.

SMITH'S DOCK LTD

SHIPYARD, South Bank   151
Smith's Dock Co Ltd until 11/1981   NZ 530216 – **Map G**

With increasing traffic on the river, the Tees Conservancy Commissioners encouraged the establishment of ship repair facilities. Smith's Dock Co Ltd was already based on the River Tyne at North Shields and, in 1906, agreed to establish two dry docks with associated quays and workshops on initially 16 acres of marsh land and muddy foreshore at South Bank. The coping stone of the first dry dock was laid by Sir Hugh Bell on 22/10/1907. It was commissioned on 2/1909 and this coincided with the last new ship to be built by Smith's Dock at North Shields. The company extended its premises at South Bank over three years to incorporate a shipbuilding yard and engine works. The first keel to be laid was for the grab hopper barge SS PRIESTMAN, which was launched on 28/2/1910. Thirty three vessels, mainly trawlers, drifters and lighters were built in 1910. Work started on constructing two more dry docks in 1913. By 1914, 174 new ships had been constructed and repair/ maintenance carried out on another 4,000 vessels. Smith's Dock was also well known for its whale-catching ships and this design was adapted by the company to produce the Flower Class corvettes for convoy escort duties in World War 2. The design evolved into the Castle Class corvettes and River Class frigates. Following World War 2, the yard diversified its production by building tankers and specialist refrigeration and container ships.

The 62 acre site was approximately triangular in shape with the longest side of about 900 yards abutting the river. Here were the building berths and four dry docks with the fitting out quay at the south west end of the site and the workshops behind. Like all shipyards, Smith's Dock relied on numerous travelling cranes – there were 15 serving the various berths – although a coal fired steam rail crane (later replaced by a diesel rail crane) was used to lower in timber supports for ships in the dry docks. The yard was connected to the Middlesbrough–Redcar line by a railway that left the sidings east of the original South Bank Station and controlled by South Bank Signal Box. It crossed the Cargo Fleet Iron Company's land, including passing by the basic slag works, to reach a set of exchange sidings before entering the shipyard at the apex of the triangle. It is then divided into individual sidings to serve the various buildings and berths. On the eastern side of the triangle was a long building housing the boilersmiths, fire station and garage. Adjoining it was the single road brick engine shed. According to an agreement dated 23/9/1947 and, with the consent of the Cargo Fleet Iron Company, a new rail-served plate stockyard was provided on the south east side of the premises with sidings for use by the LNER alongside.

Arthur Dinsdale has described operations on the railway: "Small steam cranes ran on railway lines around the shipyard. These 'Heath Robinson' cranes were cobbled together on a low four wheeled bogie using a coal fired small vertical boiler providing steam for both locomotion and to power the winch… The whole contraption was sheeted with corrugated iron to give shelter to driver, boiler, coals… A small tank locomotive engine moved heavy loads on low flat trucks, such as ships' propellers, steel plates and marine engines…" After construction, the ships' engines were partially dismantled and the parts put on wagons for a locomotive to take to the fitting out bay, where they were lowered into the hull by a rail crane. Concrete roadways were constructed during World War 2 and three-wheel mini trucks could be used then to move smaller items around the yard.

John Fowler 4200018 of 1947 at Smith's Dock South Bank shipyard on 24th July 1978 where it stood out of use for a number of years. *(Cliff Shepherd photograph)*

UK shipbuilders increasingly found it difficult to compete with foreign shipyards mass producing ships. In 1958 the tonnage launched on the Tees was Smith's Dock 5 vessels (44,816 tons), Gray 5 vessels (39,366 tons) and Furness 6 vessels (88,693 tons). The Government set up the Geddes Committee to review the overcapacity in shipbuilding and it concluded that regional groupings would be more cost effective. This led to a series of amalgamations and Smith's Dock Co Ltd was taken over by the Swan Hunter group of Wallsend on 5/7/1966. Although the intention had been to concentrate new shipbuilding on the Tyne, Smith's Dock secured orders for new vessels that were to keep it busy for the next ten years and it became a separate subsidiary of British Shipbuilders when the industry was nationalised on vesting day, 1/7/1977; it took the name Smith's Dock Ltd in 1981.

The last ship built at Smith's Dock was the NORTH ISLANDS launched on 15/10/1986 and the yard closed on 28/2/1987, ship repairing having ceased on 15/7/1982. Rail traffic finished c/1980 and the construction of the South Bank link road in 1984 between the former slag works and the shipyard severed the rail outlet. The site was later taken over by A&P Tees, which carried out some ship and offshore repair work.

References :   *Smith's Dock Co Ltd, A Shipyard Centennial*, Wilf Austen, Apprentice Press, 2011

Life at the Yard, Memories of Working at Smith's Dock, South Bank, ed. Margaret Williamson, Teesside Industrial Memories Project, 2012

SMITH'S DOCK Shipbuilders, Ian MacDonald & Len Tabener, Seaworks, 1986

The Cobbler's Nephew, Arthur Dinsdale, Whynott Direct Pub, 2012

'Smiths Dock–South Bank, Middlesbrough', Chris Twigg, *Cleveland Industrial Heritage* 29, pub Peter Tuffs, 2011, pp 2–15

Gauge : 4ft 8½in

No.1	0-4-0ST	OC	P	463	1887	(a)	(1)
No.2	0-4-0ST	OC	HL	3006	1913	New	(3)
No.1	0-4-0ST	OC	RSHN	7084	1943	New	(2)
-	0-4-0DM		JF	4200018	1947	(b)	(4)

(a)   originally Upper Forest & Worcester Steel & Tinplate Works Ltd, Morriston, Glamorgan, which sold it to J.F. Wake, Darlington (dealer) 5/1909. At Peckett, Bristol 5-7/1909 for repairs before moving to Smith's Dock.

(b)   ex Abelson & Co (Engineers) Ltd where it had been used as a hire locomotive, c/1960, by 4/1961. Originally Milford Haven Dock & Railway Co, Pembrokeshire, until 3/1957.

(1)   to Thomas Allan & Sons Ltd, Bonlea Foundry, Thornaby, /1931.
(2)   to J. Garrity, Middlesbrough, for scrap, after 4/1960; resold to Tyne-Tees Steam Shipping Co Ltd, Middlesbrough.
(3)   to J. Garrity, Middlesbrough, for scrap, c/5/1961.
(4)   to Darlington Railway Preservation Society, Darlington, Co. Durham, after 2/6/1987, by 12/7/1987.

HAVERTON HILL SHIPYARD

Haverton Hill   **152**
Smith's Dock Co Ltd until 11/1981   NZ 487226 – **Map C**
(It became a separate subsidiary of **British Shipbuilders Ltd** when the industry was nationalised on 1/7/1977)
Swan Hunter Shipbuilders Ltd until 1/7/1977
Furness Shipbuilding Co Ltd until /1968
Furness Withy Shipbuilding Co Ltd to 13/10/1917

Against the background of the German submarine menace and shipping losses in World War 1, the Admiralty gave approval in 1917 and was involved in the construction of a major new shipyard on the north bank of the River Tees at Haverton Hill; a move away from Furness Withy Shipbuilding Co's previous base at Hartlepool. The project was known as the Government's "Auxiliary Shipyard Extension No.81". A new subsidiary was formed, Furness Shipbuilding Co Ltd and a contract was let to Holloway Brothers (London) Ltd (which see) by 1/1918 to construct the shipyard. The 85 acre site had a river frontage of 2,500 feet and was laid out on modern principles of prefabrication and mass production. Eight berths were provided, together with a later group of four smaller berths on either side of a fitting out basin. The yard was connected to the NER Port Clarence Branch, south east of Haverton Hill Station. From the exchange sidings, railway lines ran into the shipyard forming an extensive system which served the stocking grounds, fabrication shops and berths. Much of the riveting took place under cover with a double railway line allowing locomotives and rail cranes to take components to the berths. The machine and fitting shops had a bay with an inspection pit allocated as a repair and running depot for the steam locomotives and rail cranes. According to J.F. Clarke, there were eight yard locomotives and nineteen travelling steam cranes, together with electric tower cranes on the main berths. The company also constructed new houses for its workers; a total of 530 had been completed by contract at Belasis village by 1920.

The keel for the first ship, a standard steamer of 10,800dwt, was laid in 5/1918, just five months after work began on the site. A total of 73,000 tons of ships was launched in 1920, but the interwar years proved difficult and the yard never operated to its full capacity. Often many berths stood empty and other steel construction work was undertaken, such as storage tanks and bridges. Examples included the gas holder at Cannon Park, Middlesbrough and the nearby Belasis Avenue (B1275) bridge over the Port Clarence Branch. By 1939, 163 ships of 655,000 gross tons had been launched by the yard. Output during World War 2 was mainly tankers and its annual total of 71,764 tons of ships in 1946 was the third highest for any British shipbuilder, but production then declined.

The yard was modernised in 1963–65 including changes to the fabrication section and the provision of a more spacious slipway to enable the yard to build large mixed cargo bulk carriers. By this date, the size of the internal railway system was much reduced with sidings mainly serving the site of the former four small slipways and the fabrication shop. Early in 1968 Furness Shipbuilding Co announced the closure of the yard, but it was bought by Swan Hunter which built a number of approximately 170,000dwt bulk carriers. After nationalisation, the yard was vested in British Shipbuilders and managed by Smith's Dock Co Ltd. Finally the last ship to be launched at Haverton Hill, NEW ZEALAND STAR, was completed in 1/1979 and the yard closed in 1980.

The use of locomotives on the site during the construction of the shipyard is confusing, but a number of contractor's locomotives appear to have come into the Furness Shipbuilding Co's possession. At the same time, the latter acquired nine new locomotives. This was far more than was needed, especially in view of the lower than expected output and the size of the fleet was gradually reduced. The article by Derek Stoyel suggests that the contractor, Norton Griffiths & Co, may also have been involved in the project and that some of the locomotives could have belonged to it. He also listed a number of other possible engines that may have been used on the site but this information is too vague and uncertain to be included in the schedule. The last locomotive, FH 3896, became the property of the Northern Machine Tools (Engineering) Ltd because it had the contract to remove heavy plant from the yard following closure. However, it was not used and was sold for scrap in 1983.

References :   *Building Ships on the North East Coast* (Part 2 c1914–c1980), J.F. Clarke, Berwick Press, 1997

Fifty Years of Furness 1919–1969, John M. Evans, pub. by author, 2002

Shipyard Memories, John M. Evans, pub. by author, 2005

'The Furness Shipbuilding Co Ltd', Derek Stoyel, *The Industrial Locomotive* 36, ILS, 1985 pp.4–8

Gauge : 4ft 8½in

-		0-6-0ST	IC	HC	347	1892	(a)	(3)
-		0-6-0T	IC	MW	1005	1887	(a)	(2)
8		0-4-0ST	OC	HL	2663	1906	(b)	(1)
6		0-6-0ST	IC	MW	1146	1890	(b)	(4)
(13)	HAVERTON HILL	0-6-0ST	IC	HE	464	1888	(c)	(5)
18		0-6-0ST	IC	MW	599	1876	(d)	s/s
-		0-4-0ST	OC	HCR	112	1871	(e)	Scr/1947
-		0-6-0T	OC	KS	3098	1918	New	(7)
-		0-6-0T	OC	KS	3099	1918	New	(6)
-		0-6-0T	OC	KS	3100	1918	New	(7)
-		0-6-0T	OC	KS	3101	1918	New	(8)
No.1 (7)		0-4-0ST	OC	MW	1969	1918	New	(11)
No.2 (6)		0-4-0ST	OC	MW	1968	1918	New	(10)
No.3		0-4-0ST	OC	KS	3125	1918	New	(13)
No.4		0-4-0ST	OC	KS	3124	1918	(f)	s/s c/1957
No.5		0-4-0ST	OC	AE	1801	1918	New	(9)
1		0-4-0DM		JF	4200023	1948	New	Scr/1970
2		0-4-0DM		JF	4200024	1948	New	Scr c/1969
3		4wDM		RH	221643	1943	(g)	s/s c/1963
4		4wDM		RH	189965	1938	(h)	Scr c/1969

5		0-4-0DM	JF	22985	1942	(i)	(12)
-		4wDM	FH	3896	1959	New	(14)
7		0-4-0ST OC	WB	2586	1938	(j)	(13)

(a) ex Holloway Brothers (London) Ltd, contractor, /1918.
(b) ex Holloway Brothers (London) Ltd, contractor.
(c) ex Holloway Brothers (London) Ltd, contractor, 10/1918.
(d) ex J.& G. Tomlinson, Pleasley, Derbyshire.
(e) ex Pease & Partners Ltd, Lackenby Iron Works, Grangetown.
(f) new. To T.J. Thomson & Son Ltd, Stockton, /1949 and returned c/1951.
(g) ex Air Ministry, British Industries Fairs Building, Castle Bromwich, Warwickshire, c/1949 (the first spares order was 17/8/1949.
(h) ex W.H. Arnott Young & Co (Shipbreakers) Ltd, Dalmuir, Dunbartonshire, c/1950. The first spares order from RH for Haverton Hill was 24/5/1950, although the last at Dalmuir was 31/5/1951.
(i) ex Abelson (Engineers) Ltd, Sheldon, Birmingham, per G.E. Simms (Machinery) Ltd, dealer, c/1952.
(j) ex London Brick Co Ltd, Calvert Brick Works, Calvert, Buckinghamshire, per George W. Bungey Ltd, dealer, sale date 5/1959, sent c6/1959.

(1) to Wm. Benson & Son Ltd, Fourstones Quarry and Colliery, near Hexham, Northumberland, c/1918, via RS (for repairs?).
(2) to Exors. of L.P. Nott, contractor, for construction of Bristol-Avonmouth Portway Road, c/1921.
(3) to Northamptonshire Ironstone Co, Byfield Quarries, Byfield, Northamptonshire, c/1923.
(4) to Rugby Portland Cement Co Ltd, New Bilton Works, Warwickshire, 1923, by c3/1924.
(5) to George Hodsman & Sons Ltd, Port Clarence, c5/1925.
(6) to Waterloo Main Colliery Co Ltd, Waterloo Main Colliery, Temple Newsham, Yorkshire (WR), via J.F. Wake, by 12/1928.
(7) to the Weardale Steel, Coal & Coke Co Ltd, Thornley Colliery, Thornley, Co. Durham, by 1/1929.
(8) to Hazlerigg & Burradon Coal Co Ltd, Northumberland, by 1/1929.
(9) to T.J. Thomson & Sons Ltd, Stockton, c/1948. Inness has /1949.
(10) to NCB Elsecar Central Workshops, Elsecar, per H. Potter & Co Ltd, dealer, 9/1948.
(11) to Davy & United Roll Foundry Ltd, Haverton Hill, via Ridley Shaw & Co Ltd, /1950.
(12) scrapped /1959; frame retained as a bogie truck, later s/s.
(13) scrapped on site by T.J. Thomson & Son Ltd, 1/1967.
(14) property of Northern Machine Tool Co Ltd from c/1980 but stayed on site; scrapped /1983, after 6/1982.

SOUTH BANK BRICK CO LTD

SOUTH BANK BRICK WORKS, South Bank   153
(Subsidiary of **Crossley Building Products Ltd**, later **Crossley & Sons Ltd**)

NZ 526206 – **Map G**

The developing iron and steel industry provided employment for many people who required houses. The northern end of the Normanby Branch traversed an area of clays that were suitable for brick making and this encouraged the establishment of a number of brick works with siding connections to the branch. Heading south from Cargo Fleet Junction, past the Cargo Fleet Iron and Steel Works (which had its own brick works), the first premises to be reached was the South Bank Brick Works located north of the later Middlesbrough Road overbridge. A siding into the works was controlled from the nearby Cargo Fleet Inner Junction Signal Box. The branch continued southwards to serve four more small brick works*.

The agreement for the South Bank Brick Co's rail connection to the branch was dated 9/3/1904. By 1913, an inclined narrow gauge railway enabled wagons of clay to be drawn up from the pit to the south and east of the main works building. The premises were often referred to as the "Station Yard" which could be misleading as it was not at the station, but had a vehicular access off Station Road. A 12 chamber coal-fired Hoffman continuous kiln operated on the north west side of the brick works. In the 1920s field drainage pipes, as well as common bricks, were produced and the clay pit had expanded on the east side of the premises. Crossley Building Products Ltd was formed in 1935 made up of a number of businesses, including the South Bank Brick Co which Crossley had acquired in 1923. Further improvements took place and Davidson records that, in 1946, a Swinney Brothers brick making machine was installed resulting in a weekly production of 118,000 common wire-cut bricks. By 1952, the original clay pit had been reclaimed leaving a pond alongside the BR branch near the signal box. In its place, a new clay pit had been excavated to the north of the works. A narrow gauge railway ran round the northern and western edge of this pit before terminating at a head shunt south of the works. From here, wagons were hauled up an incline into the building. Crossley invested in new brick works after World War 2 and the days of the South Bank Works were numbered; the agreement for the rail connection with BR was terminated 28/12/1956.

Reference :   Brickworks of the North East, Peter J. Davidson, Gateshead Libraries & Arts Service, 1986

Gauge : 2ft 6in

		4wDM	HE	1929	1938	(a)	(1)

(a)   ex Coatham Stob Estates Ltd, Eaglescliffe, /1946.

(1)   to Eaglescliffe Bricks Ltd, Eaglescliffe, /1963.

* After the old Middlesbrough Road level crossing, a siding ran off to the east to the North Eastern Brick and Tile Works, which was shown on the 1892–93 Ordnance Survey. An agreement over use of one of the works sidings to receive NER traffic was dated 26/12/1908. Opposite, another siding served Messrs Maw & Hodgson's Branch Brick and Tile Works. The agreement for the rail connection to the NER's branch was dated 2/2/1907. It had become the Central Brick and Tile Works by 1927. According to the NER, the owner was John Hodgson in 1914 and the South Bank Brick Co Ltd in 1916. The siding agreement was terminated by letter 13/5/1954. A few yards further on and, also on the east side of the branch, a siding ran across the field to the South Bank and Normanby Brick Works in the early 1890s. However, this site was then occupied by the South Bank and Normanby Gas Works which relied on the same siding for the delivery of coal; an agreement being reached with the NER on 30/12/1899. The railway company agreed, in 1902, that the "tenant" could run along the branch to the NER sidings next to the signal box just north of Middlesbrough Road. To the south were old clay pits in 1892–93 that had once been the Cleveland Brick and Tile Works.

SOUTH BANK CHEMICAL CO LTD

South Bank **154**
NZ 532214 – **Map G**

(Subsidiary of **The Cargo Fleet Iron Co Ltd** until 3/10/1953 and then **South Durham Steel & Iron Co Ltd**)

The chemical works processed tar from the Cargo Fleet Iron Co's (CFI) coke ovens; the South Bank Chemical Co (SBCC) being established as a subsidiary of CFI and the site for the plant at the east end of the slag tips leased to it on 13/10/1912. In the same year W. Carby Hall, architect, had been instructed to prepare plans for the tar distillery and Brotherton & Co Ltd was appointed to manage it. This company also had shares in the plant until 1919 when it was no longer involved. The SBCC provided the tar wagons and delivered them empty to the Cargo Fleet Works. When full, the Iron Company hauled the wagons (of about 20 tons capacity each) to SDCC's works on payment of 1½ d per ton. The CFI supplied the 10 ton wagons for creosote, pitch, grease and black varnish produced at SBCC's works and charged for hauling them between the works and Cargo Fleet Wharf. Much of the traffic was handled by the CFI's locomotives in line with the agreement; the CFI telling the NER that the "SBCC is actually a department of ours"! However, it is not known how the SBCC delivered the empty wagons to the Cargo Fleet Works. The CFI/SBCC owned a fleet of wagons for the traffic; depreciation of £275–12s–0d on it was noted on 29/12/1919 and, on 29/6/1922, the CFI agreed to purchase 106 wagons for £7,420 that had been previously loaned to it by the Ministry of Munitions. For 15 years after World War 2, the SBCC was allocated its own Fowler diesel to handle traffic. The plant closed in 1961 when CFI's coke ovens finished.

Gauge : 4ft 8½in

No.26		0-4-0DM	JF	4200006	1946	New	(1)

(1)   to South Durham Steel & Iron Co Ltd, Cargo Fleet Works, /1961.

SOUTH DURHAM STEEL & IRON CO LTD

MOOR STEEL AND IRON WORKS, Stockton **155**
Moor Steel & Iron Co Ltd until 29/12/1898 NZ 438193 – **Map D**
Johnson & Reay until /1882
Shaw, Johnson & Reay

Shaw, Johnson & Reay purchased land from Messrs Wren for the Moor Iron Works; the first sod being cut on 19/10/1871 and the initial puddled iron bar was rolled on 12/4/1872. This may have been the same James Shaw, an iron merchant, who supplied rails to many overseas governments. According to the Iron & Coal Trades Review, the works was laid out for the manufacture of puddled iron bars and plates. It contained 24 puddling furnaces, six boilers by Blair & Co, a pair of horizontal direct acting steam engines by Coupe & Co of Wigan, a 24 inch forge train consisting of three pairs of rolls and a reversing plate mill. The works was located on the west side of the NER Yarm–Norton line a short distance south of Stockton Station. In 1874 it had an order to supply 9,000 tons of rails to New Zealand. Trading conditions were difficult for the rest of the decade and, in 1882, there was a petition for the liquidation of the estate of Messrs Johnson & Reay and a statement that the iron works was restarting; the business being reconstituted as the Moor Steel & Iron Co Ltd. It was probably about this time that open hearth steel furnaces were introduced. In 1894 it was stated to have a weekly output of about 1,700 tons and employ nearly 1,000 people.

On 28/12/1898, the South Durham Steel & Iron Co Ltd was formed by the amalgamation of the West Hartlepool Steel & Iron Co Ltd, Stockton Malleable Iron Co Ltd and the Moor Steel and Iron Co Ltd. This was a move by Christopher Furness and William Gray to rationalise the supply of plates, angles and bars for their shipbuilding activities at the Hartlepools.

The Moor Works had a long frontage to the NER sidings alongside the main line and tracks entered the premises at three separate points. Inside the site, sidings served the various buildings and the slag tip on the west side. Additional sidings had been laid at the works by 1914 and a sizeable slag heap occupied the north west corner of the property. A surprisingly large fleet of Hunslet four-coupled saddle tank locomotives shunted around the works.

SDSI Co offered plant at Moor Works, including a secondhand 14 inch locomotive, for sale on 28/1/1921. The works was disused by 1922 (see T.J. Thomson & Son Ltd entry) when William Newton and James Davison White, trading as William Newton & Co, scrap merchants, had the NER working its traffic. Newton was allowed to run with its steam rail crane on to the railway company's adjacent siding.

Gauge : 4ft 8½in

No.2	0-4-0ST	OC	B	266	1880	New	(2)
2	0-4-0ST	OC	HE	177	1877	(a)	(1)
3	0-4-0ST	OC	HE	413	1887	(b)	s/s
No.4	0-4-0ST	OC	HE	608	1895	New	(3)
No.5	0-4-0ST	OC	HE	894	1905	New	(4)
No.6	0-4-0ST	OC	HE	951	1907	New	(3)
No.7	0-4-0ST	OC	HE	1086	1911	New	(3)
No.8	0-4-0ST	OC	HE	1087	1911	New	(4)
No.9	0-4-0ST	OC	HE	1108	1912	New	(3)
No.3	0-4-0ST	OC	HL	2134	1889	(c)	(5)
No.10	0-4-0ST	OC	HE	1405	1920	New	(3)

(a) ex James Muspratt & Sons Ltd, Widnes, Lancashire.
(b) ex Leeds Steel Works Ltd, Hunslet, Leeds, Yorkshire (WR).
(c) Thos W. Ward Ltd's records state that it purchased HL 2124 of 1889 from the Moor Works. However, 2124 is the number of a triple expansion engine for a ship. The sale may have referred to HL 2134 that was thought to have been scrapped at SDSI's West Hartlepool Works c/1918 but may have been transferred here about that date instead.

(1) to Hunslet Engine Co Ltd, Leeds, Yorkshire (WR), possibly in part exchange. While with Hunslet, it was loaned to Lever Bros Ltd at Port Sunlight and then sold to the Holwell Ironworks Ltd, Melton Mowbray, Leicestershire, 9/2/1900.
(2) last known spares dated 5/1919.
(3) to SDSI, West Hartlepool Works. No dates for transfers are known but they would be in the period 1898 to 1922.
(4) to SDSI, Malleable Works, Stockton.
(5) If (c) is correct, then this locomotive was sold to Thos. W. Ward Ltd on 7/10/1920 and used by them on hire before being sent to its Charlton Scrap Depot in 11/1923.

SOUTH STOCKTON IRON CO LTD (1881)

SOUTH STOCKTON IRON WORKS, Thornaby   156
North Yorkshire Iron Co Ltd until /1875   NZ 449179 – **Map E**
Richardson, Johnson & Co until /1869

Manufacturing commenced at the North Yorkshire Iron Works (later South Stockton Iron Works) belonging to Messrs Richardson, Johnson & Co on 5/4/1865. The principal partners were Joseph Richardson of the shipbuilders, Richardson Duck, and Cuthbert Johnson. The works was located south of the Darlington-Middlesbrough railway on the east bank of the Tees next to where the main line crossed the river. According to Thomas Richmond, the "works comprise 20 large puddling furnaces with powerful rolling mills for angular iron, railway and bar iron". As a demonstration of its capabilities, a 117 feet length of angle bar iron was rolled compared to the 90 feet being produced at Middlesbrough at the time. Exports included bar to Russia and rails to America. A length of rail excavated at Amberley in recent years bore the legend "N.Y.I. Co. 52lb. 1875. NZR". This was probably made by the North Yorkshire Iron Co and the rail would have been suitable for 3ft 6in gauge railways being built in New Zealand by a UK contractor. The works employed 450 people and covered 11 acres; it was operating 59 puddling furnaces in 1871.

The company also owned a brick works producing 50,000 bricks a week; a deed of indemnity for storage and carriage of a brick making machine being dated 11/1868. The site of the brick works is not known although there were clearly sources of clay close at hand. Also in 1868, the proprietors of the iron works patented a process claiming to remove the phosphorus from Cleveland iron. Bernhard Samuelson leased the North Yorkshire Iron Works in 1869 to manufacture steel rails and plates using Cleveland iron but this was not successful and it reverted to producing finished iron. A reorganisation of the company, chaired by Joseph Dodds, appears to have taken place in 1872. However the business folded in 1875 as its assets, including one tank locomotive, were taken into possession of the official liquidator. The locomotive was hired to Skinner & Walker (see Stafford Pottery) and the Bowesfield Iron Co; payments appearing in the liquidator's accounts for 11–12/1876 and 2/1877 respectively, before the locomotive was sold to Dorman Long for £500 on 2/5/1877.

The works was acquired by the South Stockton Iron Co Ltd (1881) and reopened on 28/3/1882, with a Siding Agreement being signed with the NER on 15/6/1883. In 1895, the company's railway left the main line near Tees Bridge Signal Box and ran south between a few short NER sidings and a coal depot siding in Wedgwood Street. It then divided into sets of tracks either side of the main iron works building. Slag tipping took place on the eastern part of the site up to Thornaby Road with the Ordnance Survey map indicating a possible narrow gauge line here. One of the sidings nearest the river extended into the Stafford Pottery premises. An undated note from the NER stated that this works was being demolished but that private locomotives had worked traffic between it and the NER sidings. The iron works had closed by 1892 and was mostly demolished two years later. The site was auctioned in 1897 and it was subsequently taken over by A. Bainbridge (which see).

References : *Local Records of Stockton and Neighbourhood*, Thomas Richmond, pub. William Robinson, 1868.

Pioneers of the Cleveland Iron Trade, J.S. Jeans, Middlesbrough Reid, 1875.

Correspondence in *The Industrial Locomotive*, issues 105 and 106, ILS, 2002.

Gauge : 4ft 8½in

Name	Type	Cyl	Builder	Works#	Date		Ref
CLEVELAND	0-4-0ST	OC	BH	112	1869	New	(1)
SOUTH STOCKTON No.1	0-4-0ST	OC	B			(a)	s/s
-	0-4-0ST	OC	HE	240	1880	(b)	(2)
SICO	0-4-0ST	OC	BH	935	1889	New	(3)

(a) probably built by Barclays & Co and came via the agents, Lennox Lange & Co.
(b) ex William Gradwell, Barrow-in-Furness.
(1) s/s although this may be the locomotive sold to Dorman Long in 1877.
(2) to Raisby Hill Limestone Co Ltd, near Coxhoe, Co. Durham, by 6/7/1897.
(3) to The United Alkali Co Ltd, Allhusen's Works, Gateshead, Co. Durham, by 3/1909.

STANTON & STAVELEY LTD

ORMESBY IRON WORKS (LATER COCHRANES WORKS), Cargo Fleet 157
(Subsidiary of **Stewarts & Lloyds Ltd** from 1960 and then the **British Steel Corporation** from 28/7/1967)
NZ 513207 – **Map G**
Cochrane`s (Middlesbrough) Foundry Ltd until 1/6/1962
(Subsidiary of **Stanton Ironworks Co Ltd**)
Cochrane & Co Ltd until /1933
Cochrane & Co until 27/12/1889

Note: The foundries became the responsibility of **Cochrane, Grove & Co** in the early 1860s, but the four blast furnaces stayed with Cochrane & Co (Ltd) until 1902 when the latter took overall control again. In 1887, Cochrane, Grove & Co was said to be the largest pipe manufacturer in the world. However, it appears that the works functioned as one unit and will be dealt with under the Cochrane name.

A damaged negative but it provides a good illustration of the wharf at Cochrane's works, with its stacks of pipes; one of the company's little vertical boiler locomotives appears to be not in use.

Cochrane & Co owned the Woodside Iron Works, near Dudley in Staffordshire, producing cast iron pipes and structural iron work. Alexander Brodie Cochrane had interests in a number of collieries, including some in south west Durham. He was attracted to Middlesbrough because of the publicity concerning Cleveland ironstone and the inadequate reserves of ironstone in the Midlands coal measures. He formed a partnership with Archibald Slate and Charles Geach. They approached Joseph Pease for a suitable site and established the Ormesby Iron Works on reclaimed land between the Middlesbrough-Redcar railway and the River Tees. An agreement was signed with OME on 1/7/1853 for Cochrane to purchase 17 acres of the site at a cost of £12,000. Piles were driven for the first blast furnace in 11/1853 and it was blown in a year later, to be followed by three more so that by 4/1857 a row of four blast furnaces stood on the site, together with a large building, presumably the No.1 Foundry completed in 1855, because the iron was to be used to produce pipes and castings. To the west, an area of marshland separated it from the Tees Iron Works while, to the east, the River Tees and its mudflats stretched to the old Cleveland Port. As part of the Agreement, Cochrane was allowed to reclaim the balance of the 29 acre site by infilling with slag from its blast furnaces, while OME diverted Ormesby Beck along the works' eastern boundary. Cochrane purchased the remainder of the site on 1/1/1863.

In its early years, the Ormesby Iron Works had a reputation for pioneering work. It employed E.A. Cowper to carry out experiments on regenerative hot blast stoves. Also John Gjers took over from Edwin Jones as blast furnace manager when the latter left to establish the Normanby Iron Works. Cochrane later built 17 sets of blast furnace blowing engines based on the patterns used for Gjers' Ayresome Iron Works engines.

The works went on to produce a wide range of cast iron pipes and fittings from two inch to 84 inch diameter, together with many general castings, including rail chairs, brake blocks, axle boxes, tunnel segments and tubbing. Later large quantities of ingot moulds were manufactured. The cast iron columns for the still extant Saltburn Pier were made at Ormesby Iron Works, with the first being ceremonially driven in by Thomas Vaughan on 26/1/1868.

The blast furnaces at the Ormesby Works stood in a row parallel to and close to the Middlesbrough-Redcar line and railway connections were provided into the premises from both the east and west ends. An inclined line led to a gantry over the calcining kilns and mineral stores on the south side of the blast furnaces with a "brake drop" at the opposite end. The pig beds and their sidings were on the north side. Other internal sidings served the foundries on the western half of the site, while lines delivered slag to the reclamation activities on the east side. A wharf was established on the river bank initially facilitated by OME constructing two rows of piles at Cochrane's cost. Additional sidings were laid to connect the wharf with the stockyards. The small single road engine shed was situated near the western connection with the main line opposite Whitehouse Signal Box. Unlike now with its level crossing, for a long time, there was just a pedestrian subway connecting Marsh Road to Cochrane's Works.

Given OME's involvement in the sale of the land for the works, it is not surprising that Cochrane's traffic would have to pass over the MOR lines in order to connect with the Middlesbrough-Redcar railway. The 1863 OME agreement confirmed that Cochrane had the right to pass with railway wagons "of proper construction" along the private railway belonging to Joseph Pease, provided that it paid the ¼d per ton toll. The NER possessed sidings between the works and the main line for the exchange of traffic; these were opposite the later site of Cargo Fleet Station. The above agreement also stated that "Joseph Pease" would complete and keep in good order a railway [another MOR line] from the eastern outlet of the Ormesby Iron Works heading westwards parallel to the main line before curving round to pass over the Middlesbrough Dock swing bridge to the "Steam Boat Wharfs at Middlesbrough" on or before the expiration of 18 months from 1/1/1863. The payment for Cochrane using this line for all goods was ½d per ton. A NER list dated 5/3/1897 included provision for Cochrane's private locomotives to run over the MOR to the Marsh Branch at the Ironmasters district and to Tees Union Wharf. It was noted on 19/4/1915 that slag sent by Cochrane to the NER's permanent way would not incur a toll as the NER was not paying for it! Cochrane was still using the MOR line in BR days to reach the Tyne Tees and Dent's wharves, as well as the Middlesbrough Town Goods Yard.

Cochrane obtained its first locomotive in the 1850s, but a major expansion occurred when it purchased four vertical boiler locomotives from Alexander Chaplin & Co of Glasgow in 1863. Cochrane must have been pleased with its purchases because it constructed similar locomotives to its own design in the 1860s and 1870s giving a fleet of nine vertical boiler locomotives. Two of these locomotives, at least, were fitted with unusual boilers manufactured by a company ironically called Cochran, based in Annan, which combined both vertical and horizontal features. An original photograph of vertical boiler locomotive number 6 (published in *Teesside and Old Cleveland A Further Chapter* by Robin Cook, European Library, 1995) had the following note on the back "About this time – 1900 – ten of this kind of loco, called "coffee pots", were used for pulling bogies and wagons about the Works." From the 1880s, Cochrane recognised that more powerful locomotives were required and

The view past the company's offices into Cochranes Works with green HL 2779/1907 and black AB 2279/1949 at the engine shed. On the right, Cargo Fleet Station can just be seen past the signal post with the Cargo Fleet Iron and Steel Works beyond. *(Courtesy David Webb)*

began to buy new more traditional four-coupled engines. A few of the vertical boiler locomotives lasted up to the 1930s and Ted Haigh noted number 4 (3/7/1927) and 2 and 7 (29/3/1934), but it has not been possible to associate these numbers with specific locomotives.

From 1883, Cochrane had five blast furnaces, although generally only three or four would be in blast. About 1892–94 the number of blast furnaces was reduced to four and this was the position when the Iron & Steel Institute visited the works in 1908. Each blast furnace could produce about 900 tons of Cleveland pig iron per week. The foundry comprised a substantial moulding shop capable of dealing with the largest sizes of pipes and a plant for casting pipes 8 to 24 inches in diameter. There was also a machine shop, with the locomotive repair shop measuring 74ft by 50ft. The company's wharf was 520 feet long and could accommodate vessels up to 5,000 tons; in 1926 it had a 5 ton and three 3 ton electric cranes.

Although the eastern boundary of the Ormesby Iron Works was mostly defined by the realigned Ormesby Beck, Cochrane had constructed a railway over the beck to tip slag presumably helping to reclaim land that later became part of the Deepwater Wharf Ltd sidings. Under an agreement between the NER and Cochrane & Co Ltd on 23/5/1902 (superseding a similar agreement with Cochrane, Grove & Co Ltd dated 21/7/1900), the NER leased 15,848 sq yds of land to Cochrane at the south end of the Deepwater site next to the main line and constructed a siding into the land for Cochrane. The latter used the site for storage. It was still in possession of this land, according to an agreement with the LNER dated 25/1/1930, when it was permitted to construct additional sidings there.

A short-lived umbrella organisation, the East Coast Steel Corporation mainly promoted by Benjamin Talbot of the South Durham Steel & Iron Co, acquired Cochrane's works in 1918, but later sold it to the Stanton Ironworks Co Ltd; however, the works retained the Cochrane name for many years. Only two blast furnaces were operating in 1920 and iron production appears to have ceased by 1925. They were pulled down in 1931 to make more space for pipe manufacturing. Mr E.J. Fox, managing director commented "there is plenty of pig iron available in Middlesbrough without patching up a plant which might require much money" and proceeded to order 10,000 tons of pig iron from its neighbour, Pease & Partners. Stanton acquired the Ormesby Iron Works in early 1933 and formed Cochrane's (Middlesbrough Foundry) Ltd to run it. Demand for spun pipes was growing and Ormesby Works occupied a good strategic position so a new centrifugal iron pipe plant was installed in 1935 using three cupolas for melting the iron. The iron pipe business revived with the end of World War 2 and annual output from the works increased to 34,302 tons (1934), 56,125 tons (1938) and 95,432 tons, of which 87,305 tons were pipes (1948). It appears that Cochrane was using land at the Normanby Iron Works to store pipes because BR decided to levy MOR tolls on the traffic.

As part of the nationalisation of the iron and steel industry, the Stanton & Staveley group became part of the British Steel Corporation's Northern & Tubes Group. The Ormesby Iron Works' main products at this time were spun iron pipes of 6in to 27in in diameter and other castings, such as large pipe fittings, engineering components and ingot moulds. The works was then in the habit of using modern Sentinel diesels from Dorman Long (also part of BSC); these were 39 (S10039 of 1960) and 69 (S10069 of 1961) in 3/1968. It then obtained two 'permanent' residents, Nos.1 and 2, to carry out the remaining shunting. However, the BSC was looking to rationalise its pipe making facilities and about 3pm on 1/4/1971 the doors dropped for the last time on the final cupola operating at the Ormesby Iron Works following the closure decision. The site and remaining buildings were then taken over by W.G. Readman Ltd (which see).

References :   'The Birth and Early History of a Middlesbrough Ironworks (The Ormesby Ironworks of Cochrane & Co.)', F. Jewitt, *Cleveland & Teesside Local History Society Bulletin*, 30, Winter 1975/76

Cochranes 1854-1954, published by the company to celebrate its centenary

Vertical Boiler Locomotives and Railmotors built in Great Britain, Rowland A.S. Abbott, Oakwood Press, 1989

Vertical Boiler Locomotives and Railmotors built in Great Britain Volume Two, Philip J. Ashforth and Vic Bradley, Industrial Locomotive Society, 2016

'Cochrane and Company – A Pioneering Ironworks', Jenny and Geoff Braddy, *Cleveland Industrial Heritage* 38, pub Peter Tuffs, 2016, pp. 2-9

Gauge : 4ft 8½in

DESPATCH	0-4-0ST	OC	RWH	1019	1857	New	(2)
-	0-4-0VBT		Chaplin	317	1863	New	s/s
-	0-4-0VBT		Chaplin	358	1863	New	s/s
-	0-4-0VBT		Chaplin	381	1863	New	s/s
-	0-4-0VBT		Chaplin #	382	1863	New	s/s
ORMESBY	0-4-0ST		BH	111	1869	(a)	s/s
-	0-4-0VBT	VC	Cochrane		1871	New	(1)

	Name	Type		Builder	Works No.	Date	Acquired	Disposal
6		0-4-0VBT	OC	Cochrane			New	s/s
-		0-4-0VBT		Cochrane			(b)	s/s
-		0-4-0VBT	VC	Cochrane			New	(7)
	COOMASSIE	0-4-0ST	OC	RWH	1635	1873	New	(3)
-		0-4-0VBT		Cochrane		c1880	New	s/s
	ORMESBY	0-4-0ST	OC	RWH	1820	1880	New	s/s
	WOODSIDE	0-4-0ST	OC	RWH	2010	1884	New	(11)
-		0-4-0ST	OC	RWH	1789	1879	(c)	s/s
-		0-4-0ST	OC	LE	200	1885	New	s/s
	GRETA	0-4-0ST	OC	HL	2139	1889	New	(9)
	NORTON	0-4-0ST	OC	HL	2171	1890	New	(6)
	DESPATCH	0-4-0ST	OC	HL	2358	1896	New	(4)
	DIAMOND	0-4-0ST	OC	HL	2388	1898	New	s/s
	STANGHOW	0-4-0CT	OC	HL	2516	1902	New	(10)
	POWERFULL	0-6-0ST	OC	MW	1602	1903	New	(5)
	JUBILEE	0-4-0ST	OC	MW	1641	1904	(d)	Scr c/1956
	TOGO	0-4-0ST	OC	MW	1659	1905	New	(8)
	BALKAN	0-4-0ST	OC	MW	1828	1913	New	s/s
	HOLWELL No.16	0-4-0ST	OC	HL	3304	1917	(e)	(12)
	COCHRANES No.1 (STANTON 35)	0-4-0CT	OC	AB	2040	1937	(f)	Scr /1963
No.3		0-4-0ST	OC	AE	2060	1931	(g)	Scr c/1966
	HOLWELL No.9	0-4-0ST	OC	HL	2939	1912	(h)	Scr 6/1969
No.4	(COCHRANES No.4)	0-4-0ST	OC	AB	2272	1949	New	Scr 6/1969
No.5		0-4-0ST	OC	HL	2729	1907	(i)	Scr 6/1969
-	COCHRANES No.6 (STANTON No.7)	0-4-0ST	OC	HL	3481	1920	(j)	Scr 6/1969
-		4wDM		RH	425482	1958	(k)	(16)
19		4wDH		S	10019	1960	(l)	(13)
2	(73)	4wDH		S	10073	1961	(l)	(15)
1	(36)	4wDH		S	10036	1960	(m)	(14)

\#   Chaplin's register has a note that Gabrielli & Co, a contractor on the Chatham Dockyard Extension works, writing on 1/6/1871, quoted 382 as a steam crane in its possession but the number is probably an error.

One of the vertical boiler locomotives was number 9. The locomotives in the 1950s carried what appears to have been a stock number on the side of the running plate. Examples noted include AB 2272 (5811/44), HL 2729 (5811/36), HL 3481 (5811/40), MW 1641 (5811/35) and AE 2060 (5811/37), but the cranetank AB 2040 was 5833/34.

(a)   new. Ordered 10/5/1869.
(b)   new. Locomotive with combined vertical and horizontal boiler.
(c)   ex Framwellgate Coal & Coke Co Ltd, Framwellgate Colliery, Framwellgate Moor, Co. Durham, by /1888.
(d)   new, to Stanton Ironworks Co Ltd, Market Overton Ironstone Quarries, Rutland, 2/1941 and returned 8/1948.
(e)   ex Stanton Ironworks Co Ltd, Holwell Iron Works, Melton Mowbray, Leicestershire, 12/1946; previously at Buckminster Quarries, Lincolnshire until 7/1945.
(f)   ex Stanton Ironworks Co Ltd, Holwell Iron Works, Melton Mowbray, Leicestershire, /1947.
(g)   ex Slough Estates Ltd, Slough, Berkshire, per Newman Industries Ltd, Yate, 11/1947.
(h)   ex Stanton Ironworks Co Ltd, Holwell Iron Works, Melton Mowbray, Leicestershire, /1948; returned by 9/1950. ex Holwell Iron Works, /1964, by 5/1964.
(i)   ex Babcock & Wilcox (Operations) Ltd, Dumbuck Works, Dumbarton, Dunbartonshire, 3/1948; per Abelsons & Co (Engineers) Ltd, Birmingham.
(j)   ex Stanton Ironworks Co Ltd, Stanton Iron Works, near Ilkeston, Derbyshire, 25/9/1956.
(k)   ex Sadler & Co Ltd, Middlesbrough, c/1966.
(l)   ex Dorman Long (Steel) Ltd, Cleveland Works, South Bank, 6/3/1968.
(m)   ex Dorman Long (Steel) Ltd, Redcar Works, Redcar, 29/3/1966.

(1) to Loftus Iron Co Ltd, Skinningrove Iron Works, Carlin How, c/1876.
(2) to New Brancepeth Colliery and Coke Works, New Brancepeth, Co. Durham (owned by Cochrane & Co Ltd until 12/1933, then by The New Brancepeth Colliery Co Ltd, a subsidiary of The Weardale Steel, Coal & Coke Co Ltd).
(3) to New Brancepeth Colliery and Coke Works, New Brancepeth, Co. Durham.
(4) to New Brancepeth Colliery and Coke Works, New Brancepeth, Co. Durham, 6/1905.
(5) to New Brancepeth Colliery and Coke Works, New Brancepeth, Co. Durham, /1923 (one source gives c/1928).
(6) to T.J. Thomson & Son Ltd, Stockton.
(7) to Linthorpe-Dinsdale Smelting Co Ltd, Dinsdale, Co. Durham, by 8/7/1933.
(8) to Gateshead County Borough Council, Saltmeadows Clearance Site (included former Allhusen's Chemical Works), Co. Durham, via Rolling Stock Co Ltd, 1/1936.
(9) to The North Bitchburn Fireclay Co Ltd, Rough Lea Colliery and Brickworks, New Hunswick, Co. Durham, via George Cohen, Sons & Co Ltd, /1936.
(10) to South Durham Steel & Iron Co Ltd, West Hartlepool Works, West Hartlepool, c/1940.
(11) to Stanton Ironworks Co Ltd, Stanton Iron Works, near Ilkeston, Derbyshire, c/1948.
(12) to Stanton Ironworks Co Ltd, Market Overton Ironstone Quarries, Rutland, 3/1948.
(13) to Stanton & Staveley Ltd, Holwell Foundry, Melton Mowbray, Leicestershire, 11/1968.
(14) to British Steel Corporation, Holwell Foundry, Melton Mowbray, Leicestershire, 6/1971.
(15) to Stanton & Staveley Ltd, (subsidiary of BSC), Staveley Foundry, Staveley, Derbyshire, /1972.
(16) to W.G. Readman Ltd, Cargo Fleet, c/1973.

STOCKTON HAULAGE LTD

Cargo Fleet
158
NZ 509208 – **Map G**

This road/rail transhipment depot opened in 1985 on the site of the former OME Cargo Fleet timber yard, just to the west of the Whitehouse Signal Box and crossing. A single siding led off a spur alongside the Middlesbrough–Redcar line, crossed the new Dockside Spine Road, passed through a warehouse and terminated near the River Tees. According to a notice in the signal box, use of the siding into the depot was to begin from 15/4/1985

Freshly painted AUTUMN GOLD (Moyse 1364/1976) stands outside Stockton Haulage's recently established depot at Cargo Fleet, 10th September 1985. *(Cliff Shepherd photograph)*

with BR initially shunting the wagons until the company obtained its own locomotive. A subsequent pencil note added that Stockton Haulage acquired its locomotive on 18/4/1985. To begin with rail traffic was delivered four days a week comprising Taunton cider arriving in VGA wagons, bulk powder from ICI outwards to Stranraer and steel. For example, on 10/9/1985, BR's 37013 arrived with steel on three bogie bolster wagons. It backed over the Whitehouse crossing to place the wagons on the spur. Stockton Haulage's locomotive number 2 in its sparkling yellow colour, complete with "AUTUMN GOLD" painted on the front, ran out and pulled the wagons into the warehouse for unloading. Moyse 1364 and 1464 were the working diesels with 1365 out of use on a short siding in the steel stockyard. Later traffic declined and the railway was mothballed following the cessation of BR Speedlink services (reported 28/6/1994), although 1364 was kept in running order should services resume. Stockton Haulage Ltd ceased trading c10/2003.

Gauge : 4ft 8½in

1	BESSEMER BOY	4wDE	Moyse	1464	1979	(a)	(1)
2 M43	AUTUMN GOLD	4wDE	Moyse	1364	1976	(a)	(2)
3		4wDE	Moyse	1365	1976	(a)	(3)

(a)   ex Shell (UK) Ltd, Teesport Refinery, Grangetown, by 9/1985; Stockton Haulage reported to have its own locomotive on 18/4/1985.

(1)   to Stockton Haulage Ltd, Stranraer, Dumfries & Galloway, c1/1992 and returned after 18/8/1992, by 13/5/1993. To Stockton Haulage Ltd, Stranraer c2/1994; after 13/5/1993, by 28/6/1994.

(2)   to Ed Murray & Sons Ltd, Hartlepool, 24/9/2003.

(3)   to Ed Murray & Sons Ltd, Hartlepool, 10/2003.

STOCKTON IRON FURNACE CO LTD

STOCKTON IRON WORKS, North Shore Branch, Stockton   **159**
Holdsworth, Bennington, Byers & Co until /1864   NZ 455192 – **Map D**

The original company was set up in 1854 with six shareholders providing £12,000 capital. In addition to those named above, they included George and William Fossick. The iron works was established on the bank of the River Tees to the east of the North Shore Branch near Jenny Mills Island. A start was made on erecting two blast furnaces; the first sod being turned on 15/8/1854. This was only a year after Bell Brothers had commenced its Clarence Iron Works further down the river. The first blast furnace at the Stockton Iron Works was tapped on 5/9/1855. By 1857 the company had three 50 feet high blast furnaces with a weekly capacity of 400 tons of pig iron. It also began obtaining ironstone from its Ailesbury Mine at Swainby on 3/3/1857 following the opening of the railway between Stokesley and Picton. The company had 20 puddling furnaces by 1863 to process some of the iron for making into plates and rails.

Thomas Richmond in *Local Records of Stockton and Neighbourhood*, published in 1868, states that on 6/9/1864 a new company was being formed and that the blast furnaces of Messrs Holdsworth & Co would pass into its hands, together with 28 acres of adjoining land on which rolling mills were to be erected. On 8/12/1864, the Stockton Iron Furnace Co Ltd was incorporated with a capital of £60,000 to operate the ironstone mine and blast furnaces. Only one of the furnaces was operating in 1869 but two new 80 feet high blast furnaces were completed in the following year, north of the existing structures.

A separate company, **Stockton Rail Mill Co Ltd**, was formed with a capital of £35,000, also on 8/12/1864, to construct the rolling mill (NZ 453194) on land west of the blast furnaces; its main purpose was to manufacture rails and fish plates. Among the subscribers was George Fossick of Fossick & Hackworth. However, debts on the Ailesbury Mine were increasing and money raised from shares in the Rail Mill was used to pay off the Swainby creditors. This was illegal and, as a result, George Fossick was declared bankrupt. The Rail Mill continued to function and in 12/1874 rerolled 5,000 tons of old rails for the North British Railway. The increasing popularity of steel rails and the economic downturn resulted in the Stockton Rail Mill Co ceasing to operate in 1875. All was not well with the Stockton Iron Furnace Co as well. It had purchased two new locomotives in 1870–71 and all three blast furnaces were working in 1873, but 1876 proved to be their last year of production and the company was in liquidation in 1877.

The 1859 six inch Ordnance Survey map shows sidings leaving the North Shore Branch and running due east of "brick fields" before turning south at right angles to reach the blast furnaces near to the river. With the construction of the Rail Mill and the Malleable Works (which see) to the north, the railway system on the east side of the North Shore Branch, near its termination, expanded but details are lacking because of the early demise of two of the companies. The Rail Mill probably had its own connection to the branch with a line to a jetty on the Tees; the rails for the North British Railway being delivered by ship.

There was an accident at the Stockton blast furnaces in 1870. According to the inquest on a 16 year old lad reported in the *North-Eastern Daily Gazette*, published on 18/11/1870, a vertical boiler locomotive belonging to

Head Wrightson had been hired to the Stockton Iron Furnace Co. It had broken down and was being propelled into the shed by a locomotive when its chimney hit an overhead gantry crane in the shed causing it to fall on the youth. Richard Turner, the driver of another engine, also gave evidence. This would suggest that there were two other locomotives, in addition to these listed below, including the Head Wrightson engine which may have been constructed at the Teesdale Works. There was another fatality at the Stockton blast furnaces in the following year and reported in the same newspaper on 12/12/1871. This involved No.1 locomotive from the Malleable Iron Works which occasionally worked at the Stockton Iron Furnace Co on a Sunday. On this particular day, one of the latter company's engines was under repair and No.1, with its regular crew, went to assist but the fireman was unfortunately fatally injured.

Reference : 'Fatal Accident at Stockton', Russell Wear, *The Industrial Locomotive* 141, ILS, 2011, pp.51,58.

Gauge : 4ft 8½

STOCKTON FURNACES No.3	0-4-0ST		BH	119	1870	New	s/s
No.4	0-4-0ST	OC	MW	336	1871	New	(1)

(1) to Tunnel Portland Cement Works Co Ltd, West Thurrock, Essex.

TARMAC ROADSTONE LTD

TEESPORT WORKS, Grangetown

160
NZ 539228 – **Map H**

Tarmac erected a slag crushing plant on Dorman Long's tips at South Bank on the south bank of the River Tees. The company also had use of a wharf immediately upstream from the jetties serving ICI and Shell Mex & B.P. Ltd at Teesport and near the former World War 1 submarine base. Processed slag was loaded into ships by a swinging boom conveyor. The wharf was operated by Tarmac under lease from Dorman Long and the first shipment of crushed slag left the wharf in 1957. On opening, it was claimed to be the largest roadstone plant in Europe intended to handle 15,000 tons of basic slag a week from Dorman Long's tips. In addition to the tracks on the slag heaps, a Dorman Long line also ran from its railway to Tees Dock, by the Tarmac plant and then on the side of the river to South Bank Wharf. On 14/8/1962, BR and Tarmac signed an agreement for a contractor to construct a siding for BR alongside the Dorman Long line shortly after the Tees Dock junction in order that BR could obtain permanent way ballast from Tarmac. To reach the siding, the BR locomotive would have to travel over Dorman Long's railway.

Gauge : 4ft 8½in

No.1	4wDM	RH	305301	1951	(a)	s/s c/1965
29/24	4wDM	FH	4010	1963	new	s/s c/1973

(a) ex Eccles Slag Co Ltd, Scunthorpe, Lincolnshire, c/1957.

TARSLAG (1923) LTD

TEES BRIDGE WORKS, Stockton.
Tarslag Ltd until /1923
Middlesbrough Slag Co until /1920

161
NZ 445175 – **Map D**

Blast furnace slag proved to be an excellent aggregate for road surfaces because its porosity absorbed the binding agent and it did not easily polish. The massive quantities of slag produced on Teesside became a useful raw material as tarmacadam was increasingly employed on roads in the twentieth century. In 1913, Major & Co, tar distillers of Hull, signed an agreement with the Tees Bridge Iron Co and, as the Middlesbrough Slag Co, began making tarmacadam at the iron works slag heaps. It presumably relied on the iron works locomotives to carry out any shunting. After Pease & Partners took over the Tees Bridge Iron Works, a new agreement was drawn up in 1919 and Tarslag Ltd was formed in the following year. The latter changed its name again three years later and expanded its operations, opening additional plants in the West Midlands.

Extensive use was made of 'private owner' wagons in the inter-war period. It already had a fleet of 349 railway wagons in 1923 and an additional 600 were purchased during the next four years. These were shared between the West Midlands' operations and Stockton. Keith Turton has a photograph of a five-plank wagon built by Derbyshire Carriage & Wagon Co Ltd in 1927 and lettered "TARSLAG (1923) LTD, STOCKTON, 836, WHEN EMPTY RETURN TO TEES BRIDGE IRON WORKS STOCKTON-ON-TEES". This was painted grey with white letters, shaded black and had a reinforced floor. Towards the end of the 1920s, the Tees Bridge blast furnaces were blown out and, at some stage, one of the ironworks locomotives was transferred to Tarslag. The sidings running across the slag tips on the west side of the river connected with the railway at the Tees Bridge Iron Works.

After World War 2, the fortunes of Tarslag declined, unlike its competitor Tarmac. The 4/1950 Ordnance Survey shows the branch from Bowesfield Junction running past the Tees Bridge Slag Works as a single line with just a loop serving a platform. Further south, the line divided into various sidings within Athole G. Allen's chemical works. There were no railways left on the slag tips. When Ted Haigh visited the Tees Bridge site on 19/12/1953, he found the shed disused and the rails appeared derelict. IRS records suggest that any remaining traffic had been latterly worked by Dorman Long's locomotives from its Bowesfield Works.

References :   *Private Owner Wagons*, Bill Hudson, The Oakwood Press,1996, pp.64-65

   Private Owner Wagons An Eighth Collection, Keith Turton, Lightmoor Press, 2009, pp.138-40

Gauge : 4ft 8½in

TARSLAG 4	0-4-0ST	OC	P	956	1903	(a)	(3)
-	4wDM		H	958	1931	(b)	(1)
HOWE	0-4-0ST	OC	HL	2597	1904	(c)	(2)
DAVID	0-4-0ST	OC	AB	2145	1945	New	(4)
COUNCILLOR	0-4-0ST	OC	FW	198	1873	(d)	
Rebuilt	0-4-0D*						(5)

*   Fitted with a Gardner 6LW engine.

(a)   ex Pease & Partners, Tees Bridge Iron Works, Stockton.
(b)   it was new to the South Suburban Gas Co's Lower Sydenham Gas Works in 1931 but the date of its move to Tarslag is unknown.
(c)   ex Thos W. Ward Ltd, on hire, /1942.
(d)   ex Tyne-Tees Steam Shipping Co Ltd, Stockton, c12/1952, after 30/9/1952.

(1)   to Berrisford Engineering Co Ltd, California Works, Stoke on Trent, per Thos W. Ward Ltd, sold 1/12/1936.
(2)   returned to Thos W. Ward Ltd, ex hire.
(3)   to Gjers Mills & Co Ltd, Middlesbrough; after 24/3/1947, by 30/10/1949. The Peckett had been offered for sale by Tarslag in the *Contract Journal*, 26/6/1946.
(4)   to NCB, Smithywood Coking Plant, Chapeltown, Yorkshire (WR), /1953.
(5)   sold c/1958. A 6LW diesel shunting locomotive had been offered for sale in the 23/8/1956 *Contract Journal*. Advertisements in the *Machinery Market* on 10/2/1956 and 27/4/1956 described it as a shunting locomotive with a 6LW Gardner engine for hire or sale.

Gauge : 2ft 0in

-	4wPM	MR	861	1918	(a)	(1)

(a)   former War Department locomotive, LR 2582. A possible spares order came from Tarslag (1923) Ltd, Stockton on 21/6/1924.

(1)   Tarslag advertised in the *Contract Journal* on 4/6/1959 offering for sale plant including a 2ft gauge Simplex diesel locomotive and track.

For HARTLEPOOL IRON AND STEEL WORKS, its constituents and subsequent pipe mills, see the LIBERTY PIPES (HARTLEPOOL) LTD entry.

TEES & HARTLEPOOL PORT AUTHORITY LTD

Following the report of the Rochdale Committee and under the Tees & Hartlepool Port Authority Act (authorised 9/8/1966), the Tees & Hartlepool Port Authority was established on 1/1/1967 to take over the functions of the Tees Conservancy Commissioners, Hartlepool Port and Harbour Commissioners, the local responsibilities of the British Transport Docks Board, together with Stockton Corporation's public quay and the Tyne-Tees and Dent's wharves. In 1992 the THPA was privatised and became the Tees & Hartlepool Port Authority Ltd which functioned until 1/4/2003.

HARTLEPOOL DOCKS, Hartlepool   162
NZ 518341 NZ 516342 – **Map B**

It was not until the 1990s that an industrial locomotive worked for the Port Authority at Hartlepool Docks, although ship builders and contractors on construction projects had used them.

Prior to the expansion of the coal trade, Hartlepool was a small settlement clustered around the Headland. The development of the docks at Hartlepool was due to two separate railway companies; one approaching from the north and the other from the south. In 1832 the Hartlepool Dock & Railway Company (HD&R) gained powers to construct a line from collieries in Co. Durham to the Headland and to improve the Old or Tidal Harbour. The latter was cleared of silt by operating sluices from the Slake to flush it out after high tide. The first coal was carried over the new line in 1835 and the Victoria Dock was constructed on the east side of the tidal harbour, with coal drops in between and was completed on 7/12/1840.

The Stockton & Hartlepool Railway (S&HR) opened on 12/11/1840 from the Clarence Railway at Norton and terminating at Middleton to the south of the Tidal Harbour. An extension and new incline, controlled by the Throston Engine House, enabled coal to be taken over the HD&R's tracks for shipping at the latter's dock, but at a considerable cost disadvantage. Under the forceful influence of Ralph Ward Jackson, the S&HR obtained powers in 1844 to construct its own docks south of the Slake at what was to become West Hartlepool. The 11 acre West Harbour and 7 acre West or Coal Dock opened on 1/6/1847 resulting in coal shipments being diverted from Port Clarence. Jackson formed the West Hartlepool Harbour & Railway Company (WHHR), sanctioned 30/6/1852, and this absorbed the S&HR, Clarence Railway and the docks at West Hartlepool. The new company opened the Jackson Dock on 1/6/1852, to be followed by the Swainson Dock (3/6/1856) and a timber dock and associated timber ponds next to the Slake. By 1862, the Hartlepools were ranked as England's fourth largest port and the population of West Hartlepool had grown to 14,000. With the takeover of the HD&R in 1857 and the WHHR in 1865 by the NER, the way was open to join the two railway systems again. The new railway was constructed on the west side of the docks from Church Street Junction to Cemetery North Junction and opened on 28/5/1877. It subsequently became part of the present-day Hartlepool-Sunderland line. The NER also continued the development of the port on the former Slake with the opening of Union Dock, Central Dock and a link to the Tidal Harbour via the North Basin in 1880. The shunting of the railways serving the docks, together with their associated timber yards and saw mills, was the responsibility of the NER/LNER/BR which employed numerous of their own tank locomotives. For an insight into this complex operation, the reader is referred to a 1909 NER booklet republished by the North Eastern Railway Association in 2000 entitled *Traffic Working Arrangements at the Hartlepools*. Various shipyards had also been established on the Docks estate and almost 2,000 ships were to be built here up to the 1960s. Following nationalisation, the docks became the responsibility of the British Transport Commission on 1/1/1948 and its Docks & Inland Waterways Executive was restyled the British Transport Docks Board in 1962.

With the decline in coal shipments and timber imports, together with the cessation of shipbuilding, the staithes at the Coal Dock were dismantled in 1967 and at the Old Harbour in 1972. The Swainson Dock was filled in and the land used for storage. The timber ponds were reclaimed in the early 1960s. The backfilling of the Central Dock and Graving Dock commenced in 11/1991. Today, the amalgamated Tidal Harbour and Victoria Dock form the commercial port, the Coal and Union Docks have been transformed into a marina and the Jackson Dock is the basis for the 'Historic Quay' museum. With the abandonment of much of the port's rail system, the 'Harbour Branch' now left the Hartlepool-Sunderland line and passed under Middleton Road before sweeping round to serve a set of sidings and warehouses west of the Tidal Harbour/Victoria Dock. More redevelopment resulted in the connection being moved to north of Middleton Road. In 1993–94, a THPA diesel was transferred to shunt the rail traffic which, in 1994, included steel exports from Shelton Works, Stoke on Trent. Additional alterations were made to the branch to accommodate the construction of the Marina Way Road during 1997 (see Trackwork Ltd entry). The locomotive was reported to be in regular use in 2001 but was subsequently returned to Tees Dock in 2002-03.

References : *North Eastern Railway Historical Maps*, R.A. Cook and K. Hoole, Railway and Canal Historical Society, 1991

Hartlepool Railways, George Smith, Amberley, 2013

Maritime Hartlepool, Bert Spaldin, Printability Publishing, 2005

Gauge : 4ft 8½in

| No.6 | | 0-6-0DH | RR | 10215 | 1965 | (a) | (1) |

(a) ex THPA, Tees Dock, Grangetown, c1/1994; after 23/4/1993, by 15/1/1994.

(1) returned to THPA, Tees Dock, Grangetown, after 6/7/2002, by 1/4/2003.

MIDDLESBROUGH DOCK, Middlesbrough 163
NZ 504205 – **Map G**

The inadequacies of the coal drops at "Port Darlington" on the River Tees compelled Thomas Richardson, on behalf of the OME, to propose on 18/11/1838 construction of an enclosed dock at Middlesbrough. The SDR had wished to construct the dock but was precluded by its Act and so gave an assurance to OME that all its coal traffic would pass through the dock. In 1839 land was purchased from Mr Hustler and William Cubitt, an eminent London-based engineer, prepared plans for the dock. The excavation began in spring 1840 with the contract let to Joseph Briggs who employed a number of subcontractors. The dock, which was connected to the river by an approximately ¼ mile long entrance channel, was opened with due ceremony on 12/5/1842. In addition, a ¾ mile long railway had been constructed from Old Town Junction on the Middlesbrough Branch across OME land to the south of the then town terminating at the dock in a fan of ten double sidings on a raised platform leading to the ten coal drops. These were located on the longest side of the dock which contained a water area of nine acres. The SDR had advanced £160,000 towards the cost of the works with an eye to their eventual acquisition and the dock and railway were taken over by the SDR under its 13/7/1849 Act. Earlier the Middlesbrough & Redcar Railway Company had opened its line from Dock Branch Junction to Redcar and this had been leased by the SDR in 1847.

The increasing size of ships and a greater variety of cargoes particularly based on the local iron industry, coupled with a decline in the amount of coal despatched, due to competition from West Hartlepool, resulted in successive improvements to the capacity and layout of the dock. These took place with the letting of three contracts by the NER to Messrs Hodgson & Ridley (8/1869–1873), John Jackson (1/2/1884–1889) and John Scott (1894–1906). This resulted in a water area of 25½ acres, 6,842 feet of quays, 41 hydraulic and electric cranes and 63 capstans for moving railway wagons. There were also six steam cranes at the stockyard. The coal staithes had long gone, but two 40 ton coal hoists at No.10 Quay were available for loading ships. Coal could also be shipped from the North Wharf on the entrance channel. In 1960 No.10 Quay was turned over to general cargo and two transhipment sheds were constructed there. The largest vessel to visit the dock by 1926 was the SS BERLIN (later the SS ARABIC of the White Star Line) at 17,324 tons. Occasionally locomotives were brought to Middlesbrough Dock for shipment overseas including new War Department 0-4-0 saddle tanks (Nos.50 and 51, AB 1587-88 of 1917) and a batch of ten RSH locomotives for Tasmania in 1953.

The dock was laid out in the era of railway dominance and so tracks covered most of the available land; indeed rail traffic remained heavy until the new steel terminal opened at Tees Dock in 1976. Middlesbrough Dock belonged to the railway companies (SDR, NER, LNER, British Transport Commission) until 1/1/1967 and they used their main line stock to carry out shunting. The last steam locomotives to be employed at the dock by BR were Class J71 and J72 tank engines until they were replaced by 204hp diesel mechanical shunters. Access for trains from the main line was via a connection near Guisborough Junction or along the original route by Dock Hill Signal Box. Trains from Middlesbrough Town Goods could also travel along the Vulcan Street railway. An extension of this line was linked to sidings on the north side of the dock and then continued on the swing bridge over the entrance lock before running round the end of the stockyard to reach businesses and the main line at Cargo Fleet. It had the advantage of giving access from Cargo Fleet to Middlesbrough Town Goods

THPA's number 1 (Vulcan Foundry D212 of 1953 and the former BR D2205) shunts internal wagons carrying crates of goods bound for Brazil at Middlesbrough Dock's No.9 Quay on 5th February 1980.
(Cliff Shepherd photograph)

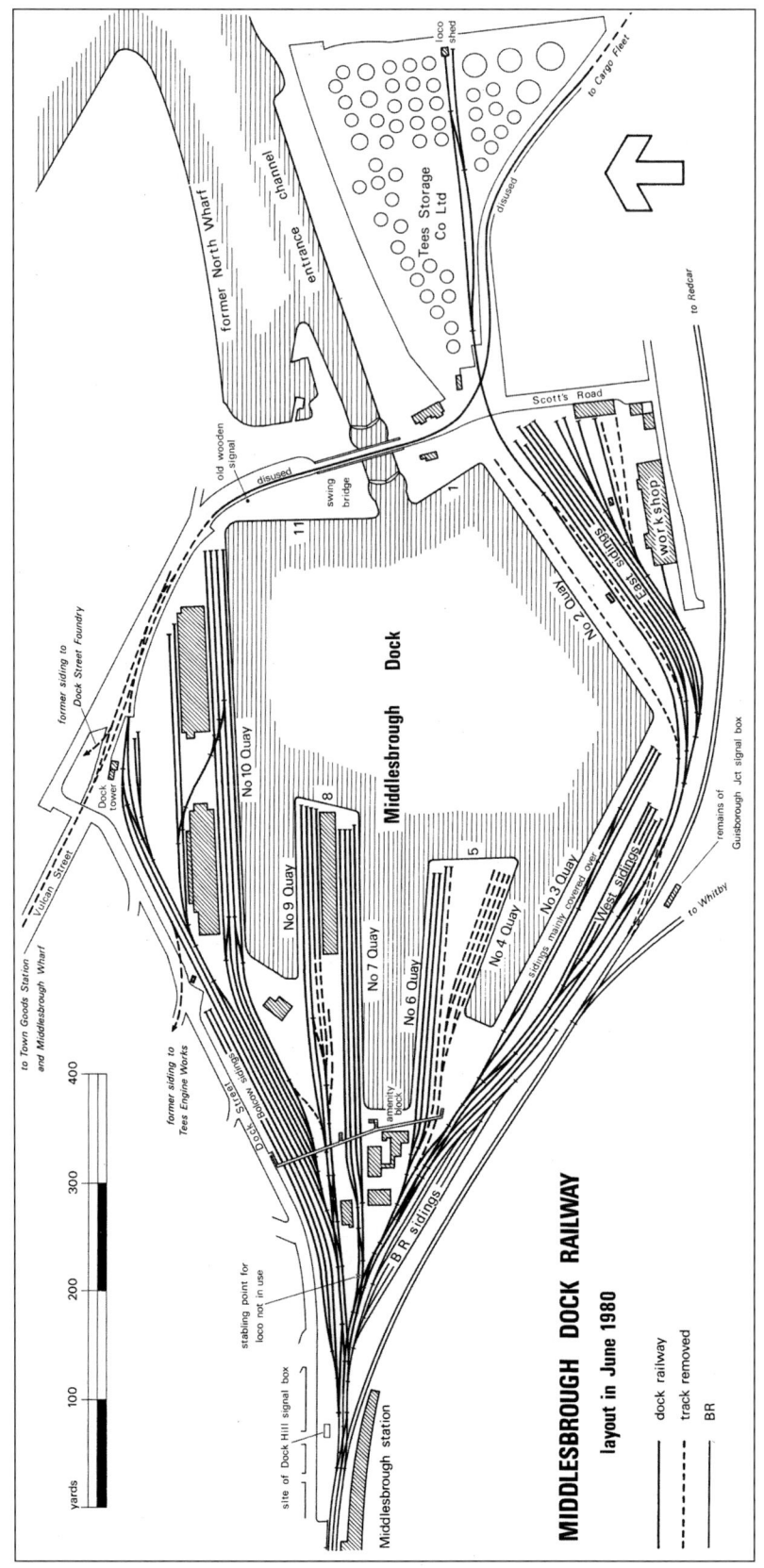

without going on the main line. Various companies' private locomotives used the line as well, subject to paying MOR tolls (except for the section round the north east side of the dock on which the tolls did not apply).

The constricted site of the dock meant that further enlargement was not possible and, in the interwar period, the LNER considered constructing a replacement dock downriver at Lackenby, but the company's finances were far from healthy and it was not until 1963 that the Tees Conservancy Commissioners' Tees Dock opened. On 1/1/1967 the THPA took over the responsibility for shunting the 21 miles of railway sidings at Middlesbrough Dock; it was once described as a 'puddle in a railway yard'. Initially it hired from BR the diesel shunters working at the dock until, in 7/1970, it purchased two of the Drewry locomotives from BR's Thornaby Depot. They were taken to the workshop, serviced and made MD1 and MD2. When Middlesbrough Wharf closed in 1971, two of its diesels were transferred to the dock although they were considered to be underpowered. After MD2 was damaged, two additional 204hp locomotives of the BR version were purchased.

The locomotives were repaired at the workshop which was situated in the south east corner of the site between the main line and Scott's Road. Latterly the Cole's rail crane (maker's number 16973) was also kept there. The THPA owned a considerable number of internal wagons and many of these were used as 'mobile warehouses' to store goods awaiting dispatch. These vehicles were ex BR or the pre-nationalisation railway companies in origin. This rolling stock was mostly kept at the East Sidings near No.2 Quay. When full, wagons had also been stabled at the stockyard on the east side of Scott's Road.

There were many complaints in the late 1920s about the poor road access because the railway lines stood proud of the quay surfaces and the LNER insisted on lorries going to the stockyard where the goods were loaded into railway wagons for the short journey to the vessels. Some shipping companies transferred to the riverside berths as a result. Later the stockyard was transformed into Tees Storage's tank farm (which see).

As the facilities at Tees Dock developed, use of Middlesbrough Dock declined. It officially closed to general cargo on 31/7/1980, although the last ship had departed on the 24th. For a short while, the dock remained tidal and could accommodate small coasters but there was no need for the railway except for the line to Tees Storage.

References : 'Middlesbrough Dock Railway', Cliff Shepherd, *Industrial Railway Record* 105, IRS, June 1986, pp.411-21

Middlesbrough Dock: 'A Few Facts and Figures', J.H. Patchett, *North Eastern Railway Magazine*, 1911

The Story of Middlesbrough Dock in Greater Middlesbrough, Norman Moorsom, Middlehaven Partners, 2006

The History of Middlesbrough Dock 1842-2000, John H. Proud, Printability Publishing Ltd, 2000. This also appeared in The *North Eastern Express* issues 159-62, 2000-01

Gauge : 4ft 8½in

1	(MD1) (D2205)	0-6-0DM	VF		D212	1953		
			DC		2486	1953	(a)	(5)
2	(MD2) (D2243)	0-6-0DM	RSHN		7862	1956		
			DC		2575	1956	(a)	(1)
2		0-4-0DM	RSHN		7925	1959		
			DC		2592	1959	(b)	(6)
5		4wDH	EEV		D908	1964	(b)	(4)
3	(D2023)	0-6-0DM	Sdn			1958	(c)	(2)
4	(D2024)	0-6-0DM	Sdn			1958	(c)	(3)

(a) ex British Rail, Thornaby Depot, 7/1970.
(b) ex Middlesbrough Wharf, /1971.
(c) ex British Rail, Lincoln Depot, 7/1972.

(1) it was damaged after running past signals at Dock Hill and colliding with a BR train coming the other way. It was dismantled /1972 and scrapped 3/1973.
(2) to Tees Dock, 15/9/1980.
(3) to Tees Dock, 25/10/1980.
(4) to Thomas Hill (Rotherham) Ltd, Kilnhurst, South Yorkshire, 18/11/1980; resold to British Aluminium Co Ltd, Falkirk, Central, 5/1/1981,
(5) the last THPA locomotive to work at Middlesbrough Dock shunting condemned wagons out of West Sidings 22/9/1980; to Tees Dock, 22/11/1980.
(6) to Darlington Railway Preservation Society, Darlington, Co. Durham, 27/3/1985.

MIDDLESBROUGH WHARF, Middlesbrough 164
NZ 494214, NZ 496213 – **Map F**

The THPA gained control of the Tyne-Tees and Dent's Wharves on the south bank of the river upstream from the Transporter Bridge on 1/1/1967 and renamed them Middlesbrough Wharf. For their earlier history, see entries under the Tyne-Tees Steam Shipping Co Ltd and T. Roddam Dent & Son Ltd. It also acquired the remaining locomotives at the two wharves, which used Roddam Dent's engine shed. BR informed the THPA that the tolls for traffic over the MOR lines would be increased from 2d to 2½d per ton (Tyne Tees) and 3d to 4d per ton (Dents) from 1/2/1968. However, the increasing size of ships and the move to containerisation meant that Middlesbrough Wharf was no longer suitable and it closed in 1971. By this date, the railway lines on the eastern part of the site had been removed and this became the THPA's engineers and craft depot; the dredgers being moored here.

Gauge : 4ft 8½in

GRADWELL	4wVBT	VCG	S	9377	1947	(a)	(1)
TYNE-TEES WHARF No.4	4wVBT	VCG	S	9562	1954	(a)	(1)
TYNE-TEES WHARF No.1	4wDM		FH	3817	1956	(a)	(3)
TYNE-TEES WHARF No.5	4wDM		FH	3949	1960	(a)	(4)
TYNE-TEES WHARF No.3	4wDM		FH	3964	1961	(a)	(4)
TYNE-TEES WHARF No.8	4wDH		EEV	D908	1964	(a)	(5)
-	0-4-0DH		EEV	D1206	1966	(a)	(2)
T. RODDAM DENT & SON LTD No.4							
	0-4-0DM		RSHN	6978	1940	(b)	Scr c/1968
D2	0-4-0DM		RSHN	7902	1958	(b)	S/S
D1	0-4-0DM		RSHN	7925			
			DC	2592	1959	(b)	(5)

(a) ex Tyne-Tees Steam Shipping Co Ltd, Tyne-Tees Wharf, Middlesbrough, with site,1/1/1967.
(b) ex T. Roddam Dent & Son Ltd, Dent's Wharf, Middlesbrough, with site, 1/1/1967.

(1) scrapped on site by Thos W. Ward Ltd, Sheffield, 4/1968.
(2) believed scrapped end of /1968.
(3) to Thomas Hill (Rotherham) Ltd, Kilnhurst, Yorkshire (WR), 28/1/1970; resold to Courtaulds Ltd, Great Coates Works, Grimsby, Lincolnshire.
(4) to Tees Dock, Grangetown, c/1970.
(5) to Middlesbrough Dock, Middlesbrough, /1971.

TEES SALT CO LTD

(Subsidiary of **ICI Ltd** from /1928)
TEES SALT WORKS, Haverton Hill 165
NZ 478220 – **Map C**

The Tees Salt Company was established on 30/6/1890 by John Wilson Watson and Robert Scrafton who were landowners and building speculators with property at Haverton Hill. The plans for the works were approved on 3/12/1890 and construction began. It comprised a panhouse containing a series of both small and large pans, a salt store, a packing warehouse, together with workshops and offices. A second panhouse had been added by 1899 and, four years later, the works was reported to have twenty pans for coarse salt and four for fine salt. The works was situated on the south side of Chilton's Lane (later named Haverton Hill Road). Railway sidings ran on either side of the two principal buildings before coming together to head north east by The Salt Union's South Durham railway line before eventually joining it. This would have given it access to the South Durham River wharf and the NER Clarence Branch. No engine shed is shown at the Tees Salt Works on the 1893 Ordnance Survey, but a revision dated 1913 and an ICI conveyance plan indicate a single road shed served by a short length of track near the convergence of the sidings.

Information on locomotives owned by the Tees Salt Co is scanty but it is known that a secondhand Andrew Barclay engine had been purchased by 1894. The company also possessed at least forty 10 ton salt vans (wagon numbers 50–69, 80–99) built by the Darlington Wagon & Engineering Co Ltd. A photograph of one appears in the Turton reference. By the mid 1920's, the number of pans at the works had not changed; brine was sold to Synthetic Ammonia & Nitrates Ltd and The United Alkali Co Ltd, as well as shipping salt from the wharf. Tees Salt Co was taken over by ICI in 1928 and negotiations began in the following year to close the works. Salt manufacture ceased on 13/12/1930, the works was demolished in 1931 and the Tees Salt Co Ltd voluntarily wound up on 28/7/1931.

References : 'Salthome A Story of Salt, Cement and Birds', Cliff Shepherd, *Industrial Railway Record* 197, IRS, 2009, pp.345-65

Private Owner Wagons A Ninth Collection, Keith Turton, Lightmoor Press, 2010, p.132

Gauge : 4ft 8½in

-		0-4-0ST	OC	AB	261	1883	(a)	(1)

(a) ex Morrison & Mason Ltd, contractors, Glasgow, by 11/1894.

(1) to Synthetic Ammonia & Nitrates Ltd, Billingham Works, Billingham, /1931.

TEES SCORIAE BRICK CO LTD

Cargo Fleet **166**
NZ 523205? – **Map G**

Joseph Woodward formed the Tees Scoriae* Brick Co in 1872 to turn slag from the blast furnaces at the Clay Lane Iron Works into blue-grey paving blocks. This involved transporting molten slag in ladles on railway bogies to a revolving brick wheel containing moulds. From here, the still hot blocks were tipped from the moulds and passed to annealing kilns where they were 'cooked' until hard. A second plant was established at the Cargo Fleet Iron Works where there was one wheel with 140 moulds and eighteen kilns capable of holding 1,000 blocks at any one time. The scoria blocks can still be seen in many locations on Teesside; in 1912, 62,881 tons were shipped from wharves on the Tees. The company had plants at various local iron works, although most relied on the latter's locomotives to handle their rail traffic.

A NER document dated 1898 listing sidings receiving and/or despatching traffic gives two works where the Tees Scoriae Brick Company was then involved; one of which was the Cargo Fleet Iron Works. The locomotive had become available after the Maryport Iron Works in Cumberland closed in 1893. Thos W. Ward purchased the Maryport premises in 1897 and offered the locomotive for sale on 23/7/1897 and again on 10/6/1898.

*This name was often simplified to Scoria, even on some of the blocks!

Gauge : 4ft 8½in

-		0-4-0ST	OC	AB	136	1872	(a)	(1)
	REDCAR No.4	0-4-0ST	OC	BH	371	1877	(b)	s/s

(a) ex Thos W. Ward Ltd, Maryport Iron Works, Cumberland after 10/6/1898, by 6/1900.
(b) ex Walker, Maynard & Co, Redcar Works, Redcar /1898. This information is from Inness but it is not known whether it came to the Cargo Fleet or Clay Lane Works.
(1) to Pease & Partners Ltd, Tees Iron Works, Cargo Fleet.

TEES SIDE IRON & ENGINE WORKS CO LTD

TEES ENGINE WORKS, Middlesbrough **167**
Tees-Side Iron & Engine Works Co Ltd until 21/9/1889   NZ 499210 – **Map G**
Hopkins, Gilkes & Co Ltd until 12/4/1880
Gilkes, Wilson & Co until 25/2/1865

A piece of land near the Port Darlington coal drops was rented by the SDR from OME on 1/5/1839 and used as a 'yard' to repair its rolling stock; presumably only a few temporary wooden sheds sufficed. Edgar Gilkes had joined the SDR as a young engineer in the same year and became manager of the yard in the early 1840s. The OME was also laying out Lower Commercial Street and Grey Street (sometimes spelt Gray) as it sought to maximise the returns from its investment on land. On 7/4/1842 Sidney, Sherwood & Smith rented just over 4 acres of grassland alongside what was to become the Vulcan Street railway with the intention of developing a foundry. The partners also acquired an access strip leading north to the river. Buildings had been erected by 1842 and the previously rented land was purchased outright in 1846, although the partnership was disbanded in that latter year.

Meanwhile Edgar Gilkes had joined with Isaac Wilson, manager of the Middlesbrough Pottery, to form Gilkes, Wilson & Co which rented 3,000 square yards of land from the OME on 30/12/1846. This was situated between Lower Commercial Street and the Vulcan Street railway with the Pottery on one side and Sydney, Sherwood & Smith's foundry on the other. Gilkes Wilson also rented the foundry from Sidney, Sherwood & Co, but not the access strip to the river. A large new foundry was constructed alongside Lower Commercial Street and investigations by J.K. Harrison suggest that this was in 1847. It probably meant that the previously rented foundry building could be released and this belonged to Sidney, Sherwood & Co until 29/5/1859 when Bolckow & Vaughan purchased it in order to gain control of the access strip (see the I. Copley & Co entry for the building's subsequent history). Gilkes Wilson soon expanded southwards beyond Lower Commercial Street by 1856 and, in 1875, its premises stretched to Lower East Street and Dock Street.

Gilkes Wilson's premises had no fewer than three connections to the Vulcan Street railway in 1856. The earliest was probably the extension of the Middlesbrough Pottery siding running parallel to the Vulcan Street railway reflecting Wilson's interest in both businesses. The second joined with a more direct curve from Vulcan Street to reach a turntable in the centre of the Tees Engine Works yard with sidings radiating to serve the various buildings. A third connection from Vulcan Street, this time coming from the west, ran into the yard and across to the fitting shop. One of the sidings in the yard crossed Grey Street to reach Sidney, Sherwood & Co's foundry and another ran over Lower Commercial Street. By 1875 the internal line to the above foundry, now occupied by the Cleveland Patent Nut and Bolt Company, had gone but Gilkes Wilson had a series of sidings serving its premises south of Lower Commercial Street and a rail connection over Dock Street to connect with the NER's lines adjoining Middlesbrough Dock. Gilkes Wilson also occupied the strip of land on the north side of Vulcan Street leading to the Pottery Wharf and established a small foundry there to produce railway chairs.

Gilkes Wilson built its first locomotive named TEES (maker's number 1) in 1847 for the York, Newcastle & Berwick Railway. It was also constructing rolling stock for the SDR and engineering products for the coal and expanding iron industries. John Fowler, aged 21, joined the company in 1847 as an apprentice before leaving in 1850 to become an agricultural engineer and eventually set up the Steam Plough Works in Leeds.

From 1859, Gilkes Wilson began to collaborate with Snowdon, Hopkins & Co, owners of the Tees Side Iron Works (see following entry) comprising blast furnaces, puddling furnaces and rolling mills. This cooperation led to the production of new lightweight bridges made of cast iron columns and wrought iron girders; examples included such well known railway bridges as the Crumlin Viaduct in South Wales and the Belah Viaduct on the Stainmore line. In 1865 the companies amalgamated to form Hopkins, Gilkes & Co Ltd.

The Tees Engine Works continued to build locomotives, together with other products for local industries such as two blowing engines for the South Durham Iron Works in Darlington, in addition to contributing to bridge construction. A list prepared by R.H. Inness in 1949 and later published by Fred Harman has 365 product numbers, of which 176 were allocated to locomotives built by the Tees Engine Works between 1847 and 1875. Although 95 were 0-6-0 tender locomotives for the SDR/NER, 60 were four-coupled shunters mostly supplied to local industries.

A locomotive almost certainly built by Hopkins Gilkes but whose identity and location is unknown. Harman had a print with the note on the back reading "No.5 or Arthur Works Nos.276 or 355" suggesting that it was either at Linthorpe or Tees Side Iron Works.

Number of locomotives reputedly made each year by the Tees Engine Works

Year	No.	Year	No.	Year	No.
1847	2	1857	4	1867	8
1848	8	1858	5	1868	1
1849	6	1859	5	1869	2
1850	3	1860	7	1870	3
1851	1	1861	9	1871	4
1852	4	1862	8	1872	1
1853	2	1863	11	1873	7
1854	5	1864	15	1874	5
1855	10	1865	14	1875	3
1856	6	1866	17		

By the mid 1870s, Hopkins Gilkes was in financial difficulties and, following the completion of the Tay Bridge contract, it went into receivership on 27/5/1879. Worse was to follow when, on 28/12/1879, some of the high girders in the bridge collapsed into the River Tay during a gale and 75 people were drowned. Hopkins Gilkes had taken over the contract to build the bridge in 1873. It supplied the cast iron columns, which were produced at a temporary foundry previously set up on site at Wormit, together with the wrought iron girders from the Middlesbrough Works (the latter were later reused in the rebuilt bridge). Even the nuts and bolts were supplied by the Cleveland Nut and Bolt Company in Vulcan Street. The subsequent enquiry found that the bridge was badly designed, constructed and maintained. Following the Tay Bridge debacle, a scheme of reconstruction for Hopkins Gilkes was agreed and the Tees-Side Iron & Engine Works Co Ltd was established on 12/4/1880 to operate the Tees Engine Works and the Tees-Side Iron Works, but no more locomotives were constructed.

For the history of the Engine Works under the Tees-Side Iron & Engine Works Co Ltd, see the following Tees Side Iron Works entry.

References :   The History and Locomotives of the Tees Engine Works, Fred W. Harman, Century Locoprints

The Industrial Heart of Old Middlesbrough, J.K. Harrison, CIAS, 2010

Gauge : 4ft 8½in

-	0-4-0tank	HG	236	1866	New	s/s

TEES SIDE IRON WORKS, Middlesbrough 168

Tees-Side Iron & Engine Works Co Ltd until 21/9/1889 NZ 491214 – **Map F**
Hopkins, Gilkes & Co Ltd until 12/4/1880
Hopkins & Co until 25/2/1865
Messrs Snowdon and Hopkins until /1861

The SDR opened its Middlesbrough Branch on 27/12/1830. After running alongside the River Tees at Newport, the line headed across the marshes to the six (later eight) new coal drops on the south bank of the river. The SDR locomotive GLOBE hauled the official train and 600 guests watched the collier SUNNISIDE being loaded with coal before enjoying a banquet. Initially given the name "Port Darlington", the enclosed coal drops extended 450 yards along the river bank with wagons being raised by lifting engines. Meanwhile The Owners of the Middlesbrough Estate (OME) had begun to develop the new town to the east. The SDR line also served the first Middlesbrough Station (1834–37) in the form of a coach shed on a siding near the coal drops and not far from the old Packet Wharf (see Tyne-Tees Steam Shipping Co entry). Unfortunately problems of silting around the coal drops compelled OME to begin work on excavating a new dock, together with laying the railway from the Middlesbrough Branch at Old Town Junction (earlier called Dock Junction). The remnant of the line to Port Darlington then became known as the Old Town Branch. Following the opening of Middlesbrough Dock, the coal drops were no longer required and were advertised for sale on 9/9/1842, but were still shown on J. Sorrenby's 1845 map.

William Randolph Hopkins moved to Middlesbrough in 1850 to manage the short lived Warlick Patent Fuel Works. This plant was located at right angles to the river and, by 1856, had five sidings connected to the Old Town Branch. Hopkins joined with Thomas Snowdon in 1853 to begin development of an iron works on the Patent Fuel Works site and the western half of the now dismantled Port Darlington coal drops. Initially a malleable iron works was established containing a rolling mill for producing bar and angle iron for shipbuilding. A rail mill was added and, by 1855, 200 tons of finished iron was being produced each week. The partners commenced building two 55 feet high blast furnaces in 1857 which were 'blown in' during 7/1859. Thomas Snowdon retired in 1861. John Gjers was appointed manager in 1862 after leaving the Ormesby Iron Works and before he set up his own business. There was increasing collaboration with Gilkes, Wilson & Co and, on 25/2/1865, the two companies' combined to form Hopkins, Gilkes & Co Ltd. In 1867, the existing blast furnaces were rebuilt and two more blown in. In addition to the blast furnaces, Tees Side Iron Works had 102 puddling

JAMES, Gilkes Wilson maker's number 130 of 1861, was supplied to Hopkins & Co's Tees Side Iron Works.
(IRS collection)

furnaces – only the Britannia Works on Teesside had more – and rolling mills producing rails, heavy angles, beams and merchant iron with a capacity of 1,200 tons of finished iron per week.

By this date, Tees Side Iron Works was a major undertaking. Rail traffic approached it along the NER (ex SDR) Old Town Branch. The latter retained its original course, but had expanded with additional sidings as it passed by the Middlesbrough Town Goods Station. The iron works was served directly by the NER branch before the latter curved to run parallel to the river. A series of sidings entered the centre of the works past the row of four blast furnaces located in the angle of the lines leading to the adjoining Linthorpe Iron Works. Other tracks led from the branch and central sidings to the puddling furnaces, mills and river wharf. Not surprisingly, several of the iron works locomotives came new from the Tees Engine Works, but little other information has survived about the fleet.

Following the depression in the mid 1870s, Hopkins Gilkes ran into financial problems and it went into voluntary liquidation on 27/5/1879. Seven months later came the Tay Bridge disaster (see the Tees Engine Works entry), which had a significant impact on the likelihood of future sales. As a result, the business was restructured as the Tees-Side Iron & Engine Works Co Ltd on 12/4/1880. The capital value was written down and financial support came from Wilsons, Pease & Co based at the Tees Iron Works at Cargo Fleet.

The next 15 years were ones of constant struggle. In 1881 it was reported that the furnaces were being kept in blast producing iron for stock because it was cheaper than taking some out of use and having to restart them. There was concern about the space for slag tipping. The General Manager commented on 9/9/1881 that the distance from the furnaces to the tip was 1½ miles and it "causes great expense in wear and tear of the locomotive power, also wear and tear of slag bogies and costly maintenance in roads". The location of the tip has not been identified, but it may have involved the company's locomotives working over the NER lines to reach a tip east of Middlesbrough Dock because it was stated on 23/6/1882 that Mr Ridley had taken a section of the slag tipping land. The alternative arrangement adopted by the company was to dump slag at sea using a secondhand tug, EXCELSIOR; orders for two barges were placed with Raylton, Dixon & Co and improvements made to the wharf. The barges were launched on 20/4/1882 and the Manager was looking to sell the slag tipping ground. The make of slag averaged 360 tons every 24 hours and it would cost 14.63d per ton to break it by hand and take it in barrows to the barges. It was proposed in 8/1882 that 130 slag bogies should be purchased representing an ongoing cost of 7.72d per ton.

In 1883 ironstone was coming from Bell Bros, Brotton Mine and Normanby or Ormesby Mines; coke from Pease & Partners, North Bitchburn Coal Co and Langley Park; limestone from Pease & Partners, Raisby Hill and the Teesdale Limestone Co, but it was reported on 21/12/1883 that there was no market for puddled iron. Most of the puddling furnaces and rolling mills had been demolished by 1883 with the works concentrating on pig iron production. Nevertheless, in that same year, the company was looking to establish a shipyard on the cleared eastern part of the premises described as "the splendid site of land we own by the riverside". Berths were laid out and workers employed but no orders came so that by the end of 1884, the only expense was the wages for a manager and watchman. The shipyard was rented to Raylton Dixon & Co Ltd in 9/1888 but handed back in 1892 when orders decreased. It had been offered to both Raylton Dixon and Ropner on 3/6/1891 to purchase for £32,000 the present yard, about ¾ acre of old mill site "and a triangular piece of land on the south side of the road, having the Railway Sidings, Engine Shed etc". The latter was probably the two-road building on the south side of Depot Road. In 1884 little profit was being made, only two blast furnaces were operating and the wharf and cranes were in a dilapidated state. The company had gained an order for the Middlesbrough Dock gate and swing bridge but missed out on a job for the Indian State Railways – "we can only conclude that the strong prejudice, in regard to the Tay Bridge, still hangs very strongly upon us".

By 1888 only one blast furnace was in blast and W.B. Peat was brought in as liquidator. The company went into liquidation and was restructured with a subtle change of name on 21/9/1889. It was reported in 1891 that two blast furnaces had been taken out of blast and the two barges let to the Tees Conservancy Commissioners for six months, although an order for two 100 inch blowing engines had been received from the Clay Lane Iron Co. This was followed in 1/1892 by orders for £38,101 of work on the extension of Newcastle Central Station and £8,990 for the iron viaduct at Beckton Gas Works. By the end of the year, three blast furnaces were operating, two using Spanish iron ore which was kept at Connal's store on the Ironmaster site and presumably was brought round by rail to the Tees Side Iron Works. The latter now comprised the blast furnaces, stoves, sidings and engine shed on the southern part of the premises, while the northern part was subdivided into two with the vacant "No.2 Dock Yard" on the east and the wharf on the west.

In 1893 the Institute of Mechanical Engineers visited the iron works and engine works. The blast furnaces were described as 75 feet high with two hoists, nine Cowper stoves and the blast supplied by five vertical non-condensing engines. Behind the furnaces was a row of nine calcining kilns with mineral bunkers surmounted by a gantry. A pneumatic hoist and drop raised full and lowered empty wagons. Slag was loaded into wagons "as shingle by Hawdon's machine and subsequently sent out to sea". The Tees Engine Works occupied 6½ acres and had two large foundries capable of producing 500 tons of castings per week. A large bridge building yard was attached to the works. It had produced the whole plant for the Leang Hoo Iron and Steel Works in China.

By 1895 the business was in serious financial trouble. The Board last met on 2/3/1896 when it was agreed that a syndicate led by Sir Christopher Furness should purchase the company with the result that:

> The order book and some of the plant at the Tees Engine Works transferred to the newly formed Tees Side Bridge & Engineering Works Ltd at Cargo Fleet.
> The Tees Engine Works premises were occupied by Furness, Westgarth & Co.
> R. Craggs took over the No.2 Dock Yard.

On 11/7/1896, W.B. Peat announced that the Tees Furnace Co Ltd was to be formed to operate two of the blast furnaces using imported iron ore via the wharf. However, *The Engineer* reported on 14/5/1897 that the blast furnaces were being dismantled and Copley Turner & Co Ltd had plant, including one 10 inch locomotive for sale.

References :   Tees Side Iron & Engine Works Co Ltd, General Manager's reports to the Board 9/1882-12/1884, 12/1890-1/1896 (BS.TSE/1/2/1, BS.TSE/1/2/2, Teesside Archives)

Gauge : 4ft 8½in

	JAMES	0-4-0tank	OC	GW	130	1861	New	(1)
No.3	ROBERT	0-4-0tank	VC	GW	165	1863	New	s/s
-		0-4-0tank		HG	235	1866	New	s/s
	NUNTHORPE	0-4-0ST	OC	HG	252	1867	New	(4)
	BULLDOG			HG	299	1873	New	(5)
No.5		0-4-0ST	OC	HG	355	1874	New	(2)
	HILDA	0-4-0ST	OC	HG	356	1874	New	(3)

Inness suggests two other locomotives were here – CASTELL 0-4-0ST with 12in x 20in cylinders and 3ft 6in diameter wheels, built by John Harris and WILLIAM, a 0-4-2 saddle tank. He adds that the Harris locomotive went to the Tees Iron Works but nothing further is known about them.

(1)   to South Cleveland Ironworks Co Ltd, Glaisdale Iron Works, Yorkshire (NR).
(2)   to William Whitwell & Co Ltd, Thornaby Iron Works, Thornaby. It is possible that this was the locomotive that later moved from Highways Construction Ltd, Port Clarence to Athole G. Allen (Stockton) Ltd in /1948.

(3)   to Wilsons, Pease & Co, Tees Iron Works, Cargo Fleet.
(4)   to The Harris Deepwater Wharf Co Ltd, Cargo Fleet.
(5)   to Gabbutt & Atkinson, contractors.

TEES STORAGE CO LTD

Middlesbrough 169
NZ 508207 – **Map G**

This chemical storage company was located at the former stockyard of Middlesbrough Dock on the south side of the entrance channel. In 1962 Sadler & Co Ltd had changed emphasis from manufacturing chemicals to storing and distributing petroleum products. It constructed storage tanks on 17½ acres of the former stockyard. In 1965 it joined with Gebr. Broere NV of Dordrecht and Europoort to form a subsidiary, Tees Storage Ltd, to store all types of liquid chemicals. A 'tank farm' was also established at Seal Sands by Tees Storage Ltd. Rail access was via the dock railway system. A single through line ran from Dock Hill Signal Box past the West and East Sidings at the dock and across Scott's Road and the former Middlesbrough Owner's railway from Vulcan Street to Cargo Fleet before entering the Tees Storage site. The line then divided into two tracks to run between the tanks and under a filling/discharge point to reach the eastern boundary. A small engine shed was located at the end of one of the tracks. The main job of the Tees Storage locomotive was to collect tank wagons from the through line by the dock's East Sidings where they had been delivered by BR. The cranes were removed from No.2 Quay in 1962 and it was converted to handle tanker traffic, being used exclusively for ships delivering chemicals to both Tees Storage and Sadlers, but the sidings serving the quay had been removed and a pipeline was used to convey chemicals from the ships. The first locomotive owned by the company was Thos Hill maker's number 101, the pioneer of the Vanguard design; it operated 2-3 times a week. When Middlesbrough Dock closed on 31/7/1980, most of its railway system was quickly dismantled leaving the through line to Tees Storage. On 31/12/1982 Tees Storage terminated the siding agreement; rail traffic having ceased and most of the track at the terminal was taken up leaving just a 200 yard spur. However, the company obtained a rail contract and so had to relay the track and bring its locomotive back into use. In 1986–87 a rake of 3 to 5 tank wagons was delivered each day by BR and left for collection by the side of the dock. Rail traffic ceased in 1992 and the locomotives were advertised for sale about August of that year. Today the site is occupied by Middlesbrough Football Club's Riverside Stadium.

Gauge : 4ft 8½in

-	4wDH	TH	101C	1960	(a)	(1)
-	4wDM	FH	3958	1961	(b)	(2)
(409)	0-4-0DH	NBQ	27644	1959	(c)	(3)

(a)   ex Upper Clyde Shipbuilders Ltd, Linthouse, Glasgow, /1969; per Leslie Sanderson Ltd, Birtley.
(b)   ex Thos Hill (Rotherham) Ltd, Kilnhurst, South Yorkshire, 6/1976; previously at Sir Hedworth Williamson's Limeworks Ltd, Kirkby Stephen, Cumbria.
(c)   ex C.F. Booth Ltd, Rotherham, South Yorkshire, 31/1/1986.

(1)   dismantled 9/1976; engine, gearbox etc to Thos Hill 9/1976, remainder to Thos. W. Ward Ltd for scrap, 11/1976.
(2)   to Yorkshire Engine Co Ltd, Rotherham, South Yorkshire, 4/8/1992.
(3)   to Ayrshire Railway Preservation Group, Dalmellington, Strathclyde, 1-2/9/1992.

T.J. THOMSON & SON LTD

MILLFIELD WORKS, Stockton 170
NZ 438193 – **Map D**

There is some uncertainty about the early years of Thomson's business. It appears to have been established as metal merchants in 1871 with an office in Middlesbrough. After resigning from Hopkins Gilkes & Co, Edgar Gilkes joined Thomas J. Thomson & Co in 1881 and Thomson & Gilkes operated from the Millfield Iron Works at a site or sites in Stockton. One of their sales was to supply wrought iron for the rebuilding of Hammersmith Bridge in London. Gilkes retired in February 1883. In 1928 the business, now titled T.J. Thomson & Son Ltd, was purchased by H.E.I. Turner although the company's name remained the same and, in 1932, it moved to occupy the former Moor Steel and Iron Works which it renamed the Millfield Works. These premises had previously belonged to the South Durham Steel & Iron Co (which see) but had closed by 1921 following which a scrap dealer, William Newton & Co, had occupied the site presumably demolishing the plant. The proprietors were William Newton and James Davison White and the NER agreed on 14/10/1922 that Newton could run its steam rail crane over the railway company's adjacent sidings. The premises were located on the west side of the LNER's Yarm-Norton line, a short distance south of Stockton Station. Approval to receive and deliver traffic on specified sidings was given by the LNER to Thomson on 16/11/1931 but Davison seems to have also occupied part of the site because he had a similar agreement with the railway company dated 29/12/1931.

From 1940, traffic was exchanged with the railway company at Phoenix Sidings situated between the main line and the scrapyard. Here, a Thomson locomotive collected wagons as well as condemned locomotives and rolling stock and propelled them into the yard where several sidings spread across the site, including some passing under overhead gantries.

In 1977, Thomson started a new business called Accredited Processed Metals Ltd (which see) to deal in non-ferrous metals at Cargo Fleet. This moved in 1981 to its own site on the southern boundary of the Millfield Works, although it was not rail connected.

Over the years Thomson purchased three new locomotives to shunt its yard, the first being RSHN 7494 in 1949 which carried "T.J. THOMSON & SON LTD" plates. At other times, one of the locomotives in for scrap would be used. The numbers of locomotives arriving increased from the 1960s with the rapid withdrawal by BR of its steam locomotives, such as the row of eight mainly class V1/V3 2-6-2 tanks seen at the yard in 3/1965. This was later followed by some main line diesels and particularly by industrial locomotives after 1990 as the number of such railway systems declined. Initially, there was less monitoring of Thomson's site and it is certain that the following schedule is incomplete but, from the 1980s, there was more scrutiny (usually from the perimeter) of the industrial locomotives being scrapped in the yard. Occasionally a locomotive owned by Staffordshire Locomotive Co Ltd would be stored by Thomson on a temporary basis. The disappearance of much heavy industry – "virtually no one on Teesside produces scrap now" – led to the rundown of the scrapyard from 2014. Most of the internal wagons were scrapped and the remaining locomotives sold to Ed Murray & Sons Ltd in 2015. The company continued to trade in scrap but stopped buying materials for processing on site in 12/2016 and the remaining track was auctioned on 1/3/2017.

EEV-AEI 4003 of 1971 stands in Thomson's scrapyard at Stockton on 29th October 2005. It was later purchased by Ed Murray and hired to become the shunter at Hartlepool Power Station. (David Love photograph)

Gauge : 4ft 8½in

\- 0-4-0ST OC I`Anson
ex R.W. Crosthwaite Ltd, Thornaby, /1927.
scrapped after 10/7/1933.

\- 0-4-0ST OC MW 399 1872
ex Gjers Mills & Co Ltd, Middlesbrough, /1927.
to Connal & Co Ltd, Middlesbrough.

0-4-0ST OC MW
origin unknown /1927.
s/s.

WEAVER 0-4-0ST OC MW 1072 1888
ex ICI Ltd, Billingham, /1934, to Richardsons Westgarth (Hartlepool) Ltd, Hartlepool Engine Works, Hartlepool, on hire, /1947 and returned.
scrapped c12/1950.

NORTON 0-4-0ST OC HL 2171 1890
ex Cochrane & Co Ltd, Cargo Fleet. Thomson offered NORTON for sale in *Machinery Market*, 16/7/1948.
s/s c/1950.

\- 0-4-0ST OC P 949 1902
ex Withnell Brick & Terra Cotta Co (1912) Ltd, Chorley, Lancashire.
to Anderston Foundry Co Ltd, Port Clarence, per Thos W. Ward Ltd, sold 12/4/1948.

\- 0-6-0T OC HL 3531 1922
ex War Department, Longmoor, Hampshire, 9/1947.
to Cargo Fleet Iron Co Ltd, Cargo Fleet, /1949.

No.5 0-4-0ST OC AE 1801 1918
ex Furness Shipbuilding Co Ltd, Haverton Hill, c/1948.
to South Durham Steel & Iron Co Ltd, West Hartlepool Works, Hartlepool, /1949.

\- 0-4-0ST OC KS 3124 1918
ex Furness Shipbuilding Co Ltd, Haverton Hill, /1949.
returned to Furness Shipbuilding Co Ltd, Haverton Hill, c/1951.

\- 0-4-0ST OC RSHN 7494 1949 New
scrapped c/1969.

SOMERTON 0-4-0ST OC HC 656 1903
ex Davy & United Roll Foundry Co Ltd, Haverton Hill, /1950.
scrapped.

No.1 0-4-0ST OC HL 3237 1917
ex Darlington & Simpson Rolling Mills Ltd, Darlington, /1963.
scrapped after 15/4/1964.

No.3 0-4-0ST OC RSHN 7160 1945
ex Darlington & Simpson Rolling Mills Ltd, Darlington, /1963.
scrapped after 15/4/1964.

No.198 ELIZABETH 0-4-0DE RH 421436 1958 New
purchased by Staffordshire Locomotive Co Ltd; to The Rugby Group plc, Barrington Cement Works, Barrington, Cambridgeshire, 19/5/1998.

| No.2 | | 0-4-0ST | OC | MW | | 1969 | 1918 |

ex Davy & United Roll Foundry Co Ltd, Haverton Hill, /1967.
scrapped 2/1967.

| 56 | | 4wDM | | RH | 338424 | 1955 |

ex British Rail, Thornaby Depot, Middlesbrough, 5/1970. These were former departmental locomotives that had been stored at Thornaby Depot since about 8/1969. The last working location for 56 was Etherley Tip, Bishop Auckland.
scrapped 10/1981.

| 82 | | 4wDM | | RH | 425485 | 1959 |

ex British Rail, Thornaby Depot, Middlesbrough, 5/1970. These were former departmental locomotives that had been stored at Thornaby Depot since about 8/1969. The last working location for 82 was Dinsdale Welded Rail Depot, near Darlington.
scrapped 10/1981.

| 87 | 4wDM | | RH | 463152 | 1961 |

ex British Rail, Thornaby Depot, Middlesbrough, 5/1970. These were former departmental locomotives that had been stored at Thornaby Depot since about 8/1969. The last working location for 87 was Geneva Yard, Darlington.
scrapped 10/1981.

| No.1 | | 0-4-0DM | | JF | 4160007 | 1952 |

ex J.D. White Ltd, Thornaby, c1/1973.
scrapped.

| No.3 | | 0-4-0DM | | JF | 4160009 | 1953 |

ex J.D. White Ltd, Thornaby, c1/1973.
scrapped.

| 2 | | 4wDM | | RH | 321735 | 1952 |

ex Head Wrightson & Co Ltd, Teesdale Iron Works, Thornaby, /1975.
scrapped c/1983

| 33 | BOYLE | 0-4-0DE | | RH | 408309 | 1957 |

ex British Steel Corporation, Hartlepool Works, Hartlepool, /1976.
scrapped c/1983.

| 34 | | 0-4-0DE | | RH | 381757 | 1955 |

ex British Steel Corporation, Hartlepool Works, Hartlepool, /1976.
scrapped c/1983.

| 01568 | HELEN | 4wDH | | TH | 264V | 1976* | New |

to Chasewater Light Railway & Museum Company, Brownhills, Staffordshire, (property of Ed Murray & Sons Ltd from c9/2015), 24/2/2017.

| No.3 | | 4wDH | | TH | 144V | 1964 |

ex British Steel Corporation, Clarence Distillation Works, Port Clarence, c7/5/1978.
to Thomas Hill (Rotherham) Ltd, Kilnhurst, South Yorkshire, for resale 29/6/1978.

| - | | 0-6-0DE | | EE | 1554 | 1948 |

ex ICI Ltd, Wilton Works, Redcar, 7/9/1983.
scrapped.

B.G.	0-4-0DH	AB	559	1970

ex British Gypsum Ltd, Thistle Plaster Works, Kirkby Thore, Cumbria, 8/1985, after 21/7/1985, by 29/8/1985; to Powell Duffryn Wagon Co Ltd, Maindy Works, Cardiff, South Glamorgan, by 18/4/1991 (last seen at Thomson on 26/2/1988); ex Staffordshire Locomotive Co Ltd, 7/12/1998; previously at Marcroft Engineering, Stoke, Staffordshire.
scrapped 22/2/2007.

J.W.H.	0-4-0DH	AB	558	1970

ex British Gypsum Ltd, Thistle Plaster Works, Kirkby Thore, Cumbria, after 28/2/1991, by 22/5/1993.
s/s c11/2013.

D3	0-6-0DH	TH	167V	1966

ex ICI Ltd, Billingham Works, Billingham, c6/1994.
purchased by Staffordshire Locomotive Co Ltd, 11/6/1996; to Parry's Transport Yard, Shawbury, Shropshire where became property of R. Prideaux, dealer.

-	6wDH	RR	10274	1968

ex Staffordshire Locomotive Co Ltd, 18/12/1995. Previously Tarmac Roadstone (Northern) Ltd, Wirksworth, Derbyshire. It had been stripped for spares at the Yorkshire Engine Co, Long Marston, Warwickshire before moving to Stockton.
scrapped.

-	0-4-0DE	RH	423659	1958

ex British Steel plc, Shapfell Quarry, Shap, Cumbria, week ending 22/12/1995 (property of Staffordshire Locomotive Co Ltd).
scrapped 9/1999.

7222/70/01	0-4-0DE	RH	323599	1953

ex British Steel plc, Shapfell Quarry, Shap, Cumbria, week ending 22/12/1995 (property of Staffordshire Locomotive Co Ltd).
scrapped 23/2/2007.

No.10 72/21/44	0-4-0DH	AB	601	1975

ex British Steel plc, Shapfell Quarry, Shap, Cumbria, week ending 22/12/1995 (property of Staffordshire Locomotive Co Ltd).
scrapped.

210	0-6-0DH	RR	10210	1964

ex British Steel plc Teesside Works, Grangetown, week ending 22/12/1995, (property of Staffordshire Locomotive Co Ltd).
scrapped 14/3/1997.

224	0-6-0DH	RR	10224	1965

ex British Steel plc, Teesside Works, Grangetown, week ending 22/12/1995, (property of Staffordshire Locomotive Co Ltd). It was then sold to Hunslet–Barclay Ltd, Kilmarnock, Strathclyde and moved there 29/3/2000. Intended for export to Portugal, it was rejected by Hunslet–Barclay and returned to Thomson, 2/2001.
scrapped 5-6/2001.

7222/70/02	0-4-0DE	RH	323605	1954

ex British Steel plc, Shapfell Quarry, Shap, Cumbria, 9/5/1996, (property of Staffordshire Locomotive Co Ltd).
scrapped 23/2/2007.

R.O.F. No.4 (01555) 9602	4wDH	TH	292V	1980

ex R. Prideaux, dealer, Chorley, Lancashire, 10/6/1996; previously British Aerospace Defence Ltd, Glascoed, Usk, Gwent. The locomotive was acquired for use.
scrapped 16/2/2016.

| - | | 0-4-0DE | YE | 2731 | 1959 |

ex Telford Steam Railway, Telford, Shropshire, 19/5/1997, (property of Staffordshire Locomotive Co Ltd).
scrapped.

| 01567 ELIZABETH | 4wDH | TH | 276V | 1977 |

ex Telford Steam Railway, Telford, Shropshire, 19/5/1998, (property of Staffordshire Locomotive Co Ltd).
to Cobra (Middlesbrough) Ltd, Middlesbrough (property of Ed Murray & Sons Ltd from c9/2015), 11/2/2017, via Tata Steel Europe, Hartlepool Pipe Mill, Hartlepool, from 9/2/2017.

| - | 0-6-0DE | YE | 2620 | 1956 |

ex Wilmott Bros (Plant Services) Ltd, Ilkeston, Derbyshire, (property of Staffordshire Locomotive Co Ltd), c20/5/1998.
scrapped w/e 20/1/2008.

| - | 0-4-0DH | RSHD | 8367 | 1962 |
| | | WB | 3212 | 1962 |

ex E.C.C. Calcium Carbonates Ltd, Quidhampton, Wiltshire (property of Staffordshire Locomotive Co Ltd), 4/6/1998. It was hired to R.F.S. Locomotives 8/1998 for use on the Croydon Metro construction, but was oou at Thomson by 19/2/2004.
scrapped w/e 21/12/2007.

| 25 | 4wDH | TH | 279V | 1978 |

ex Shell UK Oil Ltd, Stanlow & Thornton-le-Moors, Cheshire, after 5/7/1998, by 25/10/1998.
scrapped 12/2/2016.

| 01569 (27) EMMA | 4wDH | TH | 281V | 1978* |

ex Shell UK Oil Ltd, Stanlow & Thornton-le-Moors, Cheshire, after 5/7/1998, by 25/10/1998.
to British Steel Ltd, Teesside Beam Mill, Grangetown, (property of Ed Murray & Sons Ltd, from c9/2015), 8/2/2017 (initially delivered to ex SSI Lackenby Workshop).

| 28 | 4wDH | TH | 282V | 1979 |

ex Shell UK Oil Ltd, Stanlow & Thornton-le-Moors, Cheshire, after 5/7/1998, by 25/10/1998.
scrapped on site, (property of Ed Murray & Sons Ltd from c9/2015), 27/2/2017.

| 40 | 6wDE | GECT | 5478 | 1978 |

ex Wilmott Bros (Plant Services) Ltd, Ilkeston, Derbyshire, (property of Staffordshire Locomotive Co Ltd), after 5/7/1998, by 25/10/1998.
to Corus plc, Teesside Works, Grangetown, c/2000, by 4/3/2000.

| LOCO NO.9 | 4wDH | TH | 287V | 1980 |

ex Shell UK Oil Ltd, Stanlow & Thornton-le-Moors, Cheshire, 10/11/1998. To T.J. Thomson & Son Ltd, Tyne Depot, Dunston, Co. Durham, c12/2001. Ex T.J. Thomson & Son Ltd, Tyne Dock, South Shields, w/e 11/3/2015.
to Ed Murray & Sons Ltd, Casebourne Road Depot, Hartlepool, w/e 21/1/2017.

| 26 | 4wDH | TH | 280V | 1978 |

ex Shell UK Oil Ltd, Stanlow and Thornton-le-Moors, Cheshire, 13/11/1998.
scrapped on site, /2016; after 5/2016, by 11/2016. Property of Ed Murray & Sons Ltd from c9/2015.

| 19 | 4493/45 | 4wDH | S | 10019 | 1960 |

ex Staffordshire Locomotive Co Ltd, Parry's Transport Yard, Shawbury, Shropshire, 11/1/1999; previously Stanton plc, Holwell Works, Melton Mowbray, Leicestershire.
scrapped 1/1999.

No.7	0-6-0DH	S	10095	1962
	Rebuilt	AB	6007	1982
		HE	9069	1982

ex Staffordshire Locomotive Co Ltd, Telford Steam Railway, Telford, Shropshire, after 27/4/1999; previously Tees & Hartlepool Port Authority Ltd, Tees Dock, Grangetown.
scrapped 23/6/1999.

| - | 4wDH | TH | 283V | 1978 |

ex Staffordshire Locomotive Co Ltd, Parry's Transport Yard, Shawbury, Shropshire, 3/6/1999; previously Rover Group Ltd, Longbridge, West Midlands.
scrapped 14/2/2007.

| 450 | 4wDH | TH | 231V | 1971 |

ex British Steel plc, Hartlepool Pipe Mills, Hartlepool, 21/6/1999.
scrapped 9/1999.

| - | 6wDE | GECT | 5578 | 1980 |

ex Brunner Mond Ltd, Northwich, Cheshire, (property of Staffordshire Locomotive Co Ltd), 3/8/1999 for storage.
to Blue Circle Industries plc, Hope Cement Works, Hope, on hire. It was off hire 13/2/2001 but moved to Barrow Hill Engine Shed, Derbyshire.

| 8 01555 JAMES | 4wDH | TH | 288V | 1980 |

ex Staffordshire Locomotive Co Ltd 20/12/1999, previously Harry Needle Railroad Co Ltd.
to Ed Murray & Sons Ltd, Casebourne Road, Hartlepool, (property of Ed Murray & Sons Ltd from c9/2015), after 22/2/2017, by 11/6/2017.

| JANUS | 0-6-0DE | YE | 2772 | 1960 |

ex Corus plc, Shelton Works, Etruria, Staffordshire, (property of Staffordshire Locomotive Co Ltd), 13/4/2000.
scrapped w/c 10/4/2006.

| ATLAS | 0-6-0DE | YE | 2787 | 1961 |

ex Corus plc, Shelton Works, Etruria, Staffordshire, (property of Staffordshire Locomotive Co Ltd), 14/4/2000.
scrapped 10/2005.

| STANTON No.44 | 0-6-0DE | YE | 2622 | 1956 |

ex Staffordshire Locomotive Co Ltd, 10/5/2000; previously North Yorkshire Moors Railway, Pickering, North Yorkshire, until 22/12/1999.
scrapped 11/5/2000.

| 2   CATTERSTY 2 | 0-4-0DH | S | 10126 | 1963 |

ex Corus plc, Teesside Works, Grangetown, (property of Staffordshire Locomotive Co Ltd), 5/2000.
scrapped after 30/4/2001.

| 3005 | 0-6-0DH | S | 10162 | 1963 |

ex Manchester Ship Canal Co Ltd, Ellesmere Port, Cheshire, (property of Staffordshire Locomotive Co Ltd) 9/3/2001.
scrapped 4/2001.

| 3 | 0-4-0DH | S | 10127 | 1963 |

ex Corus plc, Teesside Works, Grangetown, (property of Staffordshire Locomotive Co Ltd), 23/3/2001.
to Wilmott Bros (Plant Services) Ltd, Ilkeston, Derbyshire, 6/4/2001.

| (SPRINGBOK) | 4wDH | TH | 212V | 1969 |

ex Lindsey Oil Refinery Ltd, Killingholme, Lincolnshire, (property of Staffordshire Locomotive Co Ltd), 23/4/2002.
scrapped by 14/4/2006.

| 20/110/707 | 0-6-0DH | AB | 608 | 1976 |

ex Cleveland Potash Ltd, Tees Dock, Grangetown, (property of Staffordshire Locomotive Co Ltd), 23/4/2002.
scrapped 3/2004.

| "IMPALA" | 4wDH | TH | 185V | 1967 |

ex Staffordshire Locomotive Co Ltd, Barrow Hill Engine Shed, Staveley, Derbyshire, 25/9/2003.
scrapped 9/2004, by 28/9/2004.

ENTERPRISE (PANDA)	4w-4wDH	HAB	776	1991

ex Staffordshire Locomotive Co Ltd. It had been stored at Corus plc, Lackenby and then moved for assessment to LH Group Services Ltd, Barton-under-Needwood, Staffordshire, before coming to Thomson's yard on 5/7/2007.
scrapped.

DH24	4wDH	RR	10227	1965

ex Manchester Ship Canal Co, Barton Dock, Manchester, (property of Staffordshire Locomotive Co Ltd), 2/5/2007.
scrapped 8/5/2007.

-	0-6-0DH	RR	10290	1972

ex LH Group Services Ltd, Barton-under-Needwood, Staffordshire, 12/7/2007.
scrapped c1/9/2007.

D21	0-6-0DH	RR	10270	1967

ex Staffordshire Locomotive Co Ltd, c/o LH Group Services Ltd, Barton-under-Needwood, Staffordshire where stripped for spares. Frame and bodywork to Thomson, 9/8/2007.
scrapped 21/8/2007.

No.2	0-4-0DE	RH	420142	1958

ex Chasewater Light Railway & Museum Company, Brownhills, Staffordshire, 4/9/2007.
scrapped 3/10/2007.

60 KEN	0-4-0DH	RR	10253	1966

ex Saint-Gobain Pipelines plc, Stanton, Ilkeston, Derbyshire, (property of Staffordshire Locomotive Co Ltd), 14/9/2007.
scrapped w/e 24/10/2009.

-	0-4-0DH	HC	D1346	1965

ex Ed Murray & Sons Ltd, Hartlepool, (property of Staffordshire Locomotive Co Ltd), 11/6/2008.
scrapped 11/6/2008.

D3476	0-6-0DE	Dar		1957

ex Colne Valley Railway Preservation Society Ltd, Castle Hedingham, Halstead, Essex, 3/3/2009.
scrapped c23/3/2009.

H029 DUNCAN	0-6-0DH	RR	10286	1969

ex Cleveland Potash Ltd, Tees Dock, Grangetown, (property of RMS Locotec), 10/7/2009.
scrapped w/e 19/9/2009.

H012	0-6-0DH	RR	10289	1970

ex Cleveland Potash Ltd, Tees Dock, Grangetown, (property of RMS Locotec), 7/2009; after 10/7/2009, by 18/7/2009.
scrapped w/e 26/9/2009.

No.1 ADAM	4wDH	S	10022	1959

ex LH Group Services Ltd, Barton-under-Needwood, Staffordshire, 7/10/2009.
scrapped w/e 31/10/2009.

1	0-6-0DH	S	10161	1964

ex PD Ports plc, Tees Dock, Grangetown, 7/10/2009.
scrapped 9/10/2009.

3	4wDH	S	10021	1959

ex LH Group Services Ltd, Barton-under-Needwood, Staffordshire, 9/10/2009.
scrapped by 21/11/2009.

5 20/110/710	0-6-0DH	AB	614	1977	

ex PD Ports plc, Tees Dock, Grangetown, 8/10/2009.
to LH Group Services Ltd, Barton-under-Needwood, Staffordshire, c16/10/2009.

2	0-6-0DH	RR	10239	1965	

ex PD Ports plc, Tees Dock, Grangetown, 9/10/2009.
scrapped 4/11/2009.

6	0-6-0DH	RR	10215	1965	

ex PD Ports plc, Tees Dock, Grangetown, 9/10/2009.
scrapped 21/11/2009.

KAREN	0-6-0DH	S	10147	1963	

ex LH Group Services Ltd, Barton-under-Needwood, Staffordshire, 28/10/2009.
scrapped by 7/11/2009.

13	4wDH	RR	10197	1965	

ex Weardale Railway Community Interest Co, Wolsingham, Co. Durham, 18/4/2012.
scrapped.

* To Corus plc, Lackenby, Grangetown for overhaul in 2001 (TH 264V) and 2002 (TH 281V) and returned.

THORNABY POTTERY CO LTD

STAFFORD POTTERY, Thornaby   171
Ambrose Walker & Co Ltd until c/1905   NZ 449178 – **Map E**

The Stafford Pottery was located on the east bank of the River Tees immediately south of the later North Yorkshire Iron Works. It was established by William Smith who started to produce brown ware using local clay, to be followed by blue and white ware. The pottery began operating in 11/1825 and Wm Smith & Co was formed 2/1/1826. By 1860, William Smith's share in the business had been purchased by George Skinner and it had become George Skinner & Co. When George Skinner died in 1870, his share in the pottery passed to his daughters and Ambrose Walker became the managing partner, with the company being renamed Walker, Skinner & Skinner and then Skinner & Walker. In 1876–77 Ambrose Walker gained control and it became Ambrose Walker & Co. By 1905 it was trading as the Thornaby Pottery Co Ltd.

The pottery had a wharf on the River Tees which assisted with the movement of materials and goods. However, it later had the benefit of a railway line which entered its premises from the North Yorkshire Iron Works. Once inside the pottery, this divided into four short sidings that could be easily shunted by a horse, with presumably the North Yorkshire Iron Works locomotive delivering the wagons. When the North Yorkshire Iron Co failed in 1875, the pottery company would have to arrange the movement of wagons to and from its works. The North Yorkshire Iron Co's liquidator received payments from Skinner & Walker in 11 and 12/1876 for the hire of an iron works locomotive. Also in 9/1877, the liquidator spent money on repairing the iron works railway line which was being used by Skinner & Walker.

With the restarting of the iron works in 1882, the pottery company could presumably rely on the South Stockton Iron Co to deliver its rail traffic. However, at some date, it acquired a Manning Wardle Class H locomotive with 12 inch cylinders from the Bowesfield Iron Works. The pottery closed in 1908 and the site was sold in 1912. By 1915, it was still connected by rail to the main line across the vacant iron works land, but was identified as a Corporation Depot.

Reference :   'Stafford Pottery, South Stockton: Pioneers, Pirates and Potters', Rachel Mason, *Stockton-on-Tees Local History Journal* No.1, 2001

Gauge : 4ft 8½in

-	0-4-0ST	OC	MW	447	1873	(a) s/s
-	tank		?			(b)

(a)   ex Bowesfield Iron Co Ltd, Bowesfield Iron Works, Thornaby.
(b)   in *Machinery Market* on 1/11/1894, Ambrose Walker advertised for sale a 'four-wheel tank' locomotive with 7½ inch cylinders which was geared 2:1.

TRUSTEES OF THOMAS VAUGHAN & CO

SOUTH BANK IRON WORKS, South Bank   172
Thomas Vaughan & Co until 10/1876   NZ 538215 – **Map H**
South Bank Iron Co until /1870
B. Samuelson & Co until 1862-63

Bernhard Samuelson, the son of a Liverpool merchant, was born in Hamburg on 22/11/1820. He acted as an agent on the continent for the Manchester locomotive builders, Sharp Stewart & Co, in 1841. Eight years later, he leased a small implement works and iron foundry at Banbury to manufacture agricultural equipment. It was while visiting the Cleveland Agricultural Show at Stokesley in 1853 that he gained an introduction to John Vaughan and a tour of Bolckow and Vaughan's Eston Works. As a result, Samuelson purchased land on the north side of the Middlesbrough-Redcar railway near Eston Junction and erected two 50 feet high blast furnaces immediately north of, and parallel to, the main line in 1854, with a third following in 1856. Ironstone was supplied from Bolckow & Vaughan's Eston Mine.

Samuelson operated the works until 1862–63 when he sold it to Thomas Light Elwon, son-in-law of John Vaughan and proceeded to develop the Newport Iron Works (which see) on the Ironmasters site at Middlesbrough. The new owners traded as the South Bank Iron Co and erected an additional six blast furnaces in line with the original three during 1863–64; all nine being reported in blast in 1869. Thomas Vaughan, the son of John Vaughan, had been associated with Elwon but when his father died in 1868, he acquired a fortune and set up his own business, Thomas Vaughan & Co, in 1870. In addition to purchasing the Clay Lane Iron Works (which see), Thomas Vaughan also took over the South Bank Iron Works to produce No.1 and grey foundry iron. With the economic depression of the mid to late 1870s, T. Vaughan & Co got into financial difficulties so that only two of the eight (one had been demolished circa 1874) furnaces were in blast by 1875. By the end of the following year, a newspaper report said that the business was to be liquidated with the company owing £1 million. The Trustees of Thomas Vaughan continued to operate the iron works, albeit with only two blast furnaces functioning and, in 1879, it was sold to Bolckow, Vaughan & Co.

Very little is known about the locomotives during the early stages of the works' history apart from what is stated in the engine manufacturers' records. T. Vaughan & Co also owned the nearby Clay Lane Iron Works so transfers could have taken place between the two plants. There was an auction of its locomotives from the South Bank and Clay Lane Iron Works on behalf of the Trustees of Thomas Vaughan & Co during 1-4/4/1879.

(The remaining history of the plant and its railway system will be considered under the composite Teesside Works entry.)

Gauge : 4ft 8½in

-		0-4-0VB	VC	Chaplin	305	1862	New	(1)
No.1	UNITY	0-4-0 tank		GW	179	1864	New	s/s
	CONCORD	0-4-0 tank		HG	200	1865	New	s/s
	KATE	0-4-0 tank		HG	239	1866	New	s/s
	UGTHORPE	0-4-0ST		BH	162	1871	New	s/s
	IRONMASTER	0-4-0T	OC	FJ	113	1873	(a)	s/s

(a)   new; ordered by Thomas Vaughan & Co, Middlesbrough. Inness shows it at South Bank but Union Publications has it at the Clay Lane Iron Works.

(1)   s/s. W. Wetherell of Middlesbrough, an iron, steel and machinery merchant, advertised a Chaplin locomotive with six inch cylinders for sale in the October 1883 editions of *The Engineer* and *Machinery Market*. This would indicate a 12hp locomotive and only two of this size were delivered to the local area – 305 and 1290 of 1871 to the Glaisdale Iron Works. The latter is the more likely choice as it moved to South Wales in 1885.

THOMAS TURNBULL & SONS LTD

VULCAN IRON WORKS, Thornaby   173
NZ 449182 – **Map E**

This scrap metal merchant occupied the former Vulcan Iron Works of Charles P. Kinnell & Co Ltd; a Siding Agreement with BR was dated 18/10/1968. The premises were located between the Darlington-Middlesbrough railway and the River Tees, ½ mile west of Thornaby Station. A couple of sidings enabled scrap to be sent out by rail in 16 ton mineral wagons. The site was notable for the large cranes that formed a conspicuous feature. Rail traffic ceased in 1982, the Siding Agreement being terminated on 20/9/1982 and, by 27/9/1982, the site was virtually cleared of scrap and equipment.

Gauge : 4ft 8½in

-	0-4-0ST	OC	P			(a)	s/s
-	4wDM		RH	312434	1951	(b)	(1)

(a) a photograph exists of an out of use Peckett locomotive presumably here for scrap, but the claimed date of 10/12/1961 is suspect.
(b) ex Head Wrightson & Co Ltd, Teesdale Iron Works, Thornaby, /1976.
(1) to T. Smith & Son, Hartlepool, c7/1982.

TYNE-TEES STEAM SHIPPING CO LTD

Tees Union Shipping Co Ltd until 19/10/1903
Tees Union Shipping Co to /1887
TYNE-TEES WHARF, Middlesbrough   **174**
NZ 496213 – **Map F**

Coastal shipping services for both passengers and goods between Middlesbrough and London began at an early date. In 1831 William Fallows commenced operating the paddle steamer MAJESTIC on a weekly service, although this vessel was sold in 1837. The site, which was later to become Tyne-Tees Wharf, had five jetties projecting into the river immediately downstream of the Port Darlington coal drops by 1840, together with a merchandise warehouse and wharf. All were served by sidings from the end of the SDR Old Town Branch and an embryonic Vulcan Street railway. However, by the mid-1850s, most of the jetties and some of the sidings had been removed leaving a single coal staith and the warehouse with its wharf, which was often identified as the "Packet Wharf" *. James Taylor commenced business as a wharfinger here in 1853 and took delivery of his first ship, ADVANCE, in the following year. He established the Middlesbrough & London Steam Shipping Co and had two vessels each performing a return trip to London every week. Competition came from the Stockton & London Steamship Co which had commenced a weekly service to the capital in 1857. The various Middlesbrough interests merged in the 1870s to form the London & Middlesbrough Steam Shipping Company but, in 1880, the two competing companies joined together to form the Tees Union Shipping Company with wharves at both Middlesbrough and Stockton. The managing director was James Taylor and, for several years, part of the river frontage at Middlesbrough was described as "Taylor's Wharf".

Tyne Tees Wharf at Middlesbrough with a F.C. Hibberd diesel shunter and the ship BEN LOYAL present probably on 13th March 1962.   *(Courtesy David Webb)*

The 1875 map reveals that most of the river frontage here had been developed with wharves. A number of short sidings entered the site from the Vulcan Street railway and its accompanying MOR lines to run into and alongside the warehouses and yards, while at the west end were the "old packet wharf and steam corn mill". Unlike Roddam Dent on the adjoining site, Tees Union acted as both wharfingers and shipowners so other businesses were also based on its premises. In 1897 Appleton, French & Scrafton Ltd was running the flour mill. At various dates, there were timber yards, a fuel works, an oil and manure works and three slate merchants – J.& R. Mascall Ltd, John Harrison & Son and T.S. Colbert & Son Ltd, although Mascall eventually moved to Dent's Wharf.

The locomotives owned by the Tees Union had running rights over the following NER and MOR lines subject to the payment of these charges:-

Marsh Branch	NER/MOR	Carriage and Wayleave
Vulcan St, Dock, Cargo Fleet	MOR	Wayleave
Vulcan St and Dock via Depot Road Crossing	NER/MOR	Wayleave

Matters did not always run smoothly and a meeting of the Tees Union and NER on 15/1/1895 sought to resolve a number of issues. Tees Union had claimed the right to send its locomotive to Messrs Passman and Punch's yard to collect plates for shipment at its wharf, but the NER objected as Passman's siding was adjacent to the railway company's Middlesbrough Town Goods Station and the Tees Union locomotive would have to cross over the busy Goods Yard Junction. It was agreed that such trips would only be made by special arrangement and negotiate the Goods Yard Junction in daylight. The Tees Union had also been using the MOR lines at Vulcan Street too often to stand empty and loaded wagons for its wharf, including a total of 302 trucks delivering traffic. It assured the NER that additional siding accommodation was being provided at its wharf and hoped that the railway company would treat it as an exception and forgo the charges.

Disputes with the NER continued; a claim against the railway company was outstanding for £105 8s 7d (June 1901 to 25/11/1902) for traffic in NER wagons being worked by Tees Union locomotives because NER locomotives were not available. The claim was based on 3d per ton and justified because haulage between the Middlesbrough Iron Works, Connal's wharf and Gjers Mill's Ayresome Wharf and from Dent's Wharf to the Britannia Steel Works, Cleveland Wire Mills and Ayrton Rolling Mills was 1s per ton hauled by NER locomotives and 9d by private engines. Tees Union claimed that haulage by its engines was indispensable to Ironmasters; "The nature of work at DL & Co is such that the railway company could not possibly under present conditions cope with it" and went on to assert "it is a fact that NER some 2 years ago reduced their locomotive service for this locality… thus throwing a greater obligation on Private Locomotive Owners."

In October 1903, the Tees Union Shipping Co amalgamated with Tyne Steam Shipping Co and the coastal trading activities of Furness Withy & Co to form Tyne-Tees Steam Shipping Co Ltd. This enabled it to compete more effectively for passengers and cargoes between North East England and London and the near continental ports. Although the headquarters of the new company was in Newcastle upon Tyne, the Tees component tended to operate as a separate entity.

The riverside at the Middlesbrough premises was now designated as Tyne-Tees Wharf and increasingly developed to also cater for the deep sea liner trade. In 1913 approximately eight short sidings left the Vulcan Street lines to serve the warehouses, yards and wharf. In 1926 the Tyne-Tees (or Town) Wharf was 1,048 feet long with seven electric and two steam cranes. Grabs were fitted to three of the electric cranes. The carrying of passengers on its ships to London ceased about 1925 but, by the end of that decade, the Tyne-Tees fleet consisted of 18 ships. In December 1943 Coast Lines purchased all of Tyne-Tees shares and thereafter controlled the company. Coast Lines later became part of P & O. Tyne-Tees Wharf remained busy up to the 1950s with a number of large warehouses all served by rail and a fan of five sidings for open storage. Tyne-Tees purchased T. Roddam Dent & Son Ltd, owners of Dent's Wharf; the announcement being made on 21/10/1960. However, with the formation of the Tees & Hartlepool Port Authority (which see) on 1/1/1967, both sites were taken over, together with the associated equipment including the locomotives.

*Originally "packet" boats carried Government packets of orders and despatches between England and Ireland but the term increasingly referred to any vessel regularly sailing between two ports often carrying mail and passengers.

Reference : *The Tyne-Tees Steam Shipping Company and its Associates*, Nick Robins, published by Bernard McCall, 2014

Gauge : 4ft 8½in

	ALDERMAN	0-4-0ST	OC					Scr
	ASTLEY	0-4-0ST	OC	HCR	136	1873	(a)	Scr /1927
	SWILLINGTON	0-4-0ST	OC	HCR	150	1874	(a)	(2)
	KILMARNOCK	0-4-0ST	OC	AB	814	1898	(b)	(5)
No.1		0-4-0ST	OC	HC	262	1883	(c)	(3)

PULL (WHITEHAVEN)	0-4-0ST	OC	FJ	76	1867	(d)		(1)
TYNE-TEES WHARF No.2	0-4-0ST	OC	RSHN	7333	1946	New		Scr c4/1964
SUTHERLAND	0-4-0ST	OC	HL	2439	1899			
	Rebuilt		HL	2886	1936	(e)		(4)
TYNE-TEES WHARF No.3	0-4-0ST	OC	RSHN	7358	1947	New		
						Scr /1961, after 17/4/1961		
TYNE-TEES WHARF No.1	4wDM		FH	3817	1956	New		(7)
TYNE-TEES WHARF No.5	4wDM		FH	3949	1960	New		(7)
	0-4-0ST	OC	RSHN	7084	1943	(f)		(6)
TYNE-TEES WHARF No.3	4wDM		FH	3964	1961	New		(7)
GRADWELL	4wVBT	VCG	S	9377	1947	(g)		(7)
TYNE-TEES WHARF No.4	4wVBT	VCG	S	9562	1954	(g)		(7)
TYNE-TEES WHARF No.8	4wDH		EEV	D908	1964	New		(7)
-	0-4-0DH		EEV	D1206	1966	(h)		(7)

(a) ex Hudswell, Clarke & Rodgers, Leeds, per J.F. Wake, 8/2/1888, formerly T. & R.W. Bower Ltd (subsidiary of Pease & Partners Ltd), Allerton Main Collieries, Swillington, Yorkshire (WR).
(b) new. To T. Roddam, Dent & Son Ltd, Dent's Wharf, Middlesbrough, on hire, /1961, after 23/4/1961 and returned c9/1961.
(c) ex Worthington & Co Ltd, Burton on Trent, Staffordshire, /1929.
(d) ex Stockton Wharf, Stockton, after 14/8/1930, by 16/4/1934.
(e) ex Platt Bros & Co Ltd, Moston Colliery, near Oldham, /1936 after overhauled by the makers and renamed after Sir A.M. Sutherland, Chairman of the company.
(f) ex J. Garrity, scrap merchant, Middlesbrough, c2/1961, by 14/4/1961; previously Smiths Dock Ltd, South Bank.
(g) ex Stockton Wharf, Stockton, /1965.
(h) ex Stockton Wharf, Stockton, /1966.

(1) to Stockton Wharf, Stockton, after 16/4/1934.
(2) to Stockton Wharf, Stockton, c/1951, by 22/6/1951.
(3) to Thos W. Ward Ltd, Middlesbrough, for scrap, 11/1956.
(4) scrapped on site by J. Garrity, Middlesbrough, 17/4/1961.
(5) to Stockton Wharf, Stockton, /1964, by 2/4/1964.
(6) to Stockton Wharf, Stockton, after 10/4/1965, by 13/8/1965.
(7) taken over by Tees & Hartlepool Port Authority, with site, 1/1/1967.

STOCKTON WHARF, Stockton 175
NZ 448185 – **Map D**

The railway that ran along the west bank of the River Tees at Stockton was of considerable historical importance. The SDR opened its line on 27/9/1825 terminating a short distance north of St John's Crossing on Bridge Road. Its initial purpose was to carry coal from the collieries in South West Durham for landsale at Darlington, Yarm and Stockton. The Tees Coal Co began loading coal on the ship ADAMANT at the newly erected staith in January 1826 and three more staithes had been completed on the site of the later Tyne-Tees Wharf by spring 1827. There were no formal stations to begin with, passengers being set down initially at Cottage Row, near Moat Street and later at the Custom House Tavern at the end of Finkle Street after the line had been extended northwards along the river bank. However, the difficulties for ships navigating up the river to Stockton encouraged the SDR to extend its line to Middlesbrough in 1830; this diverged from the original route at Bowesfield Junction. As a result and particularly after a replacement bridge was erected by the SDR over the River Tees in 1844, coal shipments from Stockton rapidly declined, but other businesses and warehouses had been established along the bank and they continued to trade using the river.

The branch was initially worked by the SDR and later the NER; indeed most of the lines were shown as belonging to the railway company in 1895. There were still four rail served coal staithes projecting into the river in 1857, three of which employed wagon turntables to move wagons from the sidings on to the staithes. It was also in this year that the Stockton & London Steamship Co commenced sailings to the capital from the Corporation Wharf at Stockton. In 1880 it merged with the London & Middlesbrough Steam Shipping Co to form the Tees Union Shipping Company, later Tyne-Tees Steam Shipping Co Ltd, with wharves at Stockton and Middlesbrough. The NER staithes were replaced by a Riverside wharf and the Tees Union transferred there in 1895, although this and the associated sidings were still owned by the railway company and leased to Tees Union/Tyne-Tees. It is not known when the Tees Union/Tyne-Tees took over responsibility for shunting the lines north of St John's crossing, but this was the usual arrangement for several decades, a collection of secondhand elderly steam locomotives being responsible.

It is worth examining the Stockton Goods Branch about 1900. After departing from Bowesfield Junction on the Darlington-Middlesbrough railway, the sidings for Ashmore, Benson, Pease & Co's Parkfield Works (which see) diverged to the left almost immediately. The main feature on the west side of the branch south of Bridge Road was the NER's South Stockton Goods Station opened in 1875-76, although there had been a warehouse and a single road engine shed here in 1855. On the opposite side of the branch were the railway company's coal and lime depots. Leaving the goods station behind, the single line crossed Bridge Road to reach another smaller set of sidings. From here, railways ran into Joshua Byers' timber yard and the Castle Moat shipyard; R. Craggs had reopened the shipyard in 1884 and constructed small coastal vessels and trawlers until 1909. Another line connected with Tees Union Shipping's wharf which was 450 feet long and equipped with three steam cranes in 1926. At the south end was its corrugated iron single road engine shed. To the north of the wharf, Tees Flour Mills had a couple of sidings. The branch continued alongside the river serving various small businesses to reach Stockton Council's Corporation Quay (owned since 1876). Its wharf was 650 feet long and equipped with three 5-ton and two 3-ton electric cranes, together with one 20-ton steam rail crane and a 3-ton steam crane in 1926. The 20-ton crane was disposed of in 1930 after a life of 38 years.

The formation of the Tees & Hartlepool Port Authority on 1/1/1967 resulted in the closure of both Tyne-Tees Stockton Wharf and the Corporation Quay. The *Northern Echo* on 22/8/1967 reported that the departure of the MV DORA REITH from Stockton Quay tomorrow would probably be the last before the quay closed for good on 1/9/1967; the vessel carried 700 tons of steel pipes from "Cochranes" for Amsterdam. The railway north of Bridge Road closed probably in 1969.

Reference :   *The Tyne-Tees Steam Shipping Company and its Associates*, Nick Robins, published by Bernard McCall, 2014

'Stockton's Quayside Railways', Peter H. Rigg, *Steam Days* magazine, 2011

SWILLINGTON (HCR 150/1874) was photographed in front of Tyne-Tees Steam Shipping Co's engine shed at Stockton Wharf in 1954. (ILS F. Jones photograph)

Gauge : 4ft 8½in

	Name	Type	Cyl	Builder	Works No	Date	Note	Ref
	COUNCILLOR	0-4-0ST	OC	FW	198	1873	(a)	(2)
-		0-4-0T	OC	FJ	164	1880	(b)	s/s
	WARRINGTON	0-4-0ST	OC	MW	1027	1887		
		Rebuilt		MW		1907	(c)	(3)
	BARTON	0-4-0ST	OC	HH		c1872		
		Rebuilt		Wake		1903	(d)	Scr /1927
	(PUSH)	0-4-0ST	OC	BH	423	1878	(e)	Scr c/1955
	(PULL) (WHITEHAVEN)	0-4-0ST	OC	FJ	76	1867	(f)	(1)
	SWILLINGTON	0-4-0ST	OC	HCR	150	1874	(g)	(5)
	TYNE-TEES WHARF No.4	4wVBT	VCG	S	9562	1954	New	(6)
	TYNE-TEES WHARF No.5	4wPM		FH/KC			(h)	(4)
	GRADWELL	4wVBT	VCG	S	9377	1947	(i)	(6)
	KILMARNOCK	0-4-0ST	OC	AB	814	1898	(j)	(8)
6		0-4-0ST	OC	RSHN	7084	1943	(k)	(9)
-		0-4-0DH		EEV	D1206	1966	(l)	(7)

(a) ex Hilton, Southwick & Monkwearmouth Railway, c/1876.
(b) ex Earl of Lonsdale, William Colliery, Whitehaven, Cumberland, /1895.
(c) ex J. Crosfield & Sons Ltd, Warrington, Lancashire, via Manning Wardle, /1907.
(d) ex Barton Limestone Co Ltd, Barton Quarry, Barton, Yorkshire (NR).
(e) ex Walter Scott, Contractor. Inness suggests as early as /1911. It was here by 14/8/1930.
(f) ex Dorman, Long & Co Ltd, Newport Iron Works, Middlesbrough by 5/7/1923. Inness has a note "Rebuilt from well to saddle tank at Blair's Works" but the date when this took place is not stated. To Tyne-Tees Wharf, Middlesbrough, by 16/4/1934 and returned.
(g) ex Tyne-Tees Wharf, Middlesbrough, c/1951, by 22/6/1951.
(h) ex Coast Lines Ltd, Victoria Wharf, Plymouth, Devon, after 28/1/1957, by 8/5/1957.

GRADWELL (Sentinel 9377 of 1947) hauls wagons along the Stockton Goods Branch on 29th July 1963.
(RCTS J. Faithfull photograph)

The last Husky design of locomotive was erected at English Electric's Vulcan Works using parts from the Stephenson Works at Darlington following the latter's closure. It was given maker's number D1206 and initially acted as the manufacturer's works shunter before it was despatched to Tyne-Tees wharf at Stockton. This is the locomotive on 30th November 1966, probably soon after delivery and pulling AB 814 as a test load.

(ARPT J.M. Boyes photograph)

(i) ex G. Stephenson (Builders & Contractors) Ltd, dealer, Bishop Auckland, /1958 per H. Dunn; formerly Whittingham Asylum Railway, Lancashire.
(j) ex Tyne-Tees Wharf, Middlesbrough, /1964, by 2/4/1964.
(k) ex Tyne-Tees Wharf, Middlesbrough. after 10/4/1965, by 13/8/1965.
(l) ex English Electric Co Ltd, Vulcan Works, Newton-le-Willows, Lancashire /1966. The parts for RSHD 8454 were transferred to EEV in 1964 due to the closure of the Stephenson Works. It was erected at the Vulcan Works using a secondhand Gardner engine and operated as the works shunter until despatched to Tyne-Tees at Stockton circa 30/11/1966.

(1) to Charles P. Kinnell & Co, Thornaby, c/1949.
(2) to Tarslag Ltd, Stockton, c12/1952, after 30/9/1952.
(3) to A. Bainbridge Ltd, Thornaby, for scrap, 4/1958.
(4) to A. Bainbridge Ltd, Thornaby, 5/1958; after 4/4/1958, by 1/6/1958.
(5) scrapped on site by J. Garrity, Middlesbrough, c1/1964.
(6) to Tyne-Tees Wharf, Middlesbrough, /1965.
(7) to Tyne-Tees Wharf, Middlesbrough, c/1966.
(8) scrapped on site c1-2/1968.
(9) scrapped 23/3/1968.

THE UNITED ALKALI CO LTD

HAVERTON HILL SALT WORKS, Haverton Hill 176
Newcastle Chemical Co Ltd until 1/11/1890 NZ 494219 – **Map C**

The company was founded by Christian Allhusen and owned a large works on Tyneside. It began drilling for salt near the north bank of the Tees. Although this first attempt was unsuccessful, a salt works was established here between the NER Port Clarence Branch and the river. By 1885 the company had four brine wells further north on Cowpen Marsh and this had increased to ten in 1892. Labelled as "Cowpen Salt Works" on the 1899 Ordnance Survey map, it was located to the north east of the present day A178/A1185 roundabout. A "wagonway" is shown on the map heading south in a straight line from Cowpen Salt Works to terminate at the southern end of the Port Clarence rifle range where it met a lane from other brine wells and this ran south by Bell's Iron Works. By 1913, the Cowpen Salt Works was disused and there was no sign of the wagonway. Allhusen also owned a group of eight brine wells near the Westfield Durham Co site at Haverton Hill.

The Haverton Hill Salt Works, known locally as "Ally Husen's", comprised two large buildings accommodating the panhouse and stores parallel to and on the south side of the Port Clarence Branch, together with brine reservoirs. Sidings ran along both sides of these buildings and two lines curved round from here to end at the river; that to the south east was the "Old Landing Stage" (according to an 1893–97 survey) and the other, a short distance upstream, was the operational wharf (75 feet long). Salt was loaded on to ships and taken to United Alkali's works at Gateshead where it was used to produce caustic soda. There are no surviving written records of locomotives used by the company at Haverton Hill but the above survey shows an approximately 55 feet long single road engine shed at the end of one of the sidings. It is likely that one or more of the locomotives based at the United Alkali's works on Tyneside were also used at Haverton Hill.

In 1903 United Alkali Co and the NER signed an agreement and the former hoped to send 40 to 50,000 tons of salt by rail to its Allhusen and Hebburn Works. The NER promised to supply wagons to carry about 10 tons each. The agreement stated that the "wagons when loaded shall be removed daily by the locomotive of the Alkali Company from the Salt Works siding to the Railway Company's siding, Haverton Hill." It was intended that the wagons would be conveyed in train loads of 300 tons. Latterly the brine appears to have been transported in railway tank wagons to Tyneside and the 1910 NER Working Timetable shows a goods train running from Allhusen's Haverton Hill Works three days a week. However, the manufacture of alkali by the more expensive Leblanc Process was falling out of favour and the company already had Tennant's Salt Works at Haverton Hill. The 1913 Ordnance Survey map shows that all of the railway lines at Haverton Hill Salt Works had been taken up and the company's engine shed removed, although the premises was still identified as "Allhusen's Works".

The presence of 'American' drilling rigs and the distinctive fence reveal this to be United Alkali's Haverton Hill Salt Works with Black, Hawthorn 881 of 1886 present. *(NERA R. Inness photograph)*

Gauge : 4ft 8½in

TENNANT'S No.3	0-4-0ST	OC	BH		881	1886	(a)	(1)

(a) ex The United Alkali Co Ltd, Tennant's Works, Hebburn, Co. Durham. A photograph of this locomotive was taken at Haverton Hill as revealed by the fencing and presence of American drilling rigs.

(1) returned to The United Alkali Co Ltd, Tennant's Works, Hebburn, Co.Durham.

VULCAN MATERIALS COMPANY (UK) LTD

HARTLEPOOL WORKS, West Hartlepool   **177**
Batchelor Robinson Metals & Chemicals Ltd until 1/8/1981   NZ 513308 – **Map B**
(subsidiary of **Batchelor, Robinson & Co Ltd**)
Batchelor, Robinson & Co Ltd until 1/1974

In the late 1890s, the short Longhill Branch was constructed by the NER from near the southern end of its Cliff House Branch along the north side of land used by the Seaton Carew Iron Co as a slag tip. It served a number of businesses, including the Rivet, Bolt & Nut Company's Longhill Works near the junction and J. Beach's West Hartlepool Brick Works at the end of the line. On 16/6/1899 (supplemented by an agreement dated 13/6/1901) the Hartlepools Cement Co Ltd, with the involvement of members of the Richardson family, received approval to lay down sidings for its cement works part way along the branch. In 1914 the cement works was described as disused and the site later became a small chemical plant. Between it and the Longhill Works, a short siding ran past the Seaton High Lighthouse to Longhill Farm. The NER agreed to Batchelor & Co Ltd taking over this siding on 26/7/1915. A NER memorandum dated 31/7/1917 stated that Batchelor, Robinson & Co Ltd had acquired Batchelor & Co and would conform to the agreement.

This was a metals processing business; it was described as a "Detinning Works" on the 1940 Ordnance Survey. A pair of sidings entered the premises from the Longhill Branch, one of which divided into two to pass through the main building. A notable feature in the south east corner of the site was the stone column of the no longer functioning Seaton High Lighthouse (the latter was moved to the new marina in 1997). For some time, the works mainly relied on electric winches to move wagons around but subsequently locomotives were in regular use, each being employed one to three months in turn. There were also approximately fifteen 16 ton mineral wagons available for internal use. The company purchased the former "chemical works" on its western

John Fowler 4210136 of 1958 was moving wagons at Vulcan Materials' "de-tinning" works in Hartlepool.
(IRS Malcolm Ainsworth photograph)

boundary for use as a scrap stockpile. During an IRS visit on 18/3/1983, it was stated that BR delivered scrap every weekday with the processed material going out by road or rail. The BR sidings on the branch had closed that week and a presentation made to local railwaymen. Rail traffic ceased about 1985 and, on 22/9/1986, the two locomotives stood together with the works' railway disused.

Gauge : 4ft 8½in

No.3	0-4-0DM	JF	22489	1939	(a)	(1)
-	0-4-0DM	JF	4100012	1948	(b)	(2)
-	0-4-0DM	Bg/DC	2652	1958	(c)	Scr 4/1978
-	0-4-0DM	Bg/DC	2655	1959	(d)	Scr 5/1980
-	0-4-0DM	JF	4210136	1958	(e)	Scr 10/1986
-	0-4-0DM	RSHN	7900	1958	(f)	(3)

(a)   ex Nevill's Dock & Railway Co Ltd, Llanelli, Carmarthenshire, on hire, after 9/1964.
(b)   ex Cerebos Foods Ltd, Greatham, 8/1969.
(c)   ex Bristol Mechanised Coal Ltd, Filton, Gloucestershire, 11/1971.
(d)   ex Bristol Mechanised Coal Ltd, Filton, Gloucestershire, 3/1973.
(e)   ex British Gas Corporation, Reading Gas Works, Reading, Berkshire via E.H. Bennett Ltd, Henbury, Bristol, 5/1977.
(f)   ex Whessoe Ltd, Stockton Works, Stockton, 11/1979.

(1)   returned to Nevill's Dock & Railway Co Ltd, Llanelli, Carmarthenshire, ex hire, 13/4/1965. Photographed at Worcester en route (see *Industrial Railway Record* 155).
(2)   scrapped on site by C. Herring & Son Ltd, Hartlepool, 4/1973.
(3)   to A.V. Dawson Ltd, Middlesbrough, 10/1986.

WALKER, MAYNARD & CO

KIRKLEATHAM IRONSTONE MINE, Dunsdale   178
The Kirkleatham Ironstone Co until 30/9/1881   NZ 606189 – **Map I**

The drift mine (NZ 609187) was situated on the west side of the Guisborough-Yearby road at Dunsdale. This part of the Kirkleatham estate occupied the level ground between the Eston Hills and the higher land north of Upleatham village, but shared the north-facing escarpment separating it from the marshy land bordering the Tees Estuary. The local iron industry was flourishing at the start of the 1870s and Edward Robson, formerly Bell Bros' commercial manager, joined with Arthur H.T. Newcomen of Kirkleatham Hall and James Rutherford to establish The Kirkleatham Ironstone Company on 1/7/1871.

William Walker, the company's mining engineer, advertised in the *Colliery Guardian* on 5/7/1872 for two new or secondhand 8 inch 3ft 0in gauge locomotives. QUEEN OF THE FOREST and an unknown secondhand Hunslet locomotive arrived in 7/ and 10/1872 respectively and helped with construction. According to William Walker, by 2/1873 the company had driven drifts and sunk shafts, installed stationary engines to haul ironstone to the surface and had constructed "5 miles of tramway with heavy cuttings and embankments, formed and made self-acting inclines and are now commenced working stone and have 2 locomotive engines for working tramway." In addition 71 terraced houses were constructed at Dunsdale. Peak output of ironstone was achieved in 1876 when 140,605 tons were mined. Unfortunately the extent of the Main Seam ironstone was limited due to glacial disturbance. A drift was extended to an isolated block of ironstone on the east side of the road, which was also reached by a shaft; this working being known as the New Pit.

Because of the disrupted nature of the ironstone, it was not possible to construct the mine buildings next to the main drifts. Approximately 15–20 loaded tubs were hauled out of the main drift on the 3ft 0in gauge railway by a Fowler stationary steam engine. They then passed over a wooden bridge above the beck and climbed through a brick lined tunnel up the valley side to the pit yard (NZ 606189). Here, a locomotive collected the loaded tubs and, after weighing, they were hauled about 1,000 yards to the head of a self-acting incline where they were let down about 600 yards to flatter countryside. Two locomotives were available to handle the tubs over the three miles of railway to Tod Point, Coatham. There were gates and a cabin for Wilton Crossing part way along the line. Two iron works were developed at Coatham and the ironstone was particularly destined for Robson, Maynard & Co's blast furnaces (blown in latter part of 1873) at Tod Point. Ironstone was also supplied to other local works and there were transhipment sidings with the NER Middlesbrough-Redcar railway. A calcining plant was constructed adjacent to the NER and treated stone was tipped into standard gauge wagons for movement to the iron works. The Kirkleatham Ironstone Co purchased a standard gauge locomotive to shunt the sidings at Tod Point in 1872.

Each mine tub could hold 35cwt of ironstone. The stock of wagons on 30/9/1881 comprised 189 ironstone tubs, 4 coal tubs, 9 pairs of timber bogies and 4 kibble bogies.

The late 1870s suffered a major trade depression and, in view of the substantial expenditure incurred, Messrs Robson and Rutherford were unable to meet their share of the debts and the partnership was dissolved as from 1/1/1878 with Newcomen continuing the company. Not long after, Robson withdrew from the Redcar Iron Works which became Walker, Maynard & Co and the latter purchased Kirkleatham Mine in 1881. However, reserves of ironstone at the mine were becoming rapidly depleted and it was officially abandoned on 31/12/1886.

References : 'The Kirkleatham Iron Stone [sic] Company', Peter Tuffs and Neil Parkhouse, *Archive* 22, 6/1999, pp. 31-44

Kirkleatham Ironstone Mine, Simon Chapman, pub. Peter Tuffs, 9/1999

Gauge : 4ft 8½in

REDCAR No.1	0-4-0ST	OC	HCR	123	1872	(a)	(1)
KIRKLEATHAM No.4	0-4-0ST	OC	JF	2625	1876	(b)	(2)

(a) new, purchased 12/1872 for £790.
(b) new, purchased 2/1876 for £925.

(1) to Robson, Maynard & Co, Redcar Iron Works, Coatham, prior to 30/9/1881, possibly in /1876, when the JF arrived.

(2) on 14/3/1887, it was stated to be "standing in Engine Shed at Tod Point and was in good order up to 31/12/1886. It has since been used occasionally by Walker, Maynard & Co at their furnaces and is not now in workable state, many parts having been taken away." These are not thought to have been significant and it was advertised for sale at £365 by J.F. Wake on 1/4/1889.

Loaded tubs on the 3ft gauge railway serving Kirkleatham Ironstone Mine were taken by a locomotive from the pit yard to the top of an approximately 600 yards long self-acting incline. At the foot of which, they were collected by another locomotive and hauled the three miles to the iron works at Coatham. This lovely view, probably taken in the summer of 1874, shows the foot of the incline with KIRKLEATHAM No.2 (Hudswell Clarke & Rodgers 126 of 1872) waiting for the next rake of tubs to descend. One of the sidings must have connected to the engine shed which also existed here. *(Courtesy Parkhouse/Pope Archive)*

Gauge : 3ft 0in

Name	Type		Builder	Works No.	Date		
QUEEN OF THE FOREST	0-4-0ST	OC	W.E.Bates #		1862	(a)	(3)
-			HE			(b)	(1)
KIRKLEATHAM No.2	0-4-0ST	OC	HCR	126	1872	(c)	(4)
KIRKLEATHAM No.1	0-4-0ST		HH		1874	(d)	(5)
-	0-4-0ST	OC	JF	2377	1875	(e)	(2)
KIRKLEATHAM No.3	0-4-0ST		JF	2836	1876	(f)	(6)

\# Leftwich Iron Works, Northwich, Cheshire (see *Industrial Railway Record* 222, September 2015)

(a) ex I.W. Boulton, dealer, Ashton under Lyne, Lancashire for £250, 7/1872.
(b) the existence of this secondhand loco is suggested by the Kirkleatham Ironstone Co financial ledger: "Bought from Hunslet Engine Co – Engine £240".
(c) new, costing £790, 4/12/1872; its intended name, EDWARD ROBSON, was probably never carried.
(d) new, costing £820, 2/1874.
(e) according to JF records dated 5/4/1876, its locomotive 2377, recently delivered to Palmers Shipbuilding and Iron Company's Grinkle Mine, was on loan to the Kirkleatham Ironstone Co, presumably while the latter was waiting for its new JF to arrive.
(f) new, costing £800, 22/5/1876.

(1) it must have been disposed of by 2/1875 when the Kirkleatham Ironstone Co was stated to have two locomotives (by implication, narrow gauge locomotives).
(2) returned to Palmers Shipbuilding and Iron Co's Grinkle Mine, off loan, 5(?)/1876.
(3) to I.W. Boulton, dealer, Ashton Under Lyne, Lancashire, 6(?)/1876 where rebuilt as ANT and sold to Butterworth & Brooks, Hutch Bank Quarry, Haslingden, Lancashire.
(4) on 14/3/1887, it was reported that "No.2 Loco – Standing in Engine Shed at incline Bank Foot. Is worn out." It was advertised for sale at £280 in *Machinery Market* on 1/4/1889 by J.F. Wake and subsequently sold to the Holwell Iron Co Ltd, Eaton Quarries, Leicestershire by 8/9/1890.
(5) on 14/3/1887, "No.1 Loco – This engine is standing in the Smith's Shop at Dunsdale and is [in] good working order." It was advertised for sale at £280 by J.F. Wake on 1/4/1889.
(6) on 14/3/1887, it was stated "No.3 Loco – Standing in Engine Shed at incline Bank Foot and also worn out." It was advertised for sale at £275 by J.F. Wake on 1/4/1889.

WAR DEPARTMENT

MARSKE AERODROME AND STORAGE DEPOT 179
Marske-By-The-Sea NZ 627225 – **Map I**
Royal Air Force until World War 2
Royal Flying Corps until 1/4/1918

Marske Aerodrome

The First World War had a significant impact on the small fishing village of Marske-By-The-Sea. Firstly an army reception and training camp was established on the north west side of the settlement. Several hundred soldiers were based here at any one time and training trenches were dug between Redcar and Saltburn.

In view of the heavy losses of Royal Flying Corps pilots over the Western Front in early 1916, construction started on a new training facility at Loch Doon in Scotland, but this proved to be a most unsuitable site with consequent delays. Therefore a survey was undertaken of land at Marske in 1917 as a substitute, with the aerodrome to be built on fields (NZ 627225) near Ryehills Farm on the north side of the road to Kirkleatham (old route of A174) before it crossed the NER Redcar-Saltburn line at Blacks Bridge. Support accommodation was to be located on the south side of the road. Because of the urgent need to train pilots, the RFC Air Training Brigade established the No.4 (Aux) School of Aerial Fighting and Gunnery at Marske. A "rail depot" was located next to the NER, north of Blacks Bridge and not far from the Upleatham Mine branch sidings on the opposite side of the line and this allowed people, materials and even planes to be delivered. A ground frame "Aerodrome Siding" with 4 levers controlled the connection. A row of eighteen portable Bessoneaux canvas hangars were erected on the airfield. This enabled the aerodrome to be operational by 1/11/1917 and a contemporary photograph shows about 60 aircraft parked on the adjacent field.

These temporary arrangements were to be replaced by more permanent buildings, including the construction of four large 'Double Belfast' type hangars. Tenders were invited at the end of 1917 and the contract was let to J. Gerrard & Sons Ltd of Manchester, with work reported to be well underway in early 1918. A 2ft 0in gauge railway was laid from the "rail depot" across the fields past Ryehills Farm to the work on the four hangars. The railway enabled materials to be brought in and was later used to service the aerodrome. Shale came from Upleatham Mine and was put into side-tipping wagons on the narrow gauge at the "rail depot" for use on the site. The rear roof span joists were built at Bristol and delivered by rail. Acquiring sufficient craftsmen proved difficult and Gerrard was advertising for 132 workmen in the *North Eastern Daily Gazette* on 18/6/1918. It appears that the contract was completed in early 1919, although the war had finished in November 1918. The Royal Air Force came into existence on 1/4/1918 and the site at Marske had become the No.2 School of Aerial Fighting. Unfortunately a number of pilots died during training at Marske.

Hughes, Bolckow & Co Ltd

With the end of the war, many personnel were demobilised and Marske ceased to function as a gunnery school by 1920. It then operated as a RAF Depot of Engineering Accessories, but appears to have been retained on a 'care and maintenance' basis. According to *Flight* magazine in 1/1921, Messrs Hughes, Bolckow & Co Ltd was using the four hangars and 30 acres as a concentration and disposal depot for surplus wartime material. The agreement for use of the two sidings comprising the "rail depot", between the NER and Hughes, Bolckow was dated 29/7/1921. A subsequent note says that the agreement was terminated and the sidings taken over by Dorman Long under a new agreement dated 31/10/1923. This possibly related to supplying slag for the Redcar-Marske Road contract, however, Hughes, Bolckow was continuing to use the premises. According to his obituary in the *Daily Gazette for Middlesbrough* dated 13/3/1934, Charles Bolckow had joined with R.T. Hughes to form Hughes, Bolckow & Co Ltd in 1906, it being described as "scrap merchants of Middlesbrough". The newspaper went on to state that it took over the Government Surplus Disposal at Marske Aerodrome "for scrap purposes". It is not known how much scrapping took place but, in 1923 Hughes, Bolckow was advertising a wide range of machinery for immediate disposal from Marske, including boilers, portable engines, pumps, electric motors, together with a Hudson 0-4-0 locomotive for 60cm gauge. In 1925, the Blyth shipbreaking section of the business was formed into a separate company – Hughes, Bolckow Shipbreaking Co.

J.F. Wake, at Geneva Works, Darlington had accumulated many secondhand locomotives. When Ted Haigh visited there on 7/6/1925, he recorded 22 locomotives. With the deteriorating economic situation, Wake decided that there was no market for them and he sold many to Hughes Bolckow in 1/1928 for scrap. Details are given in *The Industrial Railways & Locomotives of County Durham Part 3*. It is not known where Hughes, Bolckow cut them up but the most likely answer is at the Geneva Works, although Marske has been suggested. No substantial evidence has been found of the Marske site scrapping locomotives. With decreasing amounts of surplus equipment, Hughes, Bolckow turned the premises into a saw mill but this was advertised for sale as a going concern in the *Yorkshire Post and Leeds Intelligencer* on 23/3/1931. The sale included wood working equipment, a 5 ton Smith's travelling crane and a large stock of timber. The land, buildings and sidings were for lease.

References : *From Biggles to the Red Baron An Illustrated History of Marske Aerodrome*, John H. Watson, pub. Kirkleatham Museum

Defence of the UK Cleveland Vol 4 Part 2, North Yorkshire and Cleveland 20th Century Defence Study Group, Momentus Publications, 2011, pp.72-93

'Elusive Little Bagnalls', Allan C. Baker, *The Narrow Gauge* 156, NGRS, 1997

Gauge : 2ft 0in

KIDBROOKE	0-4-0ST	OC	WB	2043	1917	(a)	(1)
HAMPSTEAD	0-4-0ST	OC	WB	1728	1903	(b)	(2)
-	0-6-0WT	OC	HC	?		(c)	s/s
GOWRIE	0-6-4T	OC	HE	979	1908	(d)	(3)

(a) the MoM Aeronautical Department ordered three new locomotives from WB and one of these (2043) was despatched from the manufacturer on 17/12/1917. Its name suggests that it was intended to go to the Air Board, Kidbrooke in south-east London. The first order for spares was placed by the War Office (Aviation) on 23/1/1918 for delivery to Marske-by-Sea and it appears that the locomotive was actually sent to Marske for use by the contractor, Gerrard. The latter ordered spares for it at Marske on 31/7/1918 and in 1/1919.

(b) ex Swinton, Manchester, /1918. Spares were sent to Swinton Station, Manchester for War Office, A/A Park on 28/12/1917. Spares were then ordered for New Aerodrome, Marske-by-Sea on 31/7/1918 and 14/11/1918.

(c) a photograph shows what appears to be a WD 'Hudson' type 0-6-0 well tank, built by HC, on side tipping wagons at Marske.

(d) this Fairlie type locomotive was new to the North Wales Narrow Gauge Railways Company, forerunner of the Welsh Highland (Light Railway) Company. The MoM purchased it in /1916 and it was sent to J.F. Wake where it was "rebuilt" in /1918. It was demonstrated to Festiniog Railway staff in 4/1923 in steam working possibly at Marske where Hughes, Bolckow was using the storage depot to dispose of war surplus equipment. J.F. Wake continued to offer it for sale 1919-27, before apparently selling it to Hughes, Bolckow with other old locomotives.

(1) sold as Government surplus at MoM CSD Leamington-on-Tyne, 5/1920, (offered for auction, 15/4/1920), although it is not known whether it was moved to Leamington for the sale. It was purchased by J.H. Bentham & Co to transport coal at Cassop, Co Durham.
(2) to Swinton, Manchester? Spares were sent to Swinton Station, Manchester on 17/3/1921.
(3) it was offered for sale by Hughes, Bolckow at "The Aerodrome", Marske in *Machinery Market* on 6/1/1928; s/s.

Marske Storage Depot

In 3/1938, the site was considered as a bomber airfield but its position near the coast meant that it was too vulnerable. After the site had been rejected as a bomber airfield, the War Department took over the remaining buildings; some south of the Kirkleatham road had been demolished but the four large 'Belfast' hangars remained. During World War 2 these were used by the 25th Medium/Heavy Field Training Regiment of the Royal Artillery and also functioned as ordnance stores. The *Defence of the UK* reference suggests that the railway lines were a mixture of standard and narrow gauge; the standard gauge serving the hangars. No confirmation has been found but it would seem likely that the line to the "rail depot" was converted to standard gauge. Some years after World War 2, ICI Ltd leased some of the hangars and used them as a storage facility for its bulk plastic pellet production from Wilton. The pellets were bagged here and despatched by road. The depot closed prior to 1965 and the hangars were demolished in the 1970s.

Gauge :4ft 8½in

No.1	WHARTON	0-4-0ST	OC	Lewin	606	1875		
		Rebuilt				1881	(a)	(1)
801 (70232)		4wDM		RH	221648	1943	(b)	(2)
843 (72237)		0-4-0DM		AB	370	1945	(c)	(4)
800 (70231)		4wDM		RH	187075	1937	(d)	(3)

(a) ex ICI Ltd, 'Tennant's' Salt Works, Haverton Hill.
(b) ex No.4 Supply Reserve Depot, Stratton Factory, Swindon where spares were ordered 11/3/1944. Spares ordered for Marske 15 and 18/5/1944.
(c) ex Longmoor Military Railway, Hampshire, /1946.
(d) ex ICI Ltd, Mossend Works (owned by War Department but works operated by ICI), Motherwell, Lanarkshire, by 1/1952.

(1) returned to ICI Ltd, 'Tennant's' Salt Works, Haverton Hill, 1/1944.
(2) to No.4 Supply Reserve Depot, Stratton Factory, Swindon where spares ordered 17/1/1945.
(3) to Royal Engineers, Bicester Workshops, Arncott, Oxfordshire, by 6/1954.
(4) to Technical Stores Sub Depot, Wem, Shropshire, by 13/8/1956.

THOS W. WARD LTD

SCRAPYARD, South Bank   180
NZ 529205 – **Map G**

The yard was located immediately south of the Old Middlesbrough Road on the west side of the Eston Branch; this was probably the former Branch Brick and Tile Works site (see note in South Bank Brick Co). It was served by a simple siding from the branch and was probably shunted by the railway company. The branch was cut back to terminate at the yard when the coal depot at Eston finished. The remainder of the branch and Ward's yard was then shunted by a BR Class 08 diesel locomotive as required. This last remnant of the branch was lifted by 10/1982. One of the locomotives scrapped here was WHARTON (Lewin 1875, rebuilt 1881) from ICI's 'Tennant's' salt works at Haverton Hill which arrived in 1948. The yard also purchased three ex BR Class J94 0-6-0 saddle tanks, numbers 68016, 68032 and 68036, for scrap on 18/8/1964.

WARNER & CO LTD

IRON REFINERS, Cargo Fleet   181
Warner & Co until 20/10/1900   NZ 512203 – **Map G**

In 1850 Arthur Warner was an iron merchant and, during the following decade, carried out experiments to take out impurities from iron. He was subsequently granted patents from 1860 for the removal of sulphur and other impurities by the soda ash process. Having failed to interest a company to take up his inventions, in 1872, he established a small works on the north side of the Middlesbrough-Redcar railway line, west of Cargo Fleet Station, to produce refined iron as an intermediate stage between blast furnace pig iron and its use by iron foundries. The works was connected to the MOR on the opposite side to the OME's Cargo Fleet timber yard and served by the two short sidings within the premises. Although steel then became very popular, there was still a market for Warner's high quality refined iron. Arthur Warner died in 1885 and was succeeded by his sons, Arthur Maline and Herbert – their initials (AMW and HW) could be seen on the special brands of iron produced.

Warner established a new works on a six acre site south of the Middlesbrough-Redcar railway line between Marsh Road and Middle Beck and also became a public limited company; the first board meeting taking place on 23/10/1900. The site had previously been occupied by the Victoria Foundry which already had a railway connection. A NER single line swung off the Middlesbrough-Redcar railway just to the west of Cargo Fleet Station to run behind Whitehouse Signal Box passing the Navigation Inn to reach the railway company's three sidings forming one side of Marsh Road. Here two tracks reversed direction to enter the Victoria Foundry's premises.

An agreement dated 27/8/1897 between the NER and Warner identified new arrangements for "the convenience of the traffic to and from their Foundry and Works in course of erection at North Ormesby", although the siding agreement with the NER was not formally ratified by Warner until 19/2/1901. A new single line curved round from Marsh Road sidings directly into Warner's works, obviating the need for reversal. Once on the site, the siding divided with several tracks serving various platform before coming together in a short head shunt, at the end of which was the single road engine shed. The first reference to "Loco Coke" in Warner's stock book was dated 30/6/1901. A supplementary agreement with the NER on 28/8/1903 allowed Warner's locomotives to shunt its own traffic on the railway company's lines at Marsh Road, including to the nearby Tees Side Bridge & Engineering Works Ltd and William Robinson & Co Ltd (which see). From 6/11/1913 Warner was also allowed by the NER to run over the latter's sidings at Marsh Road to shunt traffic for Messrs Rundle and Campbell, F.G. Smith (coal merchant) and the Middlesbrough Co-operative Society Ltd. Warner certainly received payments from William Robinson, the Co-operative Society and Fred Smith for shunting it carried out on their behalf.

It is not known when Warner first obtained a locomotive, although it had advertised in *Machinery Market* on 1/8/1895 wanting a 9 or 10 inch standard gauge locomotive. John Thomas Junior & Co supplied scrap metal to Warner, but the latter paid £50 in 12/1901 and £26 in the week ending 31/1/1902 for the hire of a locomotive. According to a 1901 trade directory, John Thomas Junior & Co was based at Eaglescliffe Junction.

PAULA (Peckett 1506 of 1918) at Warner's iron refining works on 22nd February 1962.

(Courtesy David Webb)

Improvements continued to be made to Warner's plant. In 1934 a completely mechanised rotary furnace was added to give greater control in producing low carbon and alloy grades of refined iron. Another fully automated plant was added in 1953. Warner produced iron for a range of specialist uses and its brochure contains a photograph of a locomotive three-cylinder monobloc casting produced using its iron.

With the decline in traffic, BR and Warner agreed in 8/1978 that its rail connection with the main line could be closed, although internal rail movements continued. Both diesel shunters usually left the shed about 8am and, after a rub down, spent the day moving rusty 16 ton wagons loaded with scrap from the stockpile to the furnaces. Rail traffic finally ceased in early 1/1979 and the tracks were dismantled in 7/1980. The works did not last long after that and it closed at the end of 2/1984, when it was owned by Staveley Industries Iron Division. The site was cleared by Ward Bros of Darlington over the following five months.

Reference :   *Pioneers of Refined Pig Iron*, Warner & Co Ltd, c/1950

Gauge : 4ft 8½in

BESSIE	0-4-0ST	OC	HE	205	1878	(a)	Scr 11/1966
LOUISE	0-4-0ST	OC	P	970	1902	(b)	
	Rebuilt	Ridley Shaw			1932		
	Rebuilt	Ridley Shaw			1952		Scr c11/1970
PAULA	0-4-0ST	OC	P	1506	1918	(c)	
	Rebuilt	Ridley Shaw			1953		Scr 4/1963
JULIA	0-4-0DM		JF	4220023	1963	New	(1)
-	4wDM		RH	417894	1959	(d)	(1)

Armstrong Whitworth D22 of 1933, a 0-4-0 diesel electric, was demonstrated at one of Dorman Long's works (possibly Britannia) and then at Warner's factory but was not sold to Reyrolle & Co at Hebburn by the manufacturer until 3/1937.

(a)   ex West Hallam Coal & Iron Co Ltd, Ilkeston, Derbyshire. To Cerebos Ltd Greatham, on hire and returned. To South Medomsley Colliery Co Ltd, South Medomsley Colliery, Co Durham, on hire, /1944 and returned /1944. It was at Ridley Shaw for repairs, 11/6/1949.

(b)   new; Peckett's tender was accepted by Warner's Board on 19/11/1902. LOUISE was also the name given to a brand of iron produced by Warner.

(c)   previously Clugston Cawood Ltd, Sheffield. It was then bought by Thos W. Ward 9/4/1951 and resold to Warner /1951.

(d)   ex Thomas Allan & Sons Ltd, Bonlea Foundry, Thornaby, c/1969.

(1)   the locomotives were being scrapped on site 28-29/7/1980.

WEST HARTLEPOOL HARBOUR & RAILWAY COMPANY

North Shore Branch, Stockton   182
NZ 444204 – **Map D**

When the West Hartlepool Harbour & Railway Company took over the Stockton & Hartlepool Railway and its lease of the Clarence Railway in 1853, Ralph Ward Jackson decided to bring the production and maintenance of its locomotives in-house. It erected an engine works with a rail connection to the North Shore Branch almost opposite the junction with the Leeds Northern line. This was only a short distance north of Fossick & Hackworth's works. The WHH&R constructed about 30 locomotives at its works between 1853 and 1865 when it amalgamated with the NER. These included at least twenty three 0-6-0's, two 2-2-2's and four 0-4-0 tank locomotives. The site was later used for the NER's Stockton engine shed when a single-ended, eight road, straight shed was completed here in 1891.

WEST STOCKTON IRON CO LTD

WEST STOCKTON IRON WORKS, Stockton   183
NZ 442198 – **Map D**

The company was incorporated on 30/12/1865 with £20,000 capital and established its iron works on the west side of the Yarm-Norton railway line, a short distance north of Stockton Station (then called North Stockton) and next to the Westbourne Iron Works. The West Stockton Iron Co concentrated on producing and processing wrought iron. By 12/1866, it had 12 puddling furnaces and a mill. This had increased to 23 puddling furnaces and three rolling mills in 1871 and 33 puddling furnaces in 1873. The original works buildings stood end-on to the NER and the West Stockton Sidings where there was a connection with the iron company's railway which ran north westwards on either side of its principal building before terminating on the slag heap.

In the early 1870s, another sizeable building was constructed parallel to the main line between the existing works and the original Leeds Northern Engine Shed next to the station. This plant was also connected to the West Stockton Sidings via a headshunt adjoining Durham Road. The plate and sheet mills were stated to cover an area of 2½ acres in 1879, although the company was then in voluntary liquidation. An 1877 map gives Prosser & Co as occupiers of the iron works but, two years later, it was operated by the "Owners of West Stockton Iron Works", when it had 57 puddling furnaces and five rolling mills. Black, Hawthorn at Gateshead repaired a 0-4-0 saddle tank for the West Stockton Iron Works; the job being completed in 6/1881. In 1894 the company was making iron plates and sheets for bridges, ships, boilers and tanks. However, it closed shortly after and the business was removed from the Register of Companies in 1902. The works was dismantled and part of the site used for an extension of the NER's Stockton North marshalling yard about 1903.

Gauge : 4ft 8½in

-	'4-wheel tank'		Lill		1873	(a)	s/s
WEST STOCKTON No.2	0-4-0ST	OC	BH	547	1881	New	(1)
WEST STOCKTON No.1	0-4-0ST	OC	BH	679	1882	New	(1)

(a)   new, Lilleshall order no.181. Delivered to West Stockton Iron Co.

(1)   to Dorman, Long & Co Ltd, Britannia Works, Middlesbrough.

WHESSOE LTD

STOCKTON WORKS, Stockton.

184
NZ 436172 – **Map D**

This was the former Ashmore, Benson, Pease & Co Ltd/The Power Gas Corporation Ltd's South Works (which see for its development and railway). The premises were sold to Whessoe Ltd in 1967, which had its main works at Darlington. Whessoe gained a 64 acre site with 245,000 square feet of covered fabricating shops and 150,000 square feet of heavy and light machine shops. The premises were served by 22 overhead cranes. At this time, the expansion of the world oil and chemical industries meant that there was a sustained demand for its products, including components such as valves, control equipment, storage tanks and pressure vessels. However, the siding agreement with BR was terminated in 1970. The company's locomotive was then used about twice a week to move heavy loads within the works but a report in 1977 stated that it had seen little use in the last five years. Much of Stockton's workload had come from the UK iron and steel industry and, with this in the doldrums, the factory closed in 12/1979 with the loss of 600 jobs.

Reference :   *Whessoe Two Centuries of Engineering Distinction*, Dennis W. Hockin, The Pentland Press, 1994

Gauge : 4ft 8½in

(328)		0-4-0DM	RSHN	7900	1958	(a)	(1)

(a)   ex Ashmore, Benson, Pease & Co Ltd, with works, /1967.

(1)   to Batchelor, Robinson Metals & Chemicals Ltd, Hartlepool, 11/1979.

J.D. WHITE LTD

SCRAPYARD, Thornaby

185
NZ 458184 – **Map E**

This scrapyard was located on the south side of the Darlington-Middlesbrough railway line opposite the BR Thornaby Engine Shed; the road entrance to the shed forming part of the eastern boundary of the scrapyard. For many years, a short branch had left the main line near Thornaby Station and run south to serve J. Nimmo & Son Ltd's (later North Eastern Breweries) malting. In 1905, Stockton Race Course Committee had been given approval for a "horse dock, including booking office and porter's room" on the branch. It was from this branch that a railway was laid into White's premises where it divided into two; latterly these ran under a gantry crane. An agreement between James Davidson White and the LNER dated 7/5/1947 covered use of White's siding and identified the premises as "proposed site for new tenancy". White tended to purchase locomotives for scrap, but then use them to shunt the 16 ton mineral wagons being loaded in its yard. He did try to sell a 2ft 4in gauge 20hp Ruston & Hornsby locomotive in *Machinery Market* on 18/1/1952. Rail traffic ceased in 1972 and most of the scrap had been removed by the end of that year.

The Planet locomotive (FH 1943/1935) from Hill's Stockton factory was stored among the scrap in J.D. White's yard on 6th May 1967. A goods train on the main line is passing the roundhouse at BR's Thornaby Depot in the background. (IRS Brian Webb photograph)

Interestingly, a James Davidson White of 59 High Street, Stockton had offered for sale a 10 inch standard gauge 0-4-0 locomotive for £315 in 1904–05.

Gauge : 4ft 8½in

W. SHAW & Co LTD No.2	0-6-0ST	OC	RS	2554	1885		
		Rebuilt	Ridley Shaw		1941	(a)	Scr /1966
-	4wDM		FH	3572	1952	(b)	Scr c10/1972
W. SHAW & Co LTD No.4	4wDM		FH	3569	1951	(c)	s/s /1971
12 (MARY LOUISA)	4wDM		FH	1943	1935	(d)	Scr c10/1972
No.1	0-4-0DM		JF	4160007	1952	(e)	(1)
No.3	0-4-0DM		JF	4160009	1953	(e)	(1)

(a) ex William Shaw & Co Ltd, Middlesbrough, c/1954.
(b) ex William Gray & Co Ltd, Hartlepool, 25/7/1963.
(c) ex William Shaw & Co Ltd, Middlesbrough, c3/1966.
(d) ex F. Hills & Sons Ltd, Stockton, c/1970.
(e) ex British Steel Corporation, Malleable Works, Stockton, /1971.

(1) to T.J. Thomson & Son Ltd, Stockton, c1/1973.

WILLIAM WHITWELL & CO LTD

THORNABY IRON WORKS, Thornaby **186**
(subsidiary of **Pease & Partners Ltd** from 1923) NZ 454186 – **Map E**
William Whitwell & Co until 2/1889

The Thornaby Iron Works was situated on the north side of the Darlington-Middlesbrough railway line immediately to the east of South Stockton Station. (The latter had a complicated history, but the broad island platform existing today had opened on 1/10/1882 adjacent to the previous inadequate single platform station. The South Stockton Local Board was replaced by the Municipal Borough of Thornaby on 6/10/1892 and the station was renamed Thornaby in that year. The term 'South Stockton' may have originated when the SDR extended its line to Port Darlington (Middlesbrough) in 1830 and established a small platform with that name for passengers where the line left the suspension bridge over the Tees. Wm Whitwell & Co was established on 18/10/1859 and commenced to erect three blast furnaces on land acquired from H. Pease. There seems to have been a close relationship between the founders of the business and the Teesdale Iron Works next door as Thomas Head and Joseph Ashby were partners in the new venture, which would provide a source of iron for them. The leading partners were William and Thomas Whitwell, with the latter being responsible for the technical development of the works until his accidental death there in 1878. The site was not far from the River Tees and it was necessary to erect the blast furnaces on wooden piles. The first 60 feet high blast furnace was not blown in until 3/1862 but, by 1863, all three were operational. Towards the end of 1865, the company had 20 puddling furnaces working and had constructed rolling mills alongside, north of the blast furnaces near the river; the first iron being rolled in 5/1865. It employed 260 people and produced 10,000 tons of iron bars, angles and rails in 1865. Output had risen to 13,000 tons in 1871 when the company had 34 puddling furnaces and three rolling mills.

In 1873 the original iron plant was replaced by three 75 feet high blast furnaces which stood in a row approximately parallel to the main railway line and each serviced by Thomas Whitwell patented stoves. The company had also established a 600 feet long wharf on the River Tees and began to import richer haematite iron ore from Sweden by 1888, as well as shipping out pig and bar iron. Slag was tipped on marshy ground to the east; the purchase of the Erimus Iron Works site in 1897 extending the area available for this activity. The business had become a private limited company in 1889 dominated by the Whitwell and Pease families.

Whitwell purchased a fleet of five four-coupled locomotives from Fletcher Jennings & Co of Whitehaven in 1863–74 to shunt around the works. The latter three, makers numbers 85, 101 and 130, were members of Fletcher Jennings 'Patent' design but 101 and 130 were slightly smaller versions with 9in x 16in cylinders and 2ft 9in diameter wheels. The last two were built to order, rather than stock, and so Whitwell must have been happy with its earlier purchases.

In 1897 1,000 people were employed at the iron works which was producing 125,000 tons of haematite pig iron annually. There were already some sidings on the north side of the Darlington-Middlesbrough railway and Whitwell's private lines reversed off these to enter the works. According to Alan Betteney, in 1874 the signal box controlling the connection was called 'Whitwell's Junction' but, by 1908, it had been replaced, as part of the construction of the NER's Erimus marshalling yard by the new 'Thornaby Ironworks' box. Once inside the works, a set of reception sidings was available for incoming traffic, particularly coal and coke from County Durham and, from here, lines led to the mineral stores on the south side of the blast furnaces. Up to the finish, Whitwell relied on three large pig casting beds on the north side of the blast furnaces. The iron was lifted from here by two overhead 5 ton jib cranes and taken in railway wagons to a pig breaker supplied by the Lowca Engineering Co. Iron ore was imported from Sweden, Spain and Algeria and was unloaded at its wharf by six 3 ton cranes and taken in railway wagons to the stores; no calcining was required. Ships of up to 4,500 tons could be accommodated at the wharf. A railway line ran from the slag spouts at the blast furnaces, crossed the track to R.W. Crosthwaite's Union Foundry, continued between the rolling mills building and the wharf to reach the expanding slag tip where it divided into numerous sidings for emptying slag ladles. Other sidings within the works served the 33 puddling furnaces, two ball furnaces and six rolling mills. Ancillary facilities included the "usual fitting and roll-turning shops, pumping house, locomotive shed and testing houses" (visit by the Institute of Mechanical Engineers on 3/8/1893). The locomotive shed was extended early in the twentieth century and had three roads and two repair pits.

During World War 1, a coking plant was erected between the rolling mills and the slag tip. Sixty Otto regenerative coke ovens began operating in 10/1915 and another fifteen were added in mid 1918. These consumed 3,750 tons of coal per week which was delivered by the NER, the wagons being shunted by Whitwell's locomotives to where the contents were tipped into a hopper below rail level, from which the coal was lifted by an elevator on to a conveyor for transport to the storage bunkers. There was also a by-product plant.

The works' continued reliance on iron, rather than branching out into steel, proved to be a weakness. The business was acquired by Amalgamated Industrials by 1919, but Head Wrightson took over this interest in 1922 and, in the following year, Pease & Partners gained two thirds of the shares in William Whitwell. A coal washery and tower was erected in 1924 by the main line – the dilapidated structure was still there in 1987 – together with a wagon repair shop that was damaged by enemy action in World War 2. The works rarely operated for the last 2-3 years and it all stood idle in 4/1925, never to restart. However, it was not until 1935 that George Cohen, Sons & Co Ltd purchased the plant, including some locomotives and demolished most of the buildings. Tom Sowler identified a press cutting dated 23/10/1935 which stated "Three blast furnaces were felled today in the course of dismantling Whitwell's Ironworks, Thornaby". The site comprising just over 59 acres, including approximately 28 acres of slag, then leased to Highways Construction Ltd, was advertised for sale but not disposed of until 1942. A number of buildings had been retained, including the engine shed, together with "about four miles of [railway] track communicating with the wharf and all parts of the site". The siding arrangements were subject to an agreement under which payment of £65 per annum was due to the railway company but this had been reduced to a nominal figure during the site's inactivity.

Whitwell's Iron Works at Thornaby employed a number of elderly locomotives, including (top) this one which Inness reckoned was No.6 and built by the Grange Iron Co and (bottom) No.7 carrying a plate claiming to be built by A.E. & H. Kitching. (NERA R. Inness photograph)

Reference : 'Thomas & William Whitwell and Thornaby Iron Works'. Alan Betteney, *The Cleveland Industrial Archaeologist* 34, CIAS, 2013.

Gauge : 4ft 8½in

		0-4-0ST	OC	FJ	30	1863	New	(1)
No.1		0-4-0ST	OC	FJ	44	1864	(a)	(3)
No.4		0-4-0WT	OC	FJ	85	1870	New	(2)
No.2		0-4-0WT	OC	FJ	101	1872	(b)	(4)
No.3		0-4-0WT	OC	FJ	130	1874	New	(5)
No.6		0-4-0ST	OC	Grange		1879	(c)	
		Rebuilt		BH		1896		s/s
No.7		0-4-0ST	OC	Kitching		1881	(d)	
		Rebuilt		CF		1899		s/s
No.5	ARTHUR	0-4-0ST	OC	HG	355	1874	(e)	(6)
No.2		0-4-0ST	OC	CF	1195	1900	New	(9)
No.1		0-4-0ST	OC	HL	2684	1906	New	(8)
No.8	NORFOLK	0-4-0CT	OC	HL	2573	1904	(f)	(7)
No.3	CHARLAW	0-4-0ST	OC	BH	1037	1891	(g)	s/s
No.4		0-4-0ST	OC	HL	3175	1916	New	(8)
No.9		0-4-0ST	OC	HL	3491	1921	New	(8)

Two FJ 0-4-0s were for sale in /1907.

(a) new. To Fletcher Jennings & Co, Whitehaven for overhaul and returned, /1870.
(b) new. Patent type modified to work on slag heaps. A new firebox was fitted by Fletcher Jennings, 3/1879.
(c) origin unknown. On Inness list shown as 12 inch locomotive at Whitwell & Co. It carried a 'rebuilt' plate
(d) origin unknown. Inness has this 12 inch locomotive as No.7 at Whitwell & Co. It carried a plate reading "A.E. & H. Kitching, Engine Builder, Darlington; Chapman & Furneaux, Rebuilt 1899, Newcastle on Tyne".
(e) ex Tees Side Iron & Engine Works Co Ltd, Tees Side Iron Works, Middlesbrough.
(f) ex Thos. Firth & Sons Ltd, Sheffield, Yorkshire (WR), via Thos W. Ward Ltd.
(g) ex The Charlaw & Sacriston Collieries Co Ltd, Sacriston Colliery, Co Durham, /1912.
(1) to West Hartlepool Harbour & Railway Co where it became No.24, later being absorbed into the NER stock to become number 600 probably by 1865.
(2) to W. Blenkinsop & Co Ltd, Marsh Iron Yard, Middlesbrough for scrap.
(3) to Worth, Mackenzie & Co Ltd, Phoenix Works, Stockton, /1900.
(4) to Charles P. Kinnell & Co Ltd, Vulcan Iron Works, Thornaby.
(5) Inness has a note sold to "A.M. Terry & RWHL Co".
(6) to J. F. Wake, Slag Works, Port Clarence; after 25/8/1928, by /1931.
(7) to A. Bainbridge Ltd, North Yorkshire Iron Works, Thornaby.
(8) to ICI Ltd, Billingham Works, Billingham, c/1935. The *Colliery Guardian* on 21/6/1935 stated that Geo. Cohen, Sons & Co Ltd had acquired Whitwell's plant, including the locomotives at the Thornaby Iron Works. Hawthorn Leslie 2684, 3175 and 3491 were offered for sale in *Machinery Market* on 5/7/1935.
(9) to Irchester Ironstone Co Ltd, Irchester Quarries, Northamptonshire, /1937.

WESTBOURNE IRON WORKS, Stockton   187

John Holdsworth & Co until /1882   NZ 443199 – **Map D**

John Holdsworth had been one of the people involved in setting up the Stockton Iron Works in 1854, but he established the Westbourne Iron Works comprising 14 puddling furnaces and a merchant bar mill in 10/1866 when production of wrought iron commenced, although only billets were made during the first months. In 1870–71 the number of puddling furnaces had increased to 21. The premises were located on the west side of the NER Yarm–Norton railway and next to the West Stockton Iron Works, some 300 yards north of Stockton Station. The works occupied a site of approximately eight acres, produced mainly bar and strip iron and employed about 150 people. The main building was on the southern end of the site with a connection to the NER's North Stockton Yard. From here, a series of sidings ran south into the works and also served a slag heap to the west. No records exist of locomotives being used but it is possible that they were employed, although a full description of the premises in 1882 does not mention one. The works was purchased for a low price in 1882 by William Whitwell & Co. According to Whellan's 1894 *History and Directory of Durham*, Whitwell had about

1,000 workers in total, of which 350 were manufacturing iron bars at the Westbourne Iron Works. The premises had been demolished by 1899 and the NER's North Stockton Yard was extended over part of the site by 1914.

WORTH, MACKENZIE & CO LTD

PHOENIX WORKS, Stockton

188
NZ 440194 – **Map D**

The business was established by Messrs Worth & Mackenzie in Lucan Street, Stockton during 1880 and it became a limited company shortly after. It moved to the Phoenix Works to the south west of Stockton Station in 1889, the premises having been formerly occupied by Henry Wilson & Co. Worth, Mackenzie was described as "General Engineers and Founders, Pump Manufacturers and Colliery Engineers", but it specialised in making pumps, ranging from small wall donkey pumps of 150 gallons per hour capacity to large waterworks pumps capable of passing 150,000 gallons per hour. It also assisted with the overhaul of locomotives as T. Green & Sons Ltd of Leeds supplied three locomotive boilers to the Phoenix Works; the first for an engine named TUTHILL in 7/1900 and others in 8/1902 and 3/1903. The premises were located between the northern end of the Moor Steel & Iron Works and the NER Yarm–Norton line. A short siding entered the works building "by which coal, &c., is discharged close to the boilers, and metal direct on to the cupola stage attached to the foundry.". Eventually the company was voluntarily wound up and the works sold for scrap on 21/1/1935.

Gauge : 4ft 8½in

| No.1 | | 0-4-0ST | OC | FJ | | 44 | 1864 | (a) | | s/s |

(a)   ex William Whitwell & Co Ltd, Thornaby Iron Works, Thornaby, /1900.

CONTRACTORS' LOCOMOTIVES

ADAMSON, Leith, H. LEE & SON

HEUGH BREAKWATER CONTRACTS, Hartlepool

C1
NZ 533334 – **Map B**

The construction of the Heugh Breakwater, authorised by an Act in 1851, was the first stage in what would have been a Harbour of Refuge, but this was later seen as unnecessary with the increasing use of steam ships. Tenders for the breakwater, which was to be erected at the Headland in Hartlepool, were invited in 1853 and the contract was awarded to Mr Adamson of Leith. The first materials were unloaded at the Old Pier on the other side of the Headland during 1/6/1853 and a light railway was laid to carry them to the site of the breakwater. The foundation stone was laid on 27/8/1853. Ashlar sandstone was used in its construction and cement was provided by Trechmann & Curtis' works at Hartlepool Warren. Six hundred feet of breakwater had been completed by 12/1854. Work began on its extension in 1870 when H. Lee & Son was awarded the contract using concrete blocks. A light railway ran from the Old Pier, where a steam crane was installed to unload cargoes of building materials from coastal ships, to the stretch of sand next to the breakwater which gained its name "The Block Sands". By the end of 1880, the contractor had extended the breakwater to 1,138 feet in length, leaving 182 feet to build. A letter from J.L. Brownlie in *The Industrial Locomotive* 9 (Winter 1978) suggests that the Hartlepool Port and Harbour Commissioners employed narrow gauge locomotives on the work; according to the late Engineer in Charge, two "Naismith" [sic] locomotives were used but nothing further is known.

JOHN AIRD & CO

CONSTRUCTION OF SMITH'S DOCK LTD SHIPYARD, South Bank

C2
NZ 530216 – **Map G**

Work began on the project in 1906, the coping stone of the first dry dock was laid on 22/10/1907 and vessels began to be launched in 1910. A postcard dated 22/10/1907 shows a hive of activity with several fixed and rail cranes assisting with the construction. A locomotive is also visible in the background but no details have survived (see *Teesside and Old Cleveland re-visited in old picture postcards* by Robin Cook, page 21). Dorman Long erected the Fitting Shop.

JOHN ANDERSON

Anderson was a Middlesbrough based contractor who constructed a number of local railways from about 1834.

MIDDLESBROUGH AND REDCAR RAILWAY CONTRACT

C3
Map G

The railway was promoted by the SDR and GNE, being authorised on 21/7/1845. The 7¾ mile line following the then High Water Mark of the River Tees eastwards to Redcar was constructed by Anderson, with John Harris as Engineer. There were no major difficulties and it opened on 4/6/1846. According to Tomlinson, LOCOMOTION "with a carriage and two trucks attached to it led the way to Redcar followed by the engine "A" of the Great North of England Company."

Gauge : 4ft 8½in

GRIMSBY	0-4-0	OC	Andrew Smith	1837	(a)	s/s

(a) Anderson had earlier built the Byers Green Branch for the Clarence Railway c/1839-41. The above locomotive was used by the Clarence Railway (although owned by Anderson) and is discussed in *The Industrial Locomotive* 31, page 149 (ILS 1983). it may have moved to the Middlesbrough and Redcar Railway contract.

MIDDLESBROUGH AND GUISBOROUGH RAILWAY CONTRACT

C4
Map G

The railway, promoted by the SDR to gain access to more ironstone reserves, was authorised in 1852 and opened for mineral traffic on 11/11/1853 and passengers on 25/2/1854. The contract for the main line and associated ironstone branches was awarded to John Anderson, now of Burntisland; his tender was dated 19/3/1852. It is not known whether locomotives were involved but it would seem likely in view of the gradients encountered.

REDCAR JUNCTION – SALTBURN RAILWAY CONTRACT   C5
Map I

The Quaker proprietors of the S&DR had two main objectives in promoting an extension of their line from Redcar Junction to Saltburn. Firstly, prompted by Henry Pease, they formed the Saltburn Improvement Company to develop a seaside holiday resort and, secondly, it gave access to the ironstone reserves in East Cleveland. The 5.6 mile long line, authorised on 23/7/1858, presented no major construction obstacles to Anderson and opened on 17/8/1861.

Gauge : 4ft 8½in

NEWCASTLE	0-6-0	IC	RWH	226	1837	(a)	s/s

(a) ex Newcastle & Carlisle Railway. It had been put up for sale in 2/1858 and was sold for £185 to John Anderson, Middlesbrough. A drawing of the locomotive appears on page 163 of *A History of the Newcastle & Carlisle Railway 1824 to 1870* by Bill Fawcett (NERA, 2008).

JOHN BEST & SONS LTD

No.4 GRAVING DOCK, West Hartlepool   C6
NZ 517339 – **Map B**

A contract was let by the NER to John Best & Sons on 5/3/1906 for the reconstruction of No.4 Graving Dock at West Hartlepool. It was located between Gray's Central Shipbuilding yard and Marine Engine works. A 1909 publication gives its dimensions as 570 feet long and 60 feet wide. It is not known if any locomotives were used in the reconstruction.

CEMENTATION MINING LTD

BOULBY MINE CONTRACT, Easington   C7
NZ 763183 – **Map I**

The contract was let for the construction of the sea water pipeline at Cleveland Potash Ltd's Boulby Mine. On 4/8/1972 Wingrove & Rogers J7292 was noted on the surface and three more WR's were said to be underground. The grid reference is for the mine.

Gauge : 1ft 6in

-	0-4-0BE	WR	J7292	1969	(a)	s/s
-	0-4-0BE	WR	M7479	1972	(b)	s/s
-		WR			(c)	s/s
-		WR			(c)	s/s

(a) origin unknown.
(b) new, ex works to Boulby Mine, 20/3/1972.
(c) origin and identity unknown.

JOHN DICKSON

WHITBY, REDCAR & MIDDLESBROUGH UNION RAILWAY CONTRACT, Loftus   C8
NZ 898105 NZ 716180 – **Map I**

The Whitby, Redcar & Middlesbrough Union Railway was authorised on 16/7/1866 to build a 16½ mile line from Loftus to Whitby. To begin with, there was limited support for the scheme and it was not until 3/5/1871 that John Dickson was awarded the contract to construct the line. In 1872 it was decided to amend the course of the railway near Whitby. The revised route, authorised on 7/7/1873 ran from Bog Hall Junction, just outside Whitby Town Station, northwards along the coast to make an end on connection with the NER at Loftus; it included a shortening of the proposed Grinkle Tunnel from nearly a mile to 792 yards. Meanwhile, there had been complaints about the contractor's lack of progress and, in 9/1873 the WR&MU directors demanded that the two following locomotives stand as security and that a plate be fixed to each locomotive indicating WR&MU ownership. Three months later Dickson was relieved of the contract and, in 5/1874, the directors decided to sell off the equipment. The two locomotives were valued at £775 (PENWYLLT) and £875 (MULGRAVE) with the former stated to be in a bad state of repair and the latter as having done little work. The line was leased to the NER from 1/7/1875.

References : *The Whitby, Redcar & Middlesbrough Union Railway*, Ken Hoole, Hendon,1981

'The Whitby, Redcar & Middlesbrough Union Railway', Reg Sowler, *The North Eastern Express* 117, NERA, 2/1990

Gauge : 4ft 8½in

PENWYLLT	0-6-0ST	FW	29	1872	(a)	(1)
MULGRAVE	0-6-0ST	FW	(127	1872?)	(b)	(2)

(a) origin unknown, but the Neath & Brecon Railway's branch from Colbren Jn to Ynisygeinon Jn opened 10/11/1873, Penwyllt was 3 miles north of Colbren Junction.

(b) origin unknown. Reg Sowler says that the first locomotive to be used was MULGRAVE which arrived on 27/5/1873. (The *Railway News*, published 14/6/1873, gives the date as "On Tuesday last"). Temporary rails were laid in a zig zag from the NER line near Whitby Gas Works to take the locomotive up to the contract.

(1) a C. Morrison, with an address at Gresham Street, London, purchased PENWYLLT for £600 but wrote in 7/1875 to the WR&MU complaining that he had had to spend £335 4s 10d repairing the engine which he had now sold.

(2) the contractor, Thomas Nelson & Co, purchased MULGRAVE for £750 and, on 24/9/1874, his men started to remove the locomotive from the cliffs. It arrived at York three days later, but Nelson also complained about the poor state of the locomotive.

GEO. DOWSON
CONTRACTOR, GUISBOROUGH
C9
Map I

According to Kelly's 1893 Directory, George Dowson was a contractor based at Westgate, Guisborough. He offered for sale at £250 a standard gauge geared (3½:1 ratio) 0-4-0 'tank' locomotive with 6in x 14in cylinders and 2ft 6in diameter wheels in *Machinery Market* on 1/12/1891.

EAGRE CONSTRUCTION CO (SCUNTHORPE) LTD
RELAYING OF CARLIN HOW-BOULBY RAILWAY, LOFTUS
C10
NZ 709192 NZ 762188 – **Map I**

The sinking of Boulby Mine (see the ICL UK (CLEVELAND) Ltd entry) required the relaying of the former railway line from Carlin How to the new mine to enable the despatch of potash and salt. This also involved the construction of a new viaduct over the A174 road and the refurbishment of Grinkle Tunnel. Shaft sinking at the mine began in 1968 and Cleveland Potash Ltd let the contract for relaying to Eagre Construction. Some limited output from a single shaft began in 1973 with full production underway in 1976. Eagre Construction used a Ruston & Hornsby diesel to assist with the relaying and the reference contains two views of it on Kilton embankment and near the site of Loftus Station.

Reference : 'Boulby Potash Branch Line', *Cleveland Industrial Heritage* 23, pub by Peter Tuffs, 2008, pp. 24-25

Gauge : 4ft 8½in

-	0-4-0D	RH		(a)	s/s

(a) identity unknown. Possibly from and to Scunthorpe plant depot.

HIRST & SONS
DOCK EXTENSION WORKS & GRAVING DOCK CONSTRUCTION, Hartlepool
C11
Map B

The 1880's witnessed much construction at the docks in the Hartlepools and it has not been possible to identify the locations of these works. However, an announcement appeared in *The Engineer* on 11/7/1884 that Hirst & Sons' plant, including three four-coupled locomotives, from this contractor were to be auctioned on 30/7/1884; the sale was later postponed to 3/9/1884.

MESSRS HODGSON & RIDLEY

MIDDLESBROUGH DOCK CONTRACT, Middlesbrough   C12
NZ 504205 – **Map G**

The loss of much of the coal trade to the Hartlepools and the expansion of the iron industry meant that changes were needed at Middlesbrough Dock. Since 1860, three stationary steam cranes had been erected to handle shipments of pig iron, rails and general cargo. The first contract to enlarge the dock was carried out by Messrs Hodgson & Ridley between 8/1869 and 1874 to a design by T.E. Harrison, Chief Engineer of the NER. Three sides of the existing dock still had 1 in 3 sloping sides. New stone quays were provided on the north and south sides of the dock giving a frontage of 1,626 feet and increasing the water area from 9 to 12 acres. The resident engineer, A.H. Whipham, stated that the material excavated to make space for the walls was "raised by means of barrows, worked by a horse gin, or steam engine". Travelling gantries were erected over the quay wall construction; the front leg on timber staging and the rear leg far enough back to allow a railway line to be laid under the gantries so that stone and mortar could be brought in and lifted up by winches on top of the gantries. A new wider deeper entrance lock on the south side of the existing lock was constructed. In order to convey traffic from the iron works to the new quays, a 170 feet long manually operated swing bridge was erected across both entrance locks. This was built by Hopkins, Gilkes & Co Ltd which also supplied the ironwork for elsewhere in the contract. A railway and road crossed the 16 feet wide bridge. According to A.H. Whipham "To prevent liability to accident by collision of trains passing over the bridge, there being only a single line of railway, a signal cabin is placed at the south end, and the points and signals are interlocked". Three coal staithes were erected at the north end of the dock capable of handling 10 ton wagons instead of the chauldrons previously used. These replaced six of the former drops on the raised platform. The staithes projected into deep water and locomotives placed the wagons on a gradient of 1 in 80 falling towards the staith so that each wagon could be uncoupled and descend to the cradle. The NER's 1869 specification for the project stated that the "contractor to provide temporary railways, wagons etc for conveying materials to and carrying out the works". The Hudswell Clarke & Rogers locomotive was delivered to Cargo Fleet Station.

Reference :   'The North-Eastern Railway Company's Dock Extension Works at Middlesbrough', A.H. Whipham, *Proceedings of Cleveland Institution of Engineers*, 1874, pp.217-31

Gauge : 4ft 8½in

BRISK		0-4-0ST	OC	HCR	146	1874	New	s/s

HOLLOWAY BROTHERS (LONDON) LTD

CARGO FLEET JETTIES AND WHARVES CONTRACT, Cargo Fleet   C13
NZ 523214 – **Map G**

Following extensive slag tipping to reclaim marshland between the Middlesbrough-Redcar railway and River Tees, The Cargo Fleet Iron Co Ltd increasingly developed the area to receive and process incoming iron ore. Improvements were carried out to the wharf by Holloway Bros in about 1917–18 and Bulmer's review of Middlesbrough shipping facilities in 1926 described the 557 feet wharf as "just rebuilt". No locomotives have been identified on this contract but Holloway Bros had a number present on Teesside at this time (see below).

HAVERTON HILL SHIPYARD CONTRACT, Haverton Hill   C14
NZ 487226 – **Map C**

The Furness Shipbuilding Co Ltd, in conjunction with the Admiralty, let a contract to Holloway Bros (London) Ltd to construct a major new shipyard on the north bank of the River Tees at Haverton Hill. This was in response to the loss of merchant ships due to the German U-boat campaign. In early 2/1918 Holloway Bros advertised for navvies to work on "Auxiliary Shipyard Extension No.81 Haverton Hill". Much of the site was tidal salt marsh below the high water mark and this had to be raised by an average of 0.6 metres using blast furnace slag, with a top surface of ash and brick rubble to provide a stable level platform, on which were laid railway lines to transport the heavy construction equipment. It was estimated that a million tons of material were moved. Extensive use was made of concrete piles and reinforced concrete for the slipways. Shipping losses were becoming critical in 1917–18 and the work was carried out with great urgency, such that the first keel for a ship was laid in 5/1918 well before the contract was completed. At this date, 2,000 people were working on the site, including 500 women. Disused railway carriages provided temporary living accommodation for some of the workers. Both standard gauge and hand-worked narrow gauge railways were used in the construction, with mechanical shovels being imported from the USA to excavate the fitting out basin.

Gauge : 4ft 8½in

No.	Name	Type		Builder	Works No.	Date	Notes	Notes
No.20		0-4-0ST	OC	AB	265	1884	(a)	(1)
-		0-6-0ST	IC	HC	347	1892	(b)	(4)
2	YORKSHIRE	0-6-0T	IC	HC	1070	1914	(c)	(8)
62	GENERAL ROBERTS	0-6-0ST	IC	MW	975	1886	(d)	(9)
-		0-6-0T	IC	MW	1005	1887	(e)	(4)
-		0-6-0ST	IC	MW	1146	1890	(e)	(5)
65	IRENE	0-6-0ST	IC	MW	?	?	(f)	(3)
	PALACIOS	0-6-0ST	IC	HE	464	1888	(g)	(6)
	COWES	0-4-2ST	OC	BH	116	1870	(h)	(2)
-		0-4-0T	OC	HL	2663	1906	(i)	(5)
5	WYE	0-4-0ST	OC	HE	420	1887	(j)	(7)

The H.M. Nowell Enfield Contract entry in *Industrial Railways & Locomotives of Hertfordshire & Middlesex* (IRS, 2007) suggests that PETERHEAD 0-6-0ST IC MW 1378 of 1898 and TANFIELD 0-6-0ST IC MW 1470 of 1900 **may** have been on the Haverton Hill Shipyard contract.

(a) ex possibly a Holloway Bros housing contract at Gretna (1915-16) in connection with the MoM factory there. Previously D.J. Jardine, Corncockle Quarry, Lockerbie, Dumfriesshire; the locomotive being requisitioned by the Government c/1914.

(b) ex Baldry, Yerburgh & Hutchinson, contractor, c/1917.

(c) was new to Sir John Jackson Ltd, Bulford and Larkhill Camps and Railway contracts, Wiltshire 1914-16. To Haverton Hill Contract, c/1917.

(d) ex possibly Holloway Bros contract for Scottish National Housing Ltd, Rosyth, Fife (1917-18). To Haverton Hill Contract, c/1917.

(e) ex Pauling & Co Ltd, High Wycombe and Uxbridge Contracts, Buckinghamshire, by c/1917.

(f) this could be MW 1502 of 1900 which was used by Perry & Co (Bow) Ltd on the GWR Wombourn Branch contract until work was suspended in /1917. It was probably requisitioned by the MoM until c/1919.

(g) ex Austin Motor Co Ltd, Longbridge, Warwickshire, by 10/1917.

(h) ex Isle of Wight Central Railway, No.3, 2/1918.

(i) ex Isle of Wight Central Railway, No.1, 2/1918. This locomotive was formerly part of a 'rail motor' unit.

(j) ex S. Pearson & Son Ltd, South Metropolitan Gas Co contract for a chemical plant at East Greenwich Gas Works, London, (via MoM?), on hire, /1918

(1) s/s, although may have worked for Furness Shipbuilding Co Ltd at Haverton Hill.

(2) later to Plenmeller Collieries Ltd, Plenmeller, near Haltwhistle, Northumberland.

(3) s/s, but may have been returned to Perry.

(4) to Furness Shipbuilding Co Ltd, Haverton Hill, /1918.

(5) to Furness Shipbuilding Co Ltd, Haverton Hill.

(6) to Furness Shipbuilding Co Ltd, Haverton Hill, 10/1918.

(7) returned to S. Pearson & Son Ltd, contractor, ex hire, by 10/1919.

(8) probably for sale at Newcastle-upon-Tyne, 12/1921. It may have been involved in the contract at Graythorp prior to 1921.

(9) to Holloway Brothers (London) Ltd, Swanley Bypass contract (1921-24) for Kent County Council.

REDCAR JETTY AND WHARVES CONTRACT, Redcar   C15
NZ 548255 – Map H

This contract was carried out by Holloway Bros for Dorman, Long & Co Ltd about 1917-18. Dorman Long had acquired the Redcar Iron Works by 1916 and construction of a new steel plant began. As part of this development, the old Redcar Jetty was improved by Holloway Bros so that it became 500ft long with three electric cranes. A secondhand Manning Wardle 0-4-0ST, with 9in x 14in cylinders, standard gauge locomotive was advertised for sale at Redcar by Holloway Bros on 4/6/1919 in the *Contract Journal*.

TEES SHIPYARD CONTRACT, Graythorp   C16
NZ 522270 – Map A

Holloway Bros carried out a contract in 1919-21 to construct a graving dock and deep water quay as part of Sir William Gray & Co Ltd's new Tees Shipyard at Graythorp. The site was connected to the NER's Seaton-on-Tees Branch. It is possible that some of the locomotives used on the Haverton Hill shipyard contract moved across to work at Graythorp. Holloway Bros may also have been involved in constructing a breakwater at Seaton Carew at this time.

Gauge : 4ft 8½in

-	0-4-0ST	OC	AE +	?	?		
		Rebuilt			1907		
-	0-6-0ST	IC	MW	1665	1905	(a)	(1)
64	0-6-0ST	IC	MW	1144	1890	(b)	s/s

+   This may be AE 1055 of 1874 which was with the MoM at Birtley, Co Durham c/1916. It was seen at Billingham in /1921 and at a Derwenthaugh scrapyard in 7/1933.

(a)   ex Easton Gibb & Son Ltd, Rosyth, Fife, c/1917, after 4/1916. Holloway Bros may have used it on the Scottish National Housing Ltd contract at Rosyth (1917-18) before moving it to Graythorp.

(b)   ex South Metropolitan Gas Co, East Greenwich Gas Works, London; possibly the locomotive was for sale there 7/1919; to Graythorp, /1920.

(1)   it subsequently passed through the hands of William Jones Ltd, Greenwich, which sold it to Holland & Hannen and Cubbits Ltd for the Downham Estate contract in Kent, 10/1927.

WILLIAM HUTCHINSON

CLEVELAND RAILWAY CONTRACT   C17
Map G

William Hutchinson was both a contractor and cement manufacturer based at the West Hartlepool Cement Works. Among the contracts for which he was responsible were two involving construction of sections of the Border Counties Railway between Hexham and Riccarton Junction. One of the locomotives which may have been used on a Border Counties contract was NICHOLAS WOOD built by John Coulthard & Son and purchased by Hutchinson in 11/1859. This locomotive had once hauled coal from Westerton Colliery over the Clarence Railway to staithes on the River Tees. Hutchinson also had a contract to construct part of the Cleveland Railway which was authorised by an Act dated 22/7/1861, but work had commenced a year earlier using wayleaves. The section between Slapewath and Cargo Fleet opened on 23/11/1861. Problems with the Border Counties contract resulted in Hutchinson being declared bankrupt on 5/12/1862. His remaining assets were offered for sale at the West Hartlepool Cement Works, including the following locomotives located on the Cleveland Railway contracts according to the 5/6/1863 issue of the *Durham Chronicle*.

Reference :   'Private Locomotives on the Clarence Railway Part 2', Ken Fleming, *The Industrial Locomotive* 170, ILS, 2019, pp.10-28

Gauge : 4ft 8½in

"a six-wheeled locomotive" at Slapewath

"SWIFT, a four-wheeled locomotive" at Normanby   (1)

(1)   This may have been the same SWIFT, allegedly built by Fossick & Hackworth, that was acquired shortly afterwards by Cassop Colliery in Co. Durham.

JOHN JACKSON

MIDDLESBROUGH DOCK CONTRACT, Middlesbrough   C18
NZ 504205 – Map G

Preparations on enlarging Middlesbrough Dock to a design by T.E. Harrison, the NER's Engineer, began in 1878 and a contract for £94,447 was awarded to John Jackson for the main works on 1/1/1883. *Building News*, on 17/9/1886, reported that the expanded dock had opened last week. It had involved extending the water area from 12 to 15¼ acres with 1,400ft of new quays. The entrance channel was widened from 80ft to 200ft and deepened by 5 feet. A hydraulic power station was built and the lock gates and swing bridge converted from manual to hydraulic operation. Ten cranes and another ten capstans, all hydraulically powered, were added. The coal drops and the raised platform were dismantled and a separate coal hoist were installed for bunkering ships. The work was delayed because of poor ground conditions with 20ft of running sand encountered under the clay. The stone for the construction came from Lartington Quarries, near Barnard Castle and was transported by the NER to Middlesbrough. The total cost was approximately £200,000. Only one locomotive is known to have worked on the contract, but it seems likely that there were others.

References :   'The History of Middlesbrough Dock', D.W. Pattenden, *The Cleveland Industrial Archaeologist* 18, CIAS, 1986 pp.6-10

The History of Middlesbrough Dock 1842-2000, John H. Proud, Printability Publishing, 2000

Gauge : 4ft 8½in

	MIDDLESBROUGH	0-4-0ST	OC	B		296	1882	(a)	(1)

(a) it appears to have been built for stock in 1882 and was supplied via a dealer in the following year as it carried a plate reading "Lennox Lange & Co. 1883".

(1) later worked at Jackson's Devonport Dockyard Extension Contract where spares were sent for it by Andrew Barclay in 2/1897.

JOHN LANT LTD

CARGO FLEET IRON WORKS CONTRACT, Cargo Fleet   C19
NZ 523205 – **Map G**

J. Lant carried out a contract for the Cargo Fleet Iron Co Ltd at its works, following which a Hudswell Clarke 0-6-0 saddle tank with 12in x 24in cylinders was offered for sale on 24/5/1917 according to the *Contract Journal*. Work appears to have continued after that date because negotiations took place between the Cargo Fleet Iron Co and Middlesbrough Council on 7/5/1918 concerning the closure of a footpath due to the extension of the works.

LORD, CARTER, WADE & CO

GRANGETOWN CONTRACT, Grangetown   C20
NZ 548209 – **Map H**

The construction of the "new town" commenced on 1/4/1881 to provide accommodation for incoming workers employed at Bolckow Vaughan's recently completed steel works. The name 'Grangetown' was taken from the nearby Grange Farm. Approximately 23 acres of land were purchased and eight streets – Bessemer, Vaughan, Stapylton, Laing, Holden, Wood, Vickers and Cheetham – were laid out on a compact rectangular plan involving the construction of 768 terrace houses. The new residential development was situated next to the southern boundary of Bolckow Vaughan's Cleveland Works. The contractor, Lord, Carter, Wade & Co, had completed 418 houses by 1/10/1882 and commented in the *Daily Exchange* 1/11/1882 that "a remarkable feature in this large concern is that we have not a single horse or cart; line of rails being laid in the streets, everything is brought to the door by the steam engine." Presumably some materials could be delivered to the site by Bolckow Vaughan's locomotives over their own railway system. The contractor also made five million bricks in his brick works. The contract was completed and the *North Eastern Daily Gazette* 15/12/1883 gave notice of an auction on 20/12/1883 to sell the whole of the building plant including "locomotive engine, vertical multiplied gear" and "locomotive engine and fittings, on 4 wheels, 2ft diameter, 5in cylinder, 8[½?]in bore, 15in stroke, by H. Hughes & Co, Loughborough". Other materials included a large quantity of tram metals and an engine shed. Although the contractor had expected a railway station to be provided "on this side of the steelworks", Eston Grange Station did not open until 22/11/1885 and was linked to Grangetown by Station Road running along the eastern side of the Cleveland Works. The partnership of Joseph Lord, Robert Wade, Thomas Carter and Thos Outhwaite formed for the purpose of executing the Grangetown contract was dissolved in 1887.

Gauge : ?

		4wVB	VCG				(a)	s/s
		0-4-0		HH			(a)	s/s

(a) origin and identity unknown.

T. ARTHUR MATTHEWS

HARTLEPOOL HEADLAND PROTECTION WORKS CONTRACT, Hartlepool   C21
NZ ? – **Map B**

The contract was carried out in 1886-89, the details of which are not known. Manning Wardle 995 was a Class K locomotive but fitted with special frames, smokebox, saddle tank, footplating and brake arrangement. The manufacturer also noted that "the motion was covered in between the frames"; was this a protection against sand? The plant, including the two locomotives (10 inch and 12 inch by MW), was to be offered for sale on 18/9/1889 on completion of the contract according to the *Contract Journal*, 11/9/1889.

Gauge : 4ft 8½in

		0-4-0		MW			(a)	s/s
	MERCURY	0-6-0ST	IC	MW	995	1886	New	(1)

(a) origin and identity unknown.
(1) later at Lilleshall Co Ltd, Oakengates, Shropshire.

ROBERT McALPINE & SONS
GRAVING DOCK CONTRACT, Cargo Fleet   C22
NZ 521209 – **Map G**

The Tees Conservancy Commissioners opened its 521 feet long graving dock at Cargo Fleet on 16/11/1876. Subsequent use of the dock by ship owners was less than expected and, with the increasing size of vessels, there would be a need to enlarge the facilities. The Tees Conservancy Commissioners awarded a contract to Robert McAlpine & Sons to reconstruct the dock entrance at a cost of £7,806 in 1904-05. Presumably this involved enlargement because at some date the length of the dock was extended to 576 feet. The following two locomotives, one 12 inch and the other 13 inch, from McAlpine's contract were advertised in the *Contract Journal* on 11/1/1905 and 18/1/1905 for sale at the 27/1/1905 auction.

Gauge : 4ft 8½in

-		0-6-0 tank	MW			(a)	s/s
-		0-6-0 tank	MW			(a)	s/s

(a) origin and identity unknown.

SIR ROBERT McALPINE & SONS LTD
METAL PRODUCE RECOVERY DEPOT CONTRACT, Eaglescliffe   C23
NZ 410148 – **Map D**

McAlpine laid out the depot comprising a series of large warehouses, storage areas and railway system on the north side of the Darlington to Middlesbrough line at Eaglescliffe. The contract for £871,000 was awarded by the Ministry of Supply and ran from 21/5/1943 to 23/3/1944. The depot was then operated by Morris Motors Ltd (which see).

Gauge : 4ft 8½in

31		0-6-0ST	IC	HC	1026	1913	(a) (1)

(a) ex McAlpine's Hayes Depot, Middlesex, 21/5/1943.
(1) Latham in *The Story of McAlpine Steam Locomotives 1869-1965* states that the locomotive was later at Boscombe Down, Wiltshire from 23/3/1944.

NORMANBY IRON WORKS EXTENSION CONTRACT, Cargo Fleet   C24
NZ 518208 – **Map G**

The contract for £112,695 was let by Pease & Partners Ltd to McAlpine for alterations to be carried out at the Normanby Iron Works. This took place in 1919-22 and may have been associated with revisions to the MOR line crossing the northern part of the site.

Gauge : 4ft 8½in

30		0-6-0ST	IC	HC	1011	1912	(a) (1)

(a) the locomotive had earlier been involved in McAlpine's Turnberry Aerodrome, Ayrshire contract in 1918-19.
(1) it was used on a McAlpine contract at Watling Street, Dartford from /1922.

NORTH TEES GENERATING STATION CONTRACT, Haverton Hill   C25
NZ 478214 – **Map C**

The contract for the construction of North Tees 'A' power station on the west bank of the River Tees was awarded by the Cleveland & Durham County Electric Power Co Ltd to McAlpine in the sum of £299,763. The contract was carried out in 1918-21 but no details of any locomotives used have survived. McAlpine was also involved in the construction of the North Tees 'C' Station in 1946.

JOHN MOFFAT

PORT CLARENCE TO HARTLEPOOL NEW ROAD CONTRACT   C26
NZ 501216 NZ 525290 – **Map C**

Prior to 1914, the area of land between Port Clarence and Seaton Carew was mostly open fields and marsh punctuated by the occasional brine wellhead. A lane ran alongside the western boundary of Bell's Iron Works to serve the brine wells and Port Clarence rifle range. In 1893–97, a wagonway linked this track with more wells at Cowpen Salt Works.

John Moffat of Manchester was awarded the contract to construct the "Hartlepool New Road" comprising 6½ miles of 30 feet wide carriageway between Port Clarence and Seaton Carew, with a side road to Seaton Snook. The Tees Conservancy Commissioners accepted Moffat's tender of £38,276 on 6/10/1913 and the first sod was cut on 1/12/1913. This is the present A178 road, apart from a later deviation near Port Clarence. The agreement between John Moffat and the NER was dated 17/2/1914 and included the provision of a connection with the Greatham Creek Branch by the NER and a siding laid by the contractor capable of accommodating 18 wagons. These were to be used by Moffat to convey traffic for the new road "now under construction". The end of this siding connected with Moffat's temporary lines, but the agreement also gave approval for one of these temporary lines to cross the NER branch "by means of the said sleeper crossing and swinging rails for the purpose of conveying by his own locomotives across the company's railway empty wagons and wagons containing slag, cinders and banking material purchased by him from Bell Bros Ltd… and wagons containing coal, old bricks, plant and other materials…" A signal was to be placed there to protect the NER's line and Moffat was charged 1d per ton on all traffic conveyed by the swinging line. In a supplementary agreement of 20/8/1918 between the NER and the Furness Shipbuilding Co Ltd, the latter could use the siding and "line crossing the Company's railway" for traffic to the new shipyard being laid out at Haverton Hill. An agreement to "widen Seaton Carew Road" was not signed with the Church Commissioners until 16/3/1916. The work was eventually completed in 1918 when both locomotives were offered for sale. The siding agreement was cancelled as from 11/2/1920.

Reference :   'John Moffat – Contractor', Bob Miller, *The Industrial Locomotive* 126, ILS, 2008, p.200

Gauge : 4ft 8½in

VERNON	0-6-0ST	IC	HC	530	1899	(a)	(1)

(a)   possibly ex T. Winter & Co, Cuckoo's Nest Quarry, Hayfield, Derbyshire 1914-15. The identity of this locomotive is not confirmed.

(1)   offered for sale in *Contract Journal* 6/11/1918, although it was quoted as having 11 inch cylinders (instead of 12 inch for the above locomotive).

Gauge : 3ft 0in

MAC	0-4-0ST	OC	WB	1516	1897	(a)	(1)

(a)   ex a Moffat contract (possibly Thirlmere Viaduct), but initially to Thos D. Ridley & Sons for repairs; spares supplied 5/6/1914.

(1)   offered for sale in *Contract Journal* 6/11/1918, but not sold and moved back to Moffat's Weaste Yard, Lancashire, where it was again advertised for sale, 1/8/1919.

THOMAS NELSON & SON

CASTLE EDEN AND STOCKTON RAILWAY CONTRACT   C27
Map D

The contract for the approximately 13 mile line between Bowesfield Junction and Wellfield was awarded by the NER to Thomas Nelson & Son in 6/1874. The contract is considered in *The Industrial Railways and Locomotives of County Durham Part 3* (IRS 2021), but the first section of line to open for mineral traffic between Bowesfield Junction and Carlton South Junction on 1/5/1877 was in Stockton. Nelson offered two locomotives for sale at Stockton on 4/7/1878 – a Kitson six-wheels coupled with 13in x 20in cylinders and a R. & W. Hawthorn four-wheels coupled with 11in cylinders.

NORTH EASTERN PITWOOD ASSOCIATION LTD

WYNYARD CAMP, near Stockton   C28
NZ 403282 – **Map D**

An amendment dated 1/5/1919 to the NER's *Book of Collieries, Works and Sidings* (1918 edition) records the addition of a siding at Wynyard for the North Eastern Pitwood Association. The siding was required for the despatch of timber following tree felling on the Wynyard estate between approximately 1919 and 1921. This appears to have taken place at Tilery Wood next to Wynyard Station on the NER Stockton-Wellfield line and the siding for the camp was probably located at the Station Yard on the boundary with County Durham. The *Contract Journal* on 1/3/1922 contained notice of a sale on the following 9th March to sell, re the North Eastern Pitwood Association Ltd (in liquidation) at Wynyard Camp, plant including a 3ft gauge four-coupled saddle tank locomotive.

Gauge : 3ft 0in

-		0-4-0ST	OC	AB	297	1887	(a)	(1)

(a)   ex John Best, Delph Reservoir Contract, Bolton, Lancashire, by 3/1919.

(1)   to John Pearson Ltd, probably Portrack Lane Contract, Stockton, by 6/1922.

NORTH RIDING OF YORKSHIRE COUNTY COUNCIL

CONSTRUCTION OF BRIDGE AND IMPROVED ROAD, Dalehouse, near Staithes   C29
NZ 778181 – **Map I**

The work involved the construction of a new bridge over the beck between Dalehouse and Staithes, together with improvements to the Whitby-Loftus road (present A174) heading north past Red House Farm. A substantial embankment was required on the approach to the bridge. The scheme was considered by the Council's Highways and Bridges Committee on 12/10/1921 as a public utility for unemployment relief and approved shortly afterwards. A contract was let to J. Horsfall & Sons for the construction of the concrete culvert. Approval was given on 22/2/1922 for the construction of a "small light railway" for approximately 12 months. A month later the County Surveyor reported that a strip of land had been rented at Red House Farm for £40 pa on which to lay the temporary tramway which would transport ironstone shale from Boulby Mine tip to then run alongside the main road. The light railway plant was leased from William Jones & Co, London but later had to be purchased. The work was undertaken by the Council using direct labour with half of the cost being met by the Ministry of Transport. The committee accounts also have entries for payments to J. Wardell & Co for the hire of the locomotive; the first was approved on 20/9/1922 and the final payment on 16/1/1924. Wardell's last spares order for delivery to Staithes was dated 7/1923. The shale heap at Boulby Mine was also used to fill the Dalehouse ravine and 14,000 tons had been brought here by 1/1923.

Reference :   'Cassop and KIDBROOK', Colin Mountford, *The Industrial Locomotive* 150, ILS, 2014

Gauge : 2ft 0in

KIDBROOKE	0-4-0ST	OC	WB	2043	1917	(a)	(1)

(a)   following completion of work to remove coal from an embankment between Cassop and Thornley in Co Durham, the locomotive was advertised for sale on 2/9/1921 in *Contract Journal* by A.T. & E.A. Crow on behalf of J.H. Bentham & Co. It was presumably not sold because Bentham ordered more spares for it on 4/4/1922. In 7/1922 J. Wardell & Co of London ordered spares for delivery to the NER station at Staithes.

(1)   It appears that with the end of the locomotive hire at Staithes, J. Wardell & Co moved the locomotive to a so far unknown contract in Enfield, North London because spares were ordered in 12/1923 to be delivered to the LNER's Brimsdown Station.

H.M. NOWELL

PORTRACK LANE PLANT DEPOT, Stockton   C30
Map D

Herbert Mason Nowell was a member of the Nowell family of railway contractors who appear to have started carrying out contracts in the early 1880's. His office address was at Preston in 1887 but, in the early 1900's, he was carrying out contracts for the NER and his office was located at the Bank Chambers, High Street, Stockton in 1909. He was then living at Norton House, The Green, Norton.

In 1913, the NER gained approval for a three mile line from near its North Shore Junction to Billingham Beck where it would join end-on with its line from Haverton Hill and an agreement was signed with H.M. Nowell for the latter to construct the branch. As part of the contract, Nowell was permitted to run with his engine and crane with wagons owned or used by him along the North Shore Branch. World War 1 then intervened and a start on the contract was delayed; the agreement with the NER being terminated on 31/12/1915. The Billingham Beck Branch eventually opened in 1920. The contractor is not known but it is likely to have been H.M. Nowell. According to the NER Line Diagram showing the completed branch, there is a reference to a "Gate to Nowells Yard" between the bridge over Lustrum Beck and the Portrack Lane overbridge to the south. Nowell gave up railway contracting in 1923 when two locomotives were for sale.

Reference : *Contractors' Locomotives* Part VIII, R.W. Miller, ILS, 2008

Gauge : 4ft 8½in

BERNARD	0-6-0T	IC	HE	587	1893		(1)
	Rebuilt		HE		1906	(a)	(1)
			MW	?	?	(b)	(2)
			KS	?	?	(b)	(2)

(a) the locomotive was used by Nowell on the Hertford Loop, Enfield contract (1906-10) and it was later on a Stockton contract by 16/7/1914.

(b) identity unknown.

(1) to Nowell's Redcar Steel Works contract, Redcar, by 2/1917.

(2) A.T. & E.A. Crow was to auction on 11/4/1923 re H.M. Nowell, Portrack Lane, Stockton plant including two 'tank' locomotives, four-wheel and six-wheel, 12 inch and 14 inch Manning Wardle and Kerr, Stuart (*Machinery Market*, 30/3/1923).

REDCAR STEEL WORKS CONTRACT, Redcar　　　　　　　　　　　　　　　　C31
NZ 570254 – **Map H**

Dorman, Long & Co Ltd had acquired the Redcar Iron Works (which see) in 1916 and proceeded to erect a new open hearth steel plant. It is likely that H.M. Nowell was involved with this work.

Gauge : 4ft 8½in

BERNARD	0-6-0T	IC	HE	587	1893	
	Rebuilt		HE		1906	(a)
	Rebuilt		HE		1917	(1)

(a) following the deferment of work on the NER Billingham Beck Branch, this locomotive moved to the Redcar Works contract. Spares were ordered from Hunslet for Redcar in 2/1917 and 11/1917. The locomotive was also rebuilt as a 0-6-0ST by Hunslet in /1917.

(1) it was later at the Air Ministry, West Drayton, Middlesex, by 9/1920 (possibly by 9/1918).

JOHN PEARSON LTD

WHINSTONE QUARRY, Stainton

The company acted as a contractor, as well as operating a quarry at Stainton.

GROVE HILL HOUSING CONTRACT, Middlesbrough　　　　　　　　　　　　C32
NZ 500186 – **Map G**

Towards the end of World War 1, a number of local authorities began planning to provide new houses for the "working classes". In 1919 Pearson won the contract to lay out the sewers, road foundations and kerbing for houses at Marton Grove (Grove Hill); the first council estate to be built in Middlesbrough. Shortly after, on 22/9/1919, the Council included in the contract that Pearson could obtain railway facilities from the OME for transporting building materials. On 2/3/1920 Summerson & Son agreed to construct the railway at a cost of £2,396. The track ran from the OME Ballast line near the North Eastern Steelworks Athletics ground and is believed to have crossed Eden Road to the sidings on the housing site. Unfortunately, there is no information whether Pearson used a locomotive on the contract but, on 11/12/1922, it was reported that the line was no longer being operated because Pearson was obtaining materials locally that were being delivered by lorry.

WIDENING OF PORTRACK LANE AND HAVERTON HILL ROAD, Stockton C33
NZ 454197 NZ 471208 – Map C

The project was one of a number promoted by Stockton Council with Government grant aid to provide work for unemployed people. The section nearest Stockton town centre was awarded to Thos. Swan & Co and he appears to have used a locomotive on the contract because Hudson maker's records show that a new 20hp machine was delivered to him at Stockton being ex works on 16/2/1925. The stretch to the east was allocated to J. Pearson Ltd for £5,738 15s 3d. In 3/1923, Pearson was actively engaged laying materials for his "wagon track" and the light railway was complete in the following month. The work extended from Swan's contract eastwards over the NER Billingham Beck Branch, past Lustrum Beck to Billingham Beck Bridge. Pearson's contract involved more excavation and earthworks, hence the need for the railway. By 11/1923, Pearson had excavated 1,023 cubic yards, provided 6,470 cubic yards of fill, constructed 9,580 linear yards of foundation comprising 3in clinker and 13in slag and laid 1,240 linear yards of concrete kerb. The contract was completed 31/12/1923, apart from some surfacing of a footpath awaiting gas main work.

Gauge : 2ft 0in

-	4wPM	HU	29057	1925	New	s/s

Gauge : 3ft 0in

-	0-4-0ST	OC	AB	297	1887	(a)	s/s

(a) ex North Eastern Pitwood Association Ltd, Wynyard Camp, near Stockton; after 9/3/1922, by 6/1922. It is likely that this locomotive was used on the Portrack Lane Contract as spares were sent to John Pearson Ltd, Stockton.

DEMOLITION OF PORTRACK SLAG TIP, Stockton C34
Map D

Pearson is thought to have had this contract in 1924 and it is possible that Andrew Barclay 297 was also used on it. Later, John Pearson was in liquidation and, in 1/1936, his quarry and plant were put up for sale.

REDCAR BOROUGH COUNCIL

REDCAR-GRANGETOWN NEW TRUNK ROAD C35
Map H

Redcar Council undertook a number of construction projects to assist unemployed people in the borough following the First World War. Railways were employed on the work but no evidence has been found of locomotives being used.

Road traffic from Middlesbrough to Redcar had to travel via South Bank, Eston and Kirkleatham so, in 10/1921, it was decided to build a road between Redcar and Grangetown (the present A1085). Work began in 11/1921 and the Town Clerk was instructed to employ as many men as possible from the Relief List of the Board of Guardians. A year later, the Council's Unemployment Committee agreed to use railway track to carry out the work. As the road passed over Dorman Long's land for much of its length, the company agreed to provide materials free. By the time of its completion, 10,000 men had worked on the scheme, over £30,000 had been paid in wages and about 65,000 tons of material had been used. The road was officially opened by Sir Henry Beresford-Pierse.

EXTENSION OF REDCAR PROMENADE C36
Map I

On 29/11/1923, Redcar Council agreed to extend the promenade westwards towards Coatham. Expenditure was not to exceed £5,000 and a start was to be made at the earliest opportunity. The Town Clerk was to also approach Dorman Long with a view to securing the necessary slag from the Warrenby tips. The Council had advertised in the *Contract Journal* on 24/10/1923 wanting to hire, for six months, two miles of standard gauge railway track and wagons. In 1/1924 it agreed to hire a mile of "heavy" rails from Mr W. Oliver and to purchase secondhand rails from Hughes Bolckow & Co. The Council's Works Committee on 4/2/1924 authorised the purchase of sleepers and wagons for conveying the slag. Men were also engaged to extract material from the tips.

Although probably not connected with these works, it is interesting to note that in 1927 a siding left the LNER line at Warrenby and ran along the course of the original Middlesbrough-Redcar railway to a headshunt at Coatham, from where it reversed to pass across the Warrenby Tips terminating near Dorman Long's Redcar Steel Works.

REDCAR BOROUGH COUNCIL/GUISBOROUGH RURAL DISTRICT COUNCIL

CONSTRUCTION OF REDCAR – MARSKE COAST ROAD

C37
Map I

The Ministry of Transport offered a grant in 10/1922 towards the construction of the coast road between Redcar and Marske. Previously access had been via Redcar Lane or along the sands. Responsibility for the road was shared between Redcar Council from Granville Terrace to its eastern boundary and Guisborough RDC which administered Marske. Redcar offered to supply the District Council with slag from tips worked by the Council for 7s 6d per ton delivered by the Council's private railway to Green Lane. Redcar Council's Unemployment Committee agreed in 11/1922 that a railway should be laid from the siding serving Marske Aerodrome north along Green Lane to reach the site of the Coast Road. This line was to carry slag sent from Middlesbrough and a Mr Parsons was appointed to supervise the traffic. The Committee said it would sign an agreement with the NER and Dorman Long for the use of the siding and that as "it would be difficult to arrange for the use of a locomotive between the Siding and the Sand Banks, authority be given to engage as many horses as the circumstances may demand". Dorman Long charged 6d per ton for material it supplied. The Committee also accepted the loan of 400 yards of "light gauge railway" and four jubilee bogies and hoppers from Dorman Long for £5 per month. Guisborough RDC started its portion of the coast road project on 18/1/1923 making use of the standard gauge railway. In 10/1923 a tender for transporting slag and ashes by horses and carts from Redcar Goods Station was approved. The new road was officially opened by the Marquess of Zetland on 9/11/1923.

T.D. RIDLEY & SONS

TEESSIDE RAILWAY CONTRACT, Billingham

C38
NZ 480224, NZ 510230 – Map C

The Teesside Railway was promoted jointly by the NER and the Tees Conservancy Commissioners to open up reclaimed land on the north bank of the River Tees for industrial development. Two separate lines connecting with the NER Port Clarence Branch were authorised: (a) from Haverton Hill south westwards to Billingham Beck (NZ 480224) and (b) from Port Clarence north to Greatham Creek (NZ 510230). The contract for construction was awarded to Ridley on 18/5/1899 for £79,415 4s 6d. Work began in June 1899 and Ken Fleming in *The Industrial Locomotive* (Winter 1978) states that he had two locomotives at work by August. The Haverton Hill part of the contract was completed in 8/1900 and the line to Greatham Creek in 1901.

Gauge : 4ft 8½in

REDCAR	0-4-0ST	OC	JF	2625	1876	(a)	(1)

(a) this locomotive **may** have been used by Ridley on this contract. It had belonged to Walker, Maynard & Co's Redcar Iron Works and had been offered for sale on 1/4/1889.

(1) to Isaac Law, Shawforth Quarry, near Bacup, Lancashire, /1901. Mr Law obtained a Fowler locomotive called REDCAR following completion of "a railway contract in the Middlesbrough area"; Ridley's Teesside Railway contract was the only local one taking place at this time. It had appeared in the press for sale and may have been the 12 inch locomotive advertised by W. Blenkinsop & Co.

WILLIAM RITSON

BOOSBECK – CARLIN HOW/LOFTUS RAILWAY CONTRACTS

C39
NZ 659171 NZ 708194 – Map I

The Cleveland Railway obtained approval for a line from Belmont Junction to Skinningrove on 23/7/1858. Part of this line was opened as far as Slapewath on 23/11/1861 and had reached Boosbeck Lane by 1862. On 9/9/1863 William Ritson of Shotley Bridge submitted a tender of £25,811 9s 3d for construction of the remaining 6¾ miles. The line was opened to Brotton for mineral traffic on 23/2/1865 and to Carlin How on 21/4/1865, shortly before the line was sold to the NER.

The NER abandoned the proposal for a 1 in 6 rope-worked incline down to the Loftus Mine at Skinningrove and instead a less steeply graded line with two reversals, known as the Zig-Zag, was provided. Also the railway was extended from Carlin How across Kilton Viaduct to Loftus (earlier known as Lofthouse) where a connection could be made to the projected Whitby, Redcar & Middlesbrough Union Railway. The viaduct was 226 yards long with 12 piers supporting 13 lattice girders. It was designed by James Brunlees, Engineer, with piers for a double line although the superstructure only accommodated a single track. Later, mining subsidence resulted in the viaduct being buried in spoil to form an embankment. Ritson was again awarded the contract for the line. Mineral trains began running on 27/5/1867. Unfortunately, it is not known whether Ritson used any locomotives on the work.

JOHN SCOTT

John Scott had been the principal of his father's business, Walter Scott & Co, but then began to tender for contracts on his own account.

LONG NEWTON SERVICE RESERVOIRS CONTRACT, Long Newton C40
NZ 368138 NZ 363169 – **Map D**

John Scott's tender of £168,062 for the Long Newton Reservoirs was accepted by the Tees Valley Water Board on 8/8/1900. Scott constructed a two mile standard gauge line north from Grasspool Sidings (NZ 368138) to the site of the reservoirs (NZ 363169). The sidings were provided east of Oak Tree Junction on the Darlington-Middlesbrough line near the present Teesside Airport; the formal sidings agreement was dated 20/4/1901 (although the connection was probably made earlier). By 2/1901 Scott was laying 2ft gauge lines on the site of the reservoirs to assist with excavation and getting clay. The design had to be changed to a concrete construction because of poor ground conditions, but good progress was made and a ceremonial opening took place in 9/1905. The standard gauge line brought in nearly 250,000 tons of material comprising mainly cement and also slag from the nearby Dinsdale Iron Works

Bowtell mentions "a low built locomotive with outside cylinders and square windows in the back of the cab was photographed in June 1901 at work on standard gauge tipping wagons at the Long Newton embankments; presumably this was one of the three narrow gauge Black, Hawthorn 0-6-0 saddle tanks converted to standard gauge".

Reference : *Dam Builders' Railways from Durham's Dales to the Border,* Harold D. Bowtell, Plateway Press, 1994, pp.18-24

Gauge : 4ft 8½in

TEES VALLEY	0-6-0ST	IC	MW	1594	1903	New	(1)
MIDDLESBROUGH	0-6-0ST	IC	MW	1598	1903	New	(2)
-	0-6-0ST	OC	BH?			(a)	s/s

(a) earlier Walter Scott & Co, Baldersdale Reservoirs, near Cotherstone, contract, here by 6/1901.
(1) to John Scott, Grassholme Reservoir contract, Co. Durham.
(2) possibly to John Scott, Vittoria Dock contract, Birkenhead, Cheshire.

MIDDLESBROUGH DOCK CONTRACT, Middlesbrough C41
NZ 504205 – **Map G**

This was the third and final expansion of Middlesbrough Dock. Parliamentary authority was gained in 8/1897 and John Scott's bid of £382,559 for the contract was accepted by the NER. Work commenced in 1898 and was mostly completed in 1902 resulting in an increased water area of 25½ acres and quays totalling 6,394 feet in length. The project was carried out piecemeal to minimise disruption to maritime and rail traffic. The first part to be tackled was the creation of a middle arm of new water. According to the reference, "The excavated material was disposed of at sea by hopper barges. The steam navvy loaded spoil into end-tip wagons holding about 3 cubic yards (2.3 cubic m) which were taken, seven at a time, by locomotive up a 1:20 gradient to a temporary wooden staging on the west side of the dock where they were discharged into the hopper barges." The old locks now came within the dock area and a new entrance lock was constructed, together with a replacement 218 feet hydraulically operated swing bridge manufactured by Sir W.G. Armstrong, Whitworth & Co Ltd. It was designed to be strong enough to support the new NER E1 class (later J72) tank locomotives then coming into use. The steam cranes were replaced by electric cranes, a second coal hoist was provided and an area on the north side of the dock was filled in to give more space for wagon storage. Both COTHERSTONE and MOUNTAINEER were constructed as 3ft 0in gauge locomotives but were rebuilt before coming to Middlesbrough. In the *Contract Journal* on 1/4/1903, John Scott offered an 11 inch '6 wheel' Black, Hawthorn locomotive for sale which was working on the Middlesbrough Dock Contract.

Reference : 'The History of Middlesbrough Dock', D.W. Pattenden, *The Cleveland Industrial Archaeologist* 18, CIAS, 1986, p.p. 10-22

Gauge : 4ft 8½in

COTHERSTONE	0-6-0ST		BH	1055	1891	(a)	?
MOUNTAINEER	0-6-0ST	OC	BH	844	1885	(b)	(1)
THE CONTRACTOR	0-6-0ST	IC	BH	1115	1896	(c)	(2)
DARLINGTON	0-6-0ST	IC	MW	1309	1895	(d)	(3)

(a) ex Walter Scott & Co. The locomotive had worked on the Hury and Blackton Reservoirs contract in Baldersdale, Yorkshire (NR), until 1896-97.

(b) ex John Scott, Cat Castle Quarry, Lartington, Yorkshire (NR). The locomotive had previously been on the Baldersdale contract and moved to Cat Castle Quarry after the former opened in 1896. Stone from the quarry was used on the Middlesbrough Dock contract.

(c) the locomotive was purchased by John Scott for his "Derwentwater & Consett Railway contract" according to BH records. Harold D. Bowtell considered that it worked on Scott's Hisehope Reservoir contract near Waskerley, Co. Durham. M. Cook suggested (*The Industrial Locomotive* 28) that it then moved to a Middlesbrough contract, presumably the extension of Middlesbrough Dock.

(d) ex John Wilson & Son, Mirfield-Heaton Junction Railway contract, Yorkshire (WR); this line opened 1/10/1900.

(1) to Low Laithes Colliery Co Ltd, Low Laithes Colliery, near Dewsbury, Yorkshire (WR), c/1903.

(2) the locomotive was next shown as working for Sir Robert McAlpine & Sons Ltd, Port Rainbow Brickworks, Bromborough, Cheshire, possibly on hire, before moving to John Scott & Co's Vittoria Dock contract at Birkenhead, Cheshire.

(3) to John Scott & Co, Grassholme Reservoir Contract, Yorkshire (NR) by 4/1902.

WALTER SCOTT & CO

SALTBURN EXTENSION RAILWAY CONTRACT

C42
NX 654213 NZ 686194 – **Map I**

The 3¼ mile railway ran from Saltburn Junction (later Saltburn West) to Brotton on the former Cleveland Railway. It was authorised by an Act of 1/7/1865 and gave the NER better access to the ironstone traffic. A major feature was the impressive 783 feet long brick viaduct of eleven arches which carried the line 150 feet above Skelton Beck at Saltburn. Work commenced, the initial contractor being N.B. Fogg & Co, but Walter Scott then took over the contract in 1869 for £49,828. During 1871, the NER checked that the Hob Hill ironstone mine workings would not affect the stability of the line. The railway was opened to mineral traffic 1/6/1872, but passenger services did not commence until 1/4/1875.

The impressive viaduct over Skelton Beck valley at Saltburn was constructed by Walter Scott & Co about 1870; the line opened for mineral traffic in 1872. (Kirkleatham Museum, Redcar & Cleveland Council collection)

Reference : 'Walter Scott (1826-1910): Civil Engineering Contractor', John F. Addyman, *North Eastern Express* 186, NERA, 5/2007

Gauge : 4ft 8½in

WALTER SCOTT	0-6-0ST	IC	MW	179	1866	(a)	(1)

(a)   ex B.C. Lawton, contractor, Team Valley Railway, Gateshead, Co Durham, according to *Contractors' Locomotives Part IV*, F.D. Smith & D. Cole (Union Pub,1970).

(1)   it was later at J. & W. Stone, North Blaina Colliery, Blaina, Monmouthshire, after 8/1882, by early 1890s.

DOCK CONSTRUCTION CONTRACT, Hartlepools   C43
NZ 515333 NZ 518339 – **Map B**

Following requests from local shipowners and merchants, the NER agreed to provide more facilities at the Hartlepools and invited tenders in 1871 to construct new enclosed deep water docks with entrance locks. The contract was won by Walter Scott with work starting in 1872 and the five acre Union Dock (NZ 515333) was being excavated in 1873. By 9/1876 Scott had 400 to 500 men constructing the Union and Central Docks, together with the North Basin (NZ 518339) which would connect the two systems at West Hartlepool and Hartlepool. In order to provide the lock that would link the North Basin to the Old Harbour, a substantial coffer dam was erected which cut the NER dock line at this location and a temporary 400 feet long viaduct was built to carry the railway and a road. The contract was completed in 1880. No locomotives are known but there are likely to have been a number.

NEW RAILWAY BRIDGE ACROSS RIVER TEES CONTRACT, Thornaby   C44
NZ 447179 – **Map E**

The NER let the contract for the new bridge and associated works to Walter Scott in 10/1880 for £35,969 and, four months later, for the new South Stockton Station (£8,211). The station's name was changed to Thornaby on 1/11/1892 shortly after Thornaby was created as a separate municipal borough. The details of the contract included erection of a bridge across the River Tees on the north side of the existing bridge; lowering and widening of the railway from the Tees Bridge to a point about 240 yards from the present station; construction of two overbridges and diversion of roads over them and provision of a "carriage drive" to the intended passenger station. In the list of plans was "No.26 site for deposit of surplus excavation at Newport". Details of any locomotives used have not been located.

STOCKTON-ON-TEES BOROUGH COUNCIL

DIRECT WORKS PROJECTS, Stockton

With over 10,000 unemployed people in the borough, a number of direct works projects were carried out during the interwar period to provide relief. Each unemployed person taken on was provided with work for eight weeks. The following schemes were undertaken involving railways. Alex Bettaney obtained the information from published Council minutes.

Grangefield Road Extension Project   C45
Approximately NZ 435192 – **Map D**

The work on the west side of Stockton saw the creation of an "arterial road" running north from Hartburn Lane in the south to Durham Road in the north and forming the present A1027 road. These schemes were respectively Hartburn Avenue, Oxbridge Avenue and Bishopton Avenue. Grangefield Road left this arterial road and was extended east to Tynedale Street near Light Pipe Hall. On 18/12/1924, the Borough Engineer reported that the narrow gauge railway was expected that week. The main use of the railway was to move spoil from the Moor Iron Works tip to raise the low-lying parts of the route especially near Lustrum Beck where a culvert was to be constructed. By 20/1/1925, the narrow gauge railway was in full use with approximately 600 tons of fill having been deposited. Reports on progress continued until at least 16/6/1925.

Oxbridge Avenue Extension Project, Stockton   C46
NZ 429192 – **Map D**

In 4/1925 the Borough Engineer reported on the preparations for this project. It was intended to extend the existing narrow gauge railway on the Grangefield Road scheme to take most of the fill from the Moor Iron Works tip. The track was to run within the boundary of Rudd playing field, pass along the south side of Grangefield Road and cross it near the Stockton Cricket Ground to reach the site of the new road. He also stated "I think it would be economical to employ a petrol driven engine to haul the wagons". The 'new' locomotive, together with a "shelter", was ordered from J.C. Oliver Ltd at a cost of £480 10s 0d. This was a reconditioned 60cm gauge, 20hp Simplex petrol locomotive delivered to the LNER Stockton Station. On 15/12/1925 the locomotive was stated to be working satisfactorily; 103 tons of fill being moved daily. It had been agreed on 22/10/1925 to inspect a "Simplex petrol engine engaged on certain work near Stokesley", but this does not appear to have resulted in its acquisition.

Gauge : 60cm

-	4wPM	MR	4002*	1925	New	s/s

* This was a rebuild incorporating parts of earlier locomotives.

Bishopton Road to Durham Road Project, Stockton   C47
NZ 430201 – **Map D**

On 20/5/1931, it was reported that "Decauville" track was used from Bishopton Road to Green Lane; the latter ran to the west of Durham Road. A Ministry of Transport Inspector stated on 8/7/1931 that it would be an "advantage to employ another petrol locomotive and additional railway track and tip wagons, as some had been withdrawn from the scheme for use on Hartburn sewers". This suggests that there was already a petrol locomotive in use on this scheme. It was minuted that 700 linear yards of track and 18 wagons had been purchased. The project was completed in June 1932, although use of the railway had ceased earlier.

Resewering Lustrom Beck Project, Stockton   C48
Map D

On 22/10/1931, it was stated that the "Jubilee Wagon Track to be shortly removed" and the project was to be completed by 3/1932.

Church Row Improvement Project, Stockton   C49
NZ 446193 to NZ 454197 – **Map D**

This involved the construction of a major road from Church Row near the centre of Stockton under the North Shore Branch, along the north western boundary of the Malleable Works to connect with Portrack Lane. Equipment was being purchased/hired in 11/1934 and work started in the following month. A Ruston Bucyrus 10–RB excavator was purchased and delivered on 26/1/1935. Eight lorries were hired to move spoil and materials, however, substantial amounts of spoil had to be taken out where the new road passed under the railway. A 60cm (or 2ft 0in gauge) railway was installed using V-shaped side tipping skips and, according to the Council, this saved £1 to £2 per day compared to lorry transport. A photograph in an issue of *Steaming* shows Stockton Council's Clayton & Shuttleworth (makers' number 47750 of 1919) 5 ton steam tipper waggon mounted on timber baulks and with the rear axle acting as a winding drum for hauling skips up an incline from the excavation. A Simplex petrol locomotive also appears ready to take wagons for tipping, but this has not been identified. A feature on arterial roads in Stockton appeared in the *Contract Journal* on 16/12/1936 which stated that two petrol locomotives were used on spoil removal. On 21/5/1936, it was reported that the locomotive driver "Spackman" and the chargehand at Dog Hill Farm tip, Portrack had been laid off.

Reference :   Six photographs of the work underway appear in *Teesside and Old Cleveland A Further Chapter*, Robin Cook, European Library, 1995

Gauge : 60cm or 2ft 0in

-	4wPM				?	?

STRABAG SE

WOODSMITH MINE CONSTRUCTION OF TUNNEL WILTON-LOCKWOOD BECK   C50
NZ 580227 NZ 673142 – **Map H**

The Woodsmith Mine, located three miles south of Whitby, was initially developed by Sirius Minerals plc to extract polyhalite at a depth of approximately 1,500 metres. Polyhalite is a type of potash that is used as a fertiliser to improve crop yields. The area is reputed to contain the largest reserves of high quality polyhalite in the world. Planning approval for the mine was received in 6/2015 and a detailed feasibility study completed 17/3/2016.

All of the mined polyhalite will be transported underground on a high-capacity conveyor belt through a 37 km long and 4.9m wide tunnel at an average depth of 250m to a "material handling facility" at Wilton (NZ 580227). Here it will be processed before being conveyed to ships at a wharf in Teesport. Eventually the tunnel will also contain a railway to be used by maintenance crews. Although the mine and much of the tunnel will be dealt with later in the projected IRS North Yorkshire Handbook, the section of tunnel from Wilton to Lockwood Beck (NZ 673142) is located in Teesside. The first tunnel boring machine (TBM) arrived in 2/2019 and was officially 'launched' as STELLA ROSE in the shallow portal at Wilton in 4/2019. An Austrian company, Strabag SE, is constructing the tunnel the 13km to Lockwood Beck and is using a 900mm gauge railway to transport man-riders, grouting materials, fixing hardware and the concrete ring segments for supply to the TBM. Sirius Minerals plc was unable to secure finance for stage 2 development which resulted in a slow down on site, although the Strabag SE contract continued. From 17/3/2020, Sirius Minerals was taken over by Anglo American plc which is intent on completing the project. As at 2/2022, work on establishing the mine had slowed but construction of the tunnel continues eastwards past Lockwood Beck.

Gauge : 900mm

-	4wBE	Schöma	7084	2019	(a)
-	4wBE	Schöma	7085	2019	(a)
-	4wBE	Schöma	7086	2019	(b)
-	4wBE	Schöma	7087	2019	(b)
-	4wBE*	Schöma	6827	2019	(c)
-	4wBE	Schöma	7103	2019	(d)
-	4wBE	Schöma	7104	2019	(d)

* Also fitted with a supplementary diesel engine running as required to charge the batteries.

(a) new. These battery locomotives were the first of their type; they are thought to have been delivered about 4/2019.
(b) new, arrived between 9/4/2019 and 12/4/2019.
(c) the full height CEL 180H prototype had trials at Stuttgart in Germany but was then reworked at Schöma's Diepholz Works with a reduced height cab. It was delivered in 2-3/2019 to Strabag for running-in and staff training at the contractor's winding-down KAT2 project in Austria. Ex Strabag/Jager Bau JV, Koralmtunnel [KAT], Baulos 2 Central Section, Austria, /2019, after 21/6/2019
(d) new, arrived between 30/7/2019 and 3/8/2019.

TEES CONSERVANCY COMMISSIONERS

The mouth of the River Tees had once been a wide expanse of shallow meandering channels, salt marshes and sandbanks. In the days of sail it represented a serious hazard, particularly when a north-easterly gale blew. During one storm in February 1861 over 50 ships were lost on the seven mile long sandbank, known as the Bar. The TCC decided to construct breakwaters on both sides of the estuary to improve the navigation of the river and act as a refuge for shipping.

South Gare Breakwater Construction, Redcar C51
NZ 573247 NZ 558282 – **Map H**

There was already experience upstream of constructing training walls composed of slag to improve the river channel and reclaim land. Therefore, in 8/1861, the TCC invited tenders for the building of a slag breakwater. In the event, the TCC carried out the work assisted by its engineering consultant, John Fowler. In 8/1863 a number of ironmasters agreed to pay 2d per ton on the make of iron for the TCC to dispose of their slag. Fowler placed orders for 485 wagons at £15 each and the SDR agreed to haul them over its line to Tod Point (NZ 573247). The latter was a coastal promontory, beyond which lay Bran Sands and the South Gare sand bank. The TCC constructed a single standard gauge line from the NER at Tod Point extending north westwards as the breakwater was erected. A stable for two horses was provided at Tod Point. After some weeks, the slag trains were found to be interfering with other traffic on the line and the TCC paid the railway company to lay an additional track alongside the existing line from Cargo Fleet Junction to Tod Point Junction; this was still shown as the "slag line" on an 1895 NER plan. The iron companies initially hired the wagons from the TCC. In 12/1869 the figures were Cochrane 30, Gilkes, Wilson & Pease 60, Clay Lane Iron Co 100 and Bolckow, Vaughan 100 but, in 1871, the iron companies bought many of them. Much of the slag for use in constructing the training walls of the lower Tees and South Gare was cast into iron bogies. These had separate bodies with a removable end piece to form the mould. The latter was removed after the slag had solidified and the slag remained as a rectangular block weighing about 3½ tons on the bogie.

The foundation stone was laid on 3/11/1863 and, during one period, the breakwater was advancing 23 feet each day with 800–900 tons of slag being deposited. The breakwater was described as a "Portland cement concrete structure upon a foundation of slag with slag hearting between the exterior walls." Initially, just slag was tipped but, once deeper water was reached, a large concrete wall was erected on the seaward side with the slag being placed behind it. A gale in 1880 swept away nearly 100 yards of concrete wall. As the TCC's railway increased in length, locomotives were introduced to haul the slag bogies. Details of TCC locomotives are listed under the North Gare entry. South Gare breakwater was officially opened on 25/10/1888; it was 4,200 yards long, had absorbed 3,900,000 tons of slag and cost £220,000. The TCC's railway ran from Tod Point to the lighthouse on the end of the breakwater, a distance of two miles 51 chains, where a loop was provided for locomotives to run round. Redcar Iron Works had its own railway on the seaward side of the TCC line serving

The South Gare Breakwater was constructed by the Tees Conservancy Commissioners. The locomotive is believed to be Fletcher Jennings 111 of 1873 because it displays several of that builder's features, although the tank may not be original. *(Robin Cook collection)*

its slag tips and crossing over the TCC track (with a connection) to Redcar Jetty. Short sidings off the TCC line connected with the WD 'Submarine Miners' Depot (which see) and the powder wharf. As well as explosive materials for the WD, coasters from Belgium unloaded black powder at the wharf. The powder boxes were unloaded using a hand crane housed in a transit shed and then transferred into a railway powder van. This had been backed down an incline to the shed and the end of a long rope attached to a Redcar Iron Works locomotive. Once loaded, the van was pulled up the incline to a nearby siding. When several railway vans had been loaded, the Redcar Iron Works locomotive, suitably separated by wagons from the vans, hauled them to Tod Point sidings to await a NER locomotive for despatch to the mines. In 1895 the wharf was identified as Curtis & Harvey's powder siding but the trade probably finished with the First World War.

The railway along South Gare remained in use until about 1939 with LNER locomotives probably delivering and collecting what little traffic remained. A photograph taken circa 1930–31 shows LMS and NE (LNER) 13 ton open wagons on the line and a LNER Appendix dated 1936 specifies which of its locomotives were permitted to work along the South Gare Breakwater Branch from Tod Point. At least two wooden sail bogies, apparently converted from platelayers wagons, were used to convey small groups of people along the line. There was a block making plant for carrying out repairs to the end of the breakwater. Internal wagons were used but presumably horses were the motive power. With the outbreak of World War 2, the Government wanted the rundown railway at South Gare to be repaired, but the TCC would not contribute and so most of the line was lifted and replaced by a road paid for by the Government. One or two short sections remained connected to Dorman Long's Redcar Works slag tipping sidings.

North Gare Breakwater Construction, Seaton Carew   C52
NZ 536283 NZ 545285 – **Map A**

Even before the South Gare breakwater was finished, construction of that at North Gare began in 1882. However, there had been substantial reclamation of the intervening land before work could start. This had begun with the establishment of Bell Bros' Iron Works at Port Clarence in 1853 and slag was tipped to extend its site north and eastwards into the estuary mud flats. The TCC then embarked on a major land reclamation project at Saltholme and Greatham which, together with smaller areas at Haverton Hill and Portrack, enclosed 2,523 acres of former marsh and estuary. It required the construction of 14 miles of embankment. The work had to be undertaken during neap tides and, on occasions, up to 600 men and horses worked frantically "to beat the tide". At a meeting of the TCC's Works Committee held on 15/7/1876, its Engineer was instructed to "purchase

The reclamation of the marshes on the north side of the Tees Estuary and the construction of the North Gare Breakwater involved moving masses of slag while coping with the incoming tides. One of the locomotives involved is seen in this graphic scene. *(Courtesy Ian Macdonald)*

the locomotive engine of Mr C.G. Johnson, of Middlesbrough, for the sum of £860 delivered at Port Clarence for the use of the Saltholm [sic] and Seaton Snook reclamation work": (*Northern Echo* 8/8/1876).

The main embankment extended north from Bell Bros' Works and eventually curved round to form the north bank of what became Greatham Creek to reach Seaton Snook. From here, the embankment was constructed across North Gare sands to the site of the breakwater. A railway line laid along the top of the embankment enabled Bell Bros' locomotives to haul wagons of slag for tipping as part of the project. At Seaton Snook, a wharf was built to allow slag to be brought in on barges from other iron works which TCC's locomotives could transport to the construction sites. A rare balanced cantilever Titan crane, built by Appleby Brothers, was erected at Seaton Snook. The hot balls of slag were transported in barges, each of which could carry 40 bogies representing about 220 tons burden. A siding enabled a locomotive to collect the bogies that had been lifted out by the crane. Nearby, a row of seven cottages was provided for the workers and their families; in 1891 sixty people lived here including a man for operating the crane, a locomotive driver and fireman, together with a platelayer. The railway line ran on to the partially completed breakwater and a triangle allowed the locomotive to push its wagons along the structure for tipping. It was originally intended to construct the breakwater for 2,000 yards but, in 1891, when just over half of this length (1,110 yards) had been completed, it was decided to go no further; the final cost was £69,783. Much of the railway line along the embankment and breakwater was retained up to the First World War presumably to facilitate any remedial works.

Reference :   'South Gare', Cliff Shepherd, *Industrial Railway Record* 121, IRS, 6/1990, pp.135-40

  A History of the River Tees, M. Le Guillou, Cleveland County Libraries, 1978, pp.44-49

Gauge : 4ft 8½in

No.1	SOUTH GARE	0-4-0T	OC	FJ	111	1873	New	
	Rebuilt	0-4-0ST						s/s
No.2		0-4-0ST	OC	HE	80	1873	(a)	(1)
No.3		0-4-0ST	OC	B		1876	New	(2)
No.4		0-4-0ST	OC	HH		1876	(b)	(2)

(a) ex West Hartlepool Iron Co Ltd, Hartlepool Iron Works, Hartlepool, by 10/1875.
(b) ex Bell Brothers Ltd, Clarence Iron Works, Port Clarence, /1877.

(1)   by 9/1899, it was being used by J.T. Firbank, contractor, on the LSWR Basingstoke-Alton railway contract (1898-1901).
(2)   offered for sale 4/1900.

The above locomotives probably worked at South Gare before some transferred to assist with the construction of North Gare.

I. THOMAS

Grangetown   C53
Map H

The *Contract Journal* on 6/9/1933 contained an announcement that I. Thomas was to sell on 28/9/1933 at Trinidad Lake Asphalt Co Ltd two tank locomotives and two 10hp McP petrol locomotives, all of 2ft gauge, following completion of contracts.

TRACKWORK LTD

ALTERATIONS TO RAILWAY AT TEES DOCK, Grangetown   C54
NZ 555235 – Map H

The contract was let by the Tees & Hartlepool Port Authority Ltd to Trackwork Ltd of Doncaster to revise the railway layout at Tees Dock. This involved the relocation of the six exchange sidings, some of which now form part of the Freightliner Terminal, a revised curve to accommodate the extension of the Ro-Ro terminal and a spur to the potash unloading hopper. A temporary ballast loading siding was constructed and a 'Dogfish' 24 ton ballast open wagon and a Fairmont tamping machine used.

Gauge : 4ft 8½in

(0302) 365222   4wDM   RH   398616   1955   (a)   (1)

(a)   ex Trackwork Ltd, Plant Depot, Doncaster, South Yorkshire; the locomotive is thought to have arrived c3/1997.

(1)   to Trackwork Ltd, Hartlepool Docks contract, after 10/8/1997, by 25/8/1997.

Ruston & Hornsby 398616 of 1955 was working on the Trackwork Ltd contract at Hartlepool Dock on 31st August 1997.   *(Alex Betteney photograph)*

ALTERATIONS TO RAILWAY AT HARTLEPOOL DOCKS, Hartlepool   C55
NZ 511341 – **Map B**

These changes to the railway at Hartlepool Docks were also carried out by Trackwork Ltd on behalf of the Tees & Hartlepool Port Authority Ltd. The layout was altered to accommodate a new road (A179, Marina Way) involving a revised alignment of the 'Harbour Branch' by a bridge over the road and a new embankment/run round loop into the dock sidings. A 'Dogfish' wagon, Permaquip ballast packer and tamping machine were also present.

Gauge : 4ft 8½in

(0302) 365222	4wDM	RH	398616	1955	(a)	(1)

(a)   ex Trackwork Ltd, Tees Dock contract, after 10/8/1997, by 25/8/1997.

(1)   presumed returned to Trackwork Ltd, Doncaster, South Yorkshire, c11/1997.

JOHN WADDELL & SONS

WHITBY, REDCAR & MIDDLESBROUGH UNION RAILWAY CONTRACT   C56
John Waddell to /1879   NZ 898105 NZ 716180 – **Map I**

Following the failure of John Dickson's contract (which see), the Whitby, Redcar & Middlesbrough Union Railway Co persuaded the NER to lease the partially constructed line from Loftus to Whitby Bog Hall Junction on 1/7/1875 and an Act in the following year approved some deviations of the line. The NER undertook to complete the railway and a contract for the 1½ miles from Loftus to Easington was let to "Mr Ridley" who commenced work in 11/1876 and finished in 7/1878. The main contract was awarded to John Waddell of Edinburgh on 7/11/1878 who reported that he was ready to start work in 1/1879. The *Whitby Gazette* on 13/9/1879 reported that a team of nine powerful horses and a strong staff of Waddell's men worked from 11.30pm on Wednesday until 8am on Thursday " in getting a powerful locomotive engine in steam from the bottom of Brunswick Street to the top. After turning the corner of the street, the nine horses were detached and the engine made good

One of Waddell's locomotives was on a construction train at the south end of Easington Tunnel on the Whitby, Redcar & Middlesbrough Union Railway contract. The retaining wall was erected after some 40,000 tons of material had slipped into the cutting.

(Photograph by Frank Meadow Sutcliffe, copyright The Sutcliffe Gallery, Whitby)

progress up St. Hilda's Terrace." The contractor was reported to have an engine shed at East Row. The nature of the terrain and the need to make good the defective workmanship meant that the first official train was not able to run over the line until 3/12/1883. The WR&MU was formally vested in the NER by an Act of 5/7/1889.

References : *The Whitby, Redcar & Middlesbrough Union Railway,* Ken Hoole, Hendon, 1981

Contractors' Locomotives Part VIII, R.W. Miller, Industrial Locomotive Society, 2008

'The Whitby, Redcar and Middlesbrough Union Railway', Reg Sowler, *The North Eastern Express* 117, NERA, 2/1990

Gauge : 4ft 8½in

-		0-4-0ST	OC	AB	215	1880	(a)	(1)
	TWIZZIE GILL	0-6-0ST	OC	FW	147	1872	(b)	(2)
	SEYMOUR CLARK	0-6-0ST	OC	FW	279	1875	(c)	(3)
	MINNIE	0-6-0ST	OC	FW	358	1877-78	(d)	(4)

(a) delivered new to East Loftus on 20/5/1880.

(b) it finished on a contract for the Furness Railway at Barrow Dock, Barrow c6/1879 and was recorded on this contract by 5/1881. A pencil note in the manufacturer's records mentions "Waddell, Easington".

(c) ex W.F. Lawrence, Banbury & Cheltenham Direct Railway contract (Seymour Clark was deputy chairman of that railway c/1879), but may have gone to Waddell's Riverside Branch contract (1874-79), Newcastle on Tyne before coming here because Lawrence gave up his contract late /1876 or early /1877. It was on the WR&MU contract by 7/1881 when it was the first locomotive to pass over Staithes Viaduct.

(d) ex J. Garlick, Sheffield, c/1881.

(1) may have gone to Waddell's James Watt Dock contract (1878-85) for the Greenock Harbour Trust, Greenock.

(2) possibly to Waddell's Scarborough & Whitby Railway contract (1882-85).

(3) to Waddell's Scarborough & Whitby Railway contract (1882-85).

(4) to Skinningrove Iron Company, Carlin How, (by 5/1884?).

C.J. WILLS & SONS LTD

TEES SUBMARINE BASE CONTRACT, Grangetown   C57
NZ 540231 – Map H

The Tenth Submarine Flotilla had been created in 12/1914 at Immingham and a Submarine Base was subsequently established on reclaimed land bordering the River Tees downstream from Bolckow Vaughan's Eston Wharf. Permission was granted in 7/1916 for work to commence on site. An unfenced straight road or track headed south eastward from the base for approximately one mile to reach Grangetown railway station. A total of about 20 submarines of four different classes operated from the Tees Base between 1916 and 1918. Their duties included mine laying, protecting the Tees and operations in the North Sea. The poor conditions for submarine crews led the Admiralty to draw up contracts in late 1916 for more permanent facilities. Presumably Wills & Sons was responsible for the extra piling at the quay and reclamation of land using crushed slag, with Henry Boot & Co constructing the buildings. The Base on land comprised the main building facing the river and three rows of other buildings stretching out behind. These were used for the storage of mines, torpedoes and equipment, workshops and living quarters for officers and crews. A submarine depot ship, HMS LUCIA, was anchored in the river; this was the former Hamburg-America line SPREEWALD that had been captured in 1914 and converted in 1916. There were also jetties for the submarines and two oil storage tanks constructed near the Eston Wharf.

The contractor C.J. Wills & Sons Ltd ordered a locomotive on 3/12/1917 for the "Tees Naval Base, delivery Bolckow Vaughan sidings, Eston Wharf, South Bank (for Admiralty Works Construction)" but it was not delivered until 2/8/1918 and was advertised for sale in *Surplus* on 15/5/1920 and 1/1/1921.

In *Hansard* on 9/4/1919, the Secretary to the Admiralty was reported saying that construction of the base had cost £150,000, that no decision had been made on its future, but that the quarters were of a "semi-permanent nature adaptable for storehouses". Later in the year, the base closed and HMS LUCIA sailed to North Russia to support naval operations against the Bolshevik government. Bolckow Vaughan purchased the two oil storage tanks in 1922, although they were used by the National Benzol Co Ltd; the latter had various 10 ton cylindrical tank wagons for its South Bank business registered to run over the LNER in 1926. The buildings at the base remained and gained the name "Teesport", not to be confused with the later Tees Dock, with some being used as houses. There were no permanent railways at the Base and it seems to have become rundown. When Cyril

Fisk visited on 7/10/1934 to take a religious service, the houses had all been condemned and slag tipping continued unabated around it.

Reference: *The Silent Warriors of Teesport*, John Harland Watson, pub FORCEM

Gauge : 2ft 6in

-		0-4-0ST	OC	WB	2068	1918	New	(1)

(1) to Associated Portland Cement Manufacturers Ltd, Bevan's Works, Northfleet, Kent where it was repaired and altered to 2ft 8½in gauge (order dated 12/5/1921).

LOCOMOTIVE DEALERS, REPAIRERS, HIRERS AND BUILDERS

It is not surprising that a number of businesses operated on Teesside dealing with secondhand industrial locomotives, given the amount of heavy industry and fleets of locomotives that were employed. These companies are listed in this section. Full details of all locomotives advertised in the commercial press are not included because some descriptions are too vague to identify specific locomotives and, in many cases, there is no guarantee that they refer to engines used on Teesside. For these details, the reader is referred to contemporary publications, such as *The Contract Journal*, *The Engineer*, *Machinery Market* and *Colliery Guardian*. A flavour of the advertisements is included at the end of this section under the respective dealers, although it is stressed again that some of the locomotives will not have been present on Teesside.

BURFORD TAYLOR & CO

SCRAP DEALER, Vulcan Street, Middlesbrough and Cargo Fleet   **D1**
NZ 496212 – **Map F**

Initially Burford Taylor had use of a siding (NZ 496212) on the south side of the Vulcan Street Branch situated adjacent to Tees Bank Engineering Works, Commercial Street. A Siding Agreement was signed with the LNER on 5/9/1923, in which the railway company would deliver traffic to a nearby siding but all traffic to and from its works was to be handled by Burford Taylor or some other traders on its behalf "by means of private locomotives". Burford Taylor offered three locomotives for sale in *The Contract Journal* – 2ft gauge Hudson steam locomotive (28/4/1927), a Hudswell Clarke 0-4-0 with 13in x 20in cylinders, rebuilt /1924 (22/6/1927) and a 40hp 2ft 0in gauge Simplex that had only worked for four months (14/10/1931).

On 7/11/1945, Burford Taylor leased a site from the railway company south of the LNER Dock Stockyard Sidings and west of, and adjacent to OME's Cargo Fleet timber yard. The LNER Goods Department noted on 24/1/1947 that several wagons had been delivered to Burford Taylor's "scrap dump". These would have had to travel over MOR metals, either by Dock Hill or the Dock Lock bridge and so tolls would have been payable.

CLARKE BROS

MILTON WORKS and BALTIC WORKS, West Hartlepool   **D2**
NZ 515317 – **Map B**

Clarke Bros was variously described as Marine Stores Dealers (1917–18) and Metal and Timber Merchants (1924–25). The business was stated to be at the Milton Works according to *Machinery Market* in 1915 and 1919 but occupied the "Brunswick Siding" at West Hartlepool in the 1925 RCH *Official Handbook of Railway Stations etc*. However, by 1928, Clarke Bros had a siding at the Baltic Saw Mills (NZ 515317) located in the angle between the LNER Cliff House and Burn Road Branches. Clarke Bros advertised wanting plant, including a 12in locomotive, on 26/2/1915. In *Machinery Market* 19/9/1919, it offered for sale four Black, Hawthorn 0-4-0 locomotives suggesting that two good engines could be made out of the four. It also advertised on 4/5/1928 wanting a 15½ to 16½ inch locomotive, presumably on behalf of a client.

DOUGLAS ENGINEERING SERVICES LTD

TEES DOCK, Grangetown   **D3**
NZ 555225, NZ 547230 – **Map H**

The company acquired these redundant diesels from British Coal with a view to resale. They were kept in a compound at NZ 555225 at Teesport and then moved to Tees Dock at NZ 547230 between 1/12/1992 and 16/3/1993.

Gauge : 4ft 8½in

No.1		0-6-0DH	AB	608	1976	(a)	(3)
No.2	20/110/712	0-6-0DH	AB	616	1977	(a)	(1)
No.3		0-6-0DH	AB	614	1977	(a)	(2)

(a)   ex British Coal Corporation, Easington Colliery, Co Durham, 2/1989.

Ed Murray & Sons Ltd has been active locomotive hirers in recent years. Its number 261 (GECT 5430/1977) was stood at ICL UK (Cleveland) Ltd's Tees Dock terminal after overhaul on 5th November 2017.

(Alex Betteney photograph)

(1) repurchased by British Coal Corporation and to Hawthorn Combined Mine, Murton, Co Durham, 11/4/1990.
(2) to Tees & Hartlepool Port Authority Ltd, Tees Dock, Grangetown, c3/1993, by 16/3/1993.
(3) to Teesbulk Handling Ltd, Tees Dock, Grangetown, c1/1994, after 15/1/1994.

G.N. STEAM CO LTD

Middlesbrough, Grangetown   D4
NZ 519200, NZ 549209 – **Map G**

The company was established to mainly repair and construct miniature locomotives. It was initially based in the same premises as Graeme Walton-Binns (which see) but then moved to Teesside Power and Transmissions at Cargo Fleet (NZ 519200) and later to the Bolckow Industrial Estate, Grangetown (NZ 549209) from circa 11/2000. The company built the replica EFFIE and began work on two 1ft 7in gauge 0-4-0 tank locomotives (maker's numbers 20–21) for the Great Laxey Mine Railway. However, the company moved to a new base in Darlington from 1/12/2002 where the latter two locomotives were completed. When at Darlington, the business changed its name to Great Northern Steam Co Ltd (registered 21/1/2009).

Gauge : 1ft 3in
       EFFIE   0-4-0WT   OC   GNS   1999   (a)   (1)

(a) new, to Darlington Railway Preservation Society, Hopetown on 9/10/1999 for viewing in steam on 23/10/1999 before returning in 4/2000.
(1) to Ravenglass & Eskdale Railway, Ravenglass, Cumbria, by 26/4/2000.

W.H. LOVERIDGE & CO

IRON MERCHANT, West Hartlepool D5
Map B

Loveridge was an iron merchant who occupied part of the Milton Iron Works (see George Bower) and formed the Milton Forge & Engineering Company; a Siding Agreement was dated 2/11/1894. However, some years earlier, Loveridge had advertised in *The Contract Journal* on 12/1/1887 two tank locomotives with 9 and 10 inch cylinders for sale. It is not known whether these were from the Milton Iron Works or elsewhere. Interestingly, just over a year later, on 1/5/1888, Loveridge was seeking a cheap secondhand close-coupled tank locomotive of modern design.

Loveridge's office was located in Greatham Street, next to the Milton Works, according to directories for the 1898–1905 period. After William Gray took over the Milton Forge & Engineering Co, Loveridge occupied a siding (NZ 515307) immediately north of where the short Longhill Branch left the NER Cliff House Branch and adjacent to the Rivet, Bolt & Nut Co's Works. There was a siding agreement for Loveridge dated 20/3/1902 but this was amended by the NER to the Rivet, Bolt & Nut Co on 12/5/1906 so presumably the siding had been absorbed into the latter's premises.

ED MURRAY & SONS LTD

CASEBOURNE ROAD DEPOT, Hartlepool D6
NZ 511313 – Map B

Ed Murray is a haulier which had the contract for transporting steel pipes by road from the then Corus Pipe Mill in Hartlepool. In 2003 the company also took over responsibility for shunting rail traffic at the pipe mill, purchasing the Moyse locomotives from Stockton Haulage Ltd in Middlesbrough. For a limited period, the name 'Astontrack' was used for marketing purposes. Some repairs to the locomotives are carried out at Murray's depot at Casebourne Road on the Longhill Industrial Estate and a second depot on Brenda Road, Hartlepool with the locomotives being transported by road. Murray subsequently won other local contracts to provide locomotives for shunting, as well as purchasing the redundant GECT diesel shunters from Lackenby (see Sahaviriya Steel Industries entry).

Gauge : 4ft 8½in

MURR 1	4wDE		Moyse	1364	1976	(a)	(1)
MURR 2	4wDE		Moyse	1464	1979	(b)	(2)
3 (2)	4wDE		Moyse	1365	1976	(c)	
-	0-4-0DH		HC	D1346	1965	(d)	(4)
-	0-4-0DH		HE	7425	1981	(e)	(3)
(DH26) (01560)	4wDH		RR	10229	1965	(f)	(8)
-	4wDM	R/R	Unimog	083216	1982	(g)	(6)
-	0-6-0DH		S	10166	1963	(h)	(5)
LOCO NO.9	4wDH		TH	287V	1980	(i)	
8 0155 JAMES	4wDH		TH	288V	1980	(j)	
271	6wDE		GECT	5469	1978	(k)	(7)
-	0-6-0DH		RR	10255	1966	(l)	
306	0-6-0DH		GECT	5383	1973	(m)	
LADY POTTER	0-6-0DH		RR	10214	1964	(n)	(9)
No.4	4wDH		TH	200v	1968	(o)	

(a) ex Stockton Haulage Ltd, Middlesbrough, 24/9/2003. IRS records suggest that it went direct to the Corus plc Pipe Mill, but Murray states that it came to the yard first for repairs.
(b) ex Stockton Haulage Ltd, Stranraer, Dumfries & Galloway, c10/2003. It may have moved direct to the Corus plc Pipe Mill.
(c) ex Stockton Haulage Ltd, Middlesbrough, 10/2003. The locomotive then stood partially dismantled at the Casebourne Road depot.
(d) ex CJC Chemicals and Magnesia Ltd, Hartlepool, 25/5/2005.
(e) ex CJC Chemicals and Magnesia Ltd, Hartlepool, 5/2005.
(f) ex Bowes Railway Co Ltd, Springwell, Co Durham, (where had been the property of Staffordshire Locomotive Co Ltd), 14/6/2008. To Brenda Road Depot, Hartlepool, after 4/2009, by 18/10/2009. Ex Liberty Pipes (Hartlepool) Ltd, Hartlepool Pipe Mill, Hartlepool, w/e 4/4/2020.
(g) ex the Potter Group, Selby Storage & Freight Co Ltd, Selby Distribution Centre, Selby, North Yorkshire,

1/2014. While for subsequent sale, it was demonstrated at Tata Steel's Hartlepool Pipe Mill e.g. on 16/6/2014.
- (h) ex Lincolnshire Wolds Railway, Ludborough, Lincolnshire, c11//8/2014.
- (i) ex T.J. Thomson & Son Ltd, Millfield Works, Stockton, w/e 21/1/2017.
- (j) ex T.J. Thomson & Son Ltd, Millfield Works, Stockton, after 22/2/2017, by 11/6/2017.
- (k) ex Sahaviriya Steel Industries UK Ltd (in liquidation), Teesside Works, Lackenby, 9/6/2017.
- (l) ex W.H. Bowker Ltd, Selby Distribution Centre, Selby, North Yorkshire, w/c 15/10/2018.
- (m) ex Liberty Pipes (Hartlepool) Ltd, Hartlepool Pipe Mill, Hartlepool, c1/2020; after 9/12/2019, by 8/2/2020.
- (n) ex Liberty Pipes (Hartlepool) Ltd, Hartlepool Pipe Mill, Hartlepool, w/c 27/7/2020.
- (o) ex Chasewater Light Railway and Museum Co, Brownhills, Staffordshire, c3/2021.

- (1) to Corus plc, Hartlepool Pipe Mill, Hartlepool, on hire, 24/9/2003.
- (2) to Corus plc, Hartlepool Pipe Mill, Hartlepool, on hire, 10/2003.
- (3) to Corus plc, Hartlepool Pipe Mill, Hartlepool, for storage, by 10/1/2007.
- (4) locomotive was exchanged with RR 10229 becoming the property of Staffordshire Locomotive Co Ltd, which removed various parts and then sent it to T.J. Thomson & Son Ltd, Millfield Works, Stockton, for scrap, about 11/6/2008.
- (5) to Cleveland Potash Ltd, Boulby Mine, Easington, on hire, 13/10/2014.
- (6) to Trac International Ltd, Chesterfield, Derbyshire, 6/4/2015.
- (7) most of it scrapped on site, 30/9/2017.
- (8) to British Steel Ltd, Skinningrove Works, Carlin How, 26/6/2020.
- (9) to British Steel Ltd, Teesside Beam Mill, Lackenby, after 28/2/2021, by 4/3/2021.

BRENDA ROAD DEPOT, Hartlepool

D7
NZ 512291 – **Map A**

Ed Murray & Sons Ltd established a second depot at Brenda Road in c/2009.

Gauge : 4ft 8½in

| - | (10560) | 4wDH | RR | 10229 | 1965 | (a) | (1) |
| - | - | 0-4-0DH | HE | 7425 | 1981 | (b) | |

HUMBER (AB 1800/1923) had worked at Casebourne's cement works before becoming ICI's number 57. The photograph is thought to have been taken in Ridley Shaw's yard before the locomotive moved to ICI's Winsford works in 1953. *(E. Haigh photograph, IRS collection)*

LADY POTTER	0-6-0DH	RR	10214	1964	(c)	(2)
HARRY POTTER	0-6-0DH	RR	10220	1965	(d)	(3)

(a) ex Casebourne Road Depot, Hartlepool, after 4/2009, by 18/10/2009. To Parsons Paint Shop, Brenda Road, Hartlepool, for repainting probably w/c 14/11/2011 and returned.
(b) ex Corus plc, Hartlepool Pipe Mill, Hartlepool, c7/2007, via Bonlea Fabrications, Brenda Road, Hartlepool, for shot blasting and repainting. Arrived at Brenda Road Depot, c/2011.
(c) ex Selby Storage & Freight Co Ltd, Selby, North Yorkshire, 23/11/2011.
(d) ex Selby Storage & Freight Co Ltd, Selby, North Yorkshire, 9/12/2011.

(1) to Tata Steel Europe, Skinningrove Works, Carlin How, on hire, 21/11/2011.
(2) to Tata Steel Europe, Skinningrove Works, Carlin How, on hire, 1/12/2011.
(3) to Tata Steel Europe, Skinningrove Works, Carlin How, on hire, 21/12/2011.

NORTHERN STEAM ENGINEERING LTD

Portrack, Stockton **D8**
Adam Dalgleish Engineering Ltd until 23/5/2017   NZ 457199 – **Map D**

The company commenced trading on 4/4/2011 and concentrated on the overhaul of boilers. It took delivery of its first locomotive for overhaul in September 2015. Following a contractual dispute, Adam Dalgleish Engineering went into administration and a new company, Northern Steam Engineering, was established at the same premises and using some of the existing employees to carry on repairing boilers.

Gauge : 4ft 8½in

2253	2-8-0	OC	BLW	69496	1944	(a)	(1)
1744 (69523 4744)	0-6-2T	IC	NBH	22600	1921	(b)	
3814	2-8-0	OC	Sdn		1940	(c)	

(a) ex Locomotion, The National Railway Museum at Shildon, Co. Durham, 28/9/2015.
(b) ex Great Central Railway plc, Loughborough, Leicestershire, 15/8/2019 via North Norfolk Railway, from c7/2018 via tour of preserved railways.
(c) ex Llangollen Railway plc, Llangollen, Denbighshire, 8/2/2021.

(1) to North Yorkshire Moors Railway, Grosmont, North Yorkshire, 12/2/2019.

HENRY PORTER & CO

BOWESFIELD BOILER WORKS, Stockton   **D9**
    Map D

An 1880 directory has Henry Porter & Co running an engineering and boiler making business on Bowesfield Lane, Stockton. In *Machinery Market* on 1/6/1882, it advertised for sale a 9 inch 'tank' locomotive made by Manning Wardle. The locomotive had received a thorough overhaul three years ago but had done little work since.

RIDLEY SHAW & CO LTD

ENGINEERING WORKS, Cargo Fleet   **D10**
T.D. Ridley & Sons until c/1925-26   NZ 507206, NZ 508202 – **Map G**
T.D. Ridley until /1896

Thomas Dawson Ridley was born at Acklington, Northumberland on 4/2/1825. His father, John Ridley, had been associated as assistant engineer with the building of a number of railways, including the Newcastle & Carlisle line. He was also the Resident Engineer in 1838–40 on the erection of the fine skew viaduct at Croft, south of Darlington. Thomas' early employment involved assisting with work at Warkworth Harbour and then spending 1845 on the construction of the Whitehaven Junction Railway. In 1853 he arrived at Hartlepool to act as William Hutchinson's engineer and manager on the construction of Swainson Dock for three years before branching out as a contractor. In 1869–74, Hodgson & Ridley (which see) undertook a contract for the extension of Middlesbrough Dock and, as a consequence, Thomas Ridley settled at Redcar. He continued as a public works contractor until retiring in 1896. A sale of T.D. Ridley & Sons locomotives in 1903 suggests that the company withdrew from contracting at this date and then concentrated on engineering and the repair of locomotives.

The company's first depot (NZ 507206) occupied a narrow rectangular site north of the Middlesbrough–Redcar railway and abutting Crewdson, Hardy & Co's Yorkshire Tube Works. Sidings entered both the north and south ends of the premises from the adjoining MOR lines. Ridley may well have used one of its locomotives to move traffic over the nearby NER and MOR tracks as its name appears on a list of such owners in 1897 and it was reported to be working wagons for W.H. Campbell & Co's Dock Hill Machinery Warehouse. Although the identity of some Ridley locomotives on contracts is known, we have no information on those based at his Cargo Fleet depot. Probably as a result of the last expansion of Middlesbrough Dock (1898–1906), Ridley's premises were taken over by the NER as a dockyard workshop. Ridley then moved to a relatively small site at Louisa Street, North Ormesby. This was located on the southern boundary of the Tees Side Bridge & Engineering Works and served by a siding off that system; the line dividing into two on entering the Louisa Street premises.

When P.W.B. Semmens visited the company for his 1956 article, the managing director was Mr F.R. Shaw, a grandson of Thomas Ridley. He stated that the company had built a few four-coupled saddle tanks during the early years of the twentieth century but this proved to be not economic because of the constricted site. However, it is more likely that these locomotives were extensive rebuilds of older engines.

The locomotives involved include the following:

Ridley Number	Date Built	Name or Number	Customer
2	?		At Ord & Maddison Ltd, Co Durham by 4/1936
11	1899	AMBLE	On NER Ouston Jn-Low Fell contract 1901-03
13	1899	CLARENCE	On NER Ouston Jn-Low Fell contract 1901-03. Later at APCM Kent Works, Stone, Kent
51	1911	-	Sir W.G. Armstrong, Whitworth & Co Ltd, Elswick Works, Newcastle upon Tyne
54	1913	No.24 KENNETH	Sir W.G. Armstrong, Whitworth & Co Ltd, Scotswood Works, Newcastle upon Tyne
74	1920	ELDON No.1	Pease & Partners Ltd, Eldon Colliery, near Shildon, Co. Durham
76	1920	18	Pease & Partners Ltd, St Helen's Colliery, St Helen's Auckland, Co. Durham. May not have been to begin with but here by 1935
?	?	MABEL	Original owner unknown, but at Ord & Maddison Ltd, Crossthwaite Quarry, near Middleton-in-Teesdale, Co. Durham by April 1936
?	?		Henry Stobart & Co Ltd, Newton Cap Colliery, Toronto, Co. Durham
?	?		Hartlepool Gas & Water Company, Middleton Road Gas Works, Hartlepool

Ted Haigh visited Ridley Shaw on five occasions between 1948 and 1951 when there were usually up to seven locomotives in for repair. These were mostly from local companies (Warner's Peckett 970; Richard Hill's Hawthorn Leslie 3249) or north east collieries (Rising Sun Colliery, Wallsend's Sharp Stewart 4595; Choppington Colliery's Hawthorn Leslie 2503). A few locomotives stayed for some time, such as ex GWR 312 (Andrew Barclay 1111 of 1907) which was present on 26/3/1948, having arrived from Broomhill Colliery and Shed and did not return until after 29/9/1951! Ridley Shaw closed during 1958, the plant being auctioned on 21/5/1958. No attempt has been made to list the locomotives that passed through Ridley's hands on contracts outside Teesside, nor those coming here for repair.

References : '"Shopping" Industrial Locomotives', P.W.B. Semmens, *The Railway Magazine*, 7/1956, pp.446-48

British Steam Locomotive Builders, James W. Lowe, Guild Publishing, 2nd edition, 1989, p.549

'Ridley Shaw & Company', Cliff Shepherd, *Industrial Railway Record* 236, IRS, 2019

RUNDLE & CAMPBELL LTD

MACHINERY DEALERS, Middlesbrough and North Ormesby   **D11**
W.H. Campbell & Co until c/1903   Map G

Campbell was initially based at the Dock Hill Machinery Warehouse not for far from T.D. Ridley & Sons first premises. Ridley's locomotive appears to have delivered traffic for Campbell. The latter advertised a vertical boiler 'coffee pot' locomotive with 6in x 12in cylinders for sale at £58 (*Machinery Market*, 1/3/1893) and a John Fowler four-coupled saddle tank with 8½in x 14in cylinders, 2ft 6in diameter wheels and 4ft 8½in wheelbase for £145 (*Machinery Market*, 1/6/1895). An announcement appeared in *The Engineer* on 7/4/1899 that an auction was to take place of Campbell's plant, including a 'coffee pot' locomotive because, like Ridley, the company was having to relocate to North Ormesby to make way for the expansion of Middlesbrough Dock. In 1902 Campbell offered for sale a John Fowler locomotive with 9in x 14in cylinders for £250 and a 3ft 7in gauge locomotive with 6in cylinders. The business had become Rundle & Campbell Ltd when plant, including a tank locomotive with 9½in x 14in cylinders and a saddle tank with 7¾ x 12in cylinders, was for sale in 1904. Rundle & Campbell was located in Smeaton Street, North Ormesby according to Ward's 1917–18 Directory and probably had a siding off the NER Marsh Road lines near Robinson's timber yard. When added to the railway company's list of firms' sidings in 18/6/1900, it was commented that private engines would service Campbell's yard. This would be one of Warner's locomotives (which see).

JAS TAIT & PARTNERS LTD

ENGINEERS, Middlesbrough D12
Jas Tait Jn until ? Map G

James Tait Junior was formerly employed at T.D. Ridley & Sons before setting up as an engineer with an office at Borough Road East, Middlesbrough about 1908. He offered a number of locomotives for sale:

  9 inch OC 0-4-0ST by Black, Hawthorn (*Machinery Market*, 21/10/1910)
  14in x 22in '4 wheel' by Hawthorn Leslie (*Machinery Market*, 9/2/1912)
  13 inch 0-6-0 (*The Contract Journal*, 9/10/1912)
  Works type 12 inch 0-4-0 (*The Engineer*, 17/10/1919)
  13 inch 0-6-0 OC (*The Engineer*, 31/10/1919)

He then appears to have gone into partnership. *Industrial Railways and Locomotives of County Durham Part 1* (IRS, 2006) states that J. Tait & Partners purchased the 0-4-0 saddle tank, Hawthorn Leslie 2177 of 1890 from The Consett Iron Co Ltd, rebuilt it and allocated its number 100 of 1920 before reselling it to The Wingate Limestone Co Ltd's quarry at Wingate. No manufacturing works has been identified for Tait and it may be that the rebuilding was subcontracted, Ridley being an obvious option. No.17 GEORGE (Kitson 1705 of 1871) at the Tees Iron Works carried a similar rebuilding plate, No.90 of 1920. James B. Tait & Co Ltd, 24 Albert Road, Middlesbrough was involved in the sale of a Hudswell Clarke 0-4-0ST with 9in x 14in outside cylinders from the North-Eastern Steel Co Ltd to Thos W. Ward Ltd, Preston on 23/4/1920 but the identity of the locomotive is not known. Tait may have concluded his trading activities because it was announced in *Machinery Market* that H.W. Pilkington was to auction on 18/12/1923, re Jas Tait & Partners, plant including a 13 inch Hawthorn Leslie saddle tank.

JOSEPH TORBOCK & CO

DEALER AND MERCHANT, Middlesbrough D13
  Map G

During the last twenty years of the nineteenth century, Joseph Torbock was deeply involved in the industry and trade on Teesside with an office in the Post Office Buildings, Middlesbrough. He married into what was then regarded as high society, Florence being heiress of the Henleys of Chard in Somerset and purchased Crossrigg Hall, near Penrith, in 1912.

Torbock advertised locomotives for sale on numerous occasions and was an agent for Andrew Barclay. One of the earliest appeared in *The Colliery Guardian* on 9/5/1879 when two Hopkins Gilkes' locomotives were offered for sale – a 12 inch 0-4-0 and a 6½ inch 0-4-0 geared 'tank' locomotive. This was followed on 4/7/1879 when he claimed to have secondhand 13 inch, 12 inch, 10 inch, 9 inch, 7 inch and 6½ inch 0-4-0 tank locomotives available. In June 1879 Lloyds & Co, owners of the Lackenby Iron Works, had gone into liquidation and, on 12/12/1879, Joseph Torbock purchased that company's West Hunwick Colliery, north of Bishop Auckland. Amongst his advertisements in *Machinery Market* during the 1880s were; –

  0-4-0 'tank' locomotive with 14in x 21in outside cylinders, 3ft 4in diameter wheels on a 5ft 8in wheelbase. Only done four years work – just thoroughly overhauled (1/5/1882).

  3ft gauge 'tank' locomotive by Black, Hawthorn with 8½in x 18in cylinders and 2ft 6in diameter wheels. Only done about six months work (1/10/1882).

  'Coffee Pot' 0-4-0 vertical boiler 'tank' geared locomotive (2½:1) with 6in x 12in cylinders, 2ft 5in diameter wheels and a 6ft 6in x 3ft 4½in boiler (1/12/1882).

In *The Contract Journal* on 26/9/1888, he advertised two 3ft gauge locomotives with 8 inch cylinders by Hudswell Clarke and John Fowler. These sound similar to the Kirkleatham Ironstone Mine locomotives that Wake was offering for sale in 1889.

Torbock's advertisements tail off in the 1890's but, on 1/2/1893, he advertised the following: Dick & Stevenson 14 inch 0-4-0 locomotive £650, Hunslet 12 inch £550, John Fowler 8½in 0-4-0 with inside cylinders £325 and a 6 inch 0-4-0 'coffee pot' 'tank' locomotive £125. He also hired out locomotives, such as Barclays 208/1873 to The Woodland Collieries Co Ltd in c1886/87.

J.F. WAKE

MACHINERY BROKER, Middlesbrough

D14
Map F

John Frederick Wake (1862–1950) was born in Middlesbrough and entered business in his mid-20s. His office was at 33 Wilson Street in the centre of the town but he also had a depot at "The Marsh" in the Ironmasters District, although this may have not been rail connected. He mainly acted as a machinery and plant dealer, including some locomotives at Middlesbrough. An advertisement in the 1887–88 *Mercantile and General Directory of Middlesbrough* stated that he was also the sole agent for C.C. Marley's "Patent Improved Donkey Boiler" manufactured by T.F. Passman at the Depot Road Boiler Works in Middlesbrough. On 24/8/1889, a fire at Wake's West Marsh Store resulted in two extensive sheds and their contents being destroyed. According to the *York Herald*, "as the buildings adjoin the North-Eastern Railway, it is supposed that a spark from a passing engine had alighted on the tarred roof of one of the sheds".

Colin Mountford published his assessment of the locomotives which Wake dealt with while at Middlesbrough (see *Industrial Railways and Locomotives of County Durham Part 3*, IRS, 2021 pages 109-111). A series of sales advertisements for locomotives appeared in the press between 1888 and 1892 including the following:

    Machinery Market 1/9/1888
    9½ inch John Fowler traction engine, will work as locomotive, £100.
    Machinery Market 1/12/1888
    13 inch Fox Walker 6cpld locomotive built 1887 (probably 1877 according to the 1/8/1888 advertisement). Just overhauled by Hawthorn Leslie at a cost of £320. Practically a new engine. To be seen at Middlesbrough £300.
    15 inch (later described as 14 inch) Black, Hawthorn 6 cpld locomotive rebuilt at a cost of £600-£800.
    12 inch Hunslet 6cpld locomotive built 1877. Thoroughly overhauled and repainted. To be seen at Middlesbrough. £350.
    12 inch Hudswell Clarke 4cpld locomotive. £600.
    Machinery Market 1/4/1889
    8½ inch Hughes 3ft 0in gauge 4cpld 'tank' locomotive £280.
    8 inch Hudswell Clarke 3ft 0in gauge 4cpld 'tank' locomotive £280.
    8 inch John Fowler 3ft 0in gauge 4cpld 'tank' locomotive £275.
    10 inch John Fowler standard gauge 4cpld 'tank' locomotive £365.
    Machinery Market 1/5/1889
    12 inch Manning Wardle 6cpld locomotive, built 1886.
    Machinery Market 1/4/1890
    Two Hudswell Clarke 8in 4cpld locomotives £350 each.
    10 inch Hudswell Clarke locomotive £320.
    12 inch Hopper Radcliffe 4cpld, now being rebuilt, £600.
    Machinery Market 1/11/1890
    Small geared locomotive, 8½ inch with horizontal multi-tubular boiler, £150.
    9 inch locomotive, thoroughly overhauled, £380.
    Machinery Market 2/3/1891
    J.F. Wake advertised a "Special List of Locos in Stock at Middlesbrough" for sale:
        Std gauge coffee pot locomotive 8 inch ready for work £95.
        Std gauge 4 cpld locomotive 13 inch entirely rebuilt £800.
        Std gauge 4 cpld locomotive 8 inch by Tees Side Engine Co £120.
        Std gauge locomotive 8 inch rebuilt £275.
        Std gauge four-wheel locomotive 10 inch rebuilt £650.
        Four-wheel locomotive 12 inch, new boiler £750.
        Six-coupled Black, Hawthorn locomotive 14 inch, can be seen in South Durham £800.

Wake's last advertisement in *Machinery Market* offering locomotives for sale from Middlesbrough was dated 1/3/1892. Wake then transferred his business and head office to the Skerne Iron Works in Darlington; the first known advertisement from there was dated 14/9/1893.

List of dealers based on Teesside offering locomotives for hire or sale

ATHCO, VICTORIA BRIDGE ENGINEERING WORKS, Stockton

The works advertised, for sale or hire, 2 feet gauge 20hp and 40hp Simplex petrol locomotives in *Machinery Market* on 27/8/1943. In a subsequent issue, dated 20/10/1944, plant including a 40hp Simplex petrol locomotive, was offered for sale.

ISAAC BIGLAND, METAL BROKER, Stockton

Isaac Bigland had an office at 38 High Street, Stockton and advertised for sale in *The Engineer* on 21/10/1864 – METEOR, a 0-4-0 with 12 inch cylinders and 4ft diameter wheels and ANTELOPE, a 0-4-0 with 14 inch cylinders and 4ft diameter wheels. They were still for sale in 4/1865.

MR BRADLEY, GENERAL DEALER, MIDDLESBROUGH

A sale took place of Thomas Vaughan & Co's Whessoe Iron Works near Darlington on 28/11/1877. According to *The Colliery Guardian*, this included an ex SDR locomotive HOPETOWN with the following details – six wheels coupled, 12 inch outside cylinders, 24 tons weight and fitted with tanks and a tender, although these do not accord with the known details of HOPETOWN. It was sold to Mr Bradley for £45.

AMBROSE C. CASEBOURNE, Yarm

A four-wheel locomotive with 8½in x 12in cylinders and 25in diameter wheels was advertised for sale in *Machinery Market* on 2/5/1919.

E.R. COLWELL LTD, IRON MERCHANT, Middlesbrough

Eric Colwell's office was at Fletcher Street, Middlesbrough. The company advertised in *Machinery Market* on 12/1/1940 seeking a 14in 0-4-0 locomotive. On 28/1/1949, it placed a notice in the same publication offering a 12 inch Manning Wardle 0-4-0 saddle tank for sale.

GLADSTONE & CO, IRON MERCHANT, West Hartlepool

A 13 inch tank locomotive by Hudswell, Clarke & Rodgers was advertised for sale in *The Engineer* on 4/7/1884. In Ward's Directory 1896-97, Arthur Gladstone's company office was at Victoria Terrace, West Hartlepool Docks.

F.W.L. GRAHAM, MINERAL MERCHANT, Middlesbrough

Frances William Leland Graham was described as a mineral merchant and broker with an office in Bridge Street, Middlesbrough according to an 1880 directory. He advertised for sale plant, including a standard gauge locomotive, in *The Engineer*, 26/8/1881. The asking price was £250. On 2/1/1882, in *Machinery Market*, he was selling a tank locomotive with 8in x 15in cylinders and another with 6in x 14in cylinders.

J.H. HARRISON, Middlesbrough

In *The Engineer*, 9/12/1904, Harrison offered for sale a Black, Hawthorn 0-4-0 saddle tank with 8in x 14in cylinders, a Hudswell Clarke 0-4-0 saddle tank with 13in x 20in cylinders and a 0-4-0 saddle tank with 12in x 18in cylinders that was believed to be made by Peckett.

C.E. HUTCHINSON & CO, METAL AND MACHINERY BROKER, Stockton

Charles Hutchinson of 5 Exchange Place, Stockton offered a 10½ inch four-coupled locomotive for sale on 1/1/1889 and, possibly the same engine, stated as a 10 inch tank locomotive on 1/1/1890. The company was described as metal brokers at 15 Exchange Buildings, Stockton in Ward's 1896-97 directory. By 1914-15 he was referred to as a machinery broker at Thornaby. In *Machinery Market*, 3/3/1916, he advertised for sale a standard gauge 8 inch 0-4-0 locomotive with outside cylinders and a narrow gauge locomotive.

C.G. JOHNSON & CO, IRON MERCHANTS AND ENGINEERS, Middlesbrough

Johnson was a merchant who stated in an advertisement to deal in all types of iron sections and plates, railway plant, iron works machinery and Spanish and other ores. His office was at Exchange Place, Middlesbrough. In *The Colliery Guardian* on 1/1/1874, he claimed to have "Tank locos 8" 10" 12" 14" ready for delivery or in progress". He certainly seems to have been involved in the sale of AB 148 of 1874 to W. Jones, Middlesbrough Chemical Works.

T. LISHMAN, Stockton

In 1872 T. Lishman reported that a 7 inch tank locomotive was being repaired and this was offered for sale as a geared engine in *The Engineer* on 11/10/1872.

R. MACKAY & CO, Middlesbrough

According to *The Colliery Guardian* on 27/5/1881, Mackay was offering for sale a Hopkins Gilkes 0-4-0 'tank' locomotive with 12in x 20in outside cylinders and 3ft 10in diameter wheels and a John Fowler 0-4-0 'tank' locomotive with 8½in x 16in inside cylinders and 2ft 7in diameter wheels. R. Mackay & Co was based at the Royal Exchange, Middlesbrough and operated as accountants which suggests that a sale may have resulted from the failure of a business.

W. NEWTON & CO, IRON MERCHANT, Stockton

Ward's 1928-29 Directory states William Newton's office was at Millfield, off Dovecot Street, Stockton. On 6/12/1929 in *Machinery Market* he offered a standard gauge 13 inch 0-6-0 locomotive for sale.

WILLIAM OLIVER, ENGINEER, Guisborough

According to Kelly's Directory 1925, William Oliver was a mechanical engineer based at 68 Westgate, Guisborough. He placed two 'For Sale' advertisements in *Machinery Market*: on 10/2/1922 plant including two Manning Wardle 12 inch locomotives and, on 16/11/1923, a 2ft gauge steam locomotive and one mile of railway comprising 56lb rail on completion of a contract.

S. RICHARDSON & CO, METAL MERCHANT, West Hartlepool

The company was based at Longhill, West Hartlepool and offered for sale a 2ft gauge locomotive "as new" and equipped with a copper firebox in *The Contract Journal* on 4/5/1932.

W.A. STAINSBY LTD, Middlesbrough

A 14 inch 0-4-0 locomotive was offered for sale in *Machinery Market* on 3/6/1949.

JAS SUMMERS, Middlesborough

Jas Summers of 14 Grange Road West, Middlesbrough advertised in the *Yorkshire Post* on 12/8/1899 wanting two secondhand 0-4-0 tank locomotives with 12in x 20in outside cylinders.

W. WETHERELL, IRON AND STEEL MERCHANT, Middlesbrough

W. Wetherell had an office at 42 Albert Road, Middlesbrough and advertised locomotives for sale on various occasions during the 1880s, although it is probable that some involved the same locomotive. The following advertisements provided a little more detail:

12 inch 4-wheel tank locomotive "near Durham" £325. Also a 6 inch Chaplin-type locomotive £80 (*Machinery Market*, 1/4/1884).

12 inch four-coupled locomotive by James [sic] Harris, Darlington with 3ft 6in diameter wheels and 6ft 6in wheelbase, £325 (*Machinery Market*, 1/1/1885).

4-wheel 'tank' locomotive with 12in x 20in cylinders and 3ft 0in diameter wheels, £120 at Stockton on Tees (*The Engineer*, 21/2/1890).

PRESERVED LOCOMOTIVES

BRITISH STEEL CORPORATION

SOUTH TEESSIDE WORKS, Lackenby

P1
NZ 546211 – **Map H**

The locomotive came to the Apprentice Training Centre at South Teesside Works for some restoration to be carried out. It was then stored at another location within the works from 12/1973.

Gauge : 4ft 8½in

112	CYCLOPS	0-4-0ST OC	HL	2711	1907	(a)	(1)

(a) ex Cleveland Works, South Bank, c1/1970.

(1) to Tanfield Railway Preservation Society, Marley Hill, Co. Durham 30/12/1975.

BRITISH STEEL CORPORATION/REDPATH DORMAN LONG LTD

(Subsidiary of BSC)
BRITANNIA STEEL WORKS, Middlesbrough

P2
NZ 484212 – **Map F**

The two locomotives were the property of the Dalescroft Railfans Club and were temporarily stored here.

Gauge : 4ft 8½in

ROKER	0-4-0CT OC	RSHN	7006	1940	(a)	(1)
PALLION	0-4-0CT OC	HL	2517	1902	(b)	(2)

(a) ex Doxford & Sunderland Ltd, Pallion Dockyard, Sunderland, Co. Durham, 3/1971.
(b) ex Doxford & Sunderland Ltd, Pallion Dockyard, Sunderland, Co. Durham, 5/1971.

(1) to J.I. Blackburn & Co*, Godalming, Surrey, 27/9/1973.
(2) to J.I. Blackburn & Co*, Godalming, Surrey, 20/10/1973.

* *Industrial Railways and Locomotives of Sussex and Surrey* (IRS, 2015) has W. Lees, Farncombe, Godalming. A possible explanation is that Lees may have owned the locomotives on Blackburn's premises.

CLEVELAND IRONSTONE MINING MUSEUM

Skinningrove

P3
NZ 712912 – **Map I**

Tom Leonard was a local reporter for the *Evening Gazette* who had previously started work in the South Skelton Mine Offices. It was his ambition to open a museum to commemorate and interpret the Cleveland ironstone mining industry based on his collection of memorabilia. Tom Robinson owned the Loftus Mine site, parts of which had stood derelict since closure in 1958 and he agreed to the project to establish a mining museum. It took some five years to clear and prepare the site. Tom Leonard died before the museum opened in 1983. The first part of the museum is located in the former engine house with the remains of the Waddle fan building behind. Visitors can also enter 90 yards into the North Drift and see the 3ft gauge railway that descended on a 1 in 15 gradient to the workings. The Tom Leonard Mining Museum was renamed in 1999. Following improvements to the museum, while it was closed in 2020-21, it is intended to market it as the "Land of Iron" museum.

Gauge : 2ft 0in

(No.9)	4wDM	RH	182137	1936*	(a)

*chassis only

(a) ex J.D. Wiggins, Manless Green Mine, Skelton Green, c5/2009.

A.V. DAWSON LTD

MIDDLESBROUGH WHARF, Middlesbrough

P4
NZ 493215 – **Map F**

The locomotive was the property of the Darlington Railway Preservation Society and stored at this location. Dawson has two of its early diesel locomotives currently on display outside its Automotive Coil Store (see the 'Locomotive Worked Sites' section).

Gauge : 4ft 8½in

-		0-4-0ST	OC	P	2142	1953	(a) (1)

(a)   ex Thomas Ness Ltd, St Antony's Tar Distillation Plant, Walker, Newcastle-upon-Tyne, 14/3/1982.

(1)   to Darlington Railway Preservation Society, Darlington, Co. Durham, after 25/9/1983, by 7/7/1984.

HEAD WRIGHTSON & CO LTD,

TEESDALE IRON WORKS, Thornaby

P5
NZ 451188 – **Map E**

Three vertical boiler locomotives built by the company in the 1870s were repurchased for preservation. HW 21 was displayed with two chaldron wagons on a short length of track at the works. At one stage, it was refurbished by the apprentices with its wheels lifted off the rails so that the engine could be 'turned over' by compressed air for the entertainment of visitors!

Gauge : 4ft 8½in

(16)	0-4-0VBT	VCG	HW	21	1870	(a) (2)
-	0-4-0VBT	VCG	HW		1871	
	Rebuilt		Dorking Grey-stone Lime		9/1952	(b) (1)
(17)	0-4-0VBT	OC	HW	33	1873	
	Rebuilt		Seaham			(c) (3)

Head Wrightson re-acquired its number 21 which it had built in 1870 and put it on display with a couple of chaldron wagons at its Teesdale Works where it was seen on the 19th June 1960.

(IRS J.P. Mullett photograph)

(a)   ex Seaham Harbour Dock Co Ltd, Seaham, Co. Durham, 6/1959.
(b)   ex Dorking Lime Co Ltd, Betchworth, Surrey, 15/9/1960.
(c)   ex Seaham Harbour Dock Co Ltd, Seaham, Co. Durham, 6/1962.

(1)   to British Steel Corporation, Consett Works, Consett, Co. Durham for storage, 7/1970; then to North of England Open Air Museum, Marley Hill Shed store, Co. Durham, 9/1971. It moved to Beamish Museum in 3/1975 and was rebuilt at ICI Wilton Works c11/1982 (after 27/9/1981) to 6/1984.
(2)   to County Borough of Teesside, Preston Park Museum, Eaglescliffe, 10/1970.
(3)   to Darlington Railway Museum, North Road Station, Darlington, Co. Durham, 9/1975 after exhibition at SDR 150th anniversary, Shildon, 8/1975.

SALTBURN MINIATURE RAILWAY

Saltburn   **P6**
NZ 667216 NZ 668214 – **Map I**

The miniature railway was established in 1947 by Herbert Dunn from Co. Durham who promoted a number of 15 inch gauge lines. In its original form, the line was about 300 yards long end to end with the stock stored in the above ground 'tunnel' part way along the railway. Towards the end of 1948, work commenced to modify the line for the 1949 season. Cat Nab Station was located on the retaining wall of the boating lake and had a turntable and a run-round loop. From here, the railway ran south along the flat river meadows on the west side of Skelton Beck, passing through the tunnel en route. It then ran round a balloon loop under the Halfpenny Bridge, which carried a footpath high across the valley, allowing the train to return to the station.

About 1950 the railway was purchased by Cyril Pickering who owned Saltburn Motor Services. The balloon loop was lifted and the line extended southwards over a bridge across Skelton Beck to the Valley Gardens, where a station was provided, later called Forest Halt, with a turntable and run-round loop. Saltburn Motor Services was purchased by the joint local authorities-owned Cleveland Transit bus company in 1974 and the railway passed into the control of Langbaurgh Borough Council, which operated it for a short while until the line fell into disuse; the last operating licence was dated 1979. The Saltburn Miniature Railway Association was formed in 1983 and it started to rebuild the line in the following year. It reopened on 15/8/1987 but, following a landslip, the railway was slightly shortened and operated from a station near the wooden engine and carriage shed (NZ 667216). It was not until 1992 that the line was reopened over this section to a replacement station near the seafront.

The original miniature railway station near the sea front at Saltburn in 1955 with the steam outline Barlow locomotive of 1953 in use.   *(Hamish Stevenson collection)*

As a result of the effects of severe flooding in 11/2000 and a wish to regenerate Saltburn, the alignment of the railway was substantially altered in 2001 to run on the east side of Skelton Beck. A new Cat Nab Station was constructed at the end of the Council car park in 2002 and trains ran from here along the edge of the river meadows passing a brick-built workshop and rolling stock shed (NZ 668214, into use 12/2001) to rejoin the former route of the railway south of Skelton Beck bridge and so reach Forest Halt. Passenger services began again on Easter Sunday 2003 with a formal reopening on 17/8/2003.

The most notable locomotive to work on the railway was ELIZABETH; this had considerable claim to fame when completed as BLACOLVESLEY in 1909. It was Bassett-Lowke's only attempt at internal combustion motive power and was one of the earliest steam-outline miniature locomotives built in England. Its original German N.A.G. petrol engine was replaced by an Austin 7 petrol engine prior to arrival at Saltburn. This locomotive proved to be unsuitable for operating at the slow speeds desirable at Saltburn and it became the standby when the new Barlow diesel was acquired. Since the Association has operated the railway, other locomotives have briefly visited the line for special events, but only the resident engines are listed here.

Gauge : 1ft 3in

7		2-4-0 + 6wPM					
			s/o	N.G. Parkinson		c/1932	(a)	(1)
	ELIZABETH (YVONNE)	4-4-4TPM	s/o	BL	177	1909	(b)	(2)
	PRINCE CHARLES (PRINCE OF WALES)						
		4-6-2DE*	s/o	H.N.Barlow		1953	(c)
	GEORGE OUTHWAITE	4wDH	s/o	SMR/ICI Wilton		1994	(d)
	Rebuilt	0-4-0DH	s/o	SMR		by 1997	
(4472)	SALTBURN 150 (FLYING SCOTSMAN)						
		4-6-2 gas	s/o	Artisair		1972	(e)
				RADev+		1972	
	Rebuilt	4-6-2 DH	s/o	SMR		2011	
	BLACKLOCK R.	4-4-2	OC	Moss A.J.		2001	(f)
		Rebuilt		SMR		2015	

*   For a period in the late 1970s, it ran without a leading bogie, but regained it when overhauled by the National Engineering Training Centre at Stockton in 1987.

\+   RADev is R A Developments, Scunthorpe, Lincolnshire.

(a) ex Seaton Carew Miniature Railway, Seaton Carew, /1947, after 12/8/1947.

(b) originally supplied new to Blacolvesley Hall, Northamptonshire. Ex H. Dunn, Plant & Machinery Co, Bishop Auckland, Co. Durham, /1949, but may have spent a brief period at Dunn's Seaton Carew Miniature Railway. It carried the name YVONNE until c/1951 and was named ELIZABETH in /1953. A photograph exists of it stored at Saltburn Motor Services depot on 17/4/1968.

(c) new. During the first closure of the railway, the locomotive was stored at Langbaurgh Borough Council's Dormanstown Depot, Broadway East, Redcar (NZ 584232) from c12/1978 to before 30/7/1985. The locomotive carried the name PRINCE OF WALES from by 1965 to c/1987.

(d) new locomotive arrived as a kit of parts from ICI Wilton c8/1994 and was reassembled at Saltburn.

(e) ex Cleethorpes Coast Light Railway Ltd, Cleethorpes, Lincolnshire, 25/1/2007.

(f) ex Windmill Animal Farm Miniature Railway, Crossens, Lancashire, 29/6/2013.

(1) to Seaton Carew Miniature Railway, Seaton Carew, /1951.

(2) to Tom Tate who was constructing a new 1ft 3in gauge railway at Haswell Lodge, Co Durham, 12/12/1968.

SEATON CAREW MINIATURE RAILWAY

Seaton Carew   P7
NZ ? – **Map A**

Herbert Dunn operated the miniature railway which ran on the beach close to the South Shelter at Seaton Carew briefly in 1939 and some years from 1946. The associated fairground rides and "shuggy" boats had been established about 1946. A photograph of the locomotive on the "Seaton Express", dated 12/8/1947, shows it giving rides on the sands. YVONNE 4-4-4 PM s/o Bassett Lowke of 1909 may have spent a brief time here.

Gauge : 1ft 3in

7		2-4-0 + 6wPM				
		s/o	N.G. Parkinson	c/1932	(a)	(1)

(a) ex H. Dunn, Plant & Machinery Co, Bishop Auckland, Co Durham, /1939; previously Southend Miniature Railway, Southend. To H. Dunn, /1939 and returned to Seaton Carew /1946. To Saltburn Miniature Railway, Saltburn, /1947; after 12/8/1947 and returned, /1951.

(1) destination unknown.

STOCKTON-ON-TEES BOROUGH COUNCIL

PRESTON PARK MUSEUM, Eaglescliffe P8
County Borough of Teesside until 31/3/1974 NZ 429158 – **Map D**

The locomotive was built by and initially preserved by Head Wrightson & Co Ltd at its Teesdale Works (which see). It was then displayed at the Preston Hall Museum in the open on a short length of track. During the winter of 2001/02, it was repainted black at the BP Cats Terminal, Seal Sands before being put on the Bridge Road roundabout at Stockton (NZ 446185). It was later returned to the museum (renamed Preston Park Museum in 2012) where it stands in the open.

Gauge : 4ft 8½in

-	0-4-0VBT	VCG	HW	21	1870	(a)

(a) ex Head Wrightson & Co Ltd, Thornaby, 10/1970. To Bridge Road roundabout, Stockton, from 24/2/2002 and returned to Preston Park Museum, 12/8/2012.

R.G. WALL

Hartlepool P9
NZ 494331 – **Map B**

The locomotive was kept at the owner's home, 9 Serpentine Road, Hartlepool.

Gauge : 1ft 6in

-	0-4-0VBT		Rebuilt R.G. Wall	(a)	s/s c/1980

(a) origin unknown.

GRAEME WALTON-BINNS

Cargo Fleet, Middlesbrough P10
NZ 516197 NZ 487216 – **Map F**

Mr Binns purchased the three locomotives from South Africa and largely through his and a number of colleagues' efforts restored them over a period of 14 years. Initially the work was carried out at the rear of a Council depot at Cargo Fleet (NZ 516197) but the locomotives were then moved to a unit on Riverside Park industrial estate, located on the former Ironmasters site, Middlesbrough (NZ 487216), where the restoration was largely completed. A start was then made on restoring an ex Scottish colliery locomotive.

Reference : 'South African Steam: Bagnall Locomotives for the Tongaat Sugar Company', Allan C. Baker, *The Narrow Gauge* 270, Narrow Gauge Railway Society, September 2021.

Gauge : 2ft 0in

CHARLES WYTOCK	4-4-0T	OC	WB	2819	1946	(a) (1)
SINEMBE	4-4-0T	OC	WB	2287	1926	(a)
A. BOULLE	4-4-0T	OC	WB	2627	1940	(b)

(a) ex Welsh Highland Railway Ltd, Porthmadoc, Gwynedd, 2/1997; previously Tongaat-Hulett Sugar Mills & Estates (Pty) Ltd, Tongaat, South Africa.

(b) ex storage at Immingham, 11/1998; previously Geoff C. Titren, Himeville, Natal and before that Tongaat-Hulett Sugar Mills & Estates (Pty) Ltd, Tongaat, South Africa.

(1) to Lynton & Barnstable Railway Trust, Woody Bay Station, near Parracombe, Devon, 9/4/2014, via G. Lee, Statfold Barn Railway and arrived at Woody Bay, 11/4/2014.

Graeme Walton-Binns and his colleagues have restored three Bagnall 2ft 0in gauge locomotives from South Africa. This is A. BOULLE (WB 2627/1940) in a unit at Riverside Park, Middlesbrough on 14th November 2017. *(Cliff Shepherd photograph)*

Gauge : 4ft 8½in

| No.8 | | 0-6-0T | OC | AB | 1296 | 1912 | (a) | |

(a)   ex Ayrshire Railway Preservation Society, Waterside, Ayrshire, 11 or 12/2016.

J.D. WIGGINS, MANLESS GREEN MINE

Skelton Green P11
NZ 655181 – **Map I**

Mr Wiggins created a mock-up of a small ironstone mine in his garden including a wooden headgear, a drift into an embankment and a short length of a 2ft gauge railway with a home-built 4-wheel tub.

Reference :   'The Manless Green Mine', Jim Stancliffe, *The Narrow Gauge* 221, Narrow Gauge Railway Society, 2012/13.

Gauge : 2ft 0in

| (No.9) | | 4wDM | | RH | 202969 | 1940* | (a) | (1) |

* chassis only

(a)   ex J. Scholes, Rippingale, Lincolnshire, 3/5/2001.

(1)   to Cleveland Ironstone Mining Museum, Skinningrove, c5/2009.

NON-LOCOMOTIVE WORKED SITES

In addition to railways operated by industrial locomotives, there were many other sidings owned by companies that did not require their own engine and some are covered in this section. They may have relied on people, horses, tractors or even the main line railway companies to shunt their wagons. However, a number of the non-locomotive systems on Teesside were very extensive, especially the underground railways at ironstone mines relying on powered haulages and horses. Shipyards often employed rail cranes to both lift and move bulky loads.

SAMUEL BASTOW & CO

CLIFF HOUSE IRON WORKS, West Hartlepool   H1
NZ 516316 – **Map B**

The Cliff House Iron Works appears on the 1861 Ordnance Survey map and was located on the west side of the Norton-Hartlepool railway immediately south of the West Hartlepool Iron Works. A short siding entered the premises from the main line. Samuel Bastow had purchased the site in 1858 and erected the iron works which functioned as a foundry and engineering establishment. Robert Wood in his book on West Hartlepool claims that the Cliff House Iron Works made "locomotives and fixed steam engines, boilers, railway trucks, and cranes" but no proof has been found to support this statement. Samuel Bastow was declared bankrupt in 1869 and the premises were then acquired by Dunlop, Meredith & Co and converted into a nut and bolt works. This business went into liquidation in 4/1875 and the premises were sold in 1880 to William Smith of South Stockton who opened a pottery mainly producing white flint ware. The pottery closed in 1897 and the premises were then used by the West Hartlepool Wood Turning Co until the site was later incorporated into the South Durham Steel & Iron Co's North Works.

BELL BROTHERS & CO

SKINNINGROVE IRONSTONE MINE, Skinningrove   H2
Messrs Losh, Wilson & Bell & Co until ?
Messrs Bolckow & Vaughan until 19/10/1850
Messrs Roseby until 7/1849
NZ 714199 – **Map I**

Ironstone was collected along the coast where it had fallen from the cliffs and been quarried from beds of rock exposed at low tide. Anthony Lax Maynard from Skinningrove Hall agreed that James Burlinson could extract ironstone on his land but no action resulted and the lease was transferred to Messrs Roseby who began to dig out ironstone on 7/8/1848 from the east side of the valley near the mouth of Skinningrove Beck. Initially it was taken by horse and cart to be loaded on ships from the beach. Bolckow Vaughan took over the lease in 7/1849 and operated the mine until 19/10/1850 when the partnership began to take advantage of its more convenient ironstone reserves at Eston. Messrs Losh, Wilson & Bell then took over the mine sending ironstone to its works at Walker on the Tyne and to Bell Brothers' Clarence Iron Works which had begun operating in 1854. In order to facilitate the movement of ironstone to the ships, a wooden and iron jetty was constructed across the beach where Skinningrove Beck joins the sea and a tramway laid up the valley past the hamlet of Skinningrove, crossing the beck on a wooden bridge to reach the mine. The tramway is shown on the 1856 Ordnance Survey map, although there is a break between the landward end of the jetty and the tramway. It would almost certainly have been a narrow gauge line with mine tubs pulled by horses. There is some doubt who provided the tramway, but a wayleave for it was granted by Lord Zetland from 8/8/1850. The mine continued to operate on a small scale for some years until a Bell Brothers' plan reproduced in Issue 33 of *Cleveland Industrial Heritage* has a handwritten note in pencil which states *"Plan shewing state of Workings Oct 24, 1863 when the grounds were closed. A.L. Steavenson"*. The workings were later absorbed into the Loftus Mine.

Reference :   *Skinningrove Iron and Steel Works Its History, Railways and Locomotives*, Cliff Shepherd, IRS, 2012

BELL BROTHERS LTD

HUNTCLIFF IRONSTONE MINES, Carlin How   H3
Bell Brothers & Co until 27/11/1873
NZ 689214 NZ 697215 – **Map I**

The Cleveland Railway's line from Slapewath to Carlin How had to run round on the edge of the cliffs to circumvent the ridge of high ground that terminates at Warsett Hill and Huntcliff. The Main Seam ironstone outcrops here and the coming of the railway encouraged its exploitation. Bell Brothers obtained the lease for ironstone on 1/7/1865 and sank shallow shafts on either side of the railway to the west of Warsett Hill; the workings principally extended westwards to near Brook House Farm. A total of 10,055 tons was despatched along the railway from Cliff Mine (NZ 689214) in 1866. A visitor to the mine on 24/4/1867 noted that a 10hp Fowler stationary engine was drawing about 250 tons of ironstone out each day. According to *The Engineer,* 18/9/1874, the Fowler machine was actually a traction engine that had steamed to the mine from the station at Saltburn. The peak year of production was 1871 when 266,084 tons of ironstone was produced, but the mine closed in 1877–80 during the depression. There was an underground connection with Huntcliff Mine and later Cliff stone went out that way. By the termination of the lease in 10/1887, virtually all of the ironstone had been removed and Cliff Mine was abandoned.

Bell Brothers obtained an additional lease of ironstone to the east in 1871 and drove drifts under the railway and into the hillside where the Main Seam outcropped and Huntcliff Mine (NZ 697215) was developed. A stationary steam engine was installed on the east side of the NER's line and this hauled tubs on a railway out of the drift and up an inclined ramp. From here, the ironstone was tipped into the waiting railway wagons on the sidings below. The latter ran parallel to and connected with the NER's line. With the exhaustion of the reserves, Huntcliff Mine closed on 30/6/1906. Although the tipping gantry and sidings were quickly removed, the ramp partly faced in stone and Guibal fanhouse remain to form a striking image near the edge of the cliffs.

The curve of the railway on Huntcliff gave a good arc of fire over the sea and a 9.2 inch rail mounted gun was stationed here during World War 1, together with a 6lb anti-aircraft gun. Presumably to stable it and enhance its field of fire, a siding was laid off the NER line, passing behind West Cliff Cottages before heading northwards to the edge of the cliffs. Ken Hoole commented that a NER Class L tank engine, number 544, was allocated to this duty and fitted with condensing apparatus and a pump for lifting water from streams. Although removed, the alignment of the siding was still visible on the 1927 Ordnance Survey map.

References : "Wheels Turning and Smoke Rising" Volume 1 Mines to the north of Brotton, Simon Chapman, pub. P. Tuffs, 1997, pp.24-46

Catalogue of Cleveland Ironstone Mines, Peter Tuffs, pub. by P. Tuffs, 1997

R. BLACKETT & SONS

ST ANN'S BRICK WORKS, Stockton H4
NZ 455201 – **Map D**

The brick works was located to the north side of Portrack Lane opposite the Malleable Works. The stimulus for its establishment in 1920 was the existence of good clay deposits and the recent opening of the NER's Billingham Beck Branch. A siding ran from the branch southwards for about 500 yards terminating alongside the brick works enabling coal to be brought in and bricks taken out; the yard produced both wire-cut and special bricks. The clay pit was situated on the west side of the works and a Ruston Bucyrus steam powered excavator was employed in the 1920s to extract the clay. The mineral was then dumped into small wagons and pulled by a rope up the incline to the works. There were at least four rectangular and one circular kiln with peak production reaching 15,000 bricks per day. The works closed in 1966 but it had finally relied on road transport, with clay being brought in from Carter's brickworks at Cowpen Bewley.

Reference : *The Brickworks of the Stockton-on-Tees Area*, Alan Betteney, Tees Valley Heritage Group, 2007

CARGO FLEET IRON CO LTD, THE

NORMANBY IRONSTONE MINE, near Normanby H5
Bell Brothers Ltd until 1883 NZ 553169 NZ 550162 – **Map H**
Bell Brothers & Co until 27/11/1873
Bell Brothers until /1864

The discovery of the Main Seam ironstone at Eston encouraged Bell Bros to search for it further west along the Eston Hills. The royalty here was owned by George Ward Jackson, whose brother Ralph was chairman of the West Hartlepool Harbour & Railway Company (WHH&R) and the lease was agreed for the ironstone on the basis that Bell Bros would establish its iron works on the north bank of the River Tees. Bell Bros secured a lease of the ironstone from the Ward Jackson family on 1/10/1853. Initially the ironstone was obtained by quarrying and a drift north of Upsall Cottages (the latter face south on to the present A171 road) with production commencing in 1854. The ironstone travelled along a tramway south-westwards to a location near the present Flatts Lane/A171 road junction, where the tubs' contents were tipped into standard gauge wagons on a branch that connected with the Middlesbrough-Guisborough railway. Despite the circuitous route required to reach Bell's iron works, annual output of ironstone had risen to 165,000 tons in 1857. In 1860 Bell Bros opened a mine entrance on the north side of the Eston Hills but this ironstone had to be transported by horse-drawn tramway and an incline southwards to reach the Middlesbrough-Guisborough branch mentioned above. The ironstone was then taken by SDR train to Middlesbrough where a ferry carried it to Port Clarence, before another locomotive hauled the wagons to the blast furnaces.

The WHH&R company and interested local landowners proposed that a new railway should be constructed from Skinningrove via the royalties at Skelton and Normanby to Bell Bros iron works. The SDR saw this as a threat to its interests and there was much obstruction and not a little violence between the two parties. Much of the line west of Guisborough could be built under wayleave agreements with the landowners and construction commenced in 1860. Finally, the Cleveland Railway was authorised on 22/7/1861 and the line was opened from Slapewath to Cargo Fleet on 23/11/1861. The line ran past the northern workings of Normanby Mine (NZ 553169) and sidings were laid on the east side so that ironstone could be hauled down to the river at Normanby Jetty. A curve just before the bridge over the Middlesbrough-Redcar railway also gave access to the main line at Cargo Fleet Junction. At the jetty a barge, which could carry 24 wagons each containing 13¼ tons of ironstone, delivered it to the Clarence Iron Works Wharf. On 8/10/1861 a barge for this traffic had been launched from Pearse & Co's yard at Stockton. By late 1862 the main underground haulage road in Normanby Mine was several hundred yards long and connected with the workings north of Upsall Cottages so obviating the need to send ironstone to the Middlesbrough-Guisborough line. Tubs of ironstone were pulled out of the mine by a steam haulage engine located at the main drift entrance which was sited on top of a ridge. Here the contents of the tubs were tipped into standard gauge wagons which were let down an incline to the sidings alongside the Cleveland Railway. Maximum annual output from Normanby Mine was recorded in 1871 at 265,723 tons. By this date, both the Cleveland Railway and the SDR had become part of the NER and the redundant section of the Cleveland Railway south from the sidings for Normanby Mine to near Guisborough closed in 1873.

An area of ironstone to the west of the Cleveland Railway known as the "24 Acre" (NZ 550162) was worked, starting in 1865. Both quarrying and a drift were employed with the tubs of ironstone travelling from the drift entrance across Crabtree Walk to a headshunt before being let down a self-acting incline to a tipping shed part way along the Ormesby Mine branch line. With exhaustion of the reserves, the "24 Acre" workings had been abandoned by 5/1880.

Ironstone continued to travel down the Cleveland Railway towards Cargo Fleet, but output from Normanby Mine began to decline from the early 1880s and, in 1883, Bell Bros withdrew from it. The Cargo Fleet Iron Co secured a new lease of the royalty from 11/8/1883 and paid Bell Bros for the plant. Annual production reached 138,427 tons in 1884, but then declined as the remaining ironstone was worked out and the mine closed in 1898, being formally abandoned on 15/4/1899.

References :   *Eston and Normanby Ironstone Mines*, Richard Pepper, pub. By Peter Tuffs, 1996

The Ironstone Mines and Railways of Cleveland and Rosedale, T.E. Rounthwaite, pub. by Peter Tuffs (originally appeared in RCTS Railway Observer), 1997

ORMESBY IRONSTONE MINE, Ormesby   H6
The Cargo Fleet Iron Co until 1/1883   NZ 545167 – **Map H**
Swan, Coates & Co until /1879

The mine was established by Swan, Coates & Co in 1865 using quarrying to extract ironstone from the west end of the Eston Hills. The Bill for the Cleveland Railway had included powers to construct a ½ mile branch line on a 1 in 39 gradient from near the Normanby Mine sidings westwards to these workings. Approximately halfway along the branch, near West Farm, a siding on the south served Bell Bros' tipping shed from its "24 Acre" site. The branch terminated at Ormesby Mine and, according to T.E. Rounthwaite, two tramways fed ironstone into the standard gauge wagons at the "extremity" of the line. These were then hauled by the NER to the iron works at Cargo Fleet. Annual production was low during the 1870s, but had reached 168,909 tons in 1881. By the early 1890s, the ironstone was exhausted and the mine was abandoned on 14/1/1892, but it was not the end of the branch (see the Cleveland Magnesite & Refractory Co Ltd entry in this section).

Reference :   *The Ironstone Mines and Railways of Cleveland and Rosedale*, T.E. Rounthwaite, pub. by Peter Tuffs (originally appeared in the RCTS Railway Observer), 1997

THE CAST STEEL FOUNDRY CO LTD

ROSEBERRY STEEL WORKS, Middlesbrough   H7
Messrs Butler Bros until /1887   NZ 488214 – **Map F**

This small foundry was founded by John Theobald Butler and was located next to the MOR Old Town Branch Loop, immediately north of the Ayrton Rolling Mills (later Sheet Works) in the Ironmasters District. The Butler brothers dissolved their partnership in 1887 and the works was taken over by The Cast Steel Foundry Co Ltd. The foundry operated six small "open hearth furnaces" turning out steel castings ranging from one to ten tons.

It had a capacity of about 100 tons per week. A number of short sidings had a connection with the Loop line. A contemporary account in 1887–88 says that the foundry was constructing a "large works and offices on land adjoining the present works", however, this is probably a reference to its managing director, William Shaw, who was establishing his own business further south along Forty Foot Road. The Roseberry Steel Works was still operating in 1892–93 but, according to a note dated 24/1/1898 in the MOR records, the property had been taken into the Ayrton Sheet Works.

CLEVELAND MAGNESITE & REFRACTORY CO LTD

NORMANBY BRICK AND TILE WORKS
(later **CLEVELAND REFRACTORY WORKS** 1915-73) near Normanby   **H8**
New Normanby Brick Co until ?   NZ 551168 – **Map H**
Normanby Brick & Tile Co until c/1910

The Siding Agreement with the NER for the works was dated 3/9/1884. The brick works was situated at the end of the NER's Normanby Branch in the angle of lines leading to the Normanby Ironstone Mine to the east and Ormesby Mine to the west. In 1894 a short single track railway led from the Normanby Branch to the works sidings along the approximate course of an abandoned section of the Cleveland Railway. The shale was dug in a pit above and to the east of the works and a rope-operated incline took the loaded wagons down Ten Acre Bank to the pugging mill. By 1910 the business was named the New Normanby Brick Co and had two Hoffman kilns making pressed facing and glazed bricks. Immediately prior to World War 1, the works was taken over by the Cleveland Magnesite & Refractory Co. The Government was keen to increase steel production for the war effort and the Ministry of Munitions contracted with the company to make silica refractory bricks for the steel furnaces. Production of refractory bricks and cements continued after the war employing about 55 people. The gannister and magnesite had to be transported in by the railway company. Interestingly, the Normanby Brick Works wanted to acquire a 14 inch locomotive according to *Machinery Market* on 18/12/1925. The two Hoffman kilns were replaced by nine rectangular down draught kilns which were coal fired to 1970. Messrs Steetley Ltd took over the works in the late 1960s and ran it until closure in 1973.

Reference : *Brickworks of the North East*, Peter J. Davidson, Gateshead Libraries & Arts Service, 1986

CLEVELAND SLAG WORKING CO LTD

Cargo Fleet   **H9**
    NZ 507207 – **Map G**

According to Burnett & Hood's 1874 Directory of Middlesbrough, the Cleveland Slag Working Co Ltd was formed to make use of blast furnace slag, of which there were copious amounts. Amongst the directors were a Robert Stephenson, Isaac Wilson, Jeremiah Head and Charles Bagnall. The premises were situated on the north side of the MOR's line as it curved round to Dock Lock Bridge at Middlesbrough Dock, immediately north of the Yorkshire Tube Works. A siding is shown on an 1875 map entering the slag works from the MOR line. Slater's Directory of 1876-77 called it the Dock Gate Works and stated that it manufactured slag bricks and paving blocks, "broken" slag for road making and Portland and Selenetic cement. By 1893, the NER's stockyard for Middlesbrough Dock occupied the site of the slag works.

I. COPLEY & CO, VULCAN WORKS

Middlesbrough   **H10**
Cleveland Patent Nut & Bolt Co until c/1882   NZ 500200 – **Map G**

Following the dissolution of the Sidney, Sherwood & Smith partnership in 1846, the foundry immediately east of the Tees Engine Works and adjoining Vulcan Street was rented by Gilkes, Wilson & Co and linked to its premises by an internal railway line. In the 1850s Bolckow & Vaughan was negotiating to purchase the foundry and more importantly its access strip to the river, which impeded its plans for the expansion of Middlesbrough Iron Works. The purchase was finally completed from Sydney, Sherwood & Co on 29/5/1859. Thomas Vaughan, the son of John Vaughan, was allocated the site of the foundry as part of the establishment of Bolckow, Vaughan & Co Ltd in 1865. He had formed a new company called the Cleveland Patent Nut & Bolt Company and an article in *The Engineer* stated that this had commenced operations in 1864. The 1½ acre site was oblong in shape and served by a short siding off the Vulcan Street railway. Production mainly consisted of nuts and bolts, although some galvanising was also carried out. The company supplied most of the wrought iron nuts and bolts for the ill-fated Tay Bridge and their quality attracted criticism at the Inquiry. The business, however, survived and continued until at least 1882. Five years later the works was shown in the occupation of I. Copley & Co, marine and general engineers. John Harrison quotes a description of the Vulcan Works as it was in 1890. The 1893 Ordnance Survey plan shows a second short siding entering the premises from Vulcan Street.

Reference : *The Industrial Heart of Old Middlesbrough,* J.K. Harrison, CIAS, 2010

R. CRAGGS & SONS LTD

TEES DOCKYARD, Middlesbrough
R. Craggs & Sons until c/1903

H11
NZ 493215 – **Map F**

Robert Craggs started ship repairing in Stockton before moving to take over a small shipyard (NZ 506210) next to the Cleveland Dockyard on 26/11/1866. On 31/1/1896, R. Craggs & Sons occupied the vacant shipyard on the northern part of the Tees Side Iron & Engine Works while its former yard was absorbed into Raylton Dixon's Cleveland Dockyard. This allowed it to build bigger ships; there were four berths ranging from 400 to 600 feet in length. For the ten years from 1897, Craggs' annual output totalled almost 20,000 tons compared to less than 3,000 tons at Dock Point. The largest vessel constructed at the Tees Dockyard was the DACRE CASTLE at 5,250 tons. Many of the ships' engines came from Richardsons, Westgarth & Co Ltd, of which Henry Craggs was a director. The yard was connected to the Old Town Branch with sidings crossing Depot Road into the premises. Richardsons, Westgarth had approval from the NER for its locomotive to work between its wharf and Cragg's shipyard via Vulcan Street. Members of the Iron & Steel Institute visited Craggs' shipyard in 1908, when the yard was stated to cover about 9 acres and give employment to about 1,200 to 1,300 men. The report of the visit was fulsome in its praise but, despite this, the yard closed in 1909, the last ship being completed in July of that year. The yard was demolished but the sheerlegs crane at the fitting out quay remained in existence for a long time. In 1910, the premises were sold to T. Roddam Dent for inclusion in its Dent's Wharf (which see).

Reference : 'R. Craggs & Sons: Innovative Shipbuilders', Jenny and Geoff Braddy, *Cleveland Industrial Heritage* 15, pub. by Peter Tuffs, 2004

CRAIG, TAYLOR & CO LTD

THORNABY SHIPBUILDING YARD
Craig, Taylor & Co until /1905
Irving, Lane & Co until /1843

H12
NZ 448190 – **Map E**

This yard was situated on the south bank of the River Tees to the west of the South Stockton Shipyard and Teesdale Iron Works. It appears to have operated in 1837-43 with the first ship, SOUTH STOCKTON, being launched on 13/7/1839 by Irving, Lane & Co. The ENGLISH ROSE, completed in 1843, was believed to be the first steam ship built at Stockton and had engines made by Bolckow & Vaughan of Middlesbrough. There then follows a period of uncertainty, although an 1850 map still shows a shipyard on the site.

In 1884 Craig, Taylor & Co began reconstruction of the seven acre yard and the SAINT ANORE, completed in the following year, was the first ship to be launched by the company. It proceeded to construct a range of ships, including oil tankers and, by 1900, the site covered 12 acres with six building berths and slipways facing upstream to the Stockton turning circle on a wider part of the river, an 800 feet long wharf and an 80 feet wharf for discharging timber. The yard had no connection with the NER and only a simple internal siding in 1899 presumably for rail cranes. It also continued to rely on primitive wooden scaffolding driven into the river bank mud for constructing ships. The business became a private limited company in 1905 and three years later the number of berths had increased to eight with a 540 feet graving dock. By now, the yard could build ships up to 7,500 tons. The internal rail network had expanded by 1915 to link the wharf with the rest of the yard but there was still no connection to the main line. The years after World War 1 were difficult and between 1920 and 1931, when the yard closed, only 15 ships were built including the 11,200 ton CITY OF TOKIO for Ellerman Lines. The final ship was launched in 12/1929 with yard number 227. The site was taken over by National Shipbuilding Securities Ltd, whose remit was to reduce capacity in the industry and it was then acquired by Head Wrightson. During World War 2, the yard was used to assemble some of the 238 tank landing craft prefabricated by Head Wrightson and others under the aegis of the Stockton Construction Co.

Reference : *Shipbuilding in Stockton and Thornaby,* Alan Betteney, Tees Valley Heritage Group, 2003

CREWDSON, HARDY & CO LTD

YORKSHIRE TUBE WORKS, Cargo Fleet
Crewdson, Hardy & Co until 16/7/1893

H13
NZ 506206 – **Map G**

The works was situated between a MOR siding that ran parallel to and next to the north side of the Middlesbrough-Redcar railway and the Cargo Fleet-Vulcan Street MOR line. Crewdson, Hardy & Co started the Yorkshire Tube Works in 1873 as a wrought iron and later steel tube manufacturer. Tubes and fittings were manufactured for gas, steam, water and hydraulic functions, together with tuyeres for blast furnaces. In 1890, the works occupied a three acre site and employed about 130 people. A drawing of "proposed sidings to Messrs, Crewdson, Hardy & Co works with plan of new junction" was dated 1/6/1875. A single siding ran north to south through the centre of the site connecting with both MOR lines, although Crewdson, Hardy had some designated sidings outside its boundary for wagon standage. According to NER correspondence dated 5/8/1897, a Tees Iron Works locomotive travelled over the MOR line to carry out shunting for Crewdson, Hardy. The company still controlled the works on 27/3/1929 but, shortly after, was replaced by Cleveland Products Ltd. The latter was present in 1953 with a siding off BR's Middlesbrough Dock lines.

Reference : 'Dock Hill', Jenny and Geoff Braddy, *Cleveland Industrial Heritage* 28, pub. Peter Tuffs, 2011, pp. 31-35

JOHN CROMBIE & SON LTD

CLEVELAND SLAG AND CONCRETE WORKS, Middlesbrough

H14
NZ 486207 – **Map F**

The company was described as "Artificial stone manufacturers" in a 1914–15 directory and had been added to the MOR schedule of toll payers on 20/7/1899. The Siding Agreement with the NER was dated two days earlier. Crombie had previously rented a site from the NER called the Dock Concrete Works, but had been required to move to make way for the Middlesbrough Dock Extension. Therefore it rented some land at Ironmasters from the OME. Its premises were located alongside the Marsh Branch sidings immediately to the north west of Metz Bridge. In 1913 a line off the NER West Marsh Wharf sidings ran into Crombie's premises and divided into two, one with a passing loop. The site had been incorporated into the Britannia Works by 1927 and was occupied by sidings.

CROSSLEY & CO LTD

ORMESBY BRICK WORKS, Ormesby
Ormesby Metallic Brick Co Ltd until 29/8/1913

H15
NZ 545167 – **Map H**

The Ormesby Metallic Brick Co was registered on 19/7/1902; one of the subscribers being S.A. Sadler. It erected the brick works on the site of the Ormesby Ironstone Mine which was served by the Ormesby Mine Branch Railway. The Ormesby company was officially dissolved on 29/8/1913 and it appears that Crossley then controlled the brick works. During World War 1, the Ministry of Munitions proposed that the Cleveland Magnesite & Refractory Co lease the works from Crossley in order to increase the manufacture of silica refractory bricks. During the war, the works was operating with three Newcastle and four Hoffmann kilns. At the end of the war, Cleveland Magnesite & Refractory Co unsuccessfully tried to obtain recompense from the Government for the use and extension of Normanby Brick Works and the lease and extension of Ormesby Brick Works. The Ormesby Works was closed between 1919 and late 1921, but began making facing bricks in 1924. A new clay pit opened circa 1930. According to Davidson, four men and a boy worked in the quarry using picks and shovels to dig out the shale and load it into railway tubs. These were hooked on to a steel-ropeway fitted with a steam powered rotating drum that hauled the wagons to the mill. In late 1939 the works ceased making facing bricks and tiles and recommenced the manufacture of refractory bricks up until its closure in 1968. Latterly a class J27 steam locomotive from BR's Thornaby Shed had visited the Normanby Branch with coal and wagons for the brickworks; Ormesby used to receive 7-8 wagons, but there was often only one for Normanby. The BR branch closed on 3/10/1966 severing the link to the two brickworks.

References : 'Brickworks of South Bank, Eston, Normanby and Ormesby', David Tomlin, *Cleveland History* 71, Cleveland & Teesside Local History Society, 1996

Brickworks of the North East, Peter J. Davidson, Gateshead Libraries & Arts Service, 1986

DORMAN, LONG & CO LTD

CARLIN HOW IRONSTONE MINE, Carlin How H16
Bell Brothers Ltd until 2/5/1923 NZ 710192 – **Map I**
Bell Brothers & Co until 27/11/1873

Bell Brothers leased the ironstone royalty on 13/6/1868, but only began sinking the two shafts 151 feet to the Main Seam during 1873. Ironstone began to be taken away on the railway on 27/10/1873. The mine buildings were located part way up the west side of Skinningrove Beck Valley next to the NER's Zig Zag railway. After the Zig Zag descended from the first reversal and crossed over the Carlin Howe–Loftus road, a short siding ran northwards to the mine and the headshunt just beyond the tippler building. Wagons could then run by gravity for filling and collection by the NER. However rail traffic soon proved to be limited. In 11/1881 approval was given for an underground drift to connect the mine with the North Loftus Mine shaft. Between 1887 and 30/11/1894 almost all of the output went up that shaft to Skinningrove Iron Works. When the latter changed to relying principally on ironstone from Loftus Mine, an underground connection was put into Bell Bros' Lumpsey Mine. By 1897 all of the Ironstone was being moved in tubs by horses and then taken by rope haulage to Lumpsey. Rail traffic from the Zig Zag was probably only coal for the boilers. With the exhaustion of the reserves, Carlin How Mine closed on 20/7/1944.

Reference : *Lumpsey Mine "Flower of Cleveland" The History of the Lumpsey and Carlin How Mines*, Simon Chapman, pub. P. Tuffs, 1997

Cleveland Ironstone From Lumpsey, T. Kitching, Sotheran, 1995

SKELTON IRONSTONE MINES, Slapewath H17
Bell Brothers Ltd until 2/5/1923 NZ 644180, NZ 639158, NZ 637168 – **Map I**
Bell Brothers & Co until 27/11/1873
Bell Brothers until /1864

At Slapewath, east of Guisborough, the road to Whitby passes through a narrow valley formed by glacial action. Bell Brothers signed a lease with the landowner, J.T. Wharton, with effect from 1/1/1858, to remove ironstone from his property to the north of Slapewath. Initially a drift was made into the valley side but this was soon superseded by the establishment of **Skelton Shaft Mine** (NZ 637168). The Cleveland Railway received approval for its line to run through the valley at Slapewath, including a short standard gauge branch to reach the mine on 23/7/1858. The railway opened from the south bank of the River Tees to Slapewath, where there was a reverse junction (due to the narrowness of the valley) to give access to the branch on 23/11/1861. Sixty tons of ironstone had been despatched from the mine by the end of the month and production was up to 400–500 tons per day by 1863. In 1867 the underground workings at the Shaft Mine relied on horses to move the tubs to the Engine Plane where a haulage engine pulled sets of 10 tubs (16–17 in 1875), each weighing 10 cwt and carrying 23 cwt (28 cwt in 1875) of ironstone on the narrow gauge railway to the shaft bottom. Each tub was then raised singly to the surface where the contents were tipped into the railway company's wagons.

Bell Bros opened a drift in 1864 at Slapewath which passed under the Cleveland Railway into the north side of the valley to extract ironstone from the Hewley royalty. This was the **Spa Mine** (NZ 639158) and a haulage engine located near the Fox and Hounds Inn drew the sets of 12-15 tubs on a narrow gauge tramway with a 1 in 19 gradient out of the drift so that the contents could be tipped into standard gauge wagons standing on the sidings alongside the Cleveland Railway. This mine was sold to Gjers, Mills & Co in 1872 which operated it until 1904 when the reserves were exhausted.

Bell Bros established the larger **Skelton Park Mine** (NZ 644180) on the northern part of the Skelton royalty with the Main Seam being reached at 384 feet and production commencing in 1873. The branch railway was extended past the Shaft Mine up to Skelton Park Mine; the latter having an output of about 1,500 tons per day by 1881. The resulting railway was now 1 mile 73 chains in length (36 chains owned by the NER and the remainder by Bell Bros) and worked by the NER with control by Slapewath Junction Signal Box. The branch terminated just beyond Skelton Park Mine where there was a coal depot near the headshunt. Empty wagons gravitated from here along two lines under the tippler house for filling and were then stabled for collection. A separate siding enabled coal to be delivered to the boiler house. At the turn of the century, 55 horses were used mostly underground at Park and Shaft Pits. In 1929, Park had a 60hp electric motor powering three endless rope haulages on the main underground 'roads' (South 670 yards, South East 2,800 yards and North 1,970 yards); earlier, compressed air had been used.

Bell Bros gained approval in 9/1905 to obtain ironstone from a small area on the west side of Waterfall Beck and a 2ft 6in gauge tramway was laid from these workings to Skelton Shaft Mine. The tubs ran in sets of 21 hauled by a stationary engine at Shaft Mine until 8/1908. Skelton Shaft Mine shut on 31/7/1923, although the shafts continued to be used for ventilation and the drift reopened in 1933 to remove the ironstone from under the mine buildings. A 150hp electric hauler was transferred from Spawood Mine and this pulled a full set of tubs out of the drift, reversed direction and then raised them up a second incline to reach the tippler. Final closure came on 30/9/1938; Skelton Park Mine having finished on 19/2/1938.

There is one other railway to mention in connection with the branch. The Cargo Fleet Iron Co Ltd leased some Chaloner land at Tocketts Lythe on 24/6/1892 and established **Waterfall Mine** (NZ 628173). The first ironstone left the mine on 9/9/1893 and a narrow gauge tramway transported the stone, crossing over Waterfall Beck on a wooden bridge, to reach a tipping dock at a siding off the branch not far from Slapewath Junction. The siding agreement with the NER was dated 17/3/1893. Drainage problems and doubts about the amount of ironstone remaining led to the abandonment of the mine on 31/5/1901.

References : *Skelton Park Pit Ironstone Mine*, Simon Chapman, pub. by Peter Tuffs, 1999

'Old Shaft Mine-Skelton about 1900', Simon Chapman, *Cleveland Industrial Heritage* 9, pub. Peter Tuffs, 2001, pp. 12-16

'Visit to Skelton Park and Shaft Mines 18th May 2019', Simon Chapman, CIAS *Newsletter* No.117, 9/2019

SPAWOOD IRONSTONE MINE, Slapewath   **H18**
Sir Bernhard Samuelson & Co Ltd until /1917   NZ 638157 – **Map I**
The Weardale Iron & Coal Co Ltd until /1890
The Weardale Iron Co until 23/7/1863

The Weardale Iron Co leased a substantial royalty of ironstone south of Guisborough for its Belmont Mine and the first load was sent away over the Middlesbrough & Guisborough Railway on 5/1855. As the original lease posed too great a financial burden, part of the royalty at South Belmont was sublet, but the Weardale Iron & Coal Co began to extract ironstone from the eastern end of the royalty in the narrow defile at Slapewath during 3/1864. Sidings were laid from near the west end of Waterfall Viaduct parallel with and south of the Cleveland Railway line to the mine at Spawood. Spawood Mine was shut a number of times before it finally closed in 8/1886, being officially abandoned on 11/11/1886. The 1895 Ordnance Survey map shows that the railway sidings had been removed leaving just the earthworks and an isolated building.

Sir B. Samuelson & Co then acquired the Spawood royalty and extracted the first ironstone from it on 1/7/1890 with access probably being gained from the company's Slapewath Mine workings and the stone taken up the latter's shaft. This arrangement restricted the amount of output that could be handled and on 25/2/1897 Samuelson decided to develop Spawood Mine as the discharge point for its royalties at Spawood, Tidkinhow, Hollin Hill and Aysdalegate. The use of a drift linked to railway sidings connected to the main line meant that larger amounts of stone could be sent out. Two Lancashire boilers and a Robey steam engine were installed outside the new main drift at Spawood to haul tubs out of the mine. Standard gauge sidings were laid to connect with the NER and, according to the *Iron and Coal Trades Review* on 13/5/1898, the mine was fully operational. Ironstone ceased to be wound up the Slapewath shaft within a few years. Spa Wood Signal Box was provided with a new frame with 14 levers in 1897 and the sidings terminated at Slapewath Depot which could handle ordinary goods traffic.

The standard gauge sidings from the main line ran eastwards on a shelf cut into the hillside. As the mine workings were extended southwards, the band of shale in the ironstone thickened and so a picking belt was provided linked to the tippler house. A narrow gauge tramway took tubs from the picking belt building eastwards before making a tight turn and heading south up the hillside on a viaduct and single track incline. At the end, there was a headshunt and tubs could be let down for tipping into an old alum quarry. A Dick, Kerr 60hp motor with a double drum hauled tubs of shale to the tips. Eventually more space was required for tipping and an aerial ropeway, with its own picking belt, was installed in 1928. This took the shale south west past Old Park Farm to Cass Rock Quarry. Underground, there were principal haulages at No.1 East, No.5 West and Tidkinhow supported by a number of self-acting inclines. Spawood Mine closed on 28/6/1930, but was kept on a 'care and maintenance' basis until the decision was taken by Dorman Long in 1933 to abandon it. Nineteen miles of 2ft 6in gauge track were recovered and 1,237 tubs, of which 887 were scrapped but 350 frames and bodies were sent to Longacres Mine. The aerial ropeway was dismantled and re-erected at Dorman Long's Upton Colliery. All the equipment had been removed by 23/4/1934.

Reference : *Guisborough District Mines*, Simon Chapman, pub. by Peter Tuffs, 2001

DORMAN LONG (STEEL) LTD

LONGACRES IRONSTONE MINE, Skelton
Dorman, Long & Co Ltd until 2/10/1954
Bolckow, Vaughan & Co Ltd until 29/11/1929

H19
NZ 669195 – **Map I**

This was an extra mine sunk to the North Skelton royalty by Bolckow Vaughan, the idea being that the less deep shafts (313 feet) compared to those being sunk at North Skelton Mine, a mile to the south, would result in an earlier supply of ironstone. However, a fire and drainage problems delayed completion. Sinking had commenced on 7/3/1873 but it was not until 9/1875 that small amounts of ironstone were being extracted. Longacres Mine was situated on the east side of the NER's Saltburn–Brotton line (opened for mineral traffic 1/6/1872). The arrangements on the surface were typical for a Cleveland mine with a railway leaving the NER's line and running past the shafts to the empties sidings where the wagons were left by the railway company. These were let down by gravity under the tippler for filling and then collected by the railway company near the junction with the branch. A small coal depot was situated opposite the junction; the latter being controlled by Longacres Signal Box. Longacres Mine closed on 17/7/1915. A long underground self-acting incline was constructed enabling its ironstone to be raised at North Skelton from 1918. When extraction of the remaining ironstone pillars threatened to cut the incline, Dorman Long reopened the Longacres shaft for men and materials in 1933 and for ironstone in 1943. The mine finally closed on 27/11/1954 and the railway sidings on the surface were removed in the following year.

Reference : *Hope to Prosper A History of Ironstone Mining at North Skelton*, Simon Chapman, pub. Peter Tuffs, 1997, pp. 45-52

DURHAM PAPER MILLS LTD

HARTLEPOOLS PAPER WORKS, later DURHAM PAPER MILLS
West Hartlepool
Hartlepools Paper Mill Co Ltd until /1926
Hartlepools Pulp & Paper Co Ltd until /1920

H20
NZ 513318 – **Map B**

The short Burn Road Branch diverged from the Cliff House Branch at Mainsforth Crossing Box and ran west to the NER's Burn Road sidings. It also served various timber yards, the Council "refuse destructor" and the Hartlepools Paper Works. The Hartlepool Paper Co Ltd established the pulp works at Moreland Street in 1891 to make white paper and the Hartlepools Pulp & Paper Co Ltd was registered to operate it. The works was in a good strategic position to receive imported materials and coal, together with despatching goods by rail. A siding left the NER branch, crossed Burn Road to enter the premises where it divided to run on two sides of the main building. No private locomotives were involved and the railway company delivered wagons to the various businesses on the Burn Road Branch. The paper company went into voluntary liquidation in 1920 and the works was sold to the newly established Hartlepools Paper Mill Co Ltd. It was then acquired by Durham Paper Mills Ltd in 1926; a siding agreement being signed with the LNER on 3/12/1926. The mill was eventually taken over by the Inveresk Group via Olive & Partington Ltd and ceased production on 28/8/1965. The notable tall chimney was demolished in 1971.

Reference : *The Lost Mills A History of Papermaking in County Durham*, Jean K. Stirk, University of Sunderland, 2006

WILLIAM HARKNESS & SON LTD

SHIPYARD, DOCK POINT, Middlesborough
W. Harkness & Son until 25/3/1903

H21
NZ 507210 – **Map G**

William Harkness came to Middlesbrough in 1853 and began his small yard (NZ 507210) with three berths in 1856 on the east side of Cragg's premises near Dock Point. W. Harkness & Son specialised in small coasters and colliers; it had constructed 160 ships up to 1902, T.D. Ridley being one of the directors. The yard had three berths for ships up to 318 feet in length, a 520ft dry dock and a 130ft floating dock. The largest ship constructed was an Ellerman Lines vessel of 4,400 tons. The yard was served by a small number of sidings near the end of the Vulcan Street railway. At the conclusion of World War 1, the shipyard was sold to Monsieur Hani and it went into voluntary liquidation in 1923. It was one of the yards purchased in 1931 by the National Shipbuilders Security Ltd for dismantling.

HARTBURN BRICK & TILE CO

HARTBURN CURVE BRICK WORKS, Stockton
R.P. Dorman until?

H22
NZ 438178 – **Map D**

The brick works was located immediately north of the Castle Eden Branch shortly after that line left Bowesfield Junction. The limits of the clay pit were defined by the NER Hartburn Curve and Yarm Road. The 1899 Ordnance Survey map shows that the works had three vertical kilns and was served by two sidings from the Hartburn Curve. A separate railway track, presumably narrow gauge, connected the clay pit to the works. The Castle Eden Branch opened to goods traffic in 1877 and R.P. Dorman was operating the "Hartburn Curve Steam Brick Works" about 1885–94. The Hartburn Brick & Tile Company had taken over by 1920, but the works had been demolished according to the 1923 Ordnance Survey six inch map.

Reference : *The Brickworks of the Stockton-on-Tees Area*, Alan Betteney, Tees Valley Heritage Group, 2007

W.K. HUNTON, PORTRACK LANE BRICK AND TILE WORKS

North Shore Branch, Stockton

H23
NZ 452193 – **Map D**

According to the 1857 Ordnance Survey map, this brick works was situated on the east side of the North Shore Branch, north of Portrack Lane (NZ 449197). It was described as John Hunton's brick and tile works in 1886. Most of the yard had been quarried away by 1899 and, in the previous year, plans had been prepared on behalf of W.K. Hunton for a new Portrack Lane Brick Works (NZ 452193) again located next to the North Shore Branch but to the south between the Stockton Union Workhouse and the Malleable Works sidings. A Siding Agreement with the NER was dated 31/3/1898. The clay pit for Hunton's second site was located to the east of the brickyard and a tramway connected the pit with the works passing under a footbridge part way along its course. It later became the Portrack Lane Brick & Tile Co.

Reference : *The Brickworks of the Stockton-on-Tees Area*, Alan Betteney, Tees Valley Heritage Group, 2007, pp 32-34

IRVINE'S SHIPBUILDING & DRY DOCKS CO LTD

HARBOUR SHIPYARD, West Hartlepool
Irvine & Co until /1898
R. Irvine & Co until /1880
Irvine, Currie & Co until /1866

H24
NZ 518329 – **Map B**

Irvine's relied on travelling steam cranes to carry out any shunting at its two yards as these were essential for moving heavy and unwieldy plate and equipment. The West Hartlepool Harbour & Railway Company advertised land on the side of the West Harbour as a shipbuilding site and Robert Irvine joined with Alexander Currie to purchase it; the first vessel being launched from there in 1864. A graving dock was constructed two years later, the same year as Currie left the business and extensive ship repair work was then carried out. By the 1890s, the Harbour Yard was in need of modernisation and to permit the construction of larger ships. Local civil engineering contractor, Joseph Howe & Co, carried out this work in 1897–98. It was also at this time that Christopher Furness became the main shareholder and the business name was changed to Irvine's Shipbuilding & Dry Docks Co Ltd. The yard reopened in 1898 with increased orders for new ships and repair work. The management of the Harbour Shipyard was combined with that of the Middleton Yard in 1909. The Harbour Shipyard was situated between the Middle Pier in West Harbour and the extensive sidings serving the Coal Dock and, in 1896, a single siding from these NER lines entered the yard. Five tracks branched off this siding to run alongside the slipways; the one next to the graving dock was labelled "travelling crane". Spaldin makes reference to a 20 ton steam travelling crane, together with a number of smaller steam cranes, being used at the yard. The Harbour Yard remained busy until 1922 but only one vessel was launched in each of the following two years and it closed in 1925.

Reference : *Shipbuilders of the Hartlepools*, Bert Spaldin, Hartlepool Borough Council, 1986

IRVINE'S SHIPBUILDING & DRY DOCKS COMPANY (1930) LTD

MIDDLETON SHIPYARD, Hartlepool   H25
Irvine's Shipbuilding & Dry Docks Co Ltd until 30/1/1930   NZ 522337 – **Map B**
Furness, Withy & Co Ltd until /1909
E. Withy & Co until /1891
Withy, Alexander & Co until /1873
Denton, Gray & Co Ltd until/1869

The Middleton Shipyard was situated on the west side of the Old Harbour next to Thomas Richardson's Hartlepool Engine Works. The site had been used previously by a number of shipbuilders and repairers when John Punshon Denton took over Richardson and Parkin's yard in 1839 to build numerous wooden sailing ships.

In 1857, the then existing docks at Hartlepool and West Hartlepool were connected by a railway company's line that ran northwards past the Swainson and Jackson Docks, alongside the eastern edge of The Slake before curving round to cross the sluices at the head of the Old or Tide Harbour. As it approached the sluices, a railway trailed off to serve J.P. Denton's shipyard. John Denton formed a partnership with William Gray in 1863 to build iron ships. With increasing orders, Denton, Gray looked for larger premises and leased the vacant Pile, Spence yard in West Hartlepool.

The Middleton Yard was then taken over by the newly formed Withy, Alexander & Co in 1869. Edward Alexander retired in 1873 and Edward Withy emigrated in 1884, with the latter's interest in the company being purchased by Christopher Furness. In 1891 Christopher Furness combined his various shipping interests to form Furness, Withy & Co Ltd. The yard was modernised to construct and repair larger ships with longer slipways and a graving dock that could accommodate ships up to 7,000 tons. The creation of the North Basin had resulted in changes to the rail connection. This was now further south near Princes Street Signal Box and Richardson's locomotive delivered traffic for the Middleton Shipyard through its Hartlepool Engine Works premises. Within the yard, Furness, Withy & Co's railway served each slipway and the graving dock, together with running across Ferry Road to the buildings on the southern half of the site. Bert Spaldin states that the shipyard used nine steam travelling cranes.

In 1909, faced by falling orders, Furness amalgamated the Middleton Yard with Irvine's Harbour Yard. After World War 1 orders declined and the final three ships were launched from the Middleton Yard in 1924 and it closed in 1925, the company going into liquidation in 1930. A local syndicate took over under the title Irvine's Shipbuilding & Dry Docks Company (1930) Ltd and some ship repairing and breaking was undertaken. It operated until 1938 when the National Shipbuilders Security Corporation purchased the yard with a view to closure. Despite this, the graving dock and fitting out quay were used especially during World War 2 and to install Richardsons Westgarth's engines in ships. By the 1970s, all of this activity had ceased but the quay was later used as part of the commercial harbour.

Reference :   *Shipbuilders of the Hartlepools*, Bert Spaldin, Hartlepool Borough Council, 1986

J.G. LOWOOD & CO LTD

WHARNCLIFFE GANISTER BRICK WORKS, Middlesbrough   H26
NZ489213 – **Map F**

John Grayson Lowood had operated a brick works at Deepcar, north of Sheffield, from about 1860. The quarry at Deepcar provided fine ganister clay to manufacture fire bricks for use in kilns. With the growth of the iron industry on Teesside, Lowood wanted to tap into this market and yet keep his transport costs low. Accordingly, he established a brickworks adjoining Forty Foot Road, next to the Ayrton Rolling Mills (later Sheet Works), about 1888. It is possible that he took over the Scotch Firebrick & Ganister Co Ltd's Middlesbrough Firebrick Works that had opened about 1884 at West Marsh and was last recorded in the directories in 1887. The works used a continuous regenerative gas furnace to provide bricks which were less prone to expansion in open hearth furnaces. The silica-rich rock was quarried at Castleton and taken down a railway incline to the Esk Valley line for transport to the Ironmasters District. Here a line ran off the MOR Old Town Branch Loop, close to its junction with the Old Town Branch, dividing into three sidings that ran over Forty Foot Road at a location similar to Dawson's present crossing to enter the brick works. Directories suggest that the works continued operating into the 1930s.

Reference :   'Victoria's Youngest Child (36) Forty Foot Road: An Unnamed Thoroughfare', Jenny and Geoff Braddy, *Cleveland Industrial Heritage* 40, pub Peter Tuffs, 2017

MORRISON & CO

BROTTON IRONSTONE MINE, Brotton

H27
NZ 685201 – **Map I**

The lease of the ironstone was obtained by Robert Morrison on 11/12/1863 at the same time as work was getting underway on the construction of the Cleveland Railway from Slapeworth to Carlin How. He had previously been head of Messrs R. Morrison & Co of the Ouseborn Engine Works, Newcastle upon Tyne. Two shafts were sunk to the Main Seam ironstone with a third used for pumping. The mine was located on the west side of the railway at Brotton and a trailing connection allowed empty wagons to be shunted into the mine siding. From here, they ran by gravity under the tipping chute and into the loaded siding where a NER locomotive could collect them and rejoin the main line. Morrison was reliant on the open market to sell his ironstone; in 1867 he was sending 3,500 tons per week to Stevenson, Jaques & Co's Acklam Iron Works. When the mine was inspected in 1872, two 60hp stationary engines hauled tubs underground up the 1 in 18 north and south inclines. There were stalls underground for 22 horses to pull the mine tubs. The ironstone from Brotton Mine was well regarded and it achieved a substantial output – the fourth highest (428,716 tons) of any Cleveland mine in 1881. The mine closed on 2/2/1921 due to a sparsity of orders and the approaching exhaustion of the reserves; the tall boiler chimney being felled on 12/11/1921.

References :   *Wheels Turning and Smoke Rising*" Volume 1 Mines to the north of Brotton, Simon Chapman, pub. P. Tuffs, 1997, pp.4-25

'Morrison's Pit, Brotton about 1875', Simon Chapman, *Cleveland Industrial Heritage* 8, pub Peter Tuffs, 2001, pp.10-14

NORTHERN SMELTING & CHEMICAL CO LTD

ZINC AND SULPHURIC ACID WORKS, Seaton Carew

H28

Sulphide Corporation Ltd until 9/10/1933

NZ 535272 – **Map A**

The works was established on 52 acres of reclaimed land at the mouth of the River Tees on a relatively isolated site with a cinder track for access. It was developed by the Central Zinc Co Ltd which was closely allied to the Sulphide Corporation. The latter also established the Central Acid Co to operate the sulphuric acid plant at the works. Although close to the jetty at Seaton Snook, this was too small for ships bringing zinc ore from Broken Hill, Australia. The NER was to construct the Seaton Snook Branch from a junction on the line south of Seaton Carew Station under an agreement with the Central Zinc Co Ltd dated 21/9/1906. The Zinc Co was allowed to use it and any temporary lines. The NER let the contract for the work to John Best & Sons Ltd of Edinburgh on 1/10/1906 and the branch opened on 1/6/1907.

At the works, after passing over a weighbridge, the incoming tracks divided into various parallel sidings apparently shunted by the railway company. Three central sidings ran over a 115 feet long steel gantry provided with eight chutes for incoming coal and then continuing as two tracks serving 13 zinc ore bins. It was calculated that there was standage for 39 20 ton rail wagons. To the north was the Roaster building in which the ore was heated; the resulting sulphur dioxide gas passing to the sulphuric acid plant alongside. The latter was served by a siding and the Central Acid Co operated a fleet of rail tank wagons registered to run over the NER. The first was built by J.B. Beadman, Sons & Co Ltd of Keighley and registered on 22/4/1908; acid production began in the following June. Later wagons were built by R.Y. Pickering & Co Ltd and Chas Roberts & Co Ltd. An internal narrow gauge railway transported some of the materials between plants in side-tipping wagons, for example, between the ore bins and the Roaster furnaces. The roasted ore was then transferred to the Distillery, a long building on the south side of the central sidings, where it was heated and the zinc distilled off as a vapour before being collected in metallic form in condensing pipes, prior to being cast in moulds. Metallic zinc was first produced in 1909 and was used by the local iron industry to galvanise steel. Next to the Distillery was the Pottery, each served by a single siding, where the fireclay retorts, pipes and pots used in the smelting process were made.

Output at the works remained below expectations for many years. In 1915-16, the Sulphide Corporation Ltd took over the Central Zinc Co, followed by the Central Acid Co in 1917. The works was sold to a subsidiary of the Imperial Smelting Corporation and the Northern Smelting & Chemical Co Ltd was registered 9/10/1933 to operate it. Processing of zinc ore ceased 29/8/1936 but the Roasters continued using Canadian concentrate, later followed by imported sulphur and then 'spent oxide' (mainly ferric sulphide, a product of town gas manufacture) to keep the sulphuric acid plant running. Two zinc furnaces were restarted in 4/1941 due to war demands but zinc smelting finally ceased in 3/1945. The ore storage bins were used for 'spent oxide' and, about 1948, the narrow gauge wagons were still being employed to transfer it but appear to have been superceded circa 6/1952. From 1949, the plant was subject to various changes in ownership but the loss of 'spent oxide' supplies, due to the introduction of natural gas, resulted in the closure of the works in 1974.

Reference: 'Zinc and Sulphuric Acid Works at Seaton Snook', Alan Betteney, *The Cleveland Industrial Archaeologist* 36, 2016, pp. 1-76

PEASE & PARTNERS LTD

CRAGGS HALL IRONSTONE MINE, Carlin How   H29
J.W. Pease & Co Ltd until 19/8/1882   NZ 701197 – **Map I**
Saltburn Ironstone Company until /1873
Cragshall Ironstone Company until /1872

This was a relatively small mine situated on the north side of the NER line to the west of Carlin How. Sometimes spelt Crag Hall or Crags Hall, the surface buildings were located not far from the existing Crag Hall Signal Box. The lease for the ironstone was signed on 4/9/1867 and the sinking of the two shafts near the railway was begun by Messrs Brogden and Robson, but soon taken over by the Cragshall Ironstone Company; one of the proprietors being Thomas Vaughan. The stone was reached 1/1869 and 250 tons a day were being extracted by 7/1869. Output in 1871 was 223,984 tons. The mine does not appear to have had its own standard gauge sidings off the branch; possibly use was made of the nearby NER exchange sidings for Skinningrove Iron Works. Even a small ironstone mine employed substantial lengths of narrow gauge railway track underground. As an example, Craggs Hall in 1873 had 1,848 yards of permanent track and 3,454 yards of temporary lines. With the approaching end of the major lease in 5/1892, it was evident that Pease & Partners intended to concentrate on its Lingdale Mine, so Craggs Hall closed with the termination of the lease. After abortive attempts to sell the mine, it was formally abandoned 12/8/1893, although its remaining reserves were later removed underground by Carlin How and Lumpsey Mines.

Reference: *"Wheels Turning and Smoke Rising"* Volume 1. Mines to the north of Brotton, Simon Chapman, pub. P. Tuffs, 1997

HOB HILL IRONSTONE MINE, Saltburn   H30
J.W. Pease & Co until 19/8/1892   NZ 656205 – **Map I**

Hob Hill is situated to the west of Saltburn on either side of the road from Upleatham (the present B1268). The 9 feet thick Main Seam ironstone was of reasonable quality but limited in extent. Pease began extracting the ironstone by quarrying on the north side of the hill about 3/1865, to be supplemented by drifts into the stone further south from 1/1866. Horses initially pulled the tubs out of the main drift until a John Fowler stationary engine was installed. The engine plane eventually reached 800 yards on a gradient of 1 in 16. Horses took over again when the Fowler engine needed extensive repairs in 4/1873. Ironstone was taken away from the mine by the NER over a ½ mile branch. The latter left the Redcar–Saltburn line (opened 19/8/1861) close to where the line to Brotton also diverged. The mine branch ran south crossing the Marske Road to sidings with a headshunt on the north side of the workings. The peak year of production was 1871 when 344,016 tons of ironstone were sent away. However, by 1/1875 most of the worthwhile stone had gone and the mine closed. Pease & Partners agreed a new lease on the remaining ironstone on 29/12/1899 and an aerial ropeway with about 60 buckets was constructed by Ropeway Syndicate Ltd connecting the Hob Hill workings with the company's Upleatham Mine. Ironstone began to be transported there, principally from quarrying, in the week commencing 14/6/1902. Amounts were small and the Hob Hill Mine finally closed in early 1921. Today a golf course occupies the site.

Reference: *Hob Hill Ironstone Mine*, Simon Chapman, pub. Peter Tuffs, 2003

MIDDLESBROUGH SALT WORKS, Cargo Fleet   H31
The Owners of the Middlesbrough Estate until /1915   NZ 520209 – **Map G**

The salt works was situated on the west side of the Tees Conservancy Commissioners' Graving Dock and north of the Middlesbrough–Redcar railway line. In 6/1887 the OME had sunk a borehole near the North Ormesby Toll Bar and located the 95 feet thick salt bed at a depth of 1,436 feet below the surface. Five boreholes were present in 1892 on what later became known as the Saltwells Road site to extract salt by solution mining. A reservoir was also constructed at Saltwells Road, but the brine was pumped to the OME's Middlesbrough Salt Works at Cargo Fleet for processing. Construction of the salt pans at the works was approaching completion in 1890 and there were reported to be 12 salt pans in 1899. Rail access to Middlesbrough Salt Works was via a curve of the MOR line that ran from the main line at Cargo Fleet across the northern part of the Normanby Iron Works. This siding terminated at a headshunt in the south east corner of the Salt Works site near the Graving Dock cottages. From here, railway lines headed north to enter the building containing the furnaces and salt pans. Two more lines passed alongside this building to reach the salt works wharf on the south bank of the Tees. No records survive of a locomotive being used at the works, but evidence suggests that there may have been one. The internal railway layout was of a reasonable size and a small unidentified rail-connected building existed near the headshunt. The OME's Salt Department owned a fleet of railway vans – the NER's *Private Owner Wagons Register* records the addition of 25 8 ton salt vans, Nos.50–74 built by the Darlington & Engineering Company, about 1891. Also the NER's list, dated 5/3/1897, showing private locomotives running over the NER and MOR lines includes the OME Salt Department. It seems that one of the OME's locomotives at its timber yard may have visited the salt works. In answer to a query from OME whether a charge should be made for light engines (ownership not specified) passing over the MOR from the "Cargo Fleet timber yard to Cargo Fleet Salt Works", the NER District Goods Manager commented: "it has never been our practice to charge wayleave on light engines or empties passing over the late MO lines."

The rush to profit from salt resulted in overproduction in the UK and declining prices. It appears that Middlesbrough Salt Works was closed for some time from 1905. It was reported on 18/6/1915 that OME had sold the salt works to Pease & Partners which controlled the adjoining Normanby Iron Works. How long Pease & Partners operated the plant is questionable and, by 1927, the works was closed "by arrangement". Later the premises were taken over by Foam Slag Ltd which is covered under the Normanby Iron Works entry.

References : *The Salt Industry of the River Tees*, David M. Tomlin, De Archaeologische Pers, 1982

The Owners of the Middlesbrough Estate Business Book of Directors 1886-1901, 1902-11, 1911-21 (Teesside Archives)

J.W. PEASE & CO

HUTTON IRONSTONE MINE, Guisborough   H32
NZ 604133 – **Map I**

It was no surprise that the SDR and the Quaker businesses behind it should promote the Middlesbrough & Guisborough Railway in order to open up the ironstone known to exist in the hills south of Guisborough. The estimated cost of the line was £72,000, including two inclines to Cod Hill and Roseberry Topping. The Act was authorised in 1852 and mineral traffic was able to start along the line on 11/11/1853 from Pease's mine; the branch to the mine from Low Cross Junction having opened on the same day. According to T.E. Rounthwaite, "This branch to Codhill left the main line several hundred yards east of Pinchinthorpe Station [ran past Hutton Home Farm] and ascended the rising ground with gradients of 1 in 54, 1 in 27, culminating with a substantial incline up the hillside at 1 in 19 to the mine workings. The total length of the branch was under one and a half miles." A second incline was constructed in 1860 from the branch at "Timber House" to climb up the hillside to reach drifts which became known as Roseberry Mine. The Middlesbrough-Guisborough line was doubled in 4/1855 and ironstone accounted for 84% of its total revenue by 6/1857. Hutton Mine comprised a number of quarries where the stone outcropped in the hillside and various drifts which followed the Main Seam underground to the south. These workings had various names according to where they were located – Codhill, Roseberry, Hutton. Annual output reached 314,789 tons in 1857 but was only 39,367 tons in the final year of production when it closed in 1867 as Pease chose to concentrate on the better quality ironstone at its Upleatham Mine. The family maintained a connection with the locality because Joseph Whitwell Pease lived at Hutton Hall and a station was built at Hutton Gate on the Guisborough line for his use.

References : *The Ironstone Mines and Railways of Cleveland and Rosedale*, T.E. Rounthwaite, pub. by Peter Tuffs (originally appeared in the RCTS Railway Observer), 1997

Catalogue of Cleveland Ironstone Mines, Peter Tuffs, pub. by author, 1997

TOCKETTS IRONSTONE MINE, near Guisborough   H33
NZ 620179 – **Map I**

This was a short lived mine based on a lease from Admiral Thomas Chaloner applicable from 1/1/1874. J.W. Pease & Co constructed a standard gauge railway from Bolckow, Vaughan & Co Ltd's Chaloner Mine branch line using the contractor R. Dowson of Rothbury, Northumberland. The new railway was one mile 57 chains long and involved the construction of a 350 feet long wooden viaduct, 45 feet above Wilton Lane and another over the Dunsdale Road (450 feet long and 35 feet above the road) at Tockett's Bridge. The Guisborough Exchange on 1/7/1875 announced that the railway had just been completed at a cost of £18,000. It was single track and terminated in three sidings and a headshunt. Two shafts were sunk near Tockett's House, but no ironstone was found due to a clay washout. Another two shafts were put in near the beck adjacent to the Upleatham Mine workings. The first ironstone was raised from the shafts near the railway on 7/8/1875 and continued for the next two years, although much of it came from the Upleatham Mine. The last ironstone was 1,860 tons drawn on 13/10/1877. The mine closed in 11/1877 and the wooden viaducts were being dismantled in 1880.

References : *Catalogue of Cleveland Ironstone Mines*, Peter Tuffs, pub. by author, 1997

'*Tocketts Pit and Branch Railway*', Simon Chapman, *Cleveland Industrial Heritage* 22, pub. Peter Tuffs, pp. 14-30

PHILLIPS – IMPERIAL PETROLEUM LTD

FILLING TERMINAL, Port Clarence

H34
NZ 510223 – **Map C**

In 1964 ICI Ltd joined with Phillips Petroleum Co Ltd to form Phillips–Imperial Petroleum Ltd to construct and operate a 5 million tons pa oil refinery at Seal Sands close to the Tees Estuary with associated berths for tankers. A filling terminal was established at Port Clarence connected to the remaining section of the Greatham Branch; the siding agreement was dated 18/6/1965. The terminal comprised three, later four, parallel sidings with the last portion of the then branch acting as a run round loop outside. An electric winch and haulage wagon was used to position the rail tank wagons. An unusual visitor was BR's Class K1 locomotive, 62005, which was brought to the terminal by D3143 to act as a stationary boiler providing steam for the steam tracing pipes to the refinery between 30/12/1967 and 18/3/1968. The two locomotives had to be separated by wagons because the diesel shunter was not allowed in the terminal. Afterwards, the K1 was towed to Leeds Neville Hill Depot pending preservation. Some revisions to the layout at the terminal were reported to be almost complete on 7/1/1982. Petroplus agreed to buy the refinery on 14/12/2000 but it suspended refining on 9/11/2009, although the terminal continued to function. After Petroplus filed for insolvency, Greenergy took over and reopened the terminal as a supply depot in 11/2012. The name of the company became Navigator Terminals North Tees Ltd (incorporated 25/11/2015) which operates terminals at Seal Sands and the Port Clarence facility for storage and distribution of petroleum products, vegetable oils and chemicals to inland locations. The last train for the terminal at Port Clarence ran about 2018–19 leaving the Greatham Branch with no traffic until 11/4/2023, when a train visited the Filling Terminal with bitumen tank wagons.

SIR RAYLTON DIXON & CO LTD

CLEVELAND DOCKYARD, Middlesbrough
Raylton Dixon & Co Ltd until 14/6/1897

H35
NZ 504210 – **Map G**

The relatively narrow piece of land facing the River Tees between the Middlesbrough Iron Works and the entrance channel to Middlesbrough Dock had a varied history. Up to the early 1920s, it was the location of three shipyards which were served by a series of individual curving sidings from the eastern end of the Vulcan Street railway. After the NER took over this railway, it was sometimes referred to as an extension of the "Old Town Branch". The sidings enabled materials to be delivered to the shipyards, but the railway layout within each site was modest and no locomotives appear to have been used. Presumably rail cranes could shunt any wagons that required moving.

The largest of the three businesses was Sir Raylton Dixon & Co Ltd occupying the Cleveland Dockyard (NZ 504210). In 1862 Raylton Dixon joined with Thomas Backhouse to take over the Middlesbrough site to form the Cleveland Dockyard with six slipways. Also, about 1862, David Joy & Co opened its Cleveland Engine Works next to the Cleveland Dockyard. David Joy was a Victorian engineer arguably best known as the inventor of the Joy valve gear in 1879 which was used on a number of railway companies' locomotives. The Cleveland Engine Works failed in 1871 and was advertised in the *Railway Times* of that year as "land, buildings, railways and heavy shear legs for shipping marine boilers" and having sidings connecting to the NER. The premises were then incorporated into the Cleveland Dockyard.

Raylton Dixon & Co Ltd was formed in 1878 when Backhouse retired. The company became the leading shipbuilder on the Tees with a good worldwide reputation for building large cargo liners and, in its first 35 years, constructed over 450 vessels. It also rented the shipyard at the Tees Side Iron & Engine Works in 1888 for a few years and, in the following year, built 40,689 tons of ships, a figure only exceeded by four companies in the UK. The peak decades of the Cleveland Dockyard were the 1890s–1900s when the company was responsible for seven of the ten largest ships built on the Tees. In this period, the yard had four berths for ships of up to 550 feet in length and a 576 feet long dry dock. Raylton Dixon also had the use of Bottle House Point Wharf near Dock Point and signed an agreement with the NER on 25/11/1898 that it would pay £8 pa for three years in lieu of a wayleave for any traffic between the Cleveland Dockyard and the wharf with the stipulation "Traffic may be conveyed on Owners Bogies propelled by hand". A subsequent agreement on 18/12/1901 increased the charge to £12 pa, but allowed traffic to be conveyed "in owners wagons hauled by their own Loco Crane". During World War I, Raylton Dixon had the largest output of ships on the Tees of 132,323 tons and, in 1920, launched the HURUNUI of 9,266 tons powered by Richardsons Westgarth 5,000hp turbines. Its final vessel of 5,633 tons was built for the Paris, Lyon & Marseille Railway Company in 1922. The yard was sold to the Cleveland Shipbuilding Co (the latter formed by Parsons Marine Steam Turbine Co), but no orders were obtained in 1924 and the yard closed. It was purchased by National Shipbuilding Security Co Ltd in 1931 for dismantling.

References : *Building Ships on the North East Coast* Parts 1 and 2, J.F. Clarke, Bowick Press, 1997

The History of Middlesbrough, William Lillie, Middlesbrough CB, 1968

British Shipbuilding Yards Volume 1 North East Coast, Norman L. Middlemiss, Shield Publications, 1993

RICHARDSON, DUCK & CO LTD

SOUTH STOCKTON SHIPYARD, Thornaby H36
Richardson, Duck & Co until 3/1912 NZ 451191 – **Map E**
Stockton Iron Shipbuilding Co until 1/1854

A Mr Spence began to establish the yard on the south bank of the River Tees about 9/1836. This was almost due north of the later Teesdale Iron Works. In 1/1852 the Stockton Iron Shipbuilding Co took over the yard and on 26/1/1854 launched the first iron ship, ADVANCE (336 tons), to be constructed on the Tees. It had Fossick & Hackworth engines and was built for James Taylor, the Middlesbrough wharfinger. However, the yard was soon taken over by Richardson, Duck in late 1854. Up to 1865, this company constructed 51 steamers, 10 sailing ships and 29 barges (the latter mainly for iron companies), and 600 people were employed. The yard averaged 10,000 tons of shipping a year in 1870–72 and this increased to 21,000 tons in 1882–83. In 1912 the business became a private limited company. It had a river frontage of 1,140 feet, including 600 feet long fitting out wharf and four construction berths with slipways facing downstream. A siding entered the yard from the Teesdale Iron Works system so presumably all rail traffic with the NER had to travel over Head Wrightson's tracks. Within the shipyard, sidings led to the 'slips', wharf and main buildings in the centre of the site. There are no records of any locomotives being operated by Richardson, Duck; perhaps the company and its successors used rail cranes?

In 1919 Richardson, Duck became a public company but there was a sparsity of orders and Gould Steamship & Industrials Ltd acquired the company. The last ship was launched in 8/1924 and the yard closed shortly after, having built over 680 vessels. The site was taken over by the South Stockton Shipbuilding & Engineering Company but with apparently little activity. In 1932, South Stockton Shipbreaking Co Ltd leased the fitting out quay and ships of up to 7,000 tons were scrapped. During World War 2, tank landing craft constructed by Head Wrightson and elsewhere were assembled here by the Stockton Construction Co. Shipbreaking continued until 1954, although Stockton Shipping & Salvage Co had become responsible by 1947. Meanwhile Head Wrightson had occupied the yard and slipways which it used to 'launch' heavy fabricated objects, such as dock gates and large pressure vessels for the nuclear industry.

Reference : *Shipbuilding in Stockton and Thornaby*, Alan Betteney, Tees Valley Heritage Group, 2003

JAMES & RONALD RITCHIE LTD

ACKLAM FOUNDRY, Middlesbrough H37
James Ritchie until after /1897 NZ 489212 – **Map F**

The business was established by James Ritchie in 1878 and the foundry was located on the east side of Forty Foot Road, opposite the Ayrton Sheet Works, in the Ironmasters district. A railway line left the Old Town Branch sidings and ran through Ritchie's premises, with a couple of sidings entering the main building, before joining the MOR Old Town Branch Loop. Part of the line was rented from the NER and, in 1912, Ritchie paid 1d per ton on all traffic entering and leaving its works. According to an 1887–88 directory, the foundry was laid out for the manufacture of cast iron, gas, water and hydraulic pipes and fittings. Pig iron came to the foundry from Connal & Co's Acklam Store in 1912. The standard gauge lines within the works had not changed significantly by 1951 with the principal track passing over a weighbridge and by a loading platform. However, an extensive system of narrow gauge tramways, presumably hand-worked, connected all parts of the site and enabled heavy castings to be moved about. In 1960 James & Ronald Ritchie Ltd was a division of W. Shaw & Co Ltd and, in that year, was quoting prices for pipes to the Skinningrove Iron Company. Its name still appeared in a 1970 Directory.

ROBERT ROGER & CO LTD

STOCKTON IRON FOUNDRY, Stockton H38
 NZ 442179 – **Map D**

Robert Roger took over the Stockton Iron Foundry in West Row, central Stockton during 1844. It did not have the benefit of a siding but was quite close to the 1825 SDR line and the river. Robert died in 1869 and was eventually succeeded by his son, also Roger. The business flourished and the foundry could produce castings of ten to twenty tons weight. It manufactured steam winches, steering gear, winding engines and many types of equipment for the shipbuilding industry, together with steam rail cranes which were used by some of the local iron and steel works. It was said to employ over 400 people. During the 1880s, there was no room for expansion at West Row and an additional works was established at Bowesfield on a one acre site. This was located north of the Darlington-Middlesbrough railway and a siding left the line at Bowesfield Junction to enter the premises, with a reversal required to serve the main building. The internal railway was not large and, no doubt, a spare rail crane would be available to carry out any shunting. The business was formerly wound up 1927, the Bowesfield Lane premises being taken over by Ashmore, Benson, Pease & Co Ltd (which see) on 24/12/1926.

ROPNER SHIPBUILDING & REPAIRING CO (STOCKTON) LTD

NORTH SHORE SHIPYARD, North Shore Branch, Stockton   **H39**
Ropner & Sons Ltd until /1919   NZ 451192 – **Map D**
Ropner & Son until /1907
Pearse, Lockwood & Co until /1888

The yard was situated at the end of the North Shore Branch on the bank of the River Tees. It had a long history, the first record of commercial shipbuilding being in the 1700s. Pearse, Lockwood & Co took over the yard in 1854 and built the first iron hulled sailing ship on the Tees. The company employed 750 people in 1881. It had constructed 229 ships totalling over 180,000 tons when Ropners gained control of the yard in 1888. The yard was the third largest in the UK with an annual output of 50,000 tons of shipping in 1895 and employed about 1,500 people prior to World War 1. A series of railway lines entered the site from the west side of the North Shore Branch, but no locomotives are known to have been owned by the companies. It is likely that they relied on travelling rail cranes because of the nature of work in the shipyard. At the end of World War 1, the 9½ acre site possessed four building berths capable of accommodating ships of 8,000 to 10,000 tons. With a shortage of orders, the last ship WILLOWBROOK was launched by Ropners in 1925 and the company was liquidated 7/6/1928. The yard was taken over by Smith's Dock Co Ltd as an 'overflow' base in 5/1929, nine ships being built in 1930, but it finally closed in 6/1931.

Reference :   *Shipbuilding in Stockton and Thornaby*, Alan Betteney, Tees Valley Heritage Group, 2003, pp.23-33

SKINNINGROVE IRON CO LTD

BOULBY IRONSTONE MINE, near Staithes   **H40**
NZ 760181 (pit yard) NZ 761186 (main drift) – **Map I**

The ironstone deposits were leased by the Skinningrove Iron Company in 1903. Trial shafts located the Main Seam at 72 feet and two drifts were constructed; the first stone being taken out on 14/9/1905. A rope-worked 650 yard narrow gauge single line tramway had been constructed from the main drift in a straight line south to the NER Staithes–Loftus line. To avoid changes in gradient, parts of the tramway were constructed on wooden gantries, but soon shale was tipped to form an embankment. Wayleaves had to be paid for stone and materials travelling over the tramway. The pit yard was developed on the north side of the NER line with workshops, an engine house and picking belt. Considerable excavation was required to accommodate the standard gauge sidings, which extended west and east from the tippler and were based on an agreement with the NER dated 19/9/1903. The connection with the NER was controlled by Boulby Mines Siding Ground Frame. There was a 1 in 61 gradient on the railway company's branch to Grinkle Tunnel and NER locomotives usually took about 11 wagons at a time up to Grinkle Station (named Easington Station until 1/4/1904) before combining two loads for the short journey to Skinningrove Iron Works. There was limited space at the mine pit yard to tip the shale and so, circa 1912, it was loaded into tubs and returned along the tramway into the drift which was extended underground to the cliff face. Here, about 300 feet above the foreshore, a tippler deposited the shale on to the beach below. In 1908 the mine was producing about 15,000 tons of ironstone a month but the quality was only moderate. Pease & Partners Ltd leased the Skinningrove Iron Co Ltd from 1/10/1922 and received a report on Boulby Mine. The narrow gauge comprised 8,400 yards of 40lb per yard rails with another 650 yards of 40lb rails on the surface tramway; there were 11 horses and 405 tubs. A summary of the standard gauge sidings stated "1300 yards of permanent way 82lb per yard fitted with Walkers Patent cast-iron chains with the necessary points and crossings. Standage for 70 empty trucks and 50 full trucks." The mine had closed in 4/1921, worked again 12/1923 to 11/1925 and for a few months in 1927 before finally finishing on 16/8/1927, but the LNER's ground frame was not dispensed with until 11/1935.

Reference :   *Boulby Ironstone Mine "Window of the Earth"*, Simon Chapman, pub. P. Tuffs, 1997

LOFTUS IRONSTONE MINE, Skinningrove

H41
NZ 712193 – **Map I**

Pease & Partners Ltd until 30/9/1947
J.W. Pease & Co until 19/8/1882

J.W. Pease & Co signed the lease for 1,000 acres of ironstone on 29/5/1865. Ironstone began to be extracted later in the year using the old drift of Skinningrove Mine, until a new drift was opened further up the valley. This was driven about 20 feet above the valley floor so that the ironstone in the tubs could be tipped directly into the NER's railway wagons. The drift was nearly ready in 11/1865 with the stationary steam engine in place to work the rope hauled tubs. The first proposal by the Cleveland Railway to reach Loftus Mine in the valley was via a 1 in 6 incline, but this would have placed limits on the amount of ironstone that could be moved and the NER constructed the Skinningrove Zig Zag. Although requiring steep gradients and two reversals, it could be worked by locomotives. From the second reversal at the "Dumb End" against the valley site, trains headed along the valley to reach Loftus pit yard sidings where the empty wagons were left. These could then be run by gravity under the tipping chutes for loading prior to collection by the railway company's locomotives.

Pease initially sent its ironstone to the blast furnaces on Teesside, although rail transport was expensive because of the distance involved and cost of operating the Zig Zag. In 1/1894 Pease & Partners, with Lord Zetland, provided finance to construct two extra blast furnaces at the Skinningrove Iron Co's works and the latter agreed to take the bulk of Loftus Mine's output. A wooden viaduct was constructed across the valley to a new drift into the hillside which connected with the bottom of the North Loftus shaft. Starting in 2/1895, 3 foot gauge railway tubs were drawn up the drift out of Loftus Mine by the stationary engine and then run by gravity over the viaduct to the North Loftus shaft where they were raised to the surface. On the return, a stationary engine pulled them back across to the mine, before letting them descend to the mine workings. On 17/9/1937 this arrangement was replaced by an aerial ropeway with buckets delivering ironstone to a hopper; the wooden viaduct being dismantled in 1948. Pease secured agreement to take ironstone from Whitecliffe Mine in 1/1/1899 and this was transported underground via the Loftus Mine South Drift. Also between 1900 and 1914, ironstone from Grinkle Mine came through Loftus Mine for delivery to Skinningrove Iron Works.

With nationalisation of the coal industry, Pease & Partners withdrew from mining and the Skinningrove Iron Co took over Loftus Mine on 30/9/1947. About this time the mine possessed 44 horses and 450 mine tubs. Termination of a subsidy on home produced ore resulted in a decision to close Loftus Mine and the last tubs of ironstone rattled out of the mine on 26/9/1958. Today parts of the former Waddle Fan engine house and a short length of drift form part of the Cleveland Ironstone Mining Museum.

References :   *The Loftus Mines Skinningrove*, Simon Chapman, pub. by P. Tuffs, 1998

Skinningrove Iron and Steel Works Its History, Railways and Locomotives, Cliff Shepherd, IRS, 2012

T. SMITH

CLARENCE BRICK WORKS, Billingham

H42
NZ 468231 – **Map C**

The brick works was located on the north side of the Port Clarence Branch near to where it was crossed by the road to Cowpen Bewley. The works was being operated by Thomas Smith in 1883 and the 1897 Ordnance Survey map shows a siding off the NER running alongside the works. The latter was connected by a separate railway track to its clay pit. The works had closed by 1913 when it was shown as old clay pit on a contemporary map, although the short siding off the NER remained. The former clay pit is now Charlton's Pond nature reserve.

Reference :   *The Brickworks of the Stockton-on-Tees Area*, Alan Betteney, Tees Valley Heritage Group, 2007

STEVENSON, JAQUES & CO

BOOSBECK IRONSTONE MINE, Boosbeck

H43
NZ 658168 – **Map I**

After a series of abortive interests in establishing a mine, Stevenson, Jaques & Co began sinking two shafts in 1871. The mine was situated a short distance north east from South Skelton Mine and also alongside the ex-Cleveland Railway. The buildings were erected on the west side of High Street, 100 yards from the goods station in Boosbeck. There were two shafts that descended 288 feet to the Main Seam ironstone which was reached in 1/1872; the first stone being wound up in 1873 and despatched to the company's Acklam Iron Works. *The Weekly Exchange*, 7/12/1878, in describing a fatality, gives an insight into the underground operation of the railway: "The deceased was a runrider in Messrs. Stevenson and Jacques [sic] ironstone mine at Boosbeck and on the 19th inst. was taking a set of empty tubs to a certain part of the mine in order to be filled. At the place from which he started was a laden wagon used in drawing tubs which had been emptied at the pit shaft from a certain point to a "siding" where they remained until filling. This laden wagon was linked to a chain fixed to a drum, and on the empty tubs being sent away the deceased took up his position on the last tub and rode with his customary lights…… the laden wagon got unlinked and followed the empty set down the main way, which is at a gradient of 2½" to the yard. Just as the last of the empties, was branching into one of the working places, where the tubs were required, the laden wagon ran up so close as to knock the deceased from his position and upsetting, fell heavily upon him….".

Initially the mine produced large amounts of ironstone but, in 1882–83, approximately 200 houses in Boosbeck suffered serious damage due to subsidence and Stevenson, Jaques was ordered to pay compensation in 1885. Two years later, there was a significant inrush of water into the mine and it closed on 28/3/1887. The royalty later became part of the South Skelton Mine workings.

Reference :   *South Skelton Mine The story of mining at Boosbeck*, Simon Chapman, pub. P. Tuffs, 2010

STILLITE PRODUCTS LTD

Stillington   **H44**
NZ 373238 – **Map D**

Mr H.A. Mackay leased the land from Dorman, Long & Co Ltd approximately on the site of the former coke ovens at the old Carlton Iron Works and erected a factory in 1939 to produce slag wool as an efficient insulating cloth using material from the tip. The slag was blasted from the face of the heap at the "Cracker Hole"; an area previously quarried by the Stillington Slag Co. Loose slag was loaded into narrow gauge wagons and winched up the slope from the bottom of the "Cracker Hole" to a turntable located near the winching shed. The wagon was then turned and allowed to run down a slight incline to the furnace just inside the factory gates.

Reference :   *There was a Green Hill, The History of Stillington from its Beginnings until 1950*, J.D. Tuffs, pub. by the author, 1999

STOCKTON CHEMICAL ENGINEERS & RILEY BOILERS LTD

PERSEVERANCE BOILER WORKS, Stockton   **H45**
Riley Bros (Boilermakers) Ltd until /1932   NZ 438189 – **Map D**
Riley Bros until 13/4/1904

Two Riley brothers originally set up as boiler and ship repairers on the Stockton quayside but, in 1872, they moved to a piece of land at the corner of the splendidly named Light Pipe Hall Road and what became Riley Street. These premises were located on the west side of the Yarm–Norton railway approximately 850 yards south of Stockton Station. A siding agreement with the NER was dated 13/10/1890. In 1891 they completed the purchase of an adjacent site to the north from Thomson Gilkes & Co's Millfield Iron Works and erected the Perseverance Boiler Works. Up to 1914, the factory concentrated on the design and manufacture of Scotch marine and shell type boilers, a large proportion of which were supplied to the trawler and drifter fleets. It employed about 150 people in 1900 and completed its 5,000 boiler by 1917.

A siding off the NER main line terminated at a headshunt, next to Oxbridge Lane, enabling wagons to be reversed into the Perseverance Works. Riley's railway system was relatively simple serving its principal buildings which were situated parallel to the main line and presumably a travelling rail crane or horses could deal with the shunting, although Rowland Abbot in his book on vertical boiler locomotives suggests that Riley Bros may have built a vertical boiler locomotive but no evidence has been found. Its sidings also extended into the Oxbridge Foundry and the Stockton Steel Foundry Co Ltd. The latter was located on Light Pipe Hall Road behind the Perseverance Works and became Head Wrightson's second steel foundry in 1927. The latter appears to have used a rail crane for shunting. With declining orders, the Perseverance Works was forced to close in 1932. The company was restructured in 1934 and its name altered to Stockton Chemical Engineers & Riley Boilers Ltd (LNER has 29/8/1934) to reflect a wider product base although, up to 1945, 90% of its business was still in shell boilers. The railway layout in 1950 was similar to earlier with a railway entering the premises from BR, passing a series of coal bunkers, before running round Riley's principal building. Sidings still left its system to run under an overhead gantry crane at Head Wrightson's steel foundry. The works closed in 10/1965.

Reference :   'Riley Bros (Boilermakers) Limited. A Short History', T.H. Riley, Typed manuscript, 3/1970

The Tees Conservancy Commissioners opened a graving dock at Cargo Fleet in response to requests from ship owners for repair facilities. This photograph of the dock appeared in the Commissioners' Ports of the River Tees *booklet published in 1906.*

The Submarine Miners' Depot at South Gare reputedly about the early 1900s with mines loaded on the 18 inch gauge railway wagons. A train with open wagons stands on the standard gauge line that ran along the Gare.
(Kirkleatham Museum, Redcar & Cleveland Council collection)

TEES & HARTLEPOOL PORT AUTHORITY

GRAVING DOCK, Cargo Fleet
Tees Conservancy Commissioners until 1/1/1967

H46
NZ 521209 – **Map G**

In response to complaints about the lack of ship repairing facilities on the River Tees, the Commissioners signed an agreement with the OME (reported 13/9/1873) to acquire land on the west side of the railway to Normanby Jetty for the construction of a graving dock. Work began in 1875 at a cost of £30,000 and it was opened for public use in 11/1876. Land was allocated on the west side of the site for ancillary uses. The dock is shown on an 1884 plan at right angles to the river with a crane track round it and the nominal positions of four cranes indicated. The dock was 521 feet long, later extended to 576 feet. A footbridge over the Middlesbrough-Redcar line from the "Sailors Trod" footpath gave access to the site. Sidings ran in from the west near the footbridge – one from the NER and the other from the MOR line that traversed Normanby Iron Works. There were two main buildings on the site comprising a pump house and a large fitting shop that was served by a siding. In 1906, the machinery included a 15 ton crane and a 10 ton steam travelling crane. Despite the requests for a graving dock, subsequent use was only spasmodic and never very high. A note subsequently added to the MOR schedule dated 10/1946 recorded "There has not been any traffic to or from the graving dock via the MO's line for many years." It and the associated slipway were mainly used for servicing river craft after World War 2. Nevertheless the site still contained numerous sidings in 1952, including one on an unusual circular alignment. Tees Marine Services Ltd later used the site, but the siding agreement was terminated from wef 7/2/1970.

TEES SIDE BRIDGE & ENGINEERING WORKS LTD

CLEVELAND DOCKYARD, Middlesbrough

H47
NZ 504210 – **Map G**

During World War 2, the Tees Side Bridge & Engineering Works Ltd (TSB) took over the disused Cleveland Dockyard in 11/1940. An agreement was sealed with the LNER on 24/3/1941 for the use of a railway line into the site. The operation of a private locomotive over the LNER's tracks serving the Cleveland Dockyard was approved on 2/11/1942, but it is not known if the TSB actually had a locomotive there. TSB constructed 214 landing craft, 30 gunboats and 48 rocket firing craft between 1941 and 1944. After the war, TSB purchased the dockyard and its equipment from the Admiralty on 25/4/1946, but a project to construct barges there proved unsuccessful and so the company acquired five acres at the old Linthorpe-Dinsdale Iron Works instead.

WAR DEPARTMENT

SUBMARINE MINERS' DEPOT, South Gare

H48
NZ 556274 – **Map H**

The South Gare breakwater occupied a strategic position at the mouth of the River Tees and had gun emplacements installed at its seaward end during both World Wars. Prior to that, in 3/1887, the WD leased some land at South Gare from the Tees Conservancy Commissioners and, by 1888, had constructed a fortified depot containing barracks for the Tees Volunteer Division (Submarine Miners), Royal Engineers. The mines were probably delivered to the powder wharf immediately south of Paddy's Hole. A standard gauge siding connected the wharf with the TCC South Gare line and another siding gave entry to the depot. Here the mines were transferred to wagons on an 18 inch gauge railway which served the storage building. With an outbreak of hostilities, the mines were taken to the wharf and loaded on to a small vessel called MINER for laying in the channel. They were connected by cables to the depot, where men were responsible for detonating them. The *Defence of the UK* reference suggests that there was a narrow gauge line from the wharf to the depot, but this is not shown on the Ordnance Survey maps. The Royal Navy took over responsibility for the mine depot in 1907 and it ceased to operate as such after World War 1 ended. The North Riding 16th (Fortress) battalion maintained a searchlight company which used the depot and searchlight defence remained operational until after World War 2. The personnel manning the gun emplacements were also housed in the barracks. Although only small remnants of the railways survive, the depot buildings remain.

References : 'South Gare', Cliff Shepherd, *Industrial Railway Record* 121, IRS, 6/1990, pp.135-40

Defence of the UK Redcar Vol 1, North Yorkshire and Cleveland 20th Century Defence Study Group, Momentous Publications, 2008, pp.16-32

WARREN CEMENT WORKS LTD

WARREN CEMENT WORKS, Hartlepool
Otto Trechmann Ltd until /1919
Otto Trechmann & Co until /1892

H49
NZ 516345 – **Map B**

In 1852 Peter Otto Trechmann acquired the Roman cement, whiting and Plaster-of-Paris premises of Wilson and Richardson. The works was located on the south side of the Hartlepool Dock & Railway line at Hart Warren and converted to manufacture Portland cement from the large quantities of chalk ballast dumped by colliers returning from the Thames estuary. Otto Trechmann died in 1892 and a controlling interest was purchased by T. & W. Weekes & Co Ltd which formed a new company, Trechmann, Weekes & Co Ltd. In 1893 Otto Trechmann Ltd was established to operate the Warren Cement Works. The change from chalk to water ballast in ships deprived the business of a cheap source of raw material and it had to lease a chalk quarry at Purfleet in Essex. The railway company's line was on an embankment posing difficulties for providing a siding. Instead, the 1896 and 1914 Ordnance Survey maps show four wagon turntables on one side of the NER's lines with single tracks descending at right angles into the works. There also appears to have been an internal narrow gauge railway system which was probably hand-worked in 1914.

Trechmann lost most of its ships during World War 1 and the cement company was liquidated shortly after. However, it was known that deposits of anhydrite existed beneath the works and Warren Cement Works Ltd was established to develop them. In 1923 a contract was obtained to supply Synthetic Ammonia & Nitrates Ltd's works at Billingham with 1,000 tons of anhydrite a week for the manufacture of ammonium sulphate. The shaft was sunk in the centre of the works and the first consignment despatched in 1925. Workings were 32m to 64m below the surface. Technical problems and contamination of the anhydrite, together with Billingham opening its own mine, resulted in the closure of the Warren Works mine in 1930. It continued to produce cement using limestone from quarries west of Darlington until it finished in 1939.

Reference :   '160 Years of Cement Manufacture in Cleveland', C.H. Morris, *The Cleveland Industrial Archaeologist* 16, CIAS, 1984

J.T. WHARTON

CARRS TILERY AND BRICK WORKS, Margrove Park

H50
NZ 651161 – **Map I**

The Cleveland Railway was extended from Slapeworth to Boosbeck in 1862 and it was during the 1860s that Carrs Tilery was established on the south side of the line. The approximately one acre site was situated in the valley bottom – hence the name 'carr' meaning wet meadow – which had once been the floor of a lake that filled with alluvial clay. Initially this was a small concern producing tiles and pots. However, the sinking of various ironstone mines in the vicinity encouraged expansion to produce substantial numbers of common bricks. Large orders were obtained to supply bricks to Skelton Shaft Mine (1867-71), South Skelton Mine (1870-73) and Stanghow Mine (1871-74). The owners of the tilery installed a steam engine driving brick making machinery in 1871-72. To the north of these were the drying sheds with the kilns next to the main line. The 'L' shaped clay pit 'wrapped' round the west and south side of the works. Unfortunately, with the depression in the iron industry during the latter 1870s, output declined dramatically and the works seems to have then mainly served the needs of the Skelton Estate for tiles, pantiles and drainage tiles. J.T. Wharton, owner of the Estate, had the tilery according to Slater's 1876-77 directory. The business probably only had a short life; an audit of equipment took place in 1894, but the premises slumbered on largely undisturbed until the 1960s. The tilery was located next to the junction for the Stanghow Mine branch opposite the signal box. A siding lay by the south side of the Cleveland Railway next to the tilery. Within the site, a narrow gauge railway ran alongside this siding and connected with the engine house presumably delivering coal to the Lancashire boiler, another linked the brick making area to the drying shed and a third descended into the clay pit.

Reference :   'The Carrs Tilery, Margrove Park', J.K. Harrison, *The Cleveland Industrial Archaeologist* 16, CIAS, 1984

INDEX OF LOCOMOTIVES

NOTES: Information normally relates to the locomotive as built.
Column 1 Works Number (or original company running number for locomotives built in main line workshops without a works number).
Column 2 Date ex-works where known - this may be a later year than the year of building or the year recorded on the worksplate.
Column 3 Gauge.
Column 4 Wheel arrangement. The suffix F for diesel locomotives indicates flameproofed for underground use, otherwise the suffix F indicates that the locomotive was of fireless design.

	Steam Locomotives:	Diesel Locomotives:
Column 5	Cylinder position	Horse power
Column 6	Cylinder size	Engine type #
Column 7	Driving wheel diameter	Weight in working order
Column 8	Either weight in working order and/or Manufacturers designation	
Column 9	Page references	

Manufacturers of petrol and diesel engines:

Blackstone	Blackstone & Co Ltd, Rutland Engineering Works, Stamford, Lincolnshire.
Cat	Caterpillar Tractor Co Ltd.
Crossley	Crossley Bros Ltd, Openshaw, Manchester, Lancashire.
Cummins	Cummins Engine Co Ltd, Shotts, Lanarkshire.
Dorman	W.H. Dorman & Co Ltd, Stafford, Staffordshire.
EE	English Electric Co Ltd.
Fowler	John Fowler & Co (Leeds) Ltd, Hunslet, Leeds.
Gardner	L. Gardner & Sons Ltd, Barton Hall Engine Works, Patricroft, Manchester.
Lister	R.A. Lister & Co Ltd, Dursley, Gloucestershire.
McLaren	J. & H. McLaren Ltd, Midland Engine Works, Hunslet, Leeds.
National	National Gas & Oil Engine Co Ltd, Ashton-under-Lyne, Gtr Manchester.
Paxman	Davey, Paxman & Co Ltd, Colchester, Essex.
R-R	Rolls-Royce Ltd, Oil Engine Division, Shrewsbury, Shropshire.
Ruston	Ruston & Hornsby Ltd, Lincoln, Lincolnshire.

ANDREW BARCLAY, SONS & CO LTD, Caledonia Works, Kilmarnock — AB

Works No	Date	Gauge	Wheels	Cyl	Size	DW	Page
24	1864	4ft 8½in	0-4-0ST	OC	12x20	3ft 6in	103
136	07.03.1872	4ft 8½in	0-4-0ST	OC	12x20	3ft 6in	268,319
148	05.02.1874	4ft 8½in	0-4-0ST	OC	10x18	3ft 0in	219,288
187	22.06.1877	4ft 8½in	0-4-0ST	OC	10x18	3ft 0in	34
215	20.05.1880	4ft 8½in	0-4-0ST	OC	12x20	3ft 6in	379
232	15.06.1881	4ft 8½in	0-4-0ST	OC	11x18	-	126
239	14.12.1881	4ft 8½in	0-4-0ST	OC	13x20	3ft 6in	34,193
261	30.04.1883	4ft 8½in	0-4-0ST	OC	10x18	-	199,319
	1883	4ft 8½in	0-4-0ST	OC	[Lennox Lange plates]		279
265	20.02.1884	4ft 8½in	0-4-0ST	OC	13x20	3ft 6in	279,361
266	21.03.1884	4ft 8½in	0-4-0ST	OC	11x18	3ft 3in	38
297	13.07.1887	3ft 0in	0-4-0ST	OC	9x17	2ft 6in	366,368
299	10.01.1888	4ft 8½in	0-4-0ST	OC	13x20	3ft 6in	221,228
300	16.03.1888	4ft 8½in	0-4-0ST	OC	13x20	3ft 6in	59,233
305	03.07.1888	4ft 8½in	0-4-0ST	OC	13x20	3ft 7in	44
650	30.12.1889	4ft 8½in	0-4-0ST	OC	14x22	3ft 5in	59,160
654	03.10.1889	4ft 8½in	0-4-0ST	OC	10x18	3ft 1in	23
656	07.02.1890	4ft 8½in	0-4-0ST	OC	13x20	3ft 2in	69
662	20.05.1890	4ft 8½in	0-4-0ST	OC	13x20	3ft 7in	52
668	24.10.1890	4ft 8½in	0-4-0ST	OC	13x20	3ft 2in	205,210
671	04.04.1890	4ft 8½in	0-4-0ST	OC	14x22	3ft 5in	59,233
673	26.06.1890	4ft 8½in	0-4-0ST	OC	12x20	3ft 2in	52
721	11.10.1894	4ft 8½in	0-4-0ST	OC	13½x20	-	69

773	05.03.1897	4ft 8½in	0-4-0ST	OC	12x20	3ft 2in	113
775	31.07.1896	4ft 8½in	0-4-0ST	OC	12x20	3ft 2in	103,282
814	16.04.1898	4ft 8½in	0-4-0ST	OC	14x22	3ft 5in	286,337,340
823	18.06.1898	4ft 8½in	0-4-0ST	OC	13½x20	3ft 8in	24,279,281
857	28.12.1899	4ft 8½in	0-4-0ST	OC	16x24	3ft 7in	65
858	29.01.1900	4ft 8½in	0-4-0ST	OC	16x24	3ft 7in	65,87
868	09.04.1900	4ft 8½in	0-4-0ST	OC	12x20	3ft 2in	192,193
879	20.07.1900	4 ft 8½in	0-4-0ST	OC	16x24	3ft 7in	58,149
897	24.07.1900	4ft 8½in	0-4-0ST	OC	14x22	3ft 5in	59,283
924	22.11.1901	4ft 8½in	0-4-0ST	OC	16x24	3ft 7in	58,65,149
925	23.12.1901	4ft 8½in	0-4-0ST	OC	16x24	3ft 7in	65
926	17.01.1902	4ft 8½in	0-4-0ST	OC	16x24	3ft 7in	65,87
979	08.07.1903	4ft 8½in	0-4-0ST	OC	14x22	3ft 5in	268
998	16.12.1903	4ft 8½in	0-4-0ST	OC	16x24	3ft 7in	65,87
1008	09.01.1905	4ft 8½in	0-4-0ST	OC	16x24	3ft 7in	44
1014	07.02.1906	4ft 8½in	0-4-0ST	OC	10x18	3ft 0in	219
1018	23.01.1905	4ft 8½in	0-4-0ST	OC	12x20	3ft 2in	205
1029	13.07.1905	4ft 8½in	0-4-0ST	OC	16x24	3ft 7in	44
1040	12.05.1905	4ft 8½in	0-4-0ST	OC	16x24	3ft 7in	58,149
1047	08.09.1905	4ft 8½in	0-4-0ST	OC	14x22	3ft 5in	44
1059	17.10.1905	4ft 8½in	0-4-0ST	OC	14x22	3ft 5in	59,126,149
1061	18.12.1905	4ft 8½in	0-4-0ST	OC	16x24	3ft 7in	44
1068	27.02.1906	4ft 8½in	0-4-0ST	OC	16x24	3ft 7in	44
1076	03.05.1906	4ft 8½in	0-4-0ST	OC	16x24	3ft 7in	87
1077	11.05.1906	4ft 8½in	0-4-0ST	OC	16x24	3ft 7in	87
1095	20.12.1906	4ft 8½in	0-4-0ST	OC	14x22	3ft 5in	44
1100	31.12.1906	4ft 8½in	0-6-0ST	OC	17x24	3ft 9in	86
1102	11.04.1907	4ft 8½in	0-4-0ST	OC	14x22	3ft 5in	59,65,233
1179	18.04.1910	4ft 8½in	0-4-0ST	OC	10x18	3ft 0in	199
1196	14.06.1912	4ft 8½in	0-4-0ST	OC	12x20	3ft 2in	133,286
1224	18.10.1911	4ft 8½in	0-4-0ST	OC	14x22	3ft 5in	88
1274	17.04.1912	4ft 8½in	0-4-0ST	OC	16x24	3ft 7¼in	58,65,149
1282	18.05.1912	4ft 8½in	0-4-0ST	OC	14x22	3ft 6in	266,268
1296	13.09.1912	4ft 8½in	0-6-0T	OC	18x24	3ft 7in	396
1315	29.011913	4ft 8½in	0-4-0ST	OC	16x24	3ft 7in	86
1317	08.08.1914	4ft 8½in	0-4-0ST	OC	16x24	3ft 7in	246
1344	10.09.1913	4ft 8½in	0-4-0ST	OC	16x24	3ft 7in	86
1350	27.11.1913	3ft 0in	0-4-0T	OC	10x12	3ft 2in	50
1351	27.11.1913	3ft 0in	0-4-0T	OC	10x12	3ft 2in	50,98
1360	17.12.1913	4ft 8½in	0-4-0ST	OC	14x22	3ft 3in	87
1363	19.01.1914	4ft 8½in	0-4-0ST	OC	16x24	3ft 7in	222,228
1364	16.02.1914	4ft 8½in	0-4-0ST	OC	16x24	3ft 7in	87,167
1384	07.11.1914	4ft 8½in	0-4-0ST	OC	16x24	3ft 7in	87
1404	27.03.1915	4ft 8½in	0-6-0ST	OC	16x24	-	87,139
1408	13.10.1915	4ft 8½in	0-6-0ST	OC	14x22	3ft 5in	199
1414	10.07.1915	4ft 8½in	0-4-0ST	OC	16x24	3ft 7in	87
1464	28.04.1916	3ft 0in	0-4-0T	OC	10x12	2ft 2in	50,98
1478	08.08.1916	4ft 8½in	0-4-0ST	OC	16x24	-	58,149
1497	11.12.1916	4ft 8½in	0-6-0ST	OC	14x22	3ft 5in	52
1501	30.04.1917	4ft 8½in	0-4-0ST	OC	16x24	3ft 7in	45,222,228
1515	28.11.1917	4ft 8½in	0-4-0ST	OC	16x24	3ft 7in	58,149
1567	15.12.1917	4ft 8½in	0-4-0ST	OC	16x24	3ft 7in	87
1589	05.06.1918	4ft 8½in	0-4-0ST	OC	14x22	3ft 5in	87
1597	09.07.1918	4ft 8½in	0-4-0ST	OC	14x22	3ft 5in	246
1599	09.05.1918	4ft 8½in	0-4-0ST	OC	16x24	3ft 8in	58,149
1602	27.08.1918	4ft 8½in	0-4-0ST	OC	14x22	3ft 5in	246

1606	11.09.1918	4ft 8½in	0-4-0ST	OC	16x24	3ft 7in		246,269
1609	30.09.1918	4ft 8½in	0-4-0ST	OC	16x24	3ft 8in		222,228
1620	10.04.1919	4ft 8½in	0-4-0ST	OC	16x24	3ft 7in		59,87,167
1643	23.10.1919	4ft 8½in	0-4-0ST	OC	14x22	3ft 5in		87
1735	05.03.1921	3ft 0in	0-4-0T	OC	10x22	2ft 2in		50,98
1736	23.03.1921	4ft 8½in	0-4-0ST	OC	16x24	3ft 7in		87,167
1737	23.03.1921	4ft 8½in	0-4-0ST	OC	16x24	3ft 7in		87
1768	01.08.1922	4ft 8½in	0-4-0ST	OC	12x20	3ft 3in		114
1770	07.10.1921	4ft 8½in	0-6-0ST	OC	12x20	3ft 1in		241
1800	04.12.1923	4ft 8½in	0-4-0ST	OC	12x20	3ft 2in		109,220,205
1888	05.04.1926	4ft 8½in	0-4-0ST	OC	16x24	3ft 7in		58,149
1945	19.03.1928	4ft 8½in	0-4-0ST	OC	16x24	3ft 7in		200
1987	05.08.1930	4ft 8½in	0-4-0ST	OC	16x24	3ft 8in		70
2032	19.05.1937	4ft 8½in	0-4-0ST	OC	16x24	3ft 7in		59
2040	02.10.1937	4ft 8½in	0-4-0CT	OC	11x18	3ft 0in		309
2045	07.09.1937	4ft 8½in	0-4-0ST	OC	16x24	3ft 7in		69
2105	27.08.1940	4ft 8½in	0-4-0ST	OC	16x24	3ft 7in		228
2106	01.10.1940	4ft 8½in	0-4-0ST	OC	16x24	3ft 7in		45
2145	10.12.1942	4ft 8½in	0-4-0ST	OC	17x18	3ft 0in		313
2152	15.04.1943	4ft 8½in	0-4-0CT	OC	14x22	3ft 5in		59
2209	21.06.1946	4ft 8½in	0-4-0ST	OC	16x24	-		69
2210	01.07.1946	4ft 8½in	0-4-0ST	OC	16x24	-		69
2254	28.09.1948	4ft 8½in	0-4-0ST	OC	16x24	-		69
2269	29.05.1950	4ft 8½in	0-4-0F	OC	19x22	3ft 5in		70
2270	18.07.1950	4ft 8½in	0-4-0F	OC	19x22	3ft 5in		69
2272	09.07.1949	4ft 8½in	0-4-0ST	OC	14x22	3ft 5in		309
2323	25.02.1952	4ft 8½in	0-4-0ST	OC	16x24	3ft 7in		45,88
2324	03.03.1952	4ft 8½in	0-4-0ST	OC	16x24	3ft 7in		45,88
2328	29.10.1952	4ft 8½in	0-4-0F	OC	19x22	3ft 5in		88
2329	29.09.1952	4ft 8½in	0-4-0F	OC	21x22	3ft 5in		88
2376	24.12.1961	4ft 8½in	0-4-0F	OC	19x22	3ft 5in		89
	1948	4ft 8½in	4wWE				[MV No.1]	99
	1948	4ft 8½in	4wWE				[MV No.2]	99
352	02.04.1941	4ft 8½in	0-4-0DM	84hp	Gardner 6LW		14½T	248
370	22.06.1945	4ft 8½in	0-4-0DM	153hp	Gardner 6L3		21T	348
487	11.03.1964	4ft 8½in	0-6-0DH	311hp	R-R C8SFL			216
558	12.05.1970	4ft 8½in	0-4-0DH	280hp	Cummins NT280			328
559	20.10.1970	4ft 8½in	0-4-0DH	280hp	Cummins NT280			328
601	11.9.1975	4ft 8½in	0-4-0DH	252hp	Cummins NT855			328
608	17.11.1976	4ft 8½in	0-6-0DH	400hp	R-R C8TFL		52T	216,330,334,381
614	07.03.1977	4ft 8½in	0-6-0DH	400hp	R-R C8TFL		52T	34,261,381
616	30.04.1977	4ft 8½in	0-6-0DH	400hp	R-R C8TFL		52T	381
659	24.02.1982	4ft 8½in	0-6-0DH	400hp	R-R C8TFL		52T	216
660	10.03.1982	4ft 8½in	0-6-0DH	400hp	R-R C8TFL		52T	216

ARTISAIR LTD, Yaddlethorpe, Scunthorpe, Lincolnshire   Artisair

	1972	1ft 3in	4-6-2		394

AVONSIDE ENGINE CO LTD, Fishponds, Bristol   AE

1055	12.1874	7ft 0in	0-4-0ST	OC	14x18	3ft 0in	199
1352	1896	4ft 8½in	0-4-0ST	OC	14x20	3ft 3in	246
1593	08.12.1910	1ft 11in	0-4-0T	OC	7x10	1ft 8in	264
1701	1915	4ft 8½in	0-4-0T	OC	10x16	2ft 9in	116
1787	1917	4ft 8½in	0-6-0ST	OC	14½x20	3ft 3in	52
1793	02.1918	4ft 8½in	0-4-0ST	OC	14x20	3ft 3in	65
1801	1918	4ft 8½in	0-4-0ST	OC	14x20	3ft 3in	228,301,326

1815	1918	4ft 8½in	0-6-0ST	OC	14x20	3ft 3in		69
1830	1911	4ft 8½in	0-4-0ST	OC				200
1881	1921	4ft 8½in	0-4-0ST	OC	12x18	2ft 11in		24
2060	1931	4ft 8½in	0-4-0ST	OC	12x20	3ft 1in		309

BARCLAYS & CO, Riverbank Engine Works, Kilmarnock, Ayrshire — B

208	1873	4ft 8½in	0-4-0ST	OC	12x -			388
226	1876	4ft 8½in	0-4-0ST	OC				23
[TCC No.3]	1876	4ft 8½in	0-4-0ST	OC				376
231?		4ft 8½in	0-4-0ST	OC	[reb AB 7571/1907]			246
236	1877	4ft 8½in	0-4-0ST	OC				199
266	1880	4ft 8½in	0-4-0ST	OC	12x -			304
283	1881	4ft 8½in	0-4-0ST	OC	- x18	3ft 0½in		23
292	1882	4ft 8½in	0-4-0ST	OC	13½x -			222,228
294	1882	4ft 8½in	0-4-0ST	OC	13x -			119
295	1882	4ft 8½in	0-4-0ST	OC	11x18			119
296	1882	4ft 8½in	0-4-0ST	OC	12x -			179,363
305	1883	4ft 8½in	0-4-0ST	OC				23
[SOUTH STOCKTON No.1]		4ft 8½in	0-4-0ST	OC				305

H.N. BARLOW, Southport, Lancashire — Barlow

	1953	1ft 3in	4-6-2DE	S/O				394

W.E. BATES, Leftwich Iron Works, Northwich, Cheshire — Bates

	1862	3ft 0in	0-4-0ST	OC	[QUEEN OF THE FOREST]			346

BRUSH ELECTRICAL ENGINEERING CO LTD, Falcon Works, Loughborough — BE

	1905-6	4ft 8½in	4wBE					249

BRITISH ELECTRIC VEHICLES LTD, Southport — BEV

"BEV" branded locomotives were later built by Wingrove & Rogers - see WR

257	18.09.1920	?	4wBE					136

E.E. BAGULEY LTD, Burton-on-Trent, Staffordshire — Bg

Built by Bg for Drewry Car Co Ltd, London

2652	15.12.1958	4ft 8½in	0-4-0DM	102hp	Gardner 4L3			344
2655	29.06.1959	4ft 8½in	0-4-0DM	102hp	Gardner 4L3			344
2725	19.12.1963	4ft 8½in	0-4-0DM	195hp	Gardner 6L3B		31T 13C	133

BLACK, HAWTHORN & CO LTD, Quarry Lane, Gateshead — BH

Black, Hawthorn & Co until 23/3/1892

	order dates							
21	07.1867	4ft 8½in	0-4-0ST	OC	12x20	3ft 3in		149
23	20.08.1868	4ft 8½in	0-4-0ST	OC	10x17	3ft 0in		149
98	09.02.1869	4ft 8½in	0-4-0ST	OC	10x17	3ft 0in		44
111	10.05.1869	4ft 8½in	0-4-0ST	OC	12x19	3ft 2in		308
112	10.05.1869	4ft 8½in	0-4-0ST	OC	12x19	3ft 2in		305
116	10.08.1869	4ft 8½in	0-4-2ST	OC	10x17	3ft 3in		361
119	24.08.1869	4ft 8½in	0-4-0ST	OC	12x19	3ft 2in		312
162	08.03.1871	4ft 8½in	0-4-0ST	OC	10x17	3ft 0in		335
163	15.04.1871	4ft 8½in	0-4-0ST	OC	10x17	3ft 0in		152
166	11.12.1872	4ft 8½in	0-6-0ST	OC	14x20	3ft 6in		139
175	01.02.1872	4ft 8½in	0-4-0ST	OC	12x19	3ft 2in		152
236	21.01.1873	4ft 8½in	0-4-0ST	OC	12x19	3ft 2in		243
287	12.08.1873	4ft 8½in	0-4-0ST	OC	12x19	3ft 2in		185

298	26.04.1875	4ft 8½in	0-4-0ST	OC	12x19	3ft 2in	109
306	18.02.1875	4ft 8½in	0-4-0ST	OC	9x16	2ft 9in	23,250
307	29.01.1874	4ft 8½in	0-6-0ST	OC	14x20	3ft 6in	64
333	25.01.1876	4ft 8½in	0-4-0ST	OC	10x17	3ft 0in	166
354	25.02.1875	4ft 8½in	0-6-0ST	OC	14x20	3ft 6in	69
364	08.04.1876	4ft 8½in	0-4-0ST	OC	12x19	3ft 2in	86
365	31.10.1876	4ft 8½in	0-4-0ST	OC	12x19	3ft 2in	87
371	05.07.1875	4ft 8½in	0-4-0ST	OC	9x16	2ft 9in	166,219
396	31.05.1876	4ft 8½in	0-4-0ST	OC	13x19	3ft 2in	86
398	22.07.1882	4ft 8½in	0-4-0ST	OC	9x16	2ft 8in	177
402	12.09.1876	3ft 0in	0-4-0ST	OC	5x10	1ft 8in	97
403	12.09,1876	3ft 0in	0-4-0ST	OC	5x10	1ft 8in	97
421	20.10.1877	4ft 8½in	0-4-0ST	OC	12x19	3ft 2in	126,286
423	27.05.1878	4ft 8½in	0-4-0ST	OC	12x19	3ft 2in	340
428	05.07.1877	4ft 8½in	0-4-0ST	OC	12x19	3ft 2in	49
429	05.07.1877	4ft 8½in	0-4-0ST	OC	12x19	3ft 2in	86
431	05.07.1877	4ft 8½in	0-4-0ST	OC	12x19	3ft 2in	86
435	01.09.1877	3ft 0in	0-4-0ST	OC	5x10	1ft 8in	97
436	01.09.1877	3ft 0in	0-4-0ST	OC	5x10	1ft 8in	97
442	15.10.1877	3ft 0in	0-4-0ST	OC	5x10	1ft 8in	97
446	04.01.1878	3ft 0in	0-4-0ST	OC	5x10	1ft 8in	97
447	04.01.1878	3ft 0in	0-4-0ST	OC	5x10	1ft 8in	97
455	25.01.1878	3ft 0in	0-4-0ST	OC	5x10	1ft 8in	97
459	04.06.1878	3ft 0in	0-4-0ST	OC	5x10	1ft 8in	97
477	03.09.1878	4ft 8½in	0-4-0ST	OC	12x19	3ft 2in	57,145
478	06.01.1880	4ft 8½in	0-4-0ST	OC	12x19	3ft 2in	57,145
479	30.01.1880	4ft 8½in	0-4-0ST	OC	12x19	3ft 2in	288
487	09.01.1879	4ft 8½in	0-4-0ST	OC	12x19	3ft 2in	86
488	09.01.1879	4ft 8½in	0-4-0ST	OC	12x19	3ft 2in	86
489	04.04.1879	4ft 8½in	0-4-0ST	OC	12x19	3ft 2in	86
491	04.04.1879	4ft 8½in	0-4-0ST	OC	12x19	3ft 2in	86
492	04.04.1879	4ft 8½in	0-4-0ST	OC	12x19	3ft 2in	86
493	04.04.1879	4ft 8½in	0-4-0ST	OC	12x19	3ft 2in	86
494	03.05.1879	3ft 0in	0-4-0ST	OC	5x10	1ft 8in	97
495	03.05.1879	3ft 0in	0-4-0ST	OC	5x10	1ft 8in	97
519	12.1879	4ft 8½in	0-4-0ST	OC	13x19	3ft 2in	65,86
523	06.12.1879	4ft 8½in	0-4-0ST	OC	12x19	3ft 2in	86
524	06.12.1879	4ft 8½in	0-4-0ST	OC	12x19	3ft 2in	86
525	06.12.1879	4ft 8½in	0-4-0ST	OC	12x19	3ft 2in	49,86
527	06.12.1879	4ft 8½in	0-4-0ST	OC	12x19	3ft 2in	86
528	06.07.1880	4ft 8½in	0-4-0ST	OC	12x19	3ft 2in	126
531	05.04.1879	4ft 8½in	0-4-0ST	OC	10x17	2ft 10in	265
543	06.12.1879	4ft 8½in	0-4-0ST	OC	12x19	3ft 2in	86
544	04.01.1881	4ft 8½in	0-4-0ST	OC	12x19	3ft 2in	86
547	07.01.1879	4ft 8½in	0-4-0ST	OC	12x19	3ft 2in	57
557	21.01.1880	3ft 0in	0-4-0ST	OC	5x10	1ft 8in	97,351
558	21.01.1880	3ft 0in	0-4-0ST	OC	5x10	1ft 8in	97
559	21.01.1880	3ft 0in	0-4-0ST	OC	5x10	1ft 8in	97
561	21.01.1880	3ft 0in	0-4-0ST	OC	5x10	1ft 8in	97
562	21.01.1880	2ft 6in	0-4-0ST	OC	5x10	1ft 8in	171
563	21.01.1880	2ft 6in	0-4-0ST	OC	5x10	1ft 8in	171
595	25.11.1880	3ft 0in	0-4-0ST	OC	6x10	1ft 8in	97
596	25.11.1880	3ft 0in	0-4-0ST	OC	6x10	1ft 8in	97
597	25.11.1880	3ft 0in	0-4-0ST	OC	6x10	1ft 8in	97
598	25.11.1880	3ft 0in	0-4-0ST	OC	6x10	1ft 8in	97
599	25.11.1880	3ft 0in	0-4-0ST	OC	6x10	1ft 8in	97

601	25.11.1880	3ft 0in	0-4-0ST	OC	6x10	1ft 8in		97
605	20.12.1880	4ft 8½in	0-4-0ST	OC	14x19	3ft 2in		86
606	20.12.1880	4ft 8½in	0-4-0ST	OC	14x19	3ft 2in		86
607	20.12.1880	4ft 8½in	0-4-0ST	OC	14x19	3ft 2in		86
608	20.12.1880	4ft 8½in	0-4-0ST	OC	14x19	3ft 2in		86
609	20.12.1880	4ft 8½in	0-4-0ST	OC	14x19	3ft 2in		65,86
611	20.12.1880	4ft 8½in	0-4-0ST	OC	14x19	3ft 2in		86
613	21.12.1881	4ft 8½in	0-4-0ST	OC	12x19	3ft 2in		221
614	02.02.1882	4ft 8½in	0-4-0ST	OC	12x19	3ft 2in		59,149
679	22.10.1882	4ft 8½in	0-4-0ST	OC	12x19	3ft 2in		57,341
689	25.03.1882	4ft 8½in	0-4-0ST	OC	8x14	2ft 6in		149
691	25.03.1882	4ft 8½in	0-4-0ST	OC	8x14	2ft 6in		149
764	16.06.1885	4ft 8½in	0-4-0ST	OC	12x19	3ft 2in		149,210
844	02.04.1885	3ft 0in	0-6-0ST	OC	11x17	2ft 6in		370
852	22.07.1886	4ft 8½in	0-4-0ST	OC	12x19	3ft 2in		149
881	9.11.1886	4ft 8½in	0-4-0ST	OC	10x17	2ft 11in		205,343
883	04.02.1888	4ft 8½in	0-4-0ST	OC	12x19	3ft 3in		166
905	22.03.1887	4ft 8½in	0-4-0ST	OC	12x19	3ft 2½in		185,186
908	07.04.1887	4ft 8½in	0-4-0ST	OC	9x16	2ft 9in		177
935	1888	4ft 8½in	0-4-0ST	OC	12x19	3ft 2in		257,305
973	09.04.1889	4ft 8½in	0-4-0ST	OC	14x19	3ft 2in		64
977	12.07.1889	4ft 8½in	0-4-0ST	OC	12x19	3ft 3in		57,151
985	13.11.1889	4ft 8½in	0-4-0ST	OC	14x19	3ft 2in		64
991	12.11.1889	4ft 8½in	0-4-0ST	OC	12x19	3ft 3in		149
994	12.11.1889	4ft 8½in	0-4-0ST	OC	14x19	3ft 2in		64
1009	30.12.1889	3ft 0in	0-4-0ST	OC	6x10	1ft 10in		36,53
1011	30.12.1889	3ft 0in	0-4-0ST	OC	6x10	1ft 10in		36,53
1037	09.06.1891	4ft 8½in	0-4-0ST	OC	12x19	3ft 2in		355
1039	02.06.1891	4ft 8½in	0-4-0ST	OC	14x19	3ft 2in		86
1041	02.06.1891	4ft 8½in	0-4-0ST	OC	14x19	3ft 2in		86
1044	11.08.1891	4ft 8½in	0-4-0ST	OC	14x19	3ft 2in		64
1055	07.10.1891	3ft 0in	0-6-0ST	OC	11x17	2ft 6in		370
1066	04.05.1892	3ft 0in	0-4-0ST	OC	6x10	1ft 10in		58
1067	04.05.1892	3ft 0in	0-4-0ST	OC	6x10	1ft 10in		58
1077	14.09.1892	3ft 0in	0-4-0ST	OC	6x10	1ft 10in		58
1081	01.1893	3ft 0in	0-4-0ST	OC	6x10	1ft 10in		58
1094	08.08.1895	4ft 8½in	0-4-0ST	OC	14x19	3ft 2in		86,119
1102	06.09.1896	4ft 8½in	0-4-0ST	OC	12x19	3ft 3in		69
1115	17.05.1895	4ft 8½in	0-6-0ST	IC	12x18	3ft 0½in		370
1123	11.03.1896	4ft 8½in	0-4-0ST	OC	14x19	3ft 2in		86,119
1124	01.05.1896	4ft 8½in	0-4-0ST	OC	14x19	3ft 2in		59,149

W.J. BASSETT LOWKE, Northampton — BL

177	1909	1ft 3in	4-4-4TPM	S/O				394

BALDWIN LOCOMOTIVE WORKS, Philadelphia, USA — BLW

69496	1944	4ft 8½in	2-8-0	OC				385
		4ft 8½in	0-4-0ST	OC				191

BEYER, PEACOCK & CO LTD, Gorton Foundry, Manchester — BP

6172	31.01.1924	4ft 8½in	0-4-4-0T	4C	13½x20	3ft 4in	61½T	200

BSC HARTLEPOOL — BSC Hartlepool

	1986	4ft 8½in	4wWE					98,292
	1986	4ft 8½in	4wWE					98,292

BRUSH ELECTRICAL MACHINES LTD, Falcon Works, Loughborough — BT

	1905	4ft 8½in	4wBE					249

BUTTERLEY CO LTD, Butterley, Derbyshire — Butterley

[SADLERS No.4]		4ft 8½in	0-4-0ST	OC	reb R.Shaw, /1942			288

CLAYTON EQUIPMENT LTD, Hatton, Derbyshire — CE

B0922B	1975	1ft 6in	4wBE		215
B4618.7	2016	4ft 8½in	4wDH		213
B4618.6	2016	4ft 8½in	4wDH		213
B4618.2	2016	4ft 8½in	4wDH		213

CHAPMAN & FURNEAUX, Quarry Lane, Gateshead — CF

	order dates						
1145	12.03.1896	4ft 8½in	0-4-0ST	OC	12x19	3ft 3in	185,186,187
1146	12.03.1896	4ft 8½in	0-4-0ST	OC	12x19	3ft 3in	152
1147	12.03.1896	4ft 8½in	0-4-0ST	OC	12x19	3ft 3in	187
1156	09.12.1897	4ft 8½in	0-4-0ST	OC	11x17	2ft 11in	255
1167	09.07.1898	4ft 8½in	0-4-0ST	OC	11x17	2ft 11in	255,257
1178	10.01.1899	3ft 0in	0-4-0ST	OC	6x10	1ft 10in	58
1179	10.02.1899	4ft 8½in	0-4-0ST	OC	12x19	3ft 3in	185,186
1183	30.06.1899	4ft 8½in	0-4-0ST	OC	12x19	3ft 3in	266
1186	08.08.1899	4ft 8½in	0-4-0ST	OC	14x19	3ft 2in	86
1188	24.10.1899	4ft 8½in	0-4-0ST	OC	12x19	3ft 3in	266
1195	28.11.1899	4ft 8½in	0-4-0ST	OC	13x19	3ft 4in	355
1199	23.07.1900	4ft 8½in	0-4-0ST	OC	14x19	3ft 2in	57,59,139
1211	18.04.1901	4ft 8½in	0-4-0ST	OC	14x19	3ft 2in	86
1212	28.05.1902	4ft 8½in	0-4-0CT	OC	12x19	3ft 0in	286

ALEXANDER CHAPLIN & CO, Cranstonhill Engine Works, Glasgow — Chaplin

305	1862	4ft 8½in	0-4-0VBT	VC	335
317	1863	4ft 8½in	0-4-0VBT		308
358	1863	4ft 8½in	0-4-0VBT		308
381	1863	4ft 8½in	0-4-0VBT		308
382	1863	4ft 8½in	0-4-0VBT		308
1585	1873	4ft 8½in	0-4-0VBT		22

CLEVELAND WORKS — Cleveland

[MC1]		4ft 8½in	0-4-0CA	OC		89
[MC2]		4ft 8½in	0-4-0CA	OC		89
[MC3]	1958	4ft 8½in	0-4-0CA	OC		89
	c1967	4ft 8½in	4wWE	[reb of 4wDH TH 128V]		99
	1974	4ft 8½in	4wWE	[reb of 4wDH S 10004]		99

COCHRANE & CO LTD, Ormesby Ironworks, Middlesbrough — Cochrane

	1871	4ft 8½in	0-4-0VBT	VC	69,308
		4ft 8½in	0-4-0VBT	OC	309
		4ft 8½in	0-4-0VBT		309
		4ft 8½in	0-4-0VBT		309
	c1880	4ft 8½in	0-4-0VBT		309

CREWE WORKS, Crewe, Cheshire — Crewe

British Railways [prev LMS, LNWR]

[08576]	06.1959	4ft 8½in	0-6-0DE	350hp	EE 6KT	48T	332
[08588]	11.1959	4ft 8½in	0-6-0DE	350hp	EE 6KT	48T	261
[08743]	11.1960	4ft 8½in	0-6-0DE	350hp	EE 6KT	48T	201,216,294,295

DARLINGTON WORKS, Darlington Dar

British Railways [prev LNER, NER]

[08375]	06.1957	4ft 8½in	0-6-0DE	350hp	EE 6KT	48T	261
[D3476]	12.1957	4ft 8½in	0-6-0DE	350hp	Blackstone ER6T	47½T	333
[08871]	11.1960	4ft 8½in	0-6-0DE	350hp	EE 6KT	48T	261
[08874]	12.1960	4ft 8½in	0-6-0DE	350hp	EE 6KT	48T	216

DAVENPORT LOCOMOTIVE WORKS, Davenport, Iowa, U.S.A Davenport

2505	04.1943	4ft 8½in	0-6-0T	OC	16 ½ x23	4ft 6in	241

DREWRY CAR CO LTD, London DC

Supply agents only, locomotives built by third party.

2165	1941	see EE 1196	297
2167	1943	see VF 4859	240
2486	1953	see VF D212	260,317
2575	1956	see RSHN 7862	317
2592	1959	see RSHN 7925	286,317,318
2652	1958	see Bg 2652	344
2655	1959	see Bg 2655	344
2725	1963	see Bg 2725	133

DERBY WORKS, Derby, Derbyshire Derby

British Railways [prev LMS, MR]

[08077]	02.1955	4ft 8½in	0-6-0DE	350hp	EE 6KT	48T	216
[08308]	08.1957	4ft 8½in	0-6-0DE	350hp	EE 6KT	48T	261
[08401]	05.1958	4ft 8½in	0-6-0DE	350hp	EE 6KT	48T	243
[08410]	06.1958	4ft 8½in	0-6-0DE	350hp	EE 6KT	48T	133
[08423]	08.1958	4ft 8½in	0-6-0DE	350hp	EE 6KT	48T	261
[08598]	05.1959	4ft 8½in	0-6-0DE	350hp	EE 6KT	48T	133,213
[08600]	05.1959	4ft 8½in	0-6-0DE	350hp	EE 6KT	48T	133
[08613]	08.1959	4ft 8½in	0-6-0DE	350hp	EE 6KT	48T	261
[08622]	10.1959	4ft 8½in	0-6-0DE	350hp	EE 6KT	48T	261
[08774]	04.1960	4ft 8½in	0-6-0DE	350hp	EE 6KT	48T	121,133
[08785]	04.1960	4ft 8½in	0-6-0DE	350hp	EE 6KT	48T	29
[08788]	04.1960	4ft 8½in	0-6-0DE	350hp	EE 6KT	48T	261
[08807]	07.1960	4ft 8½in	0-6-0DE	350hp	EE 6KT	48T	133,273
[08809]	07.1960	4ft 8½in	0-6-0DE	350hp	EE 6KT	48T	261
[08816]	08.1960	4ft 8½in	0-6-0DE	350hp	EE 6KT	48T	121

DICK, KERR & CO LTD, Preston, Lancashire DK

[No.5]	1920	4ft 8½in	4wBE	112
[No.6]	1920	4ft 8½in	4wBE	112

DORMAN LONG & CO LTD DL

[43]	1949	4ft 8½in	0-4-0ST	OC	59,60
[39]	1951	4ft 8½in	0-4-0ST	OC	59,60,65
[27]	1952	4ft 8½in	0-4-0ST	OC	59,60,65,152
[35]	1952	4ft 8½in	0-4-0ST	OC	59,60
[22]	1953	4ft 8½in	0-4-0ST	OC	59,60
[36]	1953	4ft 8½in	0-4-0ST	OC	59,60,65
[1]	1959	4ft 8½in	0-4-0ST	OC	59,60,167
		4ft 8½in	0-4-0WE	98	

DONCASTER WORKS, Doncaster, Yorkshire   Don
British Railways [prev LNER, GNR]

[D2046]	12.1958	4ft 8½in	0-6-0DM	204hp	Gardner 8L3	31T [reb HE 6644]	297
[08502]	05.1958	4ft 8½in	0-6-0DE	350hp	EE 6KT	48T	208,295
[08503]	05.1958	4ft 8½in	0-6-0DE	350hp	EE 6KT	48T	208,295
[08523]	09.1958	4ft 8½in	0-6-0DE	350hp	EE 6KT	48T	261
[D2057]	05.1959	4ft 8½in	0-6-0DM	204hp	Gardner 8L3	31T [reb HE 6645]	297
[D2093]	06.1960	4ft 8½in	0-6-0DM	204hp	Gardner 8L3	31T [reb HE 6643]	298

DUBS & CO, Glasgow Locomotive Works, Polmadie, Glasgow   D

857	1875	4ft 8½in	0-4-0ST	OC			265
2051	1884	4ft 8½in	0-4-0CT	OC	12x22	3ft 6in	44

ENGLISH ELECTRIC CO LTD, Dick Kerr Works, Preston Lancashire   EE

745	1929	2ft 6in	4wWE				203
1196	21.11.1941	4ft 8½in	0-4-0DM	153hp	Gardner 6L3	22½T [DC 2165]	297
1227	1943	2ft 6in	4wWE				203
1552	1948	4ft 8½in	0-6-0DE	350hp			208
1553	1948	4ft 8½in	0-6-0DE	350hp			208
1554	1948	4ft 8½in	0-6-0DE	350hp			200,208,327
2122	1956	4ft 8½in	0-6-0DE	350hp	EE 6KT	47T [VF D312]	261
2129	1956	4ft 8½in	0-6-0DE	350hp	EE 6KT	47T [VF D319]	261
2146	1956	4ft 8½in	0-6-0DE	350hp	EE 6KT	47T [VF D336]	261

ENGLISH ELECTRIC CO LTD, Stephenson Works, Darlington, Co.Durham   EES

8431	1963	4ft 8½in	0-4-0DH	240hp	Dorman 6KUDT		332
8453	1963	4ft 8½in	0-4-0DH				253

ENGLISH ELECTRIC CO LTD, Vulcan Works, Newton-le-Willows, Lancashire   EEV

D908	1964	4ft 8½in	4wDH	200hp	Dorman 6LCT		317,318,338
D1201	1967	4ft 8½in	0-6-0DH	391hp	Cummins NT400		121
D1205	1967	4ft 8½in	0-4-0DH	305hp	Cummins NHR6		29
D1206	1966	4ft 8½in	0-4-0DH				318,332,338,340
3870	1969	4ft 8½in	0-6-0DH	274hp	Dorman 6QA		29
3994	1970	4ft 8½in	0-6-0DH	391hp	Cummins NT400		121,216,261
4003	1971	4ft 8½in	0-6-0DH	286hp	Dorman 6QA		171,231,332
5352	1971	4ft 8½in	0-6-0DH	500hp	Dorman 8QT		261
8449	1965	4ft 8½in	0-4-0DH	283hp	Dorman 6KUDT		332

ELECTROMOBILE LTD, PROSPECT WORKS, Otley, Yorkshire   Electromobile

W247	1927	4ft 8½in	4wBE				238

FALCON ENGINE & CAR WORKS LTD, Loughborough, Leicestershire   FE

117	1889	4ft 8½in	0-4-0ST	OC		[WINSFORD]	205,207
?		4ft 8½in	0-4-0ST	OC		[NEWBRIDGE]	205,207

F.C. HIBBERD & CO LTD, Park Royal Works, London   FH

1675	08.1930	4ft 8½in	4wPM	20hp	Dorman 4MAL		31
1943	09.1935	4ft 8½in	4wPM	40hp	Dorman 4JOR	8½T	193,352
3492	1951	4ft 8½in	4wDM	77hp	Dorman 4DL	18T	38
3569	1951	4ft 8½in	4wDM	77hp	Dorman 4DL	18T	296,352
3572	29.08.1952	4ft 8½in	4wDM	57hp	Dorman 3DL	11T	177,179,352
3685	30.03.1954	4ft 8½in	4wDM	77hp	Dorman 4DL	18T	208
3808	26.04.1956	4ft 8½in	4wDM	117hp	Dorman 6DL	23T	38
3817	08.10.1956	4ft 8½in	4wDM	117hp	Dorman 6DL	23T	318,337
3822	23.02.1956	4ft 8½in	4wDM	117hp	Dorman 6DL	23T	38

3832	30.04.1957	4ft 8½in	4wDM	117hp	Dorman 6DL		23T		49,89
3883	28.07.1958	4ft 8½in	4wDM		Dorman 6KUD		24T		49,60
3888	06.11.1958	4ft 8½in	4wDH		Dorman 6KUD		24T		61,89
3896	14.04.1959	4ft 8½in	4wDH		Dorman 6KUD		25T		302
3933	1960	4ft 8½in	4wDM						277
3934	26.04.1960	4ft 8½in	4wDH		Dorman 6KUD		25T		61,89
3935	29.04.1960	4ft 8½in	4wDH		Dorman 6KD		24T		89,277
3941	1960	4ft 8½in	4wDM						61
3942	1960	4ft 8½in	4wDH		Dorman 6KUD		24½T		61,298
3949	1960	4ft 8½in	4wDM						260,318,338
3958	26.05.1961	4ft 8½in	4wDH		Leyland UE680				324
3964	15.09.1961	4ft 8½in	4wDH		Dorman 6KUD				260,318,338
4010	04.06.1963	4ft 8½in	4wDM		Dorman 4LB		11T		312
4011	11.01.1965	4ft 8½in	4wDH		Dorman 6LC		23T		124,273

FLETCHER, JENNINGS & CO LTD, Lowca Engine Works, Whitehaven   FJ

30	1863	4ft 8½in	0-4-0ST	OC	10x20	4ft 0in		355
44	1864	4ft 8½in	0-4-0ST	OC	12x20	3ft 6in		355,356
75	1868	4ft 8½in	0-4-0WT	OC	10x20			160
76	1867	4ft 8½in	0-4-0T	OC	10x20	3ft 4in		160,219,338,340
79	1871	4ft 8½in	0-4-0T	OC	10x20			160
82	1869	4ft 8½in	0-4-0WT	OC	12x20			160
85	1870	4ft 8½in	0-4-0T	OC	9x16	2ft 9in		355
88	1872	4ft 8½in	0-4-0T	OC	12x20			266
90	1872	4ft 8½in	0-4-0T	OC	10x20			57,160
101	1872	4ft 8½in	0-4-0T	OC	9x16	2ft 9in		219,355
108	1872	4ft 8½in	0-4-0T	OC	12x20			254
111	1873	4ft 8½in	0-4-0T	OC	12x20	3ft 5in		376
113	1873	4ft 8½in	0-4-0T	OC	10x20			335
116	1873	4ft 8½in	0-4-0T	OC	12x20			268
130	1874	4ft 8½in	0-4-0T	OC	9x16	2ft 9in		355
134	1875	4ft 8½in	0-4-0T	OC	10x20			160
164	1880	4ft 8½in	0-4-0T	OC	10x20			340
165	1879	4ft 8½in	0-4-0T	OC	12x20			239

FOX, WALKER & CO LTD, Atlas Engine Works, Bristol   FW

29	1873	4ft 8½in	0-6-0ST		13½x -			359
127	1872	4ft 8½in	0-6-0ST					359
133	1871	4ft 8½in	0-6-0ST	OC				160
147	1871	4ft 8½in	0-6-0ST	OC	13x -			379
162	01.07.1872	4ft 8½in	0-4-0ST	OC	10x18	2ft 8in	D	69
169	12.12.1872	4ft 8½in	0-6-0ST	OC	13x20	3ft 6½in	B	160
198	30.09.1873	4ft 8½in	0-4-0ST	OC	11x18	2ft 10in	D	313,340
245	28.05.1874	4ft 8½in	0-6-0ST	OC	13x20	3ft 6in	B	44
249	18.08.1874	4ft 8½in	0-6-0ST	OC	13x20	3ft 6½in	B	85
252	06.06.1874	4ft 8½in	0-4-0ST	OC	12x18		D	69
279	20.07.1875	4ft 8½in	0-6-0ST	OC	13x20	3ft 6in	B1	379
358	1877	4ft 8½in	0-6-0ST	OC	13x20	3ft 6in	B1	69,379

GATESHEAD WORKS, Gateshead, Co.Durham   Gateshead

NER [later LNER and BR]

[945]	1888	4ft 8½in	0-4-0T	IC	13x20	3ft 6¼in	252

GREENWOOD & BATLEY LTD, Albion Works, Leeds   GB

1448	12.1936	4ft 8½in	4wWE	2x25hp 230vDC	230,232
2937	30.03.1960	4ft 8½in	0-4-0WE	2x40hp 230vDC	232

420306	06.1972	4ft 8½in	4wWE		2x45hp 250vDC			232
420355/1	12.1976	4ft 8½in	4wWE		2x75hp 415v			98,292
420355/2	12.1976	4ft 8½in	4wWE		2x75hp 415v			98,292
420408	01.1977	4ft 8½in	4wWE		2x75hp 415v			98,292

GEC TRACTION LTD, Newton-le-Willows, Lancashire — GECT

5378	1972	4ft 8½in	0-6-0DH	500hp	Dorman 8QT		121
5383	1973	4ft 8½in	0-6-0DH	500hp	Dorman 8QT		231,383
5414	1976	4ft 8½in	6wDE	760hp	Dorman 12QT	75T	91,290
5415	1976	4ft 8½in	6wDE	760hp	Dorman 12QT	75T	91,290
5416	1976	4ft 8½in	6wDE	760hp	Dorman 12QT	75T	91,100,290
5417	1976	4ft 8½in	6wDE	760hp	Dorman 12QT	75T	91
5418	1976	4ft 8½in	6wDE	760hp	Dorman 12QT	75T	91,290
5421	1977	4ft 8½in	6wDE	750hp	Dorman 12QT		92,290
5425	1977	4ft 8½in	6wDE	760hp	Dorman 12QT	75T	91,231
5426	1977	4ft 8½in	6wDE	760hp	Dorman 12QT	75T	91,100,290
5427	1977	4ft 8½in	6wDE	760hp	Dorman 12QT	75T	91,290
5428	1977	4ft 8½in	6wDE	760hp	Dorman 12QT	75T	91
5429	1977	4ft 8½in	6wDE	760hp	Dorman 12QT	75T	91,100,290
5430	1977	4ft 8½in	6wDE	760hp	Dorman 12QT	75T	91,100,216,290
5431	1977	4ft 8½in	6wDE	760hp	Dorman 12QT	75T	91,290
5432	1977	4ft 8½in	6wDE	760hp	Dorman 12QT	75T	91
5461	1977	4ft 8½in	6wDE	760hp	Dorman 12QT	75T	91,231
5462	1977	4ft 8½in	6wDE	760hp	Dorman 12QT	75T	91,100,290
5463	1977	4ft 8½in	6wDE	760hp	Dorman 12QT	75T	91,100,290
5464	1977	4ft 8½in	6wDE	760hp	Dorman 12QT	75T	91,290
5465	1977	4ft 8½in	6wDE	760hp	Dorman 12QT	75T	91,100,216,290
5466	1977	4ft 8½in	6wDE	760hp	Dorman 12QT	75T	91,100,290
5467	1977	4ft 8½in	6wDE	760hp	Dorman 12QT	75T	91,290
5468	24.08.1977	4ft 8½in	6wDE	750hp	Dorman 12QT		92,290
5469	1978	4ft 8½in	6wDE	760hp	Dorman 12QT	75T	91,290,383
5470	1978	4ft 8½in	6wDE	760hp	Dorman 12QT	75T	91
5471	1978	4ft 8½in	6wDE	760hp	Dorman 12QT	75T	91,290
5472	1978	4ft 8½in	6wDE	760hp	Dorman 12QT	75T	91,290
5473	1978	4ft 8½in	6wDE	760hp	Dorman 12QT	75T	91,100
5474	1978	4ft 8½in	6wDE	760hp	Dorman 12QT	75T	91,100,290
5475	1978	4ft 8½in	6wDE	760hp	Dorman 12QT	75T	91,216,290
5478	03.11.1978	4ft 8½in	6wDE	750hp	Dorman 12QT		231,290,329
5479	1979	4ft 8½in	6wDE	750hp	Dorman 12QT	50T	92
5578	1980	4ft 8½in	6wDE	480hp	Dorman 8QT	48T	92,330

GIBB & HOGG, Victoria Engine Works, Airdrie, Lanarkshire, Scotland — GH

[HARE]	1908	4ft 8½in	0-4-0ST	OC	143

GILKES, WILSON & CO, Tees Engine Works, Middlebrough — GW

106	11.1860	4ft 8½in	0-4-0T				64
108	1861	4ft 8½in	0-4-0ST	OC			268,272
109	02.1861	4ft 8½in	0-4-0T				64
130	12.1861	4ft 8½in	0-4-0T	OC	12x20	3ft 9in	323
165	06.1863	4ft 8½in	0-4-0T	VC			323
166	07.1863	4ft 8½in	0-4-0T		14x -		64
167	10.1863	4ft 8½in	0-4-0VBT				85
168	10.1863	4ft 8½in	0-4-0VBT				85
169	11.1863	4ft 8½in	0-4-0VBT	OC			246,268
170	11.1863	4ft 8½in	0-4-0VBT				85
171	02.1864	4ft 8½in	0-4-0VBT				85

[BEETLE]	1863	4ft 8½in	0-4-0T					268
173	11.1863	4ft 8½in	0-4-0T					119
174	03.1864	4ft 8½in	0-4-0T					160
176	07.1864	4ft 8½in	0-4-0VBT					85
177	07.1864	4ft 8½in	0-4-0VBT					85
179	06.1864	4ft 8½in	0-4-0T					335
188	06.1864	4ft 8½in	0-4-0VBT					268
189	06.1864	4ft 8½in	0-4-0T					160
190	08.1864	4ft 8½in	0-4-0VBT					85
191	09.1864	4ft 8½in	0-4-0VBT					119
192	11.1864	4ft 8½in	0-4-0T		6x14	2ft 7in		64

GRANGE IRON WORKS, Durham

	1873	4ft 8½in	0-4-0ST	OC				221,228
[No.6]	1879	4ft 8½in	0-4-0ST	OC				355

GREAT NORTHERN STEAM CO LTD, Middlesbrough   GNS

[EFFIE]	1999	1ft 3in	0-4-0WT	OC				382

JAMES & FREDK HOWARD LTD, Britannia Iron Works, Bedford   H

958	27.10.1931	4ft 8½in	4wDM		Blackstone BVH4			313

HUNSLET-BARCLAY LTD, Caledonia Works, Kilmarnock   HAB

776	03.06.1991	4ft 8½in	4w-4wDH	624hp	Cat 3412		64T	92,333

HUDSWELL, CLARKE & CO LTD, Railway Foundry, Leeds   HC

205	29.04.1880	4ft 8½in	0-4-0ST	OC	12x18	3ft 0in		149
262	27.09.1883	4ft 8½in	0-4-0ST	OC	14x20	3ft 6½in		337
292	17.10.1888	4ft 8½in	0-4-0ST	OC	13x18	3ft 0½in		149
323	22.11.1888	4ft 8½in	0-4-0ST	OC	13x18	3ft 0in		59,149
324	05.03.1889	4ft 8½in	0-4-0ST	OC	13x18	3ft 0in		59,139,149
330	26.04.1889	4ft 8½in	0-4-0ST	OC	13x18	3ft 0in		166
347	20.01.1892	4ft 8½in	0-6-0ST	IC	13x20	3ft 3in		301,361
405	08.10.1894	4ft 8½in	0-4-0ST	OC	13x18	3ft 0in		166
530	31.08.1899	4ft 8½in	0-6-0ST	IC	12x18	3ft 0in		365
636	24.03.1903	4ft 8½in	0-4-0ST	OC	14x20	3ft 3½in		166
656	17.07.1903	4ft 8½in	0-4-0ST	OC	10x16	2ft 9½in		131,326
694	22.03.1904	4ft 8½in	0-6-0T	IC	15½x20	3ft 7in		69
754	26.01.1906	4ft 8½in	0-4-0ST	OC	14x20	3ft 3½in		166
898	12.11.1909	3ft 0in	0-4-0ST	OC	9x15	2ft 6½in		182
1011	26.11.1912	4ft 8½in	0-6-0ST	IC	15x20	3ft 7in		364
1026	14.04.1913	4ft 8½in	0-6-0ST	IC	15x20	3ft 7in		364
1070	25.11.1914	4ft 8½in	0-6-0T	IC	15½x20	3ft 4in		361
1096	10.11.1914	3ft 0in	0-4-0ST	OC	9x15	2ft 6½in		182
1422	28.01.1921	4ft 8½in	0-4-0ST	OC	14x20	3ft 3½in		200
1606	27.06.1929	4ft 8½in	0-6-0ST	IC	13x20	3ft 3½in		241
1623	19.05.1928	4ft 8½in	0-4-0ST	OC	16x24	3ft 9in		200
1624	23.05.1928	4ft 8½in	0-4-0ST	OC	16x24	3ft 9in		200
1699	09.02.1938	4ft 8½in	0-6-0ST	IC	13x20	3ft 3½in		241
1734	28.12.1942	4ft 8½in	0-4-0ST	OC	14x22	3ft 3½in		116
1735	11.12.1942	4ft 8½in	0-4-0ST	OC	14x22	3ft 3½in		116
D835	02.12.1954	4ft 8½in	0-6-0DM	300hp	Crossley EST		45T.14	112
D978	28.11.1957	4ft 8½in	0-4-0DM	260hp	Nat M4AA8		37T.10	228,230
D1013	01.12.1957	4ft 8½in	0-4-0DM	260hp	Nat M4AA8		37T.10	45,230
D1032	23.12.1957	4ft 8½in	0-4-0DM	260hp	Nat M4AA8		37T.10	45,230
D1052	03.04.1958	4ft 8½in	0-4-0DM	260hp	Nat M4AA8		37T.10	228,230

D1081	03.01.1958	4ft 8½in	0-4-0DM	260hp	Nat M4AA8		37T.10	45,230
D1141	22.07.1959	4ft 8½in	0-4-0DM	260hp	Nat M4AA8		37T.10	228,230
D1191	07.04.1960	4ft 8½in	0-6-0DM	204hp	Gardner 8L3		35T.10	331
D1279	28.02.1963	4ft 8½in	0-4-0DH	252hp	Cummins NH5-6		34T.0	116
D1346	05.07.1965	4ft 8½in	0-4-0DH	191hp	Cummins NH220		27T.17	116,333,383

HUDSWELL CLARKE & ROGERS, Railway Foundry, Leeds   HCR

112	14.08.1871	4ft 8½in	0-4-0ST	OC	8x15	2ft 6in		265,301
123	24.12.1872	4ft 8½in	0-4-0ST	OC	10x16	2ft 9in		166,346
125	15.06.1873	4ft 8½in	0-4-0ST	OC	8x15	2ft 6in		194,255
126	04.12.1872	3ft 0in	0-4-0ST	OC	8x15	2ft 6in		346
136	08.07.1873	4ft 8½in	0-4-0ST	OC	13x20	3ft 6in		337
139	17.09.1874	4ft 8½in	0-4-0ST	OC	10x16	2ft 9in		38,194
146	10.08.1874	4ft 8½in	0-4-0ST	OC	8x15	2ft 6in		360
150	20.11.1874	4ft 8½in	0-4-0ST	OC	13x20	3ft 6in		337,340
151	21.12.1874	4ft 8½in	0-4-0ST	OC	13x20	3ft 6in		149,296

HUNSLET ENGINE CO LTD, Hunslet, Leeds   HE

30	03.03.1869	4ft 8½in	0-4-0ST	OC	10x18	2ft 9in		220
41	21.02.1870	4ft 8½in	0-4-0ST	OC	8x14	2ft 8½in		218
60	14.08.1871	4ft 8½in	0-4-0ST	OC	12x18	3ft 1in		220
78	10.07.1872	4ft 8½in	0-4-0ST	OC	12x18	3ft 1in		220
80	20.05.1873	4ft 8½in	0-4-0ST	OC	10x15	2ft 9in		220,376
177	28.05.1877	4ft 8½in	0-4-0ST	OC	12x18	3ft 1in		304
205	12.08.1878	4ft 8½in	0-4-0ST	OC	12x18	3ft 0in		114,350
240	09.03.1880	4ft 8½in	0-4-0ST	OC	12x18	3ft 1in		305
413	02.03.1887	4ft 8½in	0-4-0ST	OC	13x18	3ft 1in		304
420	25.05.1887	4ft 8½in	0-4-0ST	OC	10x15	2ft 9in		361
464	03.08.1888	4ft 8½in	0-6-0ST	IC	13x18	3ft 1in		189,301,361
587	05.08.1893	4ft 8½in	0-6-0T	IC	11x15	2ft 9in		367
608	24.10.1895	4ft 8½in	0-4-0ST	OC	14x20	3ft 4in		228,304
695	24.08.1899	4ft 8½in	0-6-0T	IC	14x18	3ft 3in		191
894	22.11.1905	4ft 8½in	0-4-0ST	OC	14x20	3ft 4in		52,228,304
951	04.12.1907	4ft 8½in	0-4-0ST	OC	16x22	3ft 7in		228,304
979	03.09.1908	1ft 11¼in	0-6-4T	OC	9½ x14	2ft 4in		347
1086	20.12.1911	4ft 8½in	0-4-0ST	OC	16x22	3ft 7in		228,304
1087	20.12.1911	4ft 8½in	0-4-0ST	OC	16x22	3ft 7in		52,228,304
1108	30.09.1912	4ft 8½in	0-4-0ST	OC	16x22	3ft 7in		228,304
1294	06.02.1918	4ft 8½in	0-4-0ST	OC	15x20	3ft 4in		69,265,266
1405	27.04.1920	4ft 8½in	0-4-0ST	OC	16x22	3ft 7in		228,304
3218	20.06.1945	4ft 8½in	0-6-0ST	IC	18x26	4ft 3in		88
1720	23.05.1933	3ft 0in	4wDM	20hp				98
1742	22.02.1934	1ft 5¾in	4wDM	20hp				107
1748	13.06.1934	1ft 5¾in	4wDM	20hp				107
1813	08.05.1936	3ft 0in	4wDM	20hp				98
1863	09.10.1937	2ft 0in	4wDM	20hp				120
1929	10.03.1938	2ft 6in	4wDM	20hp				169,303
2652	11.03.1943	4ft 8½in	0-4-0DM	40/44hp	Fowler			32,228,230
2653	07.04.1943	4ft 8½in	0-4-0DM	40/44hp	Fowler			45
2840	27.10.1943	4ft 8½in	0-4-0DM	40/44hp	Fowler 4B			45
3308	12.03.1946	2ft 6in	4wDM	20hp	McLaren			230
4630	01.08.1956	4ft 8½in	0-4-0DM	60hp	Gardner 4LW			65
5306	09.04.1958	4ft 8½in	4wDM	71hp	Dorman 4LB		13T	235
5680	17.07.1961	2ft 6in	4wDM	40hp	Perkins P4			154
5682	21.07.1960	2ft 6in	4wDM	40hp	Perkins P4			154
6294	13.01.1965	4ft 8½in	0-6-0DH	311hp	R-R C8SFL		40T	121,213,216,261

6295	29.01.1965	4ft 8½in	0-6-0DH		311hp	R-R C8SFL	40T	121,261
6643	05.09.1967	4ft 8½in	0-6-0DM		204hp	reb of BR Don D2093 (0-6-0DMF)		298
6644	22.09.1967	4ft 8½in	0-6-0DM		204hp	reb of BR Don D2046 (0-6-0DMF)		297
6645	10.10.1967	4ft 8½in	0-6-0DM		204hp	reb of BR Don D2057 (0-6-0DMF)		297
6662	06.12.1966	4ft 8½in	0-6-0DH		311hp	R-R C8SFL	55T	213
7041	22.11.1971	4ft 8½in	0-6-0DH		350hp	R-R C8TFL		92
7279	26.07.1972	4ft 8½in	0-6-0DH		388hp	Cummins NT400	50T	213
7425	09.03.1981	4ft 8½in	0-4-0DH		252hp	Cummins NT250	37T	116,231,383,384
7541	20.12.1976	4ft 8½in	0-6-0DH		388hp	Cummins NT400	50T	121
8977	20.05.1980	4ft 8½in	0-6-0DH		388hp	Cummins NT400	50T	201
8998	23.02.1981	4ft 8½in	0-6-0DH		450hp	Detroit 12V 71N		216
9069	18.07.1983	4ft 8½in	0-6-0DH		310hp			90,213,261,329
9307	1992	4ft 8½in	0-6-0DH					201

HOPKINS, GILKES & CO, Tees Iron Works, Middlesbrough   HG

Successors to Gilkes, Wilson & Co, 1865. GW (which see)

196	04.1865	4ft 8½in	0-4-0T		14x -	3ft 9in		119
197	09.1865	4ft 8½in	0-4-0T		12x -	3ft 3in		119
198	11.1865	2ft 1in	0-4-0T	OC	9x -			271
200	04.1865	4ft 8½in	0-4-0T		14x -	3ft 9¾in		335
211	09.1865	4ft 8½in	0-4-0T	VC				233
231	04.1866	4ft 8½in	0-4-0T		12x -			44
233	03.1866	4ft 8½in	0-4-0T		12x -			160
235	07.1866	4ft 8½in	0-4-0T		12x -			323
236	1866	4ft 8½in	0-4-0T		12x -			335
238	02.1866	4ft 8½in	0-4-0T		12x -			38,64
239	05.1866	4ft 8½in	0-4-0T		12x -			335
247	04.1867	4ft 8½in	0-4-0T		12x -			44,119
252	11.1867	4ft 8½in	0-4-0ST	OC	12x20	3ft 11½in		135,323
276	11.1870	4ft 8½in	0-4-0ST	OC	12x20	3ft 11½in		109,233
277	07.1871	4ft 8½in	0-4-0T		12x20	3ft 11½in		44
298	07.1873	4ft 8½in	0-4-0ST	OC				166
299	03.1873	4ft 8½in	?					323
355	1874	4ft 8½in	0-4-0ST	OC				26,191,323,355
356	1874	4ft 8½in	0-4-0ST	OC				68,323
[MILLIE]	1874	4ft 8½in	0-4-0ST	OC				114
[MARY IRVING]		4ft 8½in	0-4-0ST	OC				166
[No.1]		4ft 8½in	0-4-0	VC				139

HENRY HUGHES & CO, Falcon Works, Loughborough, Leicestershire   HH

[BARTON]	1872	4ft 8½in	0-4-0ST	OC	340
[No.1]	1874	3ft 0in	0-4-0T		346
[No.4]	1876	4ft 8½in	0-4-0ST		376
		4ft 8½in	0-4-0T?		106
	1876	4ft 8½in	0-4-0ST	OC	139
		4ft 8½in	0-4-0ST	OC	254

R. & W. HAWTHORN, LESLIE & CO LTD, Forth Bank Works, Newcastle upon Tyne   HL

2110	04.06.1888	4ft 8½in	0-4-0ST	OC	14x20	3ft 6in	139
2133	01.1889	4ft 8½in	0-4-0ST	OC	14x20	3ft 6in	57
2134	02.1889	4ft 8½in	0-4-0ST	OC	14x20	3ft 6in	221,228,304
2139	25.10.1889	4ft 8½in	0-4-0ST	OC	12x18	3ft 0in	309
2171	10.1890	4ft 8½in	0-4-0ST	OC	12x18	3ft 0½in	309,326
2177	29.07.1890	4ft 8½in	0-4-0ST	OC	12x19	3ft 4in	34
2247	05.09.1892	4ft 8½in	0-4-0ST	OC	12x18	3ft 0½in	269
2295	1895	4ft 8½in	0-4-0ST	OC	14x20	3ft 6in	199

2334	02.1896	4ft 8½in	0-4-0CT	OC	12x15	2ft 10in	179
2358	03.1897	4ft 8½in	0-4-0ST	OC	14x20	3ft 6in	309
2378	08.1898	4ft 8½in	0-4-0ST	OC	14x20	3ft 6in	222,228
2388	07.02.1898	4ft 8½in	0-4-0ST	OC	12x18	3ft 0½in	309
2412	03.1899	4ft 8½in	0-4-0ST	OC	14x20	3ft 6in	228
2415	04.1899	4ft 8½in	0-4-0CT	OC	12x15	2ft 10in	22
2425	09.1899	4ft 8½in	0-4-0ST	OC	14x20	3ft 6in	69
2431	08.1899	4ft 8½in	0-4-0ST	OC	12x18	3ft 0½in	243
2439	11.1899	4ft 8½in	0-4-0ST	OC	14x20	3ft 6in	338
2445	12.1899	4ft 8½in	0-4-0ST	OC	14x20	3ft 6in	222,228
2461	29.05.1900	4ft 8½in	0-4-0ST	OC	14x20	3ft 6in	45,52
2462	08.06.1900	4ft 8½in	0-4-0ST	OC	14x20	3ft 6in	52
2516	26.03.1902	4ft 8½in	0-4-0CT	OC	12x15	2ft 10in	228,309
2517	06.05.1902	4ft 8½in	0-4-0CT	OC	12x15	2ft 10in	391
2536	18.09.1902	4ft 8½in	0-4-0ST	OC	12x15	2ft 10in	24
2570	18.11.1903	4ft 8½in	0-4-0ST	OC	14x20	3ft 6in	65,87
2573	11.08.1904	4ft 8½in	0-4-0CT	OC	12x15	2ft 10in	355
2583	11.05.1904	4ft 8½in	0-4-0ST	OC	14x20	3ft 6in	200
2597	15.11.1904	4ft 8½in	0-4-0ST	OC	12x18	3ft 0½in	313
2604	06.03.1905	4ft 8½in	0-4-0ST	OC	14x22	3ft 6in	65,87
2607	03.03.1905	4ft 8½in	0-6-0ST	OC	15x22	3ft 9in	139
2615	12.05.1905	4ft 8½in	0-4-0ST	OC	14x22	3ft 6in	65
2646	02.05.1906	4ft 8½in	0-4-0ST	OC	14x22	3ft 6in	277
2662	20.02.1907	4ft 8½in	0-4-0ST	OC	14x22	3ft 6in	286
2663	30.09.1906	4ft 8½in	0-4-0ST	OC	9x14	3ft 6in	301,361
2684	11.02.1907	4ft 8½in	0-4-0ST	OC	13x19	3ft 4in	200,355
2703	01.07.1907	4ft 8½in	0-4-0ST	OC	14x22	3ft 6in	277
2711	21.06.1907	4ft 8½in	0-4-0ST	OC	14x22	3ft 6in	86,391
2712	08.08.1907	4ft 8½in	0-4-0ST	OC	14x22	3ft 6in	58,65,149
2729	07.12.1907	4ft 8½in	0-4-0ST	OC	14x22	3ft 6in	309
2730	31.03.1908	4ft 8½in	0-4-0ST	OC	14x22	3ft 6in	135,286
2732	08.09.1907	4ft 8½in	0-6-0ST	OC	14x22	3ft 7in	139
2748	15.02.1908	4ft 8½in	0-4-0ST	OC	16x24	3ft 10in	57,58
2796	03.05.1910	4ft 8½in	0-4-0ST	OC	14x22	3ft 6in	57,58,87
2797	21.03.1910	4ft 8½in	0-4-0ST	OC	14x22	3ft 6in	86
2799	04.03.1910	4ft 8½in	0-4-0ST	OC	12x18	3ft 0½in	269
2825	25.08.1910	4ft 8½in	0-4-0ST	OC	14x22	3ft 4in	86
2839	31.12.1910	4ft 8½in	0-4-0ST	OC	14x22	3ft 6in	277
2870	16.05.1911	4ft 8½in	0-4-0ST	OC	14x22	3ft 6in	246
2871	26.05.1911	4ft 8½in	0-4-0ST	OC	14x22	3ft 6in	286
2872	31.05.1911	4ft 8½in	0-4-0ST	OC	14x22	3ft 6in	266
2880	08.08.1911	4ft 8½in	0-6-0T	OC	17x24	3ft 8in	199
2891	14.09.1911	4ft 8½in	0-4-0ST	OC	14x22	3ft 6in	152
2904	24.11.1911	4ft 8½in	0-4-0ST	OC	14x22	3ft 4in	86
2905	04.01.1912	4ft 8½in	0-4-0ST	OC	14x22	3ft 6in	57,58,166
2909	31.10.1911	4ft 8½in	0-6-0ST	OC	14x22	3ft 4in	86
2913	28.02.1912	4ft 8½in	0-4-0ST	OC	14x22	3ft 4in	58,65,149
2939	12.07.1912	4ft 8½in	0-4-0ST	OC	14x22	3ft 6in	309
2940	09.08.1912	4ft 8½in	0-4-0ST	OC	14x22	3ft 6in	256
2971	27.11.1912	4ft 8½in	0-4-0ST	OC	14x22	3ft 6in	266
2993	16.05.1913	4ft 8½in	0-4-0ST	OC	16x24	3ft 10in	268
2995	30.05.1913	4ft 8½in	0-4-0ST	OC	16x24	3ft 10in	45
2997	20.08.1913	4ft 8½in	0-4-0ST	OC	16x24	3ft 10in	45
2998	25.07.1913	4ft 8½in	0-4-0ST	OC	16x24	3ft 10in	286
3006	03.09.1913	4ft 8½in	0-4-0ST	OC	14x22	3ft 6in	173,300
3025	01.12.1913	4ft 8½in	0-4-0ST	OC	15x22	3ft 8in	69

3031	10.12.1913	4ft 8½in	0-4-0ST	OC	14x22	3ft 4¾in	58,65
3032	24.12.1913	4ft 8½in	0-4-0ST	OC	14x22	3ft 6in	57,58,65
3053	10.03.1914	4ft 8½in	0-4-0ST	OC	14x22	3ft 6in	269
3082	25.11.1914	4ft 8½in	0-4-0ST	OC	14x22	3ft 4in	87
3131	13.05.1915	4ft 8½in	0-4-0ST	OC	14x22	3ft 4in	87
3132	20.05.1915	4ft 8½in	0-4-0ST	OC	14x22	3ft 4in	87
3139	20.07.1915	4ft 8½in	0-4-0ST	OC	12x18	3ft 0½in	87
3140	30.07.1915	4ft 8½in	0-4-0ST	OC	12x18	3ft 0½in	87
3157	30.11.1915	4ft 8½in	0-4-0ST	OC	14x22	3ft 6in	57,58
3169	29.02.1916	4ft 8½in	0-4-0ST	OC	14x22	3ft 4½in	58,65,139,149
3175	13.04.1916	4ft 8½in	0-4-0ST	OC	14x22	3ft 6in	200,355
3188	30.06.1916	4ft 8½in	0-4-0ST	OC	14x22	3ft 6in	166
3189	07.07.1916	4ft 8½in	0-4-0ST	OC	14x22	3ft 6in	166
3209	24.11.1916	4ft 8½in	0-4-0ST	OC	14x22	3ft 6in	49,87
3210	27.11.1916	4ft 8½in	0-4-0ST	OC	14x22	3ft 6in	166
3211	02.12.1916	4ft 8½in	0-4-0ST	OC	14x22	3ft 6in	57,58
3212	23.01.1917	4ft 8½in	0-4-0ST	OC	16x24	3ft 10in	45
3213	26.01.1917	4ft 8½in	0-4-0ST	OC	16x24	3ft 10in	45
3237	14.04.1917	4ft 8½in	0-4-0ST	OC	15x22	3ft 8in	326
3240	16.07.1917	4ft 8½in	0-4-0ST	OC	12x18	3ft 0½in	49,87
3246	14.05.1917	4ft 8½in	0-4-0ST	OC	14x22	3ft 4in	87
3248	19.06.1917	4ft 8½in	0-4-0ST	OC	16x24	3ft 8in	65,166
3249	22.10.1917	4ft 8½in	0-4-0ST	OC	12x18	3ft 1in	192
3255	18.08.1917	4ft 8½in	0-4-0ST	OC	14x22	3ft 6in	166
3256	23.08.1917	4ft 8½in	0-4-0ST	OC	14x22	3ft 6in	166
3257	29.08.1917	4ft 8½in	0-4-0ST	OC	14x22	3ft 4in	87
3267	28.09.1917	4ft 8½in	0-4-0ST	OC	14x22	3ft 6in	166
3268	10.10.1917	4ft 8½in	0-4-0ST	OC	14x22	3ft 6in	166
3304	16.11.1917	4ft 8½in	0-4-0ST	OC	14x22	3ft 6in	309
3326	15.05.1918	4ft 8½in	0-4-0ST	OC	14x22	3ft 4in	87
3334	24.05.1918	4ft 8½in	0-4-0ST	OC	14x22	3ft 6in	57,58,65
3342	23.07.1918	4ft 8½in	0-4-0ST	OC	16x24	3ft 10in	166
3343	06.08.1918	4ft 8½in	0-4-0ST	OC	16x24	3ft 10in	166
3344	21.06.1918	4ft 8½in	0-4-0ST	OC	14x22	3ft 6in	166
3346	30.07.1918	4ft 8½in	0-4-0ST	OC	14x22	3ft 6in	166
3348	10.09.1918	4ft 8½in	0-4-0ST	OC	14x22	3ft 6in	87,167
3350	12.09.1918	4ft 8½in	0-4-0ST	OC	14x22	3ft 4in	65,87
3352	17.10.1918	4ft 8½in	0-4-0ST	OC	16x24	3ft 8in	58,65,149
3353	30.10.1918	4ft 8½in	0-4-0ST	OC	16x24	3ft 8in	57,58,65
3354	15.11.1918	4ft 8½in	0-4-0ST	OC	16x24	3ft 10in	228
3355	28.11.1918	4ft 8½in	0-4-0ST	OC	16x24	3ft 10in	228
3365	28.04.1919	4ft 8½in	0-6-0ST	OC	16x24	3ft 10in	45,52
3366	26.03.1919	4ft 8½in	0-4-0ST	OC	14x22	3ft 6in	166
3369	28.05.1919	4ft 8½in	0-4-0ST	OC	14x22	3ft 4in	87
3370	03.06.1919	4ft 8½in	0-4-0ST	OC	14x22	3ft 4in	87
3385	21.08.1919	4ft 8½in	0-4-0ST	OC	14x22	3ft 6in	166
3388	25.09.1919	4ft 8½in	0-4-0ST	OC	14x22	3ft 6in	57,58
3418	16.04.1920	4ft 8½in	0-4-0ST	OC	14x22	3ft 6in	177
3420	03.06.1920	4ft 8½in	0-4-0ST	OC	14x22	3ft 6in	45
3429	27.07.1920	4ft 8½in	0-4-0ST	OC	16x24	3ft 10in	166
3430	28.08.1920	4ft 8½in	0-4-0ST	OC	16x24	3ft 10in	45
3477	06.05.1921	4ft 8½in	0-4-0ST	OC	14x22	3ft 6in	58,160
3481	20.04.1921	4ft 8½in	0-4-0ST	OC	14x22	3ft 6in	309
3482	05.05.1921	4ft 8½in	0-4-0ST	OC	14x22	3ft 6in	58,160
3491	21.02.1921	4ft 8½in	0-4-0ST	OC	14x22	3ft 6in	200,355
3501	06.05.1921	4ft 8½in	0-4-0ST	OC	14x22	3ft 4in	87

3502	13.05.1921	4ft 8½in	0-4-0ST	OC	14x22	3ft 4in	87
3503	31.05.1921	4ft 8½in	0-4-0ST	OC	14x22	3ft 4in	87
3531	29.09.1922	4ft 8½in	0-6-0T	OC	16x24	4ft 0in	45,326
3566	23.03.1927	4ft 8½in	0-4-0ST	OC	16x24	3ft 10in	57,58
3567	17.03.1928	4ft 8½in	0-4-0ST	OC	16x24	3ft 10in	199
3586	24.09.1924	4ft 8½in	0-4-0ST	OC	12x18	3ft 0½in	112
3638	20.11.1925	4ft 8½in	0-4-0ST	OC	14x22	3ft 6in	199
3639	11.02.1926	4ft 8½in	0-4-0ST	OC	14x22	3ft 6in	199
3649	26.05.1927	4ft 8½in	0-4-0ST	OC	12x18	3ft 0½in	199,205
3651	19.11.1926	4ft 8½in	0-4-0ST	OC	14x22	3ft 6in	112
3652	24.05.1927	4ft 8½in	0-4-0ST	OC	14x22	3ft 6in	199
3671	31.12.1927	4ft 8½in	0-4-0ST	OC	14x22	3ft 6in	199
3672	31.12.1927	4ft 8½in	0-4-0ST	OC	14x22	3ft 6in	199
3684	12.04.1929	4ft 8½in	0-4-0ST	OC	12x20	3ft 1in	173
3720	16.03.1928	4ft 8½in	0-4-0ST	OC	14x22	3ft 6in	199
3721	14.04.1928	4ft 8½in	0-4-0ST	OC	14x22	3ft 6in	199
3723	11.05.1928	4ft 8½in	0-4-0ST	OC	14x22	3ft 6in	200
3730	29.08.1928	4ft 8½in	0-4-0ST	OC	14x22	3ft 6in	200
3731	11.09.1928	4ft 8½in	0-4-0ST	OC	14x22	3ft 6in	200
3734	26.07.1928	4ft 8½in	0-4-0ST	OC	16x24	3ft 10in	200
3735	31.07.1928	4ft 8½in	0-4-0ST	OC	16x24	3ft 10in	200
3737	25.09.1928	4ft 8½in	0-6-0T	OC	18x24	4ft 0in	200
3738	09.10.1928	4ft 8½in	0-6-0T	OC	18x24	4ft 0in	200
3765	28.02.1930	4ft 8½in	0-4-0F	OC	17x26	2ft 11in	200
3779	22.05.1930	4ft 8½in	0-4-0ST	OC	16x24	3ft 10in	45
3861	02.06.1936	4ft 8½in	0-4-0WE		80hp	2ft 9in	98
3862	02.08.1936	4ft 8½in	0-4-0WE		80hp	2ft 9in	98
3911	22.02.1937	4ft 8½in	0-4-0ST	OC	14x22	3ft 6in	286
3914	22.03.1937	4ft 8½in	0-4-0ST	OC	16x24	3ft 8in	87
3915	05.04.1937	4ft 8½in	0-4-0ST	OC	16x24	3ft 8in	87,166
3918	12.07.1937	4ft 8½in	0-4-0ST	OC	16x24	3ft 8in	59
3919	19.07.1937	4ft 8½in	0-4-0ST	OC	16x24	3ft 8in	59,87
3935	30.12.1937	4ft 8½in	0-4-0ST	OC	16x24	3ft 8in	45,228
3936	01.02.1938	4ft 8½in	0-4-0ST	OC	16x24	3ft 8in	45

HOPPER, RATCLIFFE & CO. Britannia Iron Works, Fencehouses, Co. Durham — Hopper

[No.2]	1870	4ft 8½in	0-4-0ST	OC			139
[GEORGE LEEMAN]	1871	4ft 8½in	0-4-0ST	OC			281

HORWICH WORKS, Horwich, Lancashire — Hor

British Railways [prev LMS, L&YR]

[08648]	03.1959	4ft 8½in	0-6-0DE	350hp	EE 6KT	48T	261
[08754]	01.1961	4ft 8½in	0-6-0DE	350hp	EE 6KT	48T	261
[08764]	03.1961	4ft 8½in	0-6-0DE	350hp	EE 6KT	48T	171
[08847]	05.1961	4ft 8½in	0-6-0DE	350hp	EE 6KT	48T	261
[08885]	01.1962	4ft 8½in	0-6-0DE	350hp	EE 6KT	48T	261
[08903]	06.1962	4ft 8½in	0-6-0DE	350hp	EE 6KT	48T	201,208,284,295
[08912]	09.1962	4ft 8½in	0-6-0DE	350hp	EE 6KT	48T	133
[08913]	09.1962	4ft 8½in	0-6-0DE	350hp	EE 6KT	48T	213

HEAD, WRIGHTSON & CO LTD, Teesdale Iron Works, Thornaby-on-Tees — HW

21	1870	4ft 8½in	0-4-0VBT	VCG	6¼x12	2ft 6in	392,395
	c1870	4ft 8½in	0-4-0VBT	VCG			175
	c1870	4ft 8½in	0-4-0VBT	VCG			175
	1871	4ft 8½in	0-4-0VBT	VCG	6x12	2ft 4in	392
[BEE]	1871	4ft 8½in	0-4-0VBT	OC			246,268

33	1873	4ft 8½in	0-4-0VBT	OC	9x14	2ft 5½in		392
[No.1]	1928	4ft 8½in	4wBE					100
[BLUE GOWN]		4ft 8½in	0-4-0VBT					254
		4ft 8½in	0-4-0VBT	VCG				282

ROBERT HUDSON LTD, Leeds   HU

29057	1925	2ft 0in	4wPM					368
42701	1931	2ft 0in	4wPM					191

CHARLES I'ANSON & CO LTD, Darlington, Co.Durham   l'Anson

	1875	4ft 8½in	0-4-0ST	OC?				129,326

JOHN FOWLER & CO LTD, Leeds, Yorkshire (WR)   JF

1540	06.1871	4ft 8½in	0-6-0T	IC	8x14			272
1575	03.1873	4ft 8½in	0-4-0ST		8½x14			171
	1871	4ft 8½in	0-4-0ST	IC				239
2376	03.1875	3ft 0in	0-4-0ST	OC	9x14			182
2377	03.1875	3ft 0in	0-4-0ST	OC	9x14			182
2378	01.1876	3ft 0in	0-4-0ST	OC	9x14			182
2625	02.1876	4ft 8½in	0-4-0ST	OC	10x16			166,343,369
2836	05.1876	3ft 0in	0-4-0ST	OC	8½x			346
	1880	4ft 8½in	0-4-0ST	OC				135
21750	01.1937	4ft 8½in	0-4-0DM	40hp	Fowler 4B			120
22489	11.02.1939	4ft 8½in	0-4-0DM	80hp	Fowler 6A			344
22938	03.1941	4ft 8½in	0-4-0DM	150hp	Fowler 4C		29T	240
22939	06.05.1941	4ft 8½in	0-4-0DM	150hp	Fowler 4C		29T	65
22942	23.06.1941	4ft 8½in	0-4-0DM	150hp	Fowler 4C		29T	286
22945	09.1941	4ft 8½in	0-4-0DM	150hp	Fowler 4C		29T	240
22985	22.09.1942	4ft 8½in	0-4-0DM	150hp	Fowler 4C		29T	301
22998	1943	4ft 8½in	0-4-0DM	150hp	Fowler 4C		29T	297
4000007	04.1947	4ft 8½in	0-4-0DM	60hp	Fowler 6B			69
4100012	09.1948	4ft 8½in	0-4-0DM	80hp	Fowler 6A			114,344
4110006	14.08.1950	4ft 8½in	0-4-0DM	80hp	McLaren MR3		19T	49,88
4160007	01.1953	4ft 8½in	0-4-0DM	100hp	McLaren M4			52,327,352
4160008	03.1953	4ft 8½in	0-4-0DM	100hp	McLaren M4			45,52
4160009	04.1953	4ft 8½in	0-4-0DM	100hp	McLaren M4			52,228,327,352
4200006	12.1946	4ft 8½in	0-4-0DM	150hp	Fowler 4C		29T	45,303
4200018	08.12.1947	4ft 8½in	0-4-0DM	150hp	Fowler 4C		29T	300
4200020	11.1947	4ft 8½in	0-4-0DM	150hp	Fowler 4C		29T	38
4200023	05.1948	4ft 8½in	0-4-0DM	150hp	Fowler 4C		29T	301,350
4200024	05.1948	4ft 8½in	0-4-0DM	150hp	Fowler 4C		29T	301
4210016	01.1950	4ft 8½in	0-4-0DM	150hp	McLaren		28T	124,192
4210086	07.1953	4ft 8½in	0-4-0DM	150hp	McLaren		28T	45,52,228
4210087	09.1953	4ft 8½in	0-4-0DM	150hp	McLaren		28T	52
4210088	09.1953	4ft 8½in	0-4-0DM	150hp	McLaren		28T	52
4210089	01.1954	4ft 8½in	0-4-0DM	150hp	McLaren		28T	228,230
4210091	03.1954	4ft 8½in	0-4-0DM	150hp	McLaren		28T	228,230
4210094	08.1954	4ft 8½in	0-4-0DM	150hp	McLaren		28T	228,230
4210097	01.1955	4ft 8½in	0-4-0DM	150hp	McLaren		28T	45
4210099	03.1955	4ft 8½in	0-4-0DM	150hp	McLaren		28T	228,230
4210102	24.05.1955	4ft 8½in	0-4-0DM	150hp	McLaren		28T	228,230
4210106	10.1955	4ft 8½in	0-4-0DM	150hp	McLaren		28T	45
4210107	11.1955	4ft 8½in	0-4-0DM	150hp	McLaren		28T	228,230
4210110	04.1956	4ft 8½in	0-4-0DM	150hp	McLaren		28T	228,230
4210113	05.1956	4ft 8½in	0-4-0DM	150hp	McLaren		28T	45
4210116	07.1956	4ft 8½in	0-4-0DM	150hp	McLaren		28T	52

4210122	10.1956	4ft 8½in	0-4-0DM	150hp	McLaren		28T	45
4210125	01.1957	4ft 8½in	0-4-0DM	150hp	McLaren		28T	52
4210128	05.1957	4ft 8½in	0-4-0DM	150hp	McLaren		28T	228,230
4210135	12.1957	4ft 8½in	0-4-0DM	150hp	McLaren		28T	45
4210136	24.02.1958	4ft 8½in	0-4-0DM	150hp	McLaren		28T	344
4210139	03.1958	4ft 8½in	0-4-0DM	150hp	McLaren		28T	45
4210146	05.1958	4ft 8½in	0-4-0DM	150hp	McLaren		28T	228,230
4210147	01.1959	4ft 8½in	0-4-0DM	150hp	McLaren		28T	45,228,230
4210148	01.1959	4ft 8½in	0-4-0DM	150hp	McLaren		28T	228,230
4210149	12.1958	4ft 8½in	0-4-0DM	150hp	McLaren		28T	45
4220004	12.1959	4ft 8½in	0-4-0DH	176hp	Leyland		28T	45
4220005	01.1960	4ft 8½in	0-4-0DH	176hp	Leyland		28T	45
4220006	02.1960	4ft 8½in	0-4-0DH	176hp	Leyland		28T	45
4220019	02.1961	4ft 8½in	0-4-0DH	176hp	Leyland		28T	45
4220020	08.1961	4ft 8½in	0-4-0DH	176hp	Leyland		28T	38,45
4220023	03.1963	4ft 8½in	0-4-0DH	203hp	Leyland		28T	350
4220027	10.1963	4ft 8½in	0-4-0DH	203hp	Leyland		28T	230,248
4220028	04.1964	4ft 8½in	0-4-0DH	203hp	Leyland		28T	230
4220035	04.1965	4ft 8½in	0-4-0DH	203hp	Leyland		28T	45
4220036	06.1965	4ft 8½in	0-4-0DH	203hp	Leyland		28T	45
4220040	05.1967	4ft 8½in	0-4-0DH	230hp	Leyland		28T	45
4220041	07.1967	4ft 8½in	0-4-0DH	230hp	Leyland		28T	45
4220042	12.1967	4ft 8½in	0-4-0DH	230hp	Leyland		28T	45
4220043	01.1968	4ft 8½in	0-4-0DH	230hp	Leyland		28T	45
4240001	09.1959	4ft 8½in	0-6-0DH	230hp	Leyland EN900		35T	230
4240002	11.1959	4ft 8½in	0-6-0DH	230hp	Leyland EN900		35T	230
4240003	11.1959	4ft 8½in	0-6-0DH	230hp	Leyland EN900		35T	230
4240004	11.1959	4ft 8½in	0-6-0DH	230hp	Leyland EN900		35T	230
4240005	01.1960	4ft 8½in	0-6-0DH	230hp	Leyland EN900		35T	230
4240006	04.1960	4ft 8½in	0-6-0DH	230hp	Leyland EN900		35T	230
4240007	04.1960	4ft 8½in	0-6-0DH	230hp	Leyland EN900		35T	230
4240008	06.1960	4ft 8½in	0-6-0DH	230hp	Leyland EN900		35T	230
4240009	10.1960	4ft 8½in	0-6-0DH	230hp	Leyland EN900		35T	230
4240011	03.1961	4ft 8½in	0-6-0DH	230hp	Leyland EN900		35T	230
4240014	10.1962	4ft 8½in	0-6-0DH	275hp	Leyland EN900		35T	213,216
4240015	31.12.1962	4ft 8½in	0-6-0DH	275hp	Leyland EN900		35T	171

J. & G. JOICEY, Pottery Lane, Newcastle upon Tyne — Joicey

215	1870	4ft 8½in	0-4-0ST	OC				44
230	1872	4ft 8½in	0-4-0ST	OC				44,149
388	1890	4ft 8½in	0-4-0ST	OC				286,288

KITSON & CO LTD, Airdale Foundry, Leeds — K

1705	27.05.1871	4ft 8½in	0-4-0ST	OC	12x18	3ft 0in		265,268
1788	06.02.1872	4ft 8½in	0-4-0T	OC	12x18	3ft 0½in		119
1790	07.03.1872	4ft 8½in	0-4-0T	OC	12x18	3ft 0½in		246
2237	12.04.1879	4ft 8½in	0-4-0ST	OC	9x15	2ft 9in		246
2362	29.04.1881	4ft 8½in	0-4-0T	OC	14x21	3ft 4in		59,160
2363	03.10.1881	4ft 8½in	0-4-0T	OC	14x21	3ft 4in		160
3882	30.01.1899	4ft 8½in	0-4-0ST	OC	14x21	3ft 2½in		58,88,160
3982	24.09.1900	4ft 8½in	0-4-0ST	OC	14x21	3ft 2½in		59,160
4382	17.11.1905	4ft 8½in	0-4-0WTST	OC	14x21	3ft 2½in		160,166,175
4624	30.11.1908	4ft 8½in	0-6-0T	IC	18x26	4ft 6in		86
4737	19.04.1910	4ft 8½in	0-4-0WTST	OC	14x21	3ft 2½in		58,160
4945	02.01.1913	4ft 8½in	0-4-0WTST	OC	14x21	3ft 2½in		58,160
5020	01.10.1913	4ft 8½in	0-4-0ST	OC	14x21	3ft 2½in		87

5114	06.05.1914	4ft 8½in	0-4-0ST	OC	14x21	3ft 2½in		59,160
5115	08.05.1914	4ft 8½in	0-4-0ST	OC	14x21	3ft 2½in		59,160

KENT CONSTRUCTION & ENGINEERING CO LTD, Ashford, Kent — KC

1470	1926	4ft 8½in	4wPM	114

A.E. & H. KITCHING, Darlington, Co. Durham — Kitching

[No.7]	1881	4ft 8½in	0-4-0ST	OC	355

KERR, STUART & CO LTD, California Works, Stoke-on-Trent, Staffordshire — KS

3095	03.1918	4ft 8½in	0-4-0ST	OC	15x20	3ft 6in	Moss Bay	228
3098	26.03.1918	4ft 8½in	0-6-0T	OC	15x20	3ft 9in	Argentina	301
3099	04.1918	4ft 8½in	0-6-0T	OC	15x20	3ft 9in	Argentina	301
3100	08.04.1918	4ft 8½in	0-6-0T	OC	15x20	3ft 9in	Argentina	301
3101	04.1918	4ft 8½in	0-6-0T	OC	15x20	3ft 9in	Argentina	301
3107	07.1918	4ft 8½in	0-4-0ST	OC	12x16	2ft 9in	Huxley	31
3112	09.1918	4ft 8½in	0-4-0ST	OC	15x20	3ft 6in	Moss Bay	199
3124	10.1918	4ft 8½in	0-4-0ST	OC	15x20	3ft 6in	Moss Bay	301
3125	10.1918	4ft 8½in	0-4-0ST	OC	15x20	3ft 6in	Moss Bay	301
3126	22.10.1918	4ft 8½in	0-4-0ST	OC	15x20	3ft 6in	Moss Bay	177,179,326
4144	12.1919	4ft 8½in	0-4-0ST	OC	15x20	3ft 6in	Moss Bay	246

R.A. LISTER & CO LTD, Dursley, Gloucestershire — L

960	1929		4wPM	254
34031	1949	2ft 0in	4wPM	38

LOWCA ENGINEERING CO LTD, Whitehaven, Cumberland — LE

Successors to FJ

200	1885	4ft 8½in	0-4-0ST	OC	12x20	309
240	1900	4ft 8½in	0-4-0ST	OC	14x20	246,268

STEPHEN LEWIN, Dorset Foundry, Poole, Dorset — Lewin

606	1875	4ft 8½in	0-4-0ST	OC	10x18	1ft 2in	205,210,348

LILLISHALL CO LTD, St George's, Oakengates, Shropshire — Lill

	1875	4ft 8½in	4wT	351

MaK-SIEMENS SCHIENENFAHRZEUG TECHNIK, Kiel-Friedrichsort, Germany — MaK

1600.001	1995	4ft 8½in	Bo-BoDE	1570kW Cat 35/6	82T	290
1600.002	1996	4ft 8½in	Bo-BoDE	1570kW Cat 35/6	82T	290
1600.003	1996	4ft 8½in	Bo-BoDE	1570kW Cat 35/6	82T	290
1600.004	1996	4ft 8½in	Bo-BoDE	1570kW Cat 35/6	82T	290
1600.008	1996	4ft 8½in	Bo-BoDE	1570kW Cat 35/6	82T	290
1600.011	1996	4ft 8½in	Bo-BoDE	1570kW Cat 35/6	82T	290
1600.012	1996	4ft 8½in	Bo-BoDE	1570kW Cat 35/6	82T	290
1600.016	1996	4ft 8½in	Bo-BoDE	1570kW Cat 35/6	82T	290
1600.017	1996	4ft 8½in	Bo-BoDE	1570kW Cat 35/6	82T	290
1600.018	1996	4ft 8½in	Bo-BoDE	1570kW Cat 35/5	82T	290
1600.019	1996	4ft 8½in	Bo-BoDE	1570kW Cat 35/6	82T	290
1600.020	1996	4ft 8½in	Bo-BoDE	1570kW Cat 35/6	82T	290

LOCOTRACTEURS GASTON MOYSE, La Courneuve, France — Moyse

3539	1973	4ft 8½in	6wDE	91
1364	1976	4ft 8½in	4wDE	231,298,311,383
1365	1976	4ft 8½in	4wDE	298,311,383
1464	1979	4ft 8½in	4wDE	100,231,298,311,383

MOTOR RAIL LTD, Simplex Works, Bedford, Bedfordshire — MR

Motor Rail & Tram Co Co Ltd until 1931

861	30.04.1918	600mm	4wPM	20hp	Dorman 2JO	2½T	313
1116	10.12.1918	600mm	4wPM	20hp	Dorman 2JO	2½T	110
1848	16.05.1919	2ft 2½in	4wPM	20hp	Dorman 2JO	2½T	296
1926	15.09.1919	4ft 8½in	4wPM	40hp	Dorman 4JO	7¾T	150
1945	21.01.1920	4ft 8½in	4wPM	40hp	Dorman 4JO	7¾T	150
4002	30.10.1925	600mm	4wPM	20hp	Dorman 2JO	2½T	373
7091	08.02.1940	2ft 2½in	4wPM	20/26hp	Dorman 2JO	2½T	296
7207	02.12.1937	2ft 0in	4wDM	20/28hp	Dorman 2HW	2½T	107

METROPOLITAN-VICKERS ELECTRICAL CO LTD, Trafford Park, Manchester — MV

[No.1]	1948	4ft 8½in	4wWE	99
[No.2]	1948	4ft 8½in	4wWE	99

MANNING WARDLE & CO LTD, Boyne Engine Works, Leeds — MW

5	10.1859	4ft 8½in	0-6-0ST	IC	11x17	3ft 1½in	Old I	85
60	28.03.1862	4ft 8½in	0-4-0ST	OC	9x14	2ft 9in	E	139
84	04.07.1863	4ft 8½in	0-4-0ST	OC	6x12	2ft 6in	B	44,107,109
97	03.11.1863	4ft 8½in	0-4-0ST	OC	6x12	2ft 6in	B	49
98	11.11.1863	4ft 8½in	0-4-0ST	OC	6x12	2ft 6in	B	49
113	22.04.1864	4ft 8½in	0-4-0ST	OC	6x12	2ft 6in	B	49
123	23.09.1864	4ft 8½in	0-4-0ST	OC	8x14	2ft 8in	D	246
153	08.08.1865	4ft 8½in	0-6-0T	IC	15x22	3ft 0in	WYT	85
179	15.01.1866	4ft 8½in	0-6-0ST	IC	12x17	3ft 1¾in	K	372
191	28.02.1866	4ft 8½in	0-4-0ST	OC	8x14	2ft 8in	D	233
199	07.08.1866	4ft 8½in	0-4-0ST	OC	9½x14	2ft 9in	E	86
333	02.01.1871	4ft 8½in	0-4-0ST	OC	12x18	3ft 0in	H	175
336	28.12.1871	4ft 8½in	0-4-0ST	OC	12x18	3ft 0in	H	312
352	22.01.1872	4ft 8½in	0-4-0ST	OC	10x16	2ft 9in	F	85
385	24.09.1872	4ft 8½in	0-4-0ST	OC	9x14	2ft 9in	E	44
399	03.09.1872	4ft 8½in	0-4-0ST	OC	12x18	3ft 0in	H	126,175,326
415	18.12.1872	4ft 8½in	0-4-0ST	OC	12x18	3ft 0in	H	85
416	02.01.1873	4ft 8½in	0-4-0ST	OC	12x18	3ft 0in	H	85
438	19.04.1873	4ft 8½in	0-4-0ST	OC	12x18	3ft 0in	H	168
447	10.07.1873	4ft 8½in	0-4-0ST	OC	12x18	3ft 0in	H	152,334
456	06.11.1874	4ft 8½in	0-4-0ST	OC	10x16	2ft 9in	F	168
465	15.08.1873	4ft 8½in	0-4-0ST	OC	12x18	3ft 0in	H	119,122
498	18.05.1874	4ft 8½in	0-4-0ST	OC	12x18	3ft 0in	H	109
539	16.09.1875	4ft 8½in	0-4-0ST	OC	12x18	3ft 0in	H	252
566	25.06.1876	4ft 8½in	0-4-0ST	OC	8x14	2ft 8in	D	257
571	10.12.1875	4ft 8½in	0-6-0ST	IC	15x22	3ft 9in	O	85
591	28.02.1877	4ft 8½in	0-4-0ST	OC	12x18	3ft 0in	H	22
599	25.05.1876	4ft 8½in	0-6-0ST	IC	12x17	3ft 1⅜in	K	301
701	05.01.1880	4ft 8½in	0-6-0ST	IC	15x22	3ft 9in	O	86
756	05.10.1880	4ft 8½in	0-6-0ST	IC	13x18	3ft 0in	M	166,283
777	04.02.1881	4ft 8½in	0-4-0ST	OC	12x18	3ft 6in	Spl H	175
975	21.07.1886	4ft 8½in	0-6-0ST	IC	12x17	3ft 1⅜in	K	361
995	30.09.1886	4ft 8½in	0-6-0ST	IC	12x17	3ft 1⅜in	K	363
1005	15.06.1887	4ft 8½in	0-6-0ST	IC	13x18	3ft 0in	Spl M	301,361
1020	08.06.1887	4ft 8½in	0-4-0ST	OC	10x16	2ft 9in	F	179
1022	09.07.1887	4ft 8½in	0-4-0ST	OC	12x18	3ft 0in	H	175
1027	23.11.1887	4ft 8½in	0-4-0ST	OC	12x18	3ft 0in	H	27,340
1072	23.04.1888	4ft 8½in	0-4-0ST	OC	12x18	3ft 0in	H	199,279,326
1075	02.05.1888	4ft 8½in	0-4-0ST	OC	10x16	2ft 9in	F	22

Works No	Date	Gauge	Type		Cylinders	Wheels	Class	Page
1144	21.05.1890	4ft 8½in	0-6-0ST	IC	12x17	3ft 1⅜in	K	362
1146	02.07.1890	4ft 8½in	0-6-0ST	IC	12x17	3ft 1⅜in	K	301,361
1161	15.01.1891	4ft 8½in	0-4-0ST	OC	12x18	3ft 6in	Spl H	178
1290	29.03.1895	4ft 8½in	0-6-0ST	IC	13x18	3ft 0in	M	199
1309	10.09.1895	4ft 8½in	0-6-0ST	IC	13x18	3ft 0in	L	370
1327	19.03.1897	4ft 8½in	0-4-0ST	OC	12x18	3ft 0in	H	114,277
1328	29.03.1898	4ft 8½in	0-4-0ST	OC	12x18	3ft 0in	H	87
1367	04.06.1897	4ft 8½in	0-6-0ST	IC	16x22	3ft 6in	Spl	86
1379	01.07.1898	4ft 8½in	0-6-0ST	IC	13x18	3ft 0in	M	240
1390	19.04.1898	4ft 8½in	0-4-0ST	OC	14x18	3ft 6in	Spl	175
1418	16.09.1898	4ft 8½in	0-6-0ST	IC	12x18	3ft 0in	L	240
1457	04.08.1899	4ft 8½in	0-4-0ST	OC	14x18	3ft 0in	P	175
1469	08.01.1900	4ft 8½in	0-6-0ST	IC	16x22	3ft 6in	Spl	86
1517	19.12.1900	4ft 8½in	0-4-0ST	OC	12x18	3ft 0in	H	87
1529	02.07.1901	4ft 8½in	0-4-0ST	OC	12x18	3ft 6in	Alt H	175
1594	25.02.1903	4ft 8½in	0-6-0ST	IC	12x18	3ft 0in	L	370
1598	18.06.1903	4ft 8½in	0-6-0ST	IC	12x18	3ft 0in	L	370
1602	30.04.1903	4ft 8½in	0-6-0ST	IC	15x22	3ft 6in	O	309
1634	21.01.1905	4ft 8½in	0-4-0ST	OC	12x17	3ft 1in	K	199
1641	26.09.1904	4ft 8½in	0-4-0ST	OC	13x18	3ft 1in	I	309
1654	28.06.1905	4ft 8½in	0-6-0ST	IC	17x24	3ft 6in	Spl	86
1659	17.07.1905	4ft 8½in	0-4-0ST	OC	13½x18	3ft 0in	Alt I	309
1665	29.08.1905	4ft 8½in	0-6-0ST	IC	13x18	3ft 0in	M	362
1714	05.06.1907	4ft 8½in	0-4-0ST	OC	14x18	3ft 0in	P	175
1772	31.05.1911	4ft 8½in	0-6-0ST	IC	12x17	3ft 1in	K	199
1828	24.071913	4ft 8½in	0-4-0ST	OC	14½x20		Spl	309
1887	22.09.1915	4ft 8½in	0-4-0ST	OC	14x18	3ft 1in	P	34
1903	25.07.1916	4ft 8½in	0-4-0ST	OC	14x18	3ft 6in	Alt P	175
1911	25.01.1917	4ft 8½in	0-4-0ST	OC	14x18	3ft 1in	P	34
1967	04.11.1918	4ft 8½in	0-4-0ST	OC	16x24	3ft 8in	Spl	228
1968	24.10.1918	4ft 8½in	0-4-0ST	OC	14x20		Spl	301
1969	30.10.1918	4ft 8½in	0-4-0ST	OC	14x20		Spl	131,301,327

A.J. MOSS, Scarisbrick, Lancashire — Moss A.J.

	2001	1ft 3in	4-4-2	OC				394

NEILSON & CO, Hyde Park Works, Springburn, Glasgow, Scotland — N

2280	08.10.1877	4ft 8½in	0-4-0ST	OC	14x20	3ft 8in		44

NORTH BRITISH LOCOMOTIVE CO LTD, Glasgow, Scotland — NB

NBH Hyde Park Works, Glasgow
NBQ Queens Park Works, Glasgow

H 22600	1921	4ft 8½in	0-6-2T	IC	[69523]			385
Q 27644	1959	4ft 8½in	0-4-0DH	275hp	National M4AAU5			324

NESHAM & WELCH, Portrack, Lane Iron Works, Stockton-on-Tees — Nesham & Welch

[GORDON]	1837	4ft 8½in	0-6-0					101
[EXILE]	1837	4ft 8½in	0-6-0					101
[NER 125]	1839	4ft 8½in	0-6-0					101
[NER 122]	1840	4ft 8½in	0-6-0					101
[NER 126]	1840	4ft 8½in	0-4-0					101
[NER 127]	1840	4ft 8½in	0-4-0					101
[WITTON CASTLE]	1840	4ft 8½in	0-6-0		14x20*	4ft 0in		101

* later altered to 16x18

NINE ELMS WORKS, LONDON 9E
London & South Western Railway

[SR 81]	11.1893	4ft 8½in	0-4-0T	OC	16x22	3ft 9¾in	69

NITEQ BV, Netherlands Niteq

M2013	2007	4ft 8½in	4wBE		5000E	133

NORD NETHERLANDSCHE MACHINEFABRIEK, BV, Winschoten, Netherlands NNM

82503	1983	4ft 8½in	4wDH	R/R		201,208, 294,295
83501	1983	4ft 8½in	4wDH	R/R		201,208, 294,295
83502	1983	4ft 8½in	4wDH	R/R		201,208, 294
83503	1984	4ft 8½in	4wDH	R/R		201,208, 294,295
83504	1984	4ft 8½in	4wDH	R/R		201,208, 294,295
83505	1984	4ft 8½in	4wDH	R/R		201,208, 294

N.G. PARKINSON Parkinson

	1932	1ft 3in	2-4-0+6wPM	S/O		394

PECKETT & SONS LTD, Atlas Locomotive Works, Bristol P

462	21.09.1887	4ft 8½in	0-4-0ST	OC	14x20	3ft 2in	W4	57,58,160
463	8.11.1887	4ft 8½in	0-4-0ST	OC	14x20	3ft 2in	W4	22,300
467	28.02.1888	4ft 8½in	0-4-0ST	OC	14x20	3ft 2in	W4	222,228
582	18.07.1894	4ft 8½in	0-4-0ST	OC	14x20	3ft 2in	W4	69
583	12.12.1894	4ft 8½in	0-4-0ST	OC	14x20	3ft 2in	W4	69
629	29.04.1896	4ft 8½in	0-6-0ST	IC	16x22	3ft 10in	X	86
634	24.02.1897	4ft 8½in	0-4-0ST	OC	12x18	3ft 0in	R1	38
640	19.10.1896	4ft 8½in	0-4-0ST	OC	14x20	3ft 2in	W4	166
652	23.07.1896	4ft 8½in	0-6-0ST	IC	16x22	3ft 10in	X	86
657	01.04.1897	4ft 8½in	0-4-0ST	OC	14x20	3ft 2in	W4	222,228
667	25.06.1897	4ft 8½in	0-4-0ST	OC	14x20	3ft 2in	W4	86
668	28.07.1897	4ft 8½in	0-4-0ST	OC	14x20	3ft 2in	W4	86
669	28.07.1897	4ft 8½in	0-4-0ST	OC	14x20	3ft 2in	W4	86
736	11.04.1899	4ft 8½in	0-4-0ST	OC	14x20	3ft 2in	W4	69
845	13.09.1900	4ft 8½in	0-4-0ST	OC	14x20	3ft 2in	W4	175
877	18.04.1901	4ft 8½in	0-6-0ST	iC	16x22	3ft 10	X	45,52
880	14.05.1901	4ft 8½in	0-4-0ST	OC	14x20	3ft 2in	W4	88
904	03.10.1901	4ft 8½in	0-4-0ST	OC	12x18	3ft 0in	R1	252
919	03.03.1902	4ft 8½in	0-4-0ST	OC	14x20	3ft 2in	W4	69
921	26.01.1903	4ft 8½in	0-4-0ST	OC	14x20	3ft 2in	W4	199
949	11.08.1902	4ft 8½in	0-4-0ST	OC	12x18	3ft 0in	R1	23,326
956	13.08.1903	4ft 8½in	0-4-0ST	OC	14x20	3ft 2in	W4	175,265,266,313
970	28.11.1902	4ft 8½in	0-4-0ST	OC	12x18	3ft 0in	R1	350
1058	30.07.1906	4ft 8½in	0-4-0ST	OC	14x20	3ft 2in	W4	175
1084	01.02.1906	4ft 8½in	0-4-0ST	OC	14x20	3ft 2in	W4	69
1210	25.08.1910	4ft 8½in	0-4-0ST	OC	12x18	3ft 0in	R2	24,279
1331	25.05.1913	3ft 0in	0-4-0T	OC	8x12	3ft 0in	8in	58
1440	14.04.1916	3ft 0in	0-4-0ST	OC	8x12	3ft 0in	8in	58
1444	21.09.1916	4ft 8½in	0-6-0ST	IC	16x22	3ft 10in	X2	48,52
1506	23.05.1918	4ft 8½in	0-4-0ST	OC	14x20	3ft 2½in	W5	350
1510	27.02.1919	4ft 8½in	0-4-0ST	OC	14x20	3ft 2½in	W5	279
1513	27.06.1918	4ft 8½in	0-4-0ST	OC	15x21	3ft 7in	E	65
1589	29.03.1928	4ft 8½in	0-4-0ST	OC	15x21	3ft 7in	E	200
1605	04.12.1922	4ft 8½in	0-4-0ST	OC	12x18	3ft 0in	R2	129
1821	29.09.1931	4ft 8½in	0-4-0ST	OC	12x20	3ft 0½in	R4	288
1961	03.07.1939	4ft 8½in	0-4-0ST	OC	14x22	3ft 2½in	W7	69

2002	02.04.1941	4ft 8½in	0-4-0ST	OC	14x22	3ft 2½in	W7	70
2020	12.01.1942	4ft 8½in	0-4-0ST	OC	14x22	3ft 2½in	W7	70
2047	21.09.1943	4ft 8½in	0-4-0ST	OC	12x20	3ft 0½in	R4	241
2048	14.02.1944	4ft 8½in	0-4-0ST	OC	12x20	3ft 0½in	R4	240,241
2049	28.08.1944	4ft 8½in	0-4-0ST	OC	12x20	3ft 0½in	R4	241
2065	06.11.1945	4ft 8½in	0-4-0ST	OC	14x22	3ft 2½in	W7	286
2142	21.04.1953	4ft 8½in	0-4-0ST	OC	14x22	3ft 2½in	W7 spl	253,392

R1, R2 -Some R1 and R2 class locomotives had 3ft 0½in dia driving wheels.

WILLIAM BARNINGHAM, Pendleton Ironworks, Manchester — Pendleton

	1867	?	0-4-0WT	27

RESCO (RAILWAYS) LTD, Woolwich Works, London — Resco

L105	1978	4ft 8½in	0-6-0DE	reb of RH 480696	208
L106	1978	4ft 8½in	0-6-0DE	reb of RH 480690	208

RUSTON & HORNSBY LTD, Lincoln, Lincolnshire — RH

172908	03.12.1934	2ft 0in	4wDM	18/21	Lister 18/2	-	191
175135	22.07.1935	2ft 0in	4wDM	18/21	Lister 18/2	-	191
182137	25.11.1936	2ft 0in	4wDM	20hp	Lister 18/2	-	391
183726	05.03.1937	2ft 0in	4wDM	20hp	Lister 18/2	-	169
183764	06.05.1937	4ft 8½in	4wDM	44/48	Ruston 4VRO	7T	28
187071	22.11.1937	4ft 8½in	4wDM	44/48	Ruston 4VRO	7½T	21,248
187075	21.01.1938	4ft 8½in	4wDM	44/48	Ruston 4VRO	7½T	348
189965	19.04.1938	4ft 8½in	4wDM	44/48	Ruston 4VRO	7½T	301
198245	27.07.1939	2ft 6in	4wDM	44/48	Ruston 4VRO	6½T	169
202037	19.02.1942	2ft 6in	4wDM	33/40	Ruston 3VRO	5T	36
202969	29.10.1940	2ft 0in	4wDM	16/20	Ruston 2VSOL	2¾T	396
210479	31.12.1942	4ft 8½in	4wDM	88DS	Ruston 4VPB	17½T	179
221643	26.10.1943	4ft 8½in	4wDM	48DS	Ruston 4VROL	7½T	301
221648	01.12.1943	4ft 8½in	4wDM	48DS	Ruston 4VROL	7½T	348
224352	07.05.1944	4ft 8½in	4wDM	88DS	Ruston 4VPHL	17T	240
235512	29.08.1945	4ft 8½in	4wDM	48DS	Ruston 4VROLS	7½T	250,251,253
235514	02.09.1945	4ft 8½in	4wDM	48DS	Ruston 4VROLS	7½T	22,31,124,282
237928	04.09.1946	4ft 8½in	4wDM	48DS	Ruston 4VROL	7½T	101
242914	03.07.1946	2ft 6in	4wDM	48DL	Ruston 4VROL	6½T	169
252685	07.01.1949	4ft 8½in	0-4-0DM	165DS	Ruston 6VPHL	28T	251
265617	03.09.1948	4ft 8½in	4wDM	48DS	Ruston 4VRHL	7½T	128
275881	27.06.1949	4ft 8½in	4wDM	88DS	Ruston 4VPHL	17T	273
279593	12.09.1949	4ft 8½in	4wDM	48DS	Ruston 4VRHL	7½T	21,248
279599	12.01.1950	4ft 8½in	4wDM	48DS	Ruston 4VRHL	7½T	21
299107	18.01.1951	4ft 8½in	4wDM	88DS	Ruston 4VPHL	17½T	61
305301	24.05.1951	4ft 8½in	4wDM	48DS	Ruston 4VRHL	7½T	312
305320	02.02.1951	4ft 8½in	4wDM	88DS	Ruston 4VPHL	17T	253
305323	16.06.1951	4ft 8½in	4wDM	88DS	Ruston 4VPHL	17T	23
306088	30.12.1949	4ft 8½in	4wDM	88DS	Ruston 4VPHL	17T	28,31,124
312427	07.08.1951	4ft 8½in	4wDM	88DS	Ruston 4VPHL	20T	186,248
312429	18.09.1951	4ft 8½in	4wDM	88DS	Ruston 4VPHL	20T	185,187
312434	20.04.1951	4ft 8½in	4wDM	88DS	Ruston 4VPHL	20T	185,298,336
312989	22.12.1952	4ft 8½in	0-4-0DE	165DE	Ruston 6VPHL	28T	111
318748	11.10.1952	2ft 6in	4wDM	LBU	Ruston 3VSHL	3½T	154,156
319285	06.02.1953	4ft 8½in	0-4-0DM	165DS	Ruston 6VPHL	28T	192
321735	30.07.1952	4ft 8½in	4wDM	88DS	Ruston 4VPHL	20T	21,185,327
323599	23.01.1953	4ft 8½in	0-4-0DE	165DE	Ruston 6VPHL	28T	328
323605	04.02.1954	4ft 8½in	0-4-0DE	165DE	Ruston 6VPHL	28T	328
327966	05.03.1954	4ft 8½in	0-4-0DM	165DS	Ruston 6VPHL	28T	116

327971	13.09.1954	4ft 8½in	0-4-0DM	165DS	Ruston 6VPHL	28T		111
338424	03.02.1955	4ft 8½in	4wDM	88DS	Ruston 4VPHL	17T		327
338438	07.05.1953	2ft 6in	4wDM	LBU	Ruston 3VSHL	3½T		154
338439	12.06.1953	3ft 0in	4wDM	LBU	Ruston 3VSHL	3½T		155
353484	26.08.1953	2ft 6in	4wDM	LBU	Ruston 3VSHL	3½T		38,154,157,162
353486	13.10.1953	2ft 6in	4wDM	LBU	Ruston 3VSHL	3½T		154
353491	29.01.1954	3ft 0in	4wDM	LBU	Ruston 3VSHL	3½T		155
353492	29.01.1954	2ft 6in	4wDM	LBU	Ruston 3VSHL	3½T		154
353493	08.03.1954	2ft 6in	4wDM	LBU	Ruston 3VSHL	3½T		162
353494	08.03.1954	2ft 6in	4wDM	LBU	Ruston 3VSHL	3½T		156,157,162
371938	09.05.1954	3ft 0in	4wDM	LBU	Ruston 3VSHL	3½T		155
371942	10.05.1954	2ft 6in	4wDM	LBU	Ruston 3VSHL	3½T		162
371947	25.06.1954	2ft 6in	4wDM	LBU	Ruston 3VSHL	3½T		162
371949	06.08.1954	2ft 6in	4wDM	LBU	Ruston 3VSHL	3½T		162
375329	23.06.1954	2ft 6in	4wDM	48DLU	Ruston 4VRHL	7T		154
375693	25.08.1954	2ft 6in	4wDM	LBU	Ruston 3VSHL	3½T		154,162
375694	01.07.1954	3ft 0in	4wDM	LBU	Ruston 3VSHL	3½T		155
375717	16.05.1955	4ft 8½in	0-4-0DM	165DS	Ruston 6VPHL	28T		240
381757	27.09.1955	4ft 8½in	0-4-0DE	165DE	Ruston 6VPH	28T		230,327
386871	04.03.1955	4ft 8½in	4wDM	48DS	Ruston 4VRHL	7½T		124
392100	16.08.1955	2ft 6in	4wDM	LBU	Ruston 3VSHL	3½T		162
392119	25.08.1956	2ft 6in	4wDM	48DLU	Ruston 4VRHL	7T		154
398616	01.10.1956	4ft 8½in	4wDM	88DS	Ruston 4VPHL	17T		377,378
402808	13.10.1956	4ft 8½in	4wDM	48DS	Ruston 4YCLS	7½T		28,31,124
408309	16.04.1957	4ft 8½in	0-4-0DE	165DE	Ruston 6VPHL	28T		230,327
411318	08.04.1957	4ft 8½in	4wDM	48DS	Ruston 4YCLS	-		21,255
417891	06.05.1959	4ft 8½in	4wDM	48DS	Ruston 4YCLS	7½T		31
417894	27.11.1959	4ft 8½in	4wDM	48DS	Ruston 4YCLS	7½T		22,350
418764	04.10..1957	3ft 0in	4wDM	LBU	Ruston 3VSHL	3¾T		155
418765	03.10.1957	2ft 6in	4wDM	LBU	Ruston 3VSHL	3¾T		154
418803	31.12.1957	3ft 0in	4wDM	48DLG	Ruston 4VRHL	7T		155
420142	07.03.1958	4ft 8½in	0-4-0DE	165DE	Ruston 6VPHL	-		333
421419	29.05.1958	4ft 8½in	4wDM	88DS	Ruston 4VPHL	17T		133
421436	12.03.1958	4ft 8½in	0-4-0DE	165DE	Ruston 6VPHL	28T		326
423659	15.07.1958	4ft 8½in	0-4-0DE	165DE	Ruston 6VPHL	28T		328
425482	12.09.1958	4ft 8½in	4wDM	88DS	Ruston 4VPHL	17T		273,288,309
425485	04.12.1959	4ft 8½in	4wDM	88DS	Ruston 4VPHL	17T		327
427802	28.07.1958	3ft 0in	4wDM	LBU	Ruston 3VSHL	3¾T		155
432480	18.03.1959	4ft 8½in	4wDM	88DS	Ruston 4VPHL	17T		251
432655	29.10.1959	2ft 6in	4wDM	LBU	Ruston 2YDAL	3¾T		162
433388	19.10.1959	3ft 0in	4wDM	48DLG	Ruston 4VRHL	7T		154,156
451900	06.07.1961	3ft 0in	4wDM	48DLG	Ruston 4VRHL	7T		156
463152	31.05.1961	4ft 8½in	4wDM	88DS	Ruston 4VPHL	17T		327
466578	25.07.1961	3ft 0in	4wDM	LBU	Ruston 2YDAL	3½T		156
466579	25.07.1961	2ft 6in	4wDM	LBU	Ruston 2YDAL	3½T		156,162
480690	20.06.1962	4ft 8½in	0-6-0DE	LSSE	Ruston 6RPHL	42T		208,295
480696	13.09.1962	4ft 8½in	0-6-0DE	LSSE	Ruston 6RPHL	42T		200,208
544996	25.10.1968	4ft 8½in	4wDH	LLSH	Ruston 6Y2DAL	20T		171

ROBEL & CO, MASCHINENFABRIK, Munchen, Bayern, Germany — Robel

54.12-107 AD183	1980	4ft 8½in	4wDM			91,100,290
54.12-107 AD184	1980	4ft 8½in	4wDM			91,100,290

ROBEY & CO, Globe Works, Lincoln, Lincolnshire — Robey

LIVERTON		4ft 8½in	0-4-0ST	IC			44,106,210

RUSTON, PROCTOR & Co, Sheaf Iron Works, Lincoln, Lincolnshire — RP

969	05.1866	4ft 8½in	0-4-0ST	OC		2ft 9in	119

ROLLS-ROYCE LTD, Sentinel Works, Harlescott, Shrewsbury, Shropshire — RR

Successors to Sentinel

10186	12.06.1964	4ft 8½in	0-6-0DH	311hp	RR-C8SFL	48T	91
10197	21.10.1964	4ft 8½in	4wDH	229hp	RR C6SFL	34T	334
10208	07.07.1965	4ft 8½in	0-4-0DH	311hp	RR-C8SFL	40T	260
10210	02.10.1964	4ft 8½in	0-6-0DH	311hp	RR-C8SFL	48T	90,328
10211	02.10.1964	4ft 8½in	0-6-0DH	311hp	RR-C8SFL	48T	90
10214	27.11.1964	4ft 8½in	0-6-0DH	375hp	RR-C8TFL	48T	29,213,383,385
10215	29.01.1965	4ft 8½in	0-6-0DH	375hp	RR-C8TFL	48T	269,314,339
10217	15.04.1965	4ft 8½in	0-6-0DH	325hp	RR-C8SFL	48T	70
10220	29.06.1965	4ft 8½in	0-6-0DH	325hp	RR-C8SFL	48T	70,201,385
10224	13.04.1965	4ft 8½in	0-6-0DH	325hp	RR-C8SFL	48T	90,328
10225	20.05.1965	4ft 8½in	0-6-0DH	325hp	RR-C8SFL	48T	90,213,331
10227	28.10.1965	4ft 8½in	4wDH	229hp	RR-C6SFL	34T	333
10229	24.11.1965	4ft 8½in	4wDH	229hp	RR-C6SFL	34T	70,231,383,384
10234	01.09.1965	4ft 8½in	0-6-0DH	325hp	RR-C8SFL	48T	90,213
10239	02.12.1965	4ft 8½in	0-6-0DH	325hp	RR-C8SFL	48T	261,334
10253	23.06.1966	4ft 8½in	0-4-0DH	325hp	RR-C8SFL	40T	333
10255	25.03.1966	4ft 8½in	0-6-0DH	325hp	RR-C8SFL	48T	100,383
10257	23.02.1966	4ft 8½in	0-6-0DH	325hp	RR-C8SFL	48T	216
10260	06.10.1966	4ft 8½in	4wDH	256hp	RR-C6SFL	34T	331
10264	29.12.1966	4ft 8½in	4wDH	256hp	RR-C6SFL	34T	331
10270	26.09.1967	4ft 8½in	0-6-0DH	325hp	RR-C8SFL	48T	330
10274	09.01.1969	4ft 8½in	6wDH	445hp	RR-DV8N	52T	328
10277	04.12.1968	4ft 8½in	6wDH	608hp	RR-DV8TCA	60T	92
10280	16.05.1968	4ft 8½in	4wDH	256hp	RR-C6SFL	34T	171
10281	11.06.1968	4ft 8½in	4wDH	256hp	RR-C6SFL	34T	70
10286	16.07.1969	4ft 8½in	0-6-0DH	375hp	RR-C8TFL	48T	216,333
10289	07.09.1970	4ft 8½in	0-6-0DH	375hp	RR-C8TFL	48T	216,333
10290	23.10.1970	4ft 8½in	0-6-0DH	375hp	RR-C8TFL	48T	333
10292	26.03.1971	4ft 8½in	0-6-0DH	375hp	RR-C8TFL	48T	90

ROBERT STEPHENSON & CO LTD, Forth Street, Newcastle upon Tyne — RS

1073	08.12.1856	4ft 8½in	0-6-0	IC	16¾ x24	4ft 6in	160
2124	25.08.1873	4ft 8½in	0-4-0ST	OC	9x18	2ft 9in	199
2554	23.11.1885	4ft 8½in	0-6-0ST	OC	13x18	3ft 6in	199,296,352
2748	1892	4ft 8½in	0-4-0ST	OC	16x24	3ft 10in	246
3045	1901	4ft 8½in	0-4-0ST	OC			246
3056	1904	4ft 8½in	0-4-0ST	OC			131,296
3057	20.05.1904	4ft 8½in	0-4-0ST	OC	14x20	3ft 3in	228
3415	1910	4ft 8½in	0-4-0ST	OC			69
3673	1917	4ft 8½in	0-4-0ST	OC			69
3674	1917	4ft 8½in	0-4-0ST	OC			69

ROBERT STEPHENSON & HAWTHORNS LTD — RSH
D   Robert Stephenson & Hawthorns Ltd, Harrowgate Hill, Darlington, Co.Durham — RSHD
N   Robert Stephenson & Hawthorns Ltd, Forth Banks, Newcastle upon Tyne — RSHN

D 6948	24.05.1938	4ft 8½in	0-6-0ST	OC	16x24	3ft 8in	27
D 6973	31.10.1939	4ft 8½in	0-4-0ST	OC	16x24	3ft 8in	200
D 6974	07.11.1939	4ft 8½in	0-4-0ST	OC	16x24	3ft 3in	26

N 6978	10.01.1940	4ft 8½in	0-4-0DM	150hp	Fowler 4C	3ft 3in		286,318
N 7006	28.06.1940	4ft 8½in	0-4-0CT	OC	12x15	2ft 10in		391
D 7024	11.04.1940	4ft 8½in	0-4-0ST	OC	16x24	3ft 8in		200
D 7036	17.07.1941	4ft 8½in	0-4-0ST	OC	16x24	3ft 8in		59,87
D 7037	29.08.1941	4ft 8½in	0-4-0ST	OC	16x24	3ft 8in		59,87
D 7041	28.10.1941	4ft 8½in	0-4-0ST	OC	16x24	3ft 8in		59,88,175
N 7045	20.03.1942	4ft 8½in	0-4-0ST	OC	16x24	3ft 8in		45,228
N 7059	13.07.1942	4ft 8½in	0-4-0ST	OC	16x24	3ft 8in		69
N 7065	05.08.1942	4ft 8½in	0-4-0ST	OC	16x24	3ft 8in		59,88
N 7072	18.12.1942	4ft 8½in	0-4-0ST	OC	16x24	3ft 8in		166
N 7075	22.02.1943	4ft 8½in	0-4-0ST	OC	16x24	3ft 8in		166,175
N 7084	30.11.1943	4ft 8½in	0-4-0ST	OC	14x22	3ft 6in		173,300,338,340
N 7129	18.01.1945	4ft 8½in	0-4-0ST	OC	14x22	3ft 6in		277
N 7160	23.08.1945	4ft 8½in	0-4-0ST	OC	16x24	3ft 8in		326
N 7308	07.08.1946	4ft 8½in	0-4-0ST	OC	12x20	3ft 1in		253
N 7311	23.01.1948	4ft 8½in	0-4-0ST	OC	14x22	3ft 6in		286
N 7332	24.01.1947	4ft 8½in	0-4-0ST	OC	16x24	3ft 8in		200
N 7333	03.09.1946	4ft 8½in	0-4-0ST	OC	14x22	3ft 6in		338
N 7337	08.08.1947	4ft 8½in	0-6-0T	OC	18x24	4ft 0in		200
N 7338	05.09.1947	4ft 8½in	0-6-0T	OC	18x24	4ft 0in		200
N 7340	28.02.1947	4ft 8½in	0-4-0ST	OC	16x24	3ft 8in		59
N 7341	06.03.1947	4ft 8½in	0-4-0ST	OC	16x24	3ft 8in		59,166
N 7342	21.03.1947	4ft 8½in	0-4-0ST	OC	16x24	3ft 8in		59,166
N 7343	24.04.1947	4ft 8½in	0-4-0ST	OC	16x24	3ft 8in		166
N 7344	07.05.1947	4ft 8½in	0-4-0ST	OC	16x24	3ft 8in		83
N 7345	21.05.1947	4ft 8½in	0-4-0ST	OC	16x24	3ft 8in		59,166
N 7346	06.06.1947	4ft 8½in	0-4-0ST	OC	16x24	3ft 8in		59
N 7347	16.06.1947	4ft 8½in	0-4-0ST	OC	16x24	3ft 8in		59,88
N 7348	26.06.1947	4ft 8½in	0-4-0ST	OC	16x24	3ft 8in		59,88,175
N 7358	23.12.1947	4ft 8½in	0-4-0ST	OC	14x22	3ft 6in		338
N 7359	28.11.1947	4ft 8½in	0-4-0ST	OC	14x22	3ft 6in		112
N 7494	13.04.1949	4ft 8½in	0-4-0ST	OC	14x22	3ft 6in		326
N 7687	19.10.1951	4ft 8½in	0-6-0ST	OC	18x24	4ft 0in		88
N 7688	19.11.1951	4ft 8½in	0-6-0ST	OC	18x24	4ft 0in		88
N 7689	30.01.1952	4ft 8½in	0-6-0ST	OC	18x24	4ft 0in		88
N 7690	04.04.1952	4ft 8½in	0-6-0ST	OC	18x24	4ft 0in		88
N 7691	01.05.1952	4ft 8½in	0-6-0ST	OC	18x24	4ft 0in		88
N 7798	24.05.1954	4ft 8½in	0-4-0ST	OC	14x22	3ft 6in		112
N 7862	24.10.1956	4ft 8½in	0-6-0DM	204hp	Gardner 8L3		[DC2575]	317
N 7900	27.05.1958	4ft 8½in	0-4-0DM	107hp	Gardner 6LW	Husky		24,25,133,344,351
N 7902	1958	4ft 8½in	0-4-0DM	107hp	Gardner 6LW	Husky		286,318
N 7925	08.12.1959	4ft 8½in	0-4-0DM	153hp	Gardner 6L3		[DC 2592]	286,317,318
D 8367	1962	4ft 8½in	0-4-0DH	266hp	Dorman 6QA		[WB 3212]	329

R. & W. HAWTHORN, Forth Banks, Newcastle upon Tyne RWH

226	1837	4ft 8½in	0-6-0	IC			358
1019	09.10.1857	4ft 8½in	0-4-0ST	OC	13x18	4ft 0in	308
1635	01.12.1874	4ft 8½in	0-4-0ST	OC	13x19	3ft 6in	309
1789	1879	4ft 8½in	0-4-0ST	OC	10x15	2ft 9in	309
1819	1880	4ft 8½in	0-4-0ST	OC			166
1820	1880	4ft 8½in	0-4-0ST	OC			309
1847	1881	4ft 8½in	0-4-0ST	OC	12x18	3ft 0½in	114,160,171
2010	1884	4ft 8½in	0-4-0ST	OC			309

SENTINEL (SHREWSBURY) LTD, Harlescott, Shrewsbury, Shropshire S

Diesel locomotives later built by Rolls Royce – which see.

6154CH	1926	4ft 8½in	0-4-0VBT	VCG			44
9294	1936	4ft 8½in	4wVBT	VCG			38
9377	1947	4ft 8½in	4wVBT	VCG	100hp		318,338,340
9538	10.1951	4ft 8½in	4wVBT	VCG	200hp		88
9561	19.01.1954	4ft 8½in	4wVBT	VCG	100hp		277
9562	30.03.1954	4ft 8½in	4wVBT	VCG	100hp		318,338,340
9566	20.08.1954	4ft 8½in	4wVBT	VCG	100hp		175
9587	01.03.1956	4ft 8½in	0-6-0VBT	VCG	200hp		88
9588	15.04.1956	4ft 8½in	0-6-0VBT	VCG	200hp		88
9589	14.05.1956	4ft 8½in	0-6-0VBT	VCG	200hp		88
9590	14.06.1956	4ft 8½in	0-6-0VBT	VCG	200hp		88
9591	18.06.1956	4ft 8½in	0-6-0VBT	VCG	200hp		88
9592	23.07.1956	4ft 8½in	0-6-0VBT	VCG	200hp		88
9594	24.05.1955	4ft 8½in	4wVBT	VCG	100hp		175
9598	11.10.1955	4ft 8½in	4wVBT	VCG	100hp		175
9600	19.12.1955	4ft 8½in	4wVBT	VCG	100hp		175
9601	18.09.1956	4ft 8½in	0-6-0VBT	VCG	200hp		88
9602	12.12.1956	4ft 8½in	0-6-0VBT	VCG	200hp		88
9603	03.1957	4ft 8½in	0-6-0+0-6-0	VBT			
				VCG	400hp		88
9604	05.01,1956	4ft 8½in	0-6-0VBT	VCG	200hp		88
9605	26.03.1956	4ft 8½in	0-6-0VBT	VCG	200hp		89
9606	07.08.1956	4ft 8½in	0-6-0VBT	VCG	200hp		89
9613	28.03.1956	4ft 8½in	4wVBT	VCG	100hp		175
9619	1957	4ft 8½in	4wVBT	VCG	200hp		88
9629	25.03.1957	4ft 8½in	4wVBT	VCG	100hp		277
9633	1957	4ft 8½in	0-6-0VBT	VCG	200hp		88
9650	12.03.1958	4ft 8½in	0-6-0F	OC	200hp		88
9651	29.07.1958	4ft 8½in	0-6-0F	OC	200hp		88
9652	29.07.1958	4ft 8½in	0-6-0F	OC	200hp		88
9653	29.07.1958	4ft 8½in	0-6-0F	OC	200hp		88
10001	24.03.1959	4ft 8½in	4wDH	200hp	RR-C6SFL	34T	89
10002	09.04.1959	4ft 8½in	4wDH	200hp	RR-C6SFL	34T	89
10003	07.05.1959	4ft 8½in	4wDH	200hp	RR-C6SFL	34T	112,121
10004	08.07.1959	4ft 8½in	4wDH	200hp	RR-C6SFL	34T	89
10008	12.10.1959	4ft 8½in	4wDH	200hp	RR-C6SFL	34T	89
10009	15.10.1959	4ft 8½in	4wDH	200hp	RR-C6SFL	34T	89
10010	20.10.1959	4ft 8½in	4wDH	200hp	RR-C6SFL	34T	49,89,231
10011	28.10.1959	4ft 8½in	4wDH	200hp	RR-C6SFL	34T	60,89
10012	16.11.1959	4ft 8½in	4wDH	200hp	RR-C6SFL	34T	298
10014	08.12.1959	4ft 8½in	4wDH	200hp	RR-C6SFL	34T	60,89
10015	29.12.1959	4ft 8½in	4wDH	200hp	RR-C6SFL	34T	60,89
10016	12.01.1960	4ft 8½in	4wDH	200hp	RR-C6SFL	34T	60,89
10017	18.02.1960	4ft 8½in	4wDH	200hp	RR-C6SFL	34T	60,89,124,278
10018	29.01.1960	4ft 8½in	4wDH	200hp	RR-C6SFL	34T	61,231
10019	02.02.1960	4ft 8½in	4wDH	200hp	RR-C6SFL	34T	61,89,309,329
10021	02.12.1959	4ft 8½in	4wDH	200hp	RR-C6SFL	34T	333
10022	29.01.1960	4ft 8½in	4wDH	200hp	RR-C6SFL	34T	333
10024	26.02.1960	4ft 8½in	4wDH	200hp	RR-C6SFL	34T	61,89
10025	03.03.1960	4ft 8½in	4wDH	200hp	RR-C6SFL	34T	61,89,231
10026	10.03.1960	4ft 8½in	4wDH	200hp	RR-C6SFL	34T	89
10027	23.03.1960	4ft 8½in	4wDH	200hp	RR-C6SFL	34T	89
10028	30.03.1960	4ft 8½in	4wDH	200hp	RR-C6SFL	34T	89,231

10029	06.04.1960	4ft 8½in	4wDH	200hp	RR-C6SFL	34T	89
10030	26.04.1960	4ft 8½in	4wDH	200hp	RR-C6SFL	34T	89
10031	15.03.1960	4ft 8½in	4wDH	200hp	RR-C6SFL	34T	89
10032	15.07.1960	4ft 8½in	0-6-0DH	311hp	RR-C8SFL	50T	89
10034	08.09.1960	4ft 8½in	4wDH	200hp	RR-C6SFL	34T	89
10035	27.07.1960	4ft 8½in	4wDH	200hp	RR-C6SFL	34T	89,331
10036	27.07.1960	4ft 8½in	4wDH	200hp	RR-C6SFL	34T	89,309
10038	30.08.1960	4ft 8½in	4wDH	200hp	RR-C6SFL	34T	89,231
10039	30.08.1960	4ft 8½in	4wDH	200hp	RR-C6SFL	30T	89
10040	15.09.1960	4ft 8½in	4wDH	200hp	RR-C6SFL	34T	89
10041	21.09.1960	4ft 8½in	4wDH	200hp	RR-C6SFL	34T	89
10042	29.09.1960	4ft 8½in	4wDH	200hp	RR-C6SFL	34T	89,231
10043	18.10.1960	4ft 8½in	4wDH	200hp	RR-C6SFL	34T	89
10044	18.10.1960	4ft 8½in	4wDH	200hp	RR-C6SFL	34T	89
10048	15.12.1960	4ft 8½in	4wDH	200hp	RR-C6SFL	34T	70
10050	29.12.1960	4ft 8½in	0-6-0DH	311hp	RR-C8SFL	48T	89
10051	14.02.1961	4ft 8½in	4wDH	200hp	RR-C6SFL	34T	200
10052	20.02.1961	4ft 8½in	4wDH	200hp	RR-C6SFL	34T	200
10053	17.04.1961	4ft 8½in	0-6-0DH	311hp	RR-C8SFL	48T	89,231
10054	04.07.1961	4ft 8½in	0-6-0DH	311hp	RR-C8SFL	48T	89
10056	09.08.1961	4ft 8½in	0-6-0DH	311hp	RR-C8SFL	48T	90
10057	20.02.1961	4ft 8½in	0-4-0DH	311hp	RR-C8SFL	40T	90
10059	16.03.1961	4ft 8½in	4wDH	200hp	RR-C6SFL	34T	332
10064	06.04.1961	4ft 8½in	0-4-0DH	311hp	RR-C8SFL	40T	90
10066	18.05.1961	4ft 8½in	0-4-0DH	311hp	RR-C8SFL	40T	90
10067	04.07.1961	4ft 8½in	0-4-0DH	311hp	RR-C8SFL	40T	90
10068	18.07.1961	4ft 8½in	0-4-0DH	311hp	RR-C8SFL	40T	90
10069	22.08.1961	4ft 8½in	0-4-0DH	311hp	RR-C8SFL	40T	90
10070	16.06.1961	4ft 8½in	4wDH	200hp	RR-C6SFL	34T	32,171,200,263
10073	27.04.1961	4ft 8½in	4wDH	200hp	RR-C6SFL	34T	90,309
10074	27.04.1961	4ft 8½in	4wDH	200hp	RR-C6SFL	34T	90
10075	18.05.1961	4ft 8½in	4wDH	200hp	RR-C6SFL	34T	90
10076	25.05.1961	4ft 8½in	4wDH	200hp	RR-C6SFL	34T	90
10078	22.08.1961	4ft 8½in	0-6-0DH	311hp	RR-C8SFL	48T	90
10080	19.10.1961	4ft 8½in	0-6-0DH	311hp	RR-C8SFL	48T	70
10081	09.11.1961	4ft 8½in	0-6-0DH	311hp	RR-C8SFL	48T	70
10084	15.09.1961	4ft 8½in	0-4-0DH	311hp	RR-C8SFL	40T	90
10091	19.12.1961	4ft 8½in	0-6-0DH	311hp	RR-C8SFL	48T	90
10092	16.01.1962	4ft 8½in	0-6-0DH	311hp	RR-C8SFL	48T	90
10093	14.02.1962	4ft 8½in	0-6-0DH	311hp	RR-C8SFL	48T	90
10094	14.02.1962	4ft 8½in	0-6-0DH	311hp	RR-C8SFL	48T	90
10095	06.03.1962	4ft 8½in	0-6-0DH	311hp	RR-C8SFL	48T	90,213,261,329
10096	21.03.1962	4ft 8½in	0-4-0DH	311hp	RR-C8SFL	40T	90
10098	21.03.1962	4ft 8½in	0-4-0DH	311hp	RR-C8SFL	40T	90
10099	05.04.1962	4ft 8½in	0-4-0DH	311hp	RR-C8SFL	40T	90
10100	26.04.1962	4ft 8½in	0-4-0DH	311hp	RR-C8SFL	40T	90
10101	29.03.1962	4ft 8½in	0-6-0DH	311hp	RR-C8SFL	48T	90
10102	26.04.1962	4ft 8½in	0-6-0DH	311hp	RR-C8SFL	48T	90,216
10103	10.05.1962	4ft 8½in	0-4-0DH	311hp	RR-C8SFL	40T	90
10104	10.05.1962	4ft 8½in	0-4-0DH	311hp	RR-C8SFL	40T	90
10105	26.06.1962	4ft 8½in	0-4-0DH	311hp	RR-C8SFL	40T	70,90
10107	15.01.1963	4ft 8½in	0-6-0DH	311hp	RR-C8SFL	50T	213
10108	24.01.1963	4ft 8½in	0-6-0DH	311hp	RR-C8SFL	50T	213,216
10125	25.07.1963	4ft 8½in	0-4-0DH	311hp	RR-C8SFL	40T	70
10126	25.07.1963	4ft 8½in	0-4-0DH	311hp	RR-C8SFL	40T	70,92,330
10127	19.04.1963	4ft 8½in	0-4-0DH	272hp	RR-C8NFL	37T	70,92,230,216

10128	25.07.1963	4ft 8½in	0-4-0DH	272hp	RR-C8NFL	37T	70,91
10129	25.07.1963	4ft 8½in	0-4-0DH	272hp	RR-C8NFL	37T	70,91
10130	25.07.1963	4ft 8½in	0-4-0DH	272hp	RR-C8NFL	37T	70,91
10131	25.07.1963	4ft 8½in	0-4-0DH	272hp	RR-C8NFL	37T	70,213
10132	25.07.1963	4ft 8½in	0-4-0DH	272hp	RR-C8NFL	37T	70,91
10133	25.07.1963	4ft 8½in	0-4-0DH	272hp	RR-C8NFL	37T	70
10134	25.07.1963	4ft 8½in	0-4-0DH	272hp	RR-C8NFL	37T	70,91
10136	27.11.1962	4ft 8½in	0-8-0DH	600hp	2x RR-C8SFL	74T	91
10137	12.12.1962	4ft 8½in	0-4-0DH	311hp	RR-C8SFL	40T	260
10139	19.10.1962	4ft 8½in	0-6-0DH	311hp	RR-C8SFL	48T	332
10147	06.09.1963	4ft 8½in	0-6-0DH	311hp	RR-C8SFL	48T	213,334
10148	24.10.1963	4ft 8½in	0-6-0DH	311hp	RR-C8SFL	48T	213
10153	04.09.1963	4ft 8½in	0-6-0DH	311hp	RR-C8SFL	50T	91
10161	09.01.1964	4ft 8½in	0-6-0DH	311hp	RR-C8SFL	48T	261,333
10162	31.10.1963	4ft 8½in	0-6-0DH	311hp	RR-C8SFL	48T	330
10166	19.12.1963	4ft 8½in	0-6-0DH	311hp	RR-C8SFL	48T	90,383
10167	23.01.1964	4ft 8½in	0-6-0DH	311hp	RR-C8SFL	48T	90,213
10168	12.03.1964	4ft 8½in	0-6-0DH	311hp	RR-C8SFL	48T	90
10170	17.01.1964	4ft 8½in	0-4-0DH	311hp	RR-C8SFL	40T	269,332

CHRISTOPH SCHÖTTLER MASCHINENFABRIK GmbH, Germany — Schöma

6827	2019	900mm	4wBE	CEL180H	374
7084	2019	900mm	4wBE		374
7085	2019	900mm	4wBE		374
7086	2019	900mm	4wBE		374
7087	2019	900mm	4wBE		374
7103	2019	900mm	4wBE		374
7104	2019	900mm	4wBE		374

SOUTH DURHAM STEEL & IRON CO LTD, Hartlepool — SDSI

2ft 6in	4wRE	230

SWINDON WORKS, Swindon, Wiltshire — Sdn

GWR, British Railways

[3814]	1940	4ft 8½in	2-8-0	OC			385
[D2023]	08.1958	4ft 8½in	0-6-0DM	204hp	Gardner 8L3	30T 16c	260,317
[D2024]	08.1958	4ft 8½in	0-6-0DM	204hp	Gardner 8L3	30T 16c	260,317
[03154]	07.1960	4ft 8½in	0-6-0DM	204hp	Gardner 8L3	30T 16c	116

MARQUIS OF LONDONDERRY, Londonderry Engine Works, Seaham Seaham

1883	4ft 8½in	0-6-0ST	IC	160

SKINNINGROVE IRON CO LTD, Carlin How — Skinningrove

[MARY]	1913	4ft 8½in	0-4-0ST	OC	14x20	69

SALTBURN MINATURE RAILWAY — SMR

1994	1ft 3in	4wDH	S/O	394

ANDREW SMITH — Andrew Smith

1837	4ft 8½in	0-4-0	OC	357

THOMAS GREEN & SON LTD, Smithfield Ironworks, Leeds, Yorkshire (WR) — TG

312	03.1903	1ft 11½in	0-4-2ST	OC	7x12	191

THOS HILL (ROTHERHAM) LTD, Vanguard Works, Kilnhurst, Yorkshire (WR)   TH

No.	Date	Gauge	Type	Power	Engine	Weight	Works
101C	25.01.1960	4ft 8½in	4wDH	170hp	R-R C6NFL	26T	324
105V	22.01.1961	4ft 8½in	4wDH	178hp	R-R C6NFL	30T	277
109C	21.08.1961	4ft 8½in	0-6-0DH	311hp	R-R C8SFL	48T	90
115V	22.03.1962	4ft 8½in	4wDH	178hp	R-R C6NFL	30T	61,277
123V	31.01.1963	4ft 8½in	4wDH	157hp	R-R C6NFL	25T	331
127V	29.05.1963	4ft 8½in	4wDH	157hp	R-R C6NFL	25T	90
128V	25.06.1963	4ft 8½in	4wDH	157hp	R-R C6NFL	25T	90
144V	05.10.1964	4ft 8½in	4wDH	179hp	R-R C6NFL	30T	65,327
150C	18.03.1965	4ft 8½in	0-6-0DH	325hp	R-RC8SFL	48T	70
167V	17.06.1966	4ft 8½in	0-6-0DH	325hp	R-R C8SFL	48T	201,328
180V	14.03.1967	4ft 8½in	6wDH	272hp	R-R C8NFL	48T	331
185V	12.07.1967	4ft 8½in	4wDH	252hp	R-R C8NFL	37T	330
193V	24.05.1968	4ft 8½in	4wDH	275hp	R-R C8NFL	40T	332
200v	25.11.1968	4ft 8½in	4wDH	210hp	SF65C	34T	383
201V	28.03.1969	4ft 8½in	4wDH	278hp	R-R C6TFL	40T	90
202V	16.06.1969	4ft 8½in	4wDH	278hp	R-R C6TFL	40T	90
203V	03.09.1969	4ft 8½in	4wDH	278hp	R-R C6TFL	40T	90
204V	11.11.1969	4ft 8½in	4wDH	278hp	R-R C6TFL	40T	90
205V	28.03.1969	4ft 8½in	4wDH	278hp	R-R C6TFL	40T	90
206V	28.03.1969	4ft 8½in	4wDH	278hp	R-R C6TFL	40T	90
207V	16.06.1969	4ft 8½in	4wDH	278hp	R-R C6TFL	40T	90
208V	16.06.1969	4ft 8½in	4wDH	278hp	R-R C6TFL	40T	90
209V	03.09.1969	4ft 8½in	4wDH	278hp	R-R C6TFL	40T	90
210V	03.09.1969	4ft 8½in	4wDH	278hp	R-R C6TFL	40T	90
211V	11.11.1969	4ft 8½in	4wDH	278hp	R-R C6TFL	40T	90
212V	17.10.1969	4ft 8½in	4wDH	272hp	R-R C8NFL	40T	330
214V	17.12.1969	4ft 8½in	4wDH	272hp	R-R C8NFL	40T	332
221V	16.06.1970	4ft 8½in	4wDH	275hp	R-R C6NFL	40T	231
222V	24.07.1970	4ft 8½in	4wDH	275hp	R-R C6NFL	40T	231
223V	29.08.1970	4ft 8½in	4wDH	275hp	R-R C6NFL	40T	231
224V	10.09.1970	4ft 8½in	4wDH	275hp	R-R C6NFL	40T	231
225V	02.10.1970	4ft 8½in	4wDH	275hp	R-R C6TFL	34T	45
226V	19.11.1970	4ft 8½in	4wDH	275hp	R-R C6TFL	34T	45
231V	15.04.1971	4ft 8½in	4wDH	272hp	R-R C8NFL	42T	230,330
232V	13.05.1971	4ft 8½in	4wDH	272hp	R-R C8NFL	42T	70,230
233V	10.06.1971	4ft 8½in	4wDH	272hp	R-R C8NFL	42T	231
234V	14.07.1971	4ft 8½in	4wDH	272hp	R-R C8NFL	37T	298,332
235V	04.08.1971	4ft 8½in	4wDH	272hp	R-R C8NFL	37T	298,331
236V	10.09.1971	4ft 8½in	4wDH	272hp	R-R C8NFL	37T	298
239V	09.02.1972	4ft 8½in	0-4-0DH	272hp	R-R C8NFL	45T	332
241V	30.08.1972	4ft 8½in	4wDH	278hp	R-R C6TFL	45T	90
242V	30.08.1972	4ft 8½in	0-6-0DH	556hp	2 x R-R C6TFL	75T Titan	90
243V	28.09.1972	4ft 8½in	0-6-0DH	700hp	2 x R-R C8TFL	75T Titan	90
244V	06.12.1972	4ft 8½in	0-6-0DH	700hp	2 x R-R C8TFL	75T Titan	91
245V	01.03.1972	4ft 8½in	0-6-0DH	700hp	2 x R-R C8TFL	75T Titan	91
252V	17.12.1974	4ft 8½in	0-6-0DH	544hp	2 x R-R C8NFL	75T Titan	91
253V	16.01.1975	4ft 8½in	0-6-0DH	544hp	2 x R-R C8NFL	75T Titan	91
254V	10.03.1975	4ft 8½in	0-6-0DH	544hp	2 x R-R C8NFL	75T Titan	91
255V	21.04.1975	4ft 8½in	0-6-0DH	544hp	2 x R-R C8NFL	75T Titan	91
256V	30.07.1975	4ft 8½in	0-6-0DH	544hp	2 x R-R C8NFL	75T Titan	91
258C	16.05.1975	4ft 8½in	4wDH	230hp	R-R C6SFL	34T reb of S 10001	231
259C	28.08.1975	4ft 8½in	4wDH	230hp	R-R C6SFL	34T reb of S 10041	231
260C	22.09.1975	4ft 8½in	4wDH	230hp	R-R C6SFL	34T reb of S 10011	231
261V	06.08.1976	4ft 8½in	0-6-0DH	370hp	R-R C8TFL	48T reb YEC L124	213
264V	13.04.1976	4ft 8½in	4wDH	272hp	R-R C8NFL	40T	327

276V	06.09.1977	4ft 8½in	4wDH	300hp	R-R C6TFL	37T		121,329
279V	20.11.1978	4ft 8½in	4wDH	272hp	R-R C8NFL	40T		329
280V	14.12.1978	4ft 8½in	4wDH	272hp	R-R C8NFL	40T		329
281V	20.12.1978	4ft 8½in	4wDH	272hp	R-R C8NFL	40T		100,329
282V	10.04.1978	4ft 8½in	4wDH	272hp	R-R C8NFL	40T		329
283V	23.11.1978	4ft 8½in	4wDH	300hp	R-R C6TFL	37T		330
285V	26.03.1979	4ft 8½in	0-6-0DH	370hp	R-R C8TFL	50T		201
287V	15.01.1980	4ft 8½in	4wDH	270hp	R-R C8NFL	42T		329,383
288V	22.01.1980	4ft 8½in	4wDH	270hp	R-R C8NFL	42T		330,383
292V	03.11.1980	4ft 8½in	4wDH	210hp	R-R C6NFL	29T		328
296V	12.10.1981	4ft 8½in	6wDH	650hp	R-R DV8TCE	60T Steelman		200,331
297V	22.12.1981	4ft 8½in	6wDH	650hp	R-R DV8TCE	60T Steelman		201
V316	03.01.1987	4ft 8½in	6wDH	700hp	Cat 3412 DT	60T Steelman		91
V317	29.03.1987	4ft 8½in	6wDH	700hp	Cat 3412 DT	60T Steelman		91

THOMAS RICHARDSON & SONS, Castle Eden and Hartlepool, Co.Durham TR

208	04.1852	2ft 1in	271
209	1852	2ft 1in	271
210	1852	2ft 1in	271
232	06.1853	2ft 1in	271

THORNEWILL & WARHAM, Burton-on-Trent, Staffordshire TW

		4ft 8½in	0-4-0ST	OC	14x	131

UNIMOG – MERCEDES BENZ AG, Stuttgart, Germany Unimog

031047	1978	4ft 8½in	4wDM	R/R	240
083216	1982	4ft 8½in	4wDM	R/R	383

VULCAN FOUNDRY LTD, Newton-le-Willows, Lancashire VF

422	1858	4ft 8½in	0-6-0ST	OC	16x24	4ft 0in	139
4859	12.01.1943	4ft 8½in	0-4-0DM	153hp	Gardner 6L3	22½T [DC 2167]	240
D212	11.03.1953	4ft 8½in	0-6-0DM	204hp	Gardner 8L3	[DC2486]	260,317
D312	1956	4ft 8½in	0-6-0DE	294kW	EE 6KT	47T [EE2122 NS 625]	261
D319	1956	4ft 8½in	0-6-0DE	294kW	EE 6KT	47T [EE2129 NS 687]	261
D336	1956	4ft 8½in	0-6-0DE	294kW	EE 6KT	47T [EE2146 NS 692]	261

W.G. BAGNALL LTD, Castle Engine Works, Stafford, Staffordshire WB

1002	07.1888	2ft 6in	0-4-0IST	OC	8½x12		189
1516	08.1898	3ft 0in	0-4-0ST	OC	6x9	1ft 8in	365
1728	11.1903	1ft 9in	0-4-0ST	OC	6x9	1ft 6½in	347
1877	05.1909	2ft 6in	0-4-0ST	OC	6x9	1ft 7in	169
1910	11.1910	3ft 0in	0-4-0ST	OC	8x12	2ft 0½in	36
1917	02.06.1910	2ft 6in	0-4-0T	OC	8½x12	2ft 0½in	169,189
2043	12.1917	2ft 0in	0-4-0ST	OC	6x9	1ft 7in	347,366
2068	02.08.1918	2ft 6in	0-4-0ST	OC	8x12	2ft 0½in	380
2287	03.1926	2ft 0in	4-4-0T	OC	10x15	2ft 6in	395
2586	03.1938	4ft 8½in	0-4-0ST	OC	14x22	3ft 6½in	302
2627	02.1941	2ft 0in	4-4-0T	OC	10x15	2ft 6in	395
2819	03.1946	2ft 0in	4-4-0T	OC	10x15	2ft 6in	395
3160	10.1959	4ft 8½in	0-6-0DM	304hp			332
3212	1962	4ft 8½in	0-4-0DH	266hp		[RSHD 8367]	329

WELLMAN CRANES LTD, Darlaston, Staffordshire WC

Successors to WSO

1504	1968	4ft 8½in	4wWE	98
1505	1968	4ft 8½in	4wWE	98

WALKER BROS (WIGAN) LTD, Pagefield Ironworks, Wigan, Lancashire — WkB

[BOSTOCK]	1881	4ft 8½in	0-4-0ST	OC		205

D. WICKHAM & CO LTD, Ware, Hertfordshire — Wkm

6607	04.06.1953	4ft 8½in	2w-2PMR	Ford 10hp	27A Mk III	261
7591	09.01.1957	4ft 8½in	2w-2PMR	JAP 1323	17A	201,208,294,295
7603	21.01.1957	4ft 8½in	2w-2PMR	JAP 1323	17A	201,208,294,295
10731	28.03.1974	4ft 8½in	2w-2DMR	Ford 2262E	27A Mk IV	213
10842	20.11.1975	4ft 8½in	2w-2DMR	Ford 2712E	18A Mk VI	213

WINGROVE & ROGERS LTD, Kirkby, Liverpool, Lancashire — WR

J7292	30.09.1969	1ft 6in	0-4-0BE	W217	358
M7479	20.03.1972	1ft 6in	0-4-0BE	W217L	358
7654	31.01.1974	1ft 6in	0-4-0BE	WR5L	215

WELLMAN SMITH OWEN ENGINEERING CORPORATION LTD, Darlaston, Staffordshire — WSO

1292	1928	?	E	203
5988/1	1953	4ft 8½in	0-4-0WE	98
5988/2	1953	4ft 8½in	0-4-0WE	98

YORKSHIRE ENGINE CO LTD, Meadow Hall Works, Sheffield, Yorkshire, (WR) — YE

236	1874	4ft 8½in	0-4-0ST	OC					64,172,246
262	1875	4ft 8½in	0-4-0ST	OC					57,160,293
287	1876	4ft 8½in	0-4-0ST	OC					218
289	1877	4ft 8½in	0-4-0ST	OC					44
480	1891	4ft 8½in	0-4-0ST	OC	14x20	3ft 3in			265
2384	1937	4ft 8½in	0-4-0ST	OC					70
2414	1944	2ft 6in	4wWE						203
2595	22.8.1956	4ft 8½in	0-6-0DE	400hp	2 x R-R C6SFL			Janus	200
2596	16.01.1956	4ft 8½in	0-4-0DE	275hp	Paxman 6RPH		37T	DE2	91
2597	23.01.1956	4ft 8½in	0-4-0DE	275hp	Paxman 6RPH		37T	DE2	91
2620	26.11.1956	4ft 8½in	0-6-0DE	400hp	2 x R-R C6SFL			Janus	329
2621	12.11.1956	4ft 8½in	0-4-0DE	200hp	R-R C6SFL		30T		91
2622	24.11.1956	4ft 8½in	0-4-0DE	200hp	R-R C6SFL		30T		91,330
2624	12.01.1957	4ft 8½in	0-4-0DE	200hp	R-R C6SFL		30T		91
2625	21.01.1957	4ft 8½in	0-4-0DE	200hp	R-R C6SFL		30T		91
2629	25.11.1957	4ft 8½in	0-6-0DE	400hp	2 x R-R C6SFL			Janus	200,230
2665	03.12.1957	4ft 8½in	0-6-0DE	400hp	2 x R-R C6SFL				200
2666	11.12.1957	4ft 8½in	0-6-0DE	400hp	2 x R-R C6SFL				200
2687	02.06.1958	4ft 8½in	0-4-0DE	200hp	R-R C6SFL				331
2713	20.06.1958	4ft 8½in	0-6-0DE	400hp	2 x R-R C6SFL				213
2714	10.06.1958	4ft 8½in	0-6-0DE	400hp	2 x R-R C6SFL			Janus	200
2718	03.07.1958	4ft 8½in	0-6-0DE	400hp	2 x R-R C6SFL			Janus	200
2719	15.07.1958	4ft 8½in	0-6-0DE	400hp	2 x R-R C6SFL				200,230
2723	10.11.1958	4ft 8½in	0-6-0DE	400hp	2 x R-R C6SFL				200,208,213,216
2724	08.11.1958	4ft 8½in	0-6-0DE	400hp	2 x R-R C6SFL				200,213
2725	26.11.1958	4ft 8½in	0-6-0DE	400hp	2 x R-R C6SFL		48T	Janus	200

2731	23.09.1959	4ft 8½in	0-4-0DE	200hp	R-R C6SFL			329
2741	20.03.1959	4ft 8½in	0-6-0DE	400hp	2 x R-R C6SFL	48T	Janus	200
2742	24.03.1959	4ft 8½in	0-6-0DE	400hp	2 x R-R C6SFL			200,230
2743	25.03.1959	4ft 8½in	0-6-0DE	400hp	2 x R-R C6SFL	48T	Janus	200,230
2772	26.08.1960	4ft 8½in	0-6-0DE	440hp	2 x R-R C6SFL			330
2787	20.03.1961	4ft 8½in	0-6-0DE	440hp	2 x R-R C6SFL			330
2825	12.04.1961	4ft 8½in	0-6-0DH	300hp	R-R C8SFL			70,213
2832	19.01.1962	4ft 8½in	0-6-0DH	300hp	R-R C8SFL			70
2878	07.06.1963	4ft 8½in	0-6-0DE	440hp	2 x R-R C6SFL		Janus	121
2886	21.06.1961	4ft 8½in	0-4-0DE	300hp	R-R C8SFL			91
2911	14.02.1964	4ft 8½in	0-6-0DH	375hp	Cummins NT400			331
2938	13.04.1965	4ft 8½in	0-6-0DE	500hp	2 x Cummins NHS-6		Janus	261

YORKSHIRE ENGINE CO LTD, Rotherham, South Yorkshire — YEC

L112	1992	4ft 8½in	2w-2DHR	reb of Wkm 7591	201,208
L124	1996	4ft 8½in	0-6-0DH	reb of TH 261V	213
L147	1999	4ft 8½in	0-6-0DH	reb of S 10148	213
L180	2000	4ft 8½in	0-6-0DH	reb of EEV 3994	121,216,261

INDEX OF LOCOMOTIVE NAMES

1918	222,228
A.BOULLE	395
ABBOT	69
ACKLAM No.2	149
ACKLAM No.3	149
ACKLAM WORKS	59
ADAM	333
AILEEN	222,228
AIREDALE	86
ALBERT	87
ALDERMAN	337
ALEXANDER	87
ALEXANDRA	268
ALFRED	69
ALFRED E. ALCOCK	59,87
ALLENDALE	200,230
ALLIANCE	239
ALN	199
AMBLE	386
AMMONITE	271
ANGIE	201,208,294,295
ANGUS	87
ANNAN	200
ANNIE	69
ANTELOPE	389
ARGYLE	87
ARTHUR	69,86,233,355
ARTHUR VERNON DAWSON	121,133
ASTLEY	337
ATLAS	199,330
AUTUMN GOLD	311
AVON	199
AYRESOME	175
AYRESOME No.2	175
AYRESOME No.4	175
AYRESOME No.5	175
AYRESOME No.6	175
AYRESOME No.7	175
AYRESOME No.8	175
AYRESOME No.9	175
AYRESOME No.10	175
AYRESOME No.11	175
AYRESOME No.12	175
B.G.	328
BALKAN	309
BALNABOTH	119
BANBURY	160
BARTON	340
BEARPARK	44
BEATTY	31,49,87
BEE	246,268
BEETLE	268
BELMONT	86
BERNARD	367
BESSEMER	85,87
BESSEMER BOY	311
BESSIE	114,350
BILSDALE	200
BLACK PRINCE	69
BLACKLOCK R	394
BLUE GOWN	282
BLUEBIRD	216
BODICOTE	160
BOLCKOW	85
BONLEA No.2	22
BONLEA No.3	22
BOSTOCK	205
BOULBY	69,91,290
BOWESFIELD No.1	152
BOWESFIELD No.2	152
BOYLE	230,327
BRIDGET	69
BRISK	360
BRITANNIA	331
BROTTON	91
BRYAN TURNER	201,216,294,295
BULLDOG	323
BULLER	252
BYNG	87
CALCIUM	199
CALEDONIA	87,167
CAM	200
CARL	86
CARLIN HOW	91
CARLTON	87
CARLTON No.1	139
CARLTON No.2	139
CARLTON No.4	139
CARLTON No.7	139
CARMEL	34
CATTERSTY	69
CATTERSTY 2	70,92,330
CECIL	114
CEREBOS	113,114
CHALONER	86,91,290
CHARLAW	355
CHARLES WYTOCK	395
CHARLIE	213
CHEETHAM	86
CHELSTON	160
CHELTENHAM	160
CHRIS MOODY	89
CHURCHILL	59,88,175,231
CLARENCE	29,87,386
CLARENCE No.1	65
CLARO	191
CLEVELAND	86,88,208,305
CLEVELAND POTASH No.1	213,331
CLYDE	65,87,200
COATHAM	166
COATHAM No.1	168
COATHAM No.2	168
COBRA	121
COCHRANES No.1	309
COCHRANES No.4	309
COCHRANES No.6	309
COFFEE POT	37,69
COLONEL	45,222,228
COMET	86
COMMONDALE	200
COMPTE DE PARIS	139
COMPTON	160

Name	Page	Name	Page
CONCORD	335	GENERAL ROBERTS	361
COOMASSIE	309	GEORGE	268
COTHERSTONE	370	GEORGE OUTHWAITE	394
COUNCILLOR	313,340	GEORGE LEEMAN	281
COWES	361	GLADYS	149,222,228
CROSTHWAITE	129	GLAISDALE	91,200,208,216,290
CYCLOPS	86,391	GLENMORAG	119
DAISY	199,222,228	GORDON	101
DANBYDALE	200	GOWRIE	347
DANE	200	GRADWELL	318,338,340
DARLINGTON	370	GRANGETOWN	86,87
DAVID	313	GREENBANK	69
DAVID PAYNE	49,88	GRETA	309
DEE	199	GRIMSBY	357
DERWENT	200	GRINKLE	91,290
DESPATCH	308,309	GROSMONT	91
DIAMOND	309	GUISBOROUGH	200,208
DINSDALE No.2	88	GUY DAWNAY	44
DINSDALE No.3	175	H & H	194
DL & Co No.3	145	HAIG	49,87
DL & Co No.4	145	HAMPSTEAD	347
DON	199	HARE	143
DUKE	109	HAREHOPE	269
DUNCAN	216,333	HARRY POTTER	70,385
EAGLESCLIFFE No.1	185,186	HARWOODDALE	200,213
EARL LEOFRIC OF MERCIA	240	HAVERTON HILL	301
EAST LAYTON	286	HELEN	327
EBENEZER	189	HENRY	49,86
EDEN	199,205	HENRY CORT	86
EDGAR	69	HILDA	69,268
EDWARD WILLIAMS	86	HIPPO	261
EFFIE	382	HOLMLEA	34
ELDON No.1	386	HOLWELL No.9	309
ELEANOR DAWSON	133	HOLWELL No.16	309
ELFIN	218	HOPETOWN	389
ELIZABETH	121,326,329,394	HORNET	268
ELIZABETH II	69	HOWDEN DENE	199
ELSA	87	HOWE	313
EMILY	216	HUGO	87
EMMA	100,329	HUMBER	200,205
ENGINEERS	59	HUMMERSEA	69
ENTERPRISE	333	HUMMERSEA 1	70,92
ERIMUS	160,246	HUNTCLIFF	70
ESK	199	HURWORTH	250
ESKDALE	200	HYWELL	213
ESKDALESIDE	91,290	IBURNDALE	200
ESTATE No.1	255	ILLTYD	86
ESTATE No.2	257	IMPALA	330
ESTON	86,91,208,290	IMPERIAL	44,217
ETHEL	69	IRENE	361
ETTRICK	199	IRONMASTER	335
EVELYN JEANIE	252	ISIS	200
EXILE	101	J.W.H.	328
F.H. & Co. No.2	243	JAMES	216,323,330,383
FARNDALE	200, 230	JAMES EVANS	87
FIRELESS	200	JAMES WATT	230
FLORENCE	171	JANUS	330
FLYING SCOTSMAN	394	JEANETTE	241
FORTH	200	JEANIE	69
FRANCIS	222,228	JEANNIE	126
FRANK PALLISTER	120	JELLICOE	87
FRASER	87	JENNIE	149
FREEBROUGH	69	JERSEY	69
FRENCH	87	JESMOND	44

JOFFRE	87
JOHN	246,269
JOHN BIRD	88
JOHN EVANS	86
JOHN TURNER	86
JOHN W. ANTILL	201,208,294,295
JOSEPH STIRZAKER	59,88
JUBILEE	87,167,309
JULIA	350
JULIAN	261
JUNO	86
JUPITER	86,87
KAREN	213,334
KATE	335
KATIE	126
KELVIN	86
KEN	213,333
KENNETH	386
KIDBROOKE	347,366
KILDALE	200
KILMARNOCK	87,246,265,286,337,340
KILTON	70,91,290
KING EDWARD	87,166
KING GEORGE V	87,246
KINGSBURY	332
KIRKLEATHAM	91,290
KIRKLEATHAM No.1	346
KIRKLEATHAM No.2	346
KIRKLEATHAM No.3	346
KIRKLEATHAM No.4	345
KITCHENER	87
LACKENBY	91,290
LACKENBY No.1	265
LACKENBY No.2	265
LADY POTTER	70,100,213,231,383,385
LAING	86
LANGBAURGH	208,295
LASCELLES	268
LAWRENCE ELLIS	87
LESLIE	87
LEVEN	200
LINGDALE	91, 200, 230
LINTHORPE	233
LIVERTON	44,91,106,209,290
LIZZIE	69
LOC 1	133
LOCO 2	261
LOCO No.9	329,383
LOCO(1)	213
LOFTUS	91,290
LONGACRES	91,290
LOTTIE	139
LOUISE	350
LUMPSEY	91
MABEL	386
MAC	365
MACDONALD	265
MADDIE	213
MAGGIE	126
MAGNET II	169
MAJOR	222,228
MALLEABLE No.1	52
MALLEABLE No.2	52
MALLEABLE No.5	52
MALLEABLE No.6	52
MALLEABLE No.7	52
MARDALE	200,230
MARGARET	69
MARS	87,116
MARTON	65,85,86
MARY	69,266,268
MARY IRVING	166
MARY LOUISA	193,352
MAY	166,233
MERCURY	87,246,363
MERRYBENT	269
MERSEY	200
METEOR	389
MICKLETON	44
MIDDLESBRO'	86
MIDDLESBROUGH	87,113,363
MIDGE	69,268
MILLE	115
MINNIE	69,379
MONARCH	24
MOSELEY	44
MOUNTAINEER	370
MULGRAVE	359
MURR 1	231,383
MURR 2	100,231,383
MUSHET	86
N.E. STEEL Co No.1	149
N.E. STEEL Co No.4	149
NANCY	126,149
NELLY	44
NEWBRIDGE	205,207
NEWCASTLE	358
NEWLANDSIDE	87
NEWPORT	160
NEWPORT No.1	160
NEWPORT No.2	160
NIDD	199
NIDDERDALE	200
NILE	241
NITH	200
NORFOLK	355
NORMANBY	86,87,246
NORMANBY No.4	246
NORMANBY No.5	246
NORTH SKELTON	91,290
NORTHERN GAS BOARD No.1	253
NORTON	309,326
NUMBER ONE	177
NUNTHORPE	135,323
OLD GEOFF	261
ORMESBY	87,308,309
PALACIOS	361
PALLION	391
PANDA	92,333
PARKFIELD	24
PAT	290
PATRICIA	268
PATRIOT	85
PATTERDALE	200
PAULA	350
PENWYLLT	359
PETE GANNON LOCOMOTIVE ENGINEER 1948-2001	121

PETER	21
PIONEER No.1	109,200
PIONEER No.4	109,200
PIONEER No.5	109,200
POCHIN	86
POLLY	126
POPPY	290
PORT MULGRAVE	91,231
PRESIDENT WILSON	87
PRINCE CHARLES	394
PRINCE OF WALES	394
PRINCESS ELIZABETH	38
PRINCESS MARGARET	38
PULL	219,338,340
PUSH	340
QUEEN ELIZABETH	246
QUEEN MARY	86,246,269
QUEEN OF THE FOREST	346
R.A. LAWDAY	121
R.O.F. No.4	328
R.P.D. & Co No.1	151
R.W.C.	129
R.W.C. LTD 533	279
RAITHWAITE	91
RAKIE	69
REDCAR	91,166,208,369
REDCAR 22	100,290
REDCAR No.1	166,345
REDCAR No.3	166
REDCAR No.4	166,319
REDCAR No.5	166
REDCAR No.6	166
REDCAR No.7	166
RENNIE	69
RICHARD HILL LIMITED No.4	124
RICHARD PEASE	21,255
RISLEY	241
RISLEY YARD No.106 MED	240
ROBERT	233,323
ROF No.8	286
ROKER	391
ROOSEVELT	59,88
ROSEBERRY	69,87,91,265,266,290
ROSEDALE	32,91,171,200,263,290
ROXBY	119
RUBY	240
SALT UNION No.1	210
SALT UNION No.2	205,210
SALTBURN	87
SALTBURN 150	394
SALTUNIA	205
SAM	216
SANNOX	200
SATURN	87
SCOTT	87
SDS & I Co No.3	228
SENTINEL	44
SEVERN	200
SEYMOUR CLARK	378
SHEET WORKS No.3	151
SHERRIFFS	91,290
SICO	305
SIEMENS	86
SINEMBE	395
SIR ARTHUR DORMAN	59,87
SIR ELLIS HUNTER	59,87
SIR HENRY WILSON	87
SLAPEWATH	91,290
SOMERTON	326
SOUTH BANK	86,87
SOUTH GARE	376
SOUTH STOCKTON No.1	305
SOUTHEND	160
SOUTHFIELD	160
SPAWOOD	91,290
SPEEDY	179
SPERO	44,119
SPEY	199
SPIDER	268,272
SPRINGBOK	330
STAITHES	91,290
STANGHOW	228,309
STANTON 35	309
STANTON No.7	309
STANTON No.40	91
STANTON No.41	91
STANTON No.43	91
STANTON No.44	91,330
STANTON No.46	91
STANTON No.47	91
STANTON No.51	91
STELLA	200
STOCKBRIDGE	160
STOCKTON FORGE No.1	187
STOCKTON FURNACES No.3	312
STOKESLEY	44
SUMUS	57,160
SUTHERLAND	338
SWALE	200
SWIFT	362
SWILLINGTON	337
SYDNEY	52
T.R. DENT No.1	286
T. RODDAM DENT & SON LTD No.3	286
T. RODDAM DENT & SON LTD No.4	286,318
T. RODDAM DENT & SON LTD No.5	286
T. RODDAM DENT & SON LTD No.6	286
TAFF	199
TAME	199,208
TAMMIE	213
TARSLAG 4	313
TAW	199
TAY	199
TCC No.1	260
TCC No.2	260
TEES	49,86,199,200
TEES BRIDGE No.1	266
TEES BRIDGE No.4	266
TEES BRIDGE No.6	266
TEES BRIDGE No.7	266
TEES SIDE No.1	277
TEES SIDE No.2	114,277
TEES SIDE No.3	277
TEES SIDE No.4	277
TEES SIDE No.5	277
TEES SIDE No.6	277
TEES SIDE No.7	277

TEES SIDE No.8	61,277
TEES SIDE No.9	277
TEES VALLEY	370
TEESDALE	185
TEESDALE No.2	185,186
TEESDALE No.3	185,186, 187
TEESDALE No.4	185
TENNANT'S No.3	205,343
TEVIOT	200
THAMES	200
THE CONTRACTOR	370
THE T.H. BROWN	200
THOMAS	85
TILL	199
TOGO	309
TOM	22,119
TWEED	199
TWEETY	133
TWIZZIE GILL	378
TYNE	86,199,200
TYNE-TEES WHARF No.1	318,338
TYNE-TEES WHARF No.3	318,338
TYNE-TEES WHARF No.4	318,338,340
TYNE-TEES WHARF No.5	260,318,338,340
TYNE-TEES WHARF No.8	318,338
UGTHORPE	335
UNITY	335
UPSALL	44
VALIANT	213,216
VAUGHAN	85,86,87
VERDUN	87
VERNON	365
VICKERS	85
VICTORIA	37,246,268
VICTORY	87
VULCAN	86
W. SHAW & CO LTD No.2	296,352
W. SHAW & CO LTD No.3	131,296
W. SHAW & CO LTD No.4	296,352
WALKER No.1	44
WALTER MORRISON	222,228
WALTER SCOTT	372
WALTER URWIN	91,290
WARRINGTON	27,340
WASP	246,268
WATERFALL	91,216,290
WEAR	199,200
WEAVER	199,279,326
WEST HARTLEPOOL No.1	221
WEST HARTLEPOOL No.2	179
WEST STOCKTON No.1	351
WEST STOCKTON No.2	351
WHARTON	205,209,348
WHITECLIFFE	44
WHITEHAVEN	160,338,340
WHITWORTH	65,86
WILLIAM	70
WILLIAM ANDERSON	87
WILLIAM BROWN	252
WILLIAM GRAY No.2	177
WILLIAM HEWITT	87
WILLIAM JONES	59,88
WILTONIA	201,208,294,295
WINDSOR	86
WINSFORD	205,207
WITTON CASTLE	102
WOODSIDE	309
WYE	199,361
WYLLIE	86
YARD No.736	240
YARD No.WD 6692	240
YARD No.114	240
YARD No.115	240
YOGI	213
YORKSHIRE	361
YVONNE	394
ZURIEL	169,189

INDEX OF OWNERS AND LOCATIONS

Accredited Processed Metals Ltd	21
Acklam Coke Ovens	150
Acklam Foundry	412
Acklam Iron & Steel Works	145,146
Acklam Iron Co Ltd	145
Adamson, Leith	357
Admiralty	240
Aird, John & Co	357
Albright & Wilson Ltd	36
Allan, Thomas & Sons Ltd	21
Allan, Athole G. Ltd	25
Allied Ironfounders Ltd	21,129
Alterations to Railway at Hartlepool Docks	378
Alterations to Railway at Tees Dock	377
Anderson, John	357
Anderson Foundry	31
Anderson Foundry Co Ltd	22
Ashmore & White	24
Ashmore, Benson, Pease & Co Ltd	24
Associated Chemical Companies Ltd	36
Athco	389
Atlas Foundry	27
Atlas Steel Hoop & Wire Rod Works	239
Ayresome Iron Works	173
Ayrton Sheet Works	151
Aysdalegate Ironstone Mine	27
Bacon, W. & Co	172
Bainbridge, A. Ltd	26
Baltic Saw Mills	381
Barningham, W. & Co	27
Barton Abrasives Ltd	27
BASF Plc	28
Bastow, Samuel & Co	137,397
Batchelor Robinson Metals & Chemicals Ltd	343
Baum, Michael & Co Ltd	30
Bell & Son (Doncaster) Ltd	31
Bell Brothers	61,397,398,403
Bell Brothers Ltd	61,156,204,397,398,403
Belmont Ironstone Mines	135,136
Bessemer Blast Furnaces	99
Best, John & Sons Ltd	358
Bigland, Isaac	389
Billingham Coking Plant	203
Billingham Works	101,194,283
Bishopyon Road to Durham Road Project	373
Blackett and Foster, Messrs	126
Blackett, Hutton & Co	126
Blackett, R. & Sons	398
Blair & Co Ltd	32
Blair's (19260 Ltd, Marine Engine Works	32
Blenkinsop, W. & Co Ltd	34
Bolckow, Vaughan & Co Ltd	46,72,120,136, 140,143,162,172,264,397,405
Bonlea Foundry	21
Boosbeck-Carlin How Contract	369
Boosbeck Ironstone Mine	414
Boulby Ironstone Mine	413
Boulby Mine	211
Boulby Mine Contract	358
Bower, George	35
Bowesfield Boiler Works	385
Bowesfield Brick Co Ltd	35
Bowesfield Lane Works	25
Bowesfield Steel Co Ltd	152
Bowesfield Works	235
Bradley, Mr	389
Brenda Road Depot	384
Britannia Iron Works Co Ltd	53
Britannia Steel Works	53,391
British Chilled Roll & Engineering Co	130
British Chrome & Chemicals Ltd	36
British Electricity Authority	111
British Energy Generation Ltd	170
British Periclase Co Ltd	114
British Shipbuilders Ltd	300
British Steel Corporation	38,53,61,66,219,273, 305,391
British Steel Ltd	66,219
British Titan Products Co Ltd	101
Britmag Ltd	114
Brotton Ironstone Mine	408
Brown, G. & Bros Ltd	101
BSC (Chemicals) Ltd	61
Burford Taylor & Co	381
Butler Bros, Messrs	399
Calder, Dixon & Co Ltd	103
Campbell, W.H. & Co	356
Cargo Fleet Iron and Steel Works	39
Cargo Fleet Iron Co Ltd, The	39,103,303,398,399
Cargo Fleet Iron Works Contract	363
Cargo Fleet Jetties and Wharves Contract	360
Cargo Fleet Timber Yard	254
Carlin How Ironstone Mine	403
Carlton Iron Co Ltd	137
Carr House Iron Co	106
Carr House Iron Works	106
Carrs Tilery and Brick Works	418
Carter, G. & W.H. Ltd	106
Casebourne & Co Ltd	107,108
Casebourne & Lucas	107
Casebourne Road Depot	383
Casebourne, Ambrose C.	389
Casebourne, C. & Co Ltd	106
Cast Steel Foundry Co Ltd, The	399
Castle Cement (Ribblesdale) Ltd	110
Castle Eden and Stockton Railway Contract	365
Cementation Building Ltd	358
Central Electricity Generating Board	111,170
Central Marine Engine Works	178
Cerebos Foods Ltd	112
Chaloner Mine	140,141
Charlton, R.H. & Co	114
Charlton, T. & Co	292
Church Row Improvement Project	373
CJC Chemicals and Magnesia Ltd	114
Clarence Brick Works	414
Clarence Iron and Steel Works	61
Clarence Salt Works	204
Clarke Bros	381
Clay Lane Iron Co Ltd	117,143
Cleveland & Durham County Electric Power Co Ltd	111,249

Cleveland Branch Blast Furnaces	99
Cleveland Chemical works	287
Cleveland Coke Ovens	98
Cleveland Dockyard	30,411,417
Cleveland Ironstone Mining Museum	391
Cleveland Magnesite & Refractory Co Ltd	400
Cleveland No.3 Billet Mill	99
Cleveland Patent Nut & Bolt Co	400
Cleveland Potash Ltd	211,215
Cleveland Railway Contract	362
Cleveland Refractory Works	400
Cleveland Salt Co Ltd	120
Cleveland Slag and Concrete Works	402
Cleveland Slag Roads Ltd	120
Cleveland Slag Working Co Ltd	400
Cleveland Timber Yard	281
Cliff House Iron Works	397
Coatham Iron Works	167
Coatham Stob Brick Works	168
Coatham Stob Estates Ltd	168
Coborn Works	124
Cobra Railfreight Ltd	120
Cochrane & Co Ltd	122,305
Cochrane, Grove & Co	305
Cohen, George, Sons & Co Ltd	124
Colwell, E.R. Ltd	389
Connal & Co Ltd	125
Construction of Bridge and Improved Road, Dalehouse	366
Construction of Redcar-Marske Coast Road	369
Construction of Tunnel Wilton-Lockwood Beck	373
Cook, William, Blackett, Hutton Ltd	126
Copley, I. & Co	400
Corus Plc	66,219
County Borough of Middlesbrough Gas Department	250
County Borough of Teesside	395
Cradock, Allison & Co Ltd	168
Craggs Hall Ironstone Mine	409
Craggs, R. & Sons Ltd	401
Cragshall Ironstone Company	409
Craig, Taylor & Co Ltd	401
Crewdson, Hardy & Co Ltd	402
Crombie, John & Son Ltd	402
Crossley & Co Ltd	402
Crossley & Sons Ltd	35,168,302
Crosthwaite, R.W. Ltd	129
Dalgleish, Adam Engineering Ltd	385
Davy & United Roll Foundry Ltd	129,130
Dawson, A.V. Ltd	131,392
Deepwater Wharf Ltd, The	133
Dent's Wharf	283
Denton, Gray & Co Ltd	177,407
Derwent Iron Co	269
Dickson, John	358
Dixon, Sir Raylton & Co Ltd	411
Dock Street Foundry	46
Dorman Long (Bridge & Engineering) Ltd	53
Dorman Long (Chemicals) Ltd	61
Dorman Long (Steel) Ltd	38,46,53,72,145,405
Dorman Long & Co Ltd	46,53,61,72,135,144 146,149,151,152,153,155,156 158,162,163,403, 405
Dorman, R.P. & Co	151,406
Douglas Engineering Services Ltd	381
Downey & Co	167,264
Dowson, Geo	359
Dunlop, Meredith & Co Ltd	106
Durham Paper Mills Ltd	405
Eaglescliffe Bricks Ltd	168
Eaglescliffe Chemical Co Ltd	36
Eaglescliffe Spare Parts Distribution Depot	240
Eaglescliffe Works	36,168
Eagre Construction Co (Scunthorpe) Ltd	359
EDF Energy Plc	170
Egglescliffe Foundry	186
Egglescliffe Iron Co Ltd	186
Elwon, Malcolm & Co	117
Empire Works	129
Enron Teesside Operations Ltd	293
Erimus Iron Co Ltd	171
Eston Grange Iron Co	172
Eston Mine	140
Eston Sheet & Galvanizing Co Ltd	172
Fossick & Hackworth	32
Fossick, Blair and Co	32
Fox, Head & Co	242
Furness Withy Shipbuilding Co Ltd	300
Furness, Sir Christopher, Westgarth & Co Ltd	280
Furness, Withy & Co Ltd	407
G.N. Steam Co Ltd	382
Garrity, J. & Co	173
Gilkes, Wilson & Co	319
Gilkes, Wilson, Pease & Co	267
Gjers Mills & Co Ltd	173
Gladstone & Co	389
Gould Steamships and Industries Ltd	32
Graham, F.W.L.	389
Grangefield Road Extension Project	372
Grangetown Contract	363
Grangetown Power Station	249
Graving Dock, Cargo Fleet	364,417
Gray, William & Co Ltd	177
Greatham Brick Works	106
Greatham Salt Works	112
Greybull Capital LLP Company	66
Grinkle Ironstone Mine	180
Grinkle Park Mining Co Ltd	180
Grove Hill Housing Contract	367
Guisborough Rural District Council	369
Harkness, William & Son Ltd	405
Harris Deepwater Wharf Co Ltd, The	133
Harrison Bros (England) Ltd	27
Harrison, J.H.	389
Harrisons & Crosfeld Ltd	36
Hartburn Brick & Tile Co	406
Hartburn Curve Brick Works	406
Hartlepool Docks	313,358,359,372
Hartlepool Engine Works	278
Hartlepool Gas & Water Co Ltd	249
Hartlepool Headland Protection Works Contract	363
Hartlepool North and South Works	223,225,232
Hartlepool Pipe Mills	219
Hartlepool Power Station	170
Hartlepool Salt & Brine Co	112
Hartlepool Works	343
Haverton Hill Salt Co	209

Haverton Hill Salt Works	342	Long Newton Service Reservoirs Contract	370
Haverton Hill Shipyard	300,360	Longacres Ironstone Mine	405
Head Wrightson & Co Ltd	182,392	Longs Steel UK Ltd	66
Head, Ashby & Co	182	Lord, Carter, Wade & Co	363
Hemingway's Chilled Rolls Ltd	130	Losh, Wilson & Bell & Co, Messrs	397
Henderson Clark, M. Ltd	21	Loveridge, W.H. & Co	383
Herring, C. & Co Ltd	187	Lowood, J.G. & Co Ltd	407
Heugh Breakwater Contracts	357	Lumpsey Ironstone Mine	156
Highways Construction Ltd	187	Mackay, R.& Co	390
Hill, John & Co	242	Manless Green Mine	396
Hill, Richard & Co Ltd	191	Marsh Iron Yard	34
Hills, F. & Sons Ltd	192	Marsh Wire Works	191
Hind, E. (South Bank) Ltd	193	Marske Aerodrome and Storage Depot	346,348
Hirst & Sons	359	Marson, E. & Co Ltd	234
Hob Hill Ironstone Mine	409	Matthews, T. Arthur	363
Hodgson & Ridley, Messrs	360	McAlpine, Sir Robert & Sons Ltd	364
Hodsman, George & Sons Ltd	187	Metal Produce and Recovery Depot No.2241	
Hogg & Henderson	193	Metal Produce Recovery Depot Contract	364
Holdsworth, Bennington, Byers & Co	311	Metropolitan-Vickers-Beyer Peacock Ltd	235
Holdsworth, John & Co	355	Middlesbrough and Guisborough Railway	
Holloway Brothers (London) Ltd	360	Contract	357
Hopkins & Co	321	Middlesbrough and Redcar Railway	
Hopkins, Gilkes & Co Ltd	319,321	Contract	357
Hughes, Bolckow & Co Ltd	347	Middlesbrough Chemical Works	218
Huntcliff Ironstone Mines	397	Middlesbrough Dock Contracts	360,362,370
Hunton, W.K.	406	Middlesbrough Earthenware Co	239
Hutchinson, C.E. & Co	389	Middlesbrough Gas Works	250
Huthinson, William	362	Middlesbrough Iron Works	46
Hutton Ironstone Mine	410	Middlesbrough Pottery Co	239
ICI (Salt) Ltd	204,209	Middlesbrough Salt Works	410
ICI Ltd	194,204,205,209,318	Middlesbrough Slag Co	312
ICL UK (Cleveland) Ltd	211	Middlesbrough Steel, Strip & Hoop Co Ltd	239
Imperial Iron Works	217	Middlesbrough Wharf	125,318,392
Imperial Works, South Bank	193	Middleton Road Gas Works	249
Irvine's Shipbuilding & Dry Docks Co Ltd	406,407	Middleton Shipyard	407
Irving, Lane & Co	401	Millfield Works	324
Jackson, Gill & Co	217	Milton Iron Works	35
Jackson, John	362	Milton Works	381
Jingye Group	66	Ministry of Defence, Navy Department	240
Johnson & Reay	303	Ministry of Fuel and Power	240
Johnson, C.G. & Co	390	Ministry of Supply	114,241
Jones Bros & Co Ltd	151	Moffat, John	365
Jones, Dunning & Co Ltd	243	Monsanto Textiles Ltd	28
Jones, W. & Co	218	Moor Steel & Iron Co Ltd	303
Joseph Torbock & Co	387	Morris Motors Ltd	241
Kilton Ironstone Co	153	Morrison & Co	408
King Spark, Henry and Others	136	Murray, Ed & Sons Ltd	383
Kinnell, Charles P. & Co Ltd	219	National Power	170
Kirkleatham Ironstone Co, The	344	Nelson, Thomas & Son	365
Lackenby Iron Co	264	Nesham & Welch	101
Lackenby Slag Plant	234	New Normanby Brick Co	400
Lant, John Ltd	363	New Railway Bridge Across River Tees	
Lee, H. & Son	357	Contract	372
Liberty Pipes (Hartlepool) Ltd	219	Newcastle Chemical Co Ltd	342
Lingdale Ironstone Mine Co Ltd, The	155	Newport Iron Works	158
Linthorpe Iron Works	232	Newport Rolling Mills Ltd	242
Linthorpe-Dinsdale Smelting Co Ltd	232	Newport Wire Works	191
Lishman, T.	390	Newton, W. & Co	390
Liverton Ironstone Co Ltd	103	Normanby Brick & Tile Co	400
Lloyd & Co	232,264	Normanby Iron Works Co Ltd	243
Loftus Iron Co Ltd	67	Normanby Iron Works Extension Contract	364
Loftus Ironstone Mine	414	Normanby Ironstone Mine	398
Loftus Railway Contract	369	North Cleveland Ironstone co Ltd	272
London & North Eastern Railway	234	North Eastern Iron Refining Co Ltd	247

North Eastern Pitwood Association Ltd	366	Redcar Coke Ovens	98,292
North Eastern Railway	234	Redcar Iron and Steel Works	163,367
North Gare Breakwater Construction	375	Redcar Jetty and Wharves Contract	361
North Loftus Mine	67	Redcar Junction-Saltburn Railway Contract	358
North of England Industrial Iron & Coal Co Ltd, The	136,137	Redcar Promenade Extension	368
		Redcar-Grangetown New Trunk Road	368
North Riding of Yorkshire County Council	366	Redland Magnesia Ltd	114
North Shore Branch	32,192,350,406	Redpath Dorman Long Ltd	53,273,391
North Shore Shipyard	413	Redsteel Works	273
North Skelton Ironstone Mine	162	Relaying of Carlin How-Boulby Railway	359
North Tees Generating Station	111	Resewering Lustrom Beck Project	373
North Yorkshire Iron Co Ltd	304	Ribbledale Cement Ltd	110
North Yorkshire Iron Works	26	Richardson, Duck & Co Ltd	412
North-Eastern Electric Supply Co Ltd	111,249	Richardson, Johnson & Co	304
North-Eastern Steel Co Ltd	145,146,149	Richardson, S. & Co	390
Northern Gas Board	249	Richardsons, Westgarth & Co Ltd	278,280
Northern Smelting & Chemical Co Ltd	408	Richmond Iron & Steel Co	281
Northern Steam Engineering Ltd	385	Ridley Shaw & Co Ltd	385
Norton Iron Co Ltd	253	Ridley, T.D. & Sons	369,385
Norton Junction Sand & Gravel Co Ltd	254	Riley Boilers Ltd	415
Nowell, H.M.	366	Riley Bros (Boilermakers) Ltd	415
Nuclear Electric Plc	170	Ritchie, James & Ronald Ltd	412
Oliver, William	390	Ritson, William	369
Ormesby Brick Works	402	Robinson, W. & Co Ltd	281
Ormesby Iron Works	305	Robson, Maynard & Co	163
Ormesby Ironstone Mine	399	Roddam Dent, T. & Son Ltd	133,283
Ormesby Metallic Brick Co Ltd	402	Roger, Robert & Co Ltd	412
Owners of Clay Lane Iron Works	117,143	Ropner Shipbuilding & Repairing Co (Stockton) Ltd	413
Owners of The Middlesbrough Estate, The	250,254,409	Roseberry Steel Works	399
Oxbridge Avenue Extension Project	372	Roseby, Messrs	397
Palliser Works	114	Royal Air Force	346
Palmers Shipbuilding & Iron Co Ltd	180	Royal Flying Corps	346
Paradise Works	21	Rundle & Campbell Ltd	386
Parkfield Works	24	Sadler & Co Ltd	287
PD Ports Plc	258	Sadler, Sir Samuel. A. Ltd	287
Pearse, Lockwood & Co	413	Sahaviriya Steel Industries UK Ltd	289
Pearson, E. (Steel & Non Ferrous) Ltd	263	Salt Union Ltd, The	204,209
Pearson, John Ltd	263,367	Saltburn Extension Railway Contract	371
Pease & Partners Ltd	133,155,243,264,267 352,409,414	Saltburn Ironstone Company	409
		Saltburn Miniature Railway	393
Pease, J.F. & Co Ltd	239	Scott, John	370
Pease, J.W. & Co Ltd	155,269,272,409,410, 414	Scott, Walter & Co	371
Perseverance Boiler Works	415	Seaton Carew Iron Co Ltd	222
Phillips-Imperial Petroleum Ltd	411	Seaton Carew Miniature Railway	394
Phoenix Works	356	Sembcorp Utilities UK Ltd	293
Pile, John & Co	272	Shaw, Johnson & Reay	303
Pile, Spence & Co Ltd	220,272	Shaw, W. & Co Ltd	296
Pioneer Cement Works	108	Shell (UK) Oil Ltd	297
Port Clarence Distillation Works	61	Shell-Mex and B.P. Ltd	296
Port Clarence Foundry	22	Skelton Ironstone Mines	403
Port Clarence to Hartlepool New Road Contract	365	Skelton Park Mine	403
		Skinningrove Iron and Steel Works	67
Porter, Henry & Co	385	Skinningrove Iron Co Ltd	67,413
Portrack Lane Brick and Tile Works	406	Skinningrove Ironstone Mine	397
Portrack Lane Iron Works	101	Slag and Tarmacadam Works	187
Portrack Lane Plant Depot	366	Slapewath Ironstone Mine	292
Portrack Slag Tip Demolition	368	Smith, T.	414
Power-Gas Corporation Ltd, The	24	Smith, T. & Son	298
Preston Park Museum	395	Smith's Dock Co Ltd	298,300,357
Prosser, C.L. & Sons Ltd	273	Snowdon and Hopkins, Messrs	321
Raylton Dixon & Co Ltd	411	South Bank Brick Co Ltd	302
Readman, W.G. Ltd	273	South Bank Chemical Co Ltd	303
Redcar Borough Council	368,369	South Bank Coke Ovens	98,292

South Bank Iron Co	335
South Durham Salt Co	209
South Durham Steel & Iron Co Ltd	39,50,219 223,225,303
South Gare Breakwater Construction	374
South Skelton Ironstone Mine	143
South Stockton Iron Co (1881)	304
South Stockton Shipyard	412
South Teesside Works	391
South Works Coke Ovens	232
Spa Mine	403
Spawood Ironstone Mine	404
St Ann's Brick Works	398
Stafford Pottery	334
Stainsby, W.A. Ltd	390
Stanghow Ironstone Co	122
Stanton & Staveley Ltd	305
Steel Strip & Nail Co Ltd	239
Steetley Magnesite Co Ltd	114
Stevenson, Jaques & Co	145,414
Stewarts & Lloyds Ltd	305
Stillington Works	247
Stillite Products Ltd	415
Stockton Chemical Engineers	415
Stockton Forge Co	186
Stockton Gas Works	252
Stockton Haulage Ltd	310
Stockton Iron Foundry	412
Stockton Iron Furnace Co Ltd	311
Stockton Iron Works	311
Stockton Malleable Works	50
Stockton Rail Mill Co Ltd	311
Stockton Stone & Concrete Co Ltd, The	253
Stockton Wharf	338
Stockton-on-Tees Borough Council	372,395
Stockton-on-Tees Gas Co	252
Strabag Se	373
Stranton Iron & Steel Co Ltd	114
Submarine Miners' Depot	417
Sulphide Corporation Ltd	408
Summers, Jas	390
Sutherst & Southron, Messrs	126
Swan Hunter Shipbuilders Ltd	300
Swan, Coates & Co	39,272,399
Tait, Jas & Partners Ltd	387
Tarmac Roadstone Ltd	312
Tarslag Ltd	312
Tata Steel Europe Ltd	66,219
Tees & Hartlepool Port Authority Ltd	258, 313, 417
Tees Bridge Iron Co Ltd	266
Tees Bridge Works	312
Tees Conservancy Commissioners	258,374, 417
Tees Dock	215,258,381
Tees Dockyard	401
Tees Engine Works	319
Tees Foundries Ltd	267
Tees Furnace Co Ltd	264
Tees Iron Works	267
Tees Salt Co Ltd	318
Tees Scoriae Brick Co Ltd	319
Tees Shipyard, Graythorpe	179,361
Tees Side Bridge & Engineering Works Ltd	273,274,417
Tees Side Iron Works	321
Tees Storage Co Ltd	324
Tees Submarine Base Contract	379
Tees Union Shipping Co Ltd	336
Tees-Side Iron & Engine Works Co Ltd	319,321
Teesbulk Handling Ltd	215
Teesdale Iron Works	182,392
Teesport Refinery	297
Teesport Works	312
Teesside Engineering Works	273
Teesside Railway Contract	369
Teesside Works	72,289
Tennant, Charles & Partners Ltd	205
Tennant's' Salt Works	205
Thomas, I.	377
Thomson, T.J. & Son Ltd	324
Thornaby Iron Works	352
Thornaby Pottery Co Ltd	334
Thornaby Shipbuilding Yard	401
Tocketts Ironstone Mine	410
Toll Bar Sidings and Ballast Line	256
Trackwork Ltd	377
Trafalgar House Group	273
Trechmann, Otto Ltd	418
Trustees of Thomas Vaughan & Co	117,143,335
Turnbull, Thomas & Sons Ltd	335
Tyne-Tees Steam Shipping Co Ltd	336
Tyne-Tees Wharf	336
Union Foundry	129
United Alkali Co Ltd, The	205,342
Upleatham Ironstone Mine	269
Vaughan, Thomas & Co	117,143,335
Victoria Bridge Engineering Works	389
Vulcan Iron Works	219,335
Vulcan Materials Company (UK) Ltd	343
Vulcan Works	400
Waddell, John & Sons	378
Wake & Co Ltd	189
Wake, J.F.	388
Walker, Ambrose & Co Ltd	334
Walker, Maynard & Co Ltd	153,163,344
Wall, R.G.	395
Walton-Binns, Graeme	395
War Department	346,417
Ward, Thos W. Ltd	348
Warner & Co Ltd	348
Warren Cement Works Ltd	418
Waterfall Mine	404
Watson, George	283
Weardale Iron & Coal Co Ltd, The	135,404
Weardale Iron & Coal Ltd	135
Wellington Foundry	296
West Hartlepool Cement Works, Brick Works, and Coal Depot	107
West Hartlepool Harbour & Railway Company	350
West Hartlepool Iron & Steel Works	223
West Hartlepool Steel & Iron Co Ltd	220
West Marsh Iron Co	144
West Stockton Iron Co Ltd	350
Westbourne Iron Works	355
Wetherell, W.	390
Wharncliffe Ganister Brick Works	407
Wharton, J.T.	418
Whessoe Ltd	351
Whinstone Quarry, Stainton	263,367

Whitby, Redcar & Middlesbrough Union Railway Contract	358, 378
White J.D. Ltd	351
Whitecliffe Ironstone Mine	272
Whitwell, William & Co Ltd	352
Widening of Portrack Lane and Haverton Hill Road	368
Wiggins, J.D.	396
Williams, Edward	232
Wills, C.J. & Sons Ltd	379
Wilson & Co	36
Wilson, Isaac & Co	239
Wilsons, Pease & Co	267
Wilton International	295
Wilton Works, Redcar	207
Withy, Alexander & Co	407
Withy, E. & Co	407
Woodsmith Mine	373
Worth, Mackenzie & Co Ltd	356
Wynyard Camp	366
Yorkshire Tube Works	402
Zinc and Sulphuric Acid Works, Seaton Carew	408

Demolition of one of British Steel's Clay Lane blast furnaces on 15th May 1988.   (Courtesy British Steel)